The Mycota

Edited by
K. Esser

The Mycota

I *Growth, Differentiation and Sexuality*
1st edition ed. by J.G.H. Wessels and F. Meinhardt
2nd edition ed. by U. Kües and R. Fischer
3rd edition ed. by J. Wendland

II *Genetics and Biotechnology*
1st and 2nd edition ed. by U. Kück

III *Biochemistry and Molecular Biology*
1st and 2nd edition ed. by R. Brambl and G. Marzluf
3rd edition ed. by D. Hoffmeister

IV *Environmental and Microbial Relationships*
1st edition ed. by D. Wicklow and B. Söderström
2nd edition ed. by C.P. Kubicek and I.S. Druzhinina
3rd edition ed. by I.S. Druzhinina and C.P. Kubicek

V *Plant Relationships*
1st edition ed. by G. Carroll and P. Tudzynski
2nd edition ed. by H.B. Deising

VI *Human and Animal Relationships*
1st edition ed. by D.H. Howard and J.D. Miller
2nd edition ed. by A.A. Brakhage and P.F. Zipfel

VII *Systematics and Evolution*
1st edition ed. by D.J. McLaughlin, E.G. McLaughlin, and P.A. Lemke†
2nd edition ed. by D. McLaughlin and J.W. Spatafora

VIII *Biology of the Fungal Cell*
1st and 2nd edition ed. by R.J. Howard and N.A.R. Gow

IX *Fungal Associations*
1st and 2nd edition ed. by B. Hock

X *Industrial Applications*
1st edition ed. by H.D. Osiewacz
2nd edition ed. by M. Hofrichter

XI *Agricultural Applications*
1st and 2nd edition ed. by F. Kempken

XII *Human Fungal Pathogens*
1st edition ed. by J.E. Domer and G.S. Kobayashi
2nd edition ed. by O. Kurzai

XIII *Fungal Genomics*
1st edition ed. by A.J.P. Brown
2nd edition ed. by M. Nowrousian

XIV *Evolution of Fungi and Fungal-Like Organisms*
Ed. by S. Pöggeler and J. Wöstemeyer

XV *Physiology and Genetics: Selected Basic and Applied Aspects*
Ed. by T. Anke and D. Weber

The Mycota

A Comprehensive Treatise on Fungi as Experimental Systems for Basic and Applied Research

Edited by K. Esser

VII

Systematics and Evolution
Part A
2nd Edition

Volume Editors:
D.J. McLaughlin and J.W. Spatafora

Springer

Series Editor

Professor Dr. Dr. h.c. mult. Karl Esser
Allgemeine Botanik
Ruhr-Universität
44780 Bochum, Germany

Tel.: +49 (234)32-22211
Fax.: +49 (234)32-14211
e-mail: Karl.Esser@rub.de

Volume Editors

Professor Dr. David J. McLaughlin
Department of Plant Biology
University of Minnesota
1445 Gortner Avenue
St. Paul, MN 55108-1095, USA

Tel.: +1(612)625-5736
Fax: +1(612)625-1738
e-mail: davem@umn.edu

Professor Dr. Joseph W. Spatafora
Department of Botany and Plant Pathology
Oregon State University
Corvallis, OR 97331, USA

Tel.: +1(541)737-5304/5305
e-mail: spatafoj@science.oregonstate.edu

ISBN 978-3-662-49578-0 ISBN 978-3-642-55318-9 (eBook)
DOI 10.1007/978-3-642-55318-9
Springer Heidelberg New York Dordrecht London

Printed on acid-free paper

Springer is part of Springer Science+Business Media (www.springer.com)

Karl Esser
(born 1924) is retired Professor of General Botany and Director of the Botanical Garden at the Ruhr-Universität Bochum (Germany). His scientific work focused on basic research in classical and molecular genetics in relation to practical application. His studies were carried out mostly on fungi. Together with his collaborators he was the first to detect plasmids in higher fungi. This has led to the integration of fungal genetics in biotechnology. His scientific work was distinguished by many national and international honors, especially three honorary doctoral degrees.

David J. McLaughlin
(born 1940) is retired Professor of Plant Biology and Curator of Fungi of the Bell Museum and cofounder of the Mycological Culture Collection at the University of Minnesota, St. Paul (USA). His scientific work has focused on the evolution of subcellular structure in Fungi, and phylogeny and systematics of basidomycetes, ectomycorrhizal community structure, and the Minnesota macrofungal flora with a view to establishing baseline data for mid continental macrofungi. He was formerly Editor-in-chief of Mycologia, and has been President of the Mycological Society of America and is a Fellow and Distinguished Mycologist of that Society.

Joseph S. Spatafora
(born 1964) is Professor of Botany and Plant Pathology and Curator of Oregon State University Mycological Collection at Oregon State University, Corvallis, OR (USA). His scientific work has focused on evolution of the Fungi, with emphases on phylogenetics of Ascomycota, insect pathogenic fungi and comparative genomics. He is a Fellow of, and has served as President of, the Mycological Society of America, and he is currently a Senior Editor of the journal Fungal Biology.

Series Preface

Mycology, the study of fungi, originated as a sub discipline of botany and was a descriptive discipline, largely neglected as an experimental science until the early years of this century. A seminal paper by Blakeslee in 1904 provided evidence for self incompatibility, termed "heterothallism", and stimulated interest in studies related to the control of sexual reproduction in fungi by mating-type specificities. Soon to follow was the demonstration that sexually reproducing fungi exhibit Mendelian inheritance and that it was possible to conduct formal genetic analysis with fungi. The names Burgeff, Kniep and Lindegren are all associated with this early period of fungal genetics research.

These studies and the discovery of penicillin by Fleming, who shared a Nobel Prize in 1945, provided further impetus for experimental research with fungi. Thus began a period of interest in mutation induction and analysis of mutants for biochemical traits. Such fundamental research, conducted largely with *Neurospora crassa*, led to the one gene: one enzyme hypothesis and to a second Nobel Prize for fungal research awarded to Beadle and Tatum in 1958. Fundamental research in biochemical genetics was extended to other fungi, especially to *Saccharomyces cerevisiae*, and by the mid-1960s fungal systems were much favored for studies in eukaryotic molecular biology and were soon able to compete with bacterial systems in the molecular arena.

The experimental achievements in research on the genetics and molecular biology of fungi have benefited more generally studies in the related fields of fungal biochemistry, plant pathology, medical mycology, and systematics. Today, there is much interest in the genetic manipulation of fungi for applied research. This current interest in biotechnical genetics has been augmented by the development of DNA-mediated transformation systems in fungi and by an understanding of gene expression and regulation at the molecular level. Applied research initiatives involving fungi extend broadly to areas of interest not only to industry but to agricultural and environmental sciences as well.

It is this burgeoning interest in fungi as experimental systems for applied as well as basic research that has prompted publication of this series of books under the title *The Mycota*. This title knowingly relegates fungi into a separate realm, distinct from that of either plants, animals, or protozoa. For consistency throughout this Series of Volumes the names adopted for major groups of fungi (representative genera in parentheses) areas follows:

Pseudomycota

Division: Oomycota (*Achlya, Phytophthora, Pythium*)
Division: Hyphochytriomycota

Eumycota

Division:	Chytridiomycota (*Allomyces*)
Division:	Zygomycota (*Mucor, Phycomyces, Blakeslea*)
Division:	Dikaryomycota
Subdivision:	Ascomycotina
Class:	Saccharomycetes (*Saccharomyces, Schizosaccharomyces*)
Class:	Ascomycetes (*Neurospora, Podospora, Aspergillus*)
Subdivision:	Basidiomycotina
Class:	Heterobasidiomycetes (*Ustilago, Tremella*)
Class:	Homobasidiomycetes (*Schizophyllum, Coprinus*)

We have made the decision to exclude from *The Mycota* the slime molds which, although they have traditional and strong ties to mycology, truly represent nonfungal forms insofar as they ingest nutrients by phagocytosis, lack a cell wall during the assimilative phase, and clearly show affinities with certain protozoan taxa.

The Series throughout will address three basic questions: what are the fungi, what do they do, and what is their relevance to human affairs? Such a focused and comprehensive treatment of the fungi is long overdue in the opinion of the editors.

A volume devoted to systematics would ordinarily have been the first to appear in this Series. However, the scope of such a volume, coupled with the need to give serious and sustained consideration to any reclassification of major fungal groups, has delayed early publication. We wish, however, to provide a preamble on the nature of fungi, to acquaint readers who are unfamiliar with fungi with certain characteristics that are representative of these organisms and which make them attractive subjects for experimentation.

The fungi represent a heterogeneous assemblage of eukaryotic microorganisms. Fungal metabolism is characteristically heterotrophic or assimilative for organic carbon and some nonelemental source of nitrogen. Fungal cells characteristically imbibe or absorb, rather than ingest, nutrients and they have rigid cell walls. The vast majority of fungi are haploid organisms reproducing either sexually or asexually through spores. The spore forms and details on their method of production have been used to delineate most fungal taxa. Although there is a multitude of spore forms, fungal spores are basically only of two types: (i) asexual spores are formed following mitosis (mitospores) and culminate vegetative growth, and (ii) sexual spores are formed following meiosis (meiospores) and are borne in or upon specialized generative structures, the latter frequently clustered in a fruit body. The vegetative forms of fungi are either unicellular, yeasts are an example, or hyphal; the latter may be branched to form an extensive mycelium.

Regardless of these details, it is the accessibility of spores, especially the direct recovery of meiospores coupled with extended vegetative haploidy, that have made fungi especially attractive as objects for experimental research.

The ability of fungi, especially the saprobic fungi, to absorb and grow on rather simple and defined substrates and to convert these substances, not only into essential metabolites but into important secondary metabolites, is also noteworthy.The metabolic capacities of fungi have attracted much interest in natural products chemistry and in the production of antibiotics and other bioactive compounds. Fungi, especially yeasts, are important in fermentation processes. Other fungi are important in the production of enzymes, citric acid and other organic compounds as well as in the fermentation of foods.

Fungi have invaded every conceivable ecological niche. Saprobic forms abound, especially in the decay of organic debris. Pathogenic forms exist with both plant and animal hosts. Fungi even grow on other fungi. They are found in aquatic as well as soil environments, and their spores may pollute the air. Some are edible; others are poisonous. Many are variously associated with plants as copartners in the formation of lichens and mycorrhizae, as symbiotic endophytes or as overt pathogens. Association with animal systems varies; examples include the predaceous fungi that trap nematodes, the microfungi that grow in the anaerobic environment of the rumen, the many insect associated fungi and the medically important pathogens afflicting humans. Yes, fungi are ubiquitous and important. There are many fungi, conservative estimates are in the order of 100,000 species, and there are many ways to study them, from descriptive accounts of organisms found in nature to laboratory experimentation at the cellular and molecular level. All such studies expand our knowledge of fungi and of fungal processes and improve our ability to utilize and to control fungi for the benefit of humankind.

We have invited leading research specialists in the field of mycology to contribute to this Series. We are especially indebted and grateful for the initiative and leadership shown by the Volume Editors in selecting topics and assembling the experts. We have all been a bit ambitious in producing these Volumes on a timely basis and therein lies the possibility of mistakes and oversights in this first edition. We encourage the readership to draw our attention to any error, omission or inconsistency in this Series in order that improvements can be made in any subsequent edition.

Finally, we wish to acknowledge the willingness of Springer-Verlag to host this project, which is envisioned to require more than 5 years of effort and the publication of at least nine Volumes.

Bochum, Germany
Auburn, AL, USA
April 1994

Karl Esser
Paul A. Lemke
Series Editors

Volume Preface to the Second Edition

There have been major changes in our knowledge of the systematics and evolution of fungi since the first edition of the Mycota, Vol. VII. These changes have been driven by an outpouring of molecular phylogenetic analyses at first based on one or a few genes but now by multiple conserved genes. The Assembling the Fungal Tree of Life projects have been a major contributor to the data needed to construct the molecular phylogenies along with work from many additional labs. The resulting phylogenies have made possible a new taxonomic outline for the Fungi (Hibbett D.S. et al., 2007, Mycol. Res. 111: 509–547), which has provided a more stable systematic treatment for this kingdom, although some of the basal groups of Fungi remain incompletely resolved (Table 1). Agreement among many mycologists on nomenclature is providing a stable framework for Fungi that has been incorporated into reference works and online databases (McLaughlin D. J. et al., 2009, Trends Microbiol. 17: 488–497), and has provided an escape from the conflicting phenetic classifications of the past. These nomenclatural changes are incorporated into these volumes along with much new information on the evolution and ecology of these organisms made possible by a variety of methods, including environmental sequencing and reevaluation of character evolution using molecular phylogenies.

While there is agreement on nomenclature within Kingdom Fungi, there is less agreement on the names for groups of fungus-like organisms, although these organisms remain a major interest of those who study fungi. Some of the confusion arises from the treatment of fungus-like organisms under two nomenclatural codes (Table 1). Of special concern has been the treatment of the oomycetes and their relatives with variant spellings of the kingdom and common name. The solution adopted by Beakes (Chap. 3, Vol. VII, Part A) reserves Straminipila for the kingdom and uses the widely cited stramenopiles for the common name.

Chapters in this edition of the Mycota, Vol. VII, vary from updates of chapters published in the first edition to new chapters. All systematic chapters treat monophyletic groups; clearly polyphyletic groups, such as those based on yeasts or asexual stages (anamorphs), have been omitted. While authors have been encouraged to provide illustrations of the diversity within each group, the results are somewhat uneven. Some authors have extensively illustrated the organisms, while others for reasons of time or access have provided limited illustrations. In the interest of getting these chapters to press in a not too tardy manner, the authors have not been unduly pressed to add illustrations. The reader's understanding is requested for the omissions, which is caused in part by the difficulty of getting all of the chapters needed to cover a wide spectrum of organisms.

Table 1 Taxonomic outline for Fungi and fungus-like organisms[a]

Fungus-like organisms
Supergroup: Amoebozoa
  Phylum: Dictyosteliomycota
  Phylum: Myxomycota
Supergroup: Excavata
  Phylum: Acrasiomycota
Supergroup: Sar[b]
  Subgroup: Rhizaria
    Phylum: Phytomyxea
  Kingdom: Straminipila[c]
    Phylum: Labyrinthulomycota
    Phylum: Hyphochytriomycota
    Phylum: Oomycota
Fungi
Supergroup: Opisthokonta
  Kingdom: Fungi
    Basal fungi
      Phylum: Cryptomycota[d]
      Phylum: Microsporidia
    Traditional Chytridomycota
      Phylum: Chytridiomycota
      Phylum: Monoblepharidomycota
      Phylum: Neocallimastigomycota
      Phylum: Blastocladiomycota
    Zygomycotan (Zygomycetous) Fungi
      Phylum: Entomophthoromycota
      Phylum/a incertae sedis:
        Subphylum: Kickxellomycotina
        Subphylum: Mortierellomycotina
        Subphylum: Mucoromycotina
        Subphylum: Zoopagomycotina
      Phylum: Glomeromycota
    Subkingdom Dikarya
      Phylum: Basidiomycota
        Subphylum: Pucciniomycotina
        Subphylum: Ustilaginomycotina
        Subphylum: Agaricomycotina
      Phylum: Ascomycota
        Subphylum: Taphrinomycotina
        Subphylum: Saccharomycotina
        Subphylum: Pezizomycotina

[a]Names for Fungi and fungus-like organisms traditionally studied by botanists are governed by the *International Code for Nomenclature of algae, fungi and plants (Melbourne Code)* (McNeil J. et al., 2012, Regnum Vegetabile 154, Koeltz Scientific Books). Multiple names exist for eukaryotic microorganisms that are treated under both the Melbourne Code and the International Code of Zoological Nomenclature, except for Microsporidia, which are classified under the zoological code
[b]Sar (Stramenopiles, Alveolata, and Rhizaria)
[c]Also known as Stramenopila or Stramenopiles. The latter is used by Adl et al. (2012, J. Eukaryot. Microbiol. 59: 429–493) and as a common name, stramenopiles, for Straminipila
[d]Also known as Rozellida and Rozellomycota

The Mycota, Vol. VII, includes treatments of the systematics and related topics for Fungi and fungus-like organisms in four eukaryotic supergroups (Table 1) as well as specialized chapters on nomenclature, techniques, and evolution. Most Fungi and fungus-like organisms are covered, including the Microsporidia. Chapter 1, Vol. VII, Part A, provides an overview of fungal origins and evolution.

Chapters 2–4, Vol. VII, Part A, cover the fungus-like organisms, and Chaps. 5 to 14, Vol. VII, Part A, and Chaps. 1–6, Vol. VII, Part B, cover the Fungi. Each of these chapters covers approximately the following topics: occurrence and distribution, economic importance, morphology and ultrastructure, development of the taxonomic theory, classification, and maintenance and culture. The fungus-like organisms are distributed in three distantly related supergroups (Table 1). The basal fungi and traditional Chytridiomycota are treated as six phyla and covered in four chapters, including Chap. 1, Vol. VII, Part A. The zygomycetous fungi, whose deeper relationships remain unresolved, and Glomeromycota are covered in two chapters. The Basidiomycota and Ascomycota, the largest groups of fungi, are treated in five or six chapters each. In the Basidiomycota two chapters cover Pucciniomycotina and Ustilaginomycotina, respectively, while three chapters are devoted to classes of the Agaricomycotina. In the Ascomycota a single chapter covers Taphrinomycotina and Saccharomycotina, while eight classes of the Pezizomycotina are covered in five chapters.

The following topics are treated in Chaps. 7–11 in Vol. VII, Part B: Chap. 7 deals with the nomenclatural changes necessitated by the recent changes to the International Code for Nomenclature of algae, fungi, and plants (Table 1), including the elimination of separate names for anamorphic fungi. Chapter 8 deals with methods for preservation of cultures and specimens, while Chap. 9 reviews the phylogenetic implications of subcellular and biochemical characters and methods for ultrastructural study. Chapter 10 deals with the fungal fossil record and Chap. 11 with the impact of the availability of whole genomes on studies of Fungi.

We are entering a new era in the study of fungi with whole genomes becoming available for an increasing number of species across all the known clades of Fungi. This genome-enabled mycology will utilize large numbers of genes in phylogenomic analyses to resolve difficult to determine relationships in fungi and to provide insights into fungal biology (Hibbett D.S. et al., 2013, Mycologia 106: 1339–1349). Initial studies are already having a significant impact on our understanding of biochemical processes and their ecological impacts. In time genomic studies may shed light on the genetic processes and the genes that control the great morphological diversity in Fungi from the subcellular to the macroscopic level. Thus, there is much new information on the systematics and evolution of fungi to be expected in the future.

We thank Meredith Blackwell for sharing unpublished manuscripts and discussions on the classification system, Esther G. McLaughlin for advice throughout the work, and the U.S. National Science Foundation for support to many labs for the AFTOL 1 and AFTOL 2 projects (including DEB-0732550 to DJM, and DEB-0732993 to JWS), and numerous scientists who have contributed to the work which has made the advances in these volumes possible.

St. Paul, MN
Corvallis, OR
22 May 2014

David J. McLaughlin
Joseph W. Spatafora

Volume Preface to the First Edition

This is an exciting time to produce an overview of the systematics and evolution of the fungi. Homoplasy is evident in all lineages, e.g., those based on the gross morphology of the chytrid zoospore, the perithecium and apothecium, the smut teliospore and the agaric fruiting body, and some classifications based on light microscope morphology have been shown to be unsound. Molecular and subcellular characters, aided by new methods of phylogenetic analysis, have allowed us to see through the conflicts between various phenetic classification schemes and have given us some confidence that we are beginning to achieve a true phylogeny of the fungi. Molecular data have both supported ultrastructural characters that first began to unravel the homoplasies unrecognized at the light microscopic level, and have also revealed the relationships of fungi to other eukaryotes. They continue to enlarge the scope of the fungi, e.g., with the recent addition of the Microsporidia (see Cavalier-Smith, Chap. 1, Vol. VII, Part A), and they have shown the need for more detailed chemical, subcellular, and developmental studies for a fuller understanding of these organisms and their relationships.

This volume is a mixture of phylogenetic and more classical systematics. Progress in knowledge of species and development of taxonomic characters is mixed. Groups with few species have been studied in great detail, while in groups with large numbers of species much effort is still needed to find and determine the taxa. Classical systematics groups organisms on a phenetic basis, then sets up a classification; phylogeny is a secondary consideration. Phylogenetic systematics first determines organism relationships, then constructs a systematic classification that reflects the phylogeny. Molecular characters have made possible the establishment of a monophyletic and, hopefully, more permanent classification for the fungi. Thus, Volume VII of *The Mycota* contains both classical and phylogenetic classifications, reflecting the available data and the orientation of different authors. The incompleteness of some classifications, e.g., those for the Urediniomycetes (Swann, Frieders, and McLaughlin, Chap. 2, Vol. VII, Part B) and Homobasidiomycetes (Hibbett and Thorn, Chap. 5, Vol. VII, Part B), demonstrates that we are in the early stages of a phylogenetic systematics for these groups.

The taxonomic outline used in *The Mycota*, Vol. VII, differs somewhat from that of other volumes in the series (Table 1), reflecting current mycological systematics. There is a lack of agreement on the naming of higher taxa, and the rules of nomenclature permit more than one name for these taxa. Cavalier-Smith (Chap. 1, Vol. VII, Part A) presents an alternative view to the taxonomic outline used for the remainder of the volume (Table 2). Some of the nomenclatural problems stem from a lack of resolution of deep branches in molecular evolutionary trees, a problem that appears likely to be resolved only with additional

Table 1 Taxonomic outline at the kingdom, phylum, and class levels as used in other volumes in the series and in this volume. The classification in this volume is necessarily confusing at this time because authors are using their own classifications rather than an imposed classification

Mycota, Vol. I	Mycota, Vol. VII
PSEUDOMYCOTA	PSEUDOMYCOTA[a,b]
Oomycota	Oomycota[c]
	Peronosporomycetes
Hyphochytriomycota	Hyphochytriomycota
	Hyphochytriomycetes
	Plasmodiophoromycota
	Plasmodiophoromycetes
EUMYCOTA	EUMYCOTA
Chytridiomycota	Chytridiomycota[d]
	Chytridiomycetes
Zygomycota	Zygomycota[d]
	Zygomycetes
	Trichomycetes
Dikaryomycota	
Ascomycotina	Ascomycota[e]
Saccharomycetes	Saccharomycetes
Ascomycetes	Plectomycetes
	Hymenoascomycetes[a]
	Loculoascomycetes[a]
Basidiomycotina	Basidiomycota
Heterobasidiomycetes	Urediniomycetes
	Ustilaginomycetes
	Heterobasidiomycetes[a,f]
Homobasidiomycetes	Homobasidiomycetes[a,f]

[a]Artificial taxon

[b]For a natural classification for Oomycota and Hyphochytriomycota, kingdom Stramenopila (Stramenipila, Dick, Chap. 2, Vol. VII, Part A) or Chromista have been proposed, and for Plasmodiophoromycota, kingdom Protozoa (see Cavalier-Smith, Chap. 1, Vol. VII, Part A)

[c]Or Heterokonta (see Cavalier-Smith, Chap. 1, and Dick, Chap. 2, Vol. VII, Part A)

[d]Probably paraphyletic (see Cavalier-Smith, Chap. 1, Vol. VII, Part A, and Berbee and Taylor, Chap. 10, Vol. VII, Part B)

[e]A phylogenetic classification for Ascomycota is not available. Current thinking among ascomycete scholars is that three classes should be recognized, as follows: "Archiascomycetes", which may not be monophyletic, Hemiascomycetes (see Kurtzman and Sugiyama, Chap. 9, Vol. VII, Part A), and a filamentous group, Euascomycetes, that eventually will be subdividable, perhaps at the subclass level [M.E. Berbee and J.W. Taylor, 1995, Can J Bot 73 (Suppl. 1):S677, and Chap. 10, Vol. VII, Part B; J.W. Spatafora, 1995, Can J Bot 73 (Suppl. 1):S811]. Saccharomycetes as used here (see Barr, Chap. 8, Vol. VII, Part A) includes "Archiascomycetes" and Hemiascomycetes. See the relevant chapters for further speculation on the ultimate disposition of these groups

[f]Heterobasidiomycetes as used in Vol. VIIB cannot be separated from Homobasidiomycetes. Hymenomycetes [E.C. Swann and J.W. Taylor, 1995, Can J Bot 73 (Suppl. 1):S862] has been proposed as a class for these groups (see Berbee and Taylor, Chap. 10, Vol. VII, Part B)

data from multiple genes and the addition of missing taxa to the analysis. Problems also arise from a difference of opinion among authors. The term *fungi* has assumed an ecological meaning for all organisms with a similar nutritional mode, and therefore, Eumycota, rather than Fungi, is less confusing for the members of the phylum that encompasses a monophyletic group of these organisms. *Pseudofungi* (Cavalier-Smith, Chap. 1, Vol. VII, Part A) implies that organisms that lie outside the Eumycota but possess the fungal lifestyle are not fungi, but in an ecological sense they are fungi. P*seudomycota* is therefore used in this series for these fungal organisms that lie outside the Eumycota.

Table 2 Taxonomic outline at the kingdom, phylum, and class levels as used in the rest of this volume compared with that of Cavalier-Smith, Chap. 1, Vol. VII, Part A

Mycota, Vol. VII	Chapter 1, Vol. VII, Part A
PSEUDOMYCOTA[a]	CHROMISTA
Oomycota	Bigyra
Peronosporomycetes	Oomycetes
Hyphochytriomycota	
Hyphochytriomycetes	Hyphochytrea
	PROTOZOA
Plasmodiophoromycota	Cercozoa
Plasmodiophoromycetes	Phytomyxea
EUMYCOTA	FUNGI
Chytridiomycota	Archemycota
Chytridiomycetes	Chytridiomycetes
	Allomycetes
Zygomycota	
Zygomycetes	Zygomycetes
	Bolomycetes
	Glomomycetes[b]
Trichomycetes	Enteromycetes
	Zoomycetes[c]
	Microsporidia
	Minisporea
	Microsporea
Ascomycota	Ascomycota
Saccharomycetes	Taphrinomycetes
	Geomycetes
	Endomycetes
Plectomycetes	Plectomycetes
Hymenoascomycetes	Discomycetes
	Pyrenomycetes
Loculoascomycetes	Loculomycetes
Basidiomycota	Basidiomycota
Urediniomycetes	Septomycetes
Ustilaginomycetes	Ustomycetes
Heterobasidiomycetes	Gelimycetes[b]
Homobasidiomycetes	Homobasidiomycetes

[a]Artificial taxon
[b]Probably paraphyletic
[c]Includes Zygomycetes, Ascomycetes, and Trichomycetes

The Mycota, Vol. VII, includes treatments of the systematics and related topics of the Eumycota and Pseudomycota as well as specialized chapters on nomenclature, techniques, and evolution. Certain groups are not treated in this volume: the Labyrinthulomycetes (Pseudomycota) and the slime molds. The evolutionary position of the slime molds has been controversial. Recent evidence suggests that most slime molds are more closely related to the Eumycota than previously believed (S.L. Baldauf and W.F. Doolittle, 1997, Proc Natl Acad Sci USA 94:12007), and they should continue to be of interest to those who study fungi for both ecological and phylogenetic reasons.

Chapters 2 to 4, Vol. VII, Part A, cover the Pseudomycota, and Chaps. 5–14, Vol. VII A, and Chaps. 1–5, Vol. VII, Part B, the Eumycota. The Pseudomycota contains distantly related groups of fungi (Table 1). The Chytridiomycota and

Zygomycota are treated in one and two chapters, respectively, while the Ascomycota and Basidiomycota are treated in five or six chapters each, with separate chapters for yeasts in each phylum, although the yeasts are not monophyletic groups. Chapter 14, Vol. VII, Part A, discusses the special problems of anamorphic genera and their relationships to the teleomorphic genera and describes the attempts being made to incorporate anamorphs into modern phylogenetic systematics. In Chap. 6, Vol. VII, Part B, Hawksworth discusses the development of a unified system of biological nomenclature. Chapters 7 and 8, Vol. VII, Part B, deal with techniques for cultivation and data analysis, respectively. The final two chapters in Vol. VII, Part B, consider speciation and molecular evolution.

The Mycota, Vol. VII, was originally intended to have been Vol. I in the series. Several changes in editors and the unfortunate death of Paul Lemke delayed its production. Added to these difficulties was the fact that these are tumultuous times in systematics because of the rapid development of molecular and phylogenetic analysis techniques and the explosive accumulation of data. As these techniques and new data are more broadly incorporated into systematics, a more stable and useful classification of the fungi will result.

We thank Heather J. Olson for her substantial efforts in compiling the indices.

St. Paul, MN
Minneapolis, MN
April 2000

David J. McLaughlin
Esther G. McLaughlin
Volume Editors

Contents

List of Contributors

M. Catherine Aime
Department of Botany and Plant Pathology, Purdue University, West Lafayette, IN, 47907, USA

Robert Bauer
Fachbereich Biologie, Eberhard-Karls-Universität Tübingen, Auf der Morgenstelle 1, D-72076 Tübingen, Germany

Universität Tübingen, Organismische Botanik, Auf der Morgenstelle 1, 72076 Tübingen, Germany

Institut für Evolution und Ökologie, Universität Tübingen, Auf der Morgenstelle 1, 72076 Tübingen, Germany

G. W. Beakes
School of Biology, Newcastle University, Newcastle upon Tyne NE1 7RU, UK

James J. Becnel
USDA, ARS, CMAVE, 1600 S.W. 23rd Drive, Gainesville, FL, 32608, USA

D. Begerow
Ruhr-Universität Bochum, Geobotanik, 44801 Bochum, Germany

Gerald L. Benny
Department of Plant Pathology, University of Florida, 1453 Fifield Hall, Gainesville, FL, 32611-0680, USA

Mary L. Berbee
Department of Botany, University of British Columbia, Vancouver, BC, Canada V6T 1Z4

M. Binder
Centraalbureau voor Schimmelcultures, Uppsalalaan 8, Utrecht 3584 CT, Netherlands

James P. Braselton
Department of Environmental and Plant Biology, Ohio University, Athens, OH, 45701, USA

Simon Bulman
The New Zealand Institute for Plant and Food Research Limited, Gerald Street, Lincoln 7608, Canterbury, New Zealand

Elizabeth S. Didier
Division of Microbiology, Tulane National Primate Research Center, 18703 Three Rivers Road, Covington, LA, 70433, USA

A.J. Giachini
Departamento de Microbiologia e Parasitologia, Centro de Ciencias Biologicas, Universidade Federal de Santa Catarina, Caixa Postal 476, 88040 900 Florianopolis, SC, Brazil

D.S. Hibbett
Biology Department, Clark University, Worcester, MA, USA

D. Honda
Faculty of Science and Engineering, Department of Biology, Konan University, 8-9-1, Okamoto, Higashinada, Kobe, Hyogo, 658-8501, Japan

K. Hosaka
Department of Botany, National Museum of Nature and Science, 4-1-1 Amakubo, Tsukuba, Ibaraki, 305-0005, Japan

Richard A. Humber
USDA-ARS Biological Integrated Pest Management Unit, R.W. Holley Center for Agriculture and Health, 538 Tower Road, Ithaca, NY, 10453, USA

Timothy Y. James
Department of Ecology and Evolutionary Biology, University of Michigan, 830 N. University, Ann Arbor, MI, 48109, USA

A. Justo
Biology Department, Clark University, Worcester, MA, USA

R. Kellner
Max Planck Institute for Terrestrial Microbiology, Karl von Frisch Strasse 10, 35043 Marburg, Germany

M. Kemler
Centre of Excellence in Tree Health Biotechnology, Forestry and Agricultural Biotechnology Institute (FABI), University of Pretoria, Pretoria 0002, South Africa

Michael L. Kent
Departments of Microbiology and Biomedical Sciences, Oregon State University, 220 Nash Hall, Corvallis, OR, 97331, USA

E. Larsson
Department of Biological and Environmental Sciences, University of Gothenburg, Box 100, S-405 30, Gothenburg, Sweden

K.H. Larsson
Natural History Museum, P.O. Box 1172 Blindern, 0318 Oslo, Norway

J.D. Lawrey
Environmental Science and Policy Department, George Mason University, 4400 University Drive, Fairfax, VA, USA

Peter M. Letcher
Department of Biological Sciences, The University of Alabama, Tuscaloosa, AL, 35487, USA

W. Wallace Martin
Department of Biology, Randolph-Macon College, Ashland, VA, 23005, USA

David J. McLaughlin
Department of Plant Biology, University of Minnesota, St. Paul, MN, 55108, USA

O. Miettinen
Biology Department, Clark University, Worcester, MA, USA

Botanical Museum, University of Helsinki, P.O. Box 7, Helsinki 00014, Finland

L. Nagy
Biology Department, Clark University, Worcester, MA, USA

R.H. Nilsson
Department of Biological and Environmental Sciences, University of Gothenburg, Box 100, S-405 30, Gothenburg, Sweden

Franz Oberwinkler
Fachbereich Biologie, Eberhard-Karls-Universität Tübingen, Auf der Morgenstelle 1, D-72076 Tübingen, Germany

Universität Tübingen, Institut für Evolution und Ökologie, Auf der Morgenstelle 1, D-72076 Tübingen, Germany

Universität Tübingen, Organismische Botanik, Auf der Morgenstelle 1, 72076 Tübingen, Germany

Teresita M. Porter
Department of Biology, McMaster University, Hamilton, ON, Canada L8S 4K1

Martha J. Powell
Department of Biological Sciences, The University of Alabama, Tuscaloosa, AL, 35487, USA

D. Redecker
Université de Bourgogne, UMR1347 Agroécologie, 17 rue Sully, BP 86510, Dijon F-21000, France

José Paulo Sampaio
Centro de Recursos Microbiológicos, Departamento de Ciências da Vida, Universidade Nova de Lisboa, 2829-516 Caparica, Portugal

Justin L. Sanders
Departments of Microbiology and Biomedical Sciences, Oregon State University, 220 Nash Hall, Corvallis, OR, 97331, USA

A. M. Schäfer
Ruhr-Universität Bochum, Geobotanik, 44801 Bochum, Germany

A. Schüßler
Department Biology I, Genetics, Ludwig-Maximilians-University Munich, Großhaderner Str. 2-4, 82152 Martinsried, Germany

Steven L. Stephenson
Department of Biological Sciences, University of Arkansas, Fayetteville, AR, 72701, USA

John W. Taylor
Plant and Microbial Biology, University of California, 111 Koshland Hall, Berkeley, CA, 94720-3102, USA

M Thines
Centre for Integrative Fungal Research (IPF), Biodiversity and Climate Research Centre (BiK-F), Goethe University Frankfurt am Main, 60325 Frankfurt am Main, Germany

Department of Biological Sciences, Institute of Ecology, Evolution and Diversity, and Senckenberg Gesellschaft für Naturforschung, Senckenberganlage 25, 60325 Frankfurt am Main, Germany

R.G. Thorn
Department of Biology, University of Western Ontario, 1151 Richmond St. North, London, ON, Canada N6A 5B7

Merje Toome
Department of Botany and Plant Pathology, Purdue University, West Lafayette, IN, 47907, USA

Kerstin Voigt
Institute of Microbiology, Fungal Reference Centre, University of Jena, Neugasse 07743 Jena, Germany

Michael Weiß
Fachbereich Biologie, Eberhard-Karls-Universität Tübingen, Auf der Morgenstelle 1, D-72076 Tübingen, Germany

Fachbereich Biologie, Universität Tübingen, Auf der Morgenstelle 5, 72076 Tübingen, Germany

Louis M. Weiss
Departments of Medicine and Pathology, Albert Einstein College of Medicine, Room 504 Forchheimer, 1300 Morris Park Avenue, Bronx, NY, 10461, USA

A. Yurkov
Leibniz Institute DSMZ-German Collection of Microorganisms and Cell Cultures, Inhoffenstraße 7B, 38124 Braunschweig, Germany

1 Fungi from PCR to Genomics: The Spreading Revolution in Evolutionary Biology

John W. Taylor[1], Mary L. Berbee[2]

CONTENTS

I. PCR to Genome Sequencing and a Robust Phylogeny for Fungi

In this volume, the fruits of about two decades of molecular systematics are presented. The development of the polymerase chain reaction (PCR) in the late 1980s (Saiki et al. 1988) made it possible for systematists to become molecular phylogeneticists. Nowhere was the need for additional characters more acute than with fungi, and mycologists responded with pioneering work, none more influential than the development of ribosomal DNA primers, which have been used far beyond Kingdom Fungi (White et al. 1990). Soon thereafter were made fundamental discoveries about the extent of the monophyletic fungal kingdom and the deep divergences in the kingdom (Berbee and Taylor 1992; Bruns et al. 1992; Swann and Taylor 1993). Perhaps the most important discovery from these early investigations was that animals and fungi shared a more recent common ancestor than either did with plants (Wainright et al. 1993). **Animals and fungi,** plus related protists, **together constitute the opisthokonts** (Lang et al. 2002; Steenkamp et al. 2006) (Fig. 1.1). The single posterior flagellum that gives the opisthokonts their name is, however, retained in only a few clades of fungi that disperse in water (Fig. 1.1, *Rozella* and allied Rozellomycota=Cryptomycota [James and Berbee 2012; Jones et al. 2011; Lara et al. 2010], Chytridiomycota Monoblepharidomycota, and Neocallimastiogomycota [see Powell and Letcher 2014], Blastocladiomycota [see James et al. 2014], *Olpidium* [see Benny et al. 2014]).

Once the phylogeny is inferred, however, the biologically interesting fun begins—**unraveling the evolution of phenotype.** In fact, the inaugural use of PCR-amplified DNA sequencing was to infer the evolution of the closed fruiting body of a false truffle from its mushroom ancestors (Bruns et al. 1989). Once phylogenies became broadly inclusive, work began on the evolution

[1]Plant and Microbial Biology, University of California, 111 Koshland Hall, Berkeley, CA 94720-3102, USA; e-mail: jtaylor@berkeley.edu
[2]Department of Botany, University of British Columbia, Vancouver, BC, Canada V6T 1Z4

Systematics and Evolution, 2nd Edition
The Mycota VII Part A
D.J. McLaughlin and J.W. Spatafora (Eds.)

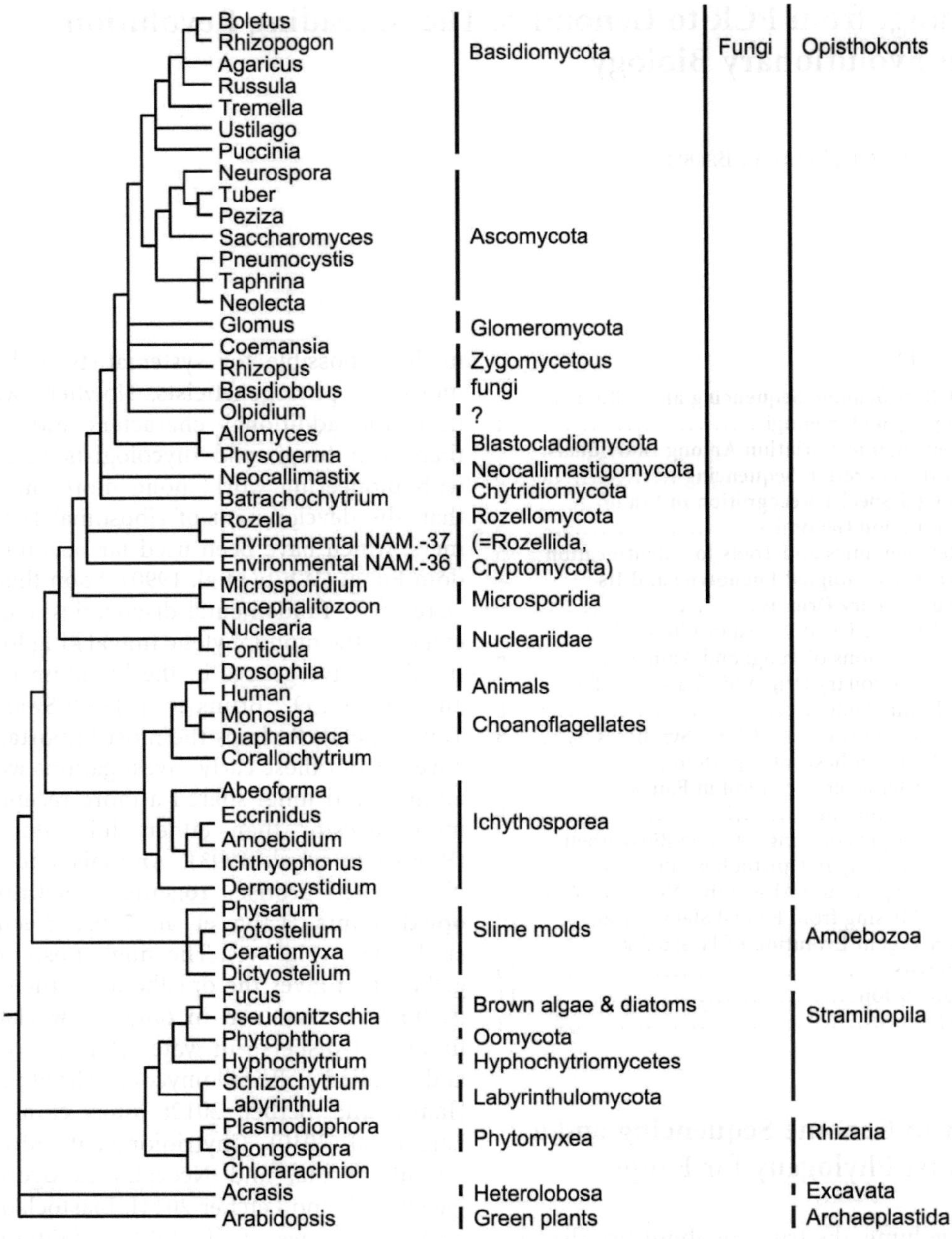

Fig. 1.1 Fungi and funguslike organisms covered in Mycota span the eukaryotic diversity diagrammed in this tree, which is based on the Tree of Life Web project (2012) and on references cited in the text

of key phenotypes, such as lichenization (Lutzoni and Pagel 1997), mycorrhizal association (Hibbett et al. 2000), parasitism (Vogler and Bruns 1998), and wood decay (Eastwood et al. 2011). Almost as soon, attempts to fit the new phylogenetic trees to the geologic time scale were made, making it possible to match evolutionary events in fungi with those in plants and animals, albeit not without controversy (Berbee and Taylor 1993, 2010; Casadevall 2005; Heckman et al. 2001; Simon et al. 1993). As with attempts to fit molecular phylogenies to geological time scales, attempts to fit phenotypes to phylogenies can be controversial, and conclusions may change as the data sets expand (Hibbett and Matheny 2009; Schoch et al. 2009).

The production of the automated sequencer accelerated the pace of molecular phylogenetics and enabled the first sequence-based studies of **fungal population genetics**. These efforts uncovered cryptic sex (Burt et al. 1996), cryptic species (Koufopanou et al. 1997), and cryptic populations (Fisher et al. 2001). The discovery that the average fungal morphospecies embraced two or more genetically differentiated phylogenetic species improved our understanding of fungal diversity and helped put an end to the notion that all microbial species have global distributions (Taylor et al. 2006). The discovery that fungi are recombining even though we mycologists have not caught them in the act put an end to the notion that a fifth of Kingdom Fungi was asexual (Taylor et al. 1999).

The first eukaryote to have a fully sequenced genome was a fungus, *Saccharomyces cerevisiae* (Goffeau et al. 1996). When the first human genomes were fully sequenced, the large centers that developed to provide the data suddenly had excess capacity and fungi became favored organisms. First, the Institute for Genomic Research (now the J. Craig Venter Institute) and the Broad Institute focused on human pathogens, but soon nonpathogens were sequenced, many of them through the Community Sequencing Project at the Joint Genome Institute. Kingdom Fungi is now the most deeply sequenced eukaryotic kingdom. It is probably impossible to keep abreast of the progress, and we hesitate to list any numbers because they will be hopelessly out of date before this volume is released. As of March 2012, more than 790 fungal genome projects were listed at GOLD (2012), none of which was zygomycetous. A shorter list at Fungal Genomes (2012) included 7 zygomycetous genera, which pushes the number to 800, a total nearly twice that of cordates, land plants, or Archaea. This wealth of data has stimulated **phylogenomics** (Fitzpatrick et al. 2006), where the task has now become to select the best genes for phylogenetic analysis (Townsend 2007) from among an embarrassing wealth of data (Rokas et al. 2005).

II. Peering into Variation Among Individuals: Next-Generation Sequencing

Technologically, the next best thing to emerge was next-generation sequencing, and again fungi are leading the way. The older Sanger sequencing allowed determination of bases of one DNA template at a time, and thousands of separate sequencing reactions were needed to completely sequence a genome. Next-generation sequencing simultaneously determines bases of a huge population of DNA fragments and can, in one lane of a single Illumina run, completely sequence a genome. The population genomics of yeast (Liti et al. 2009), human pathogens (Neafsey et al. 2010), and a model filamentous fungus, *Neurospora* (Ellison et al. 2011), have shown that genetically differentiated fungal populations can be discovered at very young ages, an order of magnitude earlier than even cryptic species, posing yet another challenge to the very definition of a species. Now that sequencing genomes of novel fungi is within the budget of an average research grant, the ultimate data for phylogenetics and molecular systematics are at hand, and the bottleneck has become the computational skills to make use of it.

Not only systematists but other biologists as well have been making use of phylogenetic results. As a result, all fields of mycology have been brought closer together because developmental biologists, industrial microbiologists, and ecologists all use the fruits of **phylogenetics and now phylogenomics as basic tools for tasks as disparate as gene cloning, improving enzyme production, and documenting the diversity of fungal communities.** No field has benefited more from the companionship with molecular systematics than fungal ecology. Mycorrhizal studies led the way, with the startling revelation that species lists from surveys of fruiting bodies bore little relationship to the species actually on the mycorrhizal roots (Horton and Bruns 2001). Even the field of fungal endophytes, which began its rapid emergence based on cultured fungi, has benefited from molecular identification (Arnold and

Lutzoni 2007) and now environmental sequencing (Jumpponen and Jones 2009). Sequencing of environmental DNA from bulk extracts from soil, air, water, or other mixed sources has greatly expanded our knowledge of clades at all taxonomic levels, from species (Suh et al. 2004) all the way up to classes in the Ascomycota (Schadt et al. 2003). For the very deep *Rozella* clade, **environmental sequencing** has shown that one cultivated member is accompanied by many other, previously unknown but highly divergent, species (Jones et al. 2011; Lara et al. 2010). Next-generation sequencing is dramatically enlarging the scope of such studies; witness the study of indoor air fungi that sampled 72 sites on all 6 habitable continents and used large subunit ribosomal DNA sequences to infer the presence of nearly 4,500 fungal species, all without a single culture (Amend et al. 2010).

Study of the evolution of phenotype in fungi has focused on ancient divergences, but next-generation sequencing is making it possible to study adaptation following the most recent divergences. In fact, next-generation sequencing of fungal populations looks to complete the amalgamation of development, evolution, and ecology through the common goal of understanding adaptive phenotypes, this time at the level of genomes.

Understanding the **mechanics of speciation** is suddenly a tractable problem. Dettman et al. (2007) provided experimental evidence for a step in the speciation process, showing that divergent selection may lead to partial reproductive isolation. They applied divergent selection to *S. cerevisiae*, experimentally creating lineages tolerant of either high salt or low glucose. After 500 generations of selection, strains' mitotic growth had improved in each selective environment. However, when crossed, the hybrids between the high-salt and low-glucose lineages had reduced meiotic efficiency. Using next-generation sequencing, Anderson et al. (2010) then tracked down genes related to increased success in each selective environment, including a gene for a proton efflux pump and for a regulator of mitochondrial protein synthesis. However, when alleles of these two genes that were favorable under opposite selective regimes were combined in the same strain under low-glucose conditions, the consequence was reduced meiotic fitness. If divergent selection leads to reduction in meiotic competence in the laboratory, it may also lead to speciation in nature.

A first stab at detecting genes associated with adaptation and speciation in natural populations has been provided by a study of *Neurospora*, where genomes of 50 individuals from one clade of *N. crassa* revealed two recently diverged populations, one tropical and the other subtropical (Ellison et al. 2011). Comparison of the genomes identified regions of exceptional divergence, in which were found candidate genes that suggested adaptation to cold temperature [an RNA helicase (Hunger et al. 2006) and prefoldin (Geissler et al. 1998)] and differences in light periods [the major circadian oscillator, *frq* (Aronson et al. 1994)]. Comparisons of fitness of wild isolates against those from which the candidate genes had been deleted (Dunlap et al. 2007) failed to reject hypotheses for adaptation to cold shock involving the RNA helicase and the prefoldin. Hypotheses linking these specific genes to adaptation can be further challenged by swapping alleles among individuals from the two populations and by examining other fungi whose populations are also separated by latitudinal gaps. This fungal study of adaptation is different from those previously conducted with animals or plants in that the genetically isolated populations were cryptic, there was no obvious candidate environmental parameter, such as light sand versus dark lava or normal soil versus serpentine soil, and there was no obvious candidate adaptive phenotype, such as mammal coat color (Nachman et al. 2003) or plant growth on serpentine soils (Turner et al. 2010). The "reverse ecological" approach to associating phenotype and genotype detailed for *Neurospora* may prove to be very powerful for natural populations of fungi where a priori identification of adaptive phenotypes can be difficult. Lacking prior bias, this approach also may offer surprises even in systems where candidate phenotypes have been selected.

III. Fungal Species Recognition in Era of Population Genomics

Population genomics has added new complexity to the task of recognizing fungal species. As noted earlier, **phylogenetic species recognition has replaced morphological species recognition** as the method of choice because fungal species typically become genetically isolated in nature long before mycologists can recognize any morphological difference (Cai et al. 2011; Giraud et al. 2008; Taylor et al. 2000). Phylogenetic species recognition has relied on the concordance of several gene genealogies, as described for *Neurospora* species (Dettman et al. 2003a) by an approach that showed good correlation with biological species recognition (Dettman et al. 2003b). However, as described in Sect. II, population genomic analysis of individuals from just one of three *N. crassa* clades revealed that it contained two, genetically distinct, populations (Ellison et al. 2011). Growth rates for the two populations showed a significant difference at low temperature, indicating that a phenotypic difference had evolved between the two populations (Ellison et al. 2011). Individuals from the two populations mate successfully in the lab, but analysis of the population genomic data indicates that intrapopulation gene flow is too low to reverse the genetic differentiation seen between the two populations (Ellison et al. 2011). Low gene flow between species that can be mated in the lab suggests that there is an extrinsic barrier to reproduction in nature. To pose the obvious rhetorical question, if genetic isolation in nature is the criterion for species recognition, should these populations be considered different species? To be sure, **species recognition by population genomics** sounds impossibly impractical today, but it might be worth remembering that species recognition by concordance of gene genealogies sounded impossibly impractical in 1997.

IV. Metagenomics and Tools for Identification

Challenges remain in interpreting the data from environmental and cultivation-independent study of fungi. No one knows how many fungal species exist, but sequencing of environmental DNA will definitely improve the accuracy of the estimate (Hawksworth 2001). To count species or to correlate fungal species across studies requires the development of **sequence identification tools** beyond GenBank. This need has arisen because databases are not populated with enough **vouchered sequences** to permit identification of a majority of environmental sequences. For example, 35 % of the ribosomal internal transcribed spacer sequences shared among the international databases GenBank, EMBL, and DDBJ were not assignable to a named taxon, and only 21 % of ITSs associated with a named taxon were also tied to a vouchered specimen (Ryberg et al. 2009). Interestingly, the rate of deposition of new fungal sequences from the environment now exceeds the deposition of sequences from fungi tied to specimens or cultures, a phenomenon that raises an important nomenclatural challenge (Hawksworth 2001; Hibbett et al. 2011). Sequences from fungi in herbaria and culture collections can be added to the database, and, even if completed for only a fraction of fungi (Hibbett et al. 2011; Jumpponen and Jones 2009), these provide an important link with environmental samples. One point to keep in mind is that, although almost all ecological studies find a very large number of total taxa, the number of commonly encountered taxa can be much smaller, e.g., only 31 common species were found among the 4,500 detected in the aforementioned study of indoor air (Amend et al. 2010). A second problem concerns the accuracy of sequences already in the international databases, of which as many as 20 % are misidentified (Bidartondo et al. 2008;

Nilsson et al. 2006). To avoid compromising its relationship with the original depositors, GenBank refuses annotation by any third party who notices a problem with a sequence. This contrasts with herbaria, which routinely welcome annotations, such as reidentifications from other researchers, contributing to the overall reliability of their data. Imagine a herbarium where no one but the original collector could annotate a specimen, and you can appreciate the problem with the genetic databases. The solution? Eliminating all so-called bad sequences may be politically impossible, but specifically designating good ones is perfectly feasible. The UNITE database provides sequences of carefully identified mycorrhizal fungi (Koljalg et al. 2005). GenBank and the Barcode of Life Database are currently creating curated, public databases of correctly identified sequences where third-party annotations will be invited and quality standards are carefully enforced (Schoch and Seifert 2010).

Having a sequence database, the next step is **developing tools for automatic identifications.** For prokaryotes, automated, reliable identification tools for environmental sequences have revolutionized microbial ecology. Identification is based on (1) a database of curated, correctly identified sequences and (2) a publicly available mechanism for matching environmental sequences to the database, then returning identifications to users. Three Web-based services (Greengenes 2012; Ribosomal Database Project 2012; Silva 2012), each with a slightly different approach (Schloss 2009), provide prokaryotic classifications and are beginning to provide classifications for fungi as well. Fungi lag behind prokaryotes due in part to the ease of aligning the 16S rRNA genes at the core of bacterial identification systems compared to the difficulty in aligning beyond the level of genus or family the more variable ribosomal internal transcribed spacer regions used to identify fungal species. Neither system is perfect. Bacteria are easier to place in a robust phylogeny, but each 16S OTU (operational taxonomic unit) harbors many genetically isolated species (Vos and Velicer 2008; Whitaker et al. 2003); fungi are more easily identified as species-level taxa, but new, divergent sequences may be impossible to link to genera or even families. Some successful fungal identification databases are therefore genus specific and targeted toward large, economically important genera including *Trichoderma* (TrichOKEY 2 2011; Druzhinina et al. 2005) and *Fusarium* (Park et al. 2011).

V. What Is a Fungus? Phenotype and Its Evolutionary Origins

A. Discoveries of Protistan Allies Affect Definitions of Fungi and Animals

Bringing us closer to the "holy grail" of understanding the evolution of complex organisms with differentiated tissues were discoveries of early diverging protists at the boundary between Kingdom Fungi (Brown et al. 2009; Steenkamp et al. 2006; Zettler et al. 2001) and Kingdom Animalia (Marshall and Berbee 2011; Mendoza et al. 2002). Because these boundary protists evolved before the origin of classical kingdom-level characters, their morphology had been a poor predictor of their relationships. In terms of morphology, perhaps the time has come to invoke Bruns' law: "There are no [expletive deleted] synapomorphies; get over it."

Molecular phylogenetics, by accommodating protists that show few fungal or animal traits, provides a simple alternative to defining taxa while offering a framework for exploring how their characters evolved (James et al. 2006). Phylogenetics linked the multicellular animals (Metazoa) and their protist allies together as Holozoa (Lang et al. 2002) within the supergroup Opisthokonta (Fig. 1.1). Protist Holozoa include choanoflagellates, or collar flagellates (Steenkamp et al. 2006), and enigmatic arthropod commensals *Amoebidium* and *Eccrinidus*, which were once considered Trichomycetes (zygomycetous fungi) and are now placed in the Ichthyosporeans or Mesomycetozoea on the animal lineage (Benny and O'Donnell 2000; Cafaro 2005; Mendoza et al. 2002).

Fungi and their unicellular relatives are classified together in the Holomycota, the sister group to the Holozoa within opisthokonts (Liu et al. 2009). Unicellular Holomycota include peculiar amoebae in *Nuclearia* and an

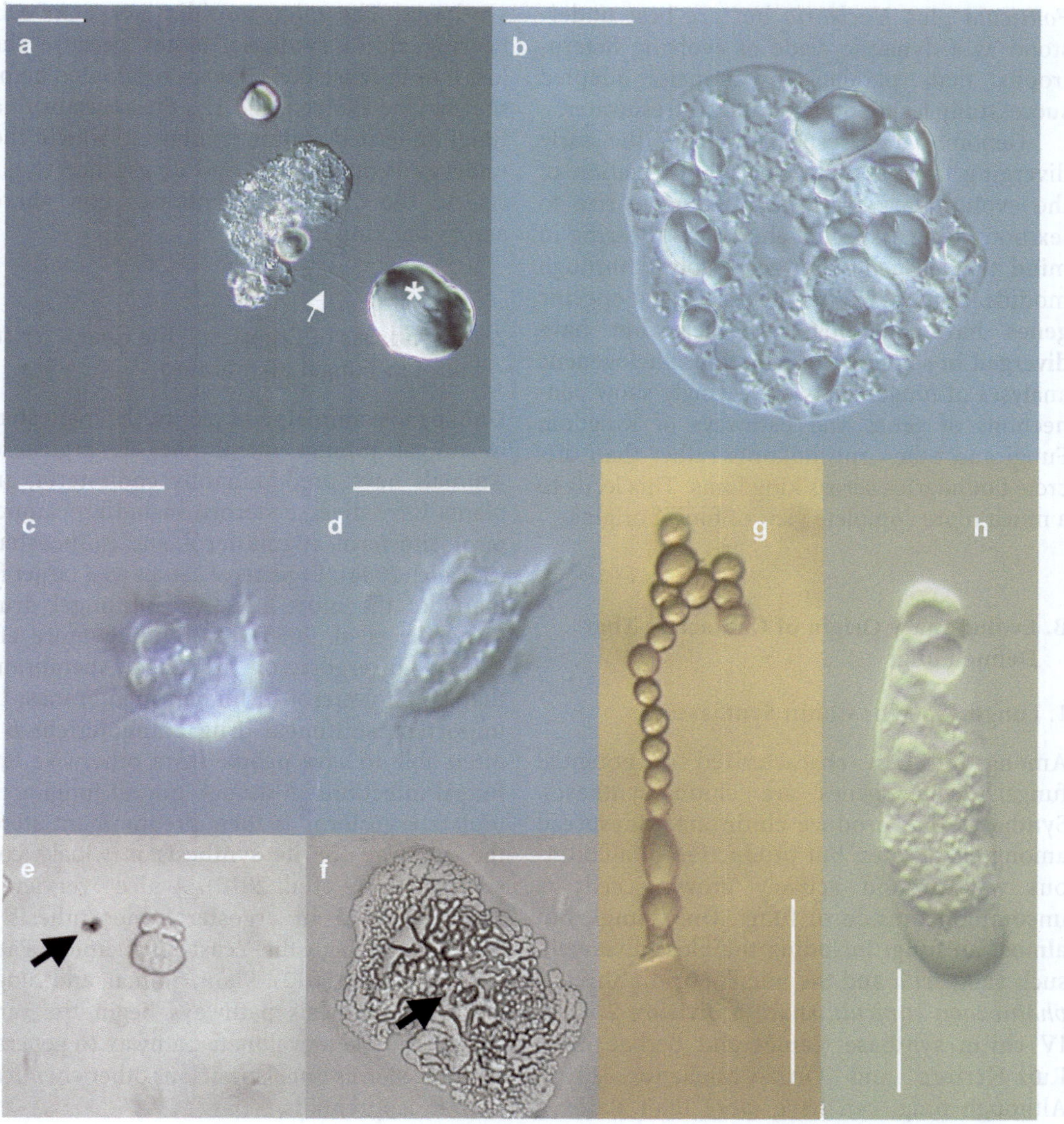

Fig. 1.2 Some of these amoeboid protists that lie phylogenetically outside of Kingdom Fungi may resemble the earliest members of the fungal lineage. *Nuclearia thermophila* (**a, b**) and *Fonticula alba* (**c, d**) are members of the sister group to Kingdom Fungi. (**a**) *N. thermophila* contacting a grain of flour (*asterisk*) using fine filose pseudopodia (*arrow*). (**b**) A *N. thermophila* amoeba that engulfed many flour particles. (**c, d**) *F. alba* amoebae with filose pseudopodia. (**e, f**) A living plasmodium of *Abeoforma whisleri* (Ichthyosporea, protist members of animal lineage) that grew from (**e**) to (**f**) in 25 h. The *arrow* indicates a particle of debris marking the same spot in both images. (**g, h**) *Acrasis helenhemmesae* is a recently discovered species in a genus once considered a social slime mold. It is in the Excavata along with the photosynthetic flagellate *Euglena* and is only distantly related to familiar social slime molds in *Dictyostelium*. (**g**) Spore production. (**h**) Amoeba. Scale bars: (**a, b**) 25 μm; (**c, d, h**) 10 μm; (**e, f**) 100 μm; (**g**) 50 μm. Photo credits: (**a**), (**b**) from Yoshida et al. (2009); (**c**), (**d**) from Brown et al. (2009); (**e**), (**f**) from Marshall and Berbee (2011); (**g**), (**h**) from Brown et al. (2010)

aberrant social slime mold, *Fonticula alba* (Figs. 1.1 and 1.2; Liu et al. 2009). Within Holomycota, phyla from Basidiomycota through Microsporidia together constitute a monophyletic group (Fig. 1.1). Fungi are therefore easily defined phylogenetically as the sister group to

Fonticula plus *Nuclearia* but are best understood as a dynamic clade of evolving heterotrophs that, parallel to animals, adapted successfully to life on land and in freshwater.

Genomic data, especially from the early diverging taxa, allow a closer appreciation of the evolutionary processes that gave rise to textbook fungal-specific characters. Bearing in mind that all genes but one evolved through modifications of earlier genes, **fungal-specific genes have homologs elsewhere but have diverged in sequence or function.** Phylogenetic analyses of fungal traits increasingly show connections of genes and pathways of Kingdom Fungi and other opisthokonts rather than discrete boundaries across kingdoms. This leads to a much more complete view of fungal origins.

B. Evolutionary Origin of Characters That Define Fungi

1. Fungus-Specific Chitin Synthases

Among the best characterized of potential fungal-specific genes are chitin synthases. Synthases that produce chitin are widespread among eukaryotes, but production of a chitinous wall around actively growing cells is uncommon outside of fungi. **Only fungi, but almost all fungi** including the most divergent, such as *Rozella* and the microsporidium *Encephalitozoon cuniculi*, **share a division 2, class IV chitin synthase** (James and Berbee 2012; Ruiz-Herrera and Ortiz-Castellanos 2010). Although fungi can have more than a dozen chitin synthases, this particular enzyme is implicated in the synthesis of the bulk of the chitin in cell walls (Munro and Gow 2001). Fungi also share one or more additional division 2 chitin synthases that bear a myosin domain at their N-terminal end (James and Berbee 2012; Ruiz-Herrera and Ortiz-Castellanos 2010). While the diatom *Thalassiosira pseudonana* once also seemed to share a myosin domain in a chitin synthase (Durkin et al. 2009), this was probably the result of an error in an early automated gene annotation. More recent gene predictions (e.g., GenBank XP_002295995) no longer show a myosin domain associated with the diatoms' chitin synthases. The microsporidia lack a myosin-bearing chitin synthase, either because they lost it or because their lineage originated before the enzyme evolved. Clearly, the ancestor of all fungi inherited chitin synthases, which then underwent duplication and divergence to give rise to the distinctive synthases now shared across the kingdom.

2. Biosynthesis of Ergosterol, the Characteristic Sterol in Fungal Membranes

Unlike most animals and plants, the main sterol in fungal plasma membranes is ergosterol. Animals have predominantly cholesterol, and plants have diverse sterols, including campesterol, sitosterol, stigmasterol, and isofucosterol (Schaller 2004). Ergosterol serves as a target for many of the most effective antifungal drugs (Francois et al. 2005). By binding more efficiently to ergosterol in fungal membranes than to cholesterol in human membranes, the important antifungal drug amphotericin B is often able to save people from otherwise fatal fungal infections. Although not all fungi accumulate ergosterol as their predominant sterol, the pathway for its synthesis is widely conserved (Weete et al. 2010). A nice overview of steps involved in ergosterol biosynthesis is available through the Yeast Biochemical Pathway Database (2012). Plant, animal, and fungal sterol biosynthesis pathways begin the same way, using the mevalonate pathway to generate not only sterols but also various other chemicals such as isoprenoids.

Although ergosterol is considered specific to fungi, the enzymes involved in biosynthesis of ergosterol all have close homologs in other organisms. To illustrate this point, we examined sterol 24-C-methyltransferase (EC 2.1.1.41) because differences at this enzymatic step help illustrate why fungi make ergosterol while plants and animals do not. In both animals and fungi, the biosynthetic pathway leading to sterol production proceeds to cyclization of squalene-2,3-oxide producing a lanosterol intermediate, which in fungi and animals is converted to zymosterol, the substrate for sterol 24-C-methyltransferase. The enzyme in *S. cerevisiae* and, presumably, other fungi (such as the chytrid *Batrachochytrium dendrobatidis* GenBank EGF84453) adds a methyl group to the tetracyclic sterol zymosterol, converting it to fecosterol (Parks et al. 1995). In most multicellular animals, the homologous enzyme lacks the sterol methyltransferase function so that animals cannot add the extra methyl group to carbon-24 of the sterol side chain, as would be necessary for ergosterol biosynthesis (Kaneshiro 2002). However, looking more deeply into early Holozoa, a homolog that seems to have sterol binding sites is present in the sponge *Amphimedon queenslandica* (GenBank XP_003387525.1). Although the differences in sterols across kingdoms are important, they sometimes resulted from loss rather than gain of function, and even then, they resulted from relatively small genetic changes.

3. Origins of Fungal Lysine Biosynthetic Pathway in Opisthokont Prehistory

While most animals must ingest the essential amino acid lysine from their diets, fungi, plants, and bacteria synthesize their own lysine. Synthesis of lysine across all domains of life takes place through one of the three or more alternative, multienzyme pathways of independent evolutionary origin. Fungi synthesize lysine using the **alpha-aminoadipate pathway** (Vogel 1965). The fungal pathway requires seven enzymatic steps (see lysine biosynthesis, Yeast Biochemical Pathway Database 2012). A pathway that also uses an alpha-aminoadipate intermediate, but is of independent evolutionary origin, leads to lysine biosynthesis in the hyperthermophilic bacterium *Thermus thermophilus* and in an anaerobic archaebacterium *Pyrococcus horikoshii* (Nishida et al. 1999). As reviewed by Zabriskie and Jackson (2000), euglenoids also use alpha-aminoadipate as an intermediate in the synthesis of lysine, and although the genes and enzymes involved have yet to be studied, this group of photosynthetic or phagotrophic protists may also have an alpha-aminoadipate lysine biosynthesis pathway of independent origin. The remaining lysine synthesis pathway, the **diaminopimelic acid pathway**, is widely distributed among prokaryotes, protists, oomycetes, and plants (Torruella et al. 2009; Vogel 1961, 1965). Like the fungal pathway, the diaminopimelic acid pathway requires seven enzymatic steps. Remarkably, however, none of the fungal enzymes for lysine biosynthesis are homologous to any of the plant diaminopimelic acid pathway enzymes, and for this reason the alpha-aminoadipate pathway for lysine biosynthesis has been considered a unifying derived characteristic that helped link chytrids to fungi rather than to oomycetes (Vogel 1961).

Vogel's insights still hold; however, closer dissection of the evolutionary relationships of individual enzymes in the pathway, coupled with comparative analysis using new protist genomes, reveals an unexpectedly complex pattern of gene duplication, functional divergence, and loss (Irvin and Bhattacharjee 1998). The first enzymes in fungal biosynthesis are, at a deep level, distant homologs of Krebs' cycle enzymes, while the two last enzymes are related to proteins involved in lysine catabolism (Irvin and Bhattacharjee 1998; Nishida and Nishiyama 2000). Nishida and Nishiyama (2000) carefully tracked the phylogeny of alpha-aminoadipate reductase (EC 1.2.1.31), the fifth gene in the pathway, and, based on sequences available at the time proposed that this enzyme was specific to fungi. However, with new genome sequences, eukaryotic homologs to alpha-aminoadipate reductase were identified in *Corallochytrium limacisporum*, which, despite its name, is not a chytrid fungus but rather a protist that diverged early in the evolutionary history of opisthokonts (Sumathi et al. 2006). Other protists in Holozoa also share the enzyme, and it is even present outside of the opisthokonts, with a homolog in *Dictyostelium discoideum* (Amoebozoa) (Torruella et al. 2009). While the functions of nonfungal homologs to fungal alpha-aminoadipate pathway enzymes have yet to be tested biochemically, it seems likely that the pathway evolved before the Amoebozoa diverged from the opisthokonts (Fig. 1.1). **For fungi, the alpha-aminoadipate pathway for lysine synthesis is a shared primitive character**, and the absence of the pathway in animals represents an evolutionary loss.

4. Hyphae and Absorptive Nutrition Were Missing from Fungal Stem Lineage

The morphology of the opisthokont protists also offers clues into the evolution of fungi. *Fonticula* and *Nuclearia* retain an ancestral habit of ingesting bacteria or algae (Cavalier-Smith 2002). The common ancestor to fungi may have done likewise before the evolution of the multicellular plants and animals that are the nutrient sources of most extant fungi. Wall-less **amoeboid dispersal phases** are common among unicellular opisthokonts and may be an ancestral characteristic of the fungal stem lineage. Amoeboid phases occur in the few Ichthyosporea that grow well in culture (Figs. 1.1 and 1.2; Marshall and Berbee 2011; Whisler 1962) as well as in *Fonticula* (Brown et al. 2009) and *Nuclearia* (Figs. 1.1 and 1.2; Liu et al. 2009; Yoshida et al. 2009; Zettler et al. 2001). Even though Chytridiomycota (Fig. 1.1) seem to have lost the capacity for extensive amoeboid motion, at least some species retain genes encoding animal cell movement proteins (Harris 2011). The movement proteins may still play a role when, for example, zoospores squeeze out of a zoosporangium or are trapped between hyphae (Gleason and Lilje 2009).

Although not ancestral in the kingdom, other classical characters of modern Kingdom Fungi, such as **hyphae, a chitinous cell wall, reproduction by spores, and absorptive nutrition**, had evolved by the time plants colonized land, by 400 million years ago, based on fossil (Taylor et al. 2004) and phylogenetic (Berbee and Taylor 2001) evidence. They must have secreted enzymes across their walls to assimilate nutrients, and they reproduced with walled spores. Most terrestrial fungi lost the flagellated, wall-less zoospore stage of their aquatic predecessors. Suggesting that **flagellar loss** may have taken place convergently and after the origin of hyphae, remnants of what may be a centriole from an ancestral flagellum remain visible in the hyphal zygomycete *Basidiobolus* (Gull and Trinci 1974) and in *Coemansia reversa* (McLaughlin et al. unpublished). Another example of convergent loss of flagella involves *Olpidium*, a flagellated unicellular fungus that disperses by zoospores, clustered phylogenetically within the terrestrial fungi and among hyphal zygomycetes (Fig. 1.1; James et al. 2006; Sekimoto et al. 2011). By implying that the ancestor to *Olpidium* was both terrestrial and a flagellate, the phylogeny suggests that early hyphal fungi on land still reproduced by motile spores. Like animals and plants, including mosses and ferns, early terrestrial fungi may have retained swimming flagellated cells as a legacy of their aquatic past.

The Microsporidia (Didier et al. 2014) present a particular challenge to the definition of fungi because their genomes had evolved so rapidly that their phylogenetic history is all but obliterated (Koestler and Ebersberger 2011). They lack shared fungal characters and have no chitin during their assimilative stage, possibly due to derived loss. As obligate parasites, microsporidia cause diseases in animals from *Daphnia* to humans, and in the past they were studied by parasitologists or medical pathologists. More recently, they are catching the attention of evolutionary biologists interested in links between parasitism and rates and modes of evolution (Gill et al. 2010; Keeling et al. 2005).

V. Convergent Evolution of Funguslike Protists

Wisely, the editors and authors have cast their net widely to include not only Kingdom Fungi but also organisms that look or behave like fungi. **Funguslike organisms are found in at least four large clades in addition to the opisthokonts**: Straminopila, Rhizaria, Excavata, and Amoebozoa. Most importantly, socially, is the Straminopila (Beakes et al. 2014), home to the Oomycota, Labyrinthulomycota, and Hyphochytriomycetes. The Oomycota harbors the plant destroyers, literally, *Phytophthora* and relatives. The Labyrinthulomycota also has some plant parasites of grasses, but the Hyphochytriomycetes is, as far as we know, innocent of phytocide. Inspired in part by DeBary's studies of *Phytophthora* species, mycologists and plant pathologists came to value hypothesis-driven experimental research.

In an era when spontaneous generation was still considered a plausible explanation for the appearance of plant disease, DeBary (1863) left a lasting legacy of a higher standard of evidence by using careful observations and experimental inoculations to prove that *Phytophthora infestans* caused late blight of potatoes.

Distinguishing straminopiles from Kingdom Fungi drew on early analysis of cell wall chemistry (Bartnicki-Garcia 1968; Von Wettstein 1921), genetics (Barksdale 1966), biochemistry (Vogel 1965), and microscopy, with molecular phylogenetics offering definitive confirmation of the deep divergence (Gunderson et al. 1987). Von Wettstein (1921) pointed out the striking division between funguslike forms that had chitinous walls versus forms with cellulose walls. He interpreted their cellulose walls as evidence that oomycetes originated relatively recently from algae, in contrast to fungi with chitinous walls, which he felt were an older group of less easily identified origin. Clearly, the nonphotosynthetic straminopiles are related to photosynthetic brown algae and diatoms, and whether the funguslike clades lost chloroplasts or never had them in the first place is still debated (Stiller et al. 2009; Tsui et al. 2009). As in the opisthokonts and plants, the straminopiles evolved into a striking diversity of body plans and ecological functions.

Slime molds, defined by their creeping plasmodium or by social amoebae, span five phylogenetic clades, three in addition to *Fonticula* from the fungal lineage and Ichthyosporea (e.g., *Abeoforma whisleri*, Fig. 1.2) on the animal lineage. The Rhizaria (Bulman and Braselton 2014) include the green, photosynthetic amoeba *Chlororachnion* and two funguslike plant pathogens, *Plasmodiophor* and *Spongospora*, scourges of Brassicaceae and potato, respectively (Cavalier-Smith and Chao 2003). The Excavata (Stephenson 2014) is home to a social slime mold, *Acrasis* (Fig. 1.2; Brown et al. 2010). Excavata also includes *Euglena*, a flagellated green photosynthetic protist, and *Naegleria fowleri*, cause of amebic meningoencephalitis, a rare human disease. If a swimmer has the bad luck to take up a nose full of water containing *Naegleria* amoebae, the amoebae can migrate to and then infect the brain (Centers for Disease Control and Prevention 2012). Finally, the Amoebozoa (Stephenson 2014) includes slime molds of both the social, cellular type (Dictyosteliomycota) and the plasmodial type (Myxomycota) and relatives, such as *Ceratiomyxa* and *Protostelium*. Dictyosteliomycota are ubiquitous and serve as model systems for research as diverse as cell migration (Ridley et al. 2003) and evolutionary cheating (Strassmann et al. 2000). Myxomycota are also model systems for research as diverse as biological oscillation (Takamatsu et al. 2000) and maze solving (Nakagaki et al. 2000) as well as being stunningly beautiful (Emoto 1977).

VII. Conclusion

The chapters in these volumes detailing phylogenetic relationships of fungi and nonfungi set the stage for future studies of phenotype and adaptation on one hand and ecological diversity on the other. We predict that over the next decade fungi will be among the most attractive targets for research associating genotype and phenotype. Fungi are eukaryotes, but simple ones and with small genomes. Fungi seem to be able to adapt to every environmental parameter. Fungi have evolved many features typical of more complex eukaryotes, including self-/non-self-recognition and even chromosomes determining sexual identity (Heitman et al. 2007; Menkis et al. 2008). Fungi can reproduce both clonally and by recombination. Fungi exchange genes within populations by mating, among populations by introgression, and even among long diverged lineages by horizontal gene transfer (Inderbitzin et al. 2005, 2011; Mehrabi et al. 2011). It would not be surprising if hybridization and introgression among recently diverged populations and species proved to be as important to fungal adaptation as horizontal gene transfer is to bacteria (Juhas et al. 2009; Lacroix et al. 2006). Cultivated fungi are immortal, so experiments that require sacrificing an individual can be replicated with the same individual, a very different situation than with most plants or animals. Cultivated fungi are often haploid, so inbred lines are unnecessary for studies aimed at associating phenotypic and genotypic variation. Where they are not haploid, fungi often are dikaryotic and function as diploids with the advantage that the dikaryons can often be broken into their haploid components and studied independently. With all of these attributes, we hope that biologists will be attracted to fungi as organisms of choice for their studies aimed at understanding the evolution of phenotype in terms of their genomes.

The phylogenies will have an equally profound effect on studies of fungi in nature. Here, fungi known only from environmental nucleic acid sequences will dominate biodiversity and will likely be among the species most important to ecosystem function. One area where such studies may prove fruitful is in documenting the biological response to global change. Given that the diversity of fungi will far outweigh that of animals or plants in any given ecosystem, it seems likely that some of these fungi will be the best sentinels for recognizing the effects of global change. One can even imagine an automated means of assessing the presence of these sentinel fungi over a region where assessing the effects of global change was a priority.

Acknowledgments Financial support to J. Taylor was provided in part by National Institutes of Health NIH Grant R24 GM081597 and National Science Foundation (NSF) Grant DEB 05-16511 and to M. Berbee by NSF DEB-0732984, Assembling the Fungal Tree of Life, and from a Natural Sciences and Engineering Research Council of Canada Discovery grant.

References

Amend AS, Seifert KA, Samson R, Bruns TD (2010) Indoor fungal composition is geographically patterned and more diverse in temperate zones than in the tropics. Proc Natl Acad Sci U S A 107:13748–13753. doi:10.1073/pnas.1000454107

Anderson JB, Funt J, Thompson DA, Prabhu S, Socha A, Sirjusingh C, Dettman JR, Parreiras L, Guttman DS, Regev A, Kohn LM (2010) Determinants of divergent adaptation and Dobzhansky-Muller interaction in experimental yeast populations. Curr Biol 20:1383–1388. doi:10.1016/j.cub.2010.06.022

Arnold AE, Lutzoni F (2007) Diversity and host range of foliar fungal endophytes: are tropical leaves biodiversity hotspots? Ecology 88:541–549

Aronson BD, Johnson KA, Loros JJ, Dunlap JC (1994) Negative feedback defining a circadian clock – autoregulation of the clock gene frequency. Science 263:1578–1584

Barksdale AW (1966) Segregation of sex in progeny of a selfed heterozygote of Achlya bisexualis. Mycologia 58:802–804

Bartnicki-Garcia S (1968) Cell wall composition and other biochemical markers in fungal phylogeny. In: Harborn JB (ed) Phytochemical phylogeny. Academic, New York, NY, pp 83–101

Beakes GW, Honda D, Thines M (2014) Systematics of the straminipila: labyrinthulomycota, hyphochytriomycota and oomycota. In: McLaughlin DJ, Spatafora JW (eds) Systematics and evolution: Part A. Springer, Heidelberg

Benny GL, O'Donnell K (2000) Amoebidium parasiticum is a protozoan, not a Trichomycete. Mycologia 92:1133–1137

Benny GL, Humber RA, Voigt K (2014) Zygomycetous fungi: phylum entomophthoromycota and subphyla kickxellomycotina, mortierellomycotina, mucoromycotina, and zoopagomycotina. In: McLaughlin DJ, Spatafora JW (eds) Systematics and evolution: Part A. Springer, Heidelberg

Berbee ML, Taylor JW (1992) Two ascomycete classes based on fruiting-body characters and ribosomal DNA sequence. Mol Biol Evol 9:278–284

Berbee ML, Taylor JW (1993) Dating the evolutionary radiations of the true fungi. Can J Bot/Rev Can Bot 71:1114–1127

Berbee ML, Taylor JW (2001) Fungal molecular evolution: gene trees and geologic time. In: McLaughlin D, McLaughlin E, Lemke P (eds) The Mycota, Part B, vol VII. Springer, Berlin, pp 229–245

Berbee ML, Taylor JW (2010) Dating the molecular clock in fungi – how close are we? Fungal Biol Rev 24:1–16

Bidartondo MI, Bruns TD, Blackwell M, Edwards I, Taylor AFS, Horton T, Zhang N, Koljalg U, May G, Kuyper TW, Bever JD, Gilbert G, Taylor JW, DeSantis TZ, Pringle A, Borneman J, Thorn G, Berbee M, Mueller GM, Andersen GL, Vellinga EC, Branco S, Anderson I, Dickie IA, Avis P, Timonen S, Kjoller R, Lodge DJ, Bateman RM, Purvis A, Crous PW, Hawkes C, Barraclough T, Burt A, Nilsson RH, Larsson KH, Alexander I, Moncalvo JM, Berube J, Spatafora J, Lumbsch HT, Blair JE, Suh SO, Pfister DH, Binder M, Boehm EW, Kohn L, Mata JL, Dyer P, Sung GH, Dentinger B, Simmons EG, Baird RE, Volk TJ, Perry BA, Kerrigan RW, Campbell J, Rajesh J, Reynolds DR, Geiser D, Humber RA, Hausmann N, Szaro T, Stajich J, Gathman A, Peay KG, Henkel T, Robinson CH, Pukkila PJ, Nguyen NH,

Villalta C, Kennedy P, Bergemann S, Aime MC, Kauff F, Porras-Alfaro A, Gueidan C, Beck A, Andersen B, Marek S, Crouch JA, Kerrigan J, Ristaino JB, Hodge KT, Kuldau G, Samuels GJ, Raja HA, Voglmayr H, Gardes M, Janos DP, Rogers JD, Cannon P, Woolfolk SW, Kistler HC, Castellano MA, Maldonado-Ramirez SL, Kirk PM, Farrar JJ, Osmundson T, Currah RS, Vujanovic V, Chen WD, Korf RP, Atallah ZK, Harrison KJ, Guarro J, Bates ST, Bonello P, Bridge P, Schell W, Rossi W, Stenlid J, Frisvad JC, Miller RM, Baker SE, Hallen HE, Janso JE, Wilson AW, Conway KE, Egerton-Warburton L, Wang Z, Eastburn D, Ho WWH, Kroken S, Stadler M, Turgeon G, Lichtwardt RW, Stewart EL, Wedin M, Li DW, Uchida JY, Jumpponen A, Deckert RJ, Beker HJ, Rogers SO, Xu JAP, Johnston P, Shoemaker RA, Liu MA, Marques G, Summerell B, Sokolski S, Thrane U, Widden P, Bruhn JN, Bianchinotti V, Tuthill D, Baroni TJ, Barron G, Hosaka K, Jewell K, Piepenbring M, Sullivan R, Griffith GW, Bradley SG, Aoki T, Yoder WT, Ju YM, Berch SM, Trappe M, Duan WJ, Bonito G, Taber RA, Coelho G, Bills G, Ganley A, Agerer R, Nagy L, Roy BA, Laessoe T, Hallenberg N, Tichy HV, Stalpers J, Langer E, Scholler M, Krueger D, Pacioni G, Poder R, Pennanen T, Capelari M, Nakasone K, Tewari JP, Miller AN, Decock C, Huhndorf S, Wach M, Vishniac HS, Yohalem DS, Smith ME, Glenn AE, Spiering M, Lindner DL, Schoch C, Redhead SA, Ivors K, Jeffers SN, Geml J, Okafor F, Spiegel FW, Dewsbury D, Carroll J, Porter TM, Pashley C, Carpenter SE, Abad G, Voigt K, Arenz B, Methven AS, Schechter S, Vance P, Mahoney D, Kang SC, Rheeder JP, Mehl J, Greif M, Ngala GN, Ammirati J, Kawasaki M, Gwo-Fang YA, Matsumoto T, Smith D, Koenig G, Luoma D, May T, Leonardi M, Sigler L, Taylor DL, Gibson C, Sharpton T, Hawksworth DL, Dianese JC, Trudell SA, Paulus B, Padamsee M, Callac P, Lima N, White M, Barreau C, Juncai MA, Buyck B, Rabeler RK, Liles MR, Estes D, Carter R, Herr JM, Chandler G, Kerekes J, Cruse-Sanders J, Marquez RG, Horak E, Fitzsimons M, Doring H, Yao S, Hynson N, Ryberg M, Arnold AE, Hughes K (2008) Preserving accuracy in GenBank. Science 319:1616

Brown MW, Spiegel FW, Silberman JD (2009) Phylogeny of the "forgotten" cellular slime mold, Fonticula alba, reveals a key evolutionary branch within Opisthokonta. Mol Biol Evol 26:2699–2709. doi:10.1093/molbev/msp185

Brown MW, Silberman JD, Spiegel FW (2010) A morphologically simple species of Acrasis (Heterolobosea, Excavata), Acrasis helenhemmesae n. sp. J Eukaryot Microbiol 57:346–353. doi:10.1111/j.1550-7408.2010.00481.x

Bruns TD, Fogel R, White TJ, Palmer JD (1989) Accelerated evolution of a false-truffle from a mushroom ancestor. Nature 339:140–142

Bruns TD, Vilgalys R, Barns SM, Gonzalez D, Hibbett DS, Lane DJ, Simon L, Stickel S, Szaro TM, Weisburg WG, Sogin ML (1992) Evolutionary relationships within the Fungi: analyses of nuclear small subunit rRNA sequences. Mol Phylogen Evol 1:231–241

Bulman S, Braselton JP (2014) Rhizaria: phytomyxea. In: McLaughlin DJ, Spatafora JW (eds) Systematics and evolution: Part A. Springer, Heidelberg

Burt A, Carter DA, Koenig GL, White TJ, Taylor JW (1996) Molecular markers reveal cryptic sex in the human pathogen Coccidioides immitis. Proc Natl Acad Sci U S A 93:770–773

Cafaro MJ (2005) Eccrinales (Trichomycetes) are not fungi, but a clade of protists at the early divergence of animals and fungi. Mol Phylogenet Evol 35:21–34. doi:10.1016/j.ympev.2004.12.019

Cai L, Giraud T, Zhang N, Begerow D, Cai GH, Shivas RG (2011) The evolution of species concepts and species recognition criteria in plant pathogenic fungi. Fungal Diver 50:121–133. doi:10.1007/s13225-011-0127-8

Casadevall A (2005) Fungal virulence, vertebrate endothermy, and dinosaur extinction: is there a connection? Fungal Genet Biol 42:98–106. doi:10.1016/j.fgb.2004.11.008

Cavalier-Smith T (2002) The phagotrophic origin of eukaryotes and phylogenetic classification of Protozoa. Int J Syst Evol Microbiol 52:297–354

Cavalier-Smith T, Chao EEY (2003) Phylogeny and classification of phylum Cercozoa (Protozoa). Protist 154:341–358

Centers for Disease Control and Prevention (2012) http://www.cdc.gov/parasites/naegleria/. Cited 13 Apr 2012

DeBary A (1863) Recherches sur le développement de quelques champignons parasites. Annales des Science Naturelles Sér 4 Botanique 20:5–148, Pl. 1–13

Dettman JR, Jacobson DJ, Taylor JW (2003a) A multilocus genealogical approach to phylogenetic species recognition in the model eukaryote Neurospora. Evolution 57:2703–2720

Dettman JR, Jacobson DJ, Turner E, Pringle A, Taylor JW (2003b) Reproductive isolation and phylogenetic divergence in Neurospora: comparing methods of species recognition in a model eukaryote. Evolution 57:2721–2741

Dettman JR, Sirjusingh C, Kohn LM, Anderson JB (2007) Incipient speciation by divergent adaptation and antagonistic epistasis in yeast. Nature 447:585–588. doi:10.1038/nature05856

Didier ES, Becnel JJ, Kent ML, Sanders JL, Weiss LM (2014) Microsporidia. In: McLaughlin DJ, Spatafora JW (eds) Systematics and evolution: Part A. Springer, Heidelberg

Druzhinina IS, Kopchinskiy AG, Komon M, Bissett J, Szakacs G, Kubicek CP (2005) An oligonucleotide barcode for species identification in Trichoderma and Hypocrea. Fungal Genet Biol 42:813–828. doi:10.1016/j.fgb.2005.06.007

Dunlap JC, Borkovich KA, Henn MR, Turner GE, Sachs MS, Glass NL, McCluskey K, Plamann M, Galagan JE, Birren BW, Weiss RL, Townsend JP, Loros JJ, Nelson MA, Lambreghts R, Colot HV, Park G, Collopy P, Ringelberg C, Crew C, Litvinkova L, DeCaprio D, Hood HM, Curilla S, Shi M, Crawford M, Koerhsen M, Montgomery P, Larson L, Pearson M, Kasuga T, Tian CG, Bastuerkmen M, Altamirano L, Xu JH (2007) Enabling a community to dissect an organism: overview of the Neurospora functional genomics project. Adv Genet 57:49–96. doi:10.1016/s0065-2660(06)57002-6

Durkin CA, Mock T, Armbrust EV (2009) Chitin in diatoms and its association with the cell wall. Eukaryot Cell 8:1038–1050. doi:10.1128/ec.00079-09

Eastwood DC, Floudas D, Binder M, Majcherczyk A, Schneider P, Aerts A, Asiegbu FO, Baker SE, Barry K, Bendiksby M, Blumentritt M, Coutinho PM, Cullen D, de Vries RP, Gathman A, Goodell B, Henrissat B, Ihrmark K, Kauserud H, Kohler A, LaButti K, Lapidus A, Lavin JL, Lee YH, Lindquist E, Lilly W, Lucas S, Morin E, Murat C, Oguiza JA, Park J, Pisabarro AG, Riley R, Rosling A, Salamov A, Schmidt O, Schmutz J, Skrede I, Stenlid J, Wiebenga A, Xie XF, Kues U, Hibbett DS, Hoffmeister D, Hogberg N, Martin F, Grigoriev IV, Watkinson SC (2011) The plant cell wall-decomposing machinery underlies the functional diversity of forest fungi. Science 333:762–765. doi:10.1126/science.1205411

Ellison CE, Hall C, Kowbel D, Welch J, Brem RB, Glass NL, Taylor JW (2011) Population genomics and local adaptation in wild isolates of a model microbial eukaryote. Proc Natl Acad Sci U S A 108:2831–2836

Emoto Y (1977) The myxomycetes of Japan. Sangyo Tosho, Tokyo

Fisher MC, Koenig GL, White TJ, San-Blas G, Negroni R, Alvarez IG, Wanke B, Taylor JW (2001) Biogeographic range expansion into South America by Coccidioides immitis mirrors New World patterns of human migration. Proc Natl Acad Sci U S A 98:4558–4562

Fitzpatrick DA, Logue ME, Stajich JE, Butler G (2006) A fungal phylogeny based on 42 complete genomes derived from supertree and combined gene analysis. BMC Evol Biol 6:99. doi:10.1186/1471-2148-6-99

Francois I, Aerts AM, Cammue BPA, Thevissen K (2005) Currently used antimycotics: spectrum, mode of action and resistance occurrence. Curr Drug Targets 6:895–907. doi:10.2174/138945005774912744

Fungal Genomes (2012) http://fungalgenomes.org/. Cited 28 Mar 2012

Geissler S, Siegers K, Schiebel E (1998) A novel protein complex promoting formation of functional a- and g-tubulin. EMBO J 17:952–966

Gill EE, Lee RCH, Corradi N, Grisdale CJ, Limpright VO, Keeling PJ, Fast NM (2010) Splicing and transcription differ between spore and intracellular life stages in the parasitic Microsporidia. Mol Biol Evol 27:1579–1584. doi:10.1093/molbev/msq050

Giraud T, Refregier G, Le Gac M, de Vienne DM, Hood ME (2008) Speciation in fungi. Fungal Genet Biol 45:791–802. doi:10.1016/j.fgb.2008.02.001

Gleason FH, Lilje O (2009) Structure and function of fungal zoospores: ecological implications. Fungal Ecol 2:53–59. doi:10.1016/j.funeco.2008.12.002

Goffeau A, Barrell BG, Bussey H, Davis RW, Dujon B, Feldmann H, Galibert F, Hoheisel JD, Jacq C, Johnston M, Louis EJ, Mewes HW, Murakami Y, Philippsen P, Tettelin H, Oliver SG (1996) Life with 6000 genes. Science 274:546–567. doi:10.1126/science.274.5287.546

GOLD (2012) http://www.genomesonline.org/cgi-bin/GOLD/phylogenetic_distribution.cgi. Cited 28 Mar 2012

Greengenes (2012) http://greengenes.lbl.gov/. Cited March 28 2012

Gull K, Trinci APJ (1974) Nuclear division in Basidiobolus ranarum. Trans Br Mycol Soc 3:457–460

Gunderson JH, Elwood H, Ingold A, Kindle K, Sogin ML (1987) Phylogenetic relationships between chlorophytes, chrysophytes, and oomycetes. Proc Natl Acad Sci U S A 84:5823–5827

Harris SD (2011) Hyphal morphogenesis: an evolutionary perspective. Fungal Biol 115:475–484. doi:10.1016/j.funbio.2011.02.002

Hawksworth DL (2001) The magnitude of fungal diversity: the 1.5 million species estimate revisited. Mycol Res 105:1422–1432

Heckman DS, Geiser DM, Eidell BR, Stauffer RL, Kardos NL, Hedges SB (2001) Molecular evidence for the early colonization of land by fungi and plants. Science 293:1129–1133

Heitman J, Kronstad JW, Taylor JW, Casselton LA (eds) (2007) Sex in fungi: molecular determination and evolutionary implications. ASM Press, Washington, DC

Hibbett DS, Matheny PB (2009) The relative ages of ectomycorrhizal mushrooms and their plant hosts estimated using Bayesian relaxed molecular clock analyses. BMC Biol 7:13. doi:10.1186/1741-7007-7-13

Hibbett DS, Gilbert LB, Donoghue MJ (2000) Evolutionary instability of ectomycorrhizal symbioses in basidiomycetes. Nature 407:506–508

Hibbett DS, Ohman A, Glotzer D, Nuhn M, Kirk P, Nilsson RH (2011) Progress in molecular and morphological taxon discovery in Fungi and options for formal classification of environmental sequences. Fungal Biol Rev 25:38–47

Horton TR, Bruns TD (2001) The molecular revolution in ectomycorrhizal ecology: peeking into the black-box. Mol Ecol 10:1855–1871

Hunger K, Beckering CL, Wiegeshoff F, Graumann PL, Marahiel MA (2006) Cold-induced putative DEAD box RNA helicases CshA and CshB are essential for cold adaptation and interact with cold shock protein B in Bacillus subtilis. J Bacteriol 188:240–248. doi:10.1128/jb.188.1.240-248.2006

Inderbitzin P, Harkness J, Turgeon BG, Berbee ML (2005) Lateral transfer of mating system in Stemphylium. Proc Natl Acad Sci U S A 102:11390–11395

Inderbitzin P, Davis RM, Bostock RM, Subbarao KV (2011) The ascomycete Verticillium longisporum is a hybrid and a plant pathogen with an expanded host range. PLoS One 6:e18260. doi:10.1371/journal.pone.0018260

Irvin SD, Bhattacharjee JK (1998) A unique fungal lysine biosynthesis enzyme shares a common ancestor with tricarboxylic acid cycle and leucine biosynthetic enzymes found in diverse organisms. J Mol Evol 46:401–408

James TY, Berbee ML (2012) No jacket required – new fungal lineage defies dress code. Bioessays 34:94–102. doi:10.1002/bies.201100110

James TY, Kauff F, Schoch CL, Matheny PB, Hofstetter V, Cox CJ, Celio G, Gueidan C, Fraker E, Miadlikowska J, Lumbsch HT, Rauhut A, Reeb V, Arnold AE, Amtoft A, Stajich JE, Hosaka K, Sung G, Johnson D, O'Rourke B, Crockett M, Binder M, Curtis JM, Slot JC, Wang Z, Wilson AW, Schuessler A, Longcore JE, O'Donnell K, Mozley-Standridge S, Porter D, Letcher PM, Powell MJ, Taylor JW, White MM, Griffith GW, Davies DR, Humber RA, Morton JB, Sugiyama J, Rossman AY, Rogers JD, Pfister DH, Hewitt D, Hansen K, Hambleton S, Shoemaker RA, Kohlmeyer J, Volkmann-Kohlmeyer B, Spotts RA, Serdani M, Crous PW, Hughes KW, Matsuura K, Langer E, Langer G, Untereiner WA, Lucking R, Buedel B, Geiser DM, Aptroot A, Diederich P, Schmitt I, Schultz M, Yahr R, Hibbett DS, Lutzoni F, McLaughlin DJ, Spatafora JW, Vilgalys R (2006) Reconstructing the early evolution of Fungi using a six-gene phylogeny. Nature 443:818–822. doi:10.1038/nature05110

James TY, Porter TM, Martin WW (2014) Blastocladiomycota. In: McLaughlin DJ, Spatafora JW (eds) Systematics and evolution: Part A. Springer, Heidelberg

Jones MDM, Forn I, Gadelha C, Egan MJ, Bass D, Massana R, Richards TA (2011) Discovery of novel intermediate forms redefines the fungal tree of life. Nature 474:200–203

Juhas M, van der Meer JR, Gaillard M, Harding RM, Hood DW, Crook DW (2009) Genomic islands: tools of bacterial horizontal gene transfer and evolution. FEMS Microbiol Rev 33:376–393. doi:10.1111/j.1574-6976.2008.00136.x

Jumpponen A, Jones KL (2009) Massively parallel 454 sequencing indicates hyperdiverse fungal communities in temperate Quercus macrocarpa phyllosphere. New Phytol 184:438–448. doi:10.1111/j.1469-8137.2009.02990.x

Kaneshiro ES (2002) Sterol biosynthesis in Pneumocystis: unique steps that define unique targets. Drug Resist Updat 5:259–298

Keeling PJ, Fast NM, Law JS, Williams BAP, Slamovits CH (2005) Comparative genomics of microsporidia. Folia Parasitol 52:8–14

Koestler T, Ebersberger I (2011) Zygomycetes, Microsporidia, and the evolutionary ancestry of sex determination. Genome Biol Evol 3:186–194

Koljalg U, Larsson KH, Abarenkov K, Nilsson RH, Alexander IJ, Eberhardt U, Erland S, Hoiland K, Kjoller R, Larsson E, Pennanen T, Sen R, Taylor AFS, Tedersoo L, Vralstad T, Ursing BM (2005) UNITE: a database providing web-based methods for the molecular identification of ectomycorrhizal fungi. New Phytol 166:1063–1068. doi:10.1111/j.1469-8137.2005.01376.x

Koufopanou V, Burt A, Taylor JW (1997) Concordance of gene genealogies reveals reproductive isolation in the pathogenic fungus Coccidioides immitis. Proc Natl Acad Sci U S A 94:5478–5482

Lacroix B, Tzfira T, Vainstein A, Citovsky V (2006) A case of promiscuity: Agrobacterium's endless hunt for new partners. Trends Genet 22:29–37. doi:10.1016/j.tig.2005.10.004

Lang BF, O'Kelly C, Nerad T, Gray MW, Burger G (2002) The closest unicellular relatives of animals. Curr Biol 12:1773–1778

Lara E, Moreira D, Lopez-Garcia P (2010) The environmental clade LKM11 and Rozella form the deepest branching clade of Fungi. Protist 161:116–121. doi:10.1016/j.protis.2009.06.005

Liti G, Carter DM, Moses AM, Warringer J, Parts L, James SA, Davey RP, Roberts IN, Burt A, Koufopanou V, Tsai IJ, Bergman CM, Bensasson D, O'Kelly MJT, van Oudenaarden A, Barton DBH, Bailes E, Ba ANN, Jones M, Quail MA, Goodhead I, Sims S, Smith F, Blomberg A, Durbin R, Louis EJ (2009) Population genomics of domestic and wild yeasts. Nature 458:337–341. doi:10.1038/nature07743

Liu Y, Steenkamp ET, Brinkmann H, Forget L, Philippe H, Lang BF (2009) Phylogenomic analyses predict sistergroup relationship of nucleariids and Fungi and paraphyly of zygomycetes with significant support. BMC Evol Biol 9:272. doi:10.1186/1471-2148-9-272

Lutzoni F, Pagel M (1997) Accelerated evolution as a consequence of transitions to mutualism. Proc Natl Acad Sci U S A 94:11422–11427

Marshall WL, Berbee ML (2011) Facing unknowns: living cultures (Pirum gemmata gen. nov., sp nov., and Abeoforma whisleri, gen. nov., sp nov.) from invertebrate digestive tracts represent an undescribed clade within the unicellular opisthokont lineage Ichthyosporea (Mesomycetozoea). Protist 162:33–57. doi:10.1016/j.protis.2010.06.002

Mehrabi R, Bahkali AH, Abd-Elsalam KA, Moslem M, Ben M'Barek S, Gohari AM, Jashni MK, Stergiopoulos I, Kema GHJ, de Wit P (2011) Horizontal gene and chromosome transfer in plant pathogenic fungi affecting host range. FEMS Microbiol Rev 35:542–554. doi:10.1111/j.1574-6976.2010.00263.x

Mendoza L, Taylor JW, Ajello L (2002) The class Mesomycetozoea: a group of microorganisms at the animal-fungal boundary. Annu Rev Microbiol 56:160950. doi:10.1146/annurev.micro.56.012302.160950

Menkis A, Jacobson DJ, Gustafsson T, Johannesson H (2008) The mating-type chromosome in the filamentous ascomycete Neurospora tetrasperma represents a model for early evolution of sex chromosomes. PLoS Genet 4:1000030. doi:10.1371/journal.pgen.1000030

Munro CA, Gow NAR (2001) Chitin synthesis in human pathogenic fungi. Med Mycol 39:41–53. doi:10.1080/744118878

Nachman MW, Hoekstra HE, D'Agostino SL (2003) The genetic basis of adaptive melanism in pocket mice. Proc Natl Acad Sci U S A 100:5268–5273. doi:10.1073/pnas.0421157100

Nakagaki T, Yamada H, Toth A (2000) Maze-solving by an amoeboid organism. Nature 407:470

Neafsey DE, Barker BM, Sharpton TJ, Stajich JE, Park DJ, Whiston E, Hung CY, McMahan C, White J, Sykes S, Heiman D, Young S, Zeng QD, Abouelleil A, Aftuck L, Bessette D, Brown A, FitzGerald M, Lui A, Macdonald JP, Priest M, Orbach MJ, Galgiani JN, Kirkland TN, Cole GT, Birren BW, Henn MR, Taylor JW, Rounsley SD (2010) Population genomic sequencing of Coccidioides fungi reveals recent hybridization and transposon control. Genome Res 20:938–946. doi:10.1101/gr.103911.109

Nilsson RH, Ryberg M, Kristiansson E, Abarenkov K, Larsson K-H, Kõljalg U (2006) Taxonomic reliability of DNA sequences in public sequence databases: a fungal perspective. PLoS One 1(1):e59

Nishida H, Nishiyama M (2000) What is characteristic of fungal lysine synthesis through the alpha-aminoadipate pathway? J Mol Evol 51:299–302

Nishida H, Nishiyama M, Kobashi N, Kosuge T, Hoshino T, Yamane H (1999) A prokaryotic gene cluster involved in synthesis of lysine through the amino adipate pathway: a key to the evollution of amino acid biosynthesis. Genome Res 9:1175–1183

Park B, Park J, Cheong KC, Choi J, Jung K, Kim D, Lee YH, Ward TJ, O'Donnell K, Geiser DM, Kang S (2011) Cyber infrastructure for Fusarium: three integrated platforms supporting strain identification, phylogenetics, comparative genomics and knowledge sharing. Nucleic Acids Res 39:D640–D646. doi:10.1093/nar/gkq1166

Parks LW, Smith SJ, Crowley SH (1995) Biochemical and physiological effects of sterol altertions in yeast–a review. Lipids 30:227–230

Powell MJ, Letcher PM (2014) Chytridiomycota, monoblepharidomycota and neocallimastigomycota. In: McLaughlin DJ, Spatafora JW (eds) Systematics and evolution: Part A. Springer, Heidelberg

Ribosomal Database Project (2012) http://rdp.cme.msu.edu/. Cited 28 Mar 2012

Ridley AJ, Schwartz MA, Burridge K, Firtel RA, Ginsberg MH, Borisy G, Parsons JT, Horwitz AR (2003) Cell migration: integrating signals from front to back. Science 302:1704–1709

Rokas A, Kruger D, Carroll SB (2005) Animal evolution and the molecular signature of radiations compressed in time. Science 310:1933–1938

Ruiz-Herrera J, Ortiz-Castellanos L (2010) Analysis of the phylogenetic relationships and evolution of the cell walls from yeasts and fungi. FEMS Yeast Res 10:225–243. doi:10.1111/j.1567-1364.2009.00589.x

Ryberg M, Kristiansson E, Sjökvist E, Nilsson RH (2009) An outlook on the fungal internal transcribed spacer sequences in GenBank and the introduction of a web-based tool for the exploration of fungal diversity. New Phytol 181:471–477. doi:10.1111/j.1469-8137.2008.02667.x

Saiki RK, Gelfand DH, Stoffel S, Scharf SJ, Higuchi R, Horn GT, Mullis KB, Erlich H (1988) Primer-directed enzymatic amplification of DNA with a thermostable DNA polymerase. Science 239:487–491

Schadt CW, Martin AP, Lipson DA, Schmidt SK (2003) Seasonal dynamics of previously unknown fungal lineages in tundra soils. Science 301:1359–1361

Schaller H (2004) New aspects of sterol biosynthesis in growth and development of higher plants. Plant Physiol Biochem 42:465–476

Schloss PD (2009) A high-throughput DNA sequence aligner for microbial ecology studies. PLoS One 4 (12):e8230

Schoch C, Seifert K (2010) The home stretch for fungal barcoding. IMA Fungus 1:2–3

Schoch CL, Sung GH, Lopez-Giraldez F, Townsend JP, Miadlikowska J, Hofstetter V, Robbertse B, Matheny PB, Kauff F, Wang Z, Gueidan C, Andrie RM, Trippe K, Ciufetti LM, Wynns A, Fraker E, Hodkinson BP, Bonito G, Groenewald JZ, Arzanlou M, de Hoog GS, Crous PW, Hewitt D, Pfister DH, Peterson K, Gryzenhout M, Wingfield MJ, Aptroot A, Suh SO, Blackwell M, Hillis DM, Griffith GW, Castlebury LA, Rossman AY, Lumbsch HT, Lucking R, Budel B, Rauhut A, Diederich P, Ertz D, Geiser DM, Hosaka K, Inderbitzin P, Kohlmeyer J, Volkmann-Kohlmeyer B, Mostert L, O'Donnell K, Sipman H, Rogers JD, Shoemaker RA, Sugiyama J, Summerbell RC, Untereiner W, Johnston PR, Stenroos S, Zuccaro A, Dyer PS, Crittenden PD, Cole MS, Hansen K, Trappe JM, Yahr R, Lutzoni F, Spatafora JW (2009) The ascomycota tree of life: a phylum-wide phylogeny clarifies the origin and evolution of fundamental reproductive and ecological traits. Syst Biol 58:224–239. doi:10.1093/sysbio/syp020

Sekimoto S, Rochon D, Long JE, Dee JM, Berbee ML (2011) A multigene phylogeny of Olpidium and its implications for early fungal evolution. BMC Evol Biol 11:331. doi:10.1186/1471-2148-11-331

Silva (2012) http://www.arb-silva.de/. Cited 28 Mar 2012

Simon L, Bousquet J, Levesque RC, Lalonde M (1993) Origin and diversification of endomycorrhizal fungi and coincidence with vascular land plants. Nature 363:67–69

Steenkamp ET, Wright J, Baldauf SL (2006) The protistan origins of animals and fungi. Mol Biol Evol 23:93–106

Stephenson SL (2014) Excavata: acrasiomycota; amoebozoa: dictyosteliomycota, myxomycota. In: McLaughlin DJ, Spatafora JW (eds) Systematics and evolution: Part A. Springer, Heidelberg

Stiller JW, Huang JL, Ding Q, Tian J, Goodwillie C (2009) Are algal genes in nonphotosynthetic protists evidence of historical plastid endosymbioses? BMC Genomics 10:484. doi:10.1186/1471-2164-10-484

Strassmann JE, Zhu Y, Queller DC (2000) Altruism and social cheating in the social amoeba Dictyostelium discoideum. Nature 408:965–967

Suh SO, McHugh JV, Blackwell M (2004) Expansion of the Candida tanzawaensis yeast clade: 16 novel Candida species from basidiocarp-feeding beetles. Int J Syst Evol Microbiol 54:2409–2429. doi:10.1099/ijs.0.63246-0

Sumathi JC, Raghukumar S, Kasbekar DP, Raghukumar C (2006) Molecular evidence of fungal signatures in the marine protist Corallochytrium limacisporum and its implications in the evolution of animals and fungi. Protist 157:363–376. doi:10.1016/j.protis.2006.05.003

Swann EC, Taylor JW (1993) Higher taxa of basidiomycetes: an 18S rRNA gene perspective. Mycologia 85:923–936

Takamatsu A, Fujii T, Endo I (2000) Time delay effect in a living coupled oscillator system with the plasmodium of Physarum polycephalum. Phys Rev Lett 85:2026–2029

Taylor JW, Jacobson DJ, Fisher MC (1999) The evolution of asexual fungi: reproduction, speciation and classification. Annu Rev Phytopathol 37:197–246

Taylor JW, Jacobson DJ, Kroken S, Kasuga T, Geiser DM, Hibbett DS, Fisher MC (2000) Phylogenetic species recognition and species concepts in fungi. Fungal Genet Biol 31:21–32. doi:10.1006/fgbi.2000.1228

Taylor TN, Klavins SD, Krings M, Taylor EL, Kerp H, Hass H (2004) Fungi from the Rhynie chert: a view from the dark side. Trans R Soc Edinb Earth Sci 94:457–473

Taylor JW, Turner E, Townsend JP, Dettman JR, Jacobson D (2006) Eukaryotic microbes, species recognition and the geographic limits of species: examples from the kingdom Fungi. Philos Trans R Soc Lond B Biol Sci 361:1947–1963. doi:10.1098/rstb.2006.1923

Torruella G, Suga H, Riutort M, Pereto J, Ruiz-Trillo I (2009) The evolutionary history of lysine biosynthesis pathways within eukaryotes. J Mol Evol 69:240–248. doi:10.1007/s00239-009-9266-x

Townsend JP (2007) Profiling phylogenetic informativeness. Syst Biol 56:222–231. doi:10.1080/10635150701311362

Tree of Life web project (2012) http://tolweb.org/tree/. Cited 13 Apr 2012

TrichOKEY 2 (2011) http://www.isth.info/tools/molkey/index.php. Cited Jun 2011

Tsui CKM, Marshall W, Yokoyama R, Honda D, Lippmeier JC, Craven KD, Peterson PD, Berbee ML (2009) Labyrinthulomycetes phylogeny and its implications for the evolutionary loss of chloroplasts and gain of ectoplasmic gliding. Mol Phylogenet Evol 50:129–140

Turner TL, Bourne EC, Von Wettberg EJ, Hu TT, Nuzhdin SV (2010) Population resequencing reveals local adaptation of Arabidopsis lyrata to serpentine soils. Nat Genet 42:260–263. doi:10.1038/ng.515

Vogel HJ (1961) Lysine synthesis and phylogeny of lower fungi – some chytrids versus Hyphochytrium. Nature 189:1026–1027

Vogel HJ (1965) Lysine biosynthesis and evolution. In: Bryson V, Vogel HJ (eds) Evolving genes and proteins. Academic, New York, NY, pp 25–40

Vogler DR, Bruns TD (1998) Phylogenetic relationships among the pine stem rust fungi (Cronartium and Peridermium spp.). Mycologia 90:244–257

Von Wettstein F (1921) Das Vorkommen von Chitin und seine Verwertung als systematisch - phylogenetisches Merkmal im Pflanzenreich. Sitzungsberichte Akademie der Wissenschaften in Wien Mathematisch-naturwissenschaftliche Klasse 130:3–20

Vos M, Velicer GJ (2008) Isolation by distance in the spore-forming soil bacterium Myxococcus xanthus. Curr Biol 18:386–391. doi:10.1016/j.cub.2008.02.050

Wainright PO, Hinkle G, Sogin ML, Stickel SK (1993) Monophyletic origins of the Metazoa: an evolutionary link with fungi. Science 260:340–342

Weete JD, Abril M, Blackwell M (2010) Phylogenetic distribution of fungal sterols. PLoS One 5(5): e10899. doi:10.1371/journal.pone.0010899

Whisler HC (1962) Culture and nutrition of Amoebidium parasiticum. Am J Bot 49:193–199

Whitaker RJ, Grogan DW, Taylor JW (2003) Geographic barriers isolate endemic populations of hyperthermophilic archaea. Science 301:976–978. doi:10.1126/science.1086909

White TJ, Bruns T, Lee S, Taylor J (1990) Amplification and direct sequencing of fungal ribosomal RNA genes for phylogenetics. In: Innis MA, Gelfand DH, Sninsky JJ, White TJ (eds) PCR protocols. A guide to methods and applications. Academic, San Diego, CA, pp 315–322

Yeast Biochemical Pathway Database (2012) http://pathway.yeastgenome.org/. Cited 10 Apr 2012

Yoshida M, Nakayama T, Inouye I (2009) Nuclearia thermophila sp nov (Nucleariidae), a new nucleariid species isolated from Yunoko Lake in Nikko (Japan). Eur J Protistol 45:147–155. doi:10.1016/j.ejop.2008.09.004

Zabriskie TM, Jackson MD (2000) Lysine biosynthesis and metabolism in fungi. Nat Prod Rep 17:85–97

Zettler LAA, Nerad TA, O'Kelly CJ, Sogin ML (2001) The nucleariid amoebae: more protists at the animal-fungal boundary. J Eukaryot Microbiol 48:293–297

Fungal-Like Organisms

2 Excavata: Acrasiomycota; Amoebozoa: Dictyosteliomycota, Myxomycota

Steven L. Stephenson[1]

CONTENTS

I. Introduction

One of the deep branches of the eukaryotic tree of life consists of an assemblage of amoeboid protists referred to as the supergroup Amoebozoa (Fiore-Donno et al. 2010). The most diverse members of the Amoebozoa are the eumycetozoans, commonly referred to as slime molds. Since their discovery, members of the Dictyosteliomycota (dictyostelids) and Myxomycota (myxomycetes or myxogastrids) have been variously classified as plants, animals, or fungi. Because they produce aerial spore-bearing structures that resemble those of certain fungi and typically occur in some of the same types of ecological situations as fungi, slime molds have been traditionally studied by mycologists (Martin and Alexopoulos 1969). However, abundant molecular data now confirm that they are amoebozoans and not fungi (Bapteste et al. 2002; Yoon et al. 2008; Baudalf 2008).

Both dictyostelids and myxomycetes are widespread and often common in the microhabitats in which they characteristically occur, where they are major predators of bacteria and other microorganisms (Stephenson and Stempen 1994). However, because of their cryptic life cycles and the fact that the number of specialists studying them is relatively small, they are among the least studied groups of terrestrial organisms in nature, although a few species, such as *Dictyostelium discoideum* (for the dictyostelids) and *Physarum polycephalum* (for the myxomycetes), have become model organisms for laboratory studies. Although once classified in the same group as the dictyostelids, the acrasid cellular slime molds (or acrasids) are not closely related to the other organisms commonly referred to as slime molds. In fact, recent studies of acrasids have revealed that as a whole they are not even closely related to one another and are more appropriately referred to as sorocarpic amoebae (Brown et al. 2011). As a group, they are much less familiar organisms than either dictyostelids or myxomycetes, and many biologists are unlikely to be aware that they even exist.

II. Acrasiomycota

Although grouped with the eumycetozoans until detailed observations of morphological features and the availability of molecular data proved otherwise, members of the phylum

[1]Department of Biological Sciences, University of Arkansas, Fayetteville, AR 72701, USA; e-mail: slsteph@uark.edu

Systematics and Evolution, 2nd Edition
The Mycota VII Part A
D.J. McLaughlin and J.W. Spatafora (Eds.)

Acrasiomycota (*sensu* Alexopoulos et al. 1996) are better placed in the supergroup Excavata (Page and Blanton 1985; Roger et al. 1996; Adl et al. 2005; Brown et al. 2009, 2011), with a few examples recently reassigned to the supergroups Opisthokonta and Amoebozoa. This small assemblage of microorganisms (usually referred to as acrasid cellular slime molds or acrasids) was recognized as the class Acrasea by Olive (1975). However, because members of the assemblage are now known to produce sorocarps that have different evolutionary origins, this taxon as circumscribed by Olive is no longer valid. As such, these microorganisms are more appropriately considered as sorocarpic amoebae.

At one time or another, the sorocarpic amoebae were thought to encompass six genera: *Acrasis*, *Pocheina*, *Copromyxa*, *Copromyxella*, *Fonticula*, and *Guttulinopsis*. However, the genus *Copromyxa* has been shown to belong to the supergroup Amoebozoa (Brown et al. 2011) and is probably closely related to *Copromyxella* (Raper 1984), whereas the genus *Fonticula* has been reassigned to the supergroup Opisthokonta and is most closely related to the nucleariid amoebae and fungi (Brown et al. 2009).

Of the taxa historically called acrasids, only members of the taxon Acrasidae, which includes the genera *Acrasis* and *Pocheina*, are currently considered valid (Adl et al. 2005). Of the Acrasidae, only *Acrasis* has been studied in any detail. The type species of the genus is *A. granulata*, described in the late nineteenth century (van Tieghem 1880). However, the taxonomic identity of *A. granulata* is somewhat controversial because the original description provided no illustrations and only limited morphological details (Olive and Stoinaovitch 1960; Olive 1975; Raper 1984). A second species (*A. rosea*) was described 80 years later by Olive and Stoinaovitch (1960). A third species (*A. helenhemmesae*) was more recently added to the genus (Brown et al. 2010). *A. rosea* is by far the best known and most widely distributed of the three species (Reinhardt 1975). Cells in all stages of the life cycle of this species have orange-pink pigmented lipid droplets in the cytoplasm. This causes the cells to have a distinctive pinkish color. In the feeding (trophic) phase, the cells are amoeboid and characterized by lobose pseudopodia (Olive 1975). When conditions are appropriate, the amoeboid cells aggregate singly or in small groups to produce an erect, spore-containing fruiting body (or sorocarp). In *A. rosea*, the sorocarp is made up of chains of spores that collectively form an arborescent-like structure (Fig. 2.1). This is borne on a thin column of living cells, one of the major features that distinguish these microorganisms from dictyostelids, in which the stalk is hollow or filled with dead cells. The sorocarps of *A. helenhemmesae* typically consist of a single chain of spores. Although *A. rosea* has been isolated from a number of localities throughout the world, relatively little is known about its ecology.

Pocheina rosea (called *Guttulina rosea* in the older literature) was first described from dead wood in Russia during the latter part of the nineteenth century and later reported from North Carolina and a number of other localities in the eastern USA by Olive (1975). The sorocarp in this species is short-stalked with an apical, rose-colored, globose structure containing the spores. The genus *Guttulinopsis*, created at the very beginning of the twentieth century by Olive (1901) to accommodate what seemed to be several species of dung-inhabiting slime-mold-like organisms that were characterized by stalked or, more rarely, sessile, globose to somewhat elongated sorocarps. Nothing is currently known about how *Guttulinopsis* is related to any of the other sorocarpic amoebae. Because they have been so poorly studied, little is known about the global distribution and ecology of any of the sorocarpic amoebae.

III. Dictyosteliomycota

The dictyostelids (also commonly called cellular slime molds) are a relatively homogeneous group of approximately 150 described species. In the single most comprehensive monograph on the group, Raper (1984) listed approximately 50 species. Hagiwara (1989) added six more in his treatment of Japanese dictyostelids.

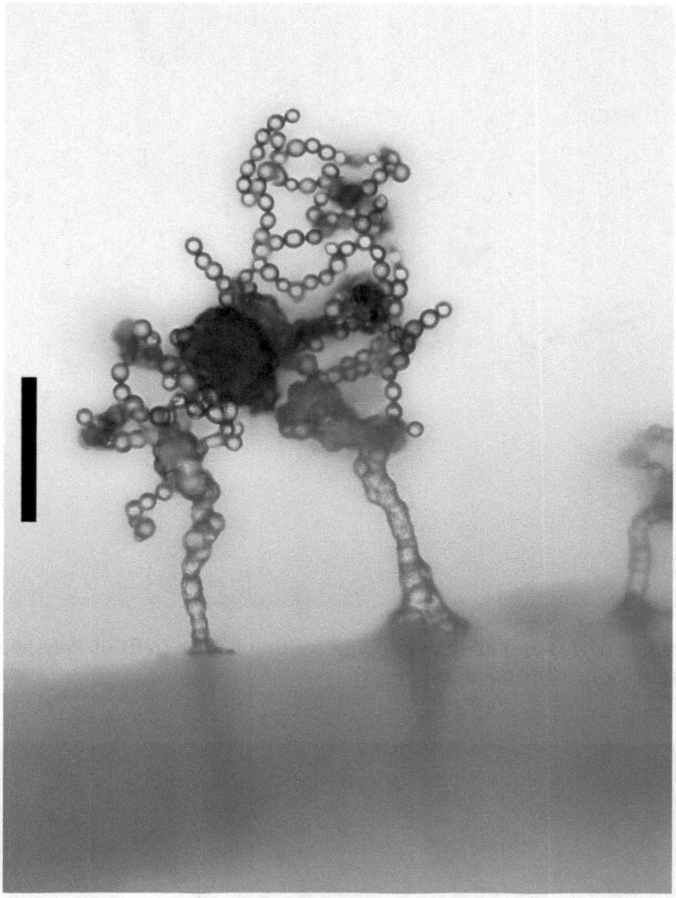

Fig. 2.1 Fruiting body of *Acrasis rosea* (photo by Matt Brown). Scale bar=0.1 mm

Since then, the number of species has more than doubled. This increase is due to the greater intensity of sampling by a larger number of individuals, sampling in regions of the world (especially the Southern Hemisphere) and habitats not previously investigated (e.g., Landolt et al. 2008; Cavender et al. 2010; Vadell et al. 2011), and evidence that some isolates previously assigned to a single species actually represent separate, distinct taxa (Romeralo et al. 2010). For example, in his treatment, Hagiwara (1989) emphasized stalk tip and base morphology, aggregation patterns, and spore morphology, which helped narrow the species concept for dictyostelids. Since then, there has been greater emphasis on the early developmental stages in delimiting species (Cavender et al. 2013). The utilization of molecular and morphological characters has also contributed to an increased understanding of the variation that exists within this group of organisms.

A. Life Cycle

All dictyostelids are characterized by having uninucleate cells with a reticulate, peripheral nucleolus (Olive 1975; Raper 1984; Cavender 1990). Amoeboid trophic cells, with acutely pointed pseudopodia, differentiate into aggregating cells that migrate in streams to an aggregation center (Fig. 2.2). The multicellular aggregation, or pseudoplasmodium, develops into one or more elongated slug-shaped structures that may migrate in some species or transform directly into a mature fruiting body (or sorocarp). The entire process is coordinated by the production of chemoattractants. The

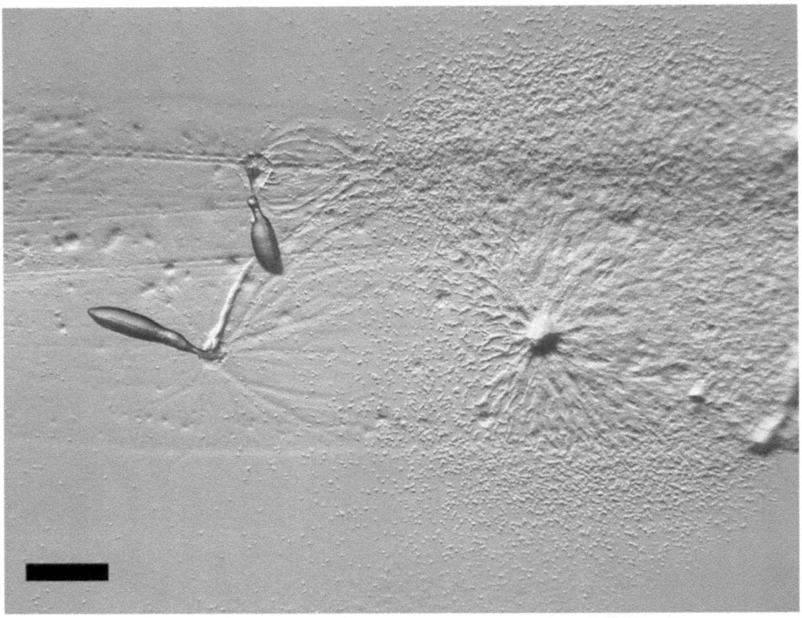

Fig. 2.2 Aggregation center (*right*) and early developing fruiting bodies (*left*) of *Polysphondylium tenuissimum* (photo by Andy Swanson). Scale bar=0.5 mm

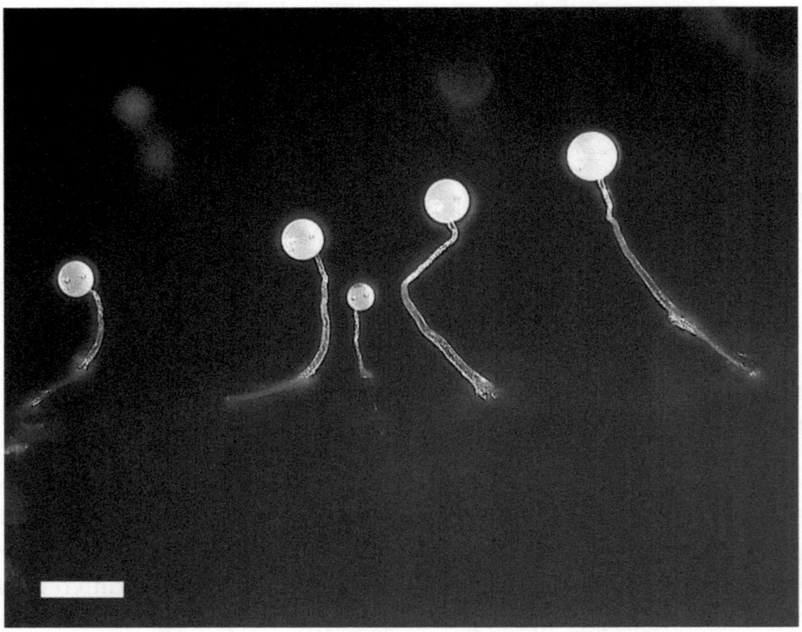

Fig. 2.3 Fruiting body of *Dictyostelium sphaerocephalum* (photo by Andy Swanson). In this species, the fruiting body consists of a stalk with a single sorus at the top. Scale bar=0.3 mm

fruiting body consists of a stalk that may display branching and one or more sori of spores (Fig. 2.3). Aggregation and fruiting represent an asexual dispersal process, but sex is known for many species (Raper 1984; Cavender 1990; Kessin 2001). The latter involves the formation of macrocysts. In brief, the process begins with the production of specific chemoattractants that cause some of the amoeboid trophic cells to aggregate. Cells of two mating types fuse under certain well-defined (Lewis and O'Day (1977) showed that a volatile sex hormone was involved) but still not completely understood environmental conditions to form a giant cell, which is essentially a diploid zygote (Chang and Raper 1981; O'Day and Keszei 2012). The giant cell then ingests the surrounding amoeboid trophic cells prior to encysting. Ultimately, meiosis takes place in the resulting macrocyst, and numerous haploid amoeboid trophic cells emerge through a rupture in the multilayered wall of the latter structure. Macrocysts were not recognized as the sexual stage of dictyostelids until the 1960s, and these structures have not yet been observed for many species. Most species of dictyostelids seem to be heterothallic, with mating types required, but homothallic strains have been reported for some species. Macrocysts also serve as a resistant stage in the life cycle, allowing the organism to survive under suboptimal conditions. Individual amoeboid trophic cells also have been observed to encyst (thus forming microcysts) in some species of dictyostelids. Microcysts thus represent yet another way that these organisms can deal with unfavorable environmental conditions (Kessin 2001).

B. Distribution and Occurrence

Dictyostelids are found in the soil microhabitat worldwide, particularly in the surface humus layers (Cavender and Raper 1965b, c; Cavender 1973, 1990; Raper 1984; Feest 1987; Hagiwara 1989; Stephenson and Landolt 1996). They are particularly abundant in the layer of leaf litter found on the forest floor and decrease in number and diversity with increasing depth (Cavender and Raper 1965b; Stephenson and Landolt 1996). Raper (1937) and Singh (1947) showed that dictyostelids can consume a variety of soil bacteria but prefer coliform bacteria if these are available. As such, dictyostelids may play a role in keeping the soil environment free of the pathogenic forms found in this group of bacteria. Dictyostelids are present in pastures and hay fields (Hammer 1984), and certain species are abundant in cultivated garden soil that is amended organically (Kauffman 1986). Moreover, the so-called canopy soil microhabitat (the mantle of soil-like dead organic matter often found at the bases of epiphytes that grow on the larger branches and trunks of trees in moist temperate and tropical forests) is now known to support an assemblage of dictyostelids (Stephenson and Landolt 1998, 2011). Interestingly, a few species were first described from these aerial microhabitats. Dictyostelids seem to be more common in forest soils than in agricultural soils, grassland soils, or deserts (Cavender and Raper 1965c; Raper 1984; Feest 1987; Cavender 1990). More species are found at lower latitudes than at higher latitudes (Cavender 1973), and at a particular latitude, more species are found at lower elevations than at higher elevations (e.g., Hagiwara 1976; Traub et al. 1981; Stephenson et al. 1999). Higher densities of dictyostelids are present in moist soils than in dry soils, although they are rare in saturated soils. Singh (1947) described the relationship that exists for fruiting ability and the level of soil moisture, while Cavender and Raper (1965c) showed that different species vary in abundance along a forest-moisture gradient and also that species abundances can be related to differences in forest composition. Horn (1971) found that there was competitive exclusion between species that depended on the same kind of bacteria.

Some species of dictyostelids seem to be strictly tropical, others are strictly temperate, and others, although cosmopolitan, are more common in either tropical or temperate regions of the world (Cavender 1973; Raper 1984; Swanson et al. 1999). The highest biodiversity of dictyostelids has been reported from neotropical rain forest soils (Vadell and Cavender 1995), but a few species can be surprisingly abundant even in tundra soils (Cavender 1978; Stephenson et al. 1991). It seems that some dictyostelids display an affinity for marginal or disturbed habitats not often sampled for these organisms previously, whereas others may be confined to a single limited geographical region of the world.

C. Isolation

Dictyostelids are usually isolated from soil (or other soil-like material) using some variation of the so-called Cavender method (Cavender and Raper 1965a; Raper 1984). In brief, this method involves collecting samples from a number of sites in a given habitat, returning these to the laboratory, and then diluting and suspending a measured mass of material from each sample in a known volume of distilled water. A small (but measured) amount of this suspension is spread evenly on a plate of a weak nutrient agar such as hay infusion agar (Raper 1984) or weak malt extract–yeast extract agar (Spiegel et al. 2004) and then overlaid with a turbid suspension of *Escherichia coli* in water. Plates are incubated at ambient temperatures for 3 or 4 days and then examined for colonies of dictyostelid fruiting bodies. Identification to species is made from direct observation of features of the fruiting bodies. When necessary, a particular isolate can be subcultured (often on water agar) to maintain it for further study. A critical consideration is to use a very weak nutrient medium that stimulates spore germination but does not promote the growth of bacteria or fungi. The *E. coli* added to such plates has an amazing ability to inhibit both soil bacteria and fungi.

Samples collected for isolation of dictyostelids should be processed in the laboratory as soon as possible because the species present gradually die off. Many of the rarer species seem to be lost within a few days or weeks (Stephenson and Cavender 1996). The reduction in numbers after 8 weeks is up to 25 % when temperate soils are refrigerated (Cavender and Raper 1965a), and this figure can be even higher when soils are exposed to fluctuating temperatures. Moreover, a number of as yet unidentified factors present in some soils inhibit the growth of dictyostelids, and some temperature-sensitive species (e.g., *Dictyostelium septentrionalis*) may not develop in culture plates even when they are present in a particular sample if the incubation temperature is above 20 °C. However, the Cavender method has yielded a considerable body of qualitative and quantitative data on the occurrence and distribution of dictyostelids throughout much of the world.

D. Taxonomy

In the taxonomic treatment traditionally used for dictyostelids, species have been assigned to three well-known genera (*Dictyostelium*, *Polysphondylium*, and *Acytostelium*) on the basis of the overall morphology and size of the fruiting body. In brief, those taxa having unbranched or laterally branched fruiting bodies have been assigned to *Dictyostelium*, those with repetitive whorls of regularly spaced side branches to *Polysphondylium*, and those characterized by fruiting bodies with acellular stalks to *Acytostelium*. However, Swanson et al. (2002) showed, using rooted cladistic analysis, that the three genera do not represent monophyletic groups. Schaap et al. (2006) developed the first molecular phylogeny of the dictyostelids) with data from the small subunit (SSU) ribosomal RNA and beta-tubulin genes. More than 100 isolates, including the majority of the species in culture at the time the study was carried out, were considered. The phylogenetic tree constructed from these data showed that the dictyostelids consist of four major groups (clades), none of which corresponds to the three traditional genera. Species of *Dictyostelium* are found in all four groups, species of *Polysphondylium* occur in two very well-separated locations in the tree, and species of *Acytostelium* form a mixed group along with species from the two other genera. Only members of the latter genus seemed to show any evidence of being monophyletic.

Romeralo et al. (2011) published an expanded phylogeny of the dictyostelids that was based on SSU ribosomal RNA data from numerous additional isolates of dictyostelids collected in various localities throughout the world. These included at least 50 species new to science. The phylogenetic tree they constructed revealed eight well-supported clades, none of which corresponds to any of the traditional genera, and also showed strong support for the four previously identified major groups (Schaap et al. 2006). In addition, three previously isolated but inconsistently resolved branches were now observed to form major divisions in their own right. These new groups have been referred to as the *polycarpum*, *polycephalum*, and *violaceum* complexes in order to retain the original groups' numbering scheme until formal names can be assigned. The new species included in the tree also expanded the range of morphological diversity found within the previously established four major groups, which suggests that the dictyostelids as a whole are in need of a major taxonomic revision. An appreciable number of the new species noted previously were characterized by small-sized fruiting bodies (i.e., an average height of no more than 2 cm), and recent data (e.g., Cavender et al. 2005, 2013) indicate that these dictyostelids with small fruiting bodies, particularly those in group 3, as reported by Schaap et al. (2006), are the most common and diverse forms found in nature. As such, most of the species remaining to be discovered are likely to be members of this assemblage.

IV. Myxomycota

Myxomycetes (also called plasmodial slime molds or myxogastrids) have been known from their fruiting bodies since at least the middle of the seventeenth century, when the first recognizable description of a member of the group (the very common species now known as *Lycogala epidendrum*) was provided by the German mycologist Thomas Panckow (Stephenson et al. 2008). Evidence from molecular studies (e.g., Baldauf and Doolittle 1997; Baldauf et al. 2000) suggests that the myxomycetes have a long evolutionary history. However, due to the fragile nature of the fruiting body, fossil records of the group are exceedingly rare. Domke (1952) described a species of *Stemonitis* and Dörfelt et al. (2003) a species of *Arcyria* from Baltic amber dating from the Eocene, whereas Waggoner and Poinar (1992) reported a rather problematic fossil of a myxomycete plasmodium in amber from Eocene–Oligocene deposits in the Dominican Republic. The maximum age that could be assigned to any of these fossils would not exceed approximately 50 million years, which is greater than that of the few records of fossil spores that seem to be those of myxomycetes, which date only from the Oligocene and Pleistocene (Graham 1971).

Although a number of early workers published recognizable descriptions of various genera and species, the first noteworthy treatment of the myxomycetes was published by de Bary in 1859. Interestingly, de Bary (1859) seems to have been the first to conclude that these organisms were more closely related to the amoeboid protozoa than to fungi. To emphasize his point, he proposed the term Mycetozoa (literally *fungus animals*) for the group. Rostafinski, who was a student of de Bary, is credited with producing the first relatively comprehensive monograph on the myxomycetes (Rostafinski 1873, 1874–1876). Unfortunately, the monograph was written in Polish and thus largely inaccessible to most of the scientific community at the time it appeared. However, much of the information contained in the monograph was made available in publications by Cooke (1877) and Massee (1892), both of which were in English. The single most significant pre-twentieth-century publication on myxomycetes was the first edition of *A Monograph of the Mycetozoa* (Lister 1894). This monograph, revised and expanded versions of which were published in 1911 and 1925 (Lister 1911, 1925), became the standard reference on the group during the early part of the twentieth century. MacBride published the first edition of *The North American Slime-Moulds* in 1899 and followed this with a greatly expanded second edition in 1922. These two works (MacBride 1899, 1922) are of particular importance because they were the basis of yet another work, *The Myxomycetes* (MacBride and Martin 1934). Several decades later, Martin collaborated with Alexopoulos to produce their comprehensive world monograph, *The Myxomycetes* (1969). This monograph is now more than 40 years old and out of print. However, it still remains the single most definitive treatment for the myxomycetes, literally representing a bible for those individuals engaged in studies of the group. Other more recent regional monographs include those by Farr (1976), Yamamoto (1998), Nannenga-Bremekamp (1991), Ing (1999), and Stephenson (2003).

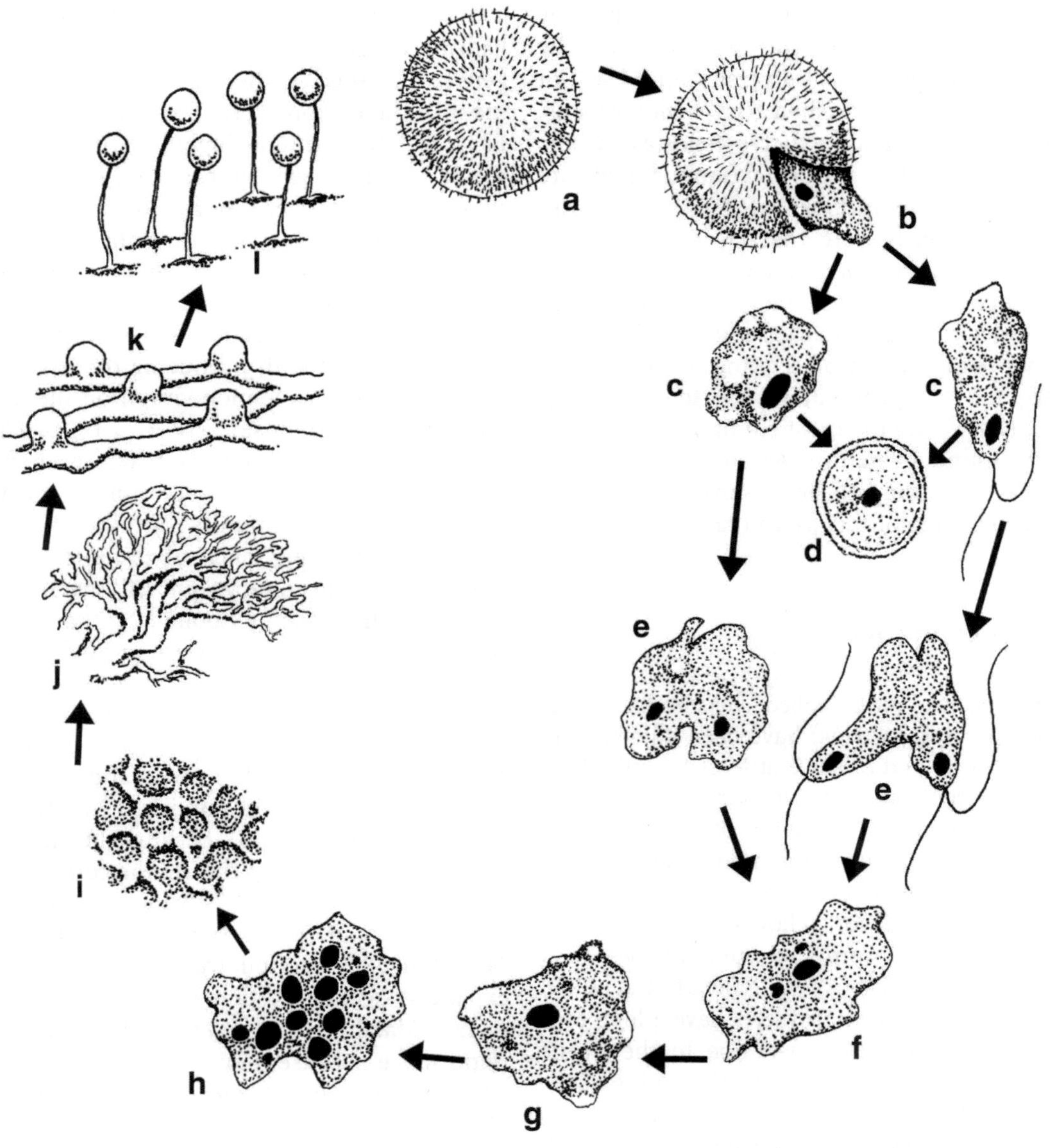

Fig. 2.4 Generalized life cycle in myxomycetes. (**a, b**) A protoplast emerges from the spore. (**c**) The protoplast can take the form of an amoeba or a flagellated cell (the term amoeboflagellate refers to both forms) during the first trophic stage. (**d**) Under dry conditions or in the absence of food, an amoeboflagellate forms a microcyst, or resting stage. (**e–g**) Compatible amoeboflagellates fuse to form a zygote (**g**). (**h–j**) The nucleus of the zygote divides by mitosis (**h**), and each subsequent nucleus also divides without being followed by cytokinesis, thereby producing a single large cell (**j**), the plasmodium, that represents the second trophic stage. Under adverse conditions, the plasmodium can form the second type of resting stage found in myxomycetes, the sclerotium (**i**). (**k, l**) Fruiting bodies are formed from the plasmodium. During fruiting body formation, spores are produced. Adapted from Stephenson (2003)

A. Life Cycle

The myxomycete life cycle (Fig. 2.4) encompasses two very different trophic stages, one consisting of uninucleate amoebae, with or without flagella, and the other consisting of a distinctive multinucleate structure, the plasmodium (Martin et al. 1983). Much of what is

Fig. 2.5 Plasmodium of a myxomycete (photo by Randy Darrah). Scale bar = 25 mm

known about the myxomycete life cycle has been derived from studies of *P. polycephalum* and *Didymium iridis*, but the life cycle of a number of other species has been observed in laboratory culture (Clark 2008). Plasmodia are motile, and those of some species can reach a size of several centimeters, with truly extraordinary examples sometimes exceeding 1 m (Fig. 2.5). A large example contains many thousands of synchronously dividing nuclei. Under favorable conditions, the plasmodium gives rise to one or more fruiting bodies (also referred to as sporocarps or sporophores) containing spores (Fig. 2.6). For practical reasons, identification of myxomycetes is based almost exclusively upon features of the fruiting body (Martin and Alexopoulos 1969). The fruiting bodies produced by myxomycetes are somewhat suggestive of those produced by certain macrofungi, although they are considerably smaller (usually no more than 1–2 mm tall). The spores of the vast majority of myxomycetes range in size from 5 to 15 μm in diameter, with most species producing spores 10 ± 2 μm in diameter. The spores are largely wind-dispersed and complete the life cycle by germinating to produce the uninucleate amoeboflagellate cells. These feed and divide by binary fission to build up large populations in the various microhabitats in which these organisms occur. Ultimately, this stage in the life cycle gives rise to the plasmodium. This process can result from gametic fusion between compatible amoeboflagellates or it can be apomictic (Collins 1980, 1981), as is described in more detail in what follows.

Most myxomycetes seem to have a basic one-locus multiple alleleic heterothallic mating system that controls syngamy between haploid amoeboflagellate cells to produce the diploid plasmodium (Clark and Haskins 2010). However, more than a single locus may be involved in some species, and three multiple alleleic loci have been reported for *P. polycephalum* (Kawano et al. 1987). Each of the morphospecies examined to date also contains a number of biological sibling species that are unable to interbreed with each other. It is not unusual for these to occur in different regions of the world. Each morphospecies generally contains numerous nonheterothallic strains that can complete the life cycle from a single isolated spore. Some of these strains are possibly homothallic (i.e., genetically identical amoeboflagellate cells fuse, resulting in a diploid

Fig. 2.6 Fruiting bodies of *Hemitrichia calyculata* (photo by Kim Fleming). In this species, the fruiting body is stalked and the lower part of the peridium persists to form a cuplike structure (or calyculus), above which the capillitium and spores are visible. Scale bar = 1.0 mm

plasmodium), but there is more evidence to suggest that they are characterized by an apomictic system derived from blockage of meiosis during spore formation. As such, these nonheterothallic strains produce diploid amoeboflagellates that can develop directly into plasmodia without the need for syngamy to take place.

In the textbook sexual life cycle outlined previously, two haploid amoeboflagellate cells fuse to form a diploid zygote, and the latter then develops into a multinucleate plasmodium in which all of the cells present are diploid. Under appropriate conditions, a plasmodium gives rise to a fruiting body, within which meiosis occurs when the spores are produced. An amoeboflagellate emerges from the spore to begin the life cycle anew. However, as already noted, some myxomycetes are known to be apomictic and thus do not follow this general pattern. Clark and Haskins (2010) listed 51 different species in which the reproductive system has been examined for one or more isolates. Of these, 14 were found to have both heterothallic and nonheterothallic (presumably apomictic) systems, 8 had only heterothallic systems, and 29 were reported to be nonheterothallic. Rather little is known about the relative proportions of heterothallic versus nonheterothalic reproduction in nature, but the latter may be more common.

The genetic structure in particular populations of myxomycetes is still largely unknown because few studies have been carried out. Fiore-Donno et al. (2011) investigated the genetic variability for three genes (SSU ribosomal, internal transcribed spacer 1, and partial elongation factor 1-alpha) in two species of *Lamproderma* associated with a spatially limited microhabitat (bryophyte-covered boulders in a series of moist ravines in Germany). Identical sequences were found to exist for a number of specimens in each of the two species, which suggests the occurrence of distinct clones that are the result of a nonheterothallic reproductive system.

Winsett and Stephenson (2010) examined the global distribution and molecular diversity (using the mitochondrial SSU gene) of *Didymium difforme*. Their data seem to support the concept of long-distance dispersal in myxomycetes since similar sequences were found to occur in widely separated regions of the world (e.g., Kenya and the central USA). However, in some instances, collections from a single region showed a very high degree of similarity, which

Fig. 2.7 Fruiting bodies of *Ceratiomyxa fruticulosa* (photo by Kim Fleming). Scale bar = 5 mm

at least suggests that some constraints to gene flow also exist.

Bacteria apparently represent the main food resource for both amoeboflagellates and plasmodia, but the latter are also known to feed upon yeasts, algae (including cyanobacteria), and fungal spores and hyphae (Stephenson and Stempen 1994). Under adverse conditions, such as drying out of the immediate environment or low temperatures, a plasmodium may convert into a hardened, resistant structure called a sclerotium, which is capable of reforming the plasmodium upon the return of favorable conditions. Moreover, amoeboflagellate cells can undergo a reversible transformation to dormant structures called microcysts. Both sclerotia and microcysts can remain viable for long periods of time and are probably very important in the continued survival of myxomycetes in some ecological situations or habitats, such as the bark surface of living trees and deserts.

B. Taxonomy

Approximately 900 species of myxomycetes have been described (Lado 2001), and in all but the most modern treatments of the group, these have been placed in six orders (Ceratiomyxales, Echinosteliales, Liceales, Physarales, Stemonitales, and Trichiales). However, members of the Ceratiomyxales, represented by the single genus *Ceratiomyxa* (Fig. 2.7), are distinctly different (e.g., their spores are produced externally on individual stalklike structures and not within a fruiting body) from any of the organisms assigned to the other orders, and modern workers have removed these organisms from the myxomycetes (Olive 1970, 1975; Olive and Stoianovitch 1979). The exact evolutionary affinities of the Ceratiomyxales are still debated, but they seem to be a sister group to the so-called true myxomycetes (Fiore-Donno et al. 2010). With the removal of the Certiomyxales, the myxomycetes constitute a well-defined and homogenous group. However, evidence from DNA sequence analysis (Baldauf and Doolittle 1997) suggests that even what seem to be closely related taxa on the basis of morphological similarity may have diverged from each other a long time ago (Clark 2000). Fiore-Donno et al. (2005) reported that phylogenetic data based on partial SSU ribosomal RNA and elongation factor-1 alpha sequences suggest a dichotomy between light-spored and dark-spored myxomycetes, with the light-spored clade consisting of the monophyletic Trichiales and the paraphyletic Liceales and the dark-spored clade consisting of the monophylectic Physarales and the paraphyletic Stemonitales.

Members of the genus *Cribraria*, traditionally assigned to the Liceales, seem to represent a sister group to both the Trichiales and Liceales (Fiore-Donno et al. 2010). These data place the genus *Echinostelium* as the sister group to the two major clades. More comprehensive analyses, based on complete SSU ribosomal RNA and elongation factor-1 alpha sequences from a wider range of taxa, indicate that *Echinostelium* branches as the sister group of the dark-spored clade (Fiore-Donno et al. 2008). The concept of this sister group has been expanded by Fiore-Donno et al. (2009), who recently provided evidence that the enigmatic organism *Semimorula liquescens* is a modified echinostelid myxomycete.

Interestingly, it has become increasingly apparent that the myxomycetes include a number of amoeboid forms that apparently do not form fruiting bodies. The latter fact prevented the true phylogenic position of these organisms from being recognized until they were subjected to the appropriate molecular-based studies. For example, Fiore-Donno et al. (2010) reported that sequences they obtained from a number of amoeboid forms previously assigned to the genus *Hyperamoeba* clearly indicated that the organisms involved should be considered as myxomycetes. Moreover, *Hyperamoeba* was found to be polyphyletic, which rendered the genus itself invalid. It seems possible that nonfruiting forms of myxomycetes are widespread in nature, sometimes occurring in certain habitats or microhabitats which would be regarded as rather extraordinary. Dyková et al. (2007) isolated an amoeboid organism from a species of sea urchin (*Sphaerechinus granularis*) that yielded SSU rDNA sequences showing a close relationship with the myxomycete genus *Didymium*. Myxomycetes have been reported from a diverse array of microhabitats, but their presence as apparent endocommensals of a sea urchin clearly indicates that the total range of potential microhabitats is even more extensive than previously realized.

Because of their small size and the limited array of morphological characters upon which their taxonomy is based, determination of what constitutes a natural biological species, in the same sense that the concept is used for many of the more familiar groups of organisms (Mayr 1970), is sometimes rather problematic. As mentioned earlier in this chapter, it is now known that a number of the more common and widespread morphospecies actually consist of complexes of geographically restricted apomictic clonal lines (El Hage et al. 2000; Clark 2000; Clark and Stephenson 2000; Irawan et al. 2000). These genetically isolated lines are capable of independent evolution, which can lead to the accumulation of minor morphological differences that reflect specific adaptations to the particular set of environmental conditions in which they occur. For example, some of the forms found in special microhabitats (e.g., the inflorescences of tropical herbs) differ in some respects (e.g., color and size of the fruiting bodies) from specimens of the same species collected from more typical habitats (Schnittler and Stephenson 2002). These almost certainly represent biotypes that are adapted to the microhabitat in question. Approximately 50 % of all described species of myxomycetes are known only from the type locality, or fewer than five localities worldwide. It seems likely that many of these so-called species are no more than morphologically distinct biotypes present in particular habitats or confined to a certain regions of the world. If so, then the criteria that need to be applied before describing a taxon as new should be reconsidered to account for this phenomenon (Schnittler and Mitchell 2000).

C. Distribution and Occurrence

Myxomycetes have been recorded from all terrestrial ecosystems examined to date. Temperature and moisture are thought to be the main factors limiting the occurrence of myxomycetes in nature (Alexopoulos 1963), and species richness tends to increase with increasing diversity and biomass of the vascular plants providing the resources (various types of detritus) that support the bacteria and other microorganisms upon which the two trophic stages in the myxomycete life cycle feed (Madelin 1984; Stephenson 1989). The pH of the substrates potentially available to myxomycetes in a particular habitat also represents an important factor influencing the distribution of these organisms (Härkönen 1977; Stephenson 1989; Wrigley de Basanta 2000; Mosquera et al. 2000).

Much of what is known about the distribution and ecology of myxomycetes in terrestrial ecosystems has been derived from studies carried out in temperate forests of the Northern Hemisphere. In such forests, myxomycetes are associated with a number of different microhabitats. These include coarse woody debris on the forest floor, the bark surface of living trees, forest floor litter, the dung of herbivorous animals, and aerial portions of dead but still standing herbaceous plants. Each of these microhabitats tends to be characterized by a distinct assemblage of species (Stephenson 1988, 1989; Stephenson and Stempen 1994). Lignicolous myxomycetes associated with coarse woody debris are the best known since the species typically occurring in this microhabitat tend to be among those characteristically producing fruiting bodies of sufficient size to be detected in the field (Martin and Alexopoulos 1969). Many of the more common and widely known myxomycete taxa, including various species of *Arcyria*, *Lycogala*, *Stemonitis*, and *Trichia*, are predominantly lignicolous. Much less is known about the myxomycetes associated with the microhabitats represented by the bark surface of living trees and forest floor litter. The primary reason for this is that many of the species involved are rather inconspicuous or sporadic in their occurrence and, thus, difficult to detect in the field. However, the moist chamber culture technique as it applies to myxomycetes (Gilbert and Martin 1933) provides a convenient and often very productive method of supplementing field collections when studying such microhabitats as bark and litter. Since its introduction, the technique has been used with considerable success by many researchers (e.g., Keller and Brooks 1976; Härkönen 1981; Blackwell and Gilbertson 1980; Stephenson 1989). More than 100 species of corticolous myxomycetes have been reported from the bark microhabitat as field or moist chamber collections (Mitchell 1980). The moist chamber culture technique is described in some detail by Stephenson and Stempen (1994).

Studies of the assemblages of myxomycetes associated with tropical forests and other major types of terrestrial ecosystems have been reviewed by Ing (1994) and Stephenson (2011). In some of these ecosystems, myxomycetes are associated with microhabitats poorly represented or absent in temperature forests. Examples include the inflorescences of large tropical herbaceous plants in tropical forests (Schnittler and Stephenson 2002) and succulent plants in deserts (Lado et al. 2007). One ecologically distinct group of myxomycetes is restricted to alpine areas of mountains, where its members are found fruiting along the margins of melting snowbanks in late spring and early summer (Ing 1999; Tamayama 2000; Stephenson and Shadwick 2009). The species that occupy this rather special and very limited habitat are usually referred to as snowbank or nivicolous myxomycetes. Interestingly, the majority of species in some genera tend to belong to this group. For example, this is the case for *Dianema* (Kowalski 1967), *Lamproderma* (Kowalski 1970), and *Lepidoderma* (Kowalski 1971).

On the whole, myxomycetes would seem to be rather opportunistic or fugitive organisms (sensu Hutchinson 1951) in that they have a high reproductive potential, seem to possess effective dispersal mechanisms, and are characterized by rapid development. These properties allow them to exploit successfully habitat islands occurring both temporally and spatially in nature. Although a particular habitat within which a species of myxomycete has been established may persist for only a short period of time, the species always survives by reestablishing itself in some new habitat (which may indeed be the very same habitat if conditions once again become favorable). Although the spores of myxomycetes would seem to have considerable potential for long-distance dispersal, there is little question that some species are more common in some regions of the world than others, and the nonavailability of certain microhabitats apparently imposes major constraints upon their occurrence even within a particular region. As such, it would seem that myxomycetes do not necessarily conform completely to the "ubiquity of small free-living eukaryotic species" concept as proposed by

Finlay (2002) and Fenchel and Finlay (2004). The very fragmented range of *Barbeyella minutissima*, a species that seems to be confined almost exclusively to a substrate complex (involving leafy liverworts, decorticated wood, and certain species of algae) found only in montane *Picea* or *Abies* forests, provides a good example (Schnittler et al. 2000).

V. Ecological Significance

A major portion of the net annual primary production in forests and other terrestrial ecosystems becomes directly or indirectly available to the decomposers in the detritus food chain. These decomposers (bacteria and fungi) are in turn an important food resource for various phagotrophic invertebrates and protozoa. For example, bacteria are preyed upon by bacterivores (e.g., protozoa and nematodes) as well as some detritivores (e.g., earthworms). Naked amoebae, which can make up 95 % of the protozoan population in some soils (Feest 1987), are the single most important group in terms of bacterial consumption. In addition to their direct influence on the structure of soil microbial communities, these amoebae play a key role in nutrient cycling. Mineralization is stimulated and decomposition enhanced by the amoebae releasing nutrients tied up in the microbial biomass. For example, amoebae are known to release ammonia to plant roots when feeding on bacteria and can produce increases in dry weight and nitrogen content (Clarholm 1981; Rosswall and Paustian 1984). It is not known what percentage of the total population of soil amoebae is made up of the amoeboid stages of dictyostelids and myxomycetes, but judging from the data available from a number of recent studies, it is significant. For example, Feest and Campbell (1986) reported that myxomycete amoebae alone represented more than 50 % of the total amoebae for some agricultural soils. Their study was based upon the use of a culture-based method (Feest and Madelin 1985), but Urich et al. (2008) used an RNA-centered megatranscriptomic approach to generate the largest data set for the entire soil protozoan community available to date. Eumycetozoans were found to represent the single largest component of total soil protozoan biodiversity, which seems to underscore the major ecological role these organisms play in the soil microhabitat.

As noted earlier in this chapter, myxomycetes are commonly associated with the microhabitats represented by the bark surface of living trees and various types of dead plant material. Although their fruiting bodies are often collected from each of these microhabitats, little is known about the exact role that myxomycetes play in each instance, although it might be assumed that it is similar to that already described for soil. For example, amoeboflagellate cells seem to be exceedingly common in decaying wood (A. Feest, unpublished data), and an individual moist chamber culture prepared with a sample of dead plant material often yields several different species, plasmodia several centimeters in total extent, and numerous fruiting bodies. Their sheer number, relative abundance, and biomass suggest that myxomycetes represent an ecologically significant component of the assemblages of organisms associated with such microhabitats.

Acknowledgments Appreciation is extended to Drs. Matt Brown, James Cavender, Jim Clark, Alan Feest, Anna Maria Fiore-Donno, Ed Haskins, John Landolt, and Maria Romeralo for reviewing portions of this chapter.

References

Adl SM, Simpson AGB, Farmer MA, Andersen RA, Anderson OR, Barta JR, Bonser SS, Brugerolle G, Fensome RA, Fredericq S, James TY, Karpov S, Kugrens P, Krug J, Lane CE, Lewis LA, Lodge J, Lynn LH, Mann DG, McCourt RM, Mendoza L, Moestrup Ø, Mozley-Standridge SE, Nerad TA, Shearer CA, Smirnov AV, Spiegel FW, Taylor MFJR (2005) The new higher level classification of eukaryotes with emphasis on the taxonomy of protists. J Eukaryot Microbiol 52:399–451

Alexopoulos CJ (1963) The myxomycetes II. Bot Rev 29:1–78

Alexopoulos CJ, Mims CW, Blackwell M (1996) Introductory mycology, 4th edn. Wiley, New York, NY

Baldauf SL, Doolittle WF (1997) Origin and evolution of the slime molds (Mycetozoa). Proc Natl Acad Sci 94:12007–12012

Baldauf SL, Roger AJ, Wenk-Siefert J, Doolittle WF (2000) A kingdom-level phylogeny of eukaryotes based on combined protein data. Science 290:972–977

Bapteste E, Brinkmann H, Lee JA, Moore DV, Sensen CW, Gordon P, Duruflé L, Gaasterland T, Lopez P, Müller M, Philippe H (2002) The analysis of 100 genes support the grouping of three highly divergent amoebae: *Dictyostelium*, *Entamoeba*, and *Mastigamoeba*. Proc Natl Acad Sci U S A 99:1414–1419

Baudalf SL (2008) An overview of the phylogeny and diversity of eukaryotes. J Syst Evol 46:263–273

Blackwell M, Gilbertson RL (1980) Sonoran desert myxomycetes. Mycotaxon 11:139–149

Brown MW, Spiegel FW, Silberman JD (2009) Phylogeny of the "forgotten" cellular slime mold, *Fonticula alba*, reveals a key evolutionary branch within Opisthokonta. Mol Biol Evol 26:2699–2709

Brown MW, Silberman JD, Spiegel FW (2010) A morphologically simple species of *Acrasis* (Heterolobosea, Excavata), *Acrasis helenhemmesae* n. sp. J Eukaryot Microbiol 57:346–353

Brown MW, Silberman JD, Spiegel FW (2011) "Slime molds" among the Tubulinea (Amoebozoa): molecular systematics and taxonomy of *Copromyxa*. Protist 162:277–287

Cavender JC (1973) Geographical distribution of Acrasiae. Mycologia 65:1044–1054

Cavender JC (1978) Cellular slime molds in tundra and forest soils of Alaska including a new species, *Dictyostelium septentrionalis*. Can J Bot 56:1326–1332

Cavender JC (1990) Phylum dictyostelida. In: Margulis L, Corliss CO, Melkonian M, Chapman DJ (eds) Handbook of proctotista. Jones and Bartlett, Boston, MA, pp 88–101

Cavender JC, Raper KB (1965a) The Acrasieae in nature. I. Isolation. Am J Bot 52:294–296

Cavender JC, Raper KB (1965b) The Acrasieae in nature. II. Forest soil as a primary habitat. Am J Bot 52:297–302

Cavender JC, Raper KB (1965c) The Acrasieae in nature. III. Occurrence and distribution in forests of eastern North America. Am J Bot 52:302–308

Cavender JC, Vadell EM, Landolt JC, Stephenson SL (2005) New species of small dictyostelids from the Great Smoky Mountains National Park. Mycologia 97:493–512

Cavender JC, Landolt JC, Ndiritu GG, Stephenson SL (2010) Dictyostelid cellular slime molds of Africa. Mycosphere 1:147–152

Cavender JC, Vadell EM, Landolt JC, Winsett KE, Stephenson SL, Rollins AW, Romeralo M (2013) Ten new small dictyostelids from seasonal rain forests of Central America. Mycologia 105(3):610–635

Chang MT, Raper KB (1981) Mating types and macrocyst formation in *Dictyostelium rosarium*. J Bacteriol 147:1049–1053

Clarholm M (1981) Protozoan grazing of bacteria in soil—impact and importance. Microb Ecol 7:343–350

Clark J (2000) The species problem in the myxomycetes. Stapfia 73:39–53

Clark J (2008) Reproductive systems and taxonomy in the myxomycetes. Syst Geogr Pt 74:209–216

Clark J, Haskins E (2010) Reproductive systems in the myxomycetes: a review. Mycosphere 1:337–353

Clark J, Stephenson SL (2000) Biosystematics of the myxomycete *Physarum melleum*. Nova Hedwigia 71:161–164

Collins OR (1980) Apomictic-heterothallic conversion in a myxomycete, *Didymium iridis*. Mycologia 72:1109–1116

Collins OR (1981) Myxomycete genetics, 1960–1981. J Elisha Mitch Sci Soc 97:101–125

Cooke MC (1877) The Myxomycetes of Great Britain arranged according to the method of Rostafinski: the characters of all the orders, families and genera, with descriptions of the British species, and original analytical tables, translated from the Polish. Williams and Norgate, London

de Bary A (1859) Die Mycetozoen. Ein Beitrag zur Kenntnis der niedersten Thiere. Z Wiss Zool 10:88–175

Domke W (1952) Der erste sichere Fund eines Myxomyceten im Baltischen Bernstein (*Stemonitis splendens* Rost. Fa. Succini fa. Nov. Foss.). Mitteilungen aus dem Gologischen Staatsinstitut in Hamburg 21:154–161

Dörfelt H, Schmidt AR, Ullmann P, Wunderlich J (2003) The oldest fossil myxogastroid slime mold. Mycol Res 107:123–126

Dyková I, Lom J, Dvořáková H, Pecková H, Fiala I (2007) *Didymium*-like myxogastrids (class Mycetozoa) as endocommensals of sea urchins (*Sphaerechinus granularis*). Folia Parasitol 54:1–12

El Hage N, Little C, Clark J, Stephenson SL (2000) Biosystematics of *Didymium squamulosum*. Mycologia 92:54–64

Farr ML (1976) Myxomycetes. Flora Neotropica, Monograph No. 16. The New York Botanical Garden, New York

Feest A (1987) The quantitative ecology of soil Mycetozoa. Prog Protistol 2:331–361

Feest A, Campbell R (1986) The microbiology of soils under successive wheat crops in relation to take-all disease. FEMS Microbiol Lett 38:99–111

Feest A, Madelin MF (1985) A method for the enumeration of myxomycetes in soils and its application to a wide range of soils. FEMS Microbiol Ecol 31:103–109

Fenchel T, Finlay BJ (2004) The ubiquity of small species: patterns of local and global diversity. BioSci 54:777–784

Finlay BJ (2002) Global dispersal of free-living microbial eukaryotic species. Science 296:1061–1063

Fiore-Donno AM, Berney C, Pawlowski J, Baldauf SL (2005) Higher-order phylogeny of plasmodial slime molds (Myxogastria) based on elongation factor 1-α and small subunit rRNA gene sequences. J Eukaryot Microbiol 52:1–10

Fiore-Donno AM, Meyer M, Baldauf SL, Pawlowski J (2008) Evolution of dark-spored Myxomycetes (slime-molds): molecules versus morphology. Mol Phylogenet Evol 46:878–889

Fiore-Donno AM, Haskins EF, Pawlowski J, Cavalier-Smith T (2009) *Semimorula liquescens* is a modified echinostelid myxomycete (Mycetozoa). Mycologia 101:773–776

Fiore-Donno AM, Nikolaev SI, Nelson M, Pawlowski J, Cavalier-Smith T, Baldauf SL (2010) Deep phylogeny and evolution of slime moulds (Mycetozoa). Protist 161:55–70

Fiore-Donno AM, Novozhilov YK, Meyer M, Schnittler M (2011) Genetic structure of two protist species (Myxogastria, Amoebozoa) suggests asexual reproduction in sexual amoebae. PLoS One. doi:10.1371/journal.pone.0022872

Gilbert HC, Martin GW (1933) Myxomycetes found on the bark of living trees. Univ Iowa Stud Nat His 15:3–8, 31

Graham A (1971) The role of Myxomyceta spores in palynology (with a brief note on the morphology of certain algal zygospores). Rev Palaeobot Palynol 11:89–99

Hagiwara H (1976) Distribution of the Dictyosteliaceae (cellular slime molds) in Mt. Ishizuchi, Shikoku. Trans Mycol Soc Jpn 17:226–237

Hagiwara H (1989) The taxonomic study of Japanese dictyostelid cellular slime molds. National Science Museum, Tokyo

Hammer CH (1984) Dictyostelids in agricultural soils. MS Thesis, Ohio University, Athens

Härkönen M (1977) Corticolous myxomycetes in three different habitats in southern Finland. Karstenia 17:19–32

Härkönen M (1981) Myxomycetes developed on litter of common Finnish trees in moist chamber cultures. Nord J Bot 1:791–794

Horn EG (1971) Food competition among the cellular slime molds. Ecology 52:475–484

Hutchinson GE (1951) Copepodology for the ornithologist. Ecology 32:571–577

Ing B (1994) The phytosociology of myxomycetes. New Phytol 126:175–201

Ing B (1999) The myxomycetes of Britain and Ireland: an identification handbook. Richmond Publishing, Slough

Irawan B, Clark JD, Stephenson SL (2000) Biosystematics of the *Physarum compressum* morphospecies. Mycologia 92:884–893

Kauffman G (1986) Effects of organic matter and moisture content on dictyostelids in agricultural soils. MS Thesis, Ohio University, Athens

Kawano S, Kuroiwa T, Anderson RW (1987) A third multialletic mating-type locus in *Physarum polycephalum*. J Gen Microbiol 133:2539–2546

Keller HW, Brooks TE (1976) Corticolous myxomycetes V: observations on the genus *Echinostelium*. Mycologia 68:1204–1220

Kessin RH (2001) *Dictyostelium*: evolution, cell biology, and the development of multicellularity. Cambridge University Press, Cambridge

Kowalski DT (1967) Observations on the Dianemaceae. Mycologia 59:1075–1084

Kowalski DT (1970) The species of *Lamproderma*. Mycologia 62:621–672

Kowalski DT (1971) The genus *Lepidoderma*. Mycologia 63:490–516

Lado C (2001) Nomenmyx. A nomenclatural taxabase of Myxomycetes. Cuadernos de Trabajo Flora Micológica Ibérica 16:1–221

Lado C, Mosquera J, Estrada-Torres A, Beltrán-Tejera E, Wrigley de Basanta D (2007) Description and culture of a new succulenticolous *Didymium* (Myxomycetes). Mycologia 99:602–611

Landolt JC, Cavender JC, Stephenson SL, Vadell EM (2008) New species of dictyostelid cellular slime moulds from Australia. Aust Syst Bot 21:50–66

Lewis KE, O'Day DH (1977) The sex hormone of *Dictyostelium discoideum* is volatile. Nature 268:730–731

Lister A (1894) A monograph of the Mycetozoa. Printed by order of the trustees, London

Lister A (1911) A monograph of the Mycetozoa. British Museum of Natural History, London (revised by G. Lister)

Lister A (1925) A monograph of the Mycetozoa. British Museum of Natural History, London (revised by G. Lister)

Macbride TH (1899) North American slime-moulds. Macmillan, New York, NY

Macbride TH (1922) North American slime-moulds, 2nd edn. Macmillan, New York, NY

Macbride TH, Martin GW (1934) The Myxomycetes. Macmillan, New York, NY

Madelin MF (1984) Myxomycetes, microorganisms and animals: a model of diversity in animal-microbial interactions. In: Anderson JN, Rayner DA, Walton DWH (eds) Invertebrate-microbial interactions. Cambridge University Press, New York, NY, pp 1–33

Martin GW, Alexopoulos CJ (1969) The myxomycetes. University of Iowa Press, Iowa City, IA

Martin GW, Alexopoulos CJ, Farr ML (1983) The genera of myxomycetes. University of Iowa Press, Iowa City, IA

Massee G (1892) A monography of the Myxogastres. Methuen, London

Mayr E (1970) Population, species, and evolution. Belknap Press, Harvard University Press, Cambridge

Mitchell DW (1980) A key to the corticolous myxomycetes. The British Mycological Society, Cambridge
Mosquera J, Lado C, Beltrán-Tejera E (2000) Morphology and ecology of *Didymium subreticulosporum*. Mycologia 92:978–983
Nannenga-Bremekamp NE (1991) A guide to temperate myxomycetes. Biopress, Bristol
O'Day DH, Keszei A (2012) Signalling and sex in the social amoebozoans. Biol Rev Camb Philos Soc 87:313–329. doi:10.1111/j.1469-185X.2011.00200.x
Olive EW (1901) A preliminary enumeration of the Sorophoreae. Proc Am Acad Arts Sci 37:333–344
Olive LS (1970) The Mycetozoa: a revised classification. Bot Rev 36:59–87
Olive LS (1975) The Mycetozoans. Academic, New York, NY
Olive LS, Stoianovitch C (1979) Observations of the mycetozoan genus *Ceratiomyxa*: description of a new species. Mycologia 71:546–555
Olive LS, Stoinaovitch C (1960) Two new members of the Acrasiales. Bull Torrey Bot Club 87:1–20
Page FC, Blanton RL (1985) The Heterolobosea (Sarcodina: Rhizopoda), a new class uniting the Schizopyrenida and the Acrasidae (Acrasida). Protistologica 21:121–132
Raper KB (1937) Growth and development of *Dictyostelium discoideum* with different bacterial associates. J Agr Res 55:289–316
Raper KB (1984) The Dictyostelids. Princeton University Press, Princeton, NJ
Reinhardt DJ (1975) Natural variants of the cellular slime mold *Acrasis rosea*. J Protozool 22:309–317
Roger AJ, Smith MW, Doolittle RF, Doolittle WF (1996) Evidence for the Heterolobosea from phylogenetic analysis of genes encoding glyceraldehyde-3-phosphate dehydrogenase. J Eukaryot Microbiol 43:475–485
Romeralo M, Spiegel FW, Baldauf SL (2010) A fully resolved phylogeny of the social amoebas (Dictyostelia) based on combined SSU and ITS rDNA sequences. Protist 161:539–548
Romeralo M, Cavender JC, Landolt JC, Stephenson SL, Baldauf S (2011) The expanding phylogeny of social amoebae defines new major lineages and emerging patterns in morphological evolution. BMC Evol Biol. doi:10.1186/1471-2148-11-84
Rosswall T, Paustian K (1984) Cycling of nitrogen in modern agricultural systems. Plant and Soil 76:3–21
Rostafinski JT (1873) Versuch eines Systems der Mycetozoen. Inaugural dissertation, University of Strassberg, Germany
Rostafinski JT (1874–1876) Sluzowce (Mycetozoa) monografia. Towarz Nauk Scis Paryzu 5:1–215 (1874); 217–432 (1875); Dodatek [appendix] 8:1–43 (1876)
Schaap P, Winckler T, Nelson M, Alvarez-Curto E, Elgie B, Hagiwara H, Cavender JC, Milano-Curto A, Rozen DE, Dingermann T, Mutzel R, Baldauf SL (2006) Molecular phylogeny and evolution of morphology in the social amoebas. Science 314:661–663
Schnittler M, Mitchell D (2000) Species diversity in myxomycetes based on the morphological species concept—a critical examination. Stapfia 73:55–62
Schnittler M, Stephenson SL (2002) Inflorescences of Neotropical herbs as a newly discovered microhabitat for myxomycetes. Mycologia 94:6–20
Schnittler M, Stephenson SL, Novozhilov YK (2000) Ecology and world distribution of *Barbeyella minutissima* (Myxomycetes). Mycol Res 104:1518–1523
Singh BN (1947) Studies of soil Acrasieae. 1. Distribution of species of *Dictyostelium* in soils of Great Britain and the effect of bacteria on their development. J Gen Microbiol 1:11–21
Spiegel FW, Stephenson SL, Keller HW, Moore DL, Cavender JC (2004) Mycetozoans. In: Mueller GM, Bills GF, Foster MS (eds) Biodiversity of fungi: inventory and monitoring methods. Academic, Amsterdam, pp 547–576
Stephenson SL (1988) Distribution and ecology of myxomycetes in temperate forests. I. Patterns of occurrence in the upland forests of southwestern Virginia. Can J Bot 66:2187–2207
Stephenson SL (1989) Distribution and ecology of myxomycetes in temperate forests. II. Patterns of occurrence on bark surface of living trees, leaf litter, and dung. Mycologia 81:608–621
Stephenson SL (2003) Myxomycetes of New Zealand. Fungal Diversity, Hong Kong
Stephenson SL (2011) From morphological to molecular: studies of myxomycetes since the publication of the Martin and Alexopoulos (1969) monograph. Fungal Divers 50:21–34
Stephenson SL, Cavender JC (1996) Dictyostelids and myxomycetes. In: Hall GS (ed) Methods for the examination of organismal diversity in soils and sediments. CAB International, Oxon, pp 91–101
Stephenson SL, Landolt JC (1996) The vertical distribution of dictyostelids and myxomycetes in the soil/litter microhabitat. Nova Hedwigia 62:105–117
Stephenson SL, Landolt JC (1998) Dictyostelid cellular slime molds in canopy soils of tropical forests. Biotropica 30:657–661
Stephenson SL, Landolt JC (2011) Dictyostelids from aerial "canopy soil" microhabitats. Fungal Ecol 4:191–195
Stephenson SL, Shadwick JD (2009) Nivicolous myxomycetes from alpine areas of south-eastern Australia. Aust J Bot 57:116–122
Stephenson SL, Stempen H (1994) Myxomycetes: a handbook of slime molds. Timber, Portland, OR
Stephenson SL, Landolt JC, Laursen GA (1991) Cellular slime molds in soils of Alaskan tundra. Arctic Alpine Res 23:104–107
Stephenson SL, Landolt JC, Moore DL (1999) Protostelids, dictyostelids, and myxomycetes in the litter

microhabitat of the Luquillo Experimental Forest, Puerto Rico. Mycol Res 103:209–214

Stephenson SL, Schnittler M, Novozhilov YK (2008) Myxomycete diversity and distribution from the fossil record to the present. Biodivers Conserv 17:285–301

Swanson AR, Vadell EM, Cavender JC (1999) Global distribution of forest soil dictyostelids. J Biogeogr 26:133–148

Swanson AR, Spiegel FW, Cavender JC (2002) Taxonomy, slime molds, and the questions we ask. Mycologia 94:968–979

Tamayama M (2000) Nivicolous taxa of the myxomycetes in Japan. Stapfia 73:121–129

Traub F, Hohl HR, Cavender JC (1981) Cellular slime molds of Switzerland. II. Distribution in forest soils. Am J Bot 68:172–182

Urich T, Lanzén A, Qi J, Huson DH, Schleper C, Schuster SC (2008) Simultaneous assessment of soil microbial community structure and function through analysis of the meta-transcriptome. PLoS One. doi:10.1371/journal.pone.0002527

Vadell EM, Cavender JC (1995) Dictyostelid cellular slime molds from forest soils of the Iguazú Falls and the Jesuitic Missions Ruins of Argentina. In: Lado C, Hernández JC (eds) Abstracts volume of the second international congress on the systematics and ecology of myxomycetes (ICSEM 2), vol 1, p 110

Vadell EM, Cavender JC, Romeralo M, Edwards SM, Stephenson SL (2011) New species of dictyostelids from Patagonia and Tierra del Fuego, Argentina. Mycologia 103:101–117

van Tieghem MP (1880) Sur quelques myxomycetes a plasmode agrege. Bull Soc Fr 27:317–322

Waggoner BM, Poinar GO (1992) A fossil myxomycete plasmodium from Eocene-Oligocene amber of the Dominican Republic. J Protozool 39:639–642

Winsett KE, Stephenson SL (2010) Global distribution and molecular diversity of *Didymium difforme*. Mycosphere 2:135–146

Wrigley de Basanta D (2000) Acid deposition in Madrid and corticolous myxomycetes. Stapfia 73:113–120

Yamamoto Y (1998) The myxomycete biota of Japan. Toyo Shorin, Tokyo

Yoon HS, Grant J, Tekle YI, Wu M, Chaon BC, Cole JC, Logsdon JM Jr, Patterson DJ, Bhattacharya D, Katz LA (2008) Broadly sampled multigene trees of eukaryotes. BMC Evol Biol 8:14. doi:10.1186/1471-2148-8-14

3 Systematics of the Straminipila: Labyrinthulomycota, Hyphochytriomycota, and Oomycota

Gordon W. Beakes[1], Daiske Honda[2], Marco Thines[3]

CONTENTS

I. Introduction

Osmotrophic fungal-like stramenopiles (as recognized by Adl et al. 2012) are characterized by their heterokont, predominantly biflagellate, zoospores (Fig. 3.16). Although well known from marine and freshwater habitats, these organisms also are widespread inhabitants of soils, and many are significant pathogens of plants and animals. It has long been recognized that these organisms have many structural and biochemical characteristics that separate them from the osmotrophic members of Kingdom Fungi (see Powell and Letcher 2014; James et al. 2014). Features such as having **cellulose** as the main structural polysaccharide in the cell wall, different pathways for **lysine biosynthesis**, mitochondria with **tubular vesiculate cristae**, largely **diploid** vegetative thalli, and often **β1-3 glucans** (laminarins) as their main storage carbohydrate (Dick 2001a) set these organisms apart from the members of Kingdom Fungi/Mycota.

There has been a recent trend to refer to the osmotrophic stramenopiles traditionally studied by mycologists as **"fungal-analogues"** or even as **"untrue fungi"** (Moore et al. 2011). However, Bartnicki-Garcia (1996), Dick (2001a), and Money (1998) have all eloquently argued that **"fungi"** should be considered a **biological lifestyle**, characterized by osmotrophic nutrition and yeast-like or hyphoid growth form, rather than a phylogenetic entity (Kingdom Mycota or Myco-

[1]School of Biology, Newcastle University, Newcastle upon Tyne NE1 7RU, United Kingdom; e-mail: gordon.beakes@ncl.ac.uk
[2]Department of Biology, Faculty of Science and Engineering, Konan University, 8-9-1, Okamoto, Higashinada, Kobe, Hyogo 658-8501, Japan; e-mail: dhonda@konan-u.ac.jp
[3]Centre for Integrative Fungal Research (IPF), Biodiversity and Climate Research Centre (BiK-F), Goethe University Frankfurt am Main (Department of Biological Sciences, Institute of Ecology, Evolution and Diversity), and Senckenberg Gesellschaft für Naturforschung, Senckenberganlage 25, 60325 Frankfurt am Main, Germany; e-mail: Marco.Thines@senckenberg.de

Systematics and Evolution, 2nd Edition
The Mycota VII Part A
D.J. McLaughlin and J.W. Spatafora (Eds.)

biota). Whichever view is taken, it is important that these groups continue to be considered an **integral part of mycology** and not be excluded or marginalized because of their different evolutionary origins.

Osmotrophic stramenopiles play important roles in **nutrient cycling** in both marine (Bongiorni 2012; Hulvey et al. 2010; Raghukumar 2002a, b) and freshwater (Dick 1976, 2001a) ecosystems. Many are also **parasites,** with the potential to wreak devastation at both the species and ecosystem levels. For example, Labyrinthulids are responsible for coastal sea grass bed dieback (Muehlstein and Porter 1991) as well as an emerging dieback disease of overirrigated (highly salinic) turf grasses (Olsen 2007). Thraustochytrids can cause serious economic losses to commercially reared shellfish such as clam (Azevedo and Corral 1997) and abalone (Bower 1987). On the positive side there is increasing interest in exploiting marine thraustochytrids as an alternative to fish as a source of polyunsaturated fatty acids such as docasahexaenoic acid, which are important dietary supplements for both animals (Miller et al. 2007) and humans (Domergue et al. 2005; Kobayashi et al. 2011; Raghukumar 2008). Oomycetes are responsible for a number of potentially species-threatening diseases in freshwater ecosystems, such as the **crayfish plague** caused by *Aphanomyces astaci* (Cerenius et al. 1988) and the salmon disease (ulcerative dermal necrosis) caused by *Saprolegnia parasitica* (Phillips et al. 2008; van West 2006). In addition to being one of the earliest documented diseases of cultivated plants, **potato blight** caused by *Phytophthora infestans* remains a major threat to potato production worldwide today (Haas et al. 2009). *Phytophthora cinnamomi* has been responsible for the widespread **dieback** of native forest in Australia and New Zealand (Hardham 2005; Newhook and Podger 1972), and over the past decade there have been a number of new and emerging *Phytophthora* diseases caused by species such as *Ph. ramorum* and *Ph. kernoviae*, which threaten both native and exotic tree species in North America (Davidson et al. 2003; Martin and Tooley 2008) and Europe (Brasier and Weber 2010).

The downy mildew *Hyaloperonospora arabidopsidis* and white blister rust *Albugo laibachii* both infect the model plant *Arabidopsis* and have provided excellent systems in which to explore **host-pathogen interactions** at the molecular level (Kemen and Jones 2012; Thines et al. 2009a). Much recent effort has been directed at unravelling the **molecular basis of pathogenicity** in economically important plant pathogenic oomycetes [see reviews by Jiang and Tyler (2012) and Thines and Kamoun (2010)]. However, compared with true Fungi, genetic manipulation of stramenopiles has generally proven difficult and frustrating. With a few exceptions, such as *Phytophthora capsici*, it has been difficult to develop routine, stable transformation systems for oomycetes—or, for that matter, any other stramenopile (Judelson and Ah-Fong 2009). In general, gene-silencing techniques often have been the only tool available to explore gene functions (Whisson et al. 2009). Representatives of the phytopathogenic genera *Albugo* (Kemen et al. 2011; Links et al. 2011), *Hyaloperonospora* (Baxter et al. 2010), several *Phytophthora species* (e.g. Haas et al. 2009; Tyler et al. 2006), *Pseudoperonospora* (Tian et al. 2011) and *Pythium ultimum* (Lévesque et al. 2010), and the fish pathogen *Saprolegnia parasitica* (Jiang et al. 2013) have had their full or partial genome sequences released. These add to genome sequences from other stramenopiles, including the human gut parasite *Blastocystis* (Denoeud et al. 2011), the ochrophytes *Aureococcus* (Gobler et al. 2011) and *Ectocarpus* (Cock et al. 2010), and the diatoms *Pheodactylum* (Bowler et al. 2008) and *Thallassiosira* (Armbrust et al. 2004) that have been published to date. Preliminary genome sequences for several members of the Labyrinthulomycota also have recently been released (Collier 2012). Comparative genomics promises to unlock many interesting secrets about these organisms (Lamour et al. 2007; Martens et al. 2008; Seidl et al. 2012). One surprising discovery seems to be the extent to which the genomes of oomycetes contain genes derived from other prokaryotes and eukaryotes, providing evidence of horizontal gene transfer from bacteria, true fungi, and red and green algal endosymbionts (Jiang and Tyler 2012; Maruyama et al. 2009; Richards et al. 2006; 2011; Soanes et al. 2007).

This account presents an updated phylogenetic and taxonomic overview of these organisms based primarily on molecular sequence data and then briefly reviews the biology and evolutionary history of these organisms in the context of our revised phylogenetic framework.

II. Molecular Phylogeny and Systematics

A. Higher-Level Relationships

Apart from the posteriorly uniflagellate chytrids (see Powell and Letcher 2014; James et al. 2014), all of the zoosporic organisms traditionally studied by mycologists can now be placed in the recently defined **stramenopile/alveolate/rhizaria (SAR)** superkingdom (Burki et al. 2008; Hackett et al. 2007; Reeb et al. 2009). The **cryptophytes** and **haptophytes** are now generally excluded from this assemblage (Dorrell and Smith 2011; Reeb et al. 2009). All of the flagellate osmotrophic organisms traditionally studied by mycologists, except the **plasmodiophorids**, fall within the **Straminipila/Heterokonta** branch of the SAR assemblage (Fig. 3.1).

Dick (2001a) argued that the kingdom name **Straminipila** (which is the **etymologically correct** version of the stamenopile name), should be adopted for the lineage, which encompasses all of the non-photosynthetic osmotrophic groups, instead of the name **Chromista**, because the name implies photosynthetic pigmentation. However, Cavalier-Smith and Chao (2006) still considered the **Straminipila** as defined by Dick (2001a) to be synonymous with their kingdom, **Chromista**. However, the original concept of the Chromista (Cavalier-Smith 1986) no longer fits with the **SAR superkingdom**, and its usage is now generally discontinued. Anderson and Cavalier-Smith (2012) consider the **Heterokonta** to be synonymous with the **stramenopiles** defined by Adl et al. (2012). In this mycological treatise, we continue to use **Straminipila** (as defined by Dick 2001a) as a formal kingdom designation but use the more widely adopted **stramenopiles** (Adl et al. 2012; Lévesque 2011) when generally referring to these organisms.

The overall relationships between the major groups within the **SAR lineage,** and the stramenopiles in particular (Fig. 3.1a), have recently been resolved using multiple protein-encoding genes (Reeb et al. 2009; Riisberg et al. 2009; Tsui et al. 2009). The statistically well-supported alveolate clade, comprised of the apicocomplexans, dinoflagellates, and ciliates, forms the sister clade to the heterokont stramenopiles (Keeling 2009). These can be divided into two main lineages (Fig. 3.1a). The first, the **BOL clade** (Fig. 3.1a), encompasses the bacteriotrophic flagellate Bicosoecida, the protistan gut-inhabiting Opalinida (plus proteromonads and *Blastocystis*), and the Labyrinthulomycota (Reeb et al. 2009; Riisberg et al. 2009; Tsui et al. 2009); the second, the **HOOf clade** (Fig. 3.1a), includes the osmotrophic Hyphochytriomycota and Oomycota, the photosynthetic Ochrophyta, and the phagotrophic flagellates *Developayella* and *Pirsonia*. ***Developayella*** is usually given as the sister clade to the Oomycota in small subunit (SSU) ribosomal RNA (rRNA) trees (Leipe et al. 1994; Tong 1995). However, recent molecular ecological studies have revealed that a number of unknown marine **picoeukaryote stramenopiles** (Massana et al. 2002, 2004, 2006) form a clade (e.g. the oomycete-related group) (Fig. 3.1b) that is sister to the Oomycota, and so their closest relatives remain a mystery. In this account the **Labyrinthulomycota, Hyphochytriomycota,** and **Oomycota** are assigned phylum rank within **Kingdom Straminipila.**

B. Systematics of Labyrinthulomycota

The **Labyrinthulomycota** consists of a relatively small group of what, at least until recently (Anderson and Cavalier-Smith 2012), were considered exclusively marine genera (Figs. 3.1b and 3.2, Table 3.1). They typically feed saprotrophically and are key players in the detrital food web, breaking down often intractable plant and animal remains and making these substrates more accessible to grazing amoebae and ciliates (Bongiorni 2012; Raghukumar 2002b). Some thraustochytrids feed bacteriotrophically (Raghukumar 2002a),

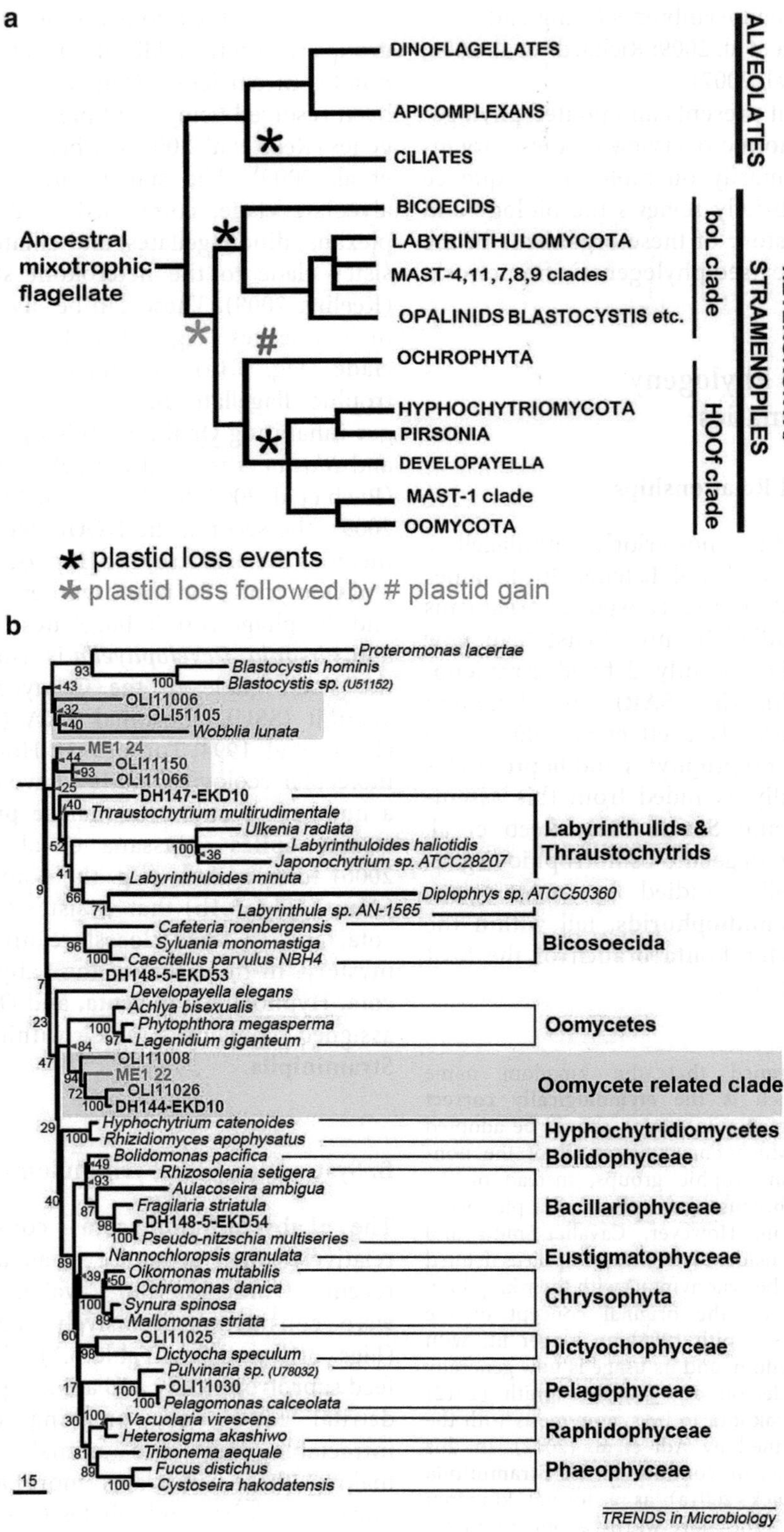

Fig. 3.1 Phylogenetic trees of stramenopiles showing the relationships between Hyphochytriomycota, Labyrinthulomycota, Oomycota, and their closest relatives. (**a**) Schematic tree based on multiple protein sequences

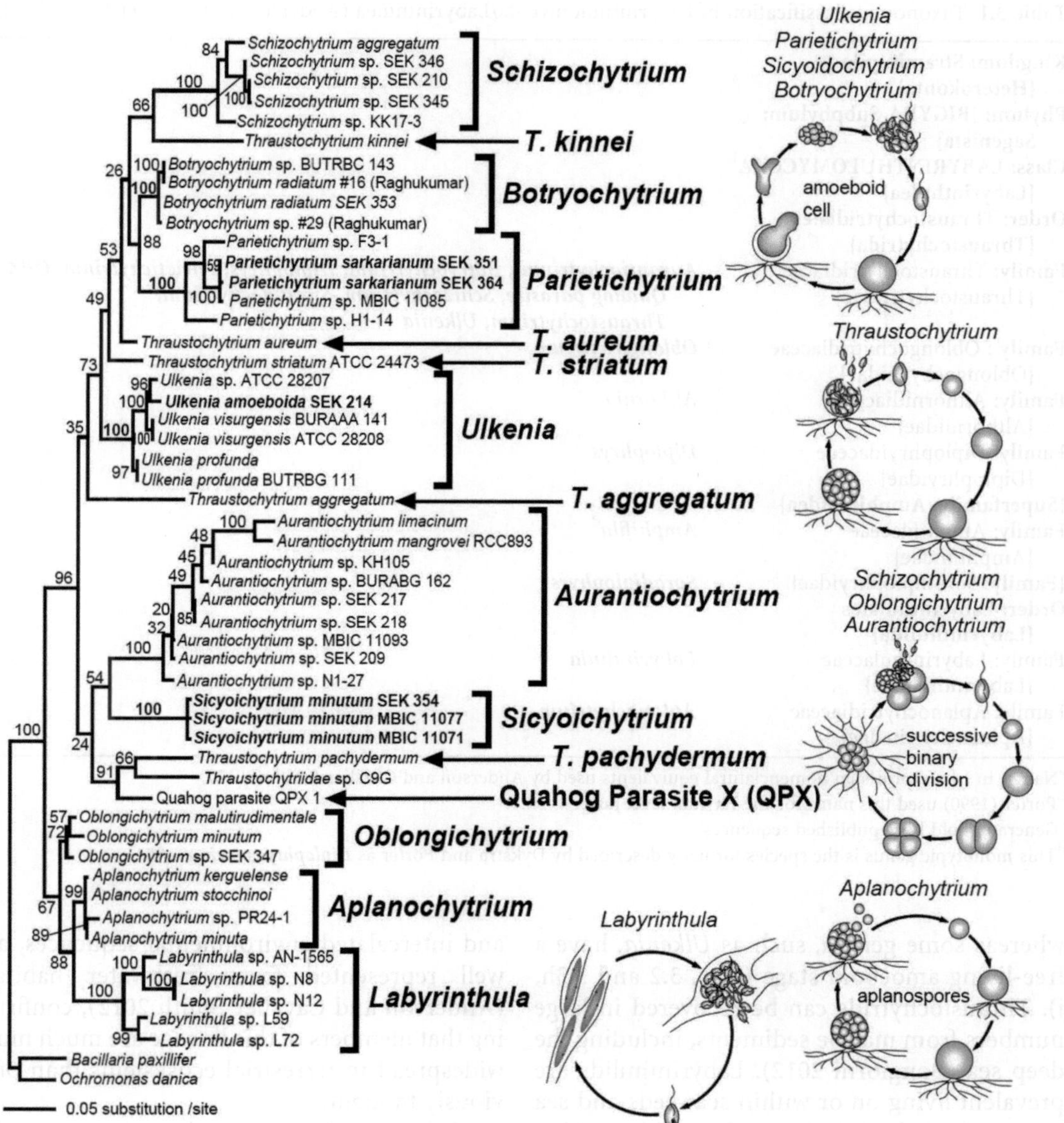

Fig. 3.2 Phylogenetic tree summarizing main phylogenetic relationships within Labyrinthulomycota based on SSU rRNA sequences, together with representative schematic life-cycle diagrams. Adapted from Yokoyama and Honda (2007) and Yokoyama et al. (2007)

Fig. 3.1 (continued) (Tsui et al. 2009) showing main clades in **alveolate and stramenopile lines.** The points where chloroplast losses may have occurred, according to Tsui et al. (2009), are indicated by *black asterisks*. An alternative hypothesis in which a single chloroplast loss event took place followed by the reacquisition of plastids by the ochrophyte line is indicated by *grey symbols* (*asterisks hash*). (**b**) Representative phylogenetic tree based on SSU rRNA sequences showing relationships between osmotrophic stramenopiles and photosynthetic ochrophytes. From Moreira and López-Garcia (2002), with permission

Table 3.1 Taxonomic classification of Labyrinthulomycota/Labyrinthulea (Anderson and Cavalier-Smith 2012)

Kingdom: Straminipila {Heterokonta}[a]	
Phylum: {BIGYRA Subphylum: Sagenista}	
Class: LABYRINTHULOMYCOTA[b] **{Labyrinthulea}**	
Order: Thraustochytridiales {Thraustochytrida}	
Family: Thraustochytridiacae {Thraustochytriidae}	***Aurantiochytrium***[c]***, Botryochytrium, Diplophrys, Parietichytrium, QPZ Quahog parasite, Schizochytrium, Sicyoidochytrium, Thraustochytrium, Ulkenia***
Family : Oblongochytridiaceae {Oblongochytriidae}	***Oblongichytrium***
Family: Althornidiaceae {Althorniidae}	*Althornia*
Family: Diplophryidaceae {Diplophryidae}	***Diplophrys***
{Superfamily: Amphifiloidea}	
Family: Amphifilaceae {Amphifilidae}	***Amphifila***[d]
{**Family:** Sorodiplophryidae}	***Sorodiplophyrs***
Order: Labyrinthulales {Labyrinthulida}	
Family: Labyrinthulaceae {Labyrinthulidae}	***Labyrinthula***
Family: Aplanochytridiaceae {Aplanochytriiadae}	***Aplanochytrium***

[a]Names in { } are protistan nomenclatural equivalents used by Anderson and Cavalier-Smith (2012)
[b]Porter (1990) used this name for the Phylum level rank
[c]Genera in bold have published sequences
[d]This monotypic genus is the species formerly described by Dykstra and Porter as ***Diplophrys marina*** (1984)

whereas some genera, such as *Ulkenia*, have a free-living amoeboid stage (Figs. 3.2 and 3.3h, j). Thraustochytrids can be recovered in large numbers from marine sediments, including the deep sea (Bongiorni 2012). Labyrinthulids are prevalent living on or within seaweeds and sea grasses, and there is increasing evidence that they can live as commensals or mutualists in plants and in other organisms, such as amoebae (Dykova et al. 2008) and mollusc tissues (Azevedo and Corral 1997). Thraustochytrids are also parasites of marine animals, such as *Mercenaria* (quahog clam; Bower 1987) and cephalopods (Jones and O'Dor 1983; Polglase 1980). A new labyrinthulid species (*Labyrinthula terrestris*) (Craven et al. 2005; Olsen 2007) has been identified as a causal agent of a serious blight disease of irrigated turf grasses (Douhan et al. 2009). Many isolates of the morphologically simple *Diplophrys*-like protists (Table 3.1), and intercalated environmental sequences, are well represented from freshwater habitats (Anderson and Cavalier-Smith 2012), confirming that members of this lineage are much more widespread in terrestrial ecosystems than previously thought.

Labyrinthulomycetes have traditionally been divided into two families, the **Labyrinthulaceae** and the **Thraustochytriaceae** (Dick 2001a; Honda et al. 1999; Porter 1990) within a single order, the **Labyrinthulales** (or **Labyrinthulida**). The former are characterized by their spindle-shaped thalli, which are enrobed in an **ectoplasmic track-like network** along which the cells freely migrate (Fig. 3.3a). The latter form ovoid or spherical thalli, which, except in the genus *Althornia* (Karling 1981; Porter 1990), are associated with a fine network of rhizoid-like **ectoplasmic threads** (Fig. 3.3c–i), which act both as anchoring and feeding struc-

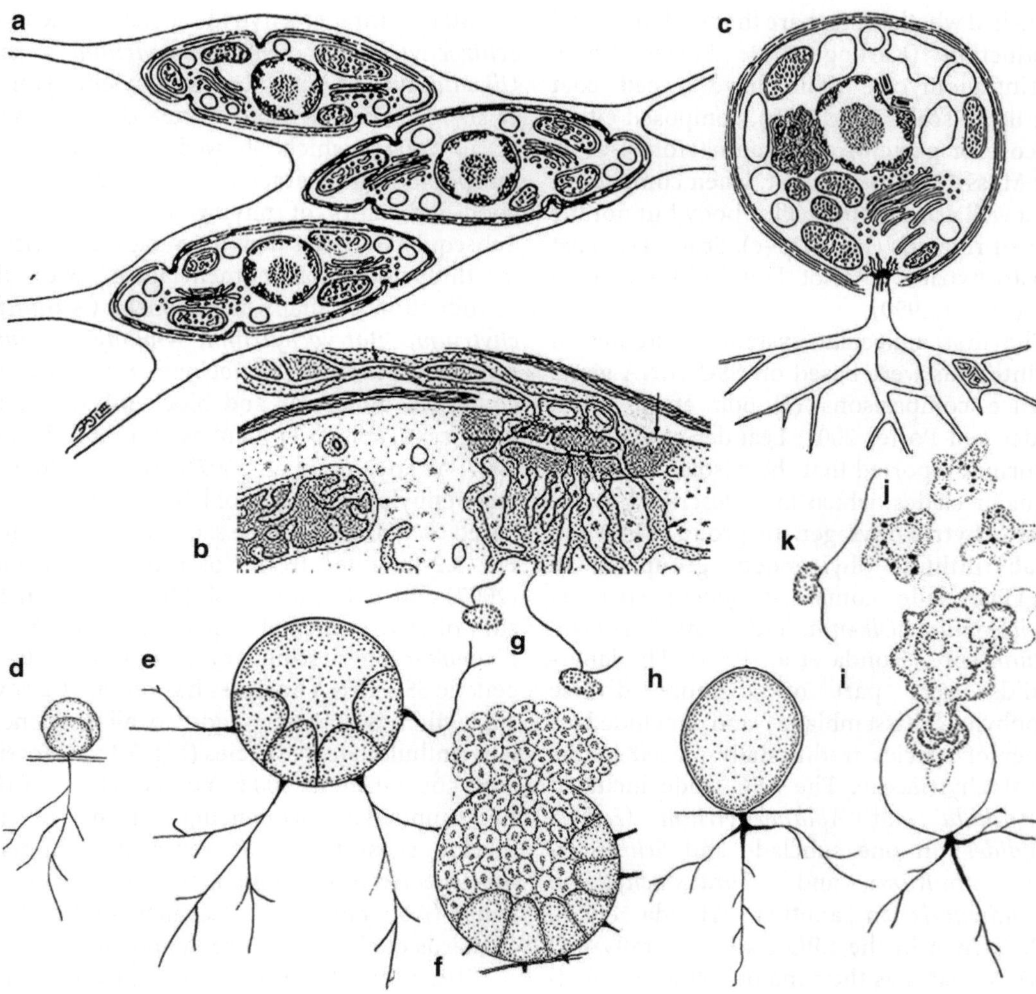

Fig. 3.3 (a–k) Morphology of Labyrinthulomycota. (**a**) Line drawing illustrating ultrastructure of uninucleate *Labyrinthula* cells enclosed by membranes of ectoplasmic network. Adapted from Porter (1990). (**b**) Ultrastructural detail of bothrosome (sagenogenetosome) complex separating labyrinthulid thallus from ectoplasmic strand in *Aplanochytrium* (formerly *Labyrinthuloides*) *minutum*, showing continuity in cisternae across complex. Adapted from Perkins (1976). (**c**) Diagram illustrating ultrastructure of immature uninucleate thallus of *Thraustochytrium* showing finely branched basal ectoplasmic network and cell coat of consolidated scales. Adapted from Porter (1990). (**d–g**) *Thraustochytrium rossi* on pollen showing internal proliferation of basal thalli (**d–f**), zoospore discharge (**f**), and biflagellate heterokont zoospores (**g**). (**h, i**) *Ulkenia amoeboida* on pollen showing mature thallus (**h**), a discharging amoeboid protoplast (**i**) that divides to form individual amoebospores (**j**) that upon settling release heterokont zoospores (**k**). Adapted from Karling (1981) based on observations of Bahnweg and Sparrow (1974)

tures (Moss 1985, 1986). A third group, the aplanochytrids, superficially resemble thraustochytrids but are able to glide slowly along ectoplasmic threads. The slime tracks and feeding rhizoids are connected to the thallus by means of a complex plug structure (Fig. 3.3b), variously known as the **sagenogenetosome** (Moss 1985) or **bothrosome** (Porter 1990). Some thraustochytrid thalli proliferate internally, as shown in *T. rossi* (Fig. 3.3d–f) (Karling 1981), whilst in genera such as *Schizochytrium*, the thalli divide by successive bipartitions (Fig. 3.2). Epibiotic resting spores are produced by some species, although it has not been established whether these are the result of sexual reproduction (Karling 1981; Porter 1990). Labyrinthulomycete thalli have a **cell coat** made up of **scales** (Fig. 3.3b), composed either of **fucose** or **galactose** derivates (Honda et al. 1999; Moss 1985, 1986), that, when compacted, form a wall around the thallus body but not the tracks or rhizoids (Fig. 3.3a–c). Scales also coat thraustochytrid, but not labyrinthulid, zoospores (Porter 1990).

The first molecular systematic studies of labyrinthulids were based on SSU rRNA gene-sequence comparisons (Honda et al. 1999; Leander and Porter 2001; Leander et al. 2004). The former reported that their isolates fell into two major clades, which they described as the **thraustochytrid phylogenetic group** (TPG) and the **labyrinthulid phylogenetic group** (LPG). The TPG clade contained genera such as *Schizochytrium*, *Ulkenia*, and many *Thraustochytrium* spp. (Honda et al. 1999). The labyrinthulids were part of a more diverse monophyletic assemblage and included a number of species traditionally placed in the Thraustochytriaceae. The LPG clade included *Labyrinthula* and *Aplanochytrium* (*Labyrinthuloides*) in one subclade and *Schizochytrium minutum* and *Thraustochytrium multirudientale* in another (Honda et al. 1999). Genera in the **LPG clade** generally had **fucose** derivates as their major cell wall constituents, whilst those in the **TPG clade** had **galactose** derivates (Honda et al. 1999). In a concurrent study, Leander and Porter (2001) recognized three major labyrinthulomycete clades (Fig. 3.2). One clade included two species, *Labyrinthuloides yorkensis* and *L. minuta*, which were subsequently transferred to genus *Aplanochytrium*, which was subsequently given family-level status, the Aplanochytriacaeae (Leander et al. 2004) as sister to the Labyrinthulaceae (Table 3.1). Most of the thraustochytrids clustered in a third clade (the thraustochytrid clade), together with the enigmatic, bothrosome-lacking, planktonic protist *Diplophrys marina* [recently transferred by Anderson and Cavalier-Smith (2012) to *Amphifila*] (Fig. 3.1b, Table 3.1) and two isolates of the *Mercenaria* (quahog clam) pathogen (QPX isolates).

Many thraustochytrid genera, such as *Schizochytrium*, *Thraustochytrium*, and *Ulkenia*, turned out to be paraphyletic (Honda et al. 1999; Leander and Porter 2001; Leander et al. 2004), which showed that traditional morphological characters (Fig. 3.2) are not good indicators of phylogenetic relatedness. Subsequent studies led to a radical revision in thraustochytrid nomenclature, with the introduction of many new genera (*Aurantiochytrium*, *Botryochytrium*, *Oblongichytrium*, *Parietichytrium*, *Sicyoidochytrium*) based on combined molecular and biochemical characteristics (Table 3.1) (Yokoyama and Honda 2007; Yokoyama et al. 2007). An example of a recent phylogenetic tree of labyrinthulomycetes based on SSU rRNA gene sequences is shown in Fig. 3.2 [adapted from Yokoyama and Honda (2007) and Yokoyama et al. (2007)]. The inclusion of environmental sequences for unknown picoeukaryote marine stramenopiles in phylogenetic SSU rRNA analyses has revealed a novel clade that is a sister clade to all sequenced labyrinthulomycote species (Fig. 3.1b) (Moreira and López-Garcia 2002; Yubuki et al. 2010). Environmental sequencing from diverse marine ecosystems has revealed a hitherto unsuspected biodiversity amongst these organisms (Anderson and Cavalier-Smith 2012; Richards et al. 2012; Worden and Not 2008).

Recently, Anderson and Cavalier-Smith (2012) proposed a revised classification of the group, summarized in Table 3.1, although we have given both the generally accepted mycological and protistological forms of their nomenclature. This scheme gives order-level status to the **Thraustochytridiales** and **Labyrinthulales** (Table 3.1). The former order contains six families, encompassing more than ten genera, whilst the latter comprises just two monotypic families, the **Labyrinthulaceae** and **Aplanochytridiaceae** (Table 3.1).

C. Systematics of Hyphochytriomycota

The **Hyphochytriomycota** (Fig. 3.4, Table 3.2) are characterized by their anteriorly flagellate, mastigonate zoospores (Figs. 3.4h, r and 3.16d) (Karling 1977; Fuller 1990, 2001). Morphologi-

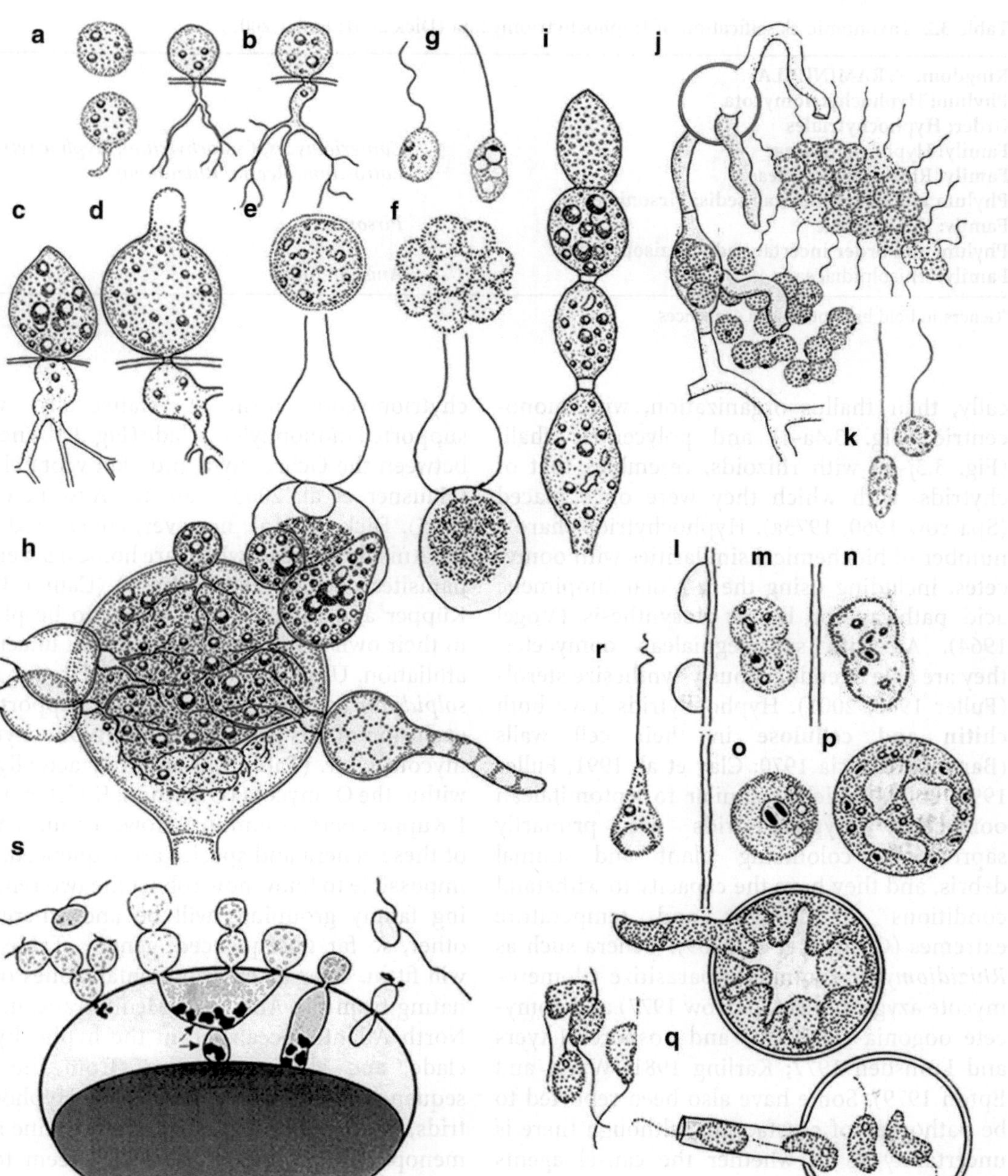

Fig. 3.4 (a–s) Morphology of Hyphochytriomycota and associated genera. (**a–h**) Drawings illustrating development of *Rhizidiomyces apophysatus* parasitizing oogonia of *Achlya racemosa*. An encysted spore germinates to produce branched rhizoids and a swollen apophysis (**a–d**) within the host tissue and an epibiontic papillate thallus (**d–e**). Mature thalli release an undifferentiated spore mass (**f**) that differentiate extrasporangially (**g**) to form anteriorly flagellate zoospores (**h**). An oogonium of *Achlya* that is heavily parasitized by *R. apophysatus* showing a range of thallus developmental stages (**i**). (**i–k**) Drawings of *Hyphochytrium catenoides* showing bead-like swellings interconnected by short hypha-like segments (**i**) and part of a mature rhizomycelial thallus showing elongate discharge tubes (**j**) that give and uniflagellate zoospores (**k**). All from Karling (1977). (**m–r**) Drawings of *Anisolpidium ectocarpi* illustrating sexual reproduction in which adjacent gametes encyst and release thalli, which immediately fuse (**m**), followed by karyogamy (**n–o**). The fused cell develops into a vacuolate sporangium (**p**), which differentiates (**q**) into uniflagellate zoospores (**r**). From Karling (1981), after Johnson (1957)

Table 3.2 Taxonomic classification of Hyphochytriomycota (Dick 2001; Fuller 2001)

Kingdom: STRAMINIPILA	
Phylum: Hyphochytriomycota	
Order: Hyphochytriales	
Family: Hyphochytriaceae	*Canteriomyces, Cystochytrium,* ***Hyphochytrium***[a]
Family: Rhizidiomycetaceae	*Latrostium, Reesia,* ***Rhizidiomyces***
Phylum and order incertae sedis: Pirsoniales	
Family: Pirsoniaceae	***Pirsonia***
Phylum and order incertae sedis: Anisopidiales	
Family: Anisolpidiaceae	*Anisolpidium*

[a]Genera in bold have published sequences

cally, their thallus organization, with monocentric (Fig. 3.4a–i) and polycentric thalli (Fig. 3.3j–k) with rhizoids, resembles that of chytrids, with which they were once placed (Sparrow 1960, 1973a). Hyphochytrids share a number of biochemical similarities with oomycetes, including using the α-ε-diaminopimelic acid pathway for lysine biosynthesis (Vogel 1964). As with saprolegnialean oomycetes, they are able to endogenously synthesize sterols (Fuller 1990, 2001). Hyphochytrids have both **chitin and cellulose** in their cell walls (Bartnicki-Garcia 1970; Clay et al. 1991; Fuller 1990, 2001), which is similar to leptomitalean oomycetes. Hyphochytrids are primarily saprotrophs colonizing plant and animal debris, and they have the capacity to withstand conditions of drought and temperature extremes (Gleason et al. 2009). Genera such as *Rhizidiomyces* commonly parasitize Glomeromycote azygospores (Sparrow 1977) and oomycete oogonia (Fig. 3.4i) and oospores (Ayers and Lumsden 1977; Karling 1981; Wynn and Epton 1979). Some have also been reported to be pathogens of crustaceans, although there is uncertainty as to whether the causal agents described were hyphochytrids (Fuller 1990, 2001).

Traditionally the hyphochytrids have been divided into three families, the monocarpic **Rhizidiomycetaceae** (Fig. 3.4a–f), the polycarpic **Hyphochytriaceae** (Fig. 3.4j–l), and the endobiotic **Anisolpidiaceae** (Fig. 3.4m–q) (Canter 1950; Fuller 1990; Karling 1981). To date, SSU rRNA sequences for two species, *Hyphochytrium catenoides* (Fig. 3.4j–l) and *Rhizidomyces inflatus* (Fig. 3.4a–h), confirm that the Hyphochytriomycota form a statistically well-supported monophyletic clade (Fig. 3.1b) nested between the Ochrophyta and Oomycota clades (Hausner et al. 2000; Van der Auwera et al. 1995). Dick (2001a), however, considered that the Anisolpidiaceae, which are holocarpic endoparasites of algae (Fig. 3.4m–r) (Canter 1950; Küpper and Müller 1999), ought to be placed in their own order (Anisopidiales) of uncertain affiliation. Unpublished sequence data for *Anisolpidium ectocarpi* (Fig. 3.4m–r) support the exclusion of this genus from the Hyphochytriomycota s. str. (Table 3.2) and may actually fall within the Oomycota (C Gachon, K Fletcher and F Kupper, pers. commun.). However, until more of these genera and species are sequenced, it is impossible to know how robust the two remaining family groupings will be and where the other, so far unsequenced, genera (Table 3.2) will fit in. Over 20 environmental clones originating from the Antarctic, Mediterranean, and North Atlantic ocean fell in the hyphochytrid clade, and all were distinct from the two sequenced species (Diéz et al. 2001). Hyphochytrids, in common with most other marine stramenopiles (Richards et al. 2012), seem to be widespread in marine ecosystems, although the niches they occupy remain to be discovered.

Astonishingly, a little-known monotypic biflagellate phagotrophic parasitoid of marine diatoms, ***Pirsonia***, apparently formed the sister clade (Fig. 3.1a) to the two sequenced hyphochytrid species (Kühn et al. 2004). However, in more recent analyses, they were shown as separate but closely related clades (Yubuki et al. 2010). For the time being, *Pirsonia* is considered *incertae sedis* but closely related to the hyphochytrids (Table 3.2).

Table 3.3 Taxonomic summary of Oomycota from Sparrow (1960, 1976)

Saprolegnialean galaxy	
Order: Saprolegniales	
Family: Saprolegniaceae	*Achlya, Aphanodictyon, Aphanomyces, Aplanes, Brevilegnia, Calyptralegnia, Dictyuchus, Geolegnia, Isoachlya, Leptolegnia, Plectospira, Protoachlya, Pythiopsis, Saprolegnia, Sommerstorffia, Thraustotheca*
Family: Leptolegniellaceae	*Aphanomycopsis, Brevilegniella, Leptolegniella*
Family: Ectrogellaceae	*Ectrogella, Pythiella*
Order: Leptomitales	
Family: Leptomitaceae	*Apodachya, Apodachyella, Leptomitus*
Order Eurychasmales	
Family: Atkinsiellaceae	*Atkinsiella*
Family: Eurychasmaceae	*Eurychasma, Eurychasmidium*
Peronosporalean Galaxy	
Order Peronosporales	
Family: Peronosporaceae	*Basidiophora, Bremia, Plasmopara, Peronospora, Peronoplasmopara, Sclerospora*
Family: Albuginaceae	*Albugo*
Family: Pythiaceae	*Diasporangium, Phytophthora, Pythium, Pythiogeton*
Order Lagenidiales	
Family: Lagenidiaceae	*Lagena, Lagenidium, Myzocytium*
Family: Olpidiopsidaceae	*Olpidiopsis, Petersenia, Pseudosphaerita, Rozellopsis*
Family: Sirolpidiacae	*Haliphthoros, Pontisma, Lagenidium, Sirolpidium*
Order Rhipidiales	
Family: Rhipidiaceae	*Araiospora, Aqualinderella, Mindeniella, Rhipidium, Sapromyces*

D. Systematics of Oomycota

The phylum Oomycota (Figs. 3.7, 3.8, 3.9, 3.10, 3.11, 3.12, 3.13, 3.14, and 3.15, Tables 3.3, 3.4, and 3.5) represents the largest group of osmotrophic stramenopiles, with well over 1,000 species, encompassing marine, freshwater, and terrestrial species (Dick 2001a, b). They are ubiquitous **saprotrophs** in freshwater and soil ecosystems (Sparrow 1960), and their ecology and roles have been extensively discussed (e.g. Dick 1976, 1990, 2001a, b; Sparrow 1960). Many oomycetes are **opportunistic necrotrophic pathogens**, and others are more specialized hemibiotrophic and **obligate biotrophic** pathogens of plants and animals. Many seem to be extremely versatile in their ability to exploit varied niches. For instance, *Pythium insidiosum* has been isolated from both plant and invertebrate material and is also an opportunistic pathogen of mammals, including humans (Mendoza 2009). Recent genomic studies on pathogenic species in genera such as *Aphanomyces* (Gaulin et al. 2008; Krajaejun et al. 2011), *Saprolegnia* (Torto-Alalibo et al. 2005; Wavra et al. 2012), and *Pythium* (Lévesque et al. 2010) have shown they possess a formidable array of **glucanase- and proteinase-encoding genes,** which have enabled them to so successfully exploit a wide range of plant and animal hosts (Jiang and Tyler 2012; Jiang et al. 2013). Genomic studies have also revealed a startling array of **pathogenicity factors and effector molecules**, which presumably have enabled *Phytophthora* species (Lamour et al. 2007; Morgan and Kamoun 2007; Qutob et al. 2002), downy mildew species (Baxter et al. 2010), and white blister rusts (Kemen et al. 2011; Links et al. 2011) to become such successful plant pathogens.

Sparrow (1960, 1973b, 1976) provided the taxonomic framework for the oomycetes that was used throughout most of the second half of the twentieth century (Table 3.3). Traditionally, the oomycetes were divided into four orders: the **Saprolegniales, Peronosporales, Lagen-**

Table 3.4 Taxonomic summary of Oomycota from Dick (2001)

Kingdom: STRAMINIPILA	
Phylum: Heterokonta	**Subphylum: Peronosporomycota**
Class: Peronosporomycetes	
Subclass: Saprolegniomycetidae	
Order: Saprolegniales	
Family: Saprolegniaceae	*Achlya, Aphanodictyon, Aplanes, Brevilegnia, Calyptralegnia, Couchia, Dictyuchus, Geolegnia, Hydatinophagus, Protoachlya, Pythiopsis, Saprolegnia, Scoliolegnia, Sommerstorffia, Thraustotheca*
Family: Leptolegniaceae	*Aphanomyces, Leptolegnia, Plectospira*
Order Sclerosporales	
Family: Sclerosporaceae	*Peronosclerospora, Sclerospora*
Family: Verrucalvaceae	*Pachymetra, Verrucalvus*
Order Salilagenidiales	
Family: Haliphthoraceae	*Atkinsiella, Haliphthorus, Halodaphnea*
Family: Salilagenidaceae	*Salilagenidium*
Order Leptomitales	
Family: Apodachlyellaceae	*Apodachlyella, Eurychasmopsis*
Family: Ducellieriaceae	*Ducellieria*
Family: Leptomitaceae	*Apodachlya, Leptomitus, Plerogone*
Family: Leptolegniellaceae	*Aphanomycopsis, Brevilegniella, Cornumyces, Leptolegniella, Nematophthora*
Subclass: Peronosporomycetidae	
Order Peronosporales	
Family: Peronosporaceae	*Basidiophora, Benua, Bremia, Bremiella, Paraperonospora, Plasmopara. Peronospora, Psudoperonospora, Peronospora*
Family: Albuginaceae	*Albugo*
Order Pythiales	
Family Pythiaceae	*Diasporangium, Cytosiphon, Halophytophthora, Lagenidium s.str., Myzocytium, Phytophthora, Pythium, Trachysphaera*
Family Pythiogetonaceae	*Medusoides, Pythiogeton*
Subclass Rhipidiomycetidae	
Order Rhipidiales	
Family Rhipidiaceae	*Araiospora, Aqualinderella, Mindeniella, Nellymyces, Rhipidium, Sapromyces*
Incertae sedis	
Order Olpidiopsidales	
Family Olpidiopsidaceae	*Olpidiopsis, Pleocystidium*
Family Sirolpidiaceae	*Haliphthoros, Pontisma, Lagenidium, Sirolpidium*
Order Myzocytiopsidales	
Family Myzocytiopsidaceae	*Chlamydomyzium, Gonimocheate, Syzgangia, Myzocytiopsis*
Family Cryptocolaceae	*Crypticola*
Order Lagenismatales	
Family Lagenismataceae	*Lagenisma*
Order Haptoglossales	
Family Haptoglossaceae	*Haptoglossa*
Order Ectrogellales	
Family Ectrogellaceae	*Ectrogella*
Incertae sedis	
Family Lagenaceae	*Lagena*
Family Sirolpidiaceae	*Sirolpidium*
Family Pontimaceae	*Petersenia, Pontisma*
Family Eurychasmaceae	*Eurychasma, Eurychasmidium*
Excluded families	
Family Rozellopsidaceae	*Rozellopsis*
Family Pseudosphaeritaceae	*Pseudosphaerita, Sphaerita, Plasmophagus*

Table 3.5 Proposed taxonomic revision of Oomycota based on molecular sequence data

Kingdom: STRAMINIPILA	
Phylum: Oomycota	
Basal orders—Class(es) incertae sedis	
Order Eurychasmales	
Family Eurychasmaceae	***Eurychasma***[a]
Family *incertae sedis* Ectrogellaceae	***Ectrogella***[b]
Family *incertae sedis* Lagenismataceae	*Lagenisma*[b]
Order Haptoglossales	
Family Haptoglossaceae	***~Haptoglossa***
Order Olpidiopsidales s. lat.	
Family Olpidiopsidacae s. lat.	***~Olpidiopsis*** s. lat.
Family Pontismataceae	*Petersenia*[b], *Pontisma*[b]
Family Sirolpidiaceae	*Sirolpidium*
Order Haliphthorales	
Family Haliphthoraceae	***Haliphthorus, Halocrusticida, Halioticida***
Order incertae sedis	
Family *incertae sedis* Pseudosphaeritacae	*Pseudosphaerita*
Family *incertae sedis* Rozellopsidaceae	*Rozellopsis*
Class: Saprolegniomycetes	
Order Atkinsiellales	
Family Atkinisellaceae	***Atkinsiella***
Family Crypticolaceae	***Crypticola***
Order Leptomitales	
Family Leptomitaceae	***Apodachlya***, *Apodachlyella*, ***Leptomitus***
Family Leptolegniellaceae s. lat.	*Aphanomycopsis*[b], *Brevilegniella*[b], ***Chlamydomyzium, Cornumyces***, *Ducellieria*[b], *Eurychasmopsis*[b], *Leptolegniella*[b], *Nematophthora*[b], *Pythiella*[b]
Order Saprolegniales	
Family Verrucalvaceae	***~Aphanomyces, Pachymetra, Plectospira***, *Sommerstorffia*[b], ***Verrucalvus***
Family Achlyaceae	***Achlya s.str, Brevilegnia, Dictyuchus, Thraustotheca***
Family Saprolegniaceae s.str.	***Aplanes, Aplanopsis, Calyptralegnia, Geolegnia, Isoachlya, Leptolegnia***[b], ***Newbya, Protoachlya, Pythiopsis, ~Saprolegnia***, *Scoliolegnia*[b]
Class: Peronosporomycetes	
Order Rhipidiales	
Family Rhipidiaceae	*Araiospora*[b], *Aqualinderella*[b], *Mindeniella*[b], *Nellymyces*[b], *Rhipidium*[b], ***Sapromyces***
Order Albuginales	
Family Albuginaceae	***Albugo, Pustula, Wilsonia***
Order Peronosporales s. lat.	
Family Salisapiliaceae	***Salisapilia***
Family Pythiaceae s. lat.	
Subclades with holocarpic "lagenidiaceous" thalli	*Gominocheate*[b], ***~Lagenidium***, *Lagena*[b], ***~Myzocytiopsis (part)***
Subclades with spp. with filamentous sporangia (*Pythium* clades A B C D)	***Pythiogeton, ~Pythium s.str,***
Subclades with more or less globose sporangia (*Pythium* Clades E F G H I J)	***Sphaerosporangium***[c], ***Elongisporangium***[c],
Family Peronosporaceae s. lat.	
Part 1 Most marine saprotrophic *Halophytophthora* sp. and *Pythium* K clade	***~Halophytophthora*** s. lat., ***Phytopythium*** *(syn. Ovatsporangium*[c]*)*
Part 2: *Phytophthora* clades 1–5 (with papillate sporangia); *Phytophthora* clades 6–10 (with non or semi-papillate sporangia)	***~Phytophthora***

(continued)

Table 3.5 (continued)

Part 3: Graminicolus downy mildews (GDM)	***Eraphthora, Graminivora, Peronosclerospora, Poakatesthia*** ***Sclerospora, Scleropthora, Viennotia***
Part 3: Brassicolous downy mildews (BDM)	***Hyaloperonospora, Perofascia***
Part 3: Downy mildews with coloured conidia (DMCC)	***Peronospora, Pseudoperonospora***
Part 3: Downy mildews with pyriform haustoria (DMPH)	***Basidiophora, Benua, Bremia, Novotelnova Paraperonospora, Plasmopara, Plasmoverna, Protobremia***

[a]Genera in **bold** have been sequenced
[b]Genus placement requires confirmation
[c]Genera introduced by Uzuhashi et al. (2010)
Tilde denotes genus which is not monophyletic and will need revision

idiales, and **Leptomitales**. In his final synopsis, Sparrow added a further two orders, the **Rhipidiales**, which he split from the **Leptomitales,** and the **Eurychasmales,** which he split from the **Saprolegniales** (Table 3.3) (Sparrow 1976). The most recent taxonomic revision of the group (summarized in Table 3.4) was made by Dick (2001a) based on a critical and scholarly re-evaluation of primarily morphological data (Table 3.4).

As part of this revision Dick (2001a) introduced the subphylum and class names **Peronosporomycotina** and **Peronosporomycetes** respectively (Table 3.4), in place of the traditional informal group/class names **oomycete/Oomycota** first introduced by Winter in 1879 (Dick 2001a). He argued that higher-level nomenclature should also reflect the type genus (*Peronospora*) rather than a morphological character (the oospore) after which the group had traditionally been named. As recommended by Lévesque (2011), the common name oomycete is retained in this account, and we use **Oomycota** as the formal phylum name.

However, since the revision of Dick (2001a) was published, many molecular taxonomic studies of oomycetes have been conducted (examples shown in Fig. 3.5), and it is now apparent that many of his changes are not supported by sequence data (cf. Tables 3.4 and 3.5). The oomycetes nevertheless form a statistically well-supported **monophyletic clade** (Figs. 3.1b and 3.5). Two single-locus [SSU rRNA and large subunit (LSU) rRNA] phylogenetic trees are given in Fig. 3.5, which include unpublished sequence data for a number of recent isolates of less-studied taxa (Fig. 3.5b). Recent molecular studies have shown that many of the traditional taxonomic paradigms used to classify these organisms require re-evaluation (Runge et al. 2011a; Thines 2009; Voglmayr 2008), and many order- and family-level circumscriptions need to be redefined. Resolving the taxonomic position of many of the smaller, less-studied orders is also problematic due to either a complete lack of sequence data or under-represented taxon sampling (Table 3.5).

In the following section on the preliminary taxonomic framework for the oomycetes, any newly introduced taxonomic names that have not been formally described are indicated by quotation marks (" ") at first mention. Orders and families that molecular data suggest are not monophyletic but that cannot at present be properly resolved are designated by a ~ symbol before their name, indicating that they will require taxonomic revision in the future. Two **monophyletic class-level clades**, the **"Saprolegniomycetes"** and **Peronosporomycetes** [but note that this name is used in a revised way from that of Dick (2001)], are recognized in this account (Fig. 3.6, Table 3.5); they correspond approximately to the two galaxies proposed by Sparrow (1976) (Table 3.3). In addition, a number of less-resolved **early-diverging clades** will almost certainly merit class-level designation (Fig. 3.6). However, until more robust data are available, they are grouped into four orders but are not assigned to classes at present and are listed as *incertae sedis* (Table 3.5). A revised

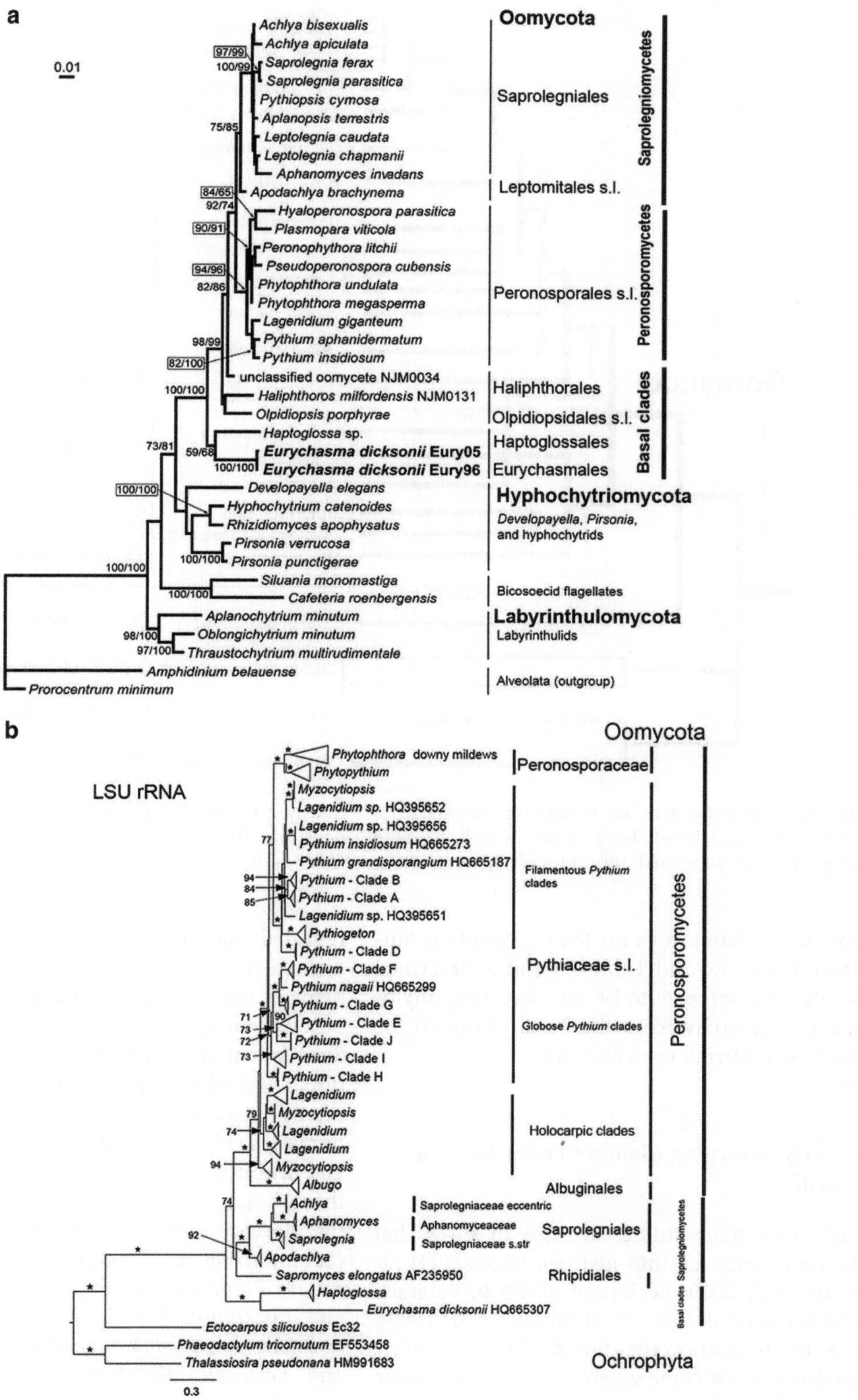

Fig. 3.5 Phylogenetic trees of Oomycota illustrating phylogenetic relationships based on representative SSU (**a**) and LSU rRNA (**b**) phylogenetic trees. The main order- and family-level clades are labelled. (**a**) Adapted from Sekimoto et al. (2008) with permission and (**b**) from Spies and Lévesque (unpublished)

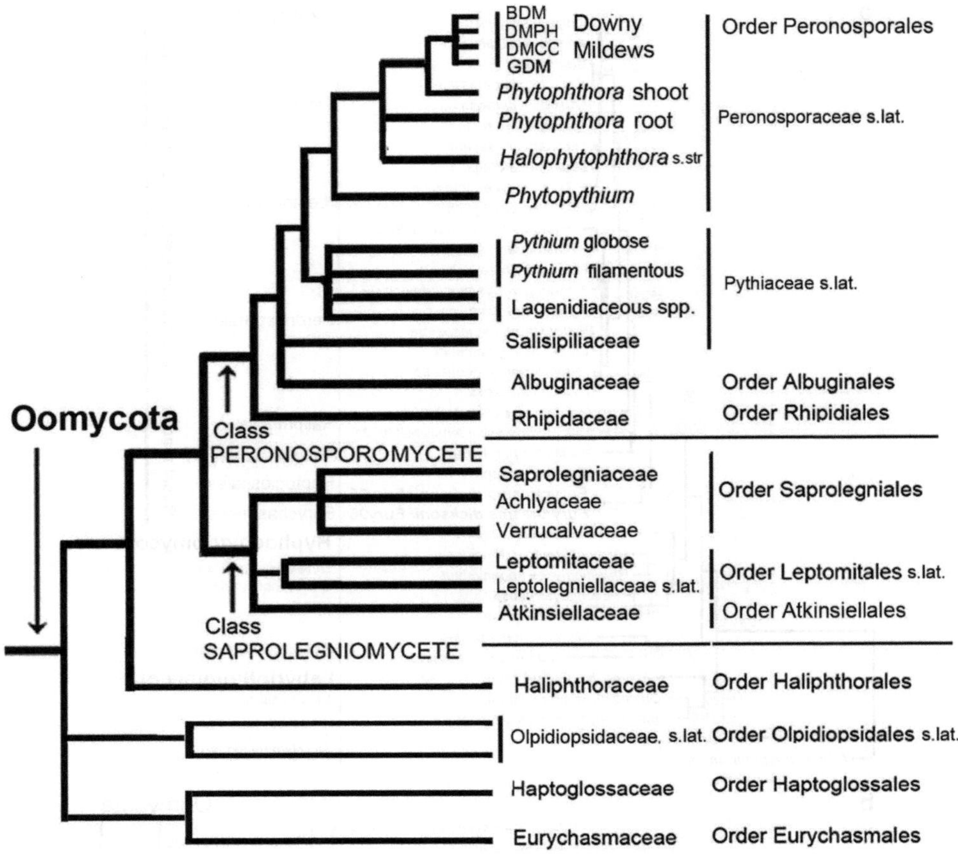

Fig. 3.6 Schematic tree of Oomycota summarizing main order- and family-level clades as well as some taxonomically undefined subclades. *BSM* brassicolus downy mildew, *DMPH* downy mildew with pyriform haustoria, *DMCC* downy mildew with coloured conidia, *GDM* graminicolus downy mildew

taxonomic framework for the Oomycota is outlined in Fig. 3.6 and Table 3.5 and is described in what we perceive to be an ascending phylogenetic hierarchy from the basal clades to the most species-rich crown groups.

E. Early Diverging Clades: Classes *Incertae Sedis*

Early molecular studies seemed to show that the oomycetes fell into just two lineages (Hudspeth et al. 2000; Léclerc et al. 2000; Petersen and Rosendahl 2000; Riethmüller et al. 1999). The first indication that this might be an overly simplistic view came when marine genera were included in analyses (Cook et al. 2001). This revealed that *Haliphthoros* and related genera formed a well-supported clade that diverged before the main Saprolegniomycete and Peronosporomycete lines separated (Fig. 3.5). Subsequent studies revealed a number of additional **early diverging genera** that are predominantly **marine organisms,** and most are **parasites** (Figs. 3.6 and 3.7, Table 3.5) (Beakes and Sekimoto 2009; Beakes et al. 2012). Many of these basal oomycetes have **holocarpic thalli** that develop initially from **unwalled plasmodia** (Sekimoto et al. 2008a). The hyperparasitic genera *Pseudosphaerita* (Anderson et al. 1995) and *Rozellopsis* (Held 1981) and the diatom parasites *Lagenisma* (Schnepf et al. 1978a, b) and *Ectrogella* (Raghukumar 1980) are also reported to form plasmodial thalli in the early stages of infection and will probably fall amongst these basal clades (Table 3.5). In the absence of sequence data they are listed as *incertae sedis*, but we place them to the extent possible where we think they are most likely to belong (Table 3.5).

1. Eurychasmales

The order Eurychasmales was first introduced by Sparrow (1976) to encompass two small monotypic marine families, the **Atkinsiellaceae** and **Eurychasmaceae**, which are parasites of crustaceans and seaweeds respectively (Table 3.3). However, molecular data have shown that *Eurychasma* and *Atkinsiella* are not closely related (Cook et al. 2001), and therefore this order is now restricted to just a single family, the **Eurychasmaceae.** ***Eurychasma dicksonii*** (Figs. 3.5a and 3.7a–d) is an obligate parasite of wide geographic distribution that infects a broad range of filamentous brown seaweeds, such as *Choristocarpus*, *Ectocarpus*, and *Pylaiella* (Gachon et al. 2009; Küpper and Müller 1999; Sekimoto et al. 2008a; Strittmatter et al. 2009). This pathogen causes host cells to become greatly enlarged without adversely affecting cytoplasmic integrity, at least during the early stages of infection (Sekimoto et al. 2008a). The flask-shaped thallus (Fig. 3.7a–c) forms one or more domed papillae that rupture the algal wall, allowing the pyriform zoospores to escape (Fig. 3.7d). The characteristic feature of this genus is that the primary cysts (aplanospores) line the sporangium wall and release their zoospores internally. This gives the mature sporangium a distinctive **net-like appearance** (Fig. 3.7c) (Sekimoto et al. 2008a; Sparrow 1960). Phylogenetic analyses based on SSU rRNA (Küpper et al. 2006), LSU rRNA (Fig. 3.5), and *cox*2 genes (Sekimoto et al. 2008a) all show that *Eurychasma* forms the **earliest diverging clade** in the oomycete tree (Fig. 3.5b). A prelimary genomic study (Grenville-Briggs et al. 2011) revealed that this pathogen has a number of unique pathogenicity factors, including many putative algal-cell-wall-degrading enzymes, which are very different from those described in higher plant pathogens. The unsequenced parasite of red seaweeds *Eurychasmidium* is also placed in this family (Dick 2001a; Karling 1981).

2. Haptoglossales

The monotypic order (family **Haptoglossaceae**) was originally included by Karling (1981) in the Lagenidiales. *Haptoglossa* (Fig. 3.7e–l) contains more than ten species that are parasites of nematodes and rotifers (Beakes and Glockling 1998, 2000, 2002; Glockling and Beakes 2000a, b, c, 2001; Hakariya et al. 2002, 2007, 2009; Karling 1981). Dick (2001a) speculated that because of functional similiarities in their infection cells (Fig. 3.7l) *Haptoglossa* might be more closely related to plasmodiophorids than to oomycetes. He placed the genus in a monotypic order, the Haptoglossales, *incertae sedis* (Table 3.4) (Dick 2001a). This genus forms unsegmented sausage-like thalli with a distinctive refractile cytoplasm (Fig. 3.7e, i). Mature thalli form exit tubes that rupture the nematode cuticle (Fig. 3.7i). The genus includes both aplanosporic (Fig. 3.7i) and zoosporic (Fig. 3.7e, f) species (Karling 1981). Some species, such as *Haptoglossa heteromorpha* (Fig. 3.7i–k) (Glockling and Beakes 2000c), *Haptoglossa polymorpha* (Glockling and Beakes 2001), and *Haptoglossa erumpens* (Beakes and Glockling 2002), produce two or three different types of infection cell (Fig. 3.7j, k), which suggests they infect multiple, as yet uncharacterized, hosts. The ultrastructure of these intricate infection cells (**gun cells**) has been described for several species (Fig. 3.7l) (Beakes and Glockling 1998, 2002; Glockling and Beakes 2000c; Robb and Barron 1982) and has revealed they contain an inverted tube containing a needle-like structure (Fig. 3.7g, h, l). Upon contacting a host the tube explosively everts, forcing the needle through the nematode cuticle and instanteously injecting an infective **sporidium** into the host body cavity (Glockling and Beakes 2000b; Robb and Barron 1982).

In *cox*2-based trees (Hakariya et al. 2007, 2009; Sekimoto et al. 2008a), *Haptoglossa* forms a monophyletic clade, which diverges immediately after *Eurychasma*, whereas in SSU and

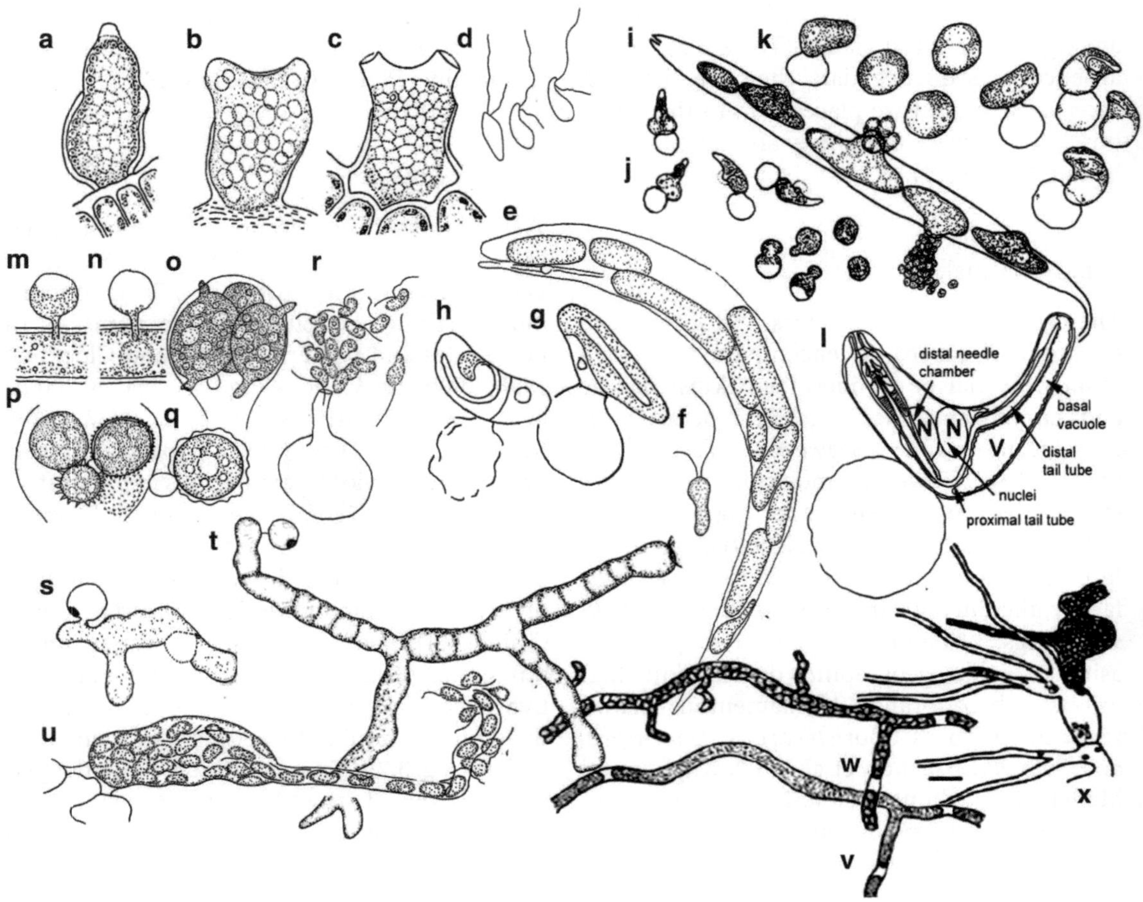

Fig. 3.7 (a–x) Morphology of genera from early-diverging clades. (**a–d**) Drawings of ***Eurychasma dicksonii*** infecting *Ectocarpus*. The mature thalli develop one (**a**) or more (**b**) exit papillae. The primary aplanospores are formed around the thallus periphery (**a**) to form a net-like array (**c**) following release of biflagellate zoospores (**d**). Adapted from Karling (1981) based on original work by Sparrow (1934). (**e–k**) Drawings of *Haptoglossa* spp. infecting rhabditid nematodes. (**e**) Nematode infected with young cylindrical thalli of ***H. heterospora*** (**e**). Biflagellate zoospores (**f**) of *H. zoospora*, which encyst and form gun cells (**g**) in which an inverted tube develops. From Karling (1981), after Drechsler (1940). A nematode infected with thalli of *H. heteromorpha* (**h**). Dimorphic aplanospores are released that form either large uninucleate gun cells (**i**) or smaller binucleate aplanopsores (**j**). From Glockling and Beakes (2000b), with permission. Drawing of a median section through a mature binucleate gun cell of ***H. erumpens***, showing inverted tube, containing needle that punctures nematode cuticle on eversion (**k**). From Glockling and Beakes (2000c), with permission (**m–q**). Drawings of ***Olpidiopsis saprolegniae*** infecting hyphae and oogonia of ***Saprolegnia ferax***. Transfer of plasmodial thallus from cyst into hyphal cytoplasm (**m, n**), which develops into a papillate thallus (**o**). Spiny resting spore stages within an infected oogonium (**p**) and a detail of a resting oospore and an adjacent antheridial thallus (**q**). Mature thallus of ***O. glenodinianum*** (**r**) showing direct release of zoospores from elongate discharge tube, which is typical of the whole genus. All from Karling (1981), from studies of Barrett (1912), Coker (1923) and Johnson (1966, 1972). (**s–u**) Drawings of ***Haliphthoros milfordensis***. Cysts germinate to produce lobed and branching thalli (**s, t**). A sporangial compartment from which zoospores are being released via a narrow discharge tube (**u**). From Karling (1981), after Sparrow (1974). (**v, w**) Drawings of young (**v**) and mature (**w**) thallus of the abalone parasite ***Halioticida noduliformans***, showing long hypha-like segments that differentiate zoospores, separated by narrow spaces. (**x**) Irregularly swollen thallus of ***Halocrusticida parasitica*** showing mature zoosporangium with slender branched discharge tubes. From Hatai (2012), with permission

LSU rRNA trees (Fig. 3.5), *Haptoglossa* seems to share a common ancestor with *Eurychasma*, although with long branch separation (Fig. 3.5b). There are few obvious features that these two genera share, which is why we have chosen to keep them in separate orders. *Haptoglossa* is also unusual amongst early-diverging genera in that it is predominantly a terrestrial genus, although *H. heterospora* has been reported to infect marine nematodes (Newell et al. 1977). The isolates of *Haptoglossa* that have so far been sequenced cluster in two or three clades (Hakariya et al. 2007, 2009), suggesting new genera will almost certainly need to be defined.

3. Olpidiopsidales s. lat.

The ~Olpidiopsidales as defined here encompass a number of relatively poorly studied groups of families [**~Olpidiopsidaceae** (Fig. 3.7m–r), **Pontismataceae** and **Sirolpidiaceae** (Tables 3.3 and 3.5)] consisting of rather small holocarpic species that had traditionally been placed in the Lagenidiales (Karling 1981). Dick (2001a) considered these families to be *incertae sedis* (Table 3.4). They are a fairly diverse assemblage of parasites of algae, protozoans, and other oomycetes that have been documented in detail by Karling (1981). A number cause economically significant losses of commercially cultivated seaweeds, such as *Olpidium porphyrae* (Sekimoto et al. 2008b), the causal agent of "red rot" disease of nori, and *Petersenia* spp., which cause "green rot" disease of *Chrondrus crispus* (Craigie and Correa 1996) and the "wasting disease" of dulse, *Palmaria palmata* (Pueschel and van der Meer 1985). Most have simple spherical or ovoid thalli that fill their infected host cells and produce one or more tube-like discharge tubes (Fig. 3.7o) to permit zoospore escape (Fig. 3.7r). The type species, ***Olpidium saprolegniae*** (Fig. 3.7m–q), parasitizes Saprolegniomycete water moulds (Karling 1981) and shows the typical characteristics of the genus. Encysted zoospores produce a fine germ tube that releases an unwalled **plasmodium** into the host cytoplasm (Fig. 3.7m, n). This ultimately develops into a small flask-shaped walled thallus, with a prominent discharge tube (Fig. 3.7o). **Sexual reproduction** has been described in *O. saprolegniae* and closely related freshwater taxa and results in the formation of spiny resting spores (Fig. 3.7p) (Karling 1981; Sparrow 1960, 1976). The adjacent thalli (Fig. 3.7q) that conjugate have been interpreted as antheridia and oogonia and the resting spores that result as oospores (Martin and Miller 1986).

To date, combined morphological, ultrastructural, and molecular sequence data have been published for just two species in the **~Olpidiopsidaceae:** *O. porphyrae* (Sekimoto et al. 2008c) and *O. bostrychiae* (Sekimoto et al. 2009), both of which infect red seaweeds (*Porphyra* and *Bostrychia*). Both diverge after the *Eurychasma* and *Haptoglossa* clades (Fig. 3.5a), but they do not form a monophyletic clade. Similarly, unpublished sequence data also show that the freshwater oomycete parasitic species fall into one or more clades that are close to, but separate from, the marine species (Inaba and Sekimoto, in preparation). This indicates that *Olpidiopsis* as currently constituted is a paraphyletic genus (Fig. 3.6) and will require substantial revision. It is assumed the Sirolpidiaceae and Pontismataceae, which include genera such as *Petersenia*, *Pontisma*, and *Sirolpidium* (Table 3.5), are likely to be closely related to the Olpidiopsidales clades (Dick 2001a). However, until there is greater taxon sampling, we retain all in the ~Olpidiosidales s. lat. for the time being (Fig. 3.6, Table 3.5).

4. Haliphthorales

The order "**Haliphthorales**" is an exclusively marine order consisting of a small number of genera belonging to a single family, the **Haliphthoraceae,** which also was originally placed in the Lagenidiales (Table 3.3) (Sparrow 1973c). The family currently contains three genera, *Haliphthoros* (Fig. 3.7s–u), *Halioticida* (Fig. 3.7v, w) (Hatai 2012; Maurosa et al. 2009), and *Halocrusticida* (syn. *Halodaphnea*) (Dick 1998) (Fig. 3.7x) (Sekimoto et al. 2007; Hatai 2012). The family was transferred by Dick

(1998) to his new order, the **Salilagenidiales** (Table 3.4), along with a number of other marine genera such as *Atkinsiella* and marine *Lagenidium* spp. The type species, *Haliphthoros milfordensis*, was described by Vishniac as an endoparasite of crustacean eggs and oyster drill, and all members of the order are parasites of molluscs, crustaceans, or their eggs (Diggles 2001; Hatai 2012; Hatai et al. 1980, 1992). They are the only group of early diverging genera to produce irregularly branched eucarpic mycelial thalli (Fig. 3.7s, t, v) that can be cultured on artificial media. Large segments of the thalli convert into zoosporangia (Fig. 3.7u, w, x), in which peripheral zoospore initials are defined by a large central vacuole, as described in the Saprolegniales (Beakes 1994). All genera form very long narrow hypha-like exit tubes (Fig. 3.7u–x) (Hatai 2012), from which the zoospores escape in uniseriate fashion, reminiscent of sporulation in *Aphanomyces*. In LSU rRNA trees (Hatai 2012) all haliphthoralean species formed a monophyletic clade, which diverged just before the peronosporomycete/saprolegniomycete split (Hatai 2012; Maurosa et al. 2009). However, in mitochondrial *cox*2 gene trees there is no good statistical support for separating or combining the Haliphthorales and Olpidiopsidales clades, and greater taxon and gene sampling is required before these basal orders can be properly resolved (Sekimoto 2008; Sekimoto et al. 2007).

F. Saprolegniomycetes

The class "**Saprolegniomycetes**" (Figs. 3.8, 3.9, 3.10, and 3.11) replaces the Saprolegniomycetidae subclass (Dick 2001) and broadly encompasses those species generally known as **water moulds**. Most have **eucarpic mycelial thalli** (Figs. 3.8d, f and 3.9b) and can usually be cultured. They are mainly freshwater saprotrophs (Willoughby 1962) or opportunist necrotrophic pathogens (Dick 1976; Dick 2001a; Sparrow 1960). Molecular studies (Hudspeth et al. 2000; Léclerc et al. 2000; Petersen and Rosendahl 2000; Riethmüller et al. 1999) have shown that this class forms a well-supported monophyletic clade (Fig. 3.5), although there is still some uncertainty about the placement of a small number of basal genera. Saprolegniomycetes are able to synthesize sterols using both lanosterol and cycloartenol pathways and can utilize ammonium and organic sulphur (Dick 2001a; Gleason 1976). They mostly form **septum-delimited zoosporangia** and can produce two morphologically distinct generations of asexual zoospores (**diplanetic**), each followed by a walled cyst stage (Fig. 3.11q) (Dick 1973b, 1990, 2001a, b; Sparrow 1960). Oospheres are formed as a result of **centrifugal cleavage** without the formation of a peripheral periplasmic layer (Beakes 1981). We take a conservative approach to their classification, recognizing three orders (Fig. 3.6, Table 3.5), the **Atkinsiellales**, the **Leptomitales**, and the **Saprolegniales.** There is no molecular support (Figs. 3.5 and 3.6) for retaining either the orders **Salilagenidiales** or **Sclerosporales** introduced by Dick (Table 3.4) and placed in his Saprolegniomycetidae subclass, and both are excluded from our revised classification (Table 3.5).

1. Atkinsiellales

We propose a new order, the "**Atkinsiellales**", for the usually earliest diverging clade in Saprolegniomycete phylogenetic trees (Fig. 3.6) (Cook et al. 2001; Sekimoto et al. 2007). This order includes the monotypic family **Atkinsiellaceae** for the crustacean parasite *Atkinsiella dubia*, which Dick (1998, 2001a) placed in the family Haliphthoraceae in the order Salilagenidiales (Table 3.4). However, *cox*2-based trees revealed that this species was not associated with *Haliphthoros* but formed a clade that diverged before the Leptomitales (Cook et al. 2001; Sekimoto et al. 2007). A second monotypic family, the **Crypticolaceae,** is also included in this order. This family contains two holocarpic species that are parasites of mosquito larvae. The first, *Crypticola entomophaga* (Dick 1998), had originally been placed in the genus *Atkinsiella* by Martin (1977), and the second, *Crypticola clavulifera*, was described by Frances et al. (1989). Their placement in this order is supported by unpublished sequence data of *C. clavulifera* (D Hudspeth and M Hudspeth pers. commun.).

2. Leptomitales s. lat. and Related Clades

In the **Leptomitales** (Fig. 3.8) only members of the family **Leptomitaceae** (Table 3.5) have been sequenced to date. This family includes the genera ***Apodachlya*** (Fig. 3.8f), ***Apodachlyella*** (Fig. 3.8g–i) and ***Leptomitus*** (Fig. 3.8d, e), which are commonly known as **sewage moulds** because they are frequently isolated from polluted or organic-rich water and have the capability of **fermentative nutrition** (Dick 1973a; Gleason 1976; Riethmüller et al. 2006). *Leptomitus* has also been reported as an opportunist necrotrophic pathogen of perch (Willoughby and Roberts 1991). They produce eucarpic mycelial thalli (Fig. 3.8d, f, g) which have constricted hyphae plugged with granules of **cellulin**, a unique **chitin-glucan** polymer (Huizar and Aronson 1986; Lee and Aronson 1975). Oogonia in the Leptomitaceae are usually uni-oosporiate (Fig. 3.8e). *Apodachlya* has a distinctive oospore with a single large fused lipid globule pushing against a large optically translucent ooplast vacuole (Fig. 3.8e) (Dick 1969). The two sequenced species, *Apodachlya* and *Leptomitus*, form a well-supported monophyletic clade deeply separated from the Saprolegniales (Fig. 3.5) (Petersen and Rosendahl 2000). The related genus *Apodachlyella* (Fig. 3.8g–i) was transferred by Dick (1986) to its own monotypic family, the **Apodachlyellaceae** (Table 3.4). This genus produces an elongate antheridium (Fig. 3.8h) containing cyst-like male gametangial units (Fig. 3.8i). Longcore et al. (1987), however, questioned whether this difference merited the reassignment of *Apodachlyella* to a family of its own and suggested it should be retained in the Leptomitaceae.

Dick (2001a) included in his revised Leptomitales two additional families (Table 3.4), the **Ducellieraceae** (Hesse et al. 1989) and **Leptolegnielliaceae** (Fig. 3.8j–x), for which there are as yet no formally published sequence data. *Ducellieria* is a monotypic genus that infects pollen grains and was originally thought to be an alga (Hesse et al. 1989). The family **Leptolegniellaceae** contains mostly holocarpic genera, such as *Aphanomycopsis*, a genus that infects algae (Fig. 3.8l) and insect eggs (Fig. 3.8m) (Karling 1981). The taxonomically problematic oomycete hyperparasite *Pythiella* (Blackwell 2010) (Fig. 3.8r–x) has centrifugal sporogenesis and an aphanomycoid pattern of zoosporogenesis similar to *Aphanomycopsis*, which suggests that this genus may also be part of the Leptolegniellaceae, despite its reportly having periplasmic oogenesis (Fig. 3.8v–x).

The holocarpic genus *Cornumyces* (as illustrated by *Lagenidium pygmaeum* which Dick transferred to *Cornumyces*) (Fig. 3.8j–k), which colonizes pollen grains, forms a separate clade with affinity to the ~Leptomitales clade (Inaba and Hariyama 2006). This supports Dick's (2001a) tentative placement of *Cornumyces* in the Leptolegniellaceae. Inaba and Beakes (unpublished data) also found that the genus *Cornumyces* formed a monophyletic clade (based on SSU rRNA sequences), together with the holocarpic nematode-infecting genus *Chlamydomyzium* (Fig. 3.8n–q). This genus, which Dick (1997) had originally placed in his newly created Myzocytiopsidales but later considered as *incertae sedis* (Dick 2001a), has been shown to be an early-diverging saprolegniomycete (Beakes et al. 2006; Glockling and Beakes 2006b). Environmental sampling has revealed that both an unknown marine stramenopile (MAST) clade (Massana et al. 2002, 2004, 2006) and a number of freshwater isolates from keratin baits (Inaba, unpublished) also fall in clades that diverge near these Leptomitales s. lat. clades. Until there is far greater taxon sampling within the **Leptomitales** s. lat., it is proposed that those genera with eucarpic constricted thalli be placed in the **Leptomitaceae** and all those with nonmycelial thalli, including *Ducellaria*, be included within an expanded ~**Leptolegniellaceae** until sequence data become available (Fig. 3.6, Table 3.5).

3. Saprolegniales

Saprolegniales includes many of the familiar and relatively well-studied genera (Figs. 3.9,

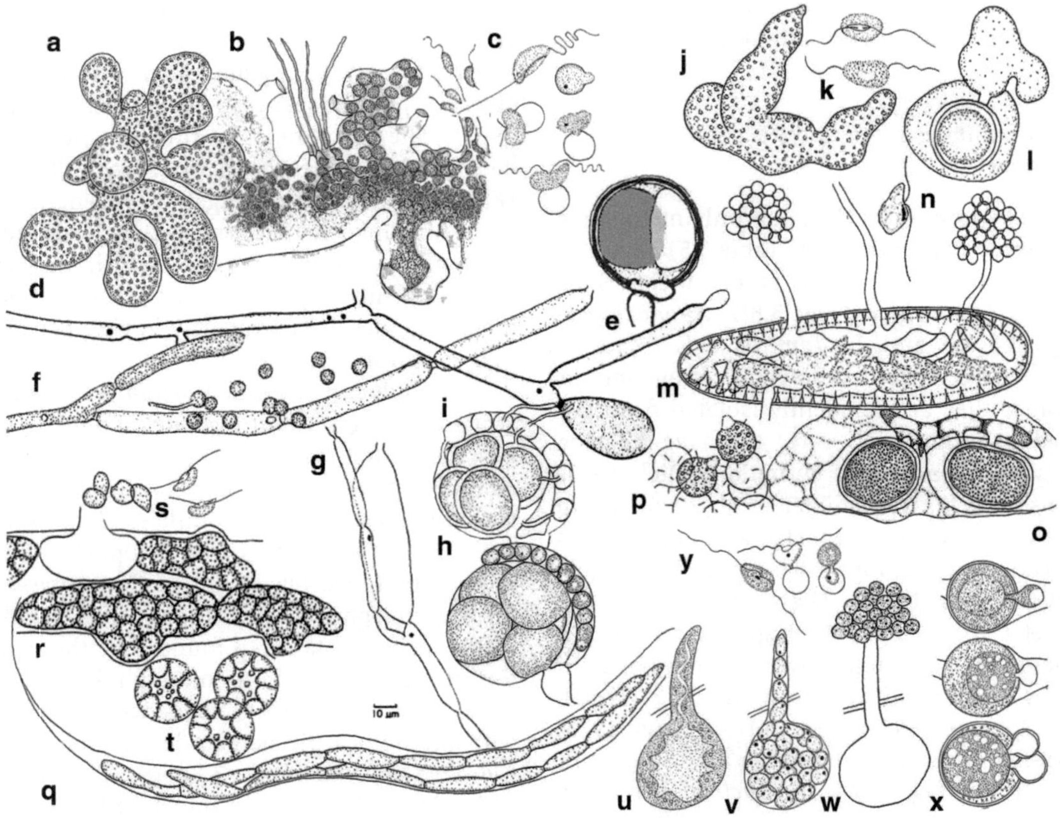

Fig. 3.8 **(a–y) Morphology of Leptomitales s. lat.** (**a–c**) Drawings of **Atkinsiellaceae.** ***Atkinsiella dubia***, showing irregularly lobate thallus (**a**) within which primary zoospores and cysts form in situ (**b**). Typical reniform **secondary zoospore** being released from a primary cyst (**c**). From Karling (1981), after Sparrow (1973). (**d–i**) Drawings of **Leptomitaceae.** Constricted hyphae of ***Leptomitus lacteus***, (**d**) and mature oogonium showing single large lipid droplet (*shaded*) and adjacent ooplast (**e**). Constricted hyphae of ***Apodachlya pyrifera*** with an empty terminal sporangium and with primary cysts nearby (**f**). From Dick (1973a) with permission. (**g–i**) Constricted hyphae of ***Apodachlyella*** showing a lateral sporangium (**g**) and young (**h**) and mature (**i**) oogonia illustrating the characteristic elongate antheridium, divided into cyst-like compartments from which fertilization tubes develop. From Longcore et al. (1987) with permission. (**j, k**) Drawings of ***Cornumyces pygmaeum*** (syn. *Lagenidium pygameum*) showing typical irregular lobate thallus (**j**) that forms within pollen grains and detail of adjacent oogonial and antheridial thallus showing developing oospore (**k**). From Karling (1981), after Zopf. (**l, m**) Drawings of ***Aphanomycopsis spp.*** *Suririella* infected with ***A. bacillarearum*** showing both elongate segmented thallus and aphanomycoid discharge of primary aplanospores (**l**). From Karling (1942), after Tokunaga (1934). Midge eggs infected with ***A. sexualis*** showing two developing oospores, associated with an elongate segmented antheridium (**m**). From Karling (1981) after Martin (1975). (**n–q**) Drawings showing development of ***Chlamydomyzium anomalum*** (syn. *Myzocytium anomalum*) within an infected nematode. Newly infected nematode showing elongate compartmentalized thalli (**n**). Each compartment swells to form an ovoid sporangium with a dome-shaped discharge papillum (**o**). Discharged aplanospores releasing secondary-type zoospores (**p**). Thick-walled resting spores that arise parthenogenetically (**q**). From Karling (1981) after Barron (1976). (**r–x**) Drawings of ***Pythiella vernalis*** infecting **hyphae** of ***Pythium***, which, although unsequenced, probably belongs in this order. Thallus with elongate discharge tube (**r**) from which a cluster of aphanomycoid primary cysts are released (**u**). These release secondary-type zoospores (**u**). Series of drawings summarizing sexual reproduction showing transfer to the oogonium of cytoplasm from an adjacent antheridia (**v, w, x**). From Karling (1981) after Couch (1935)

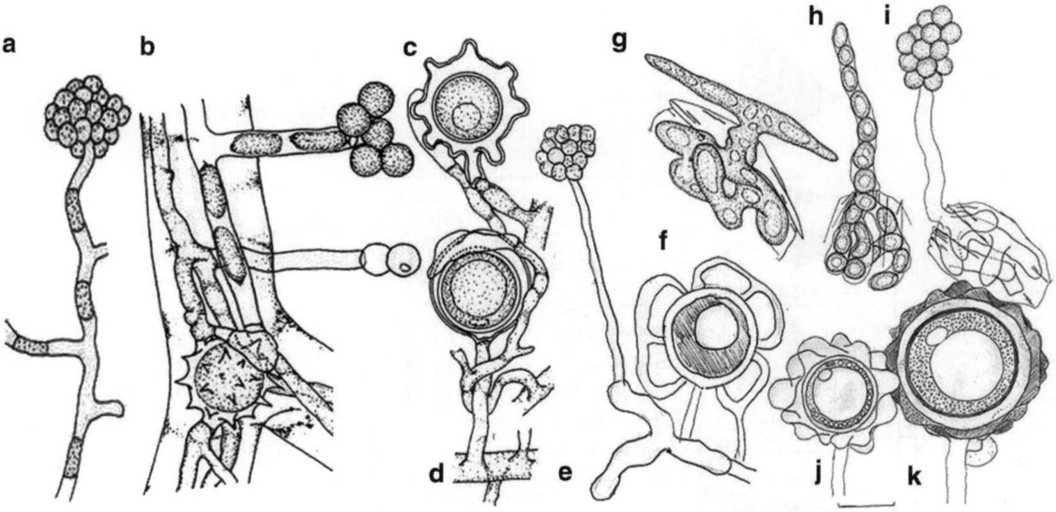

Fig. 3.9 Morphology of Verrucalvaceae. (**a–d**) Drawings of genus *Aphanomyces* spp. Eucarpic thallus of *Aphanomyces amphigynus* within infested algal filament showing an oogonium (**a**) and undifferentiated hypha-like sporangia with short lateral discharge tubes releasing naked aplanospores, which immediately encyst to form balls of spores (**b**). Detail of papillate and smooth single-oospored oogonia of *A. cladogynus* (**c**) and *A. euteiches* (**d**) showing attached antheridia and clear ooplast vacuoles of varying size. From Johnson et al. (2002). (**e, f**) Drawings of *Plectospira myriandra* a soil-borne saprotroph. Swollen sporangium segment with elongate discharge tube and cluster of primary aplanospores (**e**) and single-oospored oogonium surrounded by investing antheridial hyphae (**f**). From photographs in Watanabe (1987). (**g–i**) Drawings of *Sommerstorffia* infecting rotifer bodies. Lobate thallus and developing external trapping hypha (**g**). Thallus with differentiated aplanospores (**h**) showing aphanomycoid release (**i**). From Johnson et al. (2002). (**j, k**) Drawings of oogonia and oospores of two root-infecting graminicolus pathogens, *Pachymetra chaunorhiza* (**j**) and *Verrucalvus calvatus* (**k**), neither of which has been reported to produce asexual spores

3.10, and 3.11) of water moulds (Tables 3.3, 3.4, and 3.5) (Dick 1973b; Sparrow 1960). Most are general saprotrophs or opportunist colonizers of damaged plant and animal tissues (Dick 1976; Sparrow 1960), but they also include a number of significant pathogens of plant roots (Gaulin et al. 2007; Levenfors and Fatehi 2009), invertebrates such as crayfish (Cerenius et al. 1988) and vertebrates such as fish, and amphibians and their eggs (Lilley et al. 2003; van West 2006). The most recent taxonomic synopsis of Saprolegniales was published online (Johnson et al. 2002), replacing earlier monographs. Although the Saprolegniales forms a well-supported monophyletic clade, the formal naming and branching order of the family-level clades is not fully resolved (Fig. 3.5) (Dick et al. 1999; Inaba and Tokumasu 2002; Léclerc et al. 2000; Petersen and Rosendahl 2000; Riethmüller et al. 1999; Spencer et al. 2002). Traditionally, this order contained just a single family, the **Saprolegniaceae** (Table 3.3; Dick 1973b; Sparrow 1960). However, molecular studies suggest that the earliest diverging clade within this order encompasses those species with slender hyphae and simple hypha-like sporangia, as exemplified by the genus *Aphanomyces* (Fig. 3.9). We therefore include this genus, together with a small number of related genera, in the amended **Verrucalvaceae** family. Most remaining representatives of this order (Fig. 3.11) produce stout fast-growing hyphae on which clavate, often proliferative septum-delimited sporangia, are borne (Fig. 3.11). The dozen or more genera (Figs. 3.9, 3.10, and 3.11, Table 3.3) have been primarily defined on the basis of their different patterns of **asexual sporogenesis** (Figs. 3.10 and 3.11) (Dick 1973b; Johnson et al. 2002; Sparrow 1960). In the genus *Saprolegnia* two morphologically distinct generations of zoospores, traditionally referred

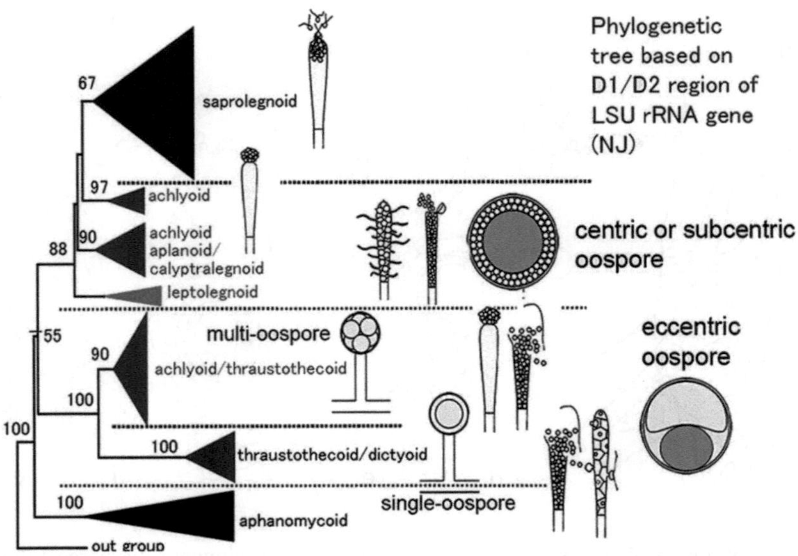

Fig. 3.10 Simplified LSU rRNA tree of members of Saprolegniales together with summary of some key morphological characters (sporogenesis and mature oospore cytology) associated with each clade. Unpublished illustration from Inaba and Tokumasu (2002)

to as **primary and secondary types,** are formed (**diplanetism**) (Fig. 3.11q). Dick (2001a) argued that these two zoospore types should be designated **auxiliary** and **principal** types, but this terminology has not been widely adopted. These zoospores in turn give rise to **primary** and **secondary cysts** (Fig. 3.11q) (Beakes 1983; Holloway and Heath 1977a, b; Sparrow 1960). In many genera, however, the first zoospore generation is suppressed, giving rise directly to **aplanospores** (equivalent to **primary cysts**), which may be retained within the sporangium or discharged. Molecular studies revealed that the large genus ***Achlya*** was paraphyletic (Inaba and Tokumasu 2002; Spencer et al. 2002), and Spencer et al. (2002) proposed transferring those species with centric or subcentric oospores to a new genus, ***Newbya*** (Figs. 3.10 and 3.11n), leaving the remainder with eccentric oospores in the genus *Achlya* s. str. (Figs. 3.10 and 3.11b). A summary of morphological characteristics that might be used to resolve taxonomic relationships within this order is given in Fig. 3.10, which was adapted from the unpublished study that was briefly reported by Inaba and Tokumasu (2002). We propose placing those genera with predominantly single-oospored oogonia and strongly **eccentric oospores** in a new family, the "**Achlyaceae**". Those genera and species with predominantly multi-oospored oogonia and **centric** or **subcentric oospores** are retained in the **Saprolegniaceae** s. str.

a) Verrucalvaceae

Dick et al. (1984) erected the family **Verrucalvaceae** to include a newly described pathogen, ***Verrucalvus flavofaciens*** (Fig. 3.9k), the causal agent of a yellowing disease of the turf grass *Pennisetum clandistinum* (Kikuyu yellows), to which they also controversially allied the graminicolous downy mildew (GDM) genus *Sclerophthora*. Subsequently, a soil-borne pathogen of sugar cane (*Saccharum* spp.) roots, ***Pachymetra chaunorhiza*** (Fig. 3.9j), was also included in this family (Dick et al. 1988). Separately, on the basis of SSU rRNA sequence data, Dick et al. (1999) created a new family, the **Leptolegniaceae,** for a clade that included the genera ***Aphanomyces, Leptolegnia,*** and ***Plectospira.*** In their final taxonomic synthesis they placed both *Pachymetra* and *Verrucalvus* in the order Sclerosporales, along with the GDM

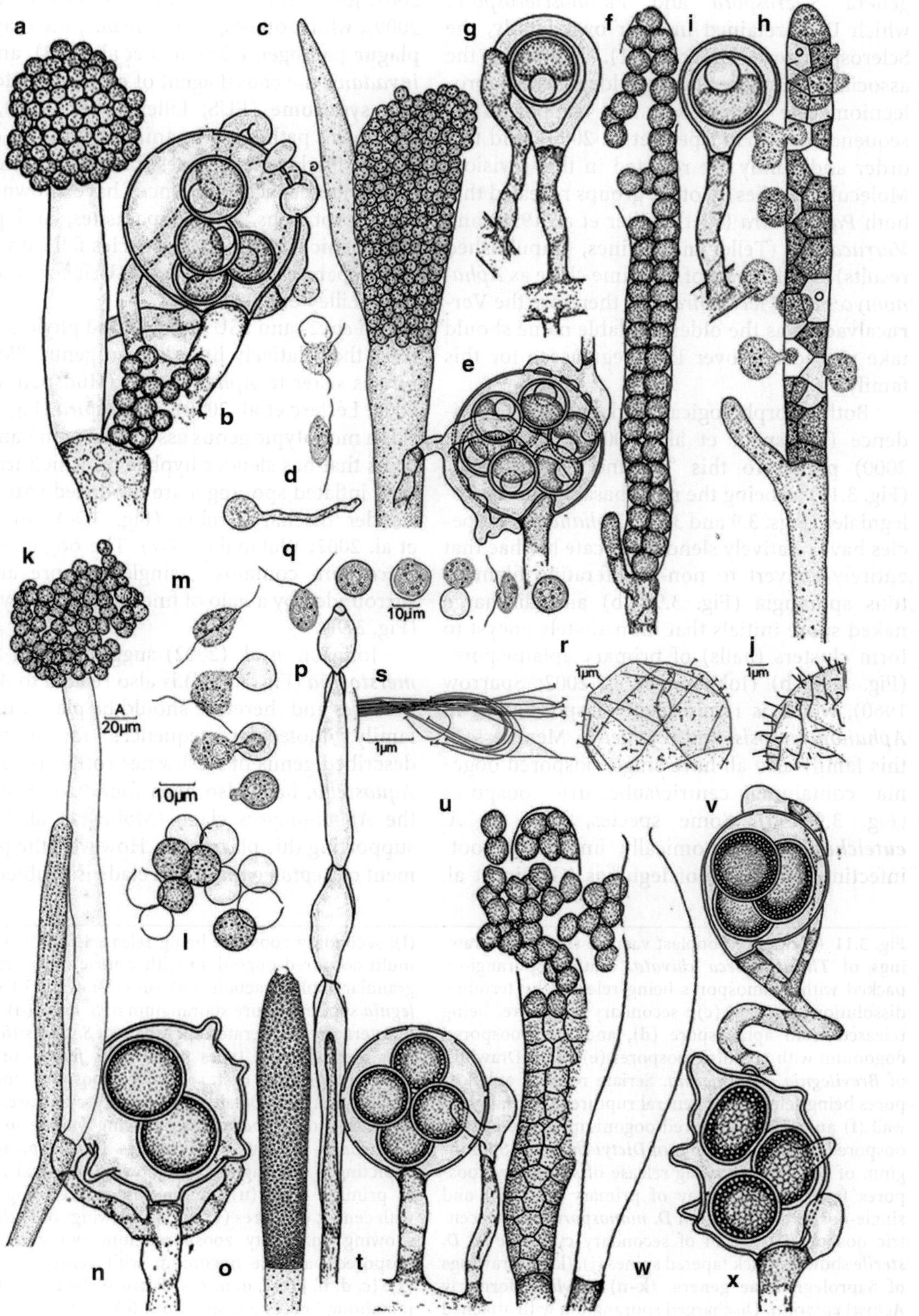

Fig. 3.11 Morphology of Achlyaceae and Saprolegniaceae. (a–i) Drawings of genera in **Achlyaceae.** Empty sporangium of *Achlya irregularis* showing terminal spore ball (a) and oogonium of *A. heterosexualis,* with diclinous antheridium and eccentric oospores containing single lipid droplet and larger crescent-shaped

genera *Sclerospora* and *Peronosclerospora*, which Dick retained in their own family, the Sclerosporaceae (Dick 2001). However, the association of these downy mildews with Saprolegniomycete genera was not supported by sequence data (Hudspeth et al. 2000), and this order and family are rejected in this revision. Molecular studies by other groups revealed that both *Pachymetra* (Riethmüller et al. 1999) and *Verrucalvus* (Telle and Thines, unpublished results) are also part of the same clade as *Aphanomyces* and *Plectospira*, and therefore the **Verrucalvaceae** as the older available name should take precedence over Leptolegniaceae for this family.

Both morphological and molecular evidence (Hudspeth et al. 2000; Léclerc et al. 2000) points to this "aphanomycoid clade" (Fig. 3.10) as being the most basal in the Saprolegniales (Figs. 3.9 and 3.10). *Aphanomyces* species have relatively slender delicate hyphae that entirely convert to non-proliferating filamentous sporangia (Fig. 3.9a, b) and discharge naked spore initials that immediately encyst to form clusters (balls) of primary aplanospores (Fig. 3.9a, b) (Johnson et al. 2002; Sparrow 1960), which is reminiscent of sporulation in *Aphanomycospsis* and *Pythiella*. Members of this family also all have single oospored oogonia containing centric/subcentric oospores (Fig. 3.9c, d). Some species, such as ***A. euteiches***, are economically important root-infecting **pathogens of legumes** (Gaulin et al. 2007; Johnson et al. 2002; Levenfors and Fatehi 2009), whilst others, like *A. astaci*, the **crayfish plague** pathogen (Cerenius et al. 1988), and *A. invadans*, the causal agent of **epizooitic ulcerative syndrome** (EUS; Lilley et al. 2003), are significant pathogens of animals (Phillips et al. 2008). Phylogenetic trees based on internal transcribed spacer sequences have shown that the saprotrophs, animal parasites, and plant pathogenic *Aphanomyces* species fall into their own separate clades (Diéguez-Uribeondo et al. 2009; Lilley et al. 2003).

In *cox*2- and SSU rRNA-based phylogenetic trees the relatively little known genus ***Plectospira*** is sister to *Aphanomyces* (Hudspeth et al. 2000; Léclerc et al. 2000). *Plectospira* (Fig. 3.9e, f) is a monotypic genus associated with bamboo roots that has slender hyphae on which irregularly inflated sporangia are produced with long slender discharge tubes (Fig. 3.9e) (Johnson et al. 2002; Watanabe 1987). The oogonium of *Plectospira* contains a single oospore and is surrounded by a halo of fine antheridial hyphae (Fig. 3.9f).

Johnson et al. (2002) suggested that ***Sommerstorffia*** (Fig. 3.9g–i) is also related to *Aphanomyces* and therefore should be placed in this family. Molecular sequences for a newly described genus of freshwater rotifer parasites, *Aquastella*, have also been found to cluster in the *Aphanomyces* clade (Molloy et al. 2014), supporting this placement. However, the placement of *Leptolegnia* in this clade is problematic

Fig. 3.11 (continued) ooplast vacuole (**b**). (**c–e**) Drawings of ***Thraustotheca clavata.*** Mature sporangium packed with aplanospores being released by terminal dissolution of wall (**c**), secondary zoospore being released from aplanospore (**d**), and multi-oospored oogonium with eccentric oospores (**e**). (**f, g**) Drawings of ***Brevilegnia megasperma.*** Seriate rows of aplanospores being released by general rupture of sporangium wall (**f**) and single-oospored oogonium with eccentric oospore (**g**). (**h–j**) Drawings of ***Dictyuchus*** spp. Sporangium of ***D. sterile*** showing release of secondary zoospores from net-like array of primary cysts (**h**) and single-oospored oogonium ***D. monosporus*** with eccentric oospore (**i**). Detail of secondary cyst case of ***D. sterile*** showing thick tapered spines (**j**). (**k–x**) **Drawings of Saprolegniaceae genera.** (**k–n**) ***Newbya*** (formerly ***Achya***) ***colorata.*** Discharged sporangium with attached cyst ball (**k**); detached spore ball with discharged cysts (**l**); secondary zoospore being released from cyst (**m**); multi-oospored oogonium with centric oospores with granular ooplast vacuoles (**n**). (**o–s**) Drawings of ***Saprolegnia*** spp. Immature sporangium of ***S. ferax*** (**o**); chain of internally proliferated sporangia of ***S. parasitica*** (**p**); four asexual spore types found in ***S. ferax***—primary zoospore, primary cyst, secondary zoospore, and secondary cyst (**q**). Detail of secondary cyst cases of ***S. ferax*** (**r**) and ***S. parasitica*** showing *bundle of boat-hook hairs* (**s**). (**u, v**) Drawings of ***Calyptralegnia*** depicting a sporangium (**u**) showing terminal release of primary cysts (**u**) and multi-oospored oogonium with centric oospores (**v**). (**w, x**) Drawings of ***Pythiopsis*** showing an empty zoosporangium (**w**) and multi-oospored papillate oogonium with centric oospores (**x**). (**c, d, h, j, k, l, o, q, r, s**) From Webster (1970). All remaining figures adapted from Johnson et al. (2002)

because in some trees, it groups with species considered by Dick (2001) to be in the **Saprolegniaceae** (Fig. 3.10) (Arcate et al. 2006; Inaba and Tokumasu 2002; Léclerc et al. 2000; Petersen and Rosendahl 2000).

b) Achlyaceae

We have introduced a new family, the "Achlyaceae" (Table 3.5), for the "achlyoid/thraustothecoid clade" with eccentric oospores (Fig. 3.10), which encompasses four genera: ***Achlya*** s. str. (Fig. 3.11a, b) (Spencer et al. 2001), ***Brevilegnia*** (Fig. 3.11f, g), ***Dictyuchus*** (Fig. 3.11h–j), and ***Thraustotheca*** (Fig. 3.11c–e). All have oogonia containing either single (Fig. 3.11g, i) or multiple (Fig. 3.11b, e) oospores with **strongly eccentric oospores**, usually with just a single coalesced lipid droplet (Figs. 3.10 and 3.11b, e, g, h). Diplanetism is not observed in this clade, although there is diversity in the pattern of zoosporogenesis (Figs. 3.10 and 3.11a, c, f, h). Species that discharge their primary aplanospores from the sporangium to form a ball of spores constitute the genus *Achlya* s. str. (Fig. 3.11a). Genera such as *Brevilegnia* (Fig. 3.11f) and *Thraustotheca* (Fig. 3.11c) retain their primary cysts (aplanospores) within the sporangium, which are released by the general or localized rupture of the sporangium wall (Fig. 3.11c, f). In contrast, in *Dictyuchus* (Fig. 3.11h), individual primary zoospores are discharged from lateral papillae, which rupture the sporangium wall and leave a distinctive net of angular aplanospore walls behind. The secondary cysts of genera within this clade are either smooth or, as in *Dictyuchus*, decorated by stout tapering spines (Fig. 3.11j).

c) Saprolegniaceae

The family **Saprolegniaceae** s. str. is now reserved for the saprolegnioid, achlyoid, aplanoid, calypralegnoid, and leptolegnoid clades (Fig. 3.10) and contains species with centric/subcentric oospores (Fig. 3.11n, t, v, x). This revised family includes the genera ***Aplanes, Aplanopsis, Calyptralegnia*** (Fig. 3.11u, v), ***Newbya*** (*Achlya*-like sporulation pattern, but with centric oospores) (Fig. 3.11k–n), ***Protoachlya, Pythiopsis*** (Fig. 3.11w, x), and the majority of described ~***Saprolegnia*** spp. (Fig. 3.11o–t, Table 3.5). All have oogonia that are predominantly **multi-oospored** with **centric/subcentric ooplast vacuoles** (Figs. 3.10 and 3.11n, t, u, v, x), which usually contain granules that are in Brownian motion (Dick 1969). Zoosporogenesis results in either the production of diplanetic zoospores (as in Fig. 3.11q) or, depending on the genus, the formation of intra- and extrasporangial primary spores (aplanospores). All primary aplanospores or cysts release reniform secondary-type zoospores with laterally inserted flagella (as in Fig. 3.11m).

The secondary cyst cases of genera in this clade are often decorated by slender boat-hook spines (Beakes 1983; Hallett and Dick 1986). Such spines are particularly large in *S. parasitica* (Fig. 3.11s), which is an important pathogen of salmonids and other fish and poses a serious threat to both wild and reared fish stocks [see reviews by van West (2006) and Phillips et al. (2008)]. This species was recently shown to produce effector-like proteins that are selectively translocated to fish cells, where they modulate the host immune response (Wavra et al. 2012). Molecular studies seem to suggest that many currently defined species of ~*Saprolegnia* are paraphyletic (Hulvey et al. 2007) but may be the result of misidentified taxon names in genetic databases. The precise placement of the genus ***Leptolegnia*** within the Saprolegniales is also problematic. LSU sequence data place it firmly in the Saprolegniaceae s. str. (Fig. 3.10) (Inaba and Tokumasu 2002; Petersen and Rosendahl 2000), although it has a number of anomalous morphological traits for this family, such as relatively slender hyphae and single oospored oogonia.

G. Peronosporomycetes

The Peronosporomycete class (Figs. 3.12, 3.13, 3.14, and 3.15) represents a major derived offshoot of the main oomycete line (Beakes and Sekimoto 2009; Beakes et al. 2012; Jiang and Tyler 2012; Thines and Kamoun 2010). Members of this class require **exogenous sterols** to complete oogenesis (Kerwin and Washino 1983). The saprotrophic members of this group are able to utilize various would be better nitrogen sources and sulphate (Dick 2001a, b; Gleason 1976). Peronosporomycetes

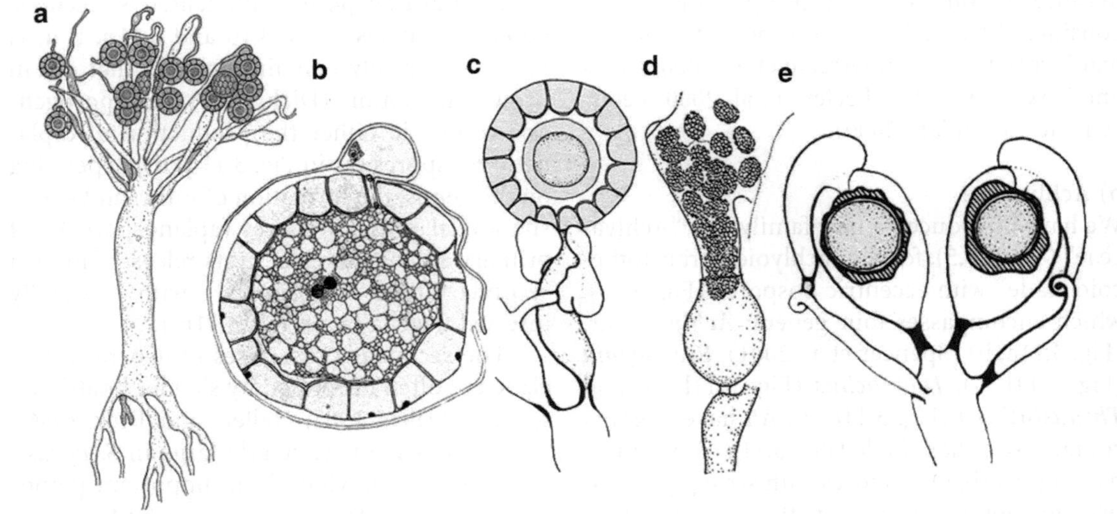

Fig. 3.12 Morphology of Rhipidiales. (**a–c**) Drawings of ***Araiospora pulchra***. Mature tree-like thallus showing fine basal hyphae and terminal branches bearing both sporangia and oogonia (**a**). Wax-embedded section of newly fertilized oosphere (**b**) showing periplasm surrounding a developing oosphere. Mature oospore with multilayered scalloped wall (**c**). (**a**) Adapted from Sparrow (1960), (**b**) from King (1903), and (**c**) from Dick (1973a) after Shanor and Olive (1943). (**d**) Sporangium of ***Rhipidium*** showing vesiculate release of zoospores. From Sparrow (1960) after Thaxter (1896). (**e**) Differentiating oospores of ***Sapromyces elongatus*** showing septate constrictions delimiting both oogonia and antheridia. From Dick (1973a), with permission

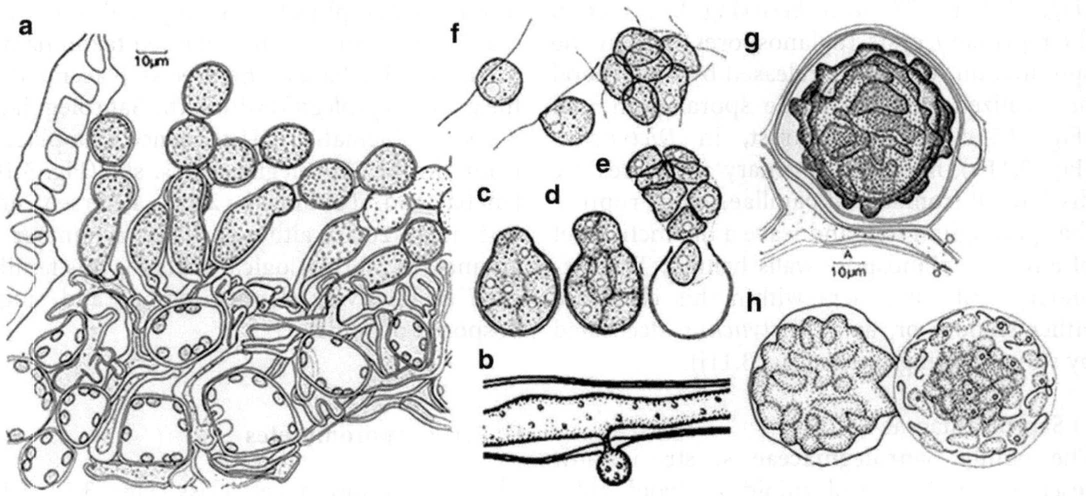

Fig. 3.13 Morphology of Albuginales. (**a–h**) Drawings of ***Albugo candida*** infecting ***Capsella bursa-pastoris.*** Section of a sporulating pustule showing chains of conidiosporangia and displaced epdidermis (**a**). Eucarpic hyphae with small globose haustorium (**b**). Sequence showing vesiculate discharge of zoospores from mature sporangium (**c–e**). The free-swimming zoospores are rather rounded (**f**). Mature oospores showing thick verrucose wall (**g**) and direct vesiculate release of zoospores upon germination (**h**). All from Webster (1970), (**h**) based on Vanterpool (1959)

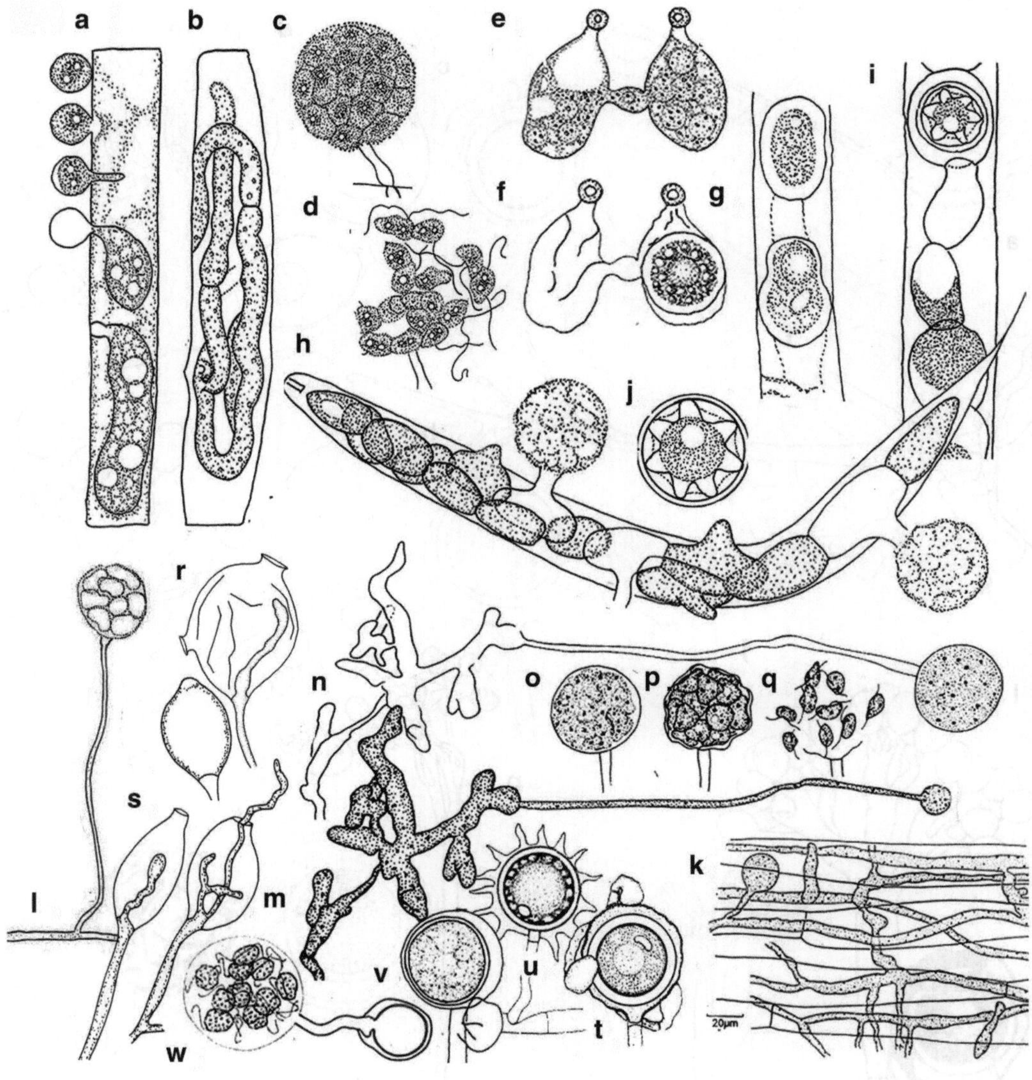

Fig. 3.14 Morphology of Pythiaceae s. lat. (**a–g**) Drawings of ***Lagena radicola*** infecting cereal root cells. Although unsequenced, it is thought to be a pythiaceous genus. Series of drawings showing infection from cysts and early (**a**) and more advanced stages (**b**) of thallus development. Vesiculate discharge of spore mass (**c**) and liberation of fully differentiated zoospores (**d**). Sexual reproduction brought about by conjugation and migration of cytoplasm from male (antheridial) to female (oogonial) thalli (**e, f**). Mature oospores within root cell (**g**). From Karling (1981) after Vanterpool and Ledingham (1930). (**h–j**) ***Myzocytiopsis lenticulare*** infecting rhabditid nematodes. Segmented thallus that has differentiated into sporangial compartments, some of which are showing vesiculate differentiation of zoospores (**h**). Adjacent antheridial and oogonial compartments (**i**). Fully mature oospore showing stellate ornamentation of thick wall (**j**). Adapted from Karling (1981) based on papers by Barron (1976). (**k–w**) Drawings of ***Pythium*** spp. Eucarpic hyphae of a *Pythium* sp. in cress-seedling hypocotyls (**k**). Filamentous sporangium of ***P. porphyrae*** (clade A) showing extra-sporangial vesicle in which zoospores differentiate (**l**). Inflated filamentous toruloid sporangium of ***P. aphanidermatum*** (clade A) before (**m**) and after (**n**) vesicle formation. Sequence illustrating formation of zoospores within extra-sporangial vesicle (**o–q**). Papillate and discharged ovoid sporangia of ***P. multisporum*** (clade E) (**r**). Internally proliferating ovoid sporangia of ***P. undulatum*** (clade H) (**s**). (**t–v**) Single-oospored oogonia of *Pythium* spp. showing both smooth and spiny wall ornamentation: ***P. debaryanum*** (clade F) (**t**), ***P. amasculinum*** (clade D) (**u**), ***P. ultimum*** (clade I) (**v**). Germinating of ***P. ultimum*** oospore showing vesiculate release of zoospores (**w**). (**l–o, t, v, w**) All adapted from Webster (1980); (**w**) based on original by Drechsler (1960). (**k, r, s, u**) Adapted from Van der Plaats-Niterink (1981)

Fig. 3.15 (a–t) **Morphology of Peronosporaceae s. lat.** (a–c) Drawings of ***Phytopythium ostracodes*** showing papillate (a) and proliferating (b) sporangia and apleurotic oogonium with thick-walled oospore (c). From Plaats-Niterink (1981). (d, e) Drawings of ***Halophytophthora vesicula.*** Mature (d) and discharging (e) sporangium showing large transient elongate vesicle characteristic of this species. (f–j) Drawings of ***Phytophthora* spp**. Non-papillate sporangium of ***Phytopthora sojae*** (clade 7) (f). From http://microbe.eolchina.org. Aerial sporangiophore of ***Ph. infestans*** (g) bearing deciduous (caducous) papillate sporangia (h) from which zoospores (i) are released. Young (j) and mature (k) oogonia of ***Ph. erythroseptica*** showing collar-like amphigynous antheridium (k) and mature oospore with central ooplast (k). Adapted from Webster (1970). (l–t) Drawing of **downy mildews in Peronosporaceae.** Branched sporangiophores with terminal sporangia of ***Peronosclerospora sorghii*** **(GDM clade)** on millet (l) and its mature thick-walled oospore (m).

produce only a single **(secondary or principal)** type of zoospore (Dick 2001a) (Figs. 3.13f and 3.14d, q, w), and all have single-oospored **oogonia** (Figs. 3.12b, c, e, 3.13g, and 3.14j, t–v) in which the cytoplasm differentiates into an egg (oosphere) surrounded by an outer **periplasmic layer** of cytoplasm (Figs. 3.12b and 3.15q). This periplasm may contribute significantly to the outer oospore wall in genera such as *Albugo* (Beakes 1981). The centric to subcentric ooplast vacuole is always homogeneous in appearance (Dick 1969).

Molecular studies (Hudspeth et al. 2003; Thines et al. 2009c) (Fig. 3.6) revealed two early diverging orders in the peronosporomycete clade, the **Rhipidiales** (Fig. 3.12) and **Albuginales** (Fig. 3.13). However, further order-level splitting and genus assignment in this class remain problematic at present. Waterhouse (1970, 1973) adopted the historical position of Rabenhorst by placing the **Pythiaceae** and **Peronosporaceae** together in a single order, **Peronosporales** (Table 3.3). Dick (2001a), however, split this order into the **Pythiales** (Fig. 3.14) and **Peronosporales** (Fig. 3.15), assigning genera as summarized in Table 3.4. However, molecular studies almost immediately revealed that some genera had been misplaced. For instance, *Phytophthora* species (Blair et al. 2008; Cooke et al. 2000) clustered with the downy mildews (Fig. 3.5) rather than amongst *Pythium* species. To avoid the proliferation of new orders that are hard to distinguish with no or few synapomorphic features, we decided to retain a **broadly defined** Peronosporales, as argued by Hulvey et al. (2010) and Runge et al. (2011a). Until more of the less-studied holocarpic representatives of the clade are included in multigene analyses, it is thought premature to split the Peronosporales s. lat. further at present. Whilst members of the family **Peronosporaceae** seem to form a well-supported monophyletic clade (Fig. 3.5), those in the family **Pythiaceae s. lat.** do not (Fig. 3.5b). However, the enormous size and diversity of Peronosporales necessitates the recognition of subgroups (Table 3.5) using informal clade names that have been previously published.

1. Rhipidiales

Rhipidiales (Fig. 3.12) contains a single family, the **Rhipidiaceae,** containing around ten genera (Table 3.3), all of which are saprotrophs, typically isolated from submerged twigs and fruit, often under fermentative conditions (Dick 1973a, 2001a; Gleason 1976). They generally have small, relatively slow-growing, determinant thalli that produce fine, rhizoid-like, basal hyphae (Fig. 3.12a) (Dick 1973a, 2001a; Sparrow 1960). Genera in this order have segmented thalli, often occluded by plugs of wall material (Fig. 3.12c–e). In *Araiospora* and *Rhipidium*, mature zoospores are released into a transient vesicle (Fig. 3.12d), which is typical of Peronosporomycete genera. They form single-oospored oogonia, and some genera, such as *Araiospora* (Fig. 3.12b), have a broad **periplasmic layer** and oospores with complex multilayered walls (Fig. 3.12c, e) (Sparrow 1960). However, it should be noted that the current phylogenetic placement of the Rhipidiales (Figs. 3.5 and 3.6) is based solely on sequence data from just one taxon, *Sapromyces elongatus.* The phylogenetic position of this taxon has proved problematic in that it may form the basal-most clade to both main classes, depending on which gene is sequenced and what other taxa are included in the analyses. In both SSU rRNA-based (Lara and

Fig. 3.15 (continued) Adapted from Weston and Uppal (1932) and (**m**) from Migula (1897). (**n**) Aerial sporangiophores of ***Plasmopara pusilla* (DMPH clade)** on *Geranium pratense* showing short terminal branches bearing relatively large conidia-like sporangiospores (**n**), together with detail of small pyriform haustoria typical of this clade in another *Plasmopara* sp. (**o**). (**n**) From Webster (1970), (**o**) from Voglmayr and Constantinescu (2008). Aerial sporangiophore of ***Bremia lactucae*** (DMPH clade) on *Senecio vulgaris* showing terminal cluster of branches bearing conidiospores (**p**). (**q, r**) Drawing of **aerial conidiophore of brassicolous downy mildew *Hyaloperonospora parasitica*** on ***Capsella bursa-pastoris*** (**q**) together with detail of hyphae within host tissue illustrating their elongate lobed haustoria (**r**). Developing oospore of ***Peronospora viciae* (in DMCC clade)** showing periplasm (**s**) and mature oospore with reticulate wall (**t**). (**p, q**) Adapted from Webster (1970), (**s–t**) from originals of Migula (1897)

Belbahri 2011) and *cox*2-based trees (Hudspeth et al. 2000), *Sapromyces* is basal to the Peronosporomycetes, whereas in LSU rRNA trees it is basal to the Saprolegniomycetes but without strong support (Fig. 3.5b) (Petersen and Rosendahl 2000; Riethmüller et al. 1999). In combined *cox*2 and nrLSU analyses it was placed basal to the remaining Peronosporomycetes with maximum support (Thines et al. 2009c). *Sapromyces*, however, has a number of anomalous features not normally associated with this order. The zoospores are released directly from the sporangium and, in *S. androgynous*, contain large structured K-bodies (Gotelli and Hanson 1987), both of which are features normally associated with Saprolegniomycetes (Beakes 1987, 1989). There is clearly an urgent need for other genera in this order to be sequenced.

The **COII amino acid** sequences derived from *cox*2 gene analyses revealed that *Sapromyces* has the same signature amino acid insertion-deletion (indel) sequence **LEF/T** as that found in other members of the Peronosporales rather than the **YTD** indel found in the Leptomitaceae (Cook et al. 2001; Hudspeth et al. 2000, 2003).

2. Albuginales

Dick (2001a) placed the **white blister rusts** (Fig. 3.13) in their own family, the Albuginaceae, in the Peronosporales (Tables 3.3 and 3.4). However, numerous phylogenetic studies using LSU rDNA and *cox*2 sequences have shown that the white blister rusts form a statistically well-supported clade that is always **early diverging** in the peronosporomycete clade (Hudspeth et al. 2003; Riethmüller et al. 2002; Thines and Spring 2005; Thines et al. 2008, 2009a) and always quite separate from other members of the Peronosporales. This led Thines and Spring (2005) to introduce the order **Albuginales** for white blister rusts containing a single family **Albuginaceae** (Table 3.5). Current multilocus phylogenies continue to support the deep branching of the Albuginales (Thines and Voglmayr 2009; Thines et al. 2009b), although their relationship to some lagenidiaceous groups remains to be explored, especially, as in single-locus SSU rRNA phylogenies, *Albugo* seems to cluster amongst the Pythiales (Lara and Belbahri 2011; Sekimoto et al. 2009), although without significant support.

The Albuginales are **obligate biotrophic pathogens** of angiosperms forming blister-like lesions on the leaves. They produce intercellular hyphae with small stalked **globose haustoria** (Fig. 3.13b) (Coffey 1975; Mims and Richardson, 2002; Soylu et al. 2003). Another feature of white blister rusts is their **subepidermal** mode of sporulation (Fig. 3.13a), with the fungus enzymatically separating the epidermis from the mesophyll to create a cavity in which sporulation takes place in rust-like fashion (Heller and Thines 2009). Dispersive sporangia are produced in a **basipetal fashion** from the apex of unbranched sporogenous hyphae (Fig. 3.13a) (Beakes 1987; Heller and Thines 2009; Mims and Richardson 2002). Zoospore formation from the deciduous conidiosporangia is frequently suppressed, although it can occur, releasing a cluster of zoospores into a transient vesicle (Fig. 3.13c–e), which quickly ruptures, releasing free-swimming zoospores (Fig. 3.13f). White blister rust species have thick, multilayered **verrucose oospore walls** (Fig. 3.13g) (Thines et al. 2009a) whose outer layers seem to be derived from the periplasm (Beakes 1981b; Tewari and Skoropad 1977). Upon germination, oospores convert into sporangia and release zoospores directly into a transient vesicle (Fig. 3.13h).

White blister rusts parasitize a broad range of mostly herbaceous flowering plants, and recent molecular studies have shown considerable genetic diversity within this order (Choi et al. 2006, 2007, 2008, 2011a, b; Mirzaee et al. 2013, Ploch et al. 2010, 2011; Thines et al. 2008, 2009c; Voglmayr and Riethmüller 2006). Currently three genera, *Albugo*, *Pustula*, and *Wilsoniana*, are parasitic to rosids, asterids, and the Caryophyllales respectively (Thines and Spring 2005; Thines and Voglmayr 2009). White blister rusts, in contrast to many oomycete plant pathogens, cause little or no visible damage to the tissues they infect (Kemen and Jones 2012; Thines and Kamoun 2010). There is increasing evidence supporting the endophytic habit of *A. candida* where asymptomatic infections are widespread in natural host populations and the pathogen can enter into the generative organs of their hosts (Ploch and Thines 2011). The **independent evolution** of **obligate biotrophy** in this group of Peronosporomycetes is also reflected by the fact that *Albugo laibachii*, the white blister rust pathogen of *Arabidopsis thaliana* (Thines et al. 2009a), has a much smaller genome compared with obligate parasites in the Peronosporales (Kemen and Jones 2012; Kemen et al. 2011) and has evolved a novel group of **CHxC/CxHC effectors** that are unique to this clade (Kemen et al. 2011). Therefore, based on the mostly restricted host range of the currently described species, it can be assumed that the 50 or so white blister rust species currently known probably represent only a fraction of the total diversity of white blister rusts (Ploch and Thines 2011; Ploch et al. 2010; Voglmayr and Thines 2009).

3. Peronosporales s. lat.

Members of the **Peronosporales s. lat.** as defined here (Fig. 3.14) are found in both marine and terrestrial ecosystems and include saprotrophs, opportunistic and obligate plant pathogens, and animal parasites (Dick 2001a; Hatai 2012; Lévesque et al. 2010; Mendoza 2009; Schurko et al. 2004; Sparrow 1973c). Thallus morphology can range from holocarpic (Fig. 3.14a–j) to eucarpic (Fig. 3.14k), but in most (with the exception of derived plant pathogens), the final differentiation of their zoospores occurs within an **external sporangial vesicle** that forms at the neck of the discharge tube or the exit pore of the sporangium (Fig. 3.14c, h, l, n). Even those genera, such as ***Halopthytophthora*** (Fig. 3.15d) and ***Phytophthora*** (Fig. 3.15h), that form their fully differentiated zoospores intrasporangially discharge their spores into a transient evanescent vesicle (as shown in Fig. 3.15e). **Peronosporales** s. lat. contains around 30 genera (Table 3.3) and 1,000 species. Dick (2001a) split his order **Pythiales** (equivalent to the ~Pythiaceae in this review) into two families, the **Pythiogetonaceae** (Voglmayr et al. 1999) and the much larger **Pythiaceae** (Table 3.4). However, there is no molecular support for separating the genus *Pythiogeton* into its own family because it forms a monophyletic clade nested within one of the *Pythium* clades (Huang et al. 2012). Until more robust molecular markers are developed, it seems unwise to create lots of new families and genera within the **Pythiaceae s. lat.**, although it may ultimately be necessary.

a) Salisapilaceae

A recent study of saprotrophs of saltmarsh litter identified a novel pythiaceous clade which was located basal to the main pythiaceous and peronosporaceaous clades (Hulvey et al. 2010). Members of this family had been originally assigned to the genus *Halophytophthora* with which they share some morphological similarities, such as **vesiculate zoospore release** (Hulvey et al. 2010). They have mycelial thalli and form **ovoid papillate sporangia** and **smooth-walled oogonia** with large translucent ooplast vacuoles (Hulvey et al. 2010). This clade was given the family designation **Salispilaceae** (Hulvey et al. 2010) and is retained as a separate family within Peronosporales s. lat. (Table 3.5). However, preliminary unpublished work suggests that some holocarpic lagenidiaceous species may diverge before the Salisapilaceae, and the delineation of families within the pythiaceous oomycetes will undoubtedly need to be reviewed.

b) Pythiaceae s. lat.

The ~**Pythiaceae s. lat.** (Fig. 3.14) is a large complex family that contains a number of well-documented subclades (Fig. 3.5b) (Lévesque and de Cock 2004). Limited molecular studies have shown that genera such as ~*Lagenidium* and ~*Myzocytiopsis* spp. are paraphyletic but share a common ancestor with ~*Pythium* (Fig. 3.5b) (Beakes et al. 2006; Schroeder et al. 2012). It is likely that the holocarpic lagenidiaceous parasites of freshwater algae, which Dick (2001a) placed in the genera *Myzocytium* s. str. and *Syzygangia* (Table 3.4), as well as the graminicolous root pathogen *Lagena radicola* (Fig. 3.14a–g), may also be part of this assemblage, although whether all of these genera (including *Lagenidium*) will remain valid is doubtful. Many holocarpic isolates form a monophyletic clade (Fig. 3.5) (Beakes et al. 2006; Hatai 2012), whilst others intercalate amongst those *Pythium* clades with predominantly filamentous sporangia (Fig. 3.14l), as indeed does the genus *Pythiogeton* (Fig. 3.5b) (Huang et al. 2012; Schroeder et al. 2012). Neither the order **Myzocytiopsidales** nor the family **Myzocytiopsidaceae** (Table 3.4) proposed by Dick (1997, 2001a) has any molecular support, and both are rejected and omitted from our revised taxonomic scheme (Table 3.5).

The largest and most comprehensively studied genus in this family is ***Pythium*** (Fig. 3.14k–w). *Pythium* was last monographed by Plaats-Niterink (1981) and will soon be updated (previewed by de Cock et al. 2012). This genus has well over 100 species, most of which have been sequenced (Bala et al. 2010; Bedard et al. 2006; Briard et al. 1995; Lévesque and de Cock 2004; Martin 2000; Schurko et al. 2004; Villa et al. 2006; Uzuhashi et al. 2010). Early molecular studies found that species assigned to *Pythium* species fell into **ten clades,** designated A–K (Lévesque and de Cock 2004) (Fig. 3.5b). Unfortunately, there are few common morphological features which characterize every species within a given clade. Although some broad overall morphological trends are recognizable, there are always exceptions to the norm, which makes it difficult to define clade synapomorphies (de Cock et al. 2012). There is a monophyletic cluster of clades (A–D: Fig. 3.5b) whose species predominantly have simple **filamentous** (Fig. 3.14l) to **irregularly inflated** (Fig. 3.14m, n) **zoosporangia** (Lévesque and de Cock 2004). These simple sporangia are found in the earliest diverging clades (if lagenidiaceous genera are excluded) and are considered to represent an ancestral state. This group includes the **type species** (clade A: ***P. monospermum***) and, as was recently proposed, represents the **genus *Pythium* s. str.** (Uzuhashi et al. 2010). This basal group contains species that are opportunist **parasites of seaweeds** (e.g. *Pythium chondricola* and *Pythium porphyrae*) and **vertebrates** (e.g. *Pythium insidiosum*; Schurko et al. 2004). However, even this redefined *Pythium* s. str. group of clades (Uzuhashi et al. 2010) has nested within it the ***Pythiogeton* subclade** (Huang et al. 2012), as well as many holocarpic isolates, previously assigned to the genera ~*Lagenidium* and *Myzocytiopsis* (as in Fig. 3.5b) (Schroeder et al. 2012), and is therefore still in need of revision.

A second group of *Pythium* clades (E–J) (Lévesque and de Cock 2004) that share a common ancestor contains species with predominantly **ovoid to globose** sporangia that frequently proliferate internally (Fig. 3.14r, s) and have oogonium walls that are often spiny (e.g. *P. amasculinum*, Fig. 3.14u). These are considered to represent a derived state and include the important root-infecting plant pathogen ***P. ultimum*** (Fig. 3.14v, w), whose genome sequence was recently published (Lévesque et al. 2010). Uzuhashi et al. (2010) also proposed that the species in clades E–G and I–J (such as *P. multisporum*, clade E, Fig. 3.14r) be transferred to a new genus, *Globisporangium*, and those in the sister clade (Clade H, e.g. *P undulatum*, Fig. 3.14s) to the genus *Elongisporangium*, although these new names have not been widely adopted.

c) Peronosporaceae s. lat.

This is a species-diverse, predominantly plant-pathogenic family that contains a significant proportion of the total number of known oomycete species (Thines and Kamoun 2010). The expanded **Peronosporaceae** proposed here contains many economically important species that account for and have been the focus of much research effort, particularly in relation to unravelling the basis of their **pathogenicity** (e.g. Kemen and Jones 2012; Lamour et al. 2007; Randall et al. 2005; Thines and Kamoun 2010; Tyler et al. 2006; Voglmayr 2008). Hulvey et al. (2010) proposed maintaining the **broad circumscription** of the family Peronosporaceae, so that it includes all downy mildews, ~*Phytophthora* and many ~*Halophytophthora* spp., in order to avoid the creation of a plethora of poorly differentiated families. **Peronosporaceae** (Fig. 3.15) as redefined here includes the genera ***Phytopythium*** (formerly the *Pythium* K-clade) (Bala et al. 2010), ~***Halophytophthora*** (Hulvey et al. 2010) and ~***Phytophthora*** (Runge et al. 2011a), all of which had previously been placed in the **Pythiaceae** (Table 3.4) (Dick 2001a). The family also includes the 19 genera of **downy mildews** (Table 3.5). Most species form ovoid, often **deciduous, sporangia** on branched **aerial sporangiophores** (Fig. 3.15). In *Halophytophthora, Phytopythium,* and *Phytophthora,* as well as a few zoosporic downy mildew species, the zoospores are usually released from the sporangium into a transient restraining **vesicle**. However, in the downy mildews (Fig. 3.15k–s) there is a trend towards the suppression or complete loss of zoospore formation. This has been confirmed by the recent finding of a complete loss of zoospore-associated genes in *Hyaloperonospora* (Baxter et al. 2010). Therefore, in many downy mildew genera, including the two largest, *Hyaloperonospora* and *Peronospora,* sporangia act as dispersive conidia that germinate solely by germ tubes (Waterhouse 1973).

There is no doubt that this large and complex family will require significant taxonomic revision, but there is at present no consensus as to how this should be achieved [see discussion by Runge et al. (2011a)]. For convenience, we divide this account of the **Peronosporaceae** into **three parts** without implying taxonomic significance but rather trophic similarities because, apart from the downy mildews, all are paraphyletic. The first section includes the largely saprotrophic genera ***Phytopythium*** **and** ~***Halophytophthora,*** the second the largely hemibiotrophic genus ~***Phytophthora,*** and the third the obligate biotrophic **downy mildews**. Recent genomic studies have shown that both ~***Phytophthora*** (Bozkurt et al. 2012; Morgan and Kamoun 2007; Raffaele et al. 2010; Tyler et al. 2006) and the **downy mildews** (Baxter et al. 2010; Stassen et al. 2011; Tian et al. 2011) produce a huge range of **RxLR-type effector** molecules, which suppress host responses to infection. The molecular basis for host–pathogen interactions is currently under intense investigation (Jiang and Tyler 2012; Thines and Kamoun 2010), and a detailed comparison of obligate biotrophy in white blister rusts and downy mildews was recently reviewed by Kemen and Jones (2012).

(iv) Part 1: *Phytopythium* and *Halophytophthora* Clades

The genera ***Phytopythium*** (e.g. *Pp. ostracodes,* Fig. 3.15a–c) and ~***Halophytophthora*** (Fig. 3.15d, e) form two relatively small clades that diverge before clades containing most of the phytopathogenic oomycetes. *Phytopythium* (Bala et al. 2010) is synonymous with the almost simultaneously named *Ovatosporangium* (Uzuhashi et al. 2010) and was formerly referred to as the K-clade *Pythiums* (Lévesque and de Cock 2004). Species in this clade have both morphological and physiological characteristics that are intermediate between *Pythium* and *Phytophthora* (Bala et al. 2010). Species have strongly **papillate** ovoid sporangia and large, thick-walled oospores (such as in *Pp. ostracodes*) (Fig. 3.15c). Furthermore, some produce **elicitin-like holoproteins** (Panabières et al. 1996) similar to those found in *Phytophthora.* This clade includes both saprotrophs (Nechwatal and Mendgen 2005) and **root/stem-infecting** pathogens such as *Pp. vexans* and the designated type species *Pp. sindham* (Bala et al. 2010), which infects bananas. Consistent with ancestral, *Pythium*-like characteristics and ecology, *Phytopythium* spp. form the earliest monophyletic clade within **Peronosporaceae** s. lat. (Fig. 3.5b) (Bala et al. 2010; Uzuhashi et al. 2010).

Also within **Peronosporaceae** s. lat. are many ~***Halophytophthora*** spp., including the species type ***H. vesicula*** (Fig. 3.15d, e). ~*Halophytophthora* as originally defined (Ho and Jong 1990; Nakagiri 2002b) is a polyphyletic genus that contains around 15 species (Hulvey et al. 2010). Species in this genus are significant saprotrophic decomposers of fallen mangrove leaves in tropical or subtropical coastal ecosystems (Ho et al. 1992; Nakagiri et al. 1994), and some species were originally described as ~*Phytophthora* species (Fell and Master 1975). Recent investigations have revealed their presence also in salt marshes in northern Europe, in the German Bight (Nigrelli and Thines 2013), providing evidence that *Halophytophthora* s. str. species might also be important for nutrient cycling in temperate marine ecosystems. ~*Halophytophthora* spp. have ovoid to elongate sporangia, often with conspicuous papillar plugs (Nakagiri 2002b), and most show a transient vesiculate discharge of their zoospores (Fig. 3.15e). Homothallic sexual reproduction has only been reported to date in approximately half the known species. All have single oospored oogonia with mostly **paragynous antheridia** (as in Fig. 3.15c) (Nakagiri 2002b).

Although extensive sequencing of species within this genus has been reported in conference abstracts or presented online (e.g. by Nakagiri 2002a; Coffey and Levesque and colleagues), only relatively limited data have actually been published in the peer-reviewed literature (e.g. Göker et al. 2007; Hulvey et al. 2010). Nakagiri (2002a) was the first to report that the *Phytophthora-Peronospora* clade was close to a group of *Halophytophthora* spp. and formed a sister clade to what he referred to as the *H. vesicula* complex. He also noted that *H. spinosa* seemed to be only distantly related to other *Halophytophthora* species in a clade close to *Sapromyces*, which makes it possible that this species belongs to the Salispilaceae described by Hulvey et al. (2010). The type species *H. vesicula* (Ho and Jong 1990) is part of a monophyletic *Halophytophthora* s. str. clade, which also includes *H. avicenniae*, *H. batemanensis*, and *H. polymorphica*. The main *Phytophthora*-downy mildew clade itself includes a number of species, such as *H. elongata*, *H. exoprolifera*, and *H. porrigovesica* (Coffey et al. 2011). Yet other species are to be found dispersed amongst *Phytopythium* and *Pythium* clades (Coffey et al. 2011). For instance, *H. kandeliae* is nested within the *Phytopythium* clade, and *H. epistomium* seems close to *Pythium monospermum*. Thus, *Halophytophthora* as previously circumscribed is a good example of how ecological preference, which might also lead to convergently evolved morphological traits, is mostly not useful for group delimitation in a taxonomic sense.

(v) Part 2: *Phytophthora* Clades

This section of the **Peronosporaceae** includes some of the most highly adapted and devastating plant-pathogenic oomycete species. The facultatively biotrophic genus ~***Phytophthora*** (Fig. 3.15f–j) is most likely paraphyletic (Cooke et al. 2000; Runge et al. 2011a) and contains over 100 species, which typically produce ovoid to ellipisoidal sporangia, often on aerial sporangiopores (Fig. 3.15f–h) and single-oospored oogonia, many of which have distinctive collar-like **amphigynous antheridia** (Fig. 3.15i, j). Many of the characters that were once thought to be important in *Phytophthora* taxonomy (Sparrow 1973c; Waterhouse 1970, 1973), such as amphigynous (as in *Ph. erythroseptica*) (Fig. 3.15i, j) versus paragynous antheridial attachment (as in *Pythium*) (Fig. 3.14t) and the genetic breeding system (homothallic versus heterothallic), have not proved to be good indicators of phylogenetic relatedness. Most *Phytophthora* species have been sequenced (Blair et al. 2008; Briard et al. 1995; Cooke et al. 2000; Förster et al. 2000; Kroon et al. 2004, 2012; Martin and Tooley 2003, 2008), and the genus has been divided into **nine or ten monophyletic subclades** (groups 1–10). The morphological characteristics and species composition of these groups were recently reviewed by Kroon et al. (2012). There does not seem to be a unique set of morphological characters (synapomorphies) that can be linked with a single clade, although developmental trends can be associated with groups of *Phytophthora* clades (Blair et al. 2008; Kroon et al. 2012; Runge et al. 2011a). Species in the **presumed basal clades** (clades 6–10) are necrotrophic soil-borne pathogens that produce non-caducous, generally non-papillate sporangia that release zoospores upon germination (as in *Ph. sojae*, clade 7, Fig. 3.15f). This group includes important root and woody stem pathogens such as ***Ph. cinnamomi*** (Hardham 2005; Newhook and Podger 1972) and ***Ph. ramorum*** (Davidson et al. 2003). Species in clades 1, 2, 4, and 5 seem to be more derived and often have an extended **biotrophic** phase, produce papillate airborne, deciduous sporangia, and are predominantly **foliage pathogens**, for example *Ph. infestans* (Fig. 3.15g, h, clade 1) and *Ph. palmivora* (clade 4) (Kroon et al. 2012; Runge et al. 2011a). Species in the more derived lineages, especially clade 4, are more difficult to isolate and culture, and some even appear to be **obli-**

gate biotrophs (Blair et al. 2008; Runge et al. 2011a).

Recently, *Phytophthora* clades were reanalysed by Runge et al. (2011a) with respect to their relationship to the **downy mildews** (Fig. 3.5b; Runge et al. 2011a). In their analysis, the monophyletic downy mildew clade form a sister clade to a group of four species that included *Ph. palmivora* and *Ph. megakarya*, which together were sister to *Ph. quercina*. These in turn form the sister clade to a group of about a dozen papillate, generally foliage pathogens, which include ***Ph. infestans*** **(the type species),** *Ph. nicotianae*, and *Ph. cactorum* and which the researchers proposed as representative of the ***Phytophthora*** **s. str. clade** (Runge et al. 2011a). Therefore, it is likely that the genus *Phytophthora* has a high degree of paraphyly with respect to the monophyletic downy mildews. Rather than the impractical solution of renaming all downy mildews and *Phytophthora* spp. as *Peronospora* spp., because this is the oldest available genus name for the group, Runge et al. (2011a) suggest that several new genera should be described within *Phytophthora* to restore the monophyly of this diverse genus. However, in a recent phylogenetic analysis based on whole genomes, albeit of the very restricted number of five taxa, Seidl et al. (2012) found that the downy mildews (represented by *Hyaloperonospora*) were sister to the *Phytophthora* clade rather than embedded within it, with the non-papillate/semi-papillate *Ph. sojae* and *Ph. ramorum* species forming a clade that was sister to the papillate *P. infestans* as in the analysis of Runge et al. (2011a).

(vi) Part 3: Downy Mildew Clades

The **downy mildews** (Fig. 3.15l–s) are a rapidly evolving, **hyperdiverse** group consisting of **19 genera** of **obligate biotrophic** pathogens of flowering plants (Table 3.5). They typically produce deciduous **conidiosporangia** that are usually borne on bifurcating persistent **aerial sporangiophores** (Fig. 3.15k, l, n, o). Downy mildew oospores are thick-walled, usually plerotic, and often have verrucose ornamentation (Fig. 3.15r, s) (Dick 2001a; Sparrow 1973c). They are by far the largest and most diverse group of oomycetes, estimated to contain **over 800 species**, accounting for more than half of all currently known oomycete species (Thines and Kamoun 2010). As with the other important plant-pathogenic peronosporomycete lineages, the downy mildews have been extensively sequenced over the past decade (García-Blázquez et al. 2008; Göker et al. 2003, 2004, 2007; Riethmüller et al. 2002; Runge et al. 2011a; Telle and Thines 2012; Telle et al. 2011; Thines et al. 2006, 2007, 2008, 2009b; Voglmayr 2003; Voglmayr and Constantinescu 2008; Voglmayr et al. 2004). An overview of their taxonomy and phylogeny was recently given by Thines et al. (2009c). Four major monophyletic subclades have been recognized within the downy mildews (Göker et al. 2007; Thines et al. 2009c), which are named on the basis of morphological characteristics or host preference (Table 3.5).

The earliest diverging group (Thines et al. 2008, 2009c) in the monophyletic downy mildew clade (Hudspeth et al. 2003; Riethmüller et al. 2002; Thines et al. 2008; Telle and Thines 2012; Telle et al. 2011) seems to be the **GDMs** (Fig. 3.15k). There is no support for placing them in their own Sclerosporales order (Table 3.4) introduced by Dick et al. (1984), and this order has been rejected in our revised classification (Table 3.5). Among the GDMs are a monophyletic group of rare monotypic genera, *Graminivora*, *Poakatesthia*, and *Viennotia*, all of which have persistent sporangiophores (Göker et al. 2003; Thines et al. 2006, 2007). *Poakatesthia* and *Viennotia* exhibit morphological characteristics that are intermediate between *Phytopththora* and the downy mildews s. str. and have been suggested as being intermediate between the two (Thines 2009). Thines (2009) suggested that the diversification of the downy mildews might have started from these *Phytophthora*-like grass parasites. The second GDM group has evanescent sporangiophores and includes the genera ***Eraphthora*** (Telle and Thines 2012), ***Peronosclerospora*** (Fig. 3.15k), ***Sclerophthora***, and ***Sclerospora*** (Telle et al.

2011; Thines et al. 2008), which contain important pathogens of cereal crops in the semi-arid tropics.

The vast majority of **downy mildew species** are pathogens of **so-called eudicot families**, although whether pathogen species are restricted to a given host species or wider host families has often been debated (Constaninescu 1991; Dick 2001; Thines et al. 2009c). However, in recent years molecular studies have helped clarify host–pathogen relationships and confirmed that most downy mildew species have very narrow host ranges (Choi et al. 2007, Riethmüller et al. 2002; Thines 2011; Thines et al. 2009b; Voglmayr 2003; Voglmayr and Thines 2007; Voglmayr et al. 2004), often encompassing only a few or even a single species within a given host genus. Only a comparatively small number of species, such as ***Pseudoperonospora cubensis*** s. lat., are able to infect more than one host genus (Choi et al. 2005; Göker et al. 2004, 2007; Runge and Thines 2011, 2012; Runge et al. 2011b, 2012). The next recognized subgroup of downy mildews are parasitic to members of the Brassicaceae; they are known as the **Brassicolous downy mildews** and include the intensively studied genus ***Hyaloperonospora*** (Fig. 3.15o, p) (Tyler et al. 2007) and the genus ***Perofascia*** (Table 3.5). The third downy mildew subgroup (Table 3.5) are those with **coloured (pigmented) conidia (DMCC)**, which includes the economically important species-rich genera ***Peronospora*** (García-Blázquez et al. 2008; Voglmayr and Constantinescu 2008) and ***Pseudoperonospora*** (Thines et al. 2009c) and has over 500 listed species. Genera such as ***Hyaloperonospora*** (Fig. 3.15o) and ***Peronospora*** often have elongate **digit-like haustoria** (as in *Hyaloperonospora*) (Fig. 3.15p), which are similar in morphology and ultrastructure (Beakes et al. 1982; Hickey and Coffey 1977) to those formed by hemibiotrophic *Phytophthora* species such as *Ph. infestans* (Coffey and Wilson 1983). The final major monophyletic lineage is a diverse group known as the downy mildews with small **pyriform haustoria (DMPH)** (Fig. 3.15m); it contains eight genera (Table 3.5), including ***Basidiophora, Bremia*** (Fig. 3.15n), ***Benua, Paraperonospora, Plasmopara*** (Fig. 3.15l), and ***Plasmoverna*** [for a review see Thines et al. (2009c)], in addition to ***Novotelnova*** and ***Protobremia,*** which are closely allied with *Bremia* (Fig. 3.15n).

III. Selected Developmental and Morphological Trends

A. Zoospore Characteristics

1. Zoospore Morphology and Flagellar Rootlet Organization

Zoospore organization and flagellum micromorphology have been extensively reviewed (Barr 1981; Barr and Désaulniers 1989; Beakes 1987, 1989; Dick 1990, 2001a, b; Fuller 1990, 2001; Perkins 1976). The stramenopile zoospore (Fig. 3.16) shows a remarkably uniform and conserved structure overall, supporting the origin of this clade from a common flagellate ancestor (Tsui et al. 2009). Stramenopile zoospores range in size between 3 and 15 μm (Dick 2001), and many are reniform with laterally inserted flagella, as in the Labyrinthulomycota and most oomycetes (Fig. 3.16a, b, l–r). In most stramenopiles the anterior flagellum is decorated with two parallel rows of proteinaceous **tripartite tubular hairs** (TTHs) (Vlk 1939) (Fig. 3.16a, b, l, o–r), which reverse the flagellum thrust, in effect pulling the zoospores through the water (Dick 2001a). Hyphochytriomycota zoospores are small (3–5 μm) and characterized by their single anterior flagellum decorated with TTHs (Fig. 3.16d) (Dick 2001a; Fuller 1990; Karling 1977; Sparrow 1960). There is some variation in the presence and distribution of TTHs within the oomycete clade. Zoospores of most *Haptoglossa* species lack TTHs (Fig. 3.16m, n) (Beakes and Glockling 1998), whilst genera such as *Crypticola* (Frances et al. 1989) and *Lagenidium giganteum* (Domas et al. 1986) have only a single row of TTHs along the anterior flagellum, and *Myzocytiopsis vermicola* has only a partial row of TTHs proximal to the zoospore body (Glockling and Beakes 2006a). Thraustochytrid zoospores are unusual amongst stramenopiles in that the cell body is coated in small scales

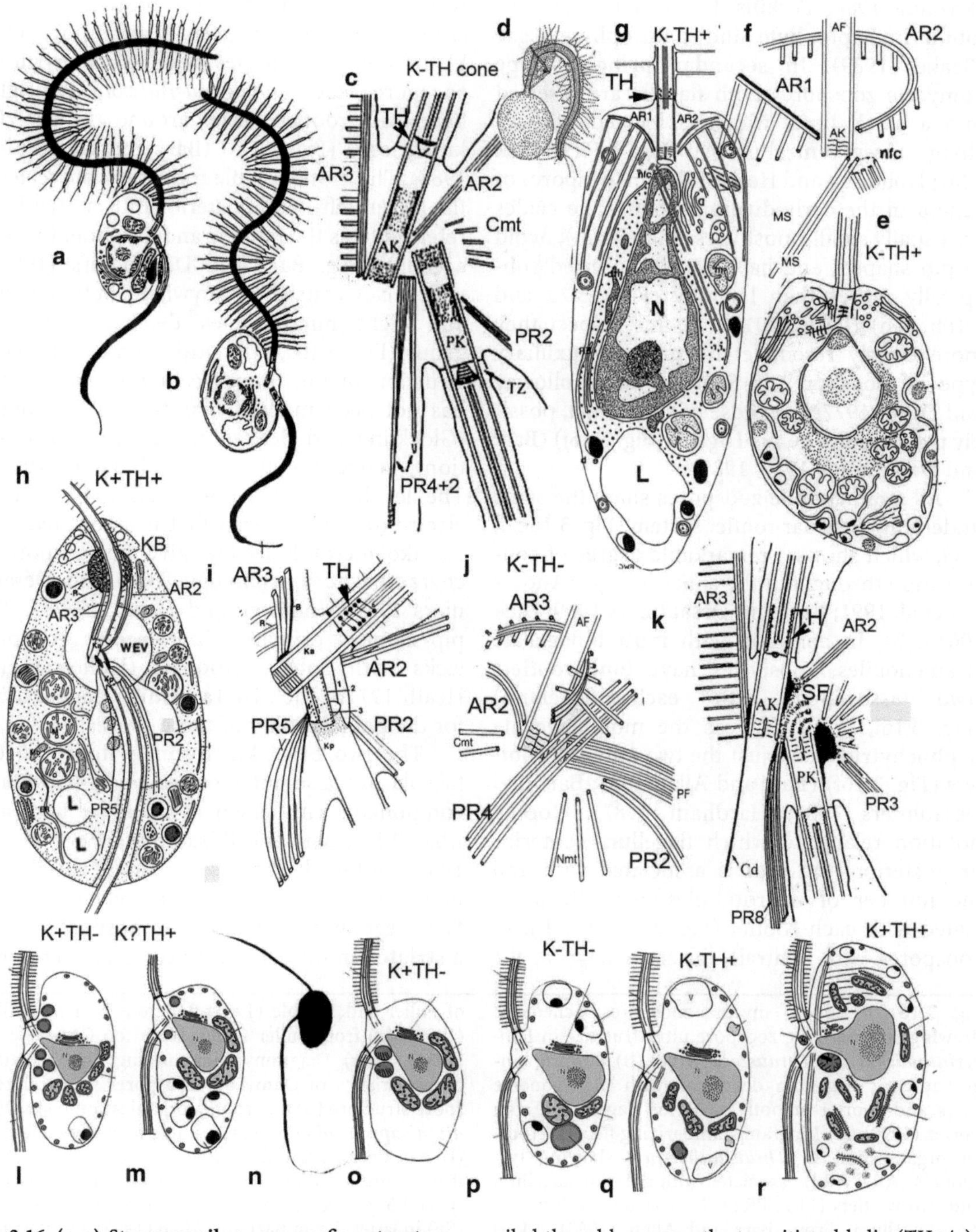

Fig. 3.16 (a–s) Stramenopile zoospore fine structure. All zoospores presented in same orientation using same rootlet notation, modified from that used by Barr and Désaulniers (1989): AK, anterior kinetosome (flagellar root); PK, posterior kientosome; AR, anterior rootlet; PR, posterior rootlet, together with number of microtubules usually recorded in each. In some species a striate fan (SF) interconnects the anterior and posterior flagella. The presence (+) or absence (-) of the coiled thread known as the transitional helix (TH+/–) (Beakes 1987) located immediately above the basal plate, which separates the flagellum from the zoospore body, is also indicated. Finally, the presence (+) or absence (-) of kinetosome associated K-bodies (Lehnen and Powell 1989) (K+/-) is also noted. Other zoospore organelles labelled include the nucleus (N) and water expulsion apparatus (WEV). **(a–c)** Diagrams showing ultrastructural characteristics of **Labrinthulo-**

(Kazama 1980; Perkins 1976), a feature also found in haptophyte and synurophyte algae (Beakes 1989). In **secondary/principal** type oomycete zoospores both flagella are inserted into a small boss, which is located midway along a deep ventral groove (Fig. 3.16r) (Dick 2001; Holloway and Heath 1977a). Zoospores of genera in the early-diverging oomycete clades are usually small (mostly less than 5 μm), ovoid to pip shaped, and have flagella inserted subapically rather than laterally (Figs. 3.7d and 3.16h, l–p) (Dick 2001a). In this respect they more closely resemble the primary/auxilliary type of zoospore of *Saprolegnia* (Holloway and Heath 1977a) or those of the unusual, possibly pythiaceous, genus *Lagena* (Fig. 3.16j) (Barr and Désaulniers 1987, 1989).

All stramenopile zoospores share the same underlying flagellar rootlet system (Fig. 3.16c, f, i–k), which shows a remarkable degree of conservation throughout the entire lineage (Andersen et al. 1991; Barr and Désaulniers 1989; Dick 2001a, b). In common with most biflagellate stramenopiles, zoospores have **four rootlets** (two associated with each flagellum) (Fig. 3.16i, j, k), whereas the monoflagellate **hyphochytrids** have just the two **anterior rootlets** (Fig. 3.16f) (Barr and Allan 1985; Barr and Désaulniers 1987; Hardham 1987). Rootlet notation relates to which flagellum (anterior or posterior) the root is associated with and the number of microtubules normally associated with each rootlet (Fig. 3.16c, f, i–k). In zoospores with ventrally inserted flagella, the two anterior and posterior rootlets align on either side of the flagellar groove (Fig. 3.16i, k), whereas in species with more pyriform zoospores, such as *Lagena radicola*, the right-hand AR3 rootlet curves around the anterior kinetosome (Fig. 3.16j) (Barr and Désaulniers 1987). The most variable rootlet in oomycetes is the larger, left-hand posterior PR8 rootlet [also referred to as the multistranded rootlet in some accounts, e.g. Barr and Désaulniers (1989)], which can consist of anywhere between four and eight microtubules depending on the genus (Fig. 3.16c, i–k). Although a full reconstruction of the *Haptoglossa* zoospore rootlet has not been made, it has the full P8 rootlet (Glockling and Beakes, unpublished observation), suggesting this is the basal condition. The left-hand AR2 rootlet has a backing of electron-dense material that is associated with rib-like microtubules that give the zoospore its characteristic morphology (Barr and Désaulniers 1989; Holloway and Heath 1977b). The pip-shaped primary *Saprolegnia* zoospore lacks both anterior rootlets (Holloway and Heath 1977b; Dick 2001a), which may account for the poor motility of these spores.

The two basal kinetosomes linked to the flagella are also interconnected by structural components (Andersen et al. 1991; Barr and Allan 1985; Barr and Désaulniers 1989; Beakes 1987), although these seem more variable in nature than the rootlets themselves. In both Saprolegniomycetes and Peronosporomycetes a **striate fan** (SF) structure connects the termi-

Fig. 3.16 (continued) mycota zoospores. Schematic drawing summarizing zoospore ultrastructure in ***Labyrinthula*** (**a**) and ***Thraustochytrium*** (**b**) showing longer anterior flagellum decorated with mastigoneme hairs and shorter smooth posterior flagellum. From Porter (1990). (**c**) Diagram summarizing flagellar rootlet organization in *Thraustochytrium* showing two rootlets (AR2, AR3) associated with anterior flagellum and two rootlets (PR2, PR4+2) associated with posterior flagellum. From Barr and Allan (1985). (**d–g**) **Hyphochytriomycota zoospores.** (**d**, **e**) Drawings made of shadowed whole mount preparation (**d**) and longitudinal section (**e**) of ***Rhizidiomyces apophysatus*** zoospores, together with a schematic diagram of flagellum base and two anterior rootlets (AR1, AR2) (**f**). Longitudinal profile zoospore of ***Hyphochytrium catenoides*** showing general arrangement of organelles and rootlets (**g**). (**d**, **e**) From Karling (1977) based on paper of Fuller and Reichle (1965), (**f**) from Barr and Allan (1985), (**g**) from Fuller (1990) based on Cooney et al. (1985). (**h–r**) Diagrams summarizing ultrastructural characteristics of **Oomycota zoospores.** General zoospore structure (**h**) and rootlet organization (**i**) in ***Olpidium saprolegniae*** in which single helically coiled TH (**i**) has been arrowed. From Bortnick et al. (1985). Rootlet organization in zoospores of ***Lagena radicola*** (**j**) and ***Saprolegnia*** sp. (**k**) Note presence of striate fan (SF) in latter. From Barr and Allan (1985) and Barr and Désaulniers (1989). Series of schematic diagrams and one whole mount preparation (**n**) comparing zoospore ultrastructure in ***Eurychasma dicksonii*** (**l**), ***Haptoglossa dickii*** (**m**, **n**), ***Olpidiopsis porphyrae*** (**o**), ***O. bostrychiae*** (**p**), ***Haliphthoros milfordensis*** (**q**) and ***Saprolegnia*** secondary-type zoospore (**r**). Diagrams adapted from Sekimoto (2008) and whole mount (**n**) from Beakes and Glockling (1998)

nal end of the posterior kinetosome to the anterior kinetosome (Fig. 3.4k) (Barr and Allan 1985; Beakes 1987). In zoospores of *Olpidiopsis saprolegniae* three electron-dense props interconnect the two kinetosomes (Fig. 3.16i), but there is no SF (Bortnick et al. 1985). However, SFs have been reported in *Haptoglossa* amongst basal oomycete genera (Beakes et al. 2012) but have not been found in either labyrinthulid or hyphochytrid zoospores (Fig. 3.16c) (Barr and Allan 1985; Porter 1990). However, an almost identical structure has been reported in bicosoecids (in *Rictus lutensis*) (Yubuki et al. 2010), which are generally considered to be the earliest diverging stramenopiles, suggesting that the origins of this structure are deeply rooted.

Another feature associated with the flagellum is the **transitional helix** (TH), which occurs in the transitional zone just above the basal plate that separates the flagellum from the main body of the zoospore (Fig. 3.16i, k). This is another structural feature that is unique to the stramenopile clade and has been the subject of much phylogenetic speculation (Beakes 1987; Cavalier-Smith and Chao 2006; Dick 2001a; Guillou et al. 1999; Patterson 1989). THs are found in slopalinids (proteromonads and opalinids as defined by Patterson 1989), some bicosecids (Patterson 1989), hyphochytrids (Barr and Désaulniers 1989), most oomycetes (Dick 2001), and, in a single helical form, in many ochrophyte groups (Cavalier-Smith and Chao 2006; Guillou et al. 1999). A single-helix variant of the TH is also reported in *Olpidiopsis saprolegniae* (Bortnick et al. 1985). Members of Labyrinthulomycota also do not have a typical TH (Fig. 3.16c), but it has been equated with a similarly placed cone-like structure [see discussion by Cavalier-Smith and Chao (2006)]. In most oomycetes the TH usually takes the form of a double-stranded helical coil with 6 to 12 turns (Dick 2001a). The TH is apparently absent from zoospores of some early-diverging species, such as *Eurychasma dicksonii* and *Olpidium porphyrae* (Sekimoto et al. 2008a, b), and from *Lagena* (Fig. 3.16j; Barr and Désaulniers 1987). So although it provides a useful general marker for stramenopiles, this structure does seem to have been lost or modified numerous times, without apparently affecting flagellar function.

2. Encystment/Adhesive Vesicles

Cortical vesicles in the peripheral cytoplasm of zoospores that are typically discharged upon encystment have also been widely discussed in relation to phylogeny (Beakes 1987, 1989; Beakes et al. 1995). In bicosoecid flagellates homologous vesicles are described as extrusosomes (Yubuki et al. 2010) but do not seem to be present in zoospores of the Labyrinthulomycota (Moss 1985, 1986; Perkins 1976; Porter 1990). Both hyphochytrid (Fuller and Reichle 1965; Fuller 1990) and oomycete (Beakes 1987, 1989) zoospores have morphologically similar cortical vesicles that upon encystment form the outermost layer of the encysted spore coat. The comparative structure of these vesicles and their homologues has already been discussed at length (Beakes 1987, 1989; Beakes et al. 1995). Electron microscopy and, later, cytochemistry revealed that the cortical vesicles in oomycetes were made up of two related vesicle populations (Gubler and Hardham 1988; Beakes 1994; Burr and Beakes 1994; Robold and Hardham 2004, 2005; Beakes et al. 1995).

In Peronosporomycetes one vesicle type [**ventral small vesicles (vsv)**] (Beakes et al. 1995; Gubler and Hardham 1988; Hardham 2005) is associated with the ventral flagellar groove, whereas in Saprolegniomycetes the homologous vesicles are **K-bodies, so** named because of their close association with kinetosomes (Holloway and Heath 1977a). K-bodies are distinguished by their relatively large size (0.3–0.5 μm) and distinctive microstructure, which includes an often crystalline matrix (Beakes 1989) and tubular inclusions (Holloway and Heath 1977a; Lehnen and Powell 1989). Upon encystment both vsv and K-bodies are discharged to form a ventral adhesive pad that helps attach the encysted spore to the substrate (Burr and Beakes 1994; Gubler and Hardham 1988; Lehnen and Powell 1989). The large structured K-bodies were considered to be a marker for saprolegnialean oomycetes because they had been observed in *Apodachlya* in the Leptomitaceae (Beakes 1987; Randolph and Powell 1992) and in genera throughout the Saprolegniales (Beakes 1987, 1989).

Morphologically similar **K-body vesicles** have now been widely found in the zoospores of basal oomycetes (Fig. 3.16h, l–r), such as *Eurychasma* (Sekimoto et al. 2008a) and both freshwater and marine *Olpidiopsis* spp. (Bortnick et al. 1985; Sekimoto et al. 2007). This is another feature that links the basal clades directly with the saprolegniomycete line. The Rhipidialean genus *Sapromyces* also has **K-bodies** (Gotelli and Hanson 1987) rather than smaller, more widely dispersed **vsv**, typical of the peronosporomycete line. This again points to this orders showing intermediate characteristics between the main classes and at the plesiomorphy of the K-bodies. The gene encoding the protein secreted by the **vsv fraction** in *Ph. cinnamomi* has been cloned and shown to contain a **thrombospondin-1 repeat**, which is a motif associated with adhesins in the malarial parasite *Plasmodium* and animals (Robold and Hardham 2005).

The second cortical vesicle fraction is more abundant and widely distributed within both hyphochytrid (Cooney et al. 1985; Fuller and Reichle 1965) and oomycete zoospores (Beakes 1983, 1987; Dick 2001) and are the source of spines that might decorate encysted spores (Fig. 3.8j, r, s). In oomycetes they were first described in *Saprolegnia* as **bar bodies** (Beakes 1987), and homologous vesicles, which have a structured periphery and hollow core (which contains any spines), have been described throughout most of the early-diverging genera, including *Eurychasma* (Sekimoto et al. 2008), *Haptoglossa* (Beakes and Glockling 1998, 2000), several *Olpidiopsis* spp. (Bortnick et al. 1986; Sekimoto et al. 2009), and *Haliphthoros* (Overton et al. 1983; Sekimoto 2008). In the Saprolegniaceae, genera with eccentric oospores, such as *Dictyuchus*, have secondary cysts decorated by broad tapered spines (Fig. 3.11j) (Beakes 1987), whereas those with centric or subcentric oospores, such as the genus *Saprolegnia*, typically have fine bifurcated boat-hook spines (Fig. 3.11r, s) (Beakes 1983; Hallett and Dick 1986; Burr and Beakes 1994; Beakes et al. 1995). In contrast, in most Peronosporomycetes, the homologous peripheral **dorsal small vesicle** (**dsv**) fraction has homogenous contents and is often morphologically indistinguishable from the ventral vesicle fraction. Although similar in appearance to **vsv**, the **dsv** fraction has different immunological properties and upon discharge forms the initial coat that protects the encysting zoospores (Beakes et al. 1995; Robold and Hardham 2004; Hardham 2005). Now that many discrete subclades have been identified in large and complex taxa such as *Pythium* and *Phytophthora*, it might be worthwhile to carry out a detailed comparative ultrastructural examination of their zoospores to see whether any structural features can be associated with particular clades, as has been so successfully done for chytrids (see Powell and Letcher 2014).

B. Life Histories and Sexual Cycle Characteristics

Most stramenopiles are **diploid organisms** that undergo gametic meiosis (Dick 2001; Sims et al. 2006), although triploid and tetraploid forms have been reported in *Phytophthora* (Tooley and Therrien 1987). However, knowledge of the precise timing of meiosis and plasmogamy in osmotrophic stramenopiles outside of Oomycota is sparse. Although the morphological changes that accompany the life cycles of different thraustochytrid genera are now well documented (Fig. 3.2), the stages where meiosis and karyogamy take place have still not been established. In labyrinthulids, evidence of meiosis is supported by the finding of synaptonemal complexes in nuclei in thalli that are forming zoospores (Perkins and Amon 1969; Porter 1990), but where syngamy takes place is still not known. In the Hyphochytriomycota s. str. it is not known where meiosis and karyogamy take place (Fuller 1990, 2001). The only documented occurrence of karyogamy is described in *Anisolpidium ectocarpii* [Karling 1943; Johnson 1957; summarized by Karling (1981)], which we now suspect may belong in the oomycete clade. In this genus, two naked endobiotic thalli, arising from two adjacent cysts (Fig. 3.4l, m), immediately fuse to produce a zygote (Fig. 3.4o, p). This fused nucleus immediately undergoes meiosis (Fig. 3.4p), which gives rise to presumably haploid zoospores (Fig. 3.4q). In oomycetes a common feature shared by almost all genera that diverge before the peronosporomycete/saprolegniomycete split is their apparent lack of a sexual cycle involving oospores (Beakes and Sekimoto 2009; Beakes et al. 2012). As noted by Sparrow (1976), sexual stages in these groups are probably not easily recognized and had been overlooked until then. This conclusion was given credence when Schnepf et al. (1977, 1978b) described an unusual life cycle in the diatom parasite *Lagenisma coscinodisci*. In this species, zoospores resulting from meiotic divisions (zoomeiospores) encyst on the surface of their host. One cyst (the presumptive male) produces a fertilization tube that fuses with the adjacent (presumptive female) cyst and plasmogamy takes place, followed by nuclear fusion and zygote formation (Schnepf et al. 1977). Recently, a remarkably similar sequence was observed in the basal species *E. dicksonii* when it infected *Choristocarpus* (Gachon, pers. commun.). Cysts settle on the host surface and fuse to give rise to an enlarged zygote cell. This fragmentary evidence suggests sexual reproduction by means of zoospores arising from meiosis (zoomeiospores), and fusion of the resulting cysts (or thalli) is the most likely form of sexual reproduction in early-diverging oomycetes. A similar pattern of sexual reproduction has also been described in *Lagena radicola* (Fig. 3.14e, f) (Barr and Désaulniers 1990). *Haptoglossa* is a diverse genus, but there are still no reports of where meiosis takes place, although fusion of adjacent thalli has been illustrated in *H. heterospora* (Karling 1981). Several *Haptoglossa* species produce both single and binucleate aplanospores, as described in *H. heteromorpha* (Fig. 3.7j, k) (Glockling and Beakes 2000c) and *H. erumpens* (Fig. 3.7l) (Beakes and Glockling 2002), but how this is related to any sexual cycle is not known.

Freshwater *Olpidiopsis* spp. are the only known basal clade representatives to form recognizable male and female thalli leading to the formation of a thick-walled resting spore (Fig. 3.7p, q) (Karling 1981; Martin and Miller 1986). In *Olpidiopsis varians*, synchronous gametangial meiosis takes place in adjacent gametangial thalli (evidenced by the presence of synaptonemal complexes) followed by nuclear transfer via a fertilization tube to the larger (presumptive female) thallus (Martin and Miller 1986). The characteristic pattern of sexual morphogenesis in oomycetes by means of egg-containing oogonia with attached antheridial cells was first observed in Leptomitales (Fig. 3.8e, m, h, i) and Rhipidiales (Fig. 3.12a, b, c, e), which are the two orders that lie at the cusp of saprolegniomycete/peronosporomycete divergence. Some genera in the ~Leptomitales s. lat., such as *Apodachlyella* (Fig. 3.8h, i) (Longore et al. 1987) and *Eurychasmopsis* (Canter and Dick 1994), produce sporangium-like antheridia in which the cysts act as individual antheridial cells, transferring their nuclei to the eggs via a germ tube. In *Aphanomycopsis sexualis* the elongate antheridium is divided into compartments, each of which forms a separate fertilization tube (Fig. 3.8m) (Martin 1975; Karling 1981). This suggests that, around the time of the divergence of the peronosporomycete line, oogonia and antheridia evolved from zoomeiosporangia as a result of the progressive suppression of cleavage and retention of the female gamete. Other basal saprolegniomycete genera, such as *Chlamydomyzium* (Fig. 3.8t) (Glockling and Beakes 2006b), *Cornumyces* (Beakes unpublished observations), and *Ducellieria* (Hesse et al. 1989), all seem to produce oospore-like resting spores parthenogenetically without any apparent involvement of antheridia. Sexual reproduction in the holocarpic lagenidiaceous members of the **Peronosporales** s. lat., such as *Myzocytiopsis vermicola*, is brought about by the transfer of a nucleus between neighbouring thallus compartments (Fig. 3.14i). The receiving oosphere is differentiated from a peripheral periplasmic layer, as expected in this order (Glockling and Beakes 2006a).

The morphology of the mature oospore has also been an important taxonomic character in oomycetes (Dick 1969, 2001a, b; Johnson et al. 2002). In Saprolegniomycetes an outer primary wall layer immediately forms around the naked differentiated oospheres (Beakes 1981b), below which the various mature oospore wall layers are later accreted. In contrast, in all Peronosporomycetes the single oosphere is cleaved from a layer of peripheral cytoplasm, the periplasm (Fig. 3.12b). This is one of the defining features of this class and is among the features indicating that members of the Rhipidiales, such as *Araiospora* (Fig. 3.12b, c) (Sparrow 1973a), should be in the peronosporomycete lineage. The mature oospores of the latter have complex outer oospore wall layers which appear to be derived from the periplasm (Fig. 3.12b, c), a situation also found in Albuginales, such as *Albugo* (Beakes 1981b; Tewari and Skoropad 1977). Many *Myzocytiopsis* species, such as *M. vermicola* (Fig. 3.14i, j) (Glockling and Beakes 2006a), have complex, scalloped, multilayered oospore walls, although these differentiate without apparent periplasmic involvement. Beakes (1987) considered complex multilayered walls to be a derived feature, but it now seems more likely that such walls are associated with early-diverging members of the peronosporomycete line.

IV. Evolutionary Timeline and the Fossil Record

Stramenopiles form a well-supported monophyletic clade that is sister to the alveolates (Keeling 2009). Hyphochytrids and oomycetes are part of the lineage that shares a common ancestor with the photosynthetic ochrophyte stramenopiles (Yubuki et al. 2010; Riisberg et al. 2009; Tsui et al. 2009). It was recently estimated that the stem origin of the Ochrophyta was around **571 million years ago** (mya) (a mean of estimates ranging from 735 to 434 mya) (Brown and Sorhannus 2010). Hyphochytrids and oomycetes presumably evolved after the ochrophyte line diverged, that is later than 570 mya. Previous molecular clock estimates had also placed the origins of the oomycetes at between 1,000 and 524 mya (Bhattacharya et al. 2009). Previously it had been claimed that fossil oomycete-like organisms were present in Precambrian deposits, but the evidence for this was considered by many to be unconvincing (Dick 2001; Johnson et al. 2002). The Labyrinthulomycota are part of a parallel clade that presumably evolved around the same time or even earlier than the other osmotrophic stramenopiles. The evidence is now overwhelming that all osmotrophic stramenopiles had their origins in the sea, even though Dick (2001a) concluded that both hyphochytrids and oomycetes were of terrestrial freshwater origin. The closest known relatives of hyphochytrids (*Pirsonia*) and oomycetes (*Developayella* and MAST-1 clade members) are marine organisms (Fig. 3.1) (Kuhn et al. 2004; Leipe et al. 1994; Moreira and López-Garcia 2002; Yubuki et al. 2010), as indeed are the majority of early diverging oomycete genera (Fig. 3.6) (Cook et al. 2001; Sekimoto 2008; Sekimoto et al. 2007, 2008a, b, 2009). The unexpectedly close relationship between the uniflagellate hyphochytrids and the phagotrophic marine biflagellate protist *Pirsonia* (Kühn et al. 2004) perhaps gives some indication of the type of ancestor that might have given rise to the osmotrophic stramenopiles. Like *Pirsonia*, many simple holocarpic oomycete genera, such as *Ectrogella* (Raghukumar 1980) and *Lagenisma* (Schnepf et al. 1978a), are also parasites of marine diatoms. Diatoms, however, evolved no earlier than 240 mya (Sims et al. 2006), which is considerably later than the oomycetes, if the timeline suggested previously is accepted. The same logic also applies to primitive oomycetes, such as *Eurychasma* (Küpper et al. 2006; Sekimoto et al. 2008), as their phaeophyte hosts did not appear until the Triassic period, approximately **220 mya**. However, it has also been reported that *Eurychasma* species infect red algae (Sparrow 1960), which are much more ancient in origin. It seems more likely that the earliest oomycetes were morphologically simple (holocarpic) necrotrophs of marine nematodes, crustaceans, and possibly algae, all of which would have been present in the Cambrian period around 550–500 mya. Thraustochytrids similarly also are parasites of molluscs and gastropods. Oomycetes, which opportunistically parasitize animal tissues, possess a huge array of proteinase and glucanase genes that were probably part of the ancestral make-up of this lineage (Jiang and Tyler 2012; Krajaejun et al. 2011).

Even after the two main eucarpic classes had diverged, there are many saprolegniomycete genera, such as *Atkinsiella*, *Crypticola*, and *Chlamydomyzium*, and peronosporomycete genera, such as ~*Lagenidium* and ~*Myzocytiopsis* (Fig. 3.6), that are parasites of arthropods and algae (Karling 1981). It is thought that the first land arthropods appeared around 500–450 mya in the Cambrian period and the first insects in the Devonian period around 400 mya. Some oomycetes could have migrated into freshwater ecosystems and onto land along with their invertebrate hosts, such as nematodes and insect larvae. It also seems likely that host-jumping from invertebrates to algae occurred multiple times. The pathogenicity factors required for successful infection of animals (Krajaejun et al. 2011; Phillips et al. 2008) and algae (Grenville-Briggs et al. 2011) are very different from those required to infect terrestrial plants (Jiang and Tyler 2012).

It is still unclear whether the saprolegniomycete/peronosporomycete schism took place before or after oomycetes had migrated to the land. The earliest diverging clade in the saprolegniomycete line is the Atkinsiellaceae, an exclusively marine family (Cook et al. 2001). However, the two earliest diverging peronosporomycete clades (the Rhipidiales and Albuginales) are wholly freshwater or terrestrial. Indeed, the evolution of peronosporomycete characteristics, like thick-walled oospores, might have been the direct result of oomycetes adapting to life on land. However, it is becoming increasingly apparent that there are many marine peronosporomycete species, such as the crustacean parasite *Lagenidium callinectes* (Hatai 2012), pythiaceous parasites of red seaweeds, such as *P. marinum* and *P. porphyrae*, and the saprotrophic saltmarsh genus *Salispilia* (Beakes et al. 2009; Cook et al. 2001; Hulvey et al. 2010). Just as is thought to have occurred in the fungi (Richards et al. 2012), there have probably been multiple transitions from land to sea (and vice versa) within the oomycete lineage. It does seem that the oomycetes were adept at exploiting estuarine ecosystems and at crossing between marine and freshwater environments.

The one out-of-place piece of the oomycete evolutionary jigsaw puzzle seems to be the Albuginales. The white blister rusts were originally considered to be a highly derived group of obligate angiosperm pathogens (Beakes 1987), and it was a surprise when they turned up amongst the early-diverging peronosporomycete clades (Hudspeth et al. 2003; Thines and Spring 2005). We now know that the earliest families to have eucarpic thalli and reproduce sexually by means of morphologically distinct oogonia and antheridia are the Leptomitaceae in the saprolegnialean line and the Rhipidiaceae and Albuginaceae in the peronosporomycete line (Beakes et al. 2009, 2011). A critical evaluation of the fossil evidence for ancient terrestrial oomycetes is given in a recent review by Krings et al. (2011). This paper describes microfossils of what appear to be oogonium and antheridium complexes in chert and coal ball samples that provide very convincing physical evidence for the appearance of terrestrial oomycetes to support the molecular clock inferences. Stidd and Consentino (1975) describe structures that they consider to be *Albugo* oospores in the megagametophyte seed tissue of the ancient gymnosperm *Nucellangium glabrum* from around 310 mya. However, the structures described were not conclusively oospores of an *Albugo* (Krings et al. 2010). A far more convincing, though still controversial, *Albugo*-like microfossil appears to be *Hassiella monosperma* from the 412 mya lower Devonian Rhynie chert (Taylor et al. 2006). Structures purported to be small oogonia in these fossils actually look much more like the small globose haustoria of *Albugo*, and the apparently amphigynous antheridium purported to be at the base of the verrucose oogonium could just be a swollen oogonium stalk. If this fossil is accepted as representing an obligately biotrophic pathogen of Rhyniophyte shoots, related to the present-day Albuginales, it implies that the evolution of obligate parasitism by oomycetes of land plants can be traced back 400 mya. This, coincidently, is also the time when the first Glomeromycota arbuscles have been identified as having formed mycorrhizal associations with similar plants (cited in Krings et al. 2010). However, all extant white blister rusts are obligate parasites of angiosperms, and the latter only diversified from a common ancestor ca. 150 mya, which is considerably later than the apparent emergence of oospores reminiscent of present-day Albuginales. However, the possibility that Albuginales-like oospores associated with fossils might be derived from an extinct organism unrelated to white blister rusts with convergent traits cannot be ruled out.

Papillate, multi-oospored oogonia, reminiscent of those found in present-day Saprolegniales, have also been described in similar Rhynie chert deposits from the same time period (Krings et al. 2010). Another fossil genus, *Combresomyces*, having spiny papillate oogonia with paragynus antheridia, resembling present-day *Pythium* species, was recently described as being associated with the remains of a 300 mya seed fern, *Lygniopteris* (Strullu-Derrien et al. 2010). *Galteriella biscalitheceae*, associated with a sporangium of a fern *Biscallitheca*, also from around ca 300 mya, has amphigynous and paragynous smooth-walled oogonia. Therefore, it seems that by the early **Mesozoic era** (ca. 300 mya) all fossils showing a range of oogonium morphologies similar to those found today in the Albuginales, Peronosporales, and Saprolegniales have been documented, which implies that much of the oomycete diversity we know of today, with most likely the exception of the hyperdiverse downy mildews, had evolved by that period.

There are other remarkable **evolutionary parallels** between the **fungi and oomycetes**. The early-diverging chytrid *Rozella* has endobiotic plasmodial thalli and microsporidia have an injecting infection mechanism (Jones et al. 2011; Lara et al. 2009) that are reminiscent of *Eurychasma* and *Haptoglossa* respectively. The clade (MAST-1) of unknown marine stramenopiles that are closest to the oomycetes (Sekimoto 2008; Yubuki et al. 2010) may be analogous to the recently described cryptofungal clade that seems to be the sister clade to the Fungi (Jones et al. 2011). This highlights that many phylogenetically critical organisms remain to be discovered and described, and we still have little idea what sort of organisms make up unknown MAST clade stramenopiles. They are probably being sampled from their zoospores, and many may be parasitoids or parasites.

V. Conclusions

Since the last account of these groups (Dick 2001b) the application of molecular systematics has greatly improved our understanding both of how individual members of the stramenopile clade relate to each other and to the wider eukaryote community. The marine origin of all the osmotrophic stramenopiles is also now evident. Labyrinthulomycota, Hyphochytriomycota, and Oomycota, together with the ochrophytes, share a common ancestor, which was most likely a photosynthetic mixotrophic flagellate (Tsui et al. 2009). The Labyrinthulomycota are part of one major stramenopile line, and the Hyphochytriomycota s. str. and Oomycota are part of another. If environmental marine sequences are included in analyses, it becomes apparent that the Hyphochytriomycota s. str. and Oomycota are probably not as closely related as originally thought (Beakes 1987). The largest and ecologically most important group are the oomycetes, which because of their economic importance have been much more extensively studied than the other groups. Since the last taxonomic synthesis it has become clear that a number of early-diverging clades lie outside of the two main galaxy clades proposed by Sparrow (1976). Nearly all early-diverging genera are marine, holocarpic parasites of invertebrates and algae, with relatively broad host ranges.

The migration of oomycetes from the sea to the land probably occurred around the time the Saprolegniomycetes and Peronosporomycetes diverged and coincided with the evolution of eucarpic thalli and sexual reproduction by means of oogonia and antheridia. The transition to freshwater and terrestrial ecosystems also seemed to bring about a broadening of the ecological niches that oomycetes occupied, from species that were general saprotrophs to those that became specialized obligate biotrophs. It is possible that these arose from ancestral species that were broad-spectrum facultative parasites of either arthropods or algae.

Finally it is clear that, compared with Fungi, oomycete systematics is still very much a work in progress. Many of the phylogenetically critical genera have yet to be sequenced. The overall taxonomic framework presented here reflects the current uncertainty over how to split many of the key groups, particularly in the Saprolegniales and Peronosporales. However, it does provide an overall taxonomic framework that is underpinned by molecular data. It is to be anticipated that the rapid advance in comparative genomics will play a major role in helping refine our understanding of the taxonomic and phylogenetic relationships in these groups in the decade to come.

Acknowledgements We would like to thank our many colleagues who have shared unpublished trees and other information, including Arthur de Cock, Mike Coffey, Frithjoff Küpper, Claire Gachon, Sally Glockling, Mike and Deborah Hudspeth, Shigeki Inaba, Tim James, Riays Jiang, Andre Lévesque, Satoshi Sekimoto, and Chris Spies, whose helpful comments and unpublished phylogenetic trees are greatly appreciated. Most of the illustrative material was adapted from the excellent line drawings of Johnson et al. (2002), John Webster and the late John Karling. MT is supported by the LOEWE program of the state of Hessen.

References

Adl SM, Simpson AGB, Lane CE, Lukes J, Bass D, Bowser SS, Brown MW, Burki F, Hampl V, Heiss A, Hoppenrath M, Lara E, Le Gall L, Lynn DH, McManus H, Mitchell EAD, Mozley-Stanridge SE, Parfrey LW, Pawlowski J, Rueckert S, Shadwick L, Schoch CL, Smirnov A, Spiegel FW (2012) A revised classification of eukaryotes. J Eukaryot Microbiol 59:429–493

Andersen RA, Barr DJS, Lynn DH, Melkonian M, Moestrup O, Sleigh MA (1991) Terminology and nomenclature of the cytoskeletal elements associated with the flagellar/ciliary apparatus in protists. Protoplasma 164:1–8

Anderson OR, Cavalier-Smith T (2012) Ultrastructure of *Diplophyrs parva*, a new small freshwater species, and a revised analysis of Labyrinthulea (Heterokonta). Acta Protozool 51:291–304

Anderson SA, Stewart A, Allen GT (1995) *Pseudosphaerita euglenae*, a fungal parasite of *Euglena* spp. in the Mangere oxidation ponds, Auckland, New Zealand. N Z J Mar Freshw Res 29:371–379

Arcate JM, Karp MA, Nelson EB (2006) Diversity of peronosporomycete (oomycete) communities associated with the rhizosphere of different plant species. Microb Ecol 51:36–50

Armbrust EV, Berges JA, Bowler C, Green BR, Martinez D, Putnam NH, Zhou S, Allen AE, Apt KE, Bechner M, Brzezinski MA, Chaal BK, Chiovitti A, Davis AK, Demarest MS, Detter JC, Glavina T, Goodstein D, Hadi MZ, Hellsten U, Hildebrand M, Jenkins BD, Jurka J, Kapitonov VV, Kröger N, Lau WW, Lane TW, Larimer FW, Lippmeier JC, Lucas S, Medina M, Montsant A, Obornik M, Parker MS, Palenik B, Pazour GJ, Richardson PM, Rynearson TA, Saito MA, Schwartz DC, Thamatrakoln K, Valentin K, Vardi A, Wilkerson FP, Rokhsar DS (2004) The genome of the diatom *Thalassiosira pseudonana*: ecology, evolution and metabolism. Science 306:79–86

Ayers WA, Lumsden RD (1977) Mycoparasitism of oospores of *Pythium* and *Aphanomyces* species by *Hyphochytrium catenoides*. Can J Microbiol 23:38–44

Azevedo C, Corral L (1997) Some ultrastructural observations of a thraustochytrid (Protoctista, Labyrinthulomycota) from clam *Ruditapes decussatus* (Mollusca, Bivalvia). Dis Aquat Organ 31:73–78

Bahnweg G, Sparrow FK (1974) Four new species of *Traustochytrium* from antarctic regions, with notes on the distribution of zoosporic fungi in the antarctic marine ecosystems. Am J Bot 61:754–766

Bala K, Robideau G, Lévesque A, de Cock WAM, Abad ZG, Lodhi AM, Coffey MD (2010) *Phytopythium*. Persoonia 24:136–137

Barr DJS (1981) The phylogenetic and taxonomic implications of flagellar rootlet morphology among zoosporic fungi. Biosystems 14:359–370

Barr DJS, Allan PME (1985) A comparison of the flagellar apparatus in *Phytophthora*, *Saprolegnia*, *Thraustochytrium*, and *Rhizidiomyces*. Can J Bot 63:138–154

Barr DJS, Désaulniers NL (1987) The ultrastructure of *Lagena radicola* zoospores, including a comparison with the primary and secondary *Saprolegnia* zoospores. Can J Bot 65:2161–2176

Barr DJS, Désaulniers NL (1989) The flagellar apparatus of the oomycetes and hyphochytriomycetes. In: Green JC, Leadbeater BSC, Diver WL (eds) The chromophyte algae: problems and perspectives. Oxford University Press, Oxford, pp 343–355

Barr DJS, Désaulniers NL (1990) The life cycle *Lagena radicola*, an oomycetous parasite of wheat roots. Can J Bot 68:2112–2118

Barrett (1912) Development and sexuality in *Olpidiopsis* (Cornu) Fisher. Ann Bot 26:209–238

Barron (1976) Nematophagous fungi: three new species of *Myzocytium*. Can J Bot 22:752–762

Bartnicki-Garcia S (1970) Cell wall composition and other biochemical markers in fungal phylogeny. In: Harborne JG (ed) Phytochemical phylogeny. Academic, New York, pp 81–103

Bartnicki-Garcia S (1996) The hypha: the unifying thread of the fungal kingdom. In: Sutton BC (ed) A century of mycology. Cambridge University Press, Cambridge, pp 105–133

Baxter L, Tripathy S, Ishaque N, Boot N, Cabral A, Kemen E, Thines M, Ah-Fong A, Anderson R,

Badejoko W, Bittner-Eddy P, Boore JL, Chibucos MC, Coates M, Dehal P, Delehaunty K, Dong S, Downton P, Dumas B, Fabro G, Fronick C, Fuerstenberg SI, Fulton L, Gaulin E, Govers F, Hughes L, Humphray S, Jiang RH, Judelson H, Kamoun S, Kyung K, Meijer H, Minx P, Morris P, Nelson J, Phuntumart V, Qutob D, Rehmany A, Rougon-Cardoso A, Ryden P, Torto-Alalibo T, Studholme D, Wang Y, Win J, Wood J, Clifton SW, Rogers J, Van den Ackerveken G, Jones JD, McDowell JM, Beynon J, Tyler BM (2010) Signatures of adaptation to obligate biotrophy in the *Hyaloperonospora* genome. Science 330:1549–1551

Beakes GW (1981) Ultrastructural aspects of oospore differentiation. In: Hohl H, Turian G (eds) The fungal spore: morphogenetic controls. Academic, London and New York, pp 71–94

Beakes GW (1983) A comparative account of cyst coat ontogeny in saprophytic and fish-lesion isolates (pathogenic) of the *Saprolegnia diclina-parasitica* complex. Can J Bot 61:603–625

Beakes GW (1987) Oomycete phylogeny: ultrastructural perspectives. In: Rayner ADM, Brasier CM, Moore D (eds) Evolutionary biology of the fungi. Cambridge University Press, Cambridge, pp 405–421

Beakes GW (1989) Oomycete fungi: their phylogeny and relationship to chromophyte algae. In: Green JC, Leadbeater BSC, Diver WL (eds) The chromophyte algae: problems and perspectives. Oxford University Press, Oxford, pp 325–342

Beakes GW (1994) Sporulation of lower fungi. In: Gow NAR, Gadd GM (eds) The growing fungus. Chapman & Hall, London, pp 337–366

Beakes GW, Glockling SL (1998) Injection tube differentiation in gun cells of a *Haptoglossa* species which infects nematodes. Fungal Genet Biol 24:45–68

Beakes GW, Glockling SL (2000) An ultrastructural analysis of organelle arrangement during gun (infection) cell differentiation in the nematode parasite *Haptoglossa dickii*. Mycol Res 104:1258–1269

Beakes GW, Glockling SL (2002) A comparative fine-structural study of dimorphic infection cells in the nematophagous parasite, *Haptoglossa erumpens*. Fungal Genet Biol 37:250–262

Beakes GW, Sekimoto S (2009) The evolutionary phylogeny of oomycetes—insights gained from studies of holocarpic parasites of algae and invertebrates. In: Lamour K, Kamoun S (eds) Oomycete genetics and genomics: diversity, interactions and research tools. Wiley, New York, pp 1–24

Beakes GW, Singh H, Dickinson CH (1982) Ultrastructure of the host-pathogen interface of *Peronospora viciae* in cultivars of pea which show different susceptibilities. Plant Pathol 31:343–354

Beakes GW, Wood SE, Burr AW, Hardham AR (1995) Application of spore surface features in defining taxonomic versus ecological groups in Oomycete fungi. Can J Bot 73(suppl):701–711

Beakes GW, Glockling SL, James TY (2006) The diversity of oomycete pathogens of nematodes and its implications to our understanding of oomycete phylogeny. In: Meyer W, Pearce C (eds) Proceedings of the 8th International Mycological Congress, Medimond, Bologna, Italy, pp 7–12

Beakes GW, Glockling SL, Sekimoto S (2012) The evolutionary phylogeny of the oomycete "fungi". Protoplasma 249:3–19

Bedard JEJ, Schurko AM, de Cock AWAM, Klassen GR (2006) Diversity and evolution of 5S rRNA gene family and organization in *Pythium*. Mycol Res 110:86–95

Bhattacharya D, Yoon HS, Hedges SB, Hackett D (2009) Eukaryotes. In: Hedges SB, Kumar S (eds) The timetree of life. Oxford University Press, New York, pp 116–120

Blackwell WH (2010) The enigmatic genus *Pythiella* (Oomycota). Phytologia 92:304–311

Blair JE, Coffey MD, Park S-Y, Geiser DM, Kang S (2008) A multi-locus phylogeny for *Phytophthora* utilizing markers derived from complete genome sequences. Fungal Genet Biol 45:266–277

Bongiorni L (2012) Thraustochytrids, a neglected component of organic matter decomposition and food webs in marine sediments. In: Raghu-Kumar C (ed) Biology of marine fungi. Prog Mol Subcell Biol 53:1–13

Bortnick RN, Powell MJ, Bangert TN (1985) Zoospore fine-structure of the parasite *Olpidiopsis saprolegniae* (Oomycetes, Lagenidiales). Mycologia 77:861–879

Bower SM (1987) *Labyrinthuloides haliotidis* n. sp. (Protozoa, Layrinthulomorpha), a pathogenic parasite of small juvenile abalone in a British Columbia mariculture facility. Can J Zool 65:2013–2020

Bowler C, Allen AE, Badger JH, Grimwood J, Jabbari K, Kuo A, Maheswari U, Martens C, Maumus F, Otillar RP, Rayko E, Salamov A, Vandepoele K, Beszteri B, Gruber A, Heijde M, Katinka M, Mock T, Valentin K, Verret F, Berges JA, Brownlee C, Cadoret JP, Chiovitti A, Choi CJ, Coesel S, De Martino A, Detter JC, Durkin C, Falciatore A, Fournet J, Haruta M, Huysman MJ, Jenkins BD, Jiroutova K, Jorgensen RE, Joubert Y, Kaplan A, Kröger N, Kroth PG, La Roche J, Lindquist E, Lommer M, Martin-Jézéquel V, Lopez PJ, Lucas S, Mangogna M, McGinnis K, Medlin LK, Montsant A, Oudot-Le Secq MP, Napoli C, Obornik M, Parker MS, Petit JL, Porcel BM, Poulsen N, Robison M, Rychlewski L, Rynearson TA, Schmutz J, Shapiro H, Siaut M, Stanley M, Sussman MR, Taylor AR, Vardi A, von Dassow P, Vyverman W, Willis A, Wyrwicz LS, Rokhsar DS, Weissenbach J, Armbrust EV, Green BR, Van de Peer Y, Grigoriev IV (2008) The *Phaeodactylum* genome reveals the evolutionary history of diatom genomes. Nature 456:239–244

Bozkurt TO, Schornack S, Banfield MJ, Kamoun S (2012) Oomycetes, effectors and all that jazz. Curr Opin Plant Biol 15:1–10

Brasier C, Weber J (2010) Sudden larch death. Nature 466:824–825
Briard M, Dutertre M, Rouxel F, Brygoo Y (1995) Ribosomal DNA sequence divergence within the Pythiaceae. Mycol Res 99:1119–1127
Brown JW, Sorhannus U (2010) A molecular genetic timescale for the diversification of autotrophic stramenopiles (Ochrophyta): substantive underestimation of putative fossil ages. PLoS One 5(9): e12759. doi:10.1371/journal.pone.0012759
Burki F, Shalchian-Tabrizi K, Pawlowski J (2008) Phylogenomics reveals a new 'megagroup' including most photosynthetic eukaryotes. Biol Lett 4:366–369
Burr AW, Beakes GW (1994) Characterization of zoospore and cyst surface structure in saprophytic and fish pathogenic *Saprolegnia* species (oomycete fungal protists). Protoplasma 181:142–163
Canter HM (1950) Studies on British chytrids IX. *Anisolpidium stigeoclonii* (De Wildeman) n. comb. Trans Br Mycol Soc 33:335–344
Canter HM, Dick MW (1994) *Eurychasmopsis multisecunda* gen. et sp. nov., a parasite of the suctorian Podophrya sp. (Ciliata). Mycol Res 98:105–117
Cavalier-Smith T (1986) The kingdom Chromista: origin and systematics. In: Round FE, Chapman DJ (eds) Progress in phycological research, vol 4. Biopress Limited, Bristol, pp 309–347
Cavalier-Smith T, Chao EEY (2006) Phylogeny and megasystematics of phagotrophic heterokonts (Kingdom Chromista). J Mol Evol 62:388–420
Cerenius L, Soderhall K, Persson M, Ajaxon R (1988) The crayfish plague fungus *Aphanomyces astaci*—diagnosis, isolation and pathobiology. Freshw Crayfish 7:131–144
Choi Y-J, Jong S-B, Shin H-D (2005) A reconsideration of *Pseudoperonospora cubensis* and *P. humuli* based on molecular and morphological data. Mycol Res 109:842–848
Choi Y-J, Jong S-B, Shin H-D (2006) Genetic diversity within the *Albugo candida* complex (Peronosporales, Oomycota) inferred from phylogenetic analysis of ITS rDNA and COX2 mt DNA sequences. Mol Pylogenet Evol 40:400–409
Choi Y-J, Shin H-D, Hong S-B, Thines M (2007) Morphological and molecular discrimination among *Albugo candida* materials infecting *Capsella bursa-pastoris* world-wide. Fungal Divers 27:11–34
Choi Y-J, Shin H-D, Ploch S, Thines M (2008) Evidence for uncharted biodiversity in the *Albugo candida* complex, with the description of a new species. Mycol Res 112:1327–1334
Choi Y-J, Thines M, Shin H-D (2011a) A new perspective on the evolution of white blister rusts: *Albugo* s.str. (Albuginales, Oomycota) is not restricted to Brassicales but also present on Fabales. Org Divers Evol 11:192–199
Choi Y-J, Shin H-D, Ploch S, Thines M (2011b) Three new phylogenetic lineages are the closest relatives of the widespread species *Albugo candida*. Fungal Biol 115:598–607
Clay RP, Benhamou N, Fuller MS (1991) Ultrastructural detection of polysaccharides in the cell walls of two members of the Hyphochytriales. Mycol Res 95:1057–1064
Cock JM, Sterck L, Rouzé P, Scornet D, Allen AE, Amoutzias G, Anthouard V, Artiguenave F, Aury JM, Badger JH, Beszteri B, Billiau K, Bonnet E, Bothwell JH, Bowler C, Boyen C, Brownlee C, Carrano CJ, Charrier B, Cho GY, Coelho SM, Collén J, Corre E, Da Silva C, Delage L, Delaroque N, Dittami SM, Doulbeau S, Elias M, Farnham G, Gachon CM, Gschloessl B, Heesch S, Jabbari K, Jubin C, Kawai H, Kimura K, Kloareg B, Küpper FC, Lang D, Le Bail A, Leblanc C, Lerouge P, Lohr M, Lopez PJ, Martens C, Maumus F, Michel G, Miranda-Saavedra D, Morales J, Moreau H, Motomura T, Nagasato C, Napoli CA, Nelson DR, Nyvall-Collén P, Peters AF, Pommier C, Potin P, Poulain J, Quesneville H, Read B, Rensing SA, Ritter A, Rousvoal S, Samanta M, Samson G, Schroeder DC, Ségurens B, Strittmatter M, Tonon T, Tregear JW, Valentin K, von Dassow P, Yamagishi T, Van de Peer Y, Wincker P (2010) The *Ectocarpus* genome and the independent evolution of multicellularity in brown algae. Nature 465:617–621
Coffey MD (1975) Ultrastructural features of the haustorial apparatus of the white blister rust *Albugo candida*. Can J Bot 53:1285–1299
Coffey MD, Wilson U (1983) An ultrastructural study of the late-blight fungus *Phytophthora infestans* and its interaction with the foliage of two potato cultivars possessing different levels of general (field) resistance. Can J Bot 61:2669–2685
Coffey MD, Huss J, Peiman M. (2011) Molecular phylogeny of the marine *Halophytophthora* species: a work in progress. http://www.phytophthoradb.org/pdf/O11Coffey230.pdf. Accessed Aug 2012
Coker WC (1923) The saprolegniaceae with notes on other water molds. University of South Carolina Press, Chapel Hill, 201 pp
Collier J (2012) Why sequence four Labrinthulomycete species? JGI DOE Joint Genome Institute. http://www.jgi.doe.gov/sequencing/why/labrinthulomycete.html. Accessed 23 Sept 2012
Cook KL, Hudspeth DSS, Hudspeth MES (2001) A cox2 phylogeny of representative marine Peronosporomycetes (Oomycetes). Nova Hedwigia 122:231–243
Cooke DEL, Drenth A, Duncan JM, Wagels G, Brasier CM (2000) A molecular phylogeny of *Phytophthora* and related oomycetes. Fungal Genet Biol 30:17–32
Cooney EW, Barr DJS, Barstow WE (1985) The ultrastructure of the zoospore of *Hyphochytrium catenoides*. Can J Bot 63:497–505
Couch J (1935) A new saprophytic species of *Lagenidium* with notes on other species. Mycologica 27:376–387

Craigie JS, Correa JA (1996) Etiology of infections diseases of cultivated *Chondrus crispus* (Gigartinales, Rhodophyta). Hydrobiologia 327:97–104

Craven KD, Peterson PD, Windham DE, Mitchell TK, Martin SB (2005) Molecular identification of the turf grass rapid blight pathogen. Mycologia 97:160–166

Davidson JM, Werres S, Garbelotto M, Hansen EM, Rizzo DM (2003) Sudden oak death and associated diseases caused by *Phytophthora ramorum*. Plant Health Prog. doi:10.1094/PHP-2003-0707-01-DG

de Cock AWAM, Abad G, Lévesque A, Robideae G, Brouwer H (2012). *Pythium*: morphological taxonomy after the molecular revision. http://www.phytopthoradb.org/pdf/O31DeCock.pdf

Denoeud F, Roussel M, Noel B, Wawrzyniak I, Da Silva C, Diogon M, Viscogliosi E, Brochier-Armanet C, Couloux A, Poulain J, Segurens B, Anthouard V, Texier C, Blot N, Poirier P, Ng GC, Tan KSW, Artiguenave F, Jaillon O, Aury J-M, Delbac F, Wincker P, Vivarès CP, El Alaoui H (2011) Genome sequence of the stramenopile *Blastocystis*, a human anaerobic parasite. Genome Biol 12:R29

Dick MW (1969) Morphology and taxonomy of the Oomycetes, with special reference to Saprolegniaceae, Leptomitaceae and Pythiaceae I. Sexual reproduction. New Phytol 68:751–755

Dick MW (1973a) Leptomitales. In: Ainsworth CG, Sparrow FK, Sussman AL (eds) The fungi, an advanced treatise, vol IVb. Academic, New York, pp 145–158

Dick MW (1973b) Saprolegniales. In: Ainsworth CG, Sparrow FK, Sussman AL (eds) The fungi, an advanced treatise, vol IVb. Academic, New York, pp 113–144

Dick MW (1976) The ecology of aquatic phycomycetes. In: Jones EBG (ed) Recent advances in aquatic mycology. Elek Press, London, pp 513–542

Dick MW (1986) A new family and genus for two taxa previously assigned to *Apodachlyella completa* (JE Humphrey) H Indoh (Peromycetidae: Leptomitales). Bot J Linn Soc 93:225–229

Dick MW (1990) Phylum Oomycota. In: Margulis L, Corliss JO, Melkonian M, Chapman D (eds) Handbook of protoctista. Jones and Bartlett, Boston, pp 661–685

Dick MW (1997) The Myzocytiopsidaceae. Mycol Res 101:878–882

Dick MW (1998) The species and systematic position of *Crypticola* in the Peronosporomycetes, and new names for the genus *Halocrusticida* and species therein. Mycol Res 102:1062–1066

Dick MW (2001a) Straminipilous fungi. Kluwer, Dordrecht

Dick MW (2001b) Oomycota. In: McLaughlin D, McLaughlin E, Lemke CA (eds) The Mycota VII Part A. Systematics and evolution. Springer, Berlin, pp 48–72

Dick MW, Wong PTW, Clark G (1984) The identity of the oomycete causing "Kikuyu Yellows", with a reclassification of the downy mildews. Bot J Linn Soc 89:171–197

Dick MW, Croft BJ, Magary RC, de Cock AWAM, Clark G (1988) A new genus of the Verrucalvaceae (Oomycetes). Bot J Linn Soc 99:97–113

Dick MW, Vick MC, Gibbings JG, Hedderson TA, Lopez Lastra CC (1999) 18S rDNA for species of *Leptolegnia* and other Peronosporomycetes: justification of the subclass taxa Saprolegniomycetidae and Peronosporomycetidae and division of the Saprolegniaceae *sensu lato* into the Leptolegniaceae and Saprolegniaceae. Mycol Res 103:1119–1125

Diéguez-Uribeondo J, Garcia MA, Cerenius LT, Kozubikova E, Ballesteros I, Windels C, Weiland J, Kator H, Soderhall K, Martn MP (2009) Phylogenetic relationships among plant and animal parasites, and saprotrophs in *Aphanomyces* (Oomycetes). Fungal Genet Biol 46:365–376

Diéz B, Pedrós-Alió C, Massana R (2001) Study of genetic diversity of eukaryotic picoplankton in different oceanic regions by small-subunit rRNA gene cloning and sequencing. Appl Environ Microbiol 67:2932–2941

Diggles BK (2001) A mycosis of juvenile spiny rock lobster, Jasus edwardsii (Hutton 1875) caused by Haliphthoros sp., and possible methods of chemical control. J Fish Dis 24:99–110

Domas A, Jaronski S, Hanton WK (1986) The zoospore and flagellar mastigonemes of Lagenidium giganteum (oomycetes, Lagenidiales). Mycologia 78:810–817

Domergue F, Abbadi A, Heinz E (2005) Relief for fish stocks: oceanic fatty acids in transgenic oilseeds synthesis. Trends Plant Sci 10:112–116

Dorrell RG, Smith AG (2011) Do red and green make brown? Perspectives on plastid acquisitions within Chromalveolates. Eukaryot Cell 10:856–868. doi:10.1128/EC.00326-10

Douhan GW, Olsen MW, Herrell A, Winder C, Wong F, Entwistle K (2009) Genetic diversity in *Labyrinthula terrestris*, a newly emergent plant pathogen, and the discovery of new Labyrinthulid organism. Mycol Res 113:1192–1199

Drechsler C (1940) Three fungi destructive to free-living terricolous nematodes. J Wash Acad Sci 30:240–254

Drechsler C (1960) Two root rot fungi closely related to *Pythium ultimum*. Sydowia 14:107–114

Dykova I, Fiala I, Dvorakova H, Peckova H (2008) Living together: the marine amoeba *Thecamoeba hilla* Schaeffer, 1926 and its endosymbiont *Labyrinthula* sp. Eur J Protistol 44:308–316

Dykstra MJ, Porter D (1984) *Diplophrys marina*, a new scale forming marine protist with labyrinthulid affinities. Mycologia 76:626–632

Fell JW, Master IM (1975) Phycomycetes (*Phytophthora* spp. nov. and *Pythium* sp. nov.) associated with degrading mangrove (*Rhizophora* mangle) leaves. Can J Bot 53:2908–2922

Förster H, Cummings MP, Coffey MD (2000) Phylogenetic relationships of *Phytophthora* species based on ribosomal ITS I DNA sequence analysis with emphasis on Waterhouse groups V and VI. Mycol Res 104:1055–1061

Frances SP, Sweeney AW, Humber RA (1989) *Crypticola clavulifera* gen. et sp. nov and *Lagenidium giganteum*: oomycetes pathogenic for dipterans infesting leaf axils in Australian rain forest. J Invertebr Pathol 54:103–111

Fuller MS (1990) Phylum Hyphochytriomycota. In: Corliss JO, Margulies L, Melkonian M (eds) Handbook of protoctista. Jones and Bartlett, New York

Fuller MS (2001) Hyphochytriomycota. In: McLaughlin D, McLaughlin E, Lemke CA (eds) The Mycota VII. Part A: systematics and evolution. Springer, Berlin, pp 74–80

Fuller MS, Reichle R (1965) The zoospore and early development of *Rhizidiomyces apophysatus*. Mycologia 57:946–961

Gachon CMM, Strittmatter M, Muller DG, Kleintech J, Kupper FC (2009) Detection of differential host susceptibility to the marine oomycete pathogen *Eurychasma dicksonii* by real-time PCR: not all algae are equal. Appl Environ Microbiol 75:322–328. doi:10.1128/AEM.01885-08

García-Blázquez G, Göker M, Voglmayr H, Martin MP, Telleria MT, Oberwinkler F (2008) Phylogeny of *Peronospora* parasitic on Fabaceae based on ITS sequences. Mycol Res 112:502–512

Gaulin E, Jacquet C, Bottin A, Dumas B (2007) Root rot disease of legumes caused by *Aphanomyces euteiches*. Mol Plant Pathol 8:539–548

Gaulin E, Madoui A-M, Bottin A, Jacquet C, Mathe C, Couloux A, Wincker P, Dumas B (2008) Transcriptome of *Aphanomyces euteiches*: new oomycete putative pathogenicity factors and metabolic pathways. PLoS One 4:e1723

Gleason F (1976) The physiology of lower freshwater fungi. In: Jones EBG (ed) Recent advances in aquatic mycology. Elek Press, London, pp 543–572

Gleason FH, Letcher PM, Evershed N, McGee PA (2009) Recovery of growth of *Hyphochytrium catenoides* after exposure to environmental stress. J Eukaryot Microbiol 55:351–354

Glockling SL, Beakes GW (2000a) A review of the biology and infection strategies of biflagellate zoosporic parasites of nematodes. Fungal Divers 4:1–20

Glockling SL, Beakes GW (2000b) An ultrastructural study of sporidium formation during infection of a rhabditid nematode by large gun cells of *Haptoglossa heteromorpha*. J Invertebr Pathol 76:208–215

Glockling SL, Beakes GW (2000c) The ultrastructure of the dimorphic infection cells of *Haptoglossa heteromorpha* illustrates the developmental plasticity of infection apparatus structures in a nematode parasite. Can J Bot 78:1095–1107

Glockling SL, Beakes GW (2001) Two new species of *Haptoglossa* from N.E. England, *H. northumbrica* and *H. polymorpha*. Bot J Linn Soc 136:329–338

Glockling SL, Beakes GW (2006a) An ultrastructural study of development and reproduction in the nematode parasite Myzocytiopsis vermicola. Mycologia 98:7–21

Glockling SL, Beakes GW (2006b) Structural and developmental studies of *Chlamydomyzium oviparasiticum* from *Rhabditis* neamotes in culture. Mycol Res 110:1119–1126

Gobler CJ, Berry DL, Dyhrman ST, Wilhelm SW, Salamov A, Lobanov AV, Zhang Y, Collier JL, Wurch LL, Kustka AB, Dill BD, Shah M, VerBerkmoes NC, Kuo A, Terry A, Pangilinan J, Lindquist EA, Lucas S, Paulsen IT, Hattenrath-Lehmann TK, Talmage SC, Walker EA, Koch F, Burson AM, Marcoval MA, Tang YZ, Lecleir GR, Coyne KJ, Berg GM, Bertrand EM, Saito MA, Gladyshev VN, Grigoriev IV (2011) Niche of harmful alga *Aureococcus anophagefferens* revealed through ecogenomics. Proc Natl Acad Sci USA 108:4352–4357

Göker M, Voglmayr H, Riethmüller A, Weiss M, Oberwinkler F (2003) Taxonomic aspects of Peronosporaceae inferred from Bayesian molecular phylogenetics. Can J Bot 81:672–683

Göker M, Riethmüller A, Voglmayr H, Weiss M, Oberwinkler F (2004) Phylogeny of *Hyaloperonospora* based on nuclear ribosomal internal transcribed spacer sequences. Mycol Prog 3:83–94

Göker M, Voglmayr H, Riethmüller A, Oberwinkler F (2007) How do obligate parasites evolve? A multigene phylogenetic analysis of downy mildews. Fungal Genet Biol 44:105–122

Gotelli D, Hanson LC (1987) An ultrastructural investigation of the zoospore of *Sapromyces androgynus* (Oomycetes, Lagenidiales). Mycologia 78:810–817

Grenville-Briggs L, Gachon CM, Strittmater M, Sterck L, Kupper FC, van West P (2011) A molecular insight into algal-oomycete warfare: cDNA analysis of Ectocarpus siliculosus infected with the basal oomycete *Eurychasma dicksonii*. PLoS One 6:e24500

Gubler F, Hardham AR (1988) Secretion of adhesive material during encystment of *Phytophthora cinnamomi* zoospores characterized by immunogold labeling with monoclonal antibodies to components of peripheral vesicles. J Cell Sci 90:225–235

Guillou L, Chrétiennot-Dinet M-J, Boulben S, Moon SY, van der Staay S, Vaulot D (1999) Symbiomonas scintillans gen. et sp. nov. and Picophagus flagellatus gen. et sp. nov. (Heterokonta): two new heterotrophic flagellates of picoplanktonic size. Protist 150:383–398

Haas BJ, Kamoun S, Zody MC, Jiang RHY, Handsaker RE, Cano LM, Grabherr M, Kodira CD, Raffaele S, Torto-Alalibo T, Bozkurt TO, Ah-Fong AMV, Alvarado L, Anderson VL, Armstrong MR, Avrova A, Baxter L, Beynon J, Boevink PC, Bollmann SR,

Bos JIB, Bulone V, Cai G, Cakir C, Carrington JC, Chawner M, Conti L, Costanzo S, Ewan R, Fahlgren N, Fischbach MA, Fugelstad J, Gilroy EM, Gnerre S, Green PJ, Grenville-Briggs LJ, Griffith J, Grünwald NJ, Horn K, Horner NR, Hu C-H, Huitema E, Jeong D-H, Jones AME, Jones JDG, Jones RW, Karlsson EK, Kunjeti SG, Lamour K, Liu Z, Ma LJ, MacLean D, Chibucos MC, McDonald H, McWalters J, Meijer HJG, Morgan W, Morris PF, Munro CA, O'Neill K, Ospina-Giraldo M, Pinzón A, Pritchard L, Ramsahoye B, Ren Q, Restrepo S, Roy S, Sadanandom A, Savidor A, Schornack S, Schwartz DC, Schumann UD, Schwessinger B, Seyer L, Sharpe T, Silvar C, Song J, Studholme DJ, Sykes S, Thines M, van de Vondervoort PJI, Phuntumart V, Wawra S, Weide R, Win J, Young C, Zhou S, Fry W, Meyers BC, van West P, Ristaino J, Govers F, Birch PRJ, Whisson SC, Judelson HS, Nusbaum C (2009) Genome sequence and analysis of the Irish potato famine pathogen *Phytophthora infestans*. Nature 461:393–398

Hackett JD, Yoon HS, Li S, Reyes-Prieto A, Rümmele SE, Bhattacharya D (2007) Phylogenomic analysis supports the monophyly of cryptophytes and haptophytes and the association of *Rhizaria* with Chromalveolates. Mol Biol Evol 24:1702–1713

Hakariya M, Masuyama N, Saikawa M (2002) Shooting of sporidium by "gun" cells in *Haptoglossa heterospora* and *H. zoospora* and secondary zoospore formation in *H. zoospora*. Mycoscience 43:119–125

Hakariya M, Hirose D, Tokumasu S (2007) A molecular phylogeny of *Haptoglossa* species, terrestrial Peronosporomycetes (oomycetes) endoparasitic on nematodes. Mycoscience 48:169–175

Hakariya M, Hirose D, Tokumasu S (2009) Molecular phylogeny of terrestrial holocarpic endoparasitic Peronosporomycetes *Haptoglossa* spp. inferred from 18S rDNA. Mycoscience 50:130–136

Hallett IC, Dick MW (1986) Fine structure of zoospore cyst ornamentation in the Saprolegniaceae and Pythiaceae. Trans Br Mycol Soc 86:457–463

Hardham AR (1987) Microtubules and the flagellar apparatus in zoospores and cysts of the fungus *Phytophthora cinnamomi*. Protoplasma 137:109–124

Hardham AR (2005) Pathogen profile: *Phytophthora cinnamomi*. Mol Plant Pathol 6:598–604

Hatai K (2012) Diseases of fish and shellfish caused by marine fungi. Raghukumar C (ed) Biology of marine fungi. Prog Mol Subcell Biol 53:15–52

Hatai K, Bian BZ, Baticados MCL, Egusa S (1980) Studies on the fungal diseases in crustaceans. II *Haliphthoros phillippinensis* sp. nov. isolated from cultivated larvae of the jumbo tiger prawn (*Penaseus monodon*). Trans Mycol Soc Jpn 21:47–55

Hatai K, Rhoobunjongde W, Wada S (1992) *Haliphthoros milfordensis* isolated from gills of juvenile kuruma prawn (*Penaeus japonicus*) with black gill disease. Trans Mycol Soc Jpn 33:185–192

Hausner G, Belkhiri A, Klassen GR (2000) Phylogenetic analysis of the small ribosomal subunit RNA gene of the hyphochytrid *Rhizidiomyces apophysatus*. Can J Bot 78:124–128

Held AA (1981) *Rozella* and *Rozellopsis*: naked endoparasitic fungi which dress-up as their hosts. Bot Rev 47:451–555

Heller A, Thines M (2009) Evidence for the importance of enzymatic digestion of epidermal walls during subepidermal sporulation and pustule opening in white blister rusts (Albuginaceae). Mycol Res 113:657–667

Hesse M, Kusel-Fetzmann E, Carniel K (1989) Life cycle and ultrastructure of *Ducellieria chodati* (Oomycetes). Plant Syst Evol 165:1–15

Hickey EL, Coffey MD (1977) A fine-structural study of the pea downy mildew fungus *Peronospora pisi* in its host *Pisum sativum*. Can J Bot 55:2845–2858

Ho HH, Jong SC (1990) *Halophytophthora* gen. nov., a new member of the family Pythiaceae. Mycotaxon 19:377–382

Ho HH, Nakagiri A, Newell SY (1992) A new species of *Halophytophthora* form Atlantic and Pacific subtropical islands. Mycologia 84:548–554

Holloway SA, Heath IB (1977a) An ultrastructural analysis of the changes in organelle arrangement and structure between the various spore types of *Saprolegnia*. Can J Bot 55:1328–1339

Holloway SA, Heath IB (1977b) Morphogenesis and the role of microtubules in synchronous populations of *Saprolegnia* zoospores. Exp Mycol 1:9–29

Honda D, Yokochi T, Nakahara T, Ragukumar S, Nakagiri A, Schaimann K, Higashirhara T (1999) Molecular phylogeny of labyrinthulids and thraustochytrids based on the sequencing of the 18S ribosomal RNA gene. J Eukaryot Microbiol 46:637–647

Huang J-H, Chen C-Y, Lin Y-H, Ann P-J, Huang H-C, Chung W-H (2012) Six new species of *Pythiogeton* in Taiwan, with an account of the molecular phylogeny of this genus. Mycoscience 54:130–147

Hudspeth DSS, Nadler SA, Hudspeth MES (2000) A cox II molecular phylogeny of the Peronosporomycetes. Mycologia 92:674–684

Hudspeth DSS, Stenger D, Hudspeth MES (2003) A cox2 phylogenetic hypothesis for the downy mildews and white rusts. Fungal Divers 13:47–57

Huizar HE, Aronson JM (1986) Aspects of cellulin deposition and chitin biosynthesis in the Leptomitaceae. Mycologia 78:489–492

Hulvey JP, Padgett DE, Bailey JC (2007) Species boundaries within the *Saprolegnia* (Saprolegniales, Oomycota) based on morphological and DNA sequence data. Mycologia 99:421–429

Hulvey JP, Telle S, Nigrelli L, Lamour K, Thines M (2010) Salisapiliaceae—a new family of oomycetes from marsh grass litter of southeastern North America. Persoonia 25:109–116

Inaba S, Hariyama S (2006) The phylogenetic studies on the genus *Cornumyces* (Oomycetes) based on the

nucleotide sequences of the nuclear large subunit ribosomal RNA and the mitochondrially-encoded cox2 genes. In: 8th International mycological congress handbook and abstracts, IMC Cairns, p 330
Inaba S, Tokumasu S (2002) Phylogenetic relationships between the genus *Saprolegnia* and related genera inferred from ITS sequences. Abstracts of the 7th international mycological congress, Oslo IMC7, p 687
James TY, Porter TM, Wallace Martin W (2014) Blastocladiomycota, chapter 7. In: McLaughlin DJ, Spatafora JW (eds) Systematics and evolution. Springer, Heidelberg
Jiang RHY, Tyler BM (2012) Mechanisms and evolution of virulence in oomycetes. Annu Rev Phytopathol 50:14.1–13.24. doi:10.1146/annrev-phyto-081211-172912
Jiang RHY, de Bruijn I, Haas BJ, Belmonte R, Lobach L, Christie J, van den Ackerveken G, Bottin A, Bulone V, D´az-Moreno SM, Dumas B, Fan L, Gaulin E, Govers F, Grenville-Briggs LJ, Horner NR, Levin JZ, Mammella M, Meijer Harold JG, Morris P, Nusbaum C, Oome S, Phillips AJ, van Rooyen D, Rzeszutek E, Saraiva M, Secombes CJ, Seidl MF, Snel B, Stassen JHM, Sykes S, Tripathy S, van den Berg H, Vega-Arreguin JC, Young SK WS, Zeng Q, Dieguez-Uribeondo J, Russ C, Tyler BM, van West P (2013) Distinctive expansion of potential virulence genes in the genome of the oomycete fish pathogen *Saprolegnia parastica*. PLoS Genet 9 (6), e1003272. doi:10.1371/journal.pgen.100327
Johnson TW (1957) Resting spore development in the marine phycomycete *Anisolpidium ectocarpii*. Am J Bot 44:875–878
Johnson TW (1966) A *Lagenidium* in the marine diatom *Coscinodiscus centralis*. Mycologica 58:131–135
Johnson TW (1972) Aquatic fungi of Iceland: *Olpidiopsis* Cornu. J Elisha Mitchell Sci Soc 88:83–91
Johnson TW, Seymour RL, Padgett DE (2002) Biology and systematics of the Saprolegniaceae. http://dl.uncw.edu/digilib/biology/fungi/taxonomy%20and%20systematics/padgett%20book/. Accessed Apr 2008
Jones GM, O'Dor RK (1983) Ultrastructural observations on a thraustochytrid fungus parasitic in the gills of squid (*Illex illecebrosus* Lesueur). J Parasitol 69:903–911
Jones MDM, Forn I, Gadelha C, Egan MJ, Bass D, Massana R, Richards TA (2011) Discovery of novel intermediate forms redefines the fungal tree of life. Nature 474:200–203
Judelson HS, Ah-Fong AMV (2009) Progress and challenges in oomycete transformation. In: Lamour K, Kamoun S (eds) Oomycete genetics and genomics: diversity, interactions and research tools. Wiley, New York, pp 435–453
Karling JS (1942) The simple holocarpic biflagellate phycomycetes. Columbia University Press, New York
Karling JS (1943) The life history of *Anisolpidium ectocarpii* gen. nov. et sp. nov., and a synopsis and classification of other fungi with anteriorly uniflagellate zoospores. Am J Bot 30:637–648
Karling JS (1977) Inconographicum iconarum, 2nd edn. Cramer, Vaduz
Karling JS (1981) Predominantly holocarpic and eucarpic simple biflagellate phycomycetes. J Cramer, Vaduz
Kazama FY (1980) The zoospore of *Schizochytrium aggregatum*. Can J Bot 58:2434–2446
Keeling PJ (2009) Chromalveolates and the evolution of plastids by secondary endosymbiosis. J Eukaryot Microbiol 56:1–8
Kemen E, Jones JDG (2012) Obligate biotroph parasitism: can we link genomes and lifestyles. Trends Plant Sci 17:448–457
Kemen E, Gardiner A, Schultz-Larsen T, Kemen AC, Balmuth AL et al (2011) Gene gain and loss during evolution of obligate parasitism in the white rust pathogen of *Arabidopsis thaliana*. PLoS Biol 9: e1001094. doi:10.1371/journal.pbio.1001094
Kerwin JL, Washino RK (1983) Sterol induction of sexual reproduction in *Lagenidium giganteum*. Exp Mycol 7:109–115
King CA (1903) Observations on the cytology of *Araiospora pulchra* Thaxter. Proc Boston Soc Nat Hist 31:11–15
Kobayashi T, Sakaguchi K, Matsuda T, Abe R, Hama Y et al (2011) Increase in eicosapentaenoic acid in thraustochytrids through thraustochytrid ubiquitin promoter-driven expression of a fatty acid 5 desaturase gene. Appl Environ Microbiol 77:3870–3876
Krajaejun T, Khositnithikul R, Lerksuthirat T, Lowhnoo T, Rujirwat T et al (2011) Expressed sequence tags reveal genetic diversity and putative virulence factors of the pathogenic oomycete *Pythium insidiosum*. Fungal Biol 115:683–696
Krings M, Dotzler N, Taylor TN, Galtier J (2010) A fungal community in plant tissue from the Lower Coal Measures (Langsettian, Lower Pennsylvanian) of Great Britain. Bull Geosci 85:679–690
Krings M, Taylor TN, Dotzler N (2011) The fossil record of the Peronosporomycetes (Oomycota). Mycologia 103:445–457
Kroon LMM, Bakker FT, van den Bosch GBM, Bonants PJM, Flier WG (2004) Phylogenetic analysis of *Phytophthora* species based on mitochondrial and nuclear DNA sequences. Fungal Genet Biol 41:766–782
Kroon LMM, Brouwer H, de Cock WAM, Govers F (2012) The genus *Phytophthora* anno 2012. Phytopathology 102:348–364
Kühn SF, Medlin LK, Eller G (2004) Phylogenetic position of the parasitoid nanoflagellate *Pirsonia* inferred from nuclear-encoded small subunit ribosomal DNA and a description of *Pseudopirsonia* n. gen. and *Pseudopirsonia mucosa* (Drebes) comb. nov. Protist 155:143–156

Küpper FC, Müller DG (1999) Massive occurrence of the heterokont and fungal parasites *Anisolpidium*, *Eurychasma* and *Chytridium* in *Pylaiella litoralis* (Ectocarpales, Phaeophyceae). Nova Hedwig 69:381–389

Küpper FC, Maier I, Müller DG, Loiseaux-de Goer S, Guillou L (2006) Phylogenetic affinities of two eukaryotic pathogens of marine macroalgae, *Eurychasma dicksonii* (Wright) Magnus and *Chytridium polysiphoniae* Cohn. Cryptogam Algol 27:165–184

Lamour KH, Win J, Kamoun S (2007) Oomycete genomics: new insights and future directions. FEMS Microbiol Lett 274:1–8

Lara E, Belbahri L (2011) SSU rRNA reveals major trends in oomycete evolution. Fungal Divers 49:93–100

Lara E, Moreira D, Lpez-García P (2009) The environmental clade LKM11 and *Rozella* form the deepest branching clade of fungi. Protist 161(1):116–21. doi:10.1016/j.protis.2009.06.005

Leander CA, Porter D (2001) The Labyrinthulomycota is comprised of three distinct lineages. Mycologia 93:459–464

Leander CA, Porter D, Leander BS (2004) Comparative morphology and molecular phylogeny of aplanochytrids (Labryrinthulomycota). Eur J Protist 40:317–328

Léclerc MC, Guillot J, Deville M (2000) Taxonomic and phylogenetic analysis of Saprolegniaceae (Oomycetes) inferred from LSU rDNA and ITS sequence comparisons. Antonie Van Leeuwenhoek 77:369–377

Lee HY, Aronson JM (1975) Composition of cellulin, a unique chitin-glucan granules of the fungus *Apodachlya* sp. Arch Microbiol 102:203–208

Lehnen LP, Powell MJ (1989) The role of kinetosome-associated organelles in the attachment of encysting secondary zoospores of *Saprolegnia ferax* to substrates. Protoplasma 149:163–174

Leipe DD, Tong SM, Goggin CL, Slemenda SB, Pieniazek NJ, Sogin ML (1994) 16S-like rDNA sequences from *Developayella elegans*, *Labyrinthuloides haliotidis*, and *Proteromonas lacertae* confirm that the stramenopiles are a primarily heterotrophic group. Eur J Protistol 33:369–377

Levenfors JP, Fatehi J (2009) Molecular characterization of *Aphanomyces* species associated with legumes. Mycol Res 108:682–689

Lévesque CA (2011) Fifty years of oomycete research—from consolidation to evolutionary and genomic exploration. Fungal Divers 50:35–46

Lévesque CA, de Cock AW (2004) Molecular phylogeny and taxonomy of the genus *Pythium*. Mycol Res 108:1363–1383

Lévesque CA, Brouwer H, Cano L, Hamilton JP, Holt C, Huitema E, Raffaele S, Robideau GP, Thines M, Win J, Zerillo MM, Beakes GW, Boore JL, Busam D, Dumas B, Ferriera S, Fuerstenberg SI, Gachon CMM, Gaulin E, Govers F, Grenville-Briggs L, Horner N, Hostetler J, Jiang RHY, Johnson J, Krajaejun T, Lin H, Meijer HJG, Moore B, Morris P, Phuntmart V, Puiu D, Shetty J, Stajich JE, Tripathy S, Wawra S, van West P, Whitty BR, Coutinho PM, Henrissat B, Martin F, Thomas PD, Tyler BM, De Vries RP, Kamoun S, Yandell M, Tisserat N, Buell CR (2010) Genome sequence of the necrotrophic plant pathogen *Pythium ultimum* reveals original pathogenicity mechanisms and effector repertoire. Genome Biol 11:R73

Lilley JH, Hart D, Panywachira V, Kanchanakhan S, Chinabut S, Soderhall K, Cerenius L (2003) Molecular characterization of the fish-pathogenic fungus *Aphanomyces invadans*. J Fish Dis 26:263–275

Links MG, Holub E, Jiang RHY, Sharpe AG, Hegedus D et al (2011) De novo sequence assembly of *Albugo candida* reveals a small genome relative to other biotrophic oomycetes. BMC Genomics 12:503

Longcore JE, Brooks JL, Homola L (1987) *Apodachyella* re-examined. Mycologia 79:621–626

Martens C, Vandepoele K, Van de Peer Y (2008) Whole-genome analysis reveals molecular innovations and evolutionary transitions in chromalveolate species. Proc Natl Acad Sci USA 105:3427–3432

Martin WW (1975) *Aphanomycopsis sexualis* a new parasite of midge eggs. Mycologia 67:923–933

Martin WW (1977) The development and possible relationships of a new *Atkinsiella* parasite in insect eggs. Am J Bot 64:766–769

Martin FN (2000) Phylogenetic relationships among some *Pythium* species inferred from sequence analysis of the mitochondrially encoded cytochrome oxidase II gene. Mycologia 92:711–727

Martin RW, Miller CE (1986) Ultrastructure of sexual reproduction in *Olpidiopsis varians*. Mycologia 78:359–370

Martin FN, Tooley PW (2003) Phylogenetic relationships among *Phytophthora* species inferred from sequence analysis of mitochondrially encoded cytochrome oxidase I and II genes. Mycologia 95:269–284

Martin FN, Tooley PW (2008) Phylogenetic relationships of *Phytophthora ramorum*, *P. nemorosa* and *P. pseudosyringae*, three species recovered from areas in California with sudden oak death. Mycol Res 107:1379–1391

Maruyama S, Matsuzaki M, Misawa K, Nozaki H (2009) Cyanobacterial contribution to the genomes of the plastid lacking protists. BMC Evol Biol 9:197. doi:10.1186/1471-2148-9-197

Massana R, Guillou L, Diez B, Pedró-Alió C (2002) Unveiling the organisms behind novel eukaryotic ribosomal DNA sequences from the ocean. Appl Environ Microbiol 68:4554–4558

Massana R, Castresana J, Balagué V, Guillou L, Romari K, Groisillier A, Valentin K, Pedró-Alió C (2004) Phylogenetic and ecological analysis of novel marine stramenopiles. Appl Environ Microbiol 70:3528–3534

Massana R, Terrado R, Forn I, Lovejoy C, Pedró-Alió C (2006) Distribution and abundance of uncultured heterotrophic flagellates in the world oceans. Environ Microbiol 8:1515–1522

Maurosa Y, Morimoto K, Sano A, NishimuraK HK (2009) A new peronosporomycete, *Halioticida noduliformans* gen.et sp. nov., isolated from white nodules in the abalone, *Haliotis* spp. from Japan. Mycoscience 50:106–115

Mendoza L (2009) *Pythium insidiosum* and mammalian hosts. In: Lamour K, Kamoun S (eds) Oomycete genetics and genomics: diversity, interactions and research tools. Wiley-Blackwell, Ann Arbor, pp 387–406

Migula M (1897) Die Characeen – Deutschlands. Osterreichs und der Schweiz. Rabenhorsts Cryptogamenflora 5. Verlag E. Kummer, Leipzig, 765 pp

Miller MR, Nichols PD, Carter CG (2007) Replacement of fish oil with thraustochytrid *Schizochytrium* sp. L oil in Atlantic salmon parr (*Salmo salar* L) diets. Comp Biochem Physiol 148:382–392

Mims CWR, Richardson EA (2002) Ultrastructure of the zoosporangia of *Albugo ipomoeae-panduratae* as revealed by conventional chemical fixation and high pressure freezing followed by freeze substitution. Mycologia 95:1–10

Mirzaee MR, Ploch S, Runge F, Telle S, Nigrelli L, Thines M (2013) A new presumably widespread species of *Albugo* parasitic to *Strigosella* spp. (Brassicaceae). Mycol Prog 12:45–52

Molloy DP, Glockling SL, Siegfried CA, Beakes GW, James TY, Mastitsky SE, Wurdak E, Giamberini L, Gaylo MJ, Nemeth MJ (2014) *Aquastella* gen.nov.: a new genus of saprolegniaceous rotifer parasites related to *Aphanomyces*, with unique sporangial outgrowths. Fungal Biol (in press). doi: 10.1016/j.funbio.2014.01.007

Money NP (1998) Why the oomycetes have not stopped being fungi. Mycol Res 102:767–768

Moore D, Robson GD, Trinci APJ (2011) 21st century guide to Fungi. Cambridge University Press, Cambridge

Moreira D, López-Garcia P (2002) The molecular ecology of microbial eukaryotes unveils a hidden world. Trends Microbiol 10:31–38

Morgan W, Kamoun S (2007) RXLR effectors of plant pathogenic oomycetes. Curr Opin Microbiol 10:332–338

Moss ST (1985) An ultrastructural study of taxonomically significant characters of the Thraustochytriales and Labyrinthulales. Bot J Linn Soc 91:329–357

Moss ST (1986) Biology and phylogeny of the Labyrinthulales and Thraustochytriales. In: Moss ST (ed) The biology of marine fungi. Cambridge University Press, Cambridge, pp 105–129

Muehlstein LK, Porter D (1991) *Labyrinthula zosterae* sp. nov., the causative agent of wasting disease of eelgrass, *Zostera marina*. Mycologia 83:180–191

Nakagiri A (2002a) Diversity and phylogeny of *Halophytophthora* (Oomycetes). Abstract of the 7th international mycological congress, Oslo, 55, p 19

Nakagiri A (2002b) *Halophytophthora* species from tropical and subtropical mangroves: a review of their characteristics. In: Hyde KD (ed) Fungi in marine environments, vol 7, Fungal diversity research series. Fungal Diversity Press, Hong Kong, pp 1–14

Nakagiri A, Newell SY, Ito T (1994) Two new *Halophytophthora* species, *H. tartarea* and *H. masteri*, from intertidal decomposing leaves in salt marsh and mangrove regions. Mycoscience 35:223–232

Nechwatal J, Mendgen K (2005) *Pythium litorale* sp. nov., a new species from the litoral of Lake Constance, Germany. FEMS Microbiol Lett 255:96–101

Newell SY, Cefalu R, Fell JW (1977) *Myzocytium*, *Haptoglossa*, and *Gonimochaete* (fungi) in littoral marine nematodes. Bull Mar Sci 27:177–207

Newhook FJ, Podger FD (1972) The role of *Phytophthora cinnamomi* in Australian and New Zealand forests. Annu Rev Phytopathol 10:299–325

Nigrelli L, Thines M (2013) Tropical oomycetes in the German Bight—climate warming or overlooked diversity? Fungal Ecol 6:152–160

Olsen MW (2007) *Labyrinthula terrestris*: a new pathogen of cool-season turfgrasses. Mol Plant Pathol 8:817–820

Overton SV, Tharp TP, Bland CE (1983) Fine structure of swimming, encysting, and germinating spores of *Haliphthoros milfordensis*. Can J Bot 61:1165–1177

Panabières F, Ponchet M, Allasia V, Cardin L, Ricci P (1996) Characterization of border species among Pythiaceae: several *Pythium* isolates produce elicitins, typical proteins from *Phytophthora* spp. Mycol Res 101:1459–1468

Patterson DJ (1989) Chromophytes from a protistan perspective. In: Green JP, Leadbeater BSC, Diver WL (eds) The chromophyte algae: problems and perspectives. Clarendon, Oxford, pp 357–379

Perkins FO (1976) Fine structure of lower marine and estuarine fungi. In: Jones EBG (ed) Recent advances in aquatic mycology. Elek Press, London, pp 513–542

Perkins FO, Amon JP (1969) Zoosporulation in *Labyrinthula* sp., an electron microscope study. J Protozool 16:235–256

Petersen AB, Rosendahl S (2000) Phylogeny of the Peronosporomycetes (Oomycota) based on partial sequences of the large ribosomal subunit (LSU rDNA). Mycol Res 104:1295–1303

Phillips AJ, Anderson VL, Robertson EJ, Secombes CJ, van West P (2008) New insights into animal pathogenic oomycetes. Trends Microbiol 16:13–19

Ploch S, Thines M (2011) Obligate biotrophic pathogens of the genus *Albugo* are widespread asymptomatic endophytes in natural populations of Brassicaceae. Mol Ecol 20:3692–3699

Ploch S, Choi Y-J, Rost C, Shin H-D, Schilling E, Thines M (2010) Evolution of diversity in *Albugo* is driven by high host specificity and multiple speciation events on closely related Brassicaceae. Mol Phylogenet Evol 57:812–820

Ploch S, Telle S, Choi Y-J, Cunnington J, Priest M, Rost C, Shin H-D, Thines M (2011) The molecular phylogeny of the white blister rust genus *Pustula* reveals a case of underestimated biodiversity with several undescribed species on ornamentals and crop plants. Fungal Biol 115:214–219

Polglase JL (1980) A preliminary report on the Thraustochytrid(s) and Labyrinthulid(s) associated with a pathological condition in the lesser octopus *Eledone cirrhosa*. Bot Mar 23:699–706

Porter D (1990) Phylum Labyrinthulomycota. In: Margulis L, Corliss JO, Melkonian M, Chapman DJ (eds) Handbook of Protoctista. Jones and Bartlett, Boston, pp 388–398

Powell MJ, Letcher PM (2014) Chytridiomycota, Monoblepharidomycota and Neocallimastigomycota, chapter 6. In: McLaughlin DJ, Spatafora JW (eds) Systematics and evolution. Springer, Heidelberg

Pueschel CM, van der Meer JP (1985) Ultrastructure of the fungus *Petersenia palmariae* (Oomycota) parasitic on the alga *Palmaria molis* (Rhodophyceae). Can J Bot 63:409–418

Qutob D, Kamoun S, Gijzen M (2002) Expression of a *Phytophthora sojae* necrosis-inducing protein occurs during the transition from biotrophy to necrotrophy. Plant J 32:361–373

Raffaele S, Win J, Cano LM, Kamoun S (2010) Analyses of genome and gene expression reveal novel virulence factors in the secretome of *Phytophthora infestans*. BMC Genomics 11:637

Raghukumar C (1980) An ultrastructural study of the marine diatom *Licmophora hyalina* and its parasite *Ectrogella perforans*. II. Development of the fungus in its host. Can J Bot 58:2557–2574

Raghukumar S (2002a) Bacterivory: a novel dual role for thraustochytrids in the sea. Mar Biol 113:165–169

Raghukumar S (2002b) Ecology of marine protists, the Labyrinthulomycetes (Thraustochytrids and Labyrinthulids). Eur J Protistol 38:127–145

Raghukumar S (2008) Thraustochytrid marine protists: production of PUFAs and other emerging technologies. Mar Biotechnol 10:631–640

Randall TA, Dwyer RA, Huitema E, Beyer K, Cvitanich C, Kelkar H, Ah Fong AMV, Gates K, Roberts S, Yatzkan E, Gaffney T, Law M, Testa A, Torto-Alalibo T, Zhang M, Zheng L, Mueller E, Windass J, Binder A, Birch PRJ, Gisi U, Govers F, Gow NA, Mauch F, van West P, Waugh ME, Yu J, Boller T, Kamoun S, Lam ST, Judelson HS (2005) Large-scale gene discovery in the oomycete *Phytophthora infestans* reveals likely components of phytopathogenicity shared with true fungi. Mol Plant Microbe Interact 18:229–243

Randolph LR, Powell MJ (1992) Ultrastructure of zoospores of the oomycete *Apodachlya pyrifera*. Mycologia 84:768–780

Reeb VC, Peglaer MT, Yoon HS, Bai JR, Wu M, Shia P et al (2009) Interrelationships of chromalveolates within a broadly sampled tree of photosynthetic protists. Mol Phylogenet Evol 53:202–211

Richards TA, Dacks JB, Jenkinson JM, Thornton CR, Talbot NJ (2006) Evolution of filamentous pathogens: gene exchange across eukaryote kingdoms. Curr Biol 16:1857–1864

Richards TA, Soanes DM, Jones MDM, Vasieva O, Leonard G, Paszkiewicz K, Foster PG, Hall N, Talbot NJ (2011) Horizontal gene transfer facilitated the evolution of plant parasitic mechanisms in the oomycetes. Proc Natl Acad Sci USA 108:15258–63

Richards TA, Jones MD, Leonard G, Bass G (2012) Marine fungi: their ecology and molecular diversity. Ann Rev Mar Sci 4:495–522

Riethmüller A, Weiss M, Oberwinkler F (1999) Phylogenetic studies of Saprolegniomycetidae and related groups based on nuclear large subunit ribosomal DNA sequences. Can J Bot 77:1790–1800

Riethmüller A, Voglmayr H, Göker M, Weiss M, Oberwinkler F (2002) Phylogenetic relationships of the downy mildews (Peronosporales) and related groups based on nuclear large subunit ribosomal DNA sequences. Mycologia 94:834–849

Riethmüller, Grundel A, Langer E (2006) The seasonal occurrence of the sewage fungus *Leptomitus lacteus* Roth C. Agard in stagnant and running waters of different chemistry of Hesse and Thuringia, Germany. Acta Hydrochem Hydrobiol 34:58–66

Riisberg I, Orr RJ, Kluge R, Shalchian-Tabrizi K, Bowers HA, Patil V, Edvardsen B, Jakobsen KS (2009) Seven gene phylogeny of heterokonts. Protist 160:191–204

Robb EJ, Barron GL (1982) Nature's ballistic missile. Science 218:1221–1222

Robold AV, Hardham AR (2004) Production of monoclonal antibodies against peripheral-vesicle proteins in zoospores of *Phytophthora nicotianae*. Protoplasma 223:121–132

Robold AV, Hardham AR (2005) During attachment *Phytophthora* spores secrete proteins containing thrombospondin type 1 repeats. Curr Genet 47:307–315

Runge F, Thines M (2011) Host matrix has major impact on morphology of *Pseudoperonospora cubensis*. Eur J Plant Pathol 129:147–156

Runge F, Thines M (2012) Re-evaluation of host specificity of the closely related species *Pseudoperonospora humuli* and *P. cubensis*. Plant Dis 96:55–61

Runge F, Telle S, Ploch S, Savory E, Day B, Sharma R, Thines M (2011a) The inclusion of downy mildews in a multi-locus-dataset and its reanalysis reveals a high degree of paraphyly in *Phytophthora*. IMA Fungus 2:163–171

Runge F, Choi Y-J, Thines M (2011b) Phylogenetic investigations in the genus *Pseudoperonospora* reveal overlooked species and cryptic diversity in the *P. cubensis* species cluster. Eur J Plant Pathol 129:135–146

Runge F, Ndambi B, Thines M (2012) Which morphological characteristics are most influenced by the host matrix in downy mildews? A case study in *Pseudoperonospora cubensis*. PLoS One 7:e44863
Schnepf E, Deichgräber G, Drebes G (1977) Development and ultrastructure of the marine, parasitic oomcete, *Lagenisma coscinodisci* (Lagenidiales): sexual reproduction. Can J Bot 56:1315–1325
Schnepf E, Deichgräber G, Drebes G (1978a) Development and ultrastructure of the marine, parasitic oomycete, *Lagenisma coscinodisci* Drebes (Lagenidiales). The infection. Arch Microbiol 116:133–139
Schnepf E, Deichgräber G, Drebes G (1978b) Development and ultrastructure of the marine parasitic oomycete *Lagenisma coscinodisci* Drebes (Lagenidiales) Thallus, zoosporangium, mitosis and meiosis. Arch Microbiol 116:121–132
Schroeder KL, Martin FN, de Cock AWAM, Levesque CA, Speis CDJ, Okubara PA, Paulitz TC (2012) Molecular detection and quantification of *Pythium* species—evolving taxonomy, new tools and challenges. Plant Dis 97:4–20
Schurko AM, Mendoza L, Lévesque CA, Désaulniers NL, de Cock AWAM, Klassen GR (2004) A molecular phylogeny of *Pythium insidiosum*. Mycol Res 107:537–544
Seidl MF, den Ackerveken V, Govers F, Snel B (2012) Reconstruction of oomycete genome evolution identifies differences in evolutionary trajectories leading to present-day large gene families. Genome Biol Evol 4:199–211
Sekimoto S (2008) The taxonomy and phylogeny of the marine holocarpic oomycetes. PhD Thesis, Graduate School of Natural Sciences, Konan University, Kobe, Japan
Sekimoto S, Hatai K, Honda D (2007) Molecular phylogeny of an unidentified *Haliphthoros*-like marine oomycete and *Haliphthoros milfordensis* inferred from nuclear-encoded small and large subunit rDNA genes and mitochondrial-encoded cox2 gene. Mycoscience 48:212–221
Sekimoto S, Beakes GW, Gachon CMM, Müller DG, Küpper FC, Honda D (2008a) The development, ultrastructural cytology, and molecular phylogeny of the basal oomycete *Eurychasma dicksonii*, infecting the filamentous phaeophyte algae *Ectocarpus siliculosus* and *Pylaiella littoralis*. Protist 159:401–412
Sekimoto S, Yokoo K, Kawamura Y, Honda D (2008b) Taxonomy, molecular phylogeny, and ultrastructural morphology of *Olpidiopsis porphyrae* sp. nov. (Oomycetes, stramenopiles), a unicellular obligate endoparasite of *Porphyra* spp. (Bangiales, Rhodophyta). Mycol Res 112:361–374
Sekimoto S, Kochkova TA, West JA, Beakes GW, Honda D (2009) *Olpidiopsis bostrychiae*: a new species endoparasitic oomycete that infects *Bostrychia* and other red algae. Phycologia 48:460–472
Shanor L, Olive LS (1943) Notes on *Araiospora streptandra*. Mycologia 34:536–542
Sims PA, Mann DG, Medlin LK (2006) Evolution of the diatoms: insights from fossil, biological and molecular data. Phycologia 45:361–402
Soanes DM, Richards TA, Talbot NJ (2007) Insights from sequencing fungal and oomycete genomes: what can we learn about plant disease and the evolution of pathogenicity. Plant Cell 19:3318–3326
Soylu EM, Soylu S, Keshavarzi M, Brown I, Mansfield JW (2003) Ultrastructural characterization of the interactions between *Arabidopsis thaliana* and *Albugo candida*. Physiol Mol Plant Pathol 63:201–211
Sparrow FK (1934) Observations on marine phycomycetes collected in Denmark. Dansk Bot Ark 8:1–24
Sparrow FK (1960) Aquatic phycomycetes, 2nd edn. University of Michigan Press, Ann Arbor
Sparrow FK (1973) The peculiar marine phycomycete *Atkinsiella dubia* from crab eggs. Arch Mikrobiol 93:137–144
Sparrow FK (1974) Observations on two uncommon fungi. Veroff Inst Meeresf Bremerh Suppl 5:9–18
Sparrow FK (1973a) Mastigomycotina. In: Ainsworth GC, Sparrow FK, Sussman AS (eds) The fungi, vol 4b. Academic, New York, pp 61–73
Sparrow FK (1973b) Chytridiomycetes. Hyphochytridiomycetes. In: Ainsworth GC, Sparrow FK, Sussman AS (eds) The fungi, vol 4b. Academic, New York, pp 85–110
Sparrow FK (1973c) Lagenidiales. In: Ainsworth GC, Sparrow FK, Sussman AS (eds) The fungi, vol 4b. Academic, New York, pp 159–164
Sparrow FK (1976) The present status of classification in biflagellate fungi. In: Gareth-Jones EB (ed) Recent advances in aquatic mycology. Elek Science, London, pp 213–222
Sparrow FK (1977) *Rhizidiomycopsis* on azygospores of *Gigaspora margarita*. Mycologia 69:1053–1058
Spencer MA, Vick MC, Dick MW (2002) Revision of *Aplanopsis*, *Pythiopsis*, and 'subcentric' *Achlya* species (Saprolegniaceae) using 18S rDNA and morphological data. Mycol Res 106:549–560
Stassen JHM, Siedl MF, Vergeer PWJ, Nijman IJ, Snel B, Cuppen E, Van den Ackerveken G (2011) Effector identification in the downy mildew *Bremia lactucae* by massively parallel transcriptome sequencing. Mol Plant Pathol 13:719–731
Stidd BM, Consentino K (1975) *Albugo*-like oogonia from the American carboniferous. Science 190:1092–1093
Strittmatter M, Gachon CMM, Kupper F (2009) Ecology of lower oomycetes. In: Lamour K, Kamoun S (eds) Oomycete genetics and genomics: diversity, interactions and research tools. Wiley-Blackwell, Ann Arbor, pp 25–46
Strullu-Derrien C, Kenrick P, Rioult JP, Strullu DG (2010) Evidence of parasitic oomycetes (Peronosporomycetes) infecting the stem cortex oft eh Carboniferous seed fern *Lygniopteris odlhamia*. Proc R Soc B. doi:10.1098/rspb.2010.1603

Taylor TN, Krings M, Keri H (2006) *Hassiella monspora* gen. et sp. nov., a microfungus from the 400 million year old Rhynie chert. Mycol Res 110:628–632

Telle S, Thines M (2012) Reclassification of an enigmatic downy mildew species on lovegrass (*Eragrostis*) to the new genus *Eraphthora*, with a key to the genera of the Peronosporaceae. Mycol Prog 11:121–129

Telle S, Shivas RG, Ryley MJ, Thines M (2011) Molecular phylogenetic analysis of *Peronosclerospora* (Oomycetes) reveals cryptic species and genetically distinct species parasitic to maize. Eur J Plant Pathol 130:521–528

Tewari JP, Skoropad WP (1977) Ultrastructure of oospore development in *Albugo candida* on rape seed. Can J Bot 55:2348–2357

Thaxter (1896) New and peculiar aquatic fungi. 4. *Rhipidium, Sapromyces* and *Araiospora*, nov. gen. Bot Gaz 21:317–330

Thines M (2009) Bridging the gulf: *Phytophthora* and downy mildews are connected by rare grass parasites. PLoS One 4:e4790

Thines M (2011) Recent outbreaks of downy mildew on grape ivy (*Parthenocissus tricuspidata* Vitaceae) in Germany are caused by a new species of *Plasmopara*. Mycol Prog 10:415–422

Thines M, Kamoun S (2010) Oomycete-plant coevolution: recent advances and future prospects. Curr Opin Biol 13:427–433

Thines M, Spring O (2005) A revision of *Albugo* (Chromista, Peronosporomycetes). Mycotaxon 92:443–458

Thines M, Voglmayr H (2009) An introduction to the white blister rusts (Albuginales). In: Lamour K, Kamoun S (eds) Oomycete genetics and genomics: diversity, interactions and research tools. Wiley, New York, pp 77–92

Thines M, Göker M, Oberwinkler F (2006) A revision of *Bremia graminicola*. Mycol Res 110:646–656

Thines M, Göker M, Oberwinkler F, Spring O (2007) A revision of *Plasmopara penniseti*, with implications for the host range of the downy mildews with pyriform haustoria. Mycol Res 111:1377–1385

Thines M, Göker M, Telle S, Ryley M, Mathur K, Narayana YD, Spring O, Thakur RP (2008) Phylogenetic relationships in graminicolous downy mildews based on cox2 sequence data. Mycol Res 112:345–351

Thines M, Choi Y-J, Kemen E, Ploch S, Holub EB, Shin H-D, Jones JDG (2009a) A new species of *Albugo* parasitic to *Arabidopsis thaliana* reveals new evolutionary patterns in white blister rusts (*Albuginaceae*). Persoonia 22:123–128

Thines M, Telle S, Ploch S, Runge F (2009b) Identity of the downy mildew pathogens of basil, coleus, and sage with implications for quarantine measures. Mycol Res 113:532–540

Thines M, Voglmayr H, Göker M (2009c) Taxonomy and phylogeny of the downy mildews. In: Lamour K, Kamoun S (eds) Oomycete genetics and genomics: diversity, interactions and research tools. Wiley, New York, pp 47–75

Tian M, Win J, Savory R, Burkhart A, Held M, Brandizzi F, Day F (2011) 454 Genome sequencing of *Pseudoperonospora cubensis* reveals effector proteins with a QXLR translocation motive. Mol Plant Microbe Interact 24:543–553

Tokunaga Y (1934) Notes on the Lagendiaceae in Japan. Trans Sapporo Nat Hist Soc 13:227–232

Tong SM (1995) *Developayella elegans* nov. gen., nov. spec., a new type of heterotrophic flagellate from marine plankton. Eur J Protistol 31:24–31

Tooley PW, Therrien CD (1987) Cytophotometric determination of the nuclear DNA content of 23 Mexican and 18 non-Mexican isolates of *Phytophthora infestans*. Exp Mycol 11:19–26

Torto-Alalibo T, Tian M, Gajendran K, Waugh ME, Van West P, Kamoun S (2005) Expressed sequence tags from the oomycete fish pathogen *Saprolegnia parasitica* reveal putative virulence factors. BMC Microbiol 5:46. doi:10.1186/1471-2180-5-46

Tsui CKM, Marshall W, Yokoyama R, Honda D, Lippmeier JC, Craven KD, Berbee ML (2009) Labyrinthulomycetes phylogeny and its implications for the evolutionary loss of chloroplasts and gain of ectoplasmic gliding. Mol Phylogenet Evol 50:129–140

Tyler BM, Tripathy S, Zhang X, Dehal P, Jiang RH, Aerts A, Arredondo FD, Baxter L, Bensasson D, Beynon JL, Chapman J, Damasceno CM, Dorrance AE, Dou D, Dickerman AW, Dubchak IL, Garbelotto M, Gijzen M, Gordon SG, Govers F, Grunwald NJ, Huang W, Ivors KL, Jones RW, Kamoun S, Krampis K, Lamour KH, Lee MK, McDonald WH, Medina M, Meijer HJ, Nordberg EK, Maclean DJ, Ospina-Giraldo MD, Morris PF, Phuntumart V, Putnam NH, Rash S, Rose JK, Sakihama Y, Salamov AA, Savidor A, Scheuring CF, Smith BM, Sobral BW, Terry A, Torto-Alalibo TA, Win J, Xu Z, Zhang H, Grigoriev IV, Rokhsar DS, Boore JL (2006) *Phytophthora* genome sequences uncover the evolutionary origins and mechanisms of pathogenesis. Science 313:1261–1266

Uzuhashi S, Tojo M, Kakishima M (2010) Phylogeny of the genus *Pythium* and description of new genera. Mycoscience 51:337–365

Van der Auwera G, Da Baere R, Van der Peer Y, Rijk PD, Van den Broeck I, De Wachter R (1995) The phylogeny of the Hyphochytriomycota as deduced from ribosomal RNA sequences of *Hyphochytrium catenoides*. Mol Biol Evol 12:671–678

Van der Plaats-Niterink AJ (1981) Monograph of the genus *Pythium*. Centraalbureau voor Schimmelcultures, Baarn

Van West P (2006) *Saprolegnia parasitica*, an oomycete pathogen with a fishy appetite; new challenges for an old problem. Mycologist 20:99–104

Vanterpool TC (1959) Oospore germination in *Albugo candida*. Can J Bot 37:169–172

Vanterpool TC, Ledingham GA (1930) Studies on browning root rot of cereals. Can J Res 2:171–194

Villa NO, Kageyama K, Asano T, Suga H (2006) Phylogenetic relationships of *Pythium* and *Phytophthora* species based on ITS, rDNA, cytochrome oxidase II and β-tubulin gene sequences. Mycologia 98:410–422

Vlk W (1939) Über die Geisselstruktur der Saprolegniaceenschwärmer. Arch Protistenkd 92:157–160

Vogel HJ (1964) Distribution of lysine pathaways among fungi: evolutionary implications. Am Nat 98:435–446

Voglmayer H, Riethmüller A (2006) Phylogenetic relationships of *Albugo* species (white blister rusts) based on LSU rDNA sequence and oospore data. Mycol Res 110:75–85

Voglmayr H (2003) Phylogenetic study of *Peronospora* and related genera based on nuclear ribosomal ITS sequences. Mycol Res 107:1132–1142

Voglmayr H (2008) Progress and challenges in systematics of downy mildews and white blister rusts: new insights from genes and morphology. Eur J Plant Pathol 122:3–18

Voglmayr H, Constantinescu O (2008) Revision and reclassification of three *Plasmopara* species based on morphological and molecular phylogenetic data. Mycol Res 112:487–501

Voglmayr H, Thines M (2007) Phylogenetic relationships and nomenclature of *Bremiella sphaerosperma* (Chromista, Peronosporales). Mycotaxon 100:11–20

Voglmayr H, Bonner L, Dick MW (1999) Taxonomy and oogonial ultrastructure of a new aero-aquatic peronosporomycete, *Medusoides* gen nov. (Pythiogetonaceae fam nov). Mycol Res 103:591–606

Voglmayr H, Riethmüller A, Göker M, Weiss M, Oberwinkler F (2004) Phylogenetic relationships of *Plasmopara*, Bremia and other genera of downy mildews with pyriform haustoria based on Bayesian analysis of partial LSU rDNA sequence data. Mycol Res 108:1011–1024

Watanabe T (1987) *Plectospira myriandra*, a rediscovered water mold in Japanese soil. Mycologia 79:77–81

Waterhouse GM (1970) The genus *Phytophthora* de Bary: Diagnoses (or descriptions) and figures from the original papers, 2nd edn. Mycological Papers 122:1–59

Waterhouse GM (1973) Peronosporales. In: Ainsworth GC, Sparrow FK, Sussman AS (eds) The fungi, vol IVb. Academic, New York, pp 165–183

Wavra S, Bain J, Durward E, Bruijn ID, Minor KL et al (2012) Host-targeting protein 1 (SpHtp1) from the oomycete *Saprolegnia parasitica* translocates specifically into fish cells in a tyrosine-O-sulphate-dependent manner. Proc Natl Acad Sci USA 109:2096–2101

Webster J (1970) Introduction to fungi. Cambridge University Press, Cambridge

Weston WH, Uppal BN (1932) The basis of *Sclerospora sorghi* as a species. Phytopathology 22:573–586

Whisson SC, Avrova A, Grenville-Briggs L, van West P (2009) Mechanisms and applications of gene silencing in oomycetes. In: Lamour K, Kamoun S (eds) Oomycete genetics and genomics: diversity, interactions and research tools. Wiley, New York, pp 493–515

Willoughby LG (1962) The occurrence of reproductive spores of Saprolegniaceae in freshwater. J Ecol 50:733–759

Willoughby LG, Roberts RJ (1991) Occurrence of the sewage fungus *Leptomitus lacteus*, a necrotroph on perch (*Perca fluviatilis*), in Windermere. Mycol Res 95:755–768

Worden AZ, Not F (2008) Ecology and diversity of piceukaryotes. In: Kirchman DL (ed) Microbial ecology of the oceans, 2nd edn. Wiley, New York, pp 159–205

Wynn AR, Epton HAS (1979) Parasitism of oospores of the *Phytophthora erythroseptica* in soil. Trans Br Mycol Soc 73:255–259

Yokoyama R, Honda D (2007) Taxonomic rearrangement of the genus *Schizochytrium* sensu lato based on morphology, chemotaxonomical characteristics, and 18S rRNA gene phylogeny (Thraustochytriaceae, Labyrinthulomycetes): emendation for *Schizochytrium* and erection of *Aurantiochytrium* and *Oblongichytrium* gen nov. Mycoscience 48:199–211

Yokoyama R, Salleh B, Honda D (2007) Taxonomic rearrangement of the genus *Ulkenia* sensu lato based on morphology, chemotaxonomical characteristics, and 18S rRNA gene phylogeny (Thraustochytriaceae, Labyrinthulomycetes): emendation for *Ulkenia* and erection of *Botrychytrium*, *Parietichytrium*, and *Sicyoidochytrium* gen. nov. Mycoscience 48:329–341

Yubuki N, Leander BS, Silberan JD (2010) Ultrastructure and molecular phylogenetic position of a novel phagotrophic position of a novel phagotrophic stramenopile from low oxygen environments: *Rictus lutensis* gen. et sp. nov. (Biocosoecida, incertae sedis). Protist 161:264–278

4 Rhizaria: Phytomyxea

Simon Bulman[1], James P. Braselton[2]

CONTENTS

I. Introduction

Phytomyxea comprises a group that historically was considered as fungi (Sparrow 1960; Waterhouse 1972) and for that reason is included here. Recent classifications place Phytomyxea in the protistan supergroup **Rhizaria** (Adl et al. 2005; Bass et al. 2009; Cavalier-Smith and Chao 2003), and molecular studies have led to the recognition of two orders within Phytomyxea (Bass et al. 2009; Cavalier-Smith and Chao 2003): **Plasmodiophorida** (the plasmodiophorids *sensu stricto*) and **Phagomyxida** (phagomyxids). Major reviews of the Phytomyxea, in addition to John Karling's monograph *The Plasmodiophorales* (1968), include Maire and Tison (1909), Cook (1933), Dylewski (1989), Dick (2001), and Neuhauser et al. (2010).

The most commonly recognized Phytomyxea are the plant pathogens *Plasmodiophora brassicae* Woronin, the causal agent of clubroot of cabbage and other brassicaceous crops worldwide (Cook and Schwartz 1930; Dixon 2009), and *Spongospora subterranea* (Wallroth) Lagerheim, the causal organism of powdery scab of potato (Kole 1954; Merz 2008; Merz and Falloon 2009). Also of economic significance are *Spongospora nasturtii* M. W. Dick, the causal agent of crook root in watercress (Tomlinson 1958), and *Polymyxa betae* Keskin, the vector for beet necrotic yellow vein virus (BNYVV), which causes rhizomania of sugar beet (McGrann et al. 2009). *Spongospora nasturtii*, *S. subterranea*, and *Polymyxa graminis* Ledingham also serve as vectors for plant-pathogenic viruses (Cooper and Asher 1988; Kanyuka et al. 2003; Rochon et al. 2004).

Karling (1981) proposed unified terminology to alleviate problems with nomenclature for stages in the life cycles of plasmodiophorids that had accumulated over the years because of contributions from researchers in a variety of disciplines. His major concern was the use of the term *cyst* for the single-celled resting structure that was not the result of a zoospore encysting on a substrate. By replacing the term cyst with **resting spore**, for consistency, the recommended term for the collection of resting spores would be ***sporosorus*** instead of ***cystosorus***. Other terms recommended by Karling included ***sporogenic***, when referring to developmental stages that lead to resting spores, and ***sporangial***, when referring to developmental stages that lead to **thin-walled sporangia** (zoosporangia) that contain zoospores. Sporogenic and sporangial phases of plasmodiophorid life cycles have also been referred to as secondary and primary, respectively.

[1]The New Zealand Institute for Plant & Food Research Limited, Gerald Street, Lincoln, 7608 Canterbury New Zealand; e-mail: simon.bulman@plantandfood.co.nz
[2]Department of Environmental & Plant Biology, Ohio University, Athens, OH 45701, USA; e-mail: braselto@ohio.edu

Systematics and Evolution, 2nd Edition
The Mycota VII Part A
D.J. McLaughlin and J.W. Spatafora (Eds.)

A unique type of nuclear division in Phytomyxea, **cruciform division**, was observed in *P. brassicae* as early as 1899 but was referred to as *promitosis* or *protomitosis* because it resembled some of the nuclear divisions in several protozoa (Cook 1933; Karling 1968). The currently used descriptive adjective *cruciform* was introduced by Blomfield and Schwartz (1910) because at metaphase the persistent nucleolus is elongated parallel to the spindle and perpendicular to the plate of chromatin, thus forming a crosslike (cruciform) configuration when viewed from the side (Figs. 4.1 and 4.2). Additional descriptive terms used for this type of nuclear division include *Saturn stage* as a synonym for cruciform and *double anchor* or *dumbbell stage* for mid to late anaphase (Blomfield and Schwartz 1910; Cook 1933) (Fig. 4.3). Features of cruciform divisions based on ultrastructural observations (Keskin 1971; Braselton et al. 1975; Dylewski et al. 1978; Garber and Aist 1979b; Dylewski and Miller 1983) (Figs. 4.1–4.3) include a persistent membrane of either nuclear envelope or endoplasmic reticulum origin, intranuclear spindle, centrioles at both poles, and a nucleolus that remains throughout nuclear division.

In addition to cruciform division, systematic features of plasmodiophorids include **multinucleate protoplasts** without walls (**plasmodia**) as growth forms (Fig. 4.1), zoospores with two anterior whiplash (lacking mastigonemes) flagella (undulipodia) of unequal lengths (Ledingham 1934; Kole and Gielink 1961), centrioles paired in an end-to-end fashion (Braselton and Miller 1973) (Fig. 4.4), environmentally resistant resting spores (Figs. 4.6 and 4.7), and intracellular, biotrophic growth forms (Dylewski 1989). Phagomyxida share these features with Plasmodiophorida, with the exception of environmentally resistant resting spores: resting spores have not been documented for *Maullinia* I. Maier, E. R. Parodi, R. Westermeier et D. G. Müller, or *Phagomyxa* Karling.

II. Life Cycle

Difficulties with describing phytomyxid life cycles arise in part because members of this group are obligate, intracellular biotrophs; no member has been shown conclusively to complete a life cycle in culture free of host cells. Dylewski's (1989) diagrammatic representation of the life cycle for members of the plasmodiophorids was in turn based on Karling's (1968) summary and serves as the basis for the life cycle presented here (Fig. 4.8). It should be emphasized that this generalized life cycle is the result of a compilation of observations made by various investigators and that variations in this scheme either have not been documented fully or are not currently understood.

Two major phases are recognized in the plasmodiophorid life cycle. The **sporogenic (secondary) phase**, which has not been observed in phagomyxids, culminates in the production of resting spores. The **sporangial (primary) phase** produces secondary zoospores within relatively thin-walled (zoo)sporangia.

In plasmodiophorids the life cycle arbitrarily may be considered to begin with a resting spore, a cell that contains a single nucleus and has an environmentally resistant cell wall. Resting spores may remain viable for several years, rendering infected soils unsuitable for susceptible hosts (Macfarlane 1952). The cell walls of *P. brassicae* (Yukawa and Tanaka 1979) and *S. subterranea* (Lahert and Kavanagh 1985) consist of three layers; *P. brassicae* cell walls contain chitin, lipids, and protein (Buczacki and Moxham 1983; Moxham and Buczacki 1983). The thickness of cell walls varies among members of the group (Figs. 4.6 and 4.7), but there has been no systematic treatment of the variations. Resting spores may occur singly, as in the genus *Plasmodiophora* Woronin, or in aggregations, sporosori, which remain the major morphological criterion for designating genera within Plasmodiophorida.

Upon germination, a resting spore releases a single, heterokont, biflagellated, uninucleate, free-swimming, primary zoospore (Kole and Gielink 1962; Macfarlane 1970; Merz 1997). When a zoospore encounters the wall of a potential host cell, the zoospore encysts and retracts its flagella (Aist and Williams 1971; Claxton et al. 1996; Merz 1997). A dense, projectile-like structure (Stachel) is within a tubular cavity (Rohr), and together these pass with the majority of the zoospore's cytoplasm into

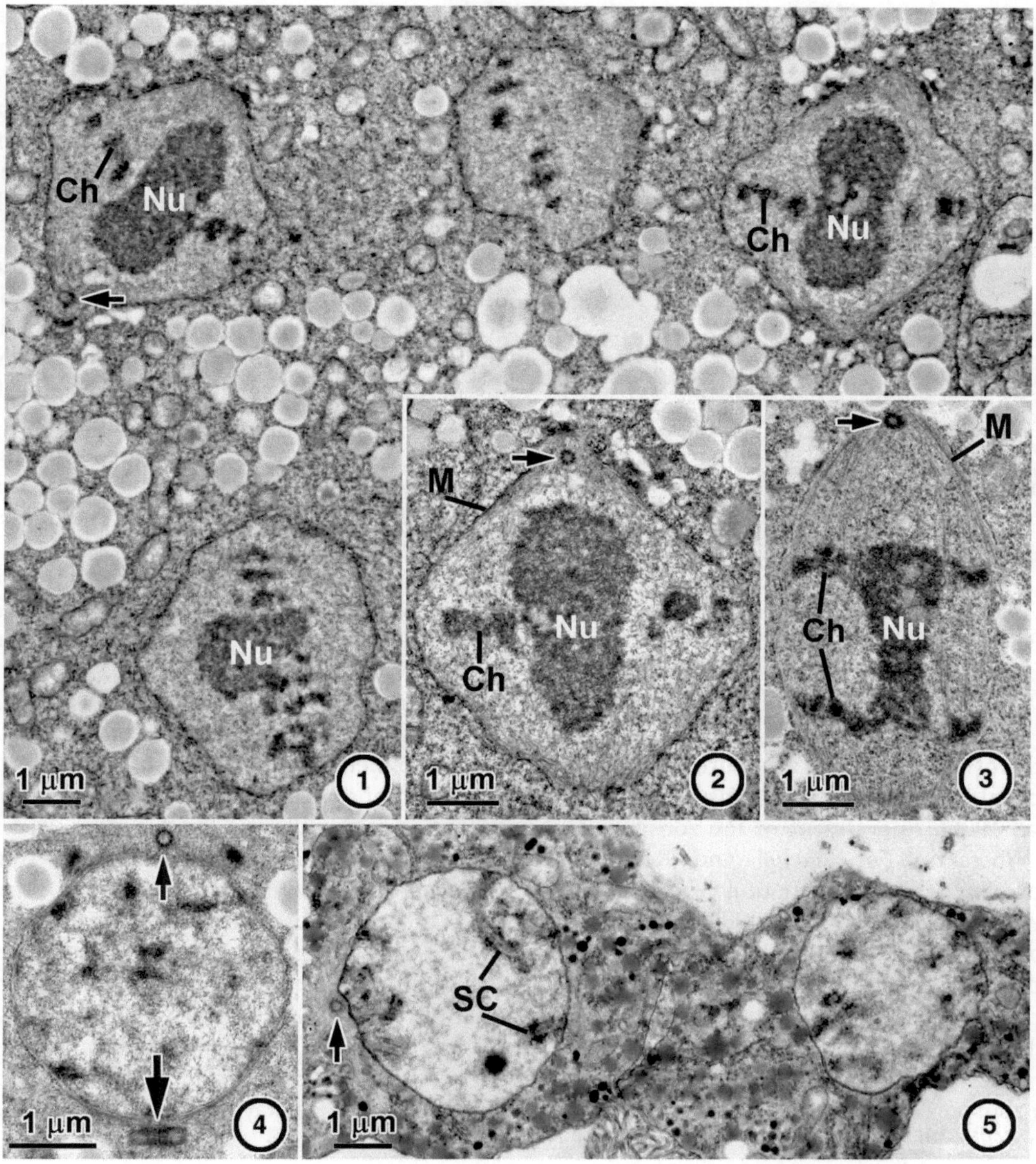

Figs. 4.1–4.5 TEMs of dividing nuclei of Phytomyxea. **Figs. 4.1–4.4** Sporangial plasmodia of *Spongospora nasturtii* on watercress. **Fig. 4.1** Survey TEM of young plasmodium with synchronous cruciform divisions. Nucleoli are elongated perpendicularly to the chromatin and centrioles are at the poles (*arrow*). **Fig. 4.2** TEM of metaphase ("Saturn stage") of cruciform nuclear division. **Fig. 4.3** TEM of anaphase of cruciform nuclear division ("double anchor stage"). **Fig. 4.4** TEM of transitional nucleus, with one pair of centrioles in the end-to-end orientation characteristic for the group shown in longitudinal view (*larger arrow*) and the other centriolar pair in transverse view (*smaller arrow*). **Fig. 4.5** *Tetramyxa parasitica* on *Ruppia maritima*. TEM of transitional sporogenic plasmodium with profiles of synaptonemal complexes in the nuclei and one centriole shown at a pole. Centriole (*arrow*), chromatin (Ch), nucleolus/ i (Nu), persistent membrane (M), synaptonemal complex (SC), and transmission electron micrograph (TEM)

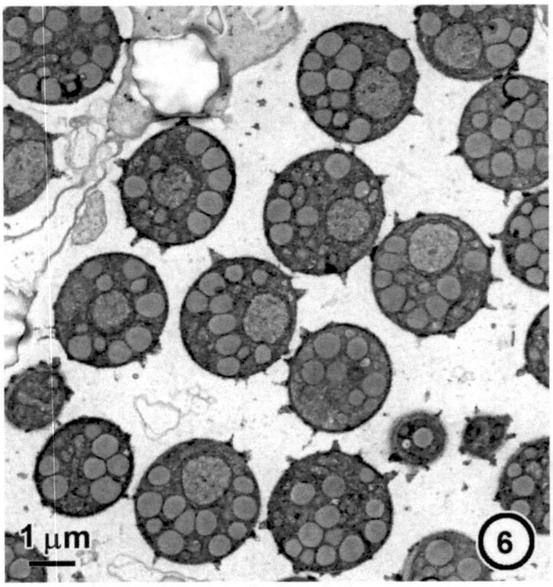

Fig. 4.6 *Plasmodiophora brassicae*. TEM of resting spores in root cell of Chinese cabbage

an outgrowth (adhesorium) from the main body of the encysted zoospore (Keskin and Fuchs 1969; Aist and Williams 1971; Claxton et al. 1996). Encystment of the zoospore with the formation of Stachel and Rohr takes approximately 2 h, formation of the adhesorium approximately 1 min, and the injection of zoospore contents through the host cell wall and plasma membrane into the host cytoplasm approximately 1 s (Aist and Williams 1971; Williams 1973).

Once within the host cell, the contents of the zoospore begin to grow by cruciform divisions (Fig. 4.1). The boundary between the plasmodium and host cytoplasm may be either a single, unit membrane for some members of the group (Braselton and Miller 1975) or a boundary thicker than a single membrane consisting of several layers for others (Williams and McNabola 1970).

What determines the path of development a plasmodium at this stage will take is not understood. For some phytomyxids, e.g., members of the genera *Polymyxa* Ledingham and *Ligniera* Maire & Tison, sporangial and sporogenic plasmodia may occur within adjacent cells of the same host tissue (Miller 1959). For others, such as *P. brassicae* and *S. subterranea*, sporangial plasmodia generally occur in root epidermal cells, particularly root hairs, whereas sporogenic plasmodia occur in cortical cells. For *Sorosphaera veronicae* Schröter,[1] sporogenic development is confined to shoots, whereas sporangial development occurs only in the roots (Miller 1958).

Conditions of the host growth medium may influence the development of the phytomyxid. For example, when *Woronina pythii* Goldie-Smith infects a *Pythium* sp. that has been growing in medium for less than a few days, the *W. pythii* will follow sporangial development. If, however, the host has been growing in medium for several days, and the medium is "stale," the *W. pythii* will follow sporogenic development (Miller and Dylewski 1983).

When a plasmodium, whether sporangial or sporogenic, reaches a stage of maturity where growth ceases, cruciform divisions no longer occur, and the nuclei become difficult to see in paraffin-sectioned specimens. The nuclei in part are difficult to recognize because the nucleoli either are reduced in size to below the resolution of optical microscopy or have disappeared altogether. Terms for this stage used by earlier microscopists included *akaryotic stage*, *enucleate stage*, *chromidial stage*, and *transitional stage*. Because nuclei are now known to be present during this stage of development (Fig. 4.5), *transitional stage* is the most appropriate term because this stage marks a change in the development of the plasmodium from a period of growth to a period of differentiation. Nuclei in this stage may be referred to as transitional nuclei.

[1] *Sorosphaera* has been used throughout this review because historically *Sorosphaera* was the name used in the literature for the genus. Neuhauser and Kirchmair (2011) noted, however, that since both Phytomyxea and Foraminifera are now recognized as members of the supergroup Rhizaria (Archibald and Keeling 2004; Bass et al. 2009; Burki et al. 2010), based on the International Code of Zoological Nomenclature (ICZN), a homonomy exists between the plasmodiophorid *Sorosphaera* J. Schröter and the foraminiferan *Sorosphaera* Brady. To resolve the homonomy, Neuhauser and Kirchmair (2011) proposed that *Sorosphaerula* nom. n. replace *Sorosphaera* J. Schröter for this genus.

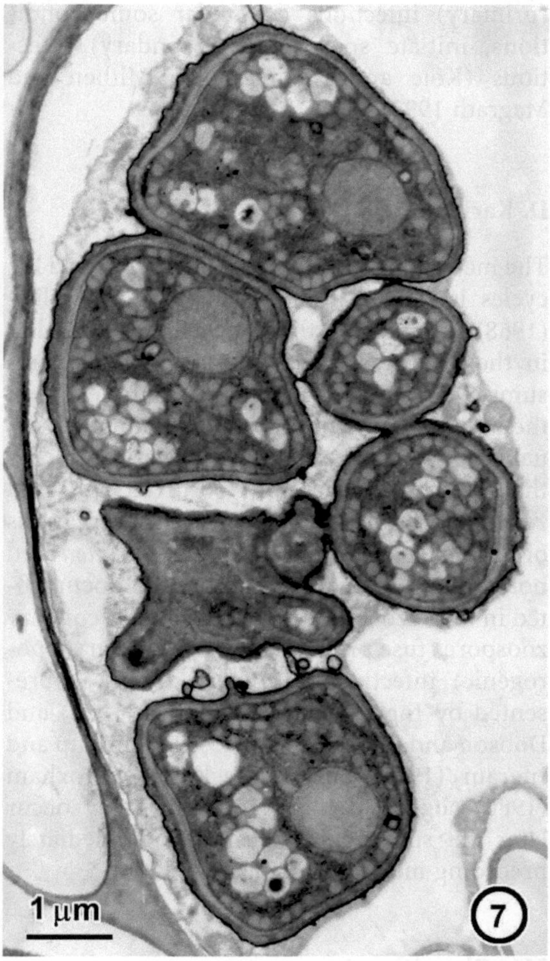

Fig. 4.7 *Polymyxa betae*. TEM of resting spores in a sporosorus in root cell of sugar beet

Up to the transitional stage, there are no obvious morphological distinctions between sporogenic or sporangial plasmodia (Miller 1959); the only time it is possible to determine definitively what type of plasmodium is present is in those situations where the two types of development occur in different host tissues as in *P. brassicae* and *S. subterranea*. Miller and Dylewski (1983) noted, however, that sporogenic plasmodia of *W. pythii* contained more lipoidal globules than sporangial plasmodia at the time cleavage is initiated. Nuclear divisions that occur in either sporogenic or sporangial transitional plasmodia are not of the cruciform type and, therefore, are referred to as *noncruciform divisions*. Although noncruciform divisions in both sporangial and sporogenic plasmodia appear similar at the level of optical microscopy, their prophases and, consequently, the type of division, may be distinguished by ultrastructural criteria.

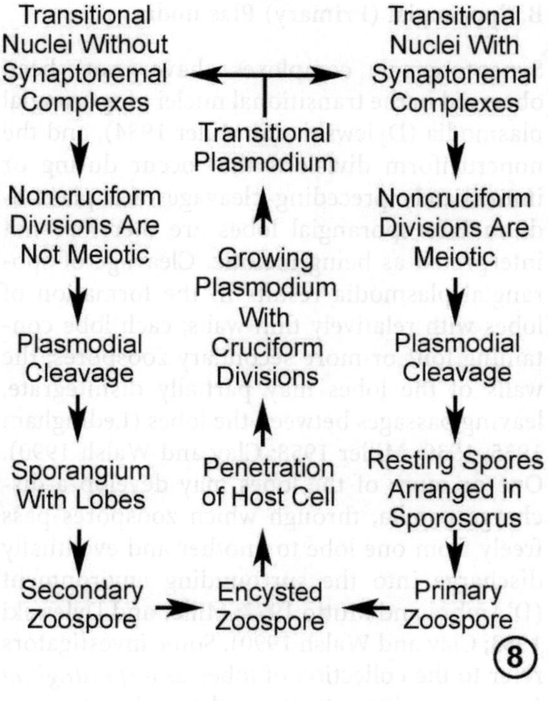

Fig. 4.8 Summary diagram of generalized life cycle for members of Plasmodiophorida

A. Sporogenic (Secondary) Plasmodia

Transmission electron microscopy of transitional nuclei in plasmodia known to be of the sporogenic type revealed **synaptonemal complexes** (Fig. 4.5), indicators of prophase I of meiosis (Garber and Aist 1979a; Braselton 1995). The noncruciform divisions that occur either immediately preceding or during cleavage of the protoplasm leading to the formation of incipient resting spores are therefore interpreted as being meiotic, as had been suspected by others (Cook 1933; Webb 1935; Heim 1955). Cleavage of cytoplasm into uninucleate cells leads to the formation of resting spores as cell walls are deposited.

B. Sporangial (Primary) Plasmodia

Synaptonemal complexes have not been observed in the transitional nuclei of sporangial plasmodia (Dylewski and Miller 1984), and the noncruciform divisions that occur during or immediately preceding cleavage of a plasmodium into sporangial lobes are therefore not interpreted as being meiotic. Cleavage of sporangial plasmodia results in the formation of lobes with relatively thin walls, each lobe containing four or more secondary zoospores; the walls of the lobes may partially disintegrate, leaving passages between the lobes (Ledingham 1935, 1939; Miller 1958; Clay and Walsh 1990). One or more of the lobes may develop a discharge papilla, through which zoospores pass freely from one lobe to another and eventually discharge into the surrounding environment (D'Ambra and Mutto 1977; Miller and Dylewski 1983; Clay and Walsh 1990). Some investigators refer to the collection of lobes as a *sporangium* (*zoosporangium*) because the collection presumably developed from one plasmodium or there are continuities between lobes once walls between them disintegrate [see Miller (1958) for a review of this terminology; Barr 1979]. Others use the term *sporangiosorus* for the collection of lobes, considering each lobe as a sporangium (Buczacki and Clay 1984).

C. Relationship of Life Cycle Phases

The relationship of the two life cycle phases is not completely understood. Dobson and Gabrielson (1983) reported that sporangial development is needed prior to sporogenic development in *P. brassicae*; sporogenic development is interpreted as being initiated by secondary zoospores produced from sporangia. Other observations for *S. subterranea* and *P. brassicae* respectively by Kole and Gielink (1963) and Mithen and Magrath (1992) have indicated that primary zoospores may give rise directly to sporogenic (secondary) infections and to sporangial infections. Secondary zoospores likewise may produce sporangial (primary) infections or, under some conditions, initiate sporogenic (secondary) infections (Kole and Gielink 1963; Mithen and Magrath 1992).

D. Karyogamy

The major unresolved aspect of phytomyxid life cycles is the location of **karyogamy**. Karling (1968) summarized the knowledge of sexuality in the group as "...largely indirect and presumptive," and the statement continues to be the best summary of our understanding of sexuality for Phytomyxea. After suggesting earlier that karyogamy possibly occurred in fused zoospores, Kole (1954) reviewed observations of fusion of zoospores of *S. subterranea* and noted that karyogamy could not be documented in fused zoospores. The idea that secondary zoospores fuse prior to initiating primary (sporogenic) infections in *P. brassicae* was presented by Ingram and Tommerup (1972) and Dobson and Gabrielson (1983). Tommerup and Ingram (1971) and Buczacki and Moxham (1980) suggested that karyogamy may occur later in sporogenic plasmodia immediately preceding meiotic divisions.

III. Classification

A. Phylogeny

Although many mycologists and plant pathologists have treated Phytomyxea as fungi (Sparrow 1960; Waterhouse 1972), others have grouped them with the protozoa (Barr 1992). Beginning with the sequencing of the *P. brassicae* ribosomal 18S gene (Castlebury and Domier 1998), DNA sequence phylogenies placed phytomyxids with a wide assemblage of protists in the Cercozoa (Cavalier-Smith and Chao 1997, 2003). Further evidence of a close relationship between phytomyxids and cercozoans came with confirmation that they shared a unique one- or two-amino-acid insertion between ubiquitin monomers (Archibald and

Keeling 2004). These insertions have been found in Cercozoa and Foraminifera but not in all other eukaryotes studied to date, including radiolarians (Archibald et al. 2003; Bass et al. 2005). Subsequently, Cercozoa was incorporated into a supergroup of diverse protists, the Rhizaria, which has been almost entirely circumscribed through molecular evidence (Bass et al. 2005; Moreira et al. 2007; Nikolaev et al. 2004) and which has an evolutionary closeness to two chromalveolate groups, stramenopiles and alveolates (Burki et al. 2007, 2008; Hackett et al. 2007; Rodriguez-Ezpeleta et al. 2007).

Although Phytomyxea is well settled in the Rhizaria, the position of the phytomyxids with respect to other rhizarians is not established. Ribosomal 18S sequences show the parasitic Phytomyxea and Ascetosporea, along with reticulose protists, solidly grouped in the subphylum Endomyxa (Bass et al. 2005, 2009; Cavalier-Smith 2003). The first phylogenomic study to include large numbers of phytomyxid gene sequences placed Phytomyxea with *Gromia* Dujardin and a clade of Acantherea and Foraminifera separate from the core Cercozoa (Burki et al. 2010). Increased density of 18S sequences from cultivated protists and anonymous sequences from environmental sources indicate that the terrestrial/freshwater Vampyrellidae in the Proteomyxidea are the closest known relatives of Phytomyxea (Bass et al. 2009). If confirmed, this will show that parasitism has arisen twice, independently of free-living ancestors in the Phytomyxea and Ascetosporea (Bass et al. 2009).

B. Genera and Species

Genera and species are based on morphological criteria; the biological species concept is not applicable for this group because sexuality has not been observed. Ten genera are recognized in the order Plasmodiophorida (Braselton 1995; Dylewski 1989; Karling 1968): *Ligniera*; *Membranosorus* Ostenfeld & Petersen; *Octomyxa* Couch, Leitner & Whiffen; *Plasmodiophora*; *Polymyxa*; *Sorodiscus* Lagerheim & Winge; *Sorosphaera*; *Spongospora* Brunchorst; *Tetramyxa* Goebel, and *Woronina*. Two genera are currently recognized in the Phagomyxida: *Maullinia* (Maier et al. 2000) and *Phagomyxa* (Schnepf 1994; Schnepf et al. 2000). Karling (1968) listed 35 recognized species in his consideration of Plasmodiophorales.

The genera of plasmodiophorids are based on the morphologies of sporosori as seen through compound optical microscopy. For several genera, sporosoral morphologies are incorporated into their generic names, such as *Tetramyxa* (four resting spores per sporosorus), *Octomyxa* (eight resting spores per sporosorus), *Membranosorus* (sporosorus consisting of resting spores primarily in a single layer), *Sorodiscus* (resting spores arranged in a disk-shaped sporosorus), *Sorosphaera* (resting spores arranged in a sphere), and *Spongospora* (resting spores arranged in a spongy-looking sporosorus). Although Palm and Burk (1933), and subsequently some reviewers of the group (e.g., Olive 1975), questioned the reliability of using sporosoral morphology, it has continued to be the main criterion for delimiting genera of plasmodiophorids. Species within genera are generally based on what hosts are infected by the given organism, with specific epithets reflecting the host name. Examples include *pythii*, *callitrichis*, *betae*, *graminis*, *subterranea*, *nasturtii*, *brassicae*, *heterantherae*, and *veronicae*.

At this time we are on the verge of a better understanding of Phytomyxea speciation based on molecular phylogenetics. While confirming the expected close relationship between the *Polymyxa* spp. and *S. veronicae*, comparisons of ribosomal DNA sequences have shown that there is considerable phylogenetic distance between *S. subterranea* and *S. nasturtii* (Bulman et al. 2001), which supported the renaming of these two members of the genus from their previously recognized formae speciales (Dick 2001).

Misidentification of some genera and species or incomplete studies have led to confusion as to whether all of the currently recognized genera are valid. Palm and Burk (1933) concluded that the presently recognized genera *Ligniera*, *Membranosorus*, and *Sorodiscus* should be considered as synonyms of *Sorosphaera*. It should be emphasized that their conclusion was based on observations of one plant of *Veronica* sp. infected with *S. veronicae*. Analyses of chromosomal numbers through serial sections of synaptonemal complexes showed that ultrastructural karyotypes of the recognized genera differ, supporting the retention of the ten recognized genera of Plasmodiophorida as valid taxa (Braselton 1995).

A paper that has led to confusion about two genera was by Wernham (1935) in which *Membranosorus heteranthera*e Ostenfeld & Petersen (Ostenfeld and Petersen 1930) was renamed *Sorodiscus heterantherae*. Wernham's misidentification created some doubt as to the validity of the genus *Membranosorus* (Karling 1968; Olive 1975), which apparently has led to its exclusion from other systematic reviews (Cavalier-Smith 1993). Ultrastructural and karyotypic studies (Braselton 1983, 1989b) supported the view that *Membranosorus* is a valid genus.

C. Molecular Applications

Molecular investigations of Phytomyxea lag behind those for other microbial groups of comparable economic significance. *P. brassicae* has been the most extensively studied phytomyxid; the progression of molecular studies in this organism was summarized by Siemens et al. (2009). A consistent driver of molecular studies for phytomyxids has been the need for rapid and accurate detection of the important plant pathogens and viral vectors. This need has led to progress toward rDNA-targeted, quantitative-PCR assays for *P. brassicae* [reviewed in Faggian and Strelkov (2009)], *S. subterranea* (Lees et al. 2008; van de Graaf et al. 2003), and *Polymyxa* spp. (Vaïanopoulos et al. 2007; Ward et al. 2005).

From the earliest studies (Buhariwalla and Mithen 1995; Buhariwalla et al. 1995; Ito et al. 1994; Möller and Harling 1996), molecular techniques have been used for detecting genetic diversity within species. Molecular techniques for differentiating the highly variable *P. brassicae* accessions remain at an exploratory phase (e.g., Manzanares-Dauleux et al. 2001), but examinations of ribosomal sequences have been successful in delimiting new variations in the genus *Polymyxa* (Legrève et al. 2002).

Large-scale genomic studies have not been completed for any phytomyxid. This is in part because of the need to sort plant from phytomyxid sequences (Burki et al. 2010). There has been progress, however, in revealing the structure of several genes from *P. brassicae* (Siemens et al. 2009) and constructing a pilot-scale DNA library for *S. subterranea* (Bulman et al. 2011). Brodmann et al. (2002) attributed an increase in trehalose in roots and hypocotyls of *Arabidopsis thaliana* (L.) Heynh. infected with *P. brassicae* to the expression of a putative trehalose synthase gene from *P. brassicae*. An in-depth characterization of a phytomyxid gene was completed for a putatively secreted proteolytic enzyme from *P. brassicae* (Feng et al. 2010). Given the plummeting cost of generating new DNA sequences, complete phytomyxid genomes are undoubtedly accessible, although correct assembly plus a full and detailed annotation of such genomic data will be more time consuming.

IV. Occurrence, Distribution, Maintenance, and Culture

Depending primarily on their respective hosts, members of the Phytomyxea occur in a variety of habitats, including terrestrial, marine, and freshwater. Hosts range from vascular plants to algae and water molds.

The commonly recognized plant pathogens *P. brassicae* and *S. subterranea* and viral vectors *P. graminis* and *P. betae* are observed on a yearly basis on crops in various parts of the world and may be obtained from crop plants grown in infected soils (Colhoun 1957).

Most investigations for maintaining Phytomyxea in the laboratory or in glasshouse conditions concern *P. brassicae* and *S. subterranea*. Clubbed roots can be stored at −20 °C and used for inoculum of *P. brassicae* for several years. *Plasmodiophora brassicae* is maintained on various *Brassica* L. (Brassicaceae) species grown in soil in the greenhouse or growth chambers by inoculating seedlings with purified resting spores or slices of infected roots (Castlebury and Glawe 1993). Root galls are visible 3–7 weeks after inoculation. Castlebury et al. (1994) described how to purify resting spores from root galls, and several reports detailed methods for initiating infections from single resting spores (Buczacki 1977; Jones et al. 1982; Scott 1985; Tinggal and Webster 1981; Voorrips 1996). Both phases of the life cycle of *P. brassicae* may be expressed on *Brassica* seedlings grown in defined, liquid, nutrient media (Crute et al. 1981; Macfarlane 1958; Williams et al. 1971). Methods for maintaining *S. subterranea* in the greenhouse on potatoes and tomatoes follow protocols similar to those used for *P. brassicae* (Kole 1954).

Polymyxa graminis may be grown on wheat in sand inoculated with infected soil samples (Barr 1987) and *P. betae* by growing sugar beet under similar conditions with sand inoculated with soils from sugar-beet-growing regions (Barr and Asher 1992). Neither *Polymyxa* species causes hypertrophy of host tissues, so localization of portions of roots that are infected must be made with optical microscopy. Both sporangial and sporogenic stages are observable in young, intact roots viewed with brightfield optical microscopy.

Collection of infected hosts from nature is the method of choice for obtaining representatives of *Ligniera*, *Membranosorus*, several species of *Plasmodiophora* other than *P. brassicae*, *Sorodiscus*, *Sorosphaera*, and *Tetramyxa*. With the exception of *Ligniera*, these parasites cause galls, which are easily identified with the unaided eye on host shoots or roots, depending on the particular host and parasite.

Membranosorus heterantherae occurs throughout the range of the host, *Heteranthera dubia* (Jacq.) MacMill. (Pontederiaceae), in freshwater lakes and rivers in the continental USA and southern Canada (Forest et al. 1986). *Sorosphaera veronicae* has been observed to cause shoot galls on various species of winter annuals in the genus *Veronica* (Plantaginaceae) in Athens, Ohio, USA (Harris et al. 1980); Chapel Hill, North Carolina, USA (Braselton and Miller 1973; Miller 1958); Sevenoaks, UK (Blomfield and Schwartz 1910); and near La Veta, Colorado, USA (Palm and Burk 1933). *Tetramyxa parasitica* Goebel is found on species of *Zannichellia* (Potamogetonaceae) and *Ruppia* (Ruppiaceae) in shallow, brackish water in Finland, Denmark, Sweden, Norway, UK, Germany, France, Italy, the USA (Luther 1949), and the Netherlands (den Hartog 1963). What was reported to be *T. parasitica* on *Halophila stipulacea* Asch. (Hydrocharitaceae) (Marziano et al. 1995) seems to be a species of *Plasmodiophora*. Two species of *Plasmodiophora* that deserve further study are widely distributed on their respective seagrass hosts, *P. diplantherae* on *Halodule* species (Cymodoceaceae) (den Hartog 1965; Walker and Campbell 2009) and *P. bicaudata* on species of *Zostera* (Zosteraceae) (den Hartog 1989). *Sorodiscus callitrichis* may be found on *Callitriche* (Plantaginaceae) species throughout Sweden in freshwater streams and ponds (Martinsson 1987).

Since species of *Ligniera* do not cause hypertrophy of host tissues, compound optical microscopy must be used to locate the various species by examining young, intact roots of hosts that have been collected from their native habitats. *Ligniera* spp. located in this manner include *L. junci* (Schwartz) Maire & Tison in roots of *Juncus triglumis* L. (Juncaceae) from englacial streams in Austria (Neuhauser and Kirchmair 2009); *L. verrucosa* Maire & Tison in roots of *Veronica* spp. collected from lawns on university campuses in Athens, Ohio, USA (Braselton 1989a; Miller et al. 1985) and Chapel Hill, North Carolina, USA (Miller 1959); and *L. pilorum* Fron & Gaillat in roots of various grasses in Ontario, Canada (Barr 1979).

Members of the genus *Woronina* are found worldwide and infect a variety of taxa of water molds and algae. *Woronina* may be located by "baiting" soil samples with hemp seeds in Emerson's (P/3) water for its hosts, primarily species of *Pythium* (Dylewski 1987; Miller and Dylewski 1983). Infected regions of hosts are enlarged and are detectable with brightfield, phase contrast, or differential interference contrast microscopy.

Location of phagomyxids has so far been largely a byproduct of research into their host species. Capture of *Phagomyxa bellerocheae* Schnepf and *P. odontellae* Kühn, Schnepf & Bulman requires close observation and expertise with phytoplankton from the Wadden Sea (Schnepf 1994; Schnepf et al. 2000). *Maullinia ectocarpii* I. Maier, E. R. Parodi, R. Westermeier et D. G. Müller has been identified as a parasite of economically important brown algae in Chile

(Maier et al. 2000). The size of infections and the culturability of its host (*Ectocarpus siliculosus* [Dillwyn] Lyngbye) mean that this phagomyxid represents the best chance for ongoing studies of these organisms.

No phytomyxid has been shown to complete a life cycle in culture free of host cells. There have been, however, successes in growing *P. brassicae* and *S. subterranea* in tissue culture with their respective plant hosts. These studies have used two approaches. First, *P. brassicae* and *S. subterranea* have been propagated successfully in hairy root cultures established by *Agrobacterium rhizogenes* (Asano et al. 2006; Qu and Christ 2007). Secondly, both *P.brassicae* and *S. subterranea* have been propagated for significant periods in plant callus cultures (Asano and Kageyama 2006; Bulman et al. 2011; Ingram 1969; Tommerup and Ingram 1971; Williams et al. 1969).

V. Conclusions and Future Prospects

Phytomyxea comprises a discrete taxonomic group that contains several members of economic importance. Despite the extensive applied literature on the control of the plant pathogens *P. brassicae* and *S. subterranea* and the viral vectors *P. graminis* and *P. betae*, several unresolved questions about the life cycles of members of the group remain. These include:

- Where in the life cycle does karyogamy occur?
- What determines when a resting spore germinates?
- How does a zoospore recognize a host cell?
- What determines whether a plasmodium will follow sporogenic or sporangial development?

It seems inevitable that Phytomyxea species are more abundant and widespread than is currently known (Neuhauser et al. 2011). Searches of potential hosts in other locations would be rewarding, and studies of environmental DNA samples may provide a new window into the group by determining the presence of undescribed species of Phytomyxea in terrestrial and aquatic environments. Further studies could include comprehensive Basic Local Alignment Search Tool searches of anonymous ribosomal RNA sequences in public databases for the presence of sequences of likely phytomyxean origin (Lesaulnier et al. 2008), multiple PCR-primer approaches (Stoeck et al. 2006), and the use of PCR primers biased toward the detection of phytomyxids (Neuhauser et al. 2011).

References

Adl SM, Simpson AGB, Farmer MA, Andersen RA, Anderson OR, Barta JR, Bowser SS, Brugerolle G, Fensome RA, Fredericq S, James TY, Karpov S, Kugrens P, Krug J, Lane CE, Lewis LA, Lodge J, Lynn DH, Mann DG, McCourt RM, Mendoza L, Moestrup Ø, Mozley-Standridge SE, Nerad TA, Shearer CA, Smirnov AV, Spiegel FW, Taylor MFJR (2005) The new higher level classification of eukaryotes with emphasis on the taxonomy of protists. J Eukaryot Microbiol 52:399–451

Aist JR, Williams PH (1971) The cytology and kinetics of cabbage root hair penetration by *Plasmodiophora brassicae* Wor. Can J Bot 49:2023–2034

Archibald JM, Keeling PJ (2004) Actin and ubiquitin protein sequences support a cercozoan/foraminiferan ancestry for the plasmodiophorid plant pathogens. J Eukaryot Microbiol 51:113–118

Archibald JM, Longet D, Pawlowski J, Keeling PJ (2003) A novel polyubiquitin structure in cercozoa and foraminifera: evidence for a new eukaryotic supergroup. Mol Biol Evol 20:62–66

Asano T, Kageyama K (2006) Growth and movement of secondary plasmodia of *Plasmodiophora brassicae* in turnip suspension-culture cells. Plant Pathol 55:145–151

Asano T, Kodama A, Kageyama K (2006) Susceptibility of hairy root lines of *Brassica* species to *Plasmodiophora brassicae* and in an *in vitro* subculture system. J Gen Plant Pathol 72:85–91

Barr DJS (1979) Morphology and host range of *Polymyxa graminis*, *Polymyxa betae*, and *Ligniera pilorum* from Ontario and some other areas. Can J Plant Pathol 1:85–94

Barr DJS (1987) Some obligate parasites in roots. In: Fuller MS, Jaworski A (eds) Zoosporic fungi in teaching and research. Southeastern Publishing Corporation, Athens, GA, pp 130–131

Barr DJS (1992) Evolution and kingdoms of organisms from the perspective of a mycologist. Mycologia 84:1–11

Barr KJ, Asher MJC (1992) The host range of *Polymyxa betae* in Britain. Plant Pathol 41:64–68

Bass D, Moreira D, Lopez-Garcia P, Polet S, Chao EE, Herden S, Pawlowski J, Cavalier-Smith T (2005)

Polyubiquitin insertions and the phylogeny of Cercozoa and Rhizaria. Protist 156:149–161

Bass D, Chao EE, Nikolaev S, Yabuki A, Ishida K, Berney C, Pakzad U, Wylezich C, Cavalier-Smith T (2009) Phylogeny of novel naked filose and reticulose Cercozoa: Granofilosea cl. n. and Proteomyxidea revised. Protist 160:75–109

Blomfield JE, Schwartz EJ (1910) Some observations on the tumours on *Veronica chamaedrys* caused by *Sorosphaera veronicae*. Ann Bot 24:35–43

Braselton JP (1983) The plasmodiophoromycete parasitic on *Heteranthera dubia*. Can J Bot 61:45–52

Braselton JP (1989a) Karyotypic analysis of *Ligniera verrucosa* (Plasmodiophoromycetes). Can J Bot 67:1216–1218

Braselton JP (1989b) Karyotypic analysis of *Membranosorus heterantherae* (Plasmodiophoromycetes). Can J Bot 67:1219–1220

Braselton JP (1995) Current status of the plasmodiophorids. Crit Rev Microbiol 21:263–275

Braselton JP, Miller CE (1973) Centrioles in *Sorosphaera*. Mycologia 65:220–226

Braselton JP, Miller CE (1975) Host-parasite interface of *Veronica persica* and *Sorosphaera veronicae* (Plasmodiophoromycetes). Arch Microbiol 104:97–99

Braselton JP, Miller CE, Pechak DG (1975) The ultrastructure of cruciform nuclear division in *Sorosphaera veronicae* (plasmodiophoromycete). Am J Bot 62:349–358

Brodmann D, Schuller A, Ludwig-Müller J, Aeschbacher RA, Wiemken A, Boller T, Wingler A (2002) Induction of trehalase in *Arabidopsis* plants infected with the trehalose-producing pathogen *Plasmodiophora brassicae*. Mol Plant Microbe Interact 15:693–700

Buczacki ST (1977) Root infections from single resting spores of *Plasmodiophora brassicae*. Trans Br Mycol Soc 69:328–329

Buczacki ST, Clay CM (1984) Some observations on secondary zoospore development in *Plasmodiophora brassicae*. Trans Br Mycol Soc 82:339–342

Buczacki ST, Moxham SE (1980) Karyogamy in *Plasmodiophora brassicae*. Trans Br Mycol Soc 75:439–444

Buczacki ST, Moxham SE (1983) Structure of the resting spore wall of *Plasmodiophora brassicae* revealed by electron microscopy and chemical digestion. Trans Br Mycol Soc 81:221–231

Buhariwalla H, Mithen R (1995) Cloning of a *Brassica* repetitive DNA element from resting spores of *Plasmodiophora brassicae*. Physiol Mol Plant Pathol 47:95–101

Buhariwalla H, Greaves S, Magrath R, Mithen R (1995) Development of specific PCR primers for the amplification of polymorphic DNA from the obligate root pathogen *Plasmodiophora brassicae*. Physiol Mol Plant Pathol 47:83–94

Bulman SR, Kühn SF, Marshall JW, Schnepf E (2001) A phylogenetic analysis of the SSU rRNA from members of the Plasmodiophorida and Phagomyxida. Protist 152:43–51

Bulman S, Candy JM, Fiers M, Lister R, Conner AJ, Eady CC (2011) Genomics of biotrophic, plant-infecting plasmodiophorids using *in vitro* dual cultures. Protist 162:449–461

Burki F, Shalchian-Tabrizi K, Minge MA, Skjaeveland A, Nikolaev SI, Jakobsen KS, Pawlowski J (2007) Phylogenomics reshuffles the eukaryotic supergroups. PLoS One 2:e790

Burki F, Shalchian-Tabrizi K, Pawlowski J (2008) Phylogenomics reveals a new 'megagroup' including most photosynthetic eukaryotes. Biol Lett 4:366–369

Burki F, Kudryavtsev A, Matz MV, Aglyamova GV, Bulman S, Fiers M, Keeling PJ, Pawlowski J (2010) Evolution of Rhizaria: new insights from phylogenomic analysis of uncultivated protists. BMC Evol Biol 10:377–394

Castlebury LA, Domier LL (1998) Small subunit ribosomal RNA phylogeny of *Plasmodiophora brassicae*. Mycologia 90:102–107

Castlebury LA, Glawe DA (1993) A comparison of three techniques for inoculating Chinese cabbage with *Plasmodiophora brassicae*. Mycologia 85:866–867

Castlebury LA, Maddox JV, Glawe DA (1994) A technique for the extraction and purification of viable *Plasmodiophora brassicae* resting spores from host tissue. Mycologia 86:458–460

Cavalier-Smith T (1993) The protozoan phylum Opalozoa. J Eukaryot Microbiol 40:609–615

Cavalier-Smith T (2003) Protist phylogeny and the high-level classification of Protozoa. Eur J Protistol 39:338–348

Cavalier-Smith T, Chao EE (1997) Sarcomonad ribosomal RNA sequences, rhizopod phylogeny, and the origin of euglyphid amoebae. Arch Protistenkd 147:227–236

Cavalier-Smith T, Chao EE (2003) Phylogeny and classification of Phylum Cercozoa (Protozoa). Protist 154:341–358

Claxton JR, Potter UJ, Blakesley D, Clarkson JM (1996) An ultrastructural study of the interaction between *Spongospora subterranea* f. sp. *nasturtii* and watercress roots. Mycol Res 100:1431–1439

Clay CM, Walsh JA (1990) Ultrastructure of sporangial development in *Spongospora subterranea* f. sp. *nasturtii* infecting watercress. Mycol Res 94:463–471

Colhoun J (1957) A technique for examining soil for the presence of *Plasmodiophora brassicae* Woron. Ann Appl Biol 45:559–565

Cook WRI (1933) A monograph of the plasmodiophorales. Arch Protistenkd 80:179–254

Cook WRI, Schwartz EJ (1930) The life-history, cytology, and method of infection of *Plasmodiophora brassicae* Woron., the cause of finger and toe disease of cabbage and other crucifers. Philos Trans R Soc Lond B Biol Sci 218:283–314

Cooper JI, Asher MJC (eds) (1988) Viruses with fungal vectors. Association of Applied Biologists, Wellesbourne, Warwick

Crute IR, Buczacki ST, Stevenson K (1981) A solution culture method for observing the development of clubroot symptoms on young brassica plants. Ann Appl Biol 99:241–245

D'Ambra V, Mutto S (1977) The ultrastructure of *Polymyxa beta* zoospore exit-tube differentiation. Can J Bot 55:831–839

den Hartog C (1963) *Tetramyxa parasitica*, een gal op *Ruppia*. Gorteria 1:138–140

den Hartog C (1965) Some notes on the distribution of *Plasmodiophora diplantherae*, a parasitic fungus on species of *Halodule*. Persoonia 4:15–18

den Hartog C (1989) Distribution of *Plasmodiophora bicaudata*, a parasitic fungus on small *Zostera* species. Dis Aquat Org 6:227–229

Dick MW (2001) Straminipilous Fungi. Systematics of the peronosporomycetes including accounts of the marine straminipilous protists, the plasmodiophorids and similar organisms. Kluwer, Dordrecht

Dixon GR (2009) The occurrence and economic impact of *Plasmodiophora brassicae* and clubroot disease. J Plant Growth Regul 28:194–202

Dobson RL, Gabrielson RL (1983) Role of primary and secondary zoospores of *Plasmodiophora brassicae* in the development of clubroot in Chinese cabbage. Phytopathology 73:559–561

Dylewski DP (1987) *Woronina pythii*. In: Fuller MS, Jaworski A (eds) Zoosporic fungi in teaching and research. Southeastern Publishing Corporation, Athens, GA, pp 108–109

Dylewski DP (1989) Phylum Plasmodiophoromycota. In: Margulis L, Corliss JO, Melkonian M, Chapman DJ (eds) Handbook of protoctista. Jones and Bartlett, Boston, pp 399–416

Dylewski DP, Miller CE (1983) Cruciform nuclear division in *Woronina pythii* (Plasmodiophoromycetes). Am J Bot 70:1325–1339

Dylewski DP, Miller CE (1984) The ultrastructure of mitosis during sporangiogenesis in *Woronina pythii* (Plasmodiophoromycetes). Protoplasma 121:42–53

Dylewski DP, Braselton JP, Miller CE (1978) Cruciform nuclear division in *Sorosphaera veronicae*. Am J Bot 65:258–267

Faggian R, Strelkov SE (2009) Detection and measurement of *Plasmodiophora brassicae*. J Plant Growth Regul 28:282–288

Feng J, Hwang R, Hwang S-F, Strelkov SE, Gossen BD, Zhou Q-X, Peng G (2010) Molecular characterization of a serine protease Pro1 from *Plasmodiophora brassicae* that stimulates resting spore germination. Mol Plant Pathol 11:503–512

Forest HS, Horn CN, Marcus BA (1986) New observations of galls on *Zosterella dubia* (Jacq.) Small (*Heteranthera dubia* (Jacq.) MacM.) infected by *Membranosorus heterantherae* Ostenfeld and Peterson [sic]. Environ Exp Bot 26:377–384

Garber RC, Aist JR (1979a) The ultrastructure of meiosis in *Plasmodiophora brassicae* (Plasmodiophorales). Can J Bot 57:2509–2518

Garber RC, Aist JR (1979b) The ultrastructure of mitosis in *Plasmodiophora brassicae* (Plasmodiophorales). J Cell Sci 40:89–110

Hackett JD, Yoon HS, Li S, Reyes-Prieto A, Rümmele SE, Bhattacharya D (2007) Phylogenomic analysis supports the monophyly of cryptophytes and haptophytes and the association of rhizaria with chromalveolates. Mol Biol Evol 24:1702–1713

Harris SE, Braselton JP, Miller CE (1980) Chromosomal number of *Sorosphaera veronicae* (Plasmodiophoromycetes) based on ultrastructural analysis of synaptonemal complexes. Mycologia 72:916–925

Heim P (1955) Le noyau dans le cycle évolutif de *Plasmodiophora brassicae* Woron. Revue de Mycologie 20:131–157

Ingram DS (1969) Growth of *Plasmodiophora brassicae* in host callus. J Gen Microbiol 55:9–18

Ingram DS, Tommerup I (1972) The life history of *Plasmodiophora brassicae* Woron. Proc R Soc Lond B Biol Sci 180:103–112

Ito S-I, Yano S, Tanaka S, Kameya-Iwaki M (1994) The use of resting spore spheroplasts in the DNA analysis of *Plasmodiophora brassicae*. Ann Phytopathol Soc Jpn 60:491–495

Jones DR, Ingram DS, Dixon GR (1982) Characterization of isolates derived from single resting spores of *Plasmodiophora brassicae* and studies of their interaction. Plant Pathol 31:239–246

Kanyuka K, Ward E, Adams MJ (2003) *Polymyxa graminis* and the cereal viruses it transmits: a research challenge. Mol Plant Pathol 4:393–406

Karling JS (1968) The Plasmodiophorales, 2nd edn. Hafner, New York

Karling JS (1981) *Woronina leptolegniae* n. sp., a plasmodiophorid parasite of *Leptolegnia*. Nova Hedwigia 35:17–24

Keskin B (1971) Beitrag zur Protomitose bei *Polymyxa betae* Keskin. Arch Mikrobiol 77:344–348

Keskin B, Fuchs WH (1969) Der Infectionsvorgang bei *Polymyxa betae*. Arch Mikrobiol 68:218–226

Kole AP (1954) A contribution to the knowledge of *Spongospora subterranea* (Wallr.) Lagerh., the cause of powdery scab of potatoes. Tijdschrift over plantenziekten 60:1–65

Kole AP, Gielink AJ (1961) Electron microscope observations on the flagella of the zoosporangial zoospores of *Plasmodiophora brassicae* and *Spongospora subterranea*. Proc Acad Sci Amsterdam Ser C 64:157–161

Kole AP, Gielink AJ (1962) Electron microscope observations on the resting-spore germination of *Plasmodiophora brassicae*. Proc Acad Sci Amsterdam Ser C 65:117–121

Kole AP, Gielink AJ (1963) The significance of the zoosporangial stage in the life cycle of the Plasmodiophorales. Neth J Plant Pathol 69:258–262

Lahert H, Kavanagh JA (1985) The fine structure of the cystosorus of *Spongospora subterranea*, the cause of powdery scab of potato. Can J Bot 63:2278–2282

Ledingham GA (1934) Zoospore ciliation in the Plasmodiophorales. Nature 133:534

Ledingham GA (1935) Occurrence of zoosporangia in *Spongospora subterranea* (Wallroth) Lagerheim. Nature 135:395

Ledingham GA (1939) Studies on *Polymyxa graminis*, n. gen. n. sp., a plasmodiophoraceous root parasite of wheat. Can J Res C 17:40–51

Lees A, Graaf P, Wale S (2008) The identification and detection of *Spongospora subterranea* and factors affecting infection and disease. Am J Potato Res 85:247–252

Legrève A, Delfosse P, Maraite H (2002) Phylogenetic analysis of *Polymyxa* species based on nuclear 5.8S and internal transcribed spacers ribosomal DNA sequences. Mycol Res 106:138–147

Lesaulnier C, Papamichail D, McCorkle S, Ollivier B, Skiena S, Taghavi S, Zak D, van der Lelie D (2008) Elevated atmospheric CO_2 affects soil microbial diversity associated with trembling aspen. Environ Microbiol 10:926–941

Luther H (1949) Beobachtungen über *Tetramyxa parasitica* Goebel. Mem Soc Fauna Flora Fenn 25:88–96

Macfarlane I (1952) Factors affecting the survival of *Plasmodiophora brassicae* Wor. in the soil and its assessment by a host test. Ann Appl Biol 39:239–256

Macfarlane I (1958) A solution culture technique for obtaining root-hair, or primary infection by *Plasmodiophora brassicae*. J Gen Microbiol 18:720–732

Macfarlane I (1970) Germination of resting spores of *Plasmodiophora brassicae*. Trans Br Mycol Soc 55:97–112

Maier I, Parodi E, Westermeier R, Müller DG (2000) *Maullinia ectocarpii* gen. et sp. nov. (Plasmodiophorea), an intracellular parasite in *Ectocarpus siliculosus* (Ectocarpales, Phaeophyceae) and other filamentous brown algae. Protist 151:225–238

Maire R, Tison A (1909) La cytologie des Plasmodiophoracées et la classe des Phytomyxinae. Ann Mycol 7:226–253

Manzanares-Dauleux MJ, Divaret I, Baron F, Thomas G (2001) Assessment of biological and molecular variability between and within field isolates of *Plasmodiophora brassicae*. Plant Pathol 50:165–173

Martinsson K (1987) Gallbildningar på *Callitriche*. Svensk Bot Tidskr 81:334–336

Marziano F, Villari R, Tripodi G (1995) A plasmodiophorid fungal parasite of the seagrass *Halophila stipulacea*. Mycotaxon 60:165–170

McGrann GRD, Grimmer MK, Mutasa-Göttens ES, Stevens M (2009) Progress towards the understanding and control of sugar beet rhizomania disease. Mol Plant Pathol 10:129–141

Merz U (1997) Microscopical observations of the primary zoospores of *Spongospora subterranea* f.sp. *subterranea*. Plant Pathol 46:670–674

Merz U (2008) Powdery scab of potato—occurrence, life cycle and epidemiology. Am J Potato Res 85:241–246

Merz U, Falloon RE (2009) Review: powdery scab of potato—increased knowledge of pathogen biology and disease epidemiology for effective disease management. Potato Res 52:17–37

Miller CE (1958) Morphology and cytology of the zoosporangia and cystosori of *Sorosphaera veronicae*. J Elisha Mitchell Sci Soc 74:49–64

Miller CE (1959) Studies on the life cycle and taxonomy of *Ligniera verrucosa*. Am J Bot 46:725–729

Miller CE, Dylewski DP (1983) Zoosporic pathogens of lower plants—what can be learned from the likes of *Woronina*? In: Buczacki ST (ed) Zoosporic plant pathogens. Academic, London, pp 249–283

Miller CE, Martin RW, Dylewski DP (1985) The ultrastructure of plasmodia, sporangia, and cystosori of *Ligniera verrucosa* (Plasmodiophoromycetes). Can J Bot 63:263–273

Mithen R, Magrath R (1992) A contribution to the life history of *Plasmodiophora brassicae*:secondary plasmodia development in root galls of *Arabidopsis thaliana*. Mycol Res 96:877–885

Möller M, Harling R (1996) Randomly amplified polymorphic DNA (RAPD) profiling of *Plasmodiophora brassicae*. Lett Appl Microbiol 22:70–75

Moreira D, von der Heyden S, Bass D, López-García P, Chao E, Cavalier-Smith T (2007) Global eukaryote phylogeny: combined small- and large-subunit ribosomal DNA trees support monophyly of Rhizaria, Retaria and Excavata. Mol Phylogenet Evol 44:255–266

Moxham SE, Buczacki ST (1983) Chemical composition of the resting sore wall of *Plasmodiophora brassicae*. Trans Br Mycol Soc 80:297–304

Neuhauser S, Kirchmair M (2009) *Ligniera junci*, a plasmodiophorid re-discovered in roots of *Juncus* in Austria. Österr Z Pilzk 18:151–157

Neuhauser S, Kirchmair M (2011) *Sorosphaerula* nom. n. for the plasmodiophorid genus *Sorosphaera* J. Schröter 1886 (Rhizaria: Endomyxa: Phytomyxea: Plasmodiophorida). J Eukaryot Microbiol 58:469–470

Neuhauser S, Bulman S, Kirchmair M (2010) Plasmodiophorids: the challenge to understand soil-borne, obligate biotrophs with a multiphasic life cycle. In: Gherbawy Y, Voight K (eds) Molecular identification of fungi. Springer, Berlin, pp 51–78

Neuhauser S, Kirchmair M, Gleason FH (2011) The ecological potentials of Phytomyxea ("plasmodiophorids") in aquatic food webs. Hydrobiologia 659:23–35

Nikolaev SI, Berney C, Fahrni J, Bolivar I, Polet S, Mylnikov AP, Aleshin VV, Petrov NB, Pawlowski J (2004) The twilight of Heliozoa and rise of Rhizaria: an emerging supergroup of amoeboid eukaryotes. Proc Natl Acad Sci USA 101:8066–8071

Olive LS (1975) The mycetozoans. Academic, New York

Ostenfeld CH, Petersen HE (1930) On a new Plasmodiophoracea found in Canada. Z Bot 23:13–18

Palm BT, Burk M (1933) The taxonomy of the Plasmodiophoraceae. Arch Protistenkd 79:263–276

Qu XS, Christ BJ (2007) *In vitro* culture of the obligate parasite *Spongospora subterranea* (Cercozoa, Plasmodiophorida) associated with root-inducing transferred-DNA transformed potato hairy roots. J Eukaryot Microbiol 54:465–467

Rochon D, Kakani K, Robbins M, Reade R (2004) Molecular aspects of plant virus transmission by *Olpidium* and plasmodiophorid vectors. Annu Rev Phytopathol 42:211–241

Rodriguez-Ezpeleta N, Brinkmann H, Burger G, Roger AJ, Gray MW, Philippe H, Lang BF (2007) Toward resolving the eukaryotic tree: the phylogenetic positions of jakobids and cercozoans. Curr Biol 17:1420–1425

Schnepf E (1994) A *Phagomyxa*-like endoparasite of the centric marine diatom *Bellerochea malleus*: a phagotrophic Plasmodiophoromycete. Bot Acta 107:374–382

Schnepf E, Kühn SF, Bulman S (2000) *Phagomyxa bellerocheae* sp. nov. and *Phagomyxa odontellae* sp. nov., Plasmodiophoromycetes feeding on marine diatoms. Helgol Mar Res 54:237–241

Scott ES (1985) Production and characterization of single-spore isolates of *Plasmodiophora brassicae*. Plant Pathol 34:287–292

Siemens J, Bulman S, Rehn F, Sundelin T (2009) Molecular biology of *Plasmodiophora brassicae*. J Plant Growth Regul 28:245–251

Sparrow FK (1960) Aquatic phycomycetes, 2nd edn. University of Michigan Press, Ann Arbor, MI

Stoeck T, Hayward B, Taylor GT, Varela R, Epstein SS (2006) A multiple PCR-primer approach to access the microeukaryotic diversity in environmental samples. Protist 157:31–43

Tinggal SH, Webster J (1981) Technique for single spore infection by *Plasmodiophora brassicae*. Trans Br Mycol Soc 76:187–190

Tomlinson JA (1958) Crook root of watercress. III: the causal organism *Spongospora subterranea* (Wallr.) Lagerh. f. sp. *nasturtii* f. sp. nov. Trans Br Mycol Soc 41:491–498

Tommerup IC, Ingram DS (1971) The life-cycle of *Plasmodiophora brassicae* Woron. in *Brassica* tissue cultures and intact roots. New Phytol 70:327–332

Vaïanopoulos C, Bragard C, Moreau V, Maraite H, Legreve A (2007) Identification and quantification of *Polymyxa graminis* f. sp. *temperata* and *P. graminis* f. sp. *tepida* on barley and wheat. Plant Dis 91:857–864

van de Graaf P, Lees AK, Cullen DW, Duncan JM (2003) Detection and quantification of *Spongospora subterranea* in soil, water and plant tissue samples using real-time PCR. Eur J Plant Pathol 109:589–597

Voorrips RE (1996) Production, characterization and interaction of single-spore isolates of *Plasmodiophora brassicae*. Eur J Plant Pathol 102:377–383

Walker AC, Campbell J (2009) First records of the seagrass parasite *Plasmodiophora diplantherae* from the northcentral Gulf of Mexico. Gulf Caribb Res 21:63–65

Ward E, Kanyuka K, Motteram J, Kornyukhin D, Adams MJ (2005) The use of conventional and quantitative real-time PCR assays for *Polymyxa graminis* to examine host plant resistance, inoculum levels and intraspecific variation. New Phytol 165:875–885

Waterhouse G (1972) Plasmodiophoromycetes. In: Ainsworth GC, Sparrow FK, Sussman AS (eds) The Fungi, vol 4B. Academic, New York, pp 75–82

Webb PCR (1935) The cytology and life-history of *Sorosphaera veronicae*. Ann Bot 49:41–52

Wernham CC (1935) A species of *Sorodiscus* on *Heteranthera*. Mycologia 27:262–273

Williams PH (1973) Penetration and infection of cabbage roots by *Plasmodiophora brassicae*. Shokubutsu Byogai Kenkyu (Forsch Gebiet Pflanzenkrankh) Kyoto 8:133–146

Williams PH, McNabola SS (1970) Fine structure of the host-parasite interface of *Plasmodiophora brassicae* in cabbage. Phytopathology 60:1557–1561

Williams PH, Reddy MN, Strandberg JO (1969) Growth of noninfected and *Plasmodiophora brassicae* infected cabbage callus in culture. Can J Bot 47:1217–1221

Williams PH, Aist SJ, Aist JR (1971) Response of cabbage root hairs to infection by *Plasmodiophora brassicae*. Can J Bot 49:41–47

Yukawa Y, Tanaka S (1979) Scanning electron microscope observations on resting sporangia of *Plasmodiophora brassicae* in clubroot tissues after alcohol cracking. Can J Bot 57:2528–2532

Fungi

5 Microsporidia

Elizabeth S. Didier[1], James J. Becnel[2], Michael L. Kent[3,4], Justin L. Sanders[3], Louis M. Weiss[5]

CONTENTS

[1]Division of Microbiology, Tulane National Primate Research Center, 18703 Three Rivers Road, Covington, LA 70433, USA; e-mail: esdnda@tulane.edu
[2]USDA, ARS, CMAVE, 1600 S.W. 23rd Drive, Gainesville, FL 32608, USA; e-mail: james.becnel@ars.usda.gov
[3]Department of Microbiology, Oregon State University, 220 Nash Hall, Corvallis, OR 97331, USA; e-mail: michael.kent@oregonstate.edu; justin.sanders@oregonstate.edu
[4]Department of Biomedical Sciences, Oregon State University, Corvallis, OR 97331, USA; e-mail: michael.kent@oregonstate.edu
[5]Departments of Medicine and Pathology, Albert Einstein College of Medicine, 1300 Morris Park Avenue, Room 504 Forchheimer, Bronx, NY 10461, USA; e-mail: louis.weiss@einstein.yu.edu

I. Introduction

The phylum Microsporidia Balbiani 1882 (Weiser 1977) is comprised of a diverse group of over 1,400 species (Didier and Weiss 2006). These organisms are obligate intracellular pathogenic protists uniquely characterized by a **specialized invasion organelle, the polar tube,** through which the cytoplasm and nucleus of these organisms pass during the infection of their host cells. Long considered early branching eukaryotes classified with the Archezoa, the microsporidia are now considered fungi based on accumulated data and more sophisticated analyses (Hibbett et al. 2007; James et al. 2006). **Microsporidia infect commercially significant animals, such as honey bees, salmon, silkworms, farm animals, and companion pets, and are of medical importance because they cause emerging opportunistic infections in humans.** A wide range of animals that are less commercially relevant can also be infected with microsporidia and thus pose a risk as environmental reservoirs of infection (Santín and Fayer 2009a). As a consequence of studies to better characterize the tree of life, the microsporidia have undergone a major transition from placement with the earliest diverging eukaryotes to

Systematics and Evolution, 2nd Edition
The Mycota VII Part A
D.J. McLaughlin and J.W. Spatafora (Eds.)

now being classified with the deep-branching fungi, closely related to the zygomycetes (Hibbett et al. 2007; James et al. 2006; Keeling 2009; Lee et al. 2008, 2010b). Questions remain, however, about the true relationship between the microsporidia and fungi (Koestler and Ebersberger 2011). Particularly noteworthy is the **evolution of gene compaction and reduction** observed among the microsporidia (Corradi and Keeling 2009; Keeling 2009; Keeling et al. 2010; Lee et al. 2010a; Texier et al. 2010; Vossbrinck et al. 2004). In this regard, microsporidia are highly efficient parasites, to the point that at least one species, *Enterocytozoon bieneusi*, even lacks genes for core carbon metabolism and depends fully on host-cell ATP import (Keeling et al. 2010).

II. Occurrence and Distribution

Microsporidia infect hosts ranging from protists to invertebrates (mainly insects) and vertebrates (mainly fish and mammals, including humans). **The only extracellular stage that survives is the spore, which is relatively resistant to environmental stress.** Because of the wide host range and the environmentally resistant spore wall, it is not surprising that microsporidia exhibit a worldwide distribution and can be found in aqueous (fresh and salt water) and terrestrial environments.

A. Arthropod Hosts

An extensive knowledge base on microsporidia isolated from terrestrial arthropod hosts has been dominated by species from insects, but other host groups include, but are not restricted to, cestodes, trematodes, nematodes, oligochaetes, isopods, myriapods, arachnids, and anaplurans (Sprague 1977). One indicator of the importance of microsporidia isolated from this group is that **nearly half of the approximately 186 described genera shown in what follows are from terrestrial arthropod hosts, with the majority of these from insects** (Becnel and Andreadis 1999). Prior to classification as fungi, the genera of microsporidia were established by the International Code of Zoological Nomenclature, and new genera are listed below in boldface type (Sprague and Becnel 1999). A request to exclude microsporidia from the International Code of Botanical Nomenclature was submitted and accepted and is now part of the revised code (Redhead et al. 2009). As such, the microsporidia are considered fungi, but descriptions of species remain subject to the International Code of Zoological Nomenclature.

The most commonly encountered microsporidia from insects are from Lepidoptera (within the *Nosema/Vairimorpha* clade) and Diptera, including the well-known genera *Amblyospora* and *Parathelohania* (Becnel and Andreadis 1999). It is likely that the large number of genera and species described from Lepidoptera and Diptera is due to the extensive studies searching for biological control agents in these groups of pest insects. Microsporidia in insects and other terrestrial arthropods have been studied extensively where they are important as natural control factors, have potential as manipulated microbial control agents for pest species, and cause chronic infections in beneficial arthropods.

Microsporidia **as important natural pest control agents** have been most extensively studied in insect pests. *Nosema pyrausta* in the European corn borer, *Ostrinia nubilalis*, is one of the most important regulators of larval populations in the USA (Andreadis 1984; Siegel et al. 1986). Populations of an important forest pest, the spruce budworm, *Choristoneura fumiferana*, are suppressed by epizootics caused by *Nosema fumiferanae* (Wilson 1973, 1981). A more recent example of natural control is in the red imported fire ant, *Solenopsis invicta*, where *Kneallhazia* (syn. *Thelohania*) *solenopsae* has become widespread in US populations of this invasive species (Oi et al. 2004).

Perhaps the best example of a species of microsporidia used as a microbial pesticide was a program for the control of rangeland grasshoppers with *Paranosema* (syns. *Nosema* and *Antonospora*) *locustae* (Henry and Oma 1981). *P. locustae* infects over 90 species of grasshoppers in the family Acrididae (Brooks 1988) and was registered by the US Environmental Protection Agency in 1989 as a microbial insecticide. Whole-organism tech-

1. *Abelspora* Azevedo 1987
2. ***Aedispora*** Kilochitskii 1997
3. *Agglomerata* Larsson & Yan 1988
4. *Agmasoma* Hazard & Oldacre 1975
5. *Alfvenia* Larsson 1983
6. *Alloglugea* Paperna & Lainson 1995
7. ***Amazonspora*** Azevedo & Matos 2003
8. *Amblyospora* Hazard & Oldacre 1975
9. *Ameson* Sprague 1977
10. *Amphiacantha* Caullery & Mesnil 1914
11. *Amphiamblys* Caullery & Mesnil 1914
12. ***Andreanna*** Simakova, Vossbrinck & Andreadis 2008
13. ***Anisofilariata*** Tokarev, Voronin, Seliverstova, Dolgikh, Pavlova, Ignatieva & Issi 2010
14. *Anncaliia* Issi, Krylova & Nicolaeva 1993
15. ***Antonospora*** Fries, Paxton, Tengo, Slemenda, da Silva & Pieniazek 1999
16. *Auraspora* Weiser & Purrini 1980
17. *Bacillidium* Janda 1928
18. *Baculea* Loubes & Akbarieh 1978
19. ***Becnelia*** Tonka & Weiser 2000
20. *Berwaldia* Larsson 1981
21. ***Binucleata*** Refardt, Decaestecker, Johnson & Vávra 2008
22. *Binucleospora* Bronnvall & Larsson 1995
23. *Bohuslavia* Larsson 1985
24. *Brachiola* Cali, Takvorian & Weiss 1998
25. ***Bryonosema*** Canning, Refardt, Vossbrinck & Curry 2002
26. *Burenella* Jouvenaz & Hazard 1978
27. *Burkea* Sprague 1977
28. *Buxtehudea* Larsson 1980
29. *Campanulospora* Issi, Radischcheva & Dolzhenko 1983
30. *Canningia* Weiser, Wegensteiner & Zizka 1995
31. *Caudospora* Weiser 1946
32. *Caulleryetta* Dogiel 1922
33. *Chapmanium* Hazard & Oldacre 1975
34. *Chytridioides* Trégouboff 1913
35. *Chytridiopsis* Schneider 1884
36. *Ciliatosporidium* Foissner & Foissner 1995
37. *Coccospora* Kudo 1925
38. *Cougourdella* Hesse 1935
39. ***Crepidulospora*** Simakova, Pankova & Issi 2004
40. ***Crispospora*** Tokarev, Voronin, Seliverstova, Pavlova & Issi 2010
41. *Cristulospora* Khodzhaeva & Issi 1989
42. *Cryptosporina* Hazard & Oldacre 1975
43. ***Cucumispora*** Ovcharenko, Bacela, Wilkinson, Ironside, Rigaud & Wattier 2009
44. *Culicospora* Weiser 1977
45. *Culicosporella* Weiser 1977
46. *Cylindrospora* Issi & Voronin 1986
47. *Cystosporogenes* Canning, Barker, Nicholas & Page 1985
48. ***Dasyatispora*** Diamant, Goren, Yokeş, Galil, Klopman, Huchon, Szitenberg & Karhan 2010
49. ***Desmozoon*** Freeman & Sommerville 2009
50. *Desportesia* Issi & Voronin 1986
51. ***Dimeispora*** Simakova, Pankova & Issi 2004
52. *Duboscqia* Perez 1908

98. *Mariona* Stempell 1909
99. *Marssoniella* Lemmermann 1900
100. *Merocinta* Pell & Canning 1993
101. *Metchnikovella* Caullery & Mesnil 1897
102. *Microfilum* Faye, Toguebaye & Bouix 1991
103. *Microgemma* Ralphs & Matthews 1986
104. *Microsporidium* Balbiani 1884
105. *Microsporidyopsis* Schereschewsky 1925
106. *Mitoplistophora* Codreanu 1966
107. ***Mockfordia*** Sokolova, Sokolov & Carlton 2010
108. *Mrazekia* Léger & Hesse 1916
109. ***Myospora*** Stentiford, Bateman, Small, Moss, Shields, Reece & Tuck 2010
110. ***Myosporidium*** Baquero, Rubio, Moura, Pieniazek & Jordana 2005
111. *Myxocystis* Mrazek 1897
112. *Nadelspora* Olson, Tiekotter & Reno 1994
113. *Napamichum* Larsson 1990
114. *Nelliemelba* Larsson 1983
115. ***Neoflabelliforma*** Morris & Freeman 2010
116. *Neoperezia* Issi & Voronin 1979
117. *Neonosemoides* Faye, Toguebaye & Bouix 1996
118. *Nolleria* Beard, Butler & Becnel 1990
119. *Norlevinea* Vávra 1984
120. *Nosema* Naegeli 1857
121. *Nosemoides* Vinckier 1975
122. ***Novothelohania*** Andreadis, Simakova, Vossbrinck, Shepard & Yurchenko 2011
123. *Nucleospora* Docker, Kent & Devlin 1996
124. *Nudispora* Larsson 1990
125. *Octosporea* Flu 1911
126. ***Octotetrasporea*** Issi, Kadyrova, Pushkar, Khodzhaeva & Krylova 1990
127. *Oligosporidium* Codreanu-Bălcescu, Codreanu & Traciuc 1981
128. *Ordospora* Larsson, Ebert & Vávra 1997
129. *Ormieresia* Vivares, Bouix & Manier 1977
130. *Orthosomella* Canning, Wigley & Barker 1991
131. *Orthothelohania* Codreanu & Bălcescu-Codreanu 1974
132. *Ovavesicula* Andreadis & Hanula 1987
133. ***Ovipleistophora*** Pekkarinen, Lom & Nilsen 2002
134. ***Pankovaia*** Simakova, Tokarev & Issi 2009
135. ***Paraepiseptum*** Hyliš, Oborník, Nebesářová & Vávra 2007
136. ***Paranosema*** Sokolova, Dolgikh, Morzhina, Nassonova, Issi, Terry, Ironside, Smith & Vossbrinck 2003
137. ***Paranucleospora*** Nylund, Nylund, Watanabe, Arnesen & Karlsbakk 2010
138. ***Parapleistophora*** Issi, Kadyrova, Pushkar, Khodzhaeva & Krylova 1990
139. ***Parastempellia*** Issi, Kadyrova, Pushkar, Khodzhaeva & Krylova 1990
140. *Parathelohania* Codreanu 1966
141. ***Paratuzetia*** Poddubnaya, Tokarev & Issi 2006
142. *Pegmatheca* Hazard &

53. *Edhazardia* Becnel, Sprague & Fukuda 1989
54. *Encephalitozoon* Levaditi, Nicolau & Schoen 1923
55. *Endoreticulatus* Brooks, Becnel & Kennedy 1988
56. *Enterocytozoon* Desportes, Le Charpentier, Galian, Bernard, Cochand-Priollet, Lavergne, Ravisse & Modigliani 1985
57. ***Enterospora*** Stentiford, Bateman, Longshaw & Feist 2007
58. *Episeptum* Larsson 1986
59. ***Euplotespora*** Fokin, Giuseppe, Erra & Dini 2008
60. *Evlachovaia* Voronin & Issi 1986
61. ***Fibrillanosema*** Galbreath, Smith, Terry, Becnel & Dunn 2004
62. *Flabelliforma* Canning, Killick-Kendrick & Killick-Kendrick 1991
63. *Geusia* Rühl & Korn 1979
64. *Glugea* Thélohan 1891
65. *Glugoides* Larsson, Ebert, Vávra & Voronin 1996
66. *Golbergia* Weiser 1977
67. *Gurleya* Doflein 1898
68. *Gurleyides* Voronin 1986
69. ***Hamiltosporidium*** Haag, Larsson, Refardt & Ebert 2010
70. *Hazardia* Weiser 1977
71. *Helmichia* Larsson 1982
72. ***Hepatospora*** Stentiford, Bateman, Dubuffet, Chambers & Stone 2011
73. *Hessea* Ormières & Sprague 1973
74. *Heterosporis* Schubert 1969
75. *Heterovesicula* Lange, Macvean, Henry & Streett 1995
76. *Hirsutusporos* Batson 1983
77. *Holobispora* Voronin 1986
78. *Hrabyeia* Lom & Dyková 1990
79. *Hyalinocysta* Hazard & Oldacre 1975
80. *lchthyosporidium* Caullery & Mesnil 1905
81. *Inodosporus* Overstreet & Weidner 1974
82. *Intexta* Larsson, Steiner & Bjørnson 1997
83. ***Intrapredatorus*** Chen, Kuo & Wu 1998
84. *Issia* Weiser 1977
85. *Janacekia* Larsson 1983
86. *Jirovecia* Weiser 1977
87. *Jiroveciana* Larsson 1980
88. *Johenrea* Lange, Becnel, Razafindratiana, Przybyszewski & Razafindrafara 1996
89. ***Kabatana*** Lom, Dyková & Tonguthai, 2000
90. *Kinorhynchospora* Adrianov & Rybakov 1991
91. ***Kneallhazia*** Sokolova & Fuxa 2008
92. ***Krishtalia*** Kilochitskii 1997
93. *Lanatospora* Voronin 1986
94. *Larssonia* Vidtman & Sokolova 1994
95. *Larssoniella* Weiser & David 1997
96. ***Liebermannia*** Sokolova, Lange & Fuxa 2006
97. *Loma* Morrison & Sprague 1981

Oldacre 1975
143. *Perezia* Léger & Duboscq 1909
144. *Pernicivesicula* Bylén & Larsson 1994
145. *Pilosporella* Hazard & Oldacre 1975
146. *Pleistophora* Gurley 1893
147. *Pleistophoridium* Codreanu-Bălcescu & Codreanu 1982
148. *Polydispyrenia* Canning & Hazard 1982
149. ***Potaspora*** Casal, Matos, Teles-Grilo & Azevedo 2008
150. ***Pseudoloma*** Matthews, Brown, Larison, Bishop-Stewart, Rogers & Kent 2001
151. ***Pseudonosema*** Canning, Refardt, Vossbrinck, Okamura & Curry 2002
152. *Pseudopleistophora* Sprague 1977
153. *Pulcispora* Vedmed, Krylova & Issi 1991
154. *Pyrotheca* Hesse 1935
155. *Rectispora* Larsson 1990
156. *Resiomeria* Larsson 1986
157. *Ringueletium* Garcia 1990
158. ***Schroedera*** Morris & Adams 2002
159. *Scipionospora* Bylén & Larsson 1996
160. *Semenovaia* Voronin & Issi 1986
161. ***Senoma*** Simakova, Pankova, Tokarev & Issi 2005
162. *Septata* Cali, Kotler & Orenstein 1993
163. ***Simuliospora*** Khodzhaeva, Krylova & Issi 1990
164. *Spherospora* Garcia 1991
165. *Spiroglugea* Léger & Hesse 1924
166. *Spraguea* Weissenberg 1976
167. *Steinhausia* Sprague, Ormières & Manier 1972
168. *Stempellia* Léger & Hesse 1910
169. *Striatospora* Issi & Voronin 1986
170. *Systenostrema* Hazard & Oldacre 1975
171. *Tardivesicula* Larsson & Bylén 1992
172. *Telomyxa* Léger & Hesse 1910
173. *Tetramicra* Matthews & Mattews 1980
174. *Thelohania* Henneguy 1892
175. *Toxoglugea* Léger & Hesse 1924
176. *Toxospora* Voronin 1993
177. *Trachipleistophora* Hollister, Canning, Weidner, Field, Kench & Marriott 1996
178. *Trichoctosporea* Larsson 1994
179. *Trichoduboscqia* Léger 1926
180. ***Trichonosema*** Canning, Refardt, Vossbrinck, Okamura & Curry 2002
181. *Trichotuzetia* Vávra, Larsson & Baker 1997
182. *Tricornia* Pell & Canning 1992
183. ***Tubulinosema*** Franzen, Fischer, Schroeder, Schölmerich & Schneuwly 2005
184. *Tuzetia* Maurand, Fize, Fenwik & Michel 1971
185. *Unikaryon* Canning, Lai & Lie 1974
186. *Vairimorpha* Pilley 1976
187. *Vavraia* Weiser 1977
188. *Vittaforma* Silveira & Canning 1995
189. *Weiseria* Doby & Saguez 1964
190. *Wittmannia* Czaker 1997

nology was used to produce large numbers of *P. locustae* spores in grasshoppers that were formulated into bait and applied by air (Henry and Oma 1981). *P. locustae* does not cause rapid mortality but has a debilitating effect on the host that can have long-term control implications when introduced.

Microsporidia are perhaps best known because several prominent species are the causative agents of **chronic disease in beneficial insects such as silkworms (*Bombyx mori*) and honey bees (*Apis melifera*).** The first named species of microsporidia was *Nosema bombycis* from *B. mori* and was the subject of landmark studies by Louis Pasteur, who established this pathogen as the etiological agent of **"pébrine" or silkworm disease** (Pasteur 1870). Pasteur proved that *N. bombycis* was transmitted from adult to progeny via the egg (transovarial transmission) and by the ingestion of spores, and he developed preventive methods that saved the silkworm industry worldwide. Adult honey bees worldwide are afflicted by nosemosis, which has been caused historically by *Nosema apis*, and more recently the Asian species, *Nosema ceranae*, has been implicated as playing a major role (Chen et al. 2008). Interactions of *Nosema* spp. with other bee pathogens have been implicated in contributing to **colony collapse disorder** and declines in honey bee colonies worldwide (Ratnieks and Carreck 2010). In addition, numerous microsporidia are implicated in reducing the effectiveness of commercially produced biological control agents. A few select examples are *Nosema muscidifuracis*, which reduced the fitness of the muscoid fly parasitoid *Muscidifurax raptor* (Geden et al. 1995), and *Oligosporidium occidentalis* from the predatory mite *Metaseiulus occidentalis*, which has a negative impact on the overall fitness of this predator (Becnel et al. 2002).

B. Aquatic Hosts (Marine and Freshwater)

Microsporidia infect a broad range of aquatic organisms, including crustaceans and amphipods, and freshwater, saltwater, and anadromous fish. The impacts of microsporidian parasites on fish in aquaculture, wild populations, and research have been documented on several occasions (Lom and Dyková 1992; Shaw and Kent 1999), and microsporidian species belonging to some 18 genera have been described in fishes (Lom 2002; Lom and Nilsen 2003). Most of these infections seem to be chronic, with minimal host mortality. Infections by some species, however, can have a profound **economic impact on wild fish and aquaculture hosts** in terms of mortality and commercial quality of fish. Several of these microsporidia have been shown to impact fish either by directly killing the host or indirectly by reducing fecundity (Ramsay et al. 2009; Wiklund et al. 1996) or decreasing the commercial quality of farmed fish. For example, Becko disease in yellowtail is caused by the formation of cysts in skeletal muscle by the microsporidium *Kabatana seriolae* Egusa 1982.

With the rapid growth of aquaculture, microsporidian pathogens in fish have increased in importance, and three species of microsporidia are the main causes of **disease in seawater-reared salmon.** *Loma salmonae* results in high mortality of salmonids reared in freshwater hatcheries and in seawater netpens in North America and Europe due to chronic gill infections (Kent and Poppe 1998). *Nucleospora salmonis*, also a microsporidian parasite of the Chinook salmon *Oncorhynchus tshawytscha*, is unique in that it infects the host cell nucleus and results in lymphoblastosis and a leukemia-like condition in fish (Chilmonczyk et al. 1991). Another intranuclear microsporidium, *Paranucleospora theridion*, infects the rainbow trout, *Oncorhynchus mykiss*, and salmonids (e.g., Atlantic salmon, *Salmo salar*), causing up to 80 % mortality in Atlantic salmon farms in Norway (Nylund et al. 2010). Additionally, *P. theridion* can infect the salmon louse *L. salmonis*, providing a potential reservoir for this parasite.

Microsporidia are also common **pathogens of baitfish**. The shiner, *Notemigonus crysoleucas*, and fathead minnow, *Pimephales promelas*, are frequently infected by *Ovipleistophora ovariae*, which generally does not result in acute mortalities but significantly affects the fecundity of spawning fish. Additionally, due to the increased use of fish in research, infections by microsporidia can have a confounding impact on experimental results using such fish (Kent et al. 2011). Zebrafish, *Danio rerio*, are also affected by microsporidia, with the first report describing infection of the spinal cord in fish purchased from a pet store for use in toxicological studies (de Kinkelin 1980). After further characterization, this microsporidian was assigned to a new genus and species, *Pseudoloma neurophilia* (Matthews et al. 2001). *P. neurophilia* infections are widespread in laboratory facilities (Kent et al. 2011) and are generally characterized as chronic and occasionally associated with spinal deformities and emaciation. *Pleistophora hyphessobryconis*, a muscle-infecting microsporidian, has also been identified in laboratory populations of zebrafish (Sanders et al. 2010). Commonly known as *neon tetra disease* for its type host, *Paracheirodon innesi*, this parasite is a frequent problem in the aquarium trade, often resulting in considerable mortality in a wide range of fishes. *P. hyphessobryconis* has a remarkably broad host range, infecting some 20 species of fishes in 4 orders (Lom and Dyková 1992; Schäperclaus 1991; Steffens 1962). As with *P. neurophilia*, *P. hyphessobryconis* can be harbored by otherwise healthy-appearing fish that may show clinical signs of the infection only after experiencing immunosuppressive events. Clinical presentation of the disease includes massive infections of myocytes resulting in liquefactive necrosis of the muscle tissue that almost invariably leads to the death of the fish. This example highlights the importance of obtaining fish used in research from reputable sources and the potential for introducing a microsporidian with a broad host range to new or accidental hosts.

In contrast to *P. hyphessobryconis*, many other microsporidia of fish are host-specific, at least at the family or genus level. One example, *L. salmonae*, infects all species of Pacific salmon, *Onchorynchus* spp., but does not infect the Atlantic salmon, *S. salar*, based on results from experimental exposure of fish to *L. salmonae*-infected gill tissue (Shaw et al. 2000). Using polymerase chain reaction to monitor infection of intestine, heart, spleen, and gill tissues, experimental exposure to *L. salmonae* showed an aberrant progression in Atlantic salmon, *S. salar*, compared to that seen in the rainbow trout, *O. mykiss* (Sanchez et al. 2001). Parasite DNA was detected in all tissues tested until week 3, at which point, rather than progressing to the gills to complete the life cycle by forming mature spores, the parasite was apparently cleared. This illustrates an abortive life cycle by *L. salmonae* infection in a nonpermissive host whereby the parasite was able to invade certain tissues and proliferate to some extent but was unable to progress to sporogony.

C. Mammalian and Avian Hosts

Members of the genus *Encephalitozoon* are the most common microsporidia infecting mammals and birds. The type species of this genus, *Enc. cuniculi*, was first identified in rabbits with motor paralysis in 1922 (Wright and Craighead 1922) and was also the first microsporidian genome to be sequenced (Katinka et al. 2001). There are now several sequenced microsporidian genomes, and the data can be found at http://microsporidiadb.org/micro/. *Enc. cuniculi* has an extraordinarily wide host range among mammals, such as rodents, lagomorphs, canines, equines, nonhuman primates, and humans. *Enc. hellem* Didier et al. 1991 and *Enc.* (syn. *Septata*) *intestinalis* Cali et al. 1993 were later isolated and identified from AIDS patients (Cali et al. 1993; Didier and Weiss 2006; Didier et al. 1991). *Enc. intestinalis* is still considered more common in humans, whereas *Enc. hellem* is more common in birds with humans believed to be zoonotic hosts. Since the *Encephalitozoon* species are indistinguishable by light microscopy, reports of *Enc. cuniculi* in birds prior to the AIDS pandemic may actually have been due to *Enc. hellem* (Didier et al. 1998; Snowden and Logan 1999; Snowden et al. 2000). *Encephalitozoon* species may infect enteric sites and contribute to diarrhea, but they more typically cause systemic infections to persist over the life of the host unless treated with effective drugs (e.g., albendazole). **Disease occurs predominantly in immune-deficient hosts** (e.g., AIDS patients, organ transplant recipients undergoing immunosuppressive therapy) and occurs sporadically in immune-competent hosts (Kotler and Orenstein 1998; Weber et al. 2000).

The most prevalent microsporidian in humans, *Ent. bieneusi* Desportes et al. 1985, was first identified in an AIDS patient in Haiti and is primarily associated with persistent and self-limiting diarrhea in immune-deficient and immune-competent humans, respectively (Desportes et al. 1985; Didier and Weiss 2006). The host range of *Ent. bieneusi* seems to be far wider than first believed and now includes wild, farm, and companion pet animals (Santín and Fayer 2009a, b). In addition, *Ent. bieneusi* has been increasingly identified in avian hosts such as chickens, pigeons, falcons, and exotic birds (Graczyk et al. 2008; Haro et al. 2005; Muller et al. 2008; Reetz et al. 2002). Currently the genus *Enterocytozoon* contains only a single species, *Ent. bieneusi.*

It is possible, however, that this organism is a species complex, and as additional information is obtained, it may be split into separate species, as was done with *Cryptosporidium parvum*. It should also be appreciated that the family Enterocytozoonidae contains the genus *Nucleospora*, which has several species, including *N. salmonis*, previously named *Ent. salmonis.*

The pathogenesis of *Ent. bieneusi* infections in immune-competent humans and nonhuman hosts has not been well characterized. For example, it is unknown whether *Ent. bieneusi* persists in otherwise healthy people and reactivates under conditions of immune deficiency. Additional species of microsporidia less frequently identified in mammals and birds include *Vittaforma corneae*, *Trachipleistophora* spp., *Anncaliia algerae*, *Pleistophora ronneaifiei*, *Nosema ocularum*, and *Microsporidium* spp. (Didier and Weiss 2006).

III. Morphology of the Microsporidian Spore

A. General Description and Common Features

Microsporidian spores are generally small and vary from 1 to 20 μm in length (Fig. 5.1). Spores of most species of microsporidia are oval in shape but may also exhibit pyriform, spherical, or rod shapes. The spore wall provides resistance to environmental influences and allows for the increase in hydrostatic pressure that causes spore discharge (see below; Frixione et al. 1997). The spore wall is surrounded by a glycoproteinaceous electron-dense exospore and electron-lucent endospore composed primarily of chitin (Vavrá and Larsson 1999). Ultrastructural studies of the genus *Encephalitozoon* using transmission electron microscopy, freeze fracture, and deep etching demonstrated that **the exospore is very complex and consists of three layers: an outer spiny layer, an intermediate electron-lucent lamina endospore, and an inner fibrous layer** (Bigliardi et al. 1996). The endospore is observed as a space crossed by bridges connecting the exospore to the plasma membrane. It has been suggested that **chitin, a major component of the endospore**, comprises the fibrils forming the bridges across the endospore and is part of the fibrillar system of the exospore (Bigliardi et al. 1996; Erickson and Blanquet 1969; Prigneau et al. 2000; Vavrá 1976). It is possible to distinguish subcompartments within the spore wall using polyclonal antisera against partially purified microsporidial proteins. A glycine- and serine-rich 51-kDa protein named SWP1 is localized to the exospore in *Enc. cuniculi* (Bohne et al. 2000) and *Enc. intestinalis* (Hayman et al. 2001). The corresponding gene, *swp1*, has been identified in *Enc. cuniculi*, *Enc. hellem*, and *Enc. intestinalis* (Bohne et al. 2000; Hayman et al. 2001). SWP1 is absent in meronts (proliferating stages) and first seen in early sporonts (stages that differentiate into spores) at a time when organisms translocate from the periphery to the center of the parasitophorous vacuole (PV) (Bohne et al. 2000). A 150-kDa glycoprotein in the spore wall named SWP2 was identified in *Enc. intestinalis* (Hayman et al. 2001). In addition, a putative glycophosphatidylinositol (GPI)-anchored chitin deacetylase has been localized to the plasmalemma endospore interface. Using proteomic techniques, a new spore wall protein, SWP3/EnP2, corresponding to ECU01_1270, was identified and localized to the endospore (Peuvel-

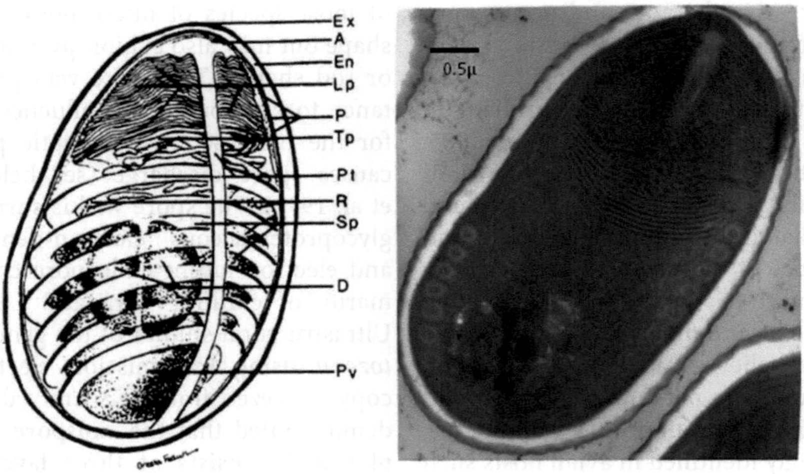

Fig. 5.1 Diagram of internal structure of a microsporidian spore (*left*) and a transmission electron micrograph of *Vavraia culicis floridensis* from *Aedes albopictus* (*right*). The spore coat has an outer electron-dense region called the exospore (Ex) and an inner thicker electron-lucent region known as the endospore (En). A unit membrane (P) separates the spore coat from the spore contents. The extrusion apparatus—anchoring disk (A), polar tubule (Pt), lamellar polaroplast (Lp), and tubular polaroplast (Tp)—dominates the spore contents and is diagnostic for microsporidian identification. The posterior vacuole (Pv) is a membrane-bound vesicle that sometimes contains a membrane whirl, a glomerular structure, flocculent material, or some combination of these structures. The spore cytoplasm is dense and contains ribosomes (R) in a tightly coiled helical array. The nucleation may consist of a single nucleus or pair of abutted nuclei, a diplokaryon (D). The size of the spore depends on the particular species and can vary from less than 1 μm to more than 10 μm. The number of polar tubule coils also varies from a few to 30 or more, again depending on the species observed. Reprinted with permission from Cali and Owen (1988)

Fanget et al. 2006; Xu et al. 2006). By immunoelectron microscopy this protein was found on the cell surface during sporogony and in the endospore in mature spores. SWP3 has several potential O-glycosylation sites and is likely a mannosylated protein like the major polar tube protein (PTP1). EnP1, corresponding to ECU01_0820, has also been localized to the endospore and demonstrated to be involved in adhesion of the spore to host cells (Peuvel-Fanget et al. 2006; Southern et al. 2007).

Under light microscopy, viable spores are refractile, and after histochemical staining (e.g., chromotrope, Gram), a **posterior vacuole** may be observed. The **unique structure that characterizes all microsporidia is the polar tube or filament** that coils within the spore and is part of the germination and infection apparatus (see below). The arrangement and number of coils of the polar filament within the spore vary among the microsporidia species. Long considered to be amitochondriate, the microsporidia have been found to possess **reduced mitochondria called mitosomes**, as well as **atypical Golgi** that lack the classical stacked dictyosome structure but instead are comprised of vesicular tubules that connect with the endoplasmic reticulum, plasma membrane, and developing polar tube. An **anchoring disk** with a membranous lamellar polaroplast is located at the anterior end of the spore and functions to anchor and fuel the extruding polar filament during germination (Keeling 2009; Vavrá and Larsson 1999). Ribosomes found along the endoplasmic reticulum are smaller than those of most eukaryotes (70S rather than 80S), being more similar to those of bacteria, and **the microsporidian nucleus may exist in a monokaryon or diplokaryon arrangement.**

B. Species (Spores) Infecting Arthropod Hosts

Studies on microsporidia in insects have been instrumental in establishing many aspects of microsporidian biology, but perhaps none more important than basic information about spores and spore types. For more than 100 years following the first microsporidian species *N. bombycis* from silk worms was named, the one spore/one species concept was almost universally accepted. Over 100 years later, Hazard and Weiser (1968) discovered that **some microsporidian species infecting mosquitoes were observed to exhibit spore dimorphism, where two spore types with distinctive morphology and function were formed over the course of the life cycle.** They reported that a binucleate spore formed in the adult female was responsible for transmitting pathogens to progeny. These studies revealed that in infected male larval progeny, uninucleate spores (meiospores) were produced, while spore development was delayed until pupation and adult emergence in infected female progeny. Binucleate spores of the original type were produced in these infected females to repeat the cycle. This proved that the two morphologically distinctive spores found in larvae and adult hosts (formerly believed to belong to two genera) may represent a single species. The means by which these microsporidia were transmitted horizontally remained a mystery until the discovery that meiospores formed in larvae were infectious to a copepod intermediate host (Sweeney et al. 1985). When ingested by mosquito larvae, the spores from the copepod intermediate host initiate a sequence of development that ends with binucleate spores in the adult female mosquito. Multiple spore types within the same species have also been documented for *K. solenopsae* infecting the fire ant *Solenopsis invicta* where four different spore types have been reported (Sokolova and Fuxa 2008), suggesting that this trait may be common in many genera.

C. Species (Spores) Infecting Aquatic Hosts

Spores of microsporidia infecting fish are generally spherical or ovoid to pyriform in shape and contain a sporoplasm that is either monokaryotic or diplokaryotic. **Most species form spores of relatively uniform size and shape, but highly variable spore sizes (macrospores and microspores) occur together in tissues of hosts infected with members of the genera *Heterosporis, Pleistophora,* and *Ichthyosporidium.*** The genus *Pleistophora* comprises numerous species that generally infect skeletal muscle of fish. Several of these species have been found to produce two and sometimes three spore types of different sizes (Canning et al. 1986). The type species of this genus, *Pleistophora typicalis*, produces elongate macrospores averaging 7.5 × 3.0 μm and, more commonly, microspores that are ovoid and average 4.4 × 2.3 μm. Spores of *Pleistophora* spp. generally have a large posterior vacuole, occupying over half the total spore volume.

Other microsporidia of fish produce polymorphic spores that vary by host and even tissue within the same host in which they develop. For example, the intranuclear microsporidian *P. theridion* develops spherical monokaryotic spores of 2.2–2.5 μm in the salmon louse, *Lepeophtheirus salmonis*. In the fish host, *P. theridion* produces spherical, diplokaryotic spores of 0.9–1.2 μm in diameter in reticuloendothelial cells and ovoid spores of 2.4–2.7 × 2.0–2.1 μm in gill and skin epithelial cells (Nylund et al. 2010). *Nucleospora salmonis*, another intranuclear microsporidian, develops similarly small, ovoid spores (1 × 2 μm), but other spore forms have not been seen for this organism (Chilmonczyk et al. 1991). Spores of *Glugea anomala* are elongate and oval, and spore size varies minimally in the same host species. However, there are some variations in sizes of spores taken from different hosts such as *Gasterosteus aculeatus* (3–6 × 1.9–1.7 μm) or *Pungitius pungitius* (3.5–5.1 × 1.9–2.6 μm), and here also the posterior vacuole is relatively large, taking up approximately half of the spore volume.

D. Species (Spores) Infecting Mammalian and Avian Hosts

Encephalitozoon spores measure approximately 1–2 × 2–4 μm and exhibit a typical microsporidian spore configuration of a glycoproteinaceous electron-dense exospore, electron-lucent endospore composed of chitin, and a plasma membrane containing the cytoplasmic organelles. The polar filament typically coils five to seven times in single row arrangement, and the nucleus is monokaryotic. Mature spores usually contain a prominent posterior vacuole that often is visible by light microscopy of histochemically stained organisms. Additional organelles include the membranous anterior anchoring disk, lamellar polaroplast with Golgi-like vesicles, endoplasmic reticulum, and ribosomes.

Spores of *Ent. bieneusi* are among the smallest of the microsporidia measuring 1 × 1.5 μm and the chitinous endospore in *Ent. bieneusi* is somewhat thinner than found in *Encephalitozoon* spores. The polar filament coils five to seven times and commonly is observed to align in two rows. A prominent posterior vacuole may be observed, and the nucleus is monokaryotic.

IV. Microsporidian Invasion Apparatus

The invasion apparatus of the microsporidia consists of a polar tube, also referred to as the polar filament prior to discharge (Lom and Vávra 1963; Takvorian and Cali 1986; Weidner 1972, 1976, 1982), that consists of two domains: an anterior straight region surrounded by a lamellar polaroplast that is attached to the inside of the anterior end of the spore by an anchoring disk and a posterior coiled region that forms from 4 to approximately 30 coils around the sporoplasm in the spore, depending on the species (Wittner and Weiss 1999). In cross section, the polar filament inside the spore is composed of electron-dense and electron-lucent concentric layers that can range from as few as 3 to as many as 20 different layers (Cali et al. 2002; Chioralia et al. 1998; Lom 1972; Sinden and Canning 1974; Vavrá 1976; Weidner 1972, 1976). **During germination the polar filament (tube) is discharged from the anterior of the spore and forms a hollow tube that remains attached to the spore and facilitates passage of its sporoplasm and nucleus (or diplokaryon) into its host cell (Frixione et al. 1992; Lom and Vávra 1963; Ohshima 1937; Walters 1958; Weidner 1972).** Electron microscopy has demonstrated elongated sporoplasm in sections of extruded polar tube and the piercing of host cell membranes by the polar tube (Lom 1972; Weidner 1976). This process serves as a unique mechanism of infection, resulting in sporoplasm transfer directly into the host cell cytoplasm (Frixione et al. 1992; Lom and Vávra 1963; Ohshima 1937; Weidner 1972). In *A. algerae*, polar tube discharge is associated with the appearance of membrane infoldings surrounding the polar tube (Cali et al. 2002). These ultrastructural observations suggest that the polar tube is actually extracytoplasmic in the spore and explains how the sporoplasm can remain intact during the explosive germination reaction.

Polar tubes range from 50 to 100 μm in length and 0.1 to 0.15 μm in diameter (Frixione et al. 1992). A germinated spore is shown in Fig. 5.2. The polar tube discharges from the anterior pole of the spore in an **explosive reaction occurring in less than 2 s** (Frixione et al. 1992; Lom and Vávra 1963; Ohshima 1937; Weidner 1972). Spore discharge occurs through phases of (1) activation, (2) increase in intrasporal osmotic pressure, (3) eversion of the polar tube, and (4) passage of sporoplasm through the polar tube. The exact mechanism of this process is not well understood. **Conditions that lead to spore germination vary widely among species, presumably reflecting the adaptation of each microsporidian to its host and external environment (Undeen and Epsky 1990; Wittner and Weiss 1999).** Since microsporidia are found in a wide range of terrestrial and aquatic hosts, different species may require unique activation conditions for spore discharge. These specific conditions are also probably important to prevent accidental discharge in the environment (Undeen and Avery 1988; Undeen and Epsky 1990). It has been theorized that, regardless of the mode of activation, microsporidia exhibit the same response to stimuli by increasing the intrasporal osmotic pressure (Lom and Vávra 1963; Ohshima 1937; Undeen and Frixione 1990, 1991). The increase in osmotic pressure results in an influx of water into the spore

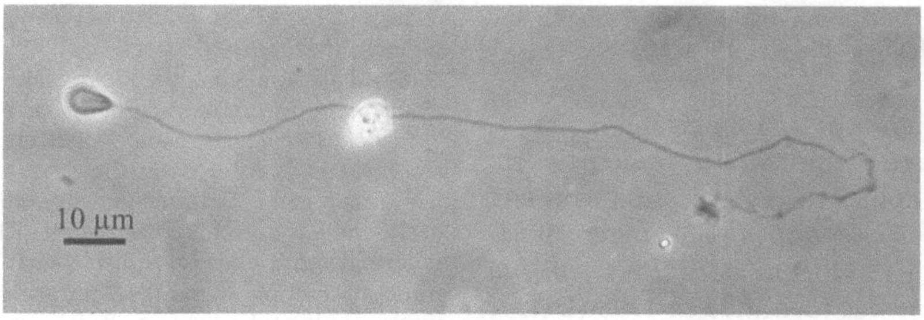

Fig. 5.2 Germinated spore of *Edhazardia aedis* from mosquito *Aedes aegypti*. The emptied spore (*left*) is shown attached to an extruded polar tube. The sporoplasm, or cytoplasmic contents of the spore, is propelled through the everting polar tube and is shown on the *right side* of the image. Under appropriate conditions, the sporoplasm is introduced into a new host cell to initiate infection. Note that the extruded polar filament is approximately 20 times the length of the spore

accompanied by swelling of the polaroplast and posterior vacuole prior to spore discharge (Frixione et al. 1992; Lom and Vávra 1963; Weidner and Byrd 1982). This pressure forces the eversion of the polar tube and expulsion of sporoplasm (Undeen 1990). In hyperosmotic solutions, polar tube discharge is inhibited or slowed down, and sporoplasm passage does not occur, thus providing indirect support for the osmotic pressure theory (Frixione et al. 1992; Lom and Vávra 1963; Ohshima 1937; Weidner 1976; Undeen and Frixione 1990).

The polar tube has flexibility, varies in diameter from 0.1 to 0.25 μm during discharge, can increase to 0.4 μm in diameter during sporoplasm passage, and shortens in length by 5–10 % after sporoplasm passage (Frixione et al. 1992; Lom and Vávra 1963; Ohshima 1937; Weidner 1972). The hollow discharged tubes appear to be two to three times as long as the dense, coiled tube inside the spore, and it has been suggested that the internal contents of the tube are incorporated at its growing tip during discharge (Frixione et al. 1992; Weidner 1972, 1976, 1982). The evagination of the polar filament has been likened to reversing a finger of a glove (Lom and Vávra 1963; Ohshima 1937; Weidner 1972, 1982; Weidner and Byrd 1982; Weidner et al. 1995). The polar tube is essentially a delivery mechanism for transversing the intestinal lumen to deliver the spore contents into intimate association with the host cell. It is not clear whether the polar tube pierces the host cell or invagination and internalization are driven by an interaction of the sporoplasm at the tip of the polar tube with host cell membranes. Although it is accepted that the sporoplasm flows through the discharged polar tube and into the host cell, the mechanisms of activation and tube formation during discharge remain unclear.

Studies have demonstrated that the polar tube has unusual solubility properties and resists dissociation in 1–3 % sodium dodecyl sulfate (SDS), 1 % Triton X-100, 1–10 % H_2O_2, 5-8 N H_2SO_4, 1-2 N HCl, chloroform, 1 % guanidine HCl, 0.1 M proteinase K, 8–10 M urea, 50 mM $NaCO_3$, and 50 mM $MgCl_2$ (Weidner 1972, 1976; Weidner and Byrd 1982). The polar tube, however, dissociates in various concentrations of 2-mercaptoethanol (2-ME) and dithiothreitol (DTT) (Keohane et al. 1994; Weidner 1972, 1976). This has allowed proteomic investigations of the composition of the polar tube (Ghosh et al. 2011). A procedure was developed for the isolation and purification of the major polar tube protein (PTP1) from the spores of microsporidia (Keohane et al. 1994; Keohane and Weiss 1998). Soluble polar tube preparation of *Glugea americanus*, *Enc. hellem*, *Enc. cuniculi*, *Enc. intestinalis*, and *A. algerae* were prepared by sequentially extracting glass-bead-disrupted spores with 1 % SDS and 9 M urea, followed by solubilization of the residual polar tubes in 2 % DTT (Keohane et al. 1994, 1999b; Weiss 2001). PTP1 in the DTT-solubilized material was then purified to homogeneity using reverse phase high-performance liquid chromatography (HPLC) (Keohane and Weiss 1998; Keohane et al. 1996a). By SDS-PAGE and silver staining, this purified fraction migrated at 43 kDa for *G. americanus*, 45 kDa for *Enc. cuniculi* and *Enc. intestinalis*, and 55 kDa for *Enc. hellem* (Keohane et al. 1999a, b). Monoclonal or polyclonal antibodies raised against the purified PTP1 demonstrated reactivity with polar tubes by immunofluorescence (IF) and immunogold electron microscopy (EM) and demonstrated cross reactivity among the species by immunoblotting and immunogold EM (Keohane and Weiss 1998; Keohane et al. 1994, 1996a, b, c).

All of the **major polar tube proteins (PTP1)** purified to date demonstrate similarities in mass, hydrophobicity, high proline content, and immunologic epitopes. The major polar tube protein, PTP1, from both *Enc. cuniculi* and *Enc. hellem* was identified in 1998 (Delbac et al. 1998a; Keohane et al. 1998). It is somewhat surprising that the translated proteins have only limited identity in amino acid sequences (Weiss 2001). Further comparisons, however, **strikingly reveal that these proteins are proline-rich and have a similar percentage of cysteine (Weiss** 2001**). PTP1 proteins have central amino acid repeat regions that are predominantly hydrophilic.** However, the repeats differ in composition and number. It is possible that this region is not important for the assembly of the polar tube and may function as an immunologic mask. In the process of evolution a similar duplication of internal sequences has been noted in malaria and other protozoan genes, and this mechanism may be operative in the microsporidia PTP gene (Rich and Ayala 2000). Analysis of PTP1 from other isolates of *Enc. hellem* supports this view, as the number of repeats is variable (Weiss 2001). Post-translational o-linked mannosylation occurs on PTP1, and this modification is probably involved in the ability of PTP1 to interact with the surface of host cells (Xu et al. 2004).

While PTP1 is the major component of the polar tube, other polar-tube-associated proteins (PTPs) are clearly present in the DTT-solubilized polar tube fraction.

For example, several putative PTPs of 23, 27, and 34 kDa have been identified in *G. americanus* using monoclonal antibodies produced to the DTT-solubilized polar tube (Keohane et al. 1994). Using two-dimensional SDS-PAGE one can also demonstrate other PTPs in DTT-solubilized *Enc. hellem* polar tube (Weiss 2001). In addition, several polyclonal and monoclonal antibodies have localized to the polar tube by IFA and immunogold EM and recognized proteins of 34, 75, and 170 kDa in *G. atherinae* and 35, 52/55, and 150 kDa in *Enc. cuniculi* and 60 and 120 kDa in *Enc. intestinalis* (Beckers et al. 1996; Delbac et al. 1998b). This resulted in the identification of PTP2, a 35-kDa protein, in *Enc. cuniculi* (Delbac et al. 2001). *Enc. cuniculi* PTP2 exists as a single copy per haploid genome and is located on the same chromosome as the *Ecptp1* gene, i.e., chromosome VI (Delbac et al. 2001), and has been found as a PTP1–PTP2 gene cluster in several other microsporidia (Delbac et al. 2001). By immunoscreening of a cDNA library of *Enc. cuniculi*, another polar tube protein, PTP3, was found (Peuvel et al. 2002). This protein, predicted to be synthesized as a 1,256-amino-acid precursor (136 kDa) with a cleavable signal peptide, is encoded by a single transcription unit (3,990 bp) located on chromosome XI of *Enc. cuniculi* (Peuvel et al. 2002). PTP3 is solubilized in the presence of SDS alone (Peuvel et al. 2002). Considering that PTP3 is extractable from *Enc. cuniculi* spores in the absence of thiol-reducing agent, lacks cysteine, but is rich in charged residues, it has been suggested that PTP3 interacts with PTP1 or PTP2 via ionic bonds and may play a role in the control of the conformational state of PTP1–PTP2 polymers (Peuvel et al. 2002). For example, when the polar tube exists as a coiled structure inside a spore, interactions with PTP3 may permit the maintenance of PTP1–PTP2 polymers in a condensed form (Peuvel et al. 2002).

It was found that DSP, a chemical cross linker that creates disulfide linkages between proteins, could mediate the purification of a large multimolecular complex from polar tubes that contained PTP1, PTP2, and PTP3 (Peuvel et al. 2002). Studies using yeast two hybrid vectors have confirmed the interaction of PTP1, PTP2, and PTP3 and determined that both the N-terminal and C-terminal regions of PTP1 are involved in these interactions, but that the central repeat region of PTP1 is not involved in these protein–protein interactions (Bouzahzah et al. 2010). It is likely that the regular multilayered organization of the microsporidian polar tube is dependent on specific interactions between its protein components.

V. Life Cycle

Microsporidia generally undergo three phases of development (Cali and Takvorian 1999). The **infective phase** occurs following the release of spores into the environment or tissues where, under suitable conditions, spores germinate and inject their spore contents through the everting polar filament to infect the host cell. Organisms then continue through a **proliferative phase** within the host cells (often referred to as *merogony*), which is followed by **sporogony, during which organisms commit to maturation and spore formation**. The modes of transmission, intracellular sites of development, number of proliferation cycles, and maturation vary widely among the species of microsporidia. Selected examples of the more common species infecting arthropod, aquatic, mammalian, and avian hosts are described in what follows.

A. Species Infecting Arthropod Hosts

The complete life cycles of many microsporidia in insects are well documented and have shown great diversity from the very simple to the complex, with some involving an intermediate host (Becnel et al. 2005). Microsporidia with **simple life cycles** are generally characterized as having a single sporulation sequence (sometimes with a second sporulation sequence) that occurs in a single host or host group. *Vavraia culicis* in mosquitoes is an example of a species that has only uninucleate stages throughout the life cycle and produces only one spore type (Vavrá and Becnel 2007). *Nosema apis* in honey bees has only binucleate (diplokaryotic) stages throughout the life cycle but is a bit more complex with the production of a primary (early) binucleate spore in the midgut epithelium that serves to spread the infection (autoinfection) to other midgut cells (de Graaf et al. 1994). These infections lead to the production of a second thick-walled (environmental) spore that can be released into the environment to infect a new host (Fries 1993).

Some species are characterized by complex life cycles involving multiple spore types responsible for horizontal and vertical transmission. They often affect two generations of the definitive host and some involve an obligate intermediate host. These microsporidia (often termed polymorphic or heterosporous) are generally very host-specific with complex developmental sequences that can be characterized by specialized stages and high levels of tissue specificity, as well. *Amblyospora californica* (Kellen and Lipa 1960) parasitizes the mosquito *Culex tarsalis* and is representative of a species with a complex life cycle that involves an intermediate host (Becnel 1992). Binucleate spores are formed in oenocytes of adult female *C. tarsalis* following a blood meal. These oenocytes invade the ovaries, where germinations occur, to infect the developing eggs. Developmental sequences in progeny are sex-dependent where females carry benign infection throughout larval development, which leads to the formation of binucleate spores in adults capable of initiating another round of transovarial transmission. Male progeny from infected adults undergo a distinctively different development where the pathogen invades fat bodies with rapid vegetative reproduction that terminates with meiosis and the production of meiospores. These male larvae die, releasing massive numbers of spores into the larval habitat. Meiospores are not infectious to mosquito larvae, but they are horizontally transmitted when ingested by females of the copepod intermediate host *Macrocyclops albidus*. Uninucleate stages replicate in ovaries of the female copepods, which terminates with the production of uninucleate spores and the death of the host. These spores are infectious when ingested by *C. tarsalis* mosquito larvae where the uninucleate stages invade larval oenocytes and remain dormant until pupation and adult emergence. In female adult mosquitoes, binucleate spores that are responsible for transovarial transmission are produced to complete the life cycle. **To date, it has been determined that the involvement of an intermediate host in the life cycles of microsporidia in insects is restricted to mosquitoes and copepod intermediate hosts** and has been documented in a number of species and genera such as *Amblyospora dyxenoides* (Sweeney et al. 1985), *Amblyospora connecticus* (Andreadis 1988), *Parathelohania anophelis* (Avery and Undeen 1990), *Culicospora magna* (Becnel et al. 1987), *Hyalinocysta chapmani* (Andreadis 2002), and *Edhazardia aedis* (Becnel et al. 1989).

B. Species Infecting Aquatic Hosts

Several hundred described species of microsporidia infect fish, but the **life cycles of only a few have been described.** Many of the fish microsporidia that have been investigated to date can **be transmitted directly by ingestion of free spores or spores from infected tissues** (Baxa-Antonio et al. 1992; Kent and Bishop-Stewart 2003; Kent and Speare 2005; McVicar 1975; Sanders et al. 2010; Weissenberg 1968). **Autoinfection** has been suggested or demonstrated for some *Loma* species, in which spores within tissues establish new infections (Matos et al. 2003; Rodriguez-Tovar et al. 2003; Shaw et al. 1998). The potential for **maternal transmission,** either transovum or transovarial, has been reported for *L. salmonae* (Sanchez et al. 2001) and *P. neurophilia* (Kent and Bishop-Stewart 2003). Phelps and Goodwin (2008) provided the most conclusive evidence for vertical transmission, showing the presence of the *Ovipleistophora ovariae* DNA by polymerase chain reaction from within the eggs of infected golden shiners, and similar results were obtained with *P. neurophilia* (Sanders and Kent 2011).

The sequential development from early infection to the site of sporulation is poorly understood for almost all microsporidia of fish. The development of *L. salmonae* has been elucidated and serves as an example for the development of microsporidia in fish (Kent and Speare 2005). The initial site of infection by *L. salmonae* is the mucosal epithelium of the stomach and intestine. Within 4 h of exposure, spores are seen in close association with the stomach epithelium and parasite DNA is present in the cytoplasm of epithelial cells and lamina propria of the small intestine by 12 h post exposure (Sprague and Hussey 1980). Two days post exposure, infected cells with dividing stages of *L. salmonae* can be visualized in the endocardium of the heart by in situ hybridization. After 2–3 weeks, uninucleate or binucleate merogonial stages can be seen developing within the endothelial cells or pillar cells of the blood vessels in the gills, the primary site of sporulation. Meronts are located at the periphery of the host cytoplasm, and the parasite (meront) cell membrane is in close proximity with the surrounding host cell membrane.

The sporogonic stages of *L. salmonae* occur in hypertrophic host cells of gills to generate a xenoparasitic complex or xenoma. **Xenomas are host cells with a radically altered structure in which the microsporidia have integrated into the host cell cytoplasm to undergo massive proliferation while isolated from the body's defense mechanisms (Lom and Dyková 2005).** Other xenoma-forming species, such as members of the genus *Glugea*, can produce very large xenomas (up to 3 mm) in many organs, especially in the subepithelium of the intestine, resulting in grossly visible tumorlike structures that are derived from a single hypertrophic host cell. Another type of xenoma develops with infection by *Ichthyosporidium giganteum*, which forms a large syncytium from the coalescence of several host cells, resulting in a large, lobular cyst (Rodriguez-Tovar et al. 2003). **In contrast to xenoma-forming species, skeletal-muscle-infecting microsporidia of the genus *Pleistophora* develop within host cells, replacing the sarcoplasm and destroying infected cells without inducing the hypertrophy characteristic of xenomas.**

Early during the first 2–3 weeks of xenoma formation, meronts of *L. salmonae* occupy the periphery of the host cell, and by 5 weeks this area begins to be occupied by mature spores (Lom and Dyková 2005). Eventually, spores can be seen throughout the xenoma that can reach a diameter of up to 0.4 mm. With other genera, sporogony occurs asynchronously throughout the xenoma (Morrison and Sprague 1983). A parasite-derived sporophorous vesicle forms prior to sporogonial division, and sporogony proceeds by binary fission, resulting in two uninucleate sporoblasts per vacuole (Chilmonczyk et al. 1991). The formation of this vacuole is absent in the genus *Spraguea*, which instead develops in direct contact with the host cytoplasm. The intranuclear microsporidia *N. salmonis* (Rodriguez-Tovar et al. 2003) and *P. theridion* (Nylund et al. 2010) develop in direct contact with the host cell nucleoplasm. Eventually, sporoblasts develop into mature spores. **The intact xenoma is surrounded by numerous inflammatory cells but**

seems to elicit little, if any, response by those cells (Rodriguez-Tovar et al. 2003). Eventual rupture of the xenoma results in the release of mature spores, elicitation of a severe proliferative inflammatory reaction, and uptake of spores by infiltrating phagocytes.

C. Species Infecting Mammalian and Avian Hosts

The most common modes of transmission of *Encephalitozoon* species in mammals, and presumably birds, are by **ingestion and inhalation of spores shed from urine, feces, or other fluids.** Transmission may occur through **direct contact (e.g., trauma), and vector-borne, sexual, and horizontal routes** have also been reported to occur in mammals. The life cycle of *Encephalitozoon* is relatively simple in comparison to that of other microsporidia. After germination and introduction of the spore cytoplasmic contents into the host cell, *Encephalitozoon* species undergo multiple cycles of binary division within a PV, the membrane of which seems to be host-cell-derived (Rönnebäumer et al. 2008). The proliferative stages or meronts tend to be larger than the mature spore and appear to adhere to the inner PV membrane. In some cases, karyokinesis occurs slightly faster than cytokinesis such that ribbons of dividing multinucleated meronts can be observed. Sporogony or spore maturation occurs as the parasite plasma membrane thickens and differentiates to form the exospore and endospore layers. These stages separate from the PV membrane and may continue to undergo a limited number of cell divisions. During this phase, the polar filament develops and the organisms become smaller and more electron-dense. Eventually, the PV becomes full of organisms leading to host cell and PV rupture and release of organisms. Among the sites of infection in mammals and birds are kidney, small intestine, and liver, so spores are commonly shed with urine and feces.

Ent. bieneusi infections typically occur in cells lining the small intestine in which organisms replicate in direct contact with the host cell cytoplasm (Cali and Owen 1990). Merogony is characterized by nuclear division without cytokinesis to generate a multinucleated plasmodium. During sporogony, electron-dense disks are observed to stack and eventually fuse to form the polar filaments in association with each nucleus. The individual nuclei become more defined, and the plasmalemma of the plasmodium begins to thicken and invaginate to surround the individual nucleus and polar filament units. Maturation continues with the thickening and differentiation of the spore wall, release of spores into the intestinal lumen, and shedding with feces.

VI. Systematics and Evolution

Microsporidia possess prokaryote-sized 70S ribosomes and lack typical mitochondria and Golgi. Early molecular biology studies also demonstrated fusion of the 5.8S and large subunit rRNAs similar to that in prokaryotes. These observations and the initial phylogenetic analyses of microsporidian small subunit rRNA genes supported divergence of the microsporidia prior to the symbiotic origin of mitochondria and placed microsporidia at the earliest and deepest branch of the eukaryotic tree (Vossbrinck et al. 1987). Evidence began to mount, however, that microsporidia are more highly evolved. Nuclear-encoded genes that target mitochondrial proteins (e.g., mHSP70, alpha and beta subunits of pyruvate dehyrdrogenase E1) were discovered and antibodies to mHSP70 identified membrane-bound organelles called mitosomes that function as mitochondrial remnants for iron–sulfur cluster assembly (Fast and Keeling 2001; Germot et al. 1997; Hirt et al. 1997; Williams and Keeling 2005; Williams et al. 2002). Genome-sequence studies of several microsporidia species (e.g., *Enc. cuniculi*, *Enc. intestinalis*, *Ent. bieneusi*, *P. locustae*, *A. algerae*) and improved phylogenetic analyses on additional genes have shed further light in demonstrating a close relationship between the microsporidia and the fungi (Akiyoshi et al. 2009; Burri et al. 2006; Cornman et al. 2009; Corradi and Slamovits 2011; Corradi et al. 2007; Katinka et al. 2001; Keeling et al.

2010; Williams et al. 2008). Microsporidia are highly efficient parasites and have undergone significant gene reduction and compaction (Keeling 2009; Keeling et al. 2010). In addition, microsporidia seem to have **evolved relatively quickly and exhibit a high degree of gene sequence divergence**, so phylogenetic analyses to address their evolution are problematic. However, a strong conservation of gene order, or synteny, among several gene clusters, including the *sex* locus, was reported among distantly related microsporidia and the zygomycetes (zygomycete mating type, *MAT*) (Dyer 2008; Lee et al. 2008, 2010b; Corradi and Keeling 2009; Corradi and Slamovits 2011). These findings were used to support a deep-branching fungal origin of the microsporidia from a zygomycete ancestor and suggest that microsporidia may have a genetically controlled sexual cycle. Concerns have been raised, however, about whether this shared syntenic relationship truly supports a microsporidial–fungal relationship because the gene cluster of the microsporidia that resemble the zygomycete sex-related loci traces back to an ancient gene cluster in the common ancestor of plants, animals, and fungi (Koestler and Ebersberger 2011).

VII. Classification

Newer molecular biology analyses and approaches being applied to better understanding evolution of the microsporidia also have impacted their taxonomy and classification. The microsporidia are now fairly well accepted for classification with Kingdom Fungi (Corradi and Keeling 2009; Hibbett et al. 2007; James et al. 2006), but some analyses question this association (Koestler and Ebersberger 2011), suggesting that further studies are required. Classification of fungi is based on the International Code of Botanical Nomenclature, but microsporidia had been described using the International Code of Zoological Nomenclature. To avoid nullification, a formal request to accept the current nomenclature of the microsporidia was presented at the last taxonomy meeting at the International Botanical Congress and was approved (Redhead et al. 2009). As a result, **microsporidia are considered fungi but remain subject to the rules of the International Code for Zoological Nomenclature.** A broad-based consensus classification (Hibbett et al. 2007) did not subdivide the microsporidia within the fungi due to a lack of well-sampled multilocus analyses at that time. More recently, gene order (i.e., synteny) between several unrelated microsporidia and the zygomycetes was highly conserved (Corradi and Keeling 2009; Dyer 2008; Lee et al. 2008), but again, others suggest that synteny was not more similar between microsporidia and the zygomycetes than with any other fungal taxon (Koestler and Ebersberger 2011). The phylum name Zygomycota is considered invalid because the interrelationships among the major clades are still unresolved, and it was named without a Latin description, so further classification of the basal fungi to relate microsporidia to the zygomycetes or another fungal group is still in progress (Hibbett et al. 2007). Based on the complexity of microsporidian evolution, it is also possible that the microsporidia might represent a sister group to the fungi. As newer analytical tools incorporate additional genomic and proteomic information, a better picture will emerge regarding the classification of the microsporidia.

Primary classification of organisms into the phylum Microsporidia was based on the presence of the polar tube. Further classification was based on morphological and ultrastructural features, as well as host and habitat (Larsson 1986; Sprague et al. 1992). More specific characteristics used to classify the microsporidia include host cell, spore size, nucleus configuration (i.e., monokaryon, diplokaryon), number and configuration of the polar filament coils, type of nuclear and cellular division (e.g., binary division, plasmotomy), interface with the host cell (e.g., replication within a PV, direct contact with host cell cytoplasm), and whether a sporophorous vesicle is formed. Microsporidia initially fell into two groups based on the presence or absence of a sporoblast vesicle (Pansporoblastina and Apansporoblastina, respectively) and then were divided into groups based on

nuclear configuration as single (Haplophasea) or double (Dihaplophasea) nuclear arrangement, the latter being grouped on the basis of diplokaryon formation through meiosis or nuclear dissociation.

With the advancement of molecular biology technology, classifications within the phylum incorporated phylogenetic analyses (Vossbrinck and Debrunner-Vossbrinck 2005), and genera accepted to date are found in Sect. II.A. A comparative molecular phylogenetic analysis using ssrDNA sequences of 125 species in relation to host and habitat led to a proposal for grouping microsporidia into five clades among three new classes: the Aquasporidia (clades I, II, and V), the Marinosporidia (clade III), and the Terresporidia (clade IV). This new classification, however, is considered to be under development due to the relatively small representation for analyzing only 125 of over 1,200 species of microsporidia, the as yet undescribed microsporidia that are likely to be found, and a need to account for features related to morphology, life cycle, and host–parasite relationship (Larsson 2005; Vossbrinck and Debrunner-Vossbrinck 2005).

VIII. Maintenance and Culture

A. Species Infecting Arthropod Hosts

Brooks (1988) presents an excellent review of spore storage and maintenance of microsporidia infecting arthropods, but the optimal storage conditions must be determined experimentally for each isolate. **There are no standard guidelines on the best practices to preserve spore viability.** There is general agreement, however, that many microsporidian spores from terrestrial hosts will tolerate freezing or desiccation, whereas spores from aquatic hosts do not but in some cases can be maintained long term under other conditions. Spores to be stored are most commonly handled as **intact infected cadavers or as purified suspensions.**

Many species of microsporidia can be maintained for extended periods (months to years) as highly purified spores held in deionized water at 5 °C (±3). Antibiotics and fungicides are routinely added to the suspensions to retard microbial growth, which can reduce spore viability. Highly purified spores of *A. algerae* cannot survive freezing but have maintained viability after being held at 5 °C (±3) for more than 10 years. Many terrestrial species of microsporidia can be frozen (–30 to –20 °C) as cadavers, or purified spores can be placed into liquid nitrogen for long-term storage. The addition of 50 % glycerol to the pure spore suspensions as a cryoprotectant is often beneficial. Some spores can also be stored in the dried host cadaver for extended periods, such as *Nosema whitei*, a pathogen of flour beetles. Spores that can be dried can often be successfully lyophilized. If information on storage parameters for a species is not available, it is suggested that highly purified spores be held in deionized water at 5 °C (±3).

In vitro culture of microsporidia in insects has a long history and began with the successful infection of a *B. mori* cell line with *N. bombycis* Trager 1937, but few additional species have been established in cell culture. Until the mid to late 1980s, only about eight species of microsporidia from insects had been cultured in insect cell systems that included *A. algerae*, *N. apis*, *N. bombycis*, *N. disstriae*, *N. heliothidis*, *N. mesnili*, *Vairimorpha necatrix*, and *Vavraia culicis* (Brooks 1988; Jaronski 1984). More recently, a few additional species have been cultured, including *Cystosporogenes operophterae*, *N. furnacalis*, and a *Vairimorpha* sp. (Becnel and Andreadis 1999). The species with the broadest host range and ability to grow in both invertebrate and vertebrate cell lines is *A. algerae*. It has been grown in many insect cell lines (Brooks 1988) and in pig kidney cells (Undeen 1975), rabbit kidney cells (Lowman et al. 2000), several warm-water fish cell lines (Monaghan et al. 2011), and human muscle fibroblasts (Trammer et al. 1999). *A. algerae* has also been grown at elevated temperatures (37 °C), which is unique for insect microsporidia (Lowman et al. 2000).

B. Species Infecting Aquatic Hosts

Few microsporidia that infect aquatic organisms have been successfully propagated in long-term cell culture, and these generally depend upon maintaining groups of infected hosts or obtaining infected hosts from the wild. **The presence of microsporidia in non-mammalian model organisms, such as the zebrafish and the nematode *Caenorhabditis elegans*, provides researchers the opportunity to study these parasites in well-described systems with numerous genetic tools (Troemel**

2011) to elucidate host responses to microsporidia infections. The in vitro **propagation of microsporidia infecting fish has proven difficult,** and the use of fish cell cultures in the long-term maintenance of fish microsporidia was recently reviewed (Monaghan et al. 2009).

For example, the intranuclear microsporidium *N. salmonis* has been successfully maintained in a long-term primary culture of salmonid mononuclear leukocytes grown in supplemented Iscove's modified Dulbecco's medium by adding small numbers of infected leukocytes to uninfected leukocytes (Wongtavatchai et al. 1994). Infected cultures can be preserved long term by freezing in liquid nitrogen with cryoprotectant.

A **continuous cell line, EP-1,** derived from the Japanese eel, *Anguilla japonica*, is persistently infected with *Heterosporis anguillarum*. Whereas this cell line was passaged over 223 times in vitro for maintaining intracellular merogonic stages of the parasite, no spore stages were observed to develop, yet eels inoculated with cells from this culture system became infected and exhibited intramuscular cysts consistent with *H. anguillarum* infection (Kou et al. 1995). To date, this **remains the only cell line developed to be persistently infected with a microsporidian parasite of fish.**

Four fish cell lines—channel catfish ovary, zebrafish caudal fin fibroblast, carp epithelioma, and fathead minnow—have been shown to support limited growth of the microsporidian parasite of zebrafish, *P. neurophilia*. Whereas sporogony occurs in all cell lines, development to the spore stages is limited, and the parasites could not be passaged into new cultures (Watral et al. 2006). Similarly, spores of *Glugea* spp. were internalized by Chinook salmon embryo cells and, while meronts were detected, development ceased by 48 h and no sporogony was observed (Lores et al. 2003). The same parasite did develop in a mosquito cell line (ECACC 90100401), producing spores within 72 h post inoculation, illustrating the potential for insect cell lines in the propagation of fish microsporidia in vitro.

C. Species Infecting Mammalian and Avian Hosts

Enc. cuniculi was the first mammalian microsporidian to be isolated from a rabbit and grown in long-term tissue culture (Shadduck 1969). Since then, *Enc. hellem*, *Enc. intestinalis*, *A. algerae*, *V. corneae*, and *Trachipleistophora hominis* isolates from humans have been grown in culture, but unfortunately long-term culture of *Ent. bieneusi* still has not been accomplished (Braunfuchsová et al. 1999; Didier et al. 1991, 1996; Juarez et al. 2005; Lafranchi-Tristem et al. 2001; Monaghan et al. 2009; Trammer et al. 1999; Visvesvara 2002).

Cultures are typically initiated via coculture of source specimen (tissue biopsy or fluids such as urine, feces, or sputum) and host cells such as Vero, RK-13, MDCK, and other epithelial cells. Examples of tissue culture media that support the growth of the host cells and facilitate propagation of the microsporidia include RPMI 1604 or D-MEM supplemented with 2 mML-glutamine, 5–10 % fetal bovine serum, and antibiotics (e.g., penicillin, streptomycin, and amphotericin B). The medium is typically changed twice a week and the supernatants can be collected in sterile bottles for short-term storage at 4 °C.

Encephalitozoon-infected cells appear to contain vacuoles filled with organisms. *V. corneae* replicates in the cytoplasm of the host cell, and infected cells may appear larger and multinucleated when filled with organisms. Individual microsporidia suspended in the supernatants after release from ruptured host cells can be observed approximately 2–4 weeks after initiation of coculture, but sometimes longer periods of time are required if the initial inoculum dose of organisms is low. In the case of *V. corneae*, large aggregates of parasite-laden host cells are often also observed in the culture supernatants and can be separated by vortexing or washing the collected culture supernatants. Host cells tend to replicate and replace the ruptured infected cells, but if overgrowth of microsporidia occurs, fresh host cells can be added to the culture flasks.

To enrich microsporidia from host cell debris, the sedimented culture supernatants can be washed sequentially in distilled water, tris-buffered saline (TBS)\ containing 0.3 % Tween 20 (TBS-TW), and TBS (400× *g* for 15 min). The pellets can be further enriched by centrifugation through 50 % Percoll (i.e., mixing equal volumes of spores in TBS and 100 % Percoll) at 400×*g* for 30–45 min. Extraneous host cell debris remains in the top layers, and the spores centrifuge to the pellet (Didier et al. 1996). Microsporidia to be used for extracting DNA or RNA require additional washing with mild ionic detergent (e.g., 0.5–1 % sodium dodecyl sulfate) to remove host cell DNA that can adhere to the spore surface (Corradi et al. 2010).

Cryopreservation of mammalian microsporidia can be accomplished most efficiently by "scraping" or trypsinizing infected host cells, centrifuging at 400×*g* (15 min at 4 °C), and resuspending in fetal bovine serum (FBS) containing 10 % dimethyl sulfoxide (DMSO). The vials are then frozen slowly (1 °C per minute) using commercially available cryopreservation containers, followed by final storage of vials in liquid nitrogen. To reestablish culture from cryopreserved spores, flasks of host cells at approximately 50 % confluence should be prepared. The frozen vial of microsporidia should be thawed quickly and added directly to the host cells; a few hours later, after the microsporidia have had an opportunity to infect the host cells, the culture medium should be changed to remove the DMSO. Alternatively, the inoculum of cryopreserved spores can be washed (i.e., centrifuged) and the pellet resuspended in a small volume of medium to remove the DMSO prior to inoculation of the culture flasks. If host cells other than those used to generate the cyropreserved spores are used, the spore inoculum will need to be washed with mild detergent (e.g., 0.5 % SDS) to prevent growth of cryopreserved host cells in the new culture.

IX. Conclusions

The microsporidia comprise a fascinating group of organisms that infect their hosts through an unusual spore germination process of polar filament extrusion and direct inoculation of the sporoplasm into the cell. They are an extremely successful group of organisms that are widespread in both vertebrate and invertebrate hosts and are highly efficient parasites, as noted by their gene compaction and reduction. Over the past 10 years, molecular studies have reshaped our understanding of phylogeny and led to the classification of microsporidia as fungi, although knowledge of the exact relationship is still in flux (Hibbett et al. 2007; Koestler and Ebersberger 2011). Whereas previously the microsporidia had been recognized pathogens of agriculturally and commercially relevant insects, fish, companion pets, domestic animals, and food-producing animals, it was only recently, during the AIDS pandemic, that these organisms came to be seen as common causes of opportunistic and emerging infections in humans. The tremendous increase in the recognition of new species of microsporidia in such a wide host range and the application of newer molecular tools will now need to be applied to improving diagnostics, developing intervention and chemotherapeutic strategies, and learning more about the basic biology and phylogeny of the microsporidia.

Acknowledgements The authors gratefully acknowledge funding from the US National Institutes of Health (AI37188 to LMW, RR017386 to JLS and MLK, and OD011104 and AI071778 subcontract to ESD) that supported research results reported in this chapter. We also recognize the excellent technical assistance of Neil Sanscrainte (to JJB) and Lisa Bowers (to ESD).

References

Akiyoshi DE, Morrison HG, Lei S, Feng X, Zhang Q, Corradi N, Mayanja H, Tumwine JK, Keeling PJ, Weiss LM, Tzipori S (2009) Genomic survey of the non-cultivatable opportunistic human pathogen, *Enterocytozoon bieneusi*. PLoS Pathog 5:e1000261

Andreadis TG (1984) Epizootiology of *Nosema pyrausta* in field populations of the European corn borer. Environ Entomol 13:882–887

Andreadis TG (1988) Comparative susceptibility of the copepod *Acanthocyclops vernalis* to a microsporidian parasite, *Amblyospora connecticus*, from the mosquito *Aedes cantator*. J Invertebr Pathol 52:73–77

Andreadis TG (2002) Epizootiology of *Hyalinocysta chapmani* (Microsporidia: Thelohaniidae) infections in field populations of *Culiseta melanura* (Diptera: Culicidae) and *Orthocyclops modestus* (Copepoda: Cyclopidae): a three-year investigation. J Invertebr Pathol 81:114–121

Avery SW, Undeen AH (1990) Horizontal transmission of *Parathelohania anophelis* to the copepod, *Microcyclops varicans*, and the mosquito, *Anopheles quadrimaculatus*. J Invertebr Pathol 56:98–105

Baxa-Antonio D, Groff JM, Hedrick RP (1992) Experimental horizontal transmission of *Enterocytozoon salmonis* to chinook salmon, *Oncorhynchus tshawytscha*. J Protozool 39:699–702

Beckers PJ, Derks GJ, van Gool T, Rietveld FJ, Sauerwein RW (1996) *Encephalocytozoon intestinalis*-specific monoclonal antibodies for laboratory diagnosis of microsporidiosis. J Clin Microbiol 34:282–285

Becnel JJ (1992) Horizontal transmission and subsequent development of *Amblyospora californica* (Microsporida: Amblyosporiae) in the intermediate and definitive hosts. Dis Aquat Org 13:17–28

Becnel JJ, Andreadis TG (1999) Microsporidia in insects. In: Wittner M, Weiss LM (eds) The microsporidia and microsporidiosis. American Society for Microbiology Press, Washington, DC, pp 447–501

Becnel JJ, Hazard EI, Fukuda T, Sprague V (1987) Life cycle of *Culicospora magna* (Kudo, 1920) (Microsporida: Culicosporidae) in *Culex restuans* Theobald with special reference to sexuality. J Eukaryot Microbiol 34:313–322

Becnel JJ, Sprague V, Fukuda T, Hazard EI (1989) Development of *Edhazardia aedis* (Kudo, 1930) n. g., n. comb. (Microsporida: Amblyosporidae) in the mosquito *Aedes aegypti* (L.) (Diptera: Culicidae). J Protozool 36:119–130

Becnel JJ, Jeyaprakash A, Hoy MA, Shapiro A (2002) Morphological and molecular characterization of a new microsporidian species from the predatory mite *Metaseiulus occidentalis* (Nesbitt) (Acari, Phytoseiidae). J Invertebr Pathol 79:163–172

Becnel JJ, White SE, Shapiro AM (2005) Review of microsporidia-mosquito relationships: from the simple to the complex. Folia Parasitol (Praha) 52:41–50

Bigliardi E, Selmi MG, Lupetti P, Corona S, Gatti S, Scaglia M, Sacchi L (1996) Microsporidian spore wall: ultrastructural findings on *Encephalitozoon hellem* exospore. J Eukaryot Microbiol 43:181–186

Bohne W, Ferguson DJ, Kohler K, Gross U (2000) Developmental expression of a tandemly repeated, glycine- and serine-rich spore wall protein in the microsporidian pathogen *Encephalitozoon cuniculi*. Infect Immun 68:2268–2275

Bouzahzah B, Nagajyothi F, Ghosh K, Takvorian PM, Cali A, Tanowitz HB, Weiss LM (2010) Interactions of *Encephalitozoon cuniculi* polar tube proteins. Infect Immun 78:2745–2753

Braunfuchsová P, Kopecký J, Ditrich O, Koudela B (1999) Cytokine response to infection with the microsporidian, *Encephalitozoon cuniculi*. Folia Parasitol (Praha) 46:91–95

Brooks WM (1988) Entomogenous Protozoa. In: Ignoffo CM (ed) Handbook of natural pesticides, Vol. V, microbial insecticides Part A, entomogenous protozoa and fungi. CRC, Boca Raton, FL, pp 1–149

Burri L, Williams BA, Bursac D, Lithgow T, Keeling PJ (2006) Microsporidian mitosomes retain elements of the general mitochondrial targeting system. Proc Natl Acad Sci USA 103:15916–15920

Cali A, Owen RL (1988) Microsporidiosis. In: Balows WJ, Hausler WJ, Ohashi M, Turano A (eds) Laboratory diagnosis of infectious diseases: principles and practice. Springer, New York, pp 929–946

Cali A, Owen RL (1990) Intracellular development of *Enterocytozoon*, a unique microsporidian found in the intestine of AIDS patients. J Protozool 37:145–155

Cali A, Takvorian P (1999) Developmental morphology and life cycles of the microsporidia. In: Wittner M, Weiss LM (eds) The microsporidia and microsporidiosis. American Society for Microbiology, Washington, DC, pp 85–128

Cali A, Kotler DP, Orenstein JM (1993) *Septata intestinalis* N. G., N. Sp., an intestinal microsporidian associated with chronic diarrhea and dissemination in AIDS patients. J Eukaryot Microbiol 40:101–112

Cali A, Weiss LM, Takvorian PM (2002) *Brachiola algerae* spore membrane systems, their activity during extrusion, and a new structural entity, the multilayered interlaced network, associated with the polar tube and the sporoplasm. J Eukaryot Microbiol 49:164–174

Canning EU, Lom J, Dyková I (1986) The microsporidia of vertebrates. Academic, London

Chen Y, Evans JD, Smith IB, Pettis JS (2008) *Nosema ceranae* is a long-present and wide-spread microsporidian infection of the European honey bee (*Apis mellifera*) in the United States. J Invertebr Pathol 97:186–188

Chilmonczyk S, Cox WT, Hedrick RP (1991) *Enterocytozoon salmonis* n. sp.: an intranuclear microsporidium from salmonid fish. J Protozool 38:264–269

Chioralia G, Trammer T, Maier WA, Seitz HM (1998) Morphologic changes in Nosema algerae (Microspora) during extrusion. Parasitol Res 84:123–131

Cornman RS, Chen YP, Schatz MC, Street C, Zhao Y, Desany B, Egholm M, Hutchison S, Pettis JS, Lipkin WI, Evans JD (2009) Genomic analyses of the microsporidian *Nosema ceranae*, an emergent pathogen of honey bees. PLoS Pathog 5:e1000466

Corradi N, Keeling PJ (2009) Microsporidia: a journey through radical taxonomical revisions. Fungal Biol Rev 23:1–8

Corradi N, Slamovits CH (2011) The intriguing nature of microsporidian genomes. Brief Funct Genomics 10:115–124

Corradi N, Akiyoshi DE, Morrison HG, Feng X, Weiss LM, Tzipori S, Keeling PJ (2007) Patterns of genome evolution among the microsporidian parasites *Encephalitozoon cuniculi*, *Antonospora* locustae and *Enterocytozoon bieneusi*. PLoS One 2: e1277

Corradi N, Pombert JF, Farinelli L, Didier ES, Keeling PJ (2010) The complete sequence of the smallest known nuclear genome from the microsporidian *Encephalitozoon intestinalis*. Nat Commun 1:77. doi:10.1038/ncomms1082

de Graaf DC, Raes H, Jacobs FJ (1994) Spore dimorphism in *Nosema apis* (Microsporida, Nosematidae) developmental cycle. J Invertebr Pathol 63:92–94

de Kinkelin P (1980) Occurrence of a microsporidian infection in zebra danio *Brachydanio rerio* (Hamilton-Buchanan). J Fish Dis 3:71–73

Delbac F, Duffieux F, David D, Méténier G, Vivarès CP (1998a) Immunocytochemical identification of spore proteins in two microsporidia, with emphasis on extrusion apparatus. J Eukaryot Microbiol 45:224–231

Delbac F, Peyret P, Méténier G, David D, Danchin A, Vivarès CP (1998b) On proteins of the microsporidian invasive apparatus: complete sequence of a polar tube protein of *Encephalitozoon cuniculi*. Mol Microbiol 29:825–834

Delbac F, Peuvel I, Méténier G, Peyretaillade E, Vivarès CP (2001) Microsporidian invasion apparatus: identification of a novel polar tube protein and evidence for clustering of ptp1 and ptp2 genes in three *Encephalitozoon species*. Infect Immun 69:1016–1024

Desportes I, Le Charpentier Y, Galian A, Bernard F, Cochand-Priollet B, Lavergne A, Ravisse P, Modigliani R (1985) Occurrence of a new microsporidan: *Enterocytozoon bieneusi* n.g., n. sp., in the enterocytes of a human patient with AIDS. J Protozool 32:250–254

Didier ES, Weiss LM (2006) Microsporidiosis: current status. Curr Opin Infect Dis 19:485–492. doi:10.1097/01.qco.0000244055.46382.23

Didier ES, Didier PJ, Friedberg DN, Stenson SM, Orenstein JM, Yee RW, Tio FO, Davis RM, Vossbrinck C, Millichamp N et al (1991) Isolation and characterization of a new human microsporidian, *Encephalitozoon hellem* (n. sp.), from three AIDS patients with keratoconjunctivitis. J Infect Dis 163:617–621

Didier ES, Rogers LB, Orenstein JM, Baker MD, Vossbrinck CR, Van Gool T, Hartskeerl R, Soave R, Beaudet LM (1996) Characterization of *Encephalitozoon (Septata) intestinalis* isolates cultured from nasal mucosa and bronchoalveolar lavage fluids of two AIDS patients. J Eukaryot Microbiol 43:34–43

Didier ES, Snowden KF, Shadduck JA (1998) Biology of microsporidian species infecting mammals. Adv Parasitol 40:283–320

Dyer PS (2008) Evolutionary biology: microsporidia sex–a missing link to fungi. Curr Biol 18:R1012–R1014

Erickson BW Jr, Blanquet RS (1969) The occurrence of chitin in the spore wall of *Glugea weissenbergi*. J Invertebr Pathol 14:358–364

Fast NM, Keeling PJ (2001) Alpha and beta subunits of pyruvate dehydrogenase E1 from the microsporidian *Nosema locustae*: mitochondrion-derived carbon metabolism in microsporidia. Mol Biochem Parasitol 117:201–209

Fries I (1993) *Nosema apis*—a parasite in the honey bee colony. Bee World 74:5–19

Frixione E, Ruiz L, Santillán M, de Vargas LV, Tejero JM, Undeen AH (1992) Dynamics of polar filament discharge and sporoplasm expulsion by microsporidian spores. Cell Motil Cytoskeleton 22:38–50

Frixione E, Ruiz L, Cerbon J, Undeen AH (1997) Germination of *Nosema algerae* (Microspora) spores: conditional inhibition by D2O, ethanol and Hg2+ suggests dependence of water influx upon membrane hydration and specific transmembrane pathways. J Eukaryot Microbiol 44:109–116

Geden CJ, Long SJ, Rutz DA, Becnel JJ (1995) Nosema disease of the parasitoid *Muscidifurax raptor* (Hymenoptera: Pteromalidae): prevalence, patterns of transmission, management, and impact. Biol Control 5:607–614

Germot A, Philippe H, Le Guyader H (1997) Evidence for loss of mitochondria in Microsporidia from a mitochondrial-type HSP70 in *Nosema locustae*. Mol Biochem Parasitol 87:159–168

Ghosh K, Nieves E, Keeling P, Pombert JF, Henrich PP, Cali A, Weiss LM (2011) Branching network of proteinaceous filaments within the parasitophorous vacuole of *Encephalitozoon cuniculi* and *Encephalitozoon hellem*. Infect Immun 79:1374–1385

Graczyk TK, Majewska AC, Schwab KJ (2008) The role of birds in dissemination of human waterborne enteropathogens. Trends Parasitol 24:55–59

Haro M, Izquierdo F, Henriques-Gil N, Andrés I, Alonso F, Fenoy S, del Aguila C (2005) First detection and genotyping of human-associated microsporidia in pigeons from urban parks. Appl Environ Microbiol 71:3153–3157

Hayman JR, Hayes SF, Amon J, Nash TE (2001) Developmental expression of two spore wall proteins during maturation of the microsporidian *Encephalitozoon intestinalis*. Infect Immun 69:7057–7066

Hazard EI, Weiser J (1968) Spores of *Thelohania* in adult female anopheles: development and transovarial transmission, and redescriptions of *T. legeri* Hesse and *T. obesa* Kudo. J Protozool 15:817–823

Henry JE, Oma EA (1981) Pest control by *Nosema locustae*, a pathogen of grasshoppers and crickets. In: Burges HD (ed) Microbial control of pests and plant diseases. Academic, London, pp 573–586

Hibbett DS, Binder M, Bischoff JF, Blackwell M, Cannon PF, Eriksson OE, Huhndorf S, James T, Kirk PM, Lucking R, Thorsten Lumbsch H, Lutzoni F, Matheny PB, McLaughlin DJ, Powell MJ, Redhead S, Schoch CL, Spatafora JW, Stalpers JA, Vilgalys R, Aime MC, Aptroot A, Bauer R, Begerow D, Benny GL, Castlebury LA, Crous PW, Dai YC, Gams W, Geiser DM, Griffith GW, Gueidan C, Hawksworth DL, Hestmark G, Hosaka K, Humber RA, Hyde KD, Ironside JE, Koljalg U, Kurtzman

CP, Larsson KH, Lichtwardt R, Longcore J, Miadlikowska J, Miller A, Moncalvo JM, Mozley-Standridge S, Oberwinkler F, Parmasto E, Reeb V, Rogers JD, Roux C, Ryvarden L, Sampaio JP, Schussler A, Sugiyama J, Thorn RG, Tibell L, Untereiner WA, Walker C, Wang Z, Weir A, Weiss M, White MM, Winka K, Yao YJ, Zhang N (2007) A higher-level phylogenetic classification of the Fungi. Mycol Res 111:509–547

Hirt RP, Healy B, Vossbrinck CR, Canning EU, Embley TM (1997) A mitochondrial Hsp70 orthologue in *Vairimorpha necatrix*: molecular evidence that microsporidia once contained mitochondria. Curr Biol 7:995–998

James TY, Kauff F, Schoch CL, Matheny PB, Hofstetter V, Cox CJ, Celio G, Gueidan C, Fraker E, Miadlikowska J, Lumbsch HT, Rauhut A, Reeb V, Arnold AE, Amtoft A, Stajich JE, Hosaka K, Sung GH, Johnson D, O'Rourke B, Crockett M, Binder M, Curtis JM, Slot JC, Wang Z, Wilson AW, Schussler A, Longcore JE, O'Donnell K, Mozley-Standridge S, Porter D, Letcher PM, Powell MJ, Taylor JW, White MM, Griffith GW, Davies DR, Humber RA, Morton JB, Sugiyama J, Rossman AY, Rogers JD, Pfister DH, Hewitt D, Hansen K, Hambleton S, Shoemaker RA, Kohlmeyer J, Volkmann-Kohlmeyer B, Spotts RA, Serdani M, Crous PW, Hughes KW, Matsuura K, Langer E, Langer G, Untereiner WA, Lucking R, Budel B, Geiser DM, Aptroot A, Diederich P, Schmitt I, Schultz M, Yahr R, Hibbett DS, Lutzoni F, McLaughlin DJ, Spatafora JW, Vilgalys R (2006) Reconstructing the early evolution of Fungi using a six-gene phylogeny. Nature 443:818–822

Jaronski ST (1984) Microsporidia in cell culture. Adv Cell Culture 18:183–229

Juarez SI, Putaporntip C, Jongwutiwes S, Ichinose A, Yanagi T, Kanbara H (2005) In vitro cultivation and electron microscopy characterization of *Trachipleistophora anthropophthera* isolated from the cornea of an AIDS patient. J Eukaryot Microbiol 52:179–190

Katinka MD, Duprat S, Cornillot E, Méténier G, Thomarat F, Prensier G, Barbe V, Peyretaillade E, Brottier P, Wincker P, Delbac F, El Alaoui H, Peyret P, Saurin W, Gouy M, Weissenbach J, Vivarès CP (2001) Genome sequence and gene compaction of the eukaryote parasite *Encephalitozoon cuniculi*. Nature 414:450–453

Keeling P (2009) Five questions about microsporidia. PLoS Pathog 5:e1000489

Keeling PJ, Corradi N, Morrison HG, Haag KL, Ebert D, Weiss LM, Akiyoshi DE, Tzipori S (2010) The reduced genome of the parasitic microsporidian *Enterocytozoon bieneusi* lacks genes for core carbon metabolism. Genome Biol Evol 2:304–309

Kellen WR, Lipa JJ (1960) *Thelohania californica* n.sp., a microsporidian parasite of *Culex tarsalis* Coquillet. J Invertebr Pathol 2:1–12

Kent ML, Bishop-Stewart JK (2003) Transmission and tissue distribution of *Pseudoloma neurophilia* (Microsporidia) of zebrafish, *Danio rerio* (Hamilton). J Fish Dis 26:423–426

Kent ML, Poppe TT (1998) Diseases of seawater netpen-reared salmonid fishes. Pacific Biological Station, Department of Fisheries and Oceans, Nanimo, BC

Kent ML, Speare DJ (2005) Review of the sequential development of *Loma salmonae* (Microsporidia) based on experimental infections of rainbow trout (*Oncorhynchus mykiss*) and Chinook salmon (O. tshawytscha). Folia Parasitol (Praha) 52:63–68

Kent ML, Buchner C, Watral VG, Sander JL, LaDu J, Peterson TS, Tanguay RL (2011) Development and maintenance of a specific pathogen-free (SPF) zebrafish research facility for *Pseudoloma neurophilia*. Dis Aquat Org 95:73–79

Keohane EM, Weiss LM (1998) Characterization and function of the microsporidian polar tube: a review. Folia Parasitol (Praha) 45:117–127

Keohane E, Takvorian PM, Cali A, Tanowitz HB, Wittner M, Weiss LM (1994) The identification and characterization of a polar tube reactive monoclonal antibody. J Eukaryot Microbiol 41:48S

Keohane EM, Orr GA, Takvorian PM, Cali A, Tanowitz HB, Wittner M, Weiss LM (1996a) Purification and characterization of a microsporidian polar tube protein. Mol Biochem Parasitol 79:255–2559

Keohane EM, Orr GA, Takvorian PM, Cali A, Tanowitz HB, Wittner M, Weiss LM (1996b) Purification and characterization of human microsporidian polar tube proteins. J Eukaryot Microbiol 43:100S

Keohane EM, Takvorian PM, Cali A, Tanowitz HB, Wittner M, Weiss LM (1996c) Identification of a microsporidian polar tube protein reactive monoclonal antibody. J Eukaryot Microbiol 43:26–31

Keohane EM, Orr GA, Zhang HS, Takvorian PM, Cali A, Tanowitz HB, Wittner M, Weiss LM (1998) The molecular characterization of the major polar tube protein from *Encephalitozoon hellem*, a microsporidian parasite of humans. Mol Biochem Parasitol 94:227–2236

Keohane EM, Orr GA, Takvorian PM, Cali A, Tanowitz HB, Wittner M, Weiss LM (1999a) Analysis of the major microsporidian polar tube proteins. J Eukaryot Microbiol 46:29S–30S

Keohane EM, Orr GA, Takvorian PM, Cali A, Tanowitz HB, Wittner M, Weiss LM (1999b) Polar tube proteins of microsporidia of the family encephalitozoonidae. J Eukaryot Microbiol 46:1–5

Koestler T, Ebersberger I (2011) Zygomycetes, microsporidia, and the evolutionary ancestry of sex determination. Genome Biol Evol 3:186–194

Kotler DP, Orenstein JM (1998) Clinical syndromes associated with microsporidiosis. Adv Parasitol 40:321–349

Kou G-H, Wang C-H, Hung H-W, Jang Y-S, Chou C-M, Lo C-F (1995) A cell line (EP-1 cell line) derived from "Beko disease" affected Japanese eel elver

(*Anguilla japonica*) persistently infected with *Pleistophora anguillarum.* Aquaculture 132:161–173

Lafranchi-Tristem NJ, Curry A, Cheney SA, Canning EU (2001) Growth of Trachipleistophora hominis (Microsporidia: Pleistophoridae) in C2, C12 mouse myoblast cells and response to treatment with albendazole. Folia Parasitol (Praha) 48:192–200

Larsson R (1986) Ultrastructure, function, and classification of microsporidia. Prog Protistol 1:325–390

Larsson JI (2005) Molecular versus morphological approach to microsporidian classification. Folia Parasitol (Praha) 52:143–144

Lee SC, Corradi N, Byrnes EJ 3rd, Torres-Martinez S, Dietrich FS, Keeling PJ, Heitman J (2008) Microsporidia evolved from ancestral sexual fungi. Curr Biol 18:1675–1679

Lee RC, Gill EE, Roy SW, Fast NM (2010a) Constrained intron structures in a microsporidian. Mol Biol Evol 27:1979–1982

Lee SC, Corradi N, Doan S, Dietrich FS, Keeling PJ, Heitman J (2010b) Evolution of the sex-related locus and genomic features shared in microsporidia and fungi. PLoS One 5:e10539

Lom J (1972) On the structure of the extruded microsporidian polar filament. Parasitol Res 38:200–213

Lom J (2002) A catalogue of described genera and species of microsporidians parasitic in fish. Syst Parasitol 53:81–99

Lom J, Dyková I (1992) Protozoan parasites of fishes. Elsevier, Amsterdam

Lom J, Dyková I (2005) Microsporidian xenomas in fish seen in wider perspective. Folia Parasitol (Praha) 52:69–81

Lom J, Nilsen F (2003) Fish microsporidia: fine structural diversity and phylogeny. Int J Parasitol 33:107–127

Lom J, Vávra J (1963) The mode of sporoplasm extrusion in microsporidian spores. Acta Protozool 1:81–89

Lores B, Rosales MJ, Mascaro C, Osuna A (2003) In vitro culture of Glugea sp. Vet Parasitol 112:185–196

Lowman PM, Takvorian PM, Cali A (2000) The effects of elevated temperatures and various time-temperature combinations on the development of *Brachiola (Nosema) algerae* N. Comb in mammalian cell culture. J Eukaryot Microbiol 47:221–234

Matos E, Corral L, Azevedo C (2003) Ultrastructural details of the xenoma of *Loma myrophis* (phylum Microsporidia) and extrusion of the polar tube during autoinfection. Dis Aquat Org 54:203–207

Matthews JL, Brown AM, Larison K, Bishop-Stewart JK, Rogers P, Kent ML (2001) *Pseudoloma neurophilia* n. g., n. sp., a new microsporidium from the central nervous system of the zebrafish (*Danio rerio*). J Eukaryot Microbiol 48:227–233

McVicar AH (1975) Infection of plaice *Pleuronectes platessa* L. with *Glugea (Nosema) stephani* (Hagenmüller 1899) (Protozoa: Microsporidia) in a fish farm and under experimental conditions. J Fish Biol 7:611–619

Monaghan SR, Kent ML, Watral VG, Kaufman RJ, Lee LE, Bols NC (2009) Animal cell cultures in microsporidial research: their general roles and their specific use for fish microsporidia. In Vitro Cell Dev Biol Anim 45:135–147

Monaghan SR, Rumney RL, Vo NT, Bols NC, Lee LE (2011) In vitro growth of microsporidia *Anncaliia algerae* in cell lines from warm water fish. In Vitro Cell Dev Biol Anim 47:104–113

Morrison CM, Sprague V (1983) *Loma salmonae* (Putz, Hoffman and Dunbar, 1965) in the rainbow trout, *Salmo gairdneri* Richardson, and *L. fontinalis* sp. nov. (Microsporida) in the brook trout, *Salvelinus fontinalis* (Mitchill). J Fish Dis 6:345–353

Muller MG, Kinne J, Schuster RK, Walochnik J (2008) Outbreak of microsporidiosis caused by *Enterocytozoon bieneusi* in falcons. Vet Parasitol 152:67–78

Nylund S, Nylund A, Watanabe K, Arnesen CE, Karlsbakk E (2010) *Paranucleospora theridion* n. gen., n. sp. (Microsporidia, *Enterocytozoonidae*) with a life cycle in the salmon louse (*Lepeophtheirus salmonis*, Copepoda) and Atlantic salmon (*Salmo salar*). J Eukaryot Microbiol 57:95–114

Ohshima K (1937) On the function of the polar filament of *Nosema bombycis*. Parasitology 29:220–224

Oi DH, Valles SM, Pereira RM (2004) Prevalence of *Thelohania solenopsae* (Microsporidia: Thelohaniidae) infection in monogyne and polygyne red imported fire ants (Hymenoptera: Formicidae). Environ Entomol 33:340–345

Pasteur L (1870) Etudes sur la Maladie des vers a Soie. Gauthier-Villars, Paris

Peuvel I, Peyret P, Méténier G, Vivarès CP, Delbac F (2002) The microsporidian polar tube: evidence for a third polar tube protein (PTP3) in *Encephalitozoon cuniculi.* Mol Biochem Parasitol 122:69–80

Peuvel-Fanget I, Polonais V, Brosson D, Texier C, Kuhn L, Peyret P, Vivarès C, Delbac F (2006) EnP1 and EnP2, two proteins associated with the *Encephalitozoon cuniculi* endospore, the chitin-rich inner layer of the microsporidian spore wall. Int J Parasitol 36:309–318

Phelps NBD, Goodwin AE (2008) Vertical transmission of *Ovipleistophora ovariae* (Microspora) within the eggs of the golden shiner. J Aquat Anim Health 20:45–53

Prigneau O, Achbarou A, Bouladoux N, Mazier D, Desportes-Livage I (2000) Identification of proteins in *Encephalitozoon intestinalis*, a microsporidian pathogen of immunocompromised humans:

an immunoblotting and immunocytochemical study. J Eukaryot Microbiol 47:48–56

Ramsay JM, Watral V, Schreck CB, Kent ML (2009) *Pseudoloma neurophilia* infections in zebrafish *Danio rerio*: effects of stress on survival, growth, and reproduction. Dis Aquat Org 88:69–84

Ratnieks FL, Carreck NL (2010) Ecology. Clarity on honey bee collapse? Science 327:152–153

Redhead SA, Kirk P, Keeling PJ, Weiss LM (2009) Proposals to exclude the phylum Microsporidia from the code. Taxon 58:10–11

Reetz J, Rinder H, Thomschke A, Manke H, Schwebs M, Bruderek A (2002) First detection of the microsporidium *Enterocytozoon bieneusi* in non-mammalian hosts (chickens). Int J Parasitol 32:785–787

Rich SM, Ayala FJ (2000) Population structure and recent evolution of *Plasmodium falciparum*. Proc Natl Acad Sci USA 97:6994–7001

Rodriguez-Tovar LE, Wadowska DW, Wright GM, Groman DB, Speare DJ, Whelan DS (2003) Ultrastructural evidence of autoinfection in the gills of Atlantic cod *Gadus morhua* infected with *Loma* sp. (phylum Microsporidia). Dis Aquat Org 57:227–230

Rönnebäumer K, Gross U, Bohne W (2008) The nascent parasitophorous vacuole membrane of *Encephalitozoon cuniculi* is formed by host cell lipids and contains pores which allow nutrient uptake. Eukaryot Cell 7:1001–1008

Sanchez JG, Speare DJ, Markham RJ, Wright GM, Kibenge FS (2001) Localization of the initial developmental stages of *Loma salmonae* in rainbow trout (*Oncorhynchus mykiss*). Vet Pathol 38:540–546

Sanders JL, Kent ML (2011) Development of a sensitive assay for the detection of *Pseudoloma neurophilia* and laboratory populations of the zebrafish, *Danio rerio*. Dis Aquat Org 96:145–156

Sanders JL, Lawrence C, Nichols DK, Brubaker JF, Peterson TS, Murray KN, Kent ML (2010) *Pleistophora hyphessobryconis* (Microsporidia) infecting zebrafish *Danio rerio* in research facilities. Dis Aquat Org 91:47–56

Santín M, Fayer R (2009a) *Enterocytozoon bieneusi* genotype nomenclature based on the internal transcribed spacer sequence: a consensus. J Eukaryot Microbiol 56:34–38

Santín M, Fayer R (2009b) A longitudinal study of *Enterocytozoon bieneusi* in dairy cattle. Parasitol Res 105:141–144

Schäperclaus W (1991) Plistophora disease of neon and other aquarium fish (Pleistophorosis). In: Schäperclaus W, Kulow H, Schreckenbach K (eds) Fish Diseases (vol 2, 5th ed.), Amerind, New Delh, 1,398 pp

Shadduck JA (1969) *Nosema cuiculi*: in vitro isolation. Science 166:516–517

Shaw R, Kent ML (1999) Fish microsporidia. In: Wittner M, Weiss LM (eds) The microsporidia and microsporidiosis. American Society for Microbiology Press, Washington, DC, pp 418–444

Shaw RW, Kent ML, Adamson ML (1998) Modes of transmission of *Loma salmonae* (Microsporidia). Dis Aquat Org 33:151–156

Shaw RW, Kent ML, Brown AM, Whipps CM, Adamson ML (2000) Experimental and natural host specificity of *Loma salmonae* (Microsporidia). Dis Aquat Org 40:131–136

Siegel JP, Maddox JV, Ruesink WG (1986) Lethal and sublethal effects of *Nosema pyrausta* on the European corn borer (*Ostrinia nubilalis*) in central Illinois. J Invertebr Pathol 48:167–173

Sinden RE, Canning EU (1974) The ultrastructure of the spore of *Nosema algerae* (Protozoa, Microsporida), in relation to the hatching mechanism of microsporidian spores. J Gen Microbiol 85:350–357

Snowden K, Logan K (1999) Molecular identification of *Encephalitozoon hellem* in an ostrich. Avian Dis 43:779–782

Snowden KF, Logan K, Phalen DN (2000) Isolation and characterization of an avian isolate of *Encephalitozoon hellem*. Parasitology 121(Pt 1):9–14

Sokolova YY, Fuxa JR (2008) Biology and life-cycle of the microsporidium *Kneallhazia solenopsae* Knell Allan Hazard 1977 gen. n., comb. n., from the fire ant *Solenopsis invicta*. Parasitology 135:903–929

Southern TR, Jolly CE, Lester ME, Hayman JR (2007) EnP1, a microsporidian spore wall protein that enables spores to adhere to and infect host cells in vitro. Eukaryot Cell 6:1354–1362

Sprague V (1977) Systematics of the Microsporidia. In: Bulla LA Jr, Cheng TC (eds) Comparative pathobiology. Plenum, New York

Sprague V, Becnel JJ (1999) Appendix: checklist of available generic names for microsporidia with type species and type hosts. In: Wittner M, Weiss LM (eds) The microsporidia and microsporidiosis. American Society for Microbiology Press, Washington, DC, pp 531–539

Sprague V, Hussey KL (1980) Observations on *Ichthyosporidium giganteum* (Microsporida) with particular reference to the host-parasite relations during merogony. J Protozool 27:169–175

Sprague V, Becnel JJ, Hazard EI (1992) Taxonomy of phylum microspora. Crit Rev Microbiol 18:285–395

Steffens W (1962) The current status of distribution of *Plistophora hyphessobryconis* Schaperclaus 1941 (Sporozoa, Microsporidia). Z Parasitenkd 21:535–541

Sweeney AW, Hazard EI, Graham MF (1985) Intermediate host for an *Amblyospora* sp. (microspora) infecting the mosquito, *Culex annulirostris*. J Invertebr Pathol 46:98–102

Takvorian PM, Cali A (1986) The ultrastructure of spores (Protozoa: Microsporida) from *Lophius americanus*, the angler fish. J Protozool 33:570–575

Texier C, Vidau C, Vigues B, El Alaoui H, Delbac F (2010) Microsporidia: a model for minimal parasite-host interactions. Curr Opin Microbiol 13:443–449

Trammer T, Chioralia G, Maier WA, Seitz HM (1999) In vitro replication of *Nosema algerae* (Microsporidia), a parasite of anopheline mosquitoes, in human cells above 36 degrees C. J Eukaryot Microbiol 46:464–468

Troemel ER (2011) New models of microsporidiosis: infections in zebrafish, *C. elegans*, and honey bees. PLoS Pathog 7:e1001243

Undeen AH (1975) Growth of *Nosema algerae* in pig kidney cell cultures. J Protozool 22:107–110

Undeen AH (1990) A proposed mechanism for the germination of microsporidian (Protozoa: Microspora) spores. J Theor Biol 142:223–235

Undeen AH, Avery SW (1988) Ammonium chloride inhibition of the germination of spores of *Nosema algerae* (Microspora: Nosematidae). J Invertebr Pathol 52:326–334

Undeen AH, Epsky ND (1990) In vitro and in vivo germination of *Nosema locustae* (Microspora: Nosematidae) spores. J Invertebr Pathol 56:371–379

Undeen AH, Frixione E (1990) The role of osmotic pressure in the germination of *Nosema algerae* spores. J Protozool 37:561–567

Undeen AH, Frixione E (1991) Structural alteration of the plasma membrane in spores of the microsporidium Nosema algerae on germination. J Protozool 38:511–518

Vavrá J (1976) Structure of the microsporidia. In: Bulla LA Jr, Cheng TC (eds) Comparative pathology. Plenum, New York, pp 1–85

Vavrá J, Becnel JJ (2007) *Vavraia culicis* (Weiser, 1947) Weiser, 1977 revisited: cytological characterisation of a *Vavraia culicis*-like microsporidium isolated from mosquitoes in Florida and the establishment of *Vavraia culicis floridensis* subsp. n. Folia Parasitol (Praha) 54:259–271

Vavrá J, Larsson JIR (1999) Structure of the microsporidia. In: Wittner M, Weiss LM (eds) The microsporidia and microsporidiosis. American Society for Microbiology Press, Washington, DC, pp 7–84

Visvesvara GS (2002) In vitro cultivation of microsporidia of clinical importance. Clin Microbiol Rev 15:401–413

Vossbrinck CR, Debrunner-Vossbrinck BA (2005) Molecular phylogeny of the Microsporidia: ecological, ultrastructural and taxonomic considerations. Folia Parasitol (Praha) 52:131–142, discussion p130

Vossbrinck CR, Maddox JV, Friedman S, Debrunner-Vossbrinck BA, Woese CR (1987) Ribosomal RNA sequence suggests microsporidia are extremely ancient eukaryotes. Nature 326:411–414

Vossbrinck CR, Andreadis TG, Weiss LM (2004) Phylogenetics: taxonomy and the microsporidia as derived fungi. In: Lindsay DS, Weiss LM (eds) Opportunistic infections: toxoplasma, sarcocystis and microsporidia. Springer, New York, pp 189–213

Walters VA (1958) Structure, hatching and size variation of the spores in a species of *Nosema* (Microsporidia) found in *Hyalophora cecropia* (Lepidoptera). Parasitology 48:113–120

Watral VG, Kauffmann RB, Kent ML (2006) In vitro culture of *Pseudoloma neurophilia*, a common microsporidian of zebrafish (*Danio rerio*). In: IX International workshops on opportunistic protists and the international society of protistologists 57th annual meeting, Lisbon, Portugal

Weber R, Deplazes P, Schwartz D (2000) Diagnosis and clinical aspects of human microsporidiosis. Contrib Microbiol 6:166–192

Weidner E (1972) Ultrastructural study of microsporidian invasion into cells. Z Parasitenkd 40:227–242

Weidner E (1976) The microsporidian spore invasion tube. The ultrastructure, isolation, and characterization of the protein comprising the tube. J Cell Biol 71:23–34

Weidner E (1982) The microsporidian spore invasion tube. III. Tube extrusion and assembly. J Cell Biol 93:976–979

Weidner E, Byrd W (1982) The microsporidian spore invasion tube. II. Role of calcium in the activation of invasion tube discharge. J Cell Biol 93:970–975

Weidner E, Manale SB, Halonen SK, Lynn JW (1995) Protein-membrane interaction is essential to normal assembly of the microsporidian spore invasion tube. Biol Bull 188:128–135

Weiser J (1977) Contribution to the classification of Microsporidia. Vestnik Ceskoslovenske Spolecnosti Zoologicke 41:308–320

Weiss LM (2001) Microsporidia: emerging pathogenic protists. Acta Trop 78:89–102

Weissenberg R (1968) Intracellular development of the Microsporidan *Glugea anomala* Moniez in hypertrophying migratory cells of the fish *Gasterosteus aculeatus* L., an example of the formation of "xenoma" tumors. J Eukaryot Microbiol 15:44–57

Wiklund T, Lounasheimo L, Lom J, Bylund G (1996) Gonadal impairment in roach *Rutilus rutilus* from Finnish coastal areas of the northern Baltic Sea. Dis Aquat Org 26:163–171

Williams BA, Keeling PJ (2005) Microsporidian mitochondrial proteins: expression in *Antonospora locustae* spores and identification of genes coding for two further proteins. J Eukaryot Microbiol 52:271–276

Williams BA, Hirt RP, Lucocq JM, Embley TM (2002) A mitochondrial remnant in the microsporidian *Trachipleistophora hominis*. Nature 418:865–869

Williams BA, Lee RC, Becnel JJ, Weiss LM, Fast NM, Keeling PJ (2008) Genome sequence surveys of *Brachiola algerae* and *Edhazardia aedis* reveal microsporidia with low gene densities. BMC Genomics 9:200

Wilson GG (1973) Incidence of microsporidia in a field population of spruce budworm. Can For Serv Bi-Mon Res Notes 29:35

Wilson GG (1981) *Nosema fumiferanae*, a natural pathogen of forest pests: potential for pest management. In: Burges HD (ed) Microbial control of pests and plant diseases. Academic, London, pp 595–602

Wittner M, Weiss LM (1999) The microsporidia and microsporidiosis. American Society for Microbiology Press, Washington, DC

Wongtavatchai J, Conrad PA, Hedrick RP (1994) In vitro cultivation of the microsporidian: *Enterocytozoon salmonis* using a newly developed medium for salmonid lymphocytes. J Tissue Cult Methods 16:125

Wright JH, Craighead EM (1922) Infectious motor paralysis in young rabbits. J Exp Med 36:135–140

Xu Y, Takvorian PM, Cali A, Orr G, Weiss LM (2004) Glycosylation of the major polar tube protein of *Encephalitozoon hellem*, a microsporidian parasite that infects humans. Infect Immun 72:6341–6350

Xu Y, Takvorian P, Cali A, Wang F, Zhang H, Orr G, Weiss LM (2006) Identification of a new spore wall protein from *Encephalitozoon cuniculi*. Infect Immun 74:239–247

6 Chytridiomycota, Monoblepharidomycota, and Neocallimastigomycota

Martha J. Powell[1], Peter M. Letcher[1]

CONTENTS

I. Introduction

Phylogenetic analyses of molecular sequences (James et al. 2000, 2006a, b) have generated monumental growth in our understanding of evolutionary relationships among zoosporic fungi since Barr's (2001) review of the morphology, life history, and occurrence of Chytridiomycota over a decade ago. New understandings of evolutionary relationships among zoosporic fungi have sharpened our focus on the value of zoospore ultrastructural characters in systematic analyses (Letcher et al. 2008a, c; Simmons 2011) and have given us insights into the convergence of thallus features (Letcher et al. 2005; Mozley-Standridge et al. 2009) once used as primary taxonomic characters (Sparrow 1960). Molecular techniques allow us to detect uncultured and unseen chytrids in environmental samples and to demonstrate that zoosporic fungi are essentially ubiquitous and abundant in a wide range of habitats, including temperate soils and aquatic environments (Chen et al. 2008; Lefèvre et al. 2008, 2012; Lepère et al. 2008; Miki et al. 2011; Monchy et al. 2011; Sime-Ngando et al. 2011) as well as especially stressful environments such as anoxic deep-sea cold seeps and hydrothermal vent ecosystems (LeCalvez et al. 2009; Nagahama et al. 2011; Stoeck and Epstein 2003), exposed soils at high elevations (Freeman et al. 2009; Schmidt et al. 2012), and soils at Arctic latitudes (Stoeck et al. 2007). A renaissance of interest in zoosporic fungi is occurring because, as basal members in the evolution of fungi, they hold the key to reconstructing ancestral forms and forces that may have driven the evolutionary radiation of fungi (Amaral Zettler et al. 2001; Stajich et al. 2009; Steenkamp et al. 2006). Moreover, their roles as parasites of phytoplankton (Bruning et al. 1992; Holfeld 2000) and amphibians (Longcore et al. 1999, 2007; Piotrowski et al. 2004; Voyles 2011) cause concern among conservationists (Bai et al. 2010; Rosenblum et al. 2008; Schloegel et al. 2012; Weldon et al. 2004); their recognition as key players in food webs alerts ecologists of their potential impact on aquatic and terrestrial sustainability (Gleason et al. 2008; Kagami et al.

[1]Department of Biological Sciences, The University of Alabama, Tuscaloosa, AL 35487, USA; e-mail: mpowell@biology.as.ua.edu; letch006@bama.ua.edu

Systematics and Evolution, 2nd Edition
The Mycota VII Part A
D.J. McLaughlin and J.W. Spatafora (Eds.)

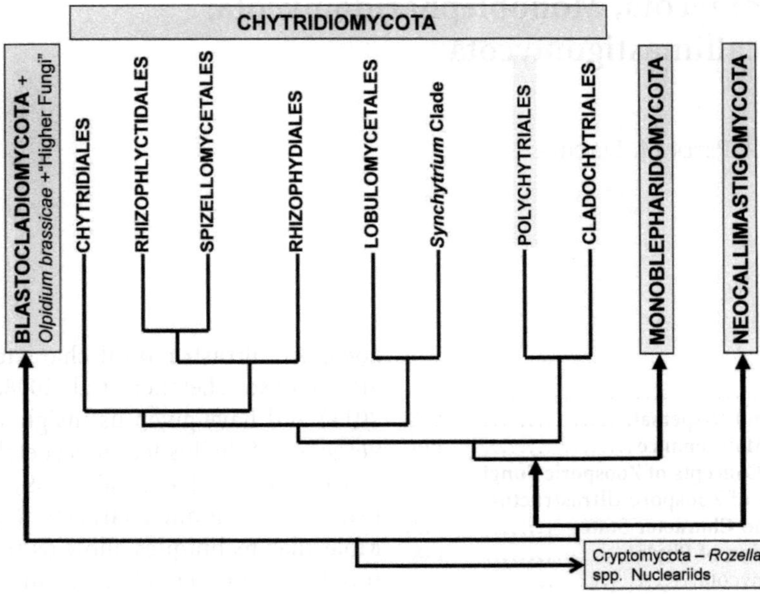

Fig. 6.1 Phylogenetic hypothesis for zoosporic fungi based on James et al. (2006b). Four phyla have been formally circumscribed: Blastocladiomycota, Chytridiomycota, Monoblepharidomycota, and Neocallimastigomycota. The phylum Cryptomycota, which is considered either a fungal or protistian phylum, includes the zoosporic genus *Rozella* spp., earlier classified in the Chytridiomycota

2007, 2011, 2012; Miki et al. 2011; Sime-Ngando et al. 2011).

This chapter updates Barr's (2001) review of zoosporic fungi, with an emphasis on how molecular and ultrastructural phylogenetic analyses have revolutionized the taxonomy of zoosporic fungi. As we rapidly learn more about the biology, diversity, and global distribution of zoosporic fungi, we are revising their systematics, which, as a consequence, is in a state of flux (Powell and Letcher 2012). The phylum Chytridiomycota as circumscribed by Barr (2001) has now been separated into three additional validly published phyla: Blastocladiomycota (James et al. 2006b), Monoblepharidomycota (Doweld 2001), and Neocallimastigomycota (Hibbett et al. 2007), each circumscribing a monophyletic lineage (Fig. 6.1, Table 6.1). Moreover, *Rozella* species, once classified in the Chytridiomycota, are now amalgamated with filose pseudopodiate (pseudociliate) organisms in the **phylum Cryptomycota** (Jones et al. 2011; Karpov et al. 2013). **Blastocladiomycota** is commonly placed as the sister group of zygomycetous fungi (Fig. 6.1) and diverges from other groups of zoosporic fungi [James et al. 2006b; see James et al. (2014)]. However, in different analyses, other placements may be found (Ebersberger et al. 2012; Sekimoto et al. 2011). We summarize progress in the systematics of **Chytridiomycota** (chytrids), **Monoblepharidomycota** (monoblephs), and **Neocallimastigomycota** (neocallimastigos).

II. Occurrence and Dispersal

Zoosporic fungi are common members of aquatic and soil microbial communities and can be isolated from or detected on leaf litter and tree-canopy detritus (Bandoni and Barr 1976; Bills et al. 2004; Letcher and Powell 2001, 2002b; Longcore 2005; Nikolcheva and Barlocher 2004; Powell 1993; Shearer et al. 2004). **Chytrids** are microscopic, and their thalli may be observed from environmental samples of algae, other hosts, or organic substrates. **Monoblephs** typically occur on totally submerged waterlogged twigs, decorticated

Table 6.1 Classification of Zoosporic Fungi from 1990 to 2013

Barr (1990)	Barr (2001)[a]	This volume
Chytridiomycota	**Chytridiomycota**	**Chytridiomycota**
Chytridiomycetes	Chytridiomycetes	Chytridiomycetes
Chytridiales	Chytridiales	Chytridiales
		Rhizophydiales
		Lobulomycetales
		Cladochytriales
		Polychytriales
Spizellomycetales	Spizellomycetales	Spizellomycetales
		Rhizophlyctidales
Monoblepharidales	Monoblepharidales	**Monoblepharidomycota**
		Monoblepharidomycetes
		Monoblepharidales
		Harpochytriales
		Hyaloraphidiomycetes
		Hyaloraphidiales
	Neocallimastigales	**Neocallimastigomycota**
		Neocallimastigomycetes
		Neocallimastigales
Blastocladiales	Blastocladiales	**Blastocladiomycota**
		Blastocladiomycetes
		Blastocladiales

[a]*Olpidium* and *Rozella*, previously classified in the Spizellomycetales and Olpidiaceae, place outside of the Chytridiomycota in molecular phylogenetic studies. Gene sequences of the type species of these two genera have not been obtained; consequently, the genera and family are *incertae sedis*. *Rozella allomycis* has been classified in a nomenclaturally validly described phylum, Cryptomycota (Jones et al. 2011)

twigs, fruits, or insect material in shallow freshwater habitats, and as filamentous growth, often as tufts with a slimy texture. **Neocallimastigos** are adapted for growth in the rumen and digestive tracts of animals, including sheep, goats, cows, horses, deer, elephants, and buffalo. They may be even more widespread among herbivores than previously recognized; they have recently been found associated with the digestive system of the green iguana, a herbivorous reptile (Liggenstoffer et al. 2010). Molecular techniques have detected them outside of host animals in anoxic landfills rich in cellulosic materials (Lockhart et al. 2006), and resistant spores can survive outside of their hosts in dried feces (Milne et al. 1989; Wubah et al. 1991).

The notion that zoosporic fungi are strictly aquatic fungi has been dispelled because there are essentially terrestrial groups such as Spizellomycetales and Rhizophlyctidales (Letcher et al. 2008a; Wakefield et al. 2010). Their **adaptations** for dispersal and survival are more complex than generally recognized. Clearly, chytrids require water or humidity to trigger zoospore release from sporangia and for zoospores to disperse. Although the zoospore is covered with a cell coat of varying prominence (Dorward and Powell 1983; Powell 1994; Shields and Fuller 1996), the zoospore is unwalled and becomes desiccated if left out of water for any extended period of time before it encysts. Zoospores of many chytrids, especially those of the Spizellomycetales, are capable of a squirming amoeboid-type motion that can advance the zoospore in a thin water film, but some moisture is still required for zoospore motility.

How far zoospores (Fig. 6.2A, E) can swim under their own power is not known, but distant dispersal by individual zoospores seems to be limited to a few centimeters (Hampson and Coombes 1989). Zoospores can remain motile after release from sporangia for a few seconds, hours, or even (rarely) days; but typically zoospore motility is ultimately limited in time because of their dependence on endogenous

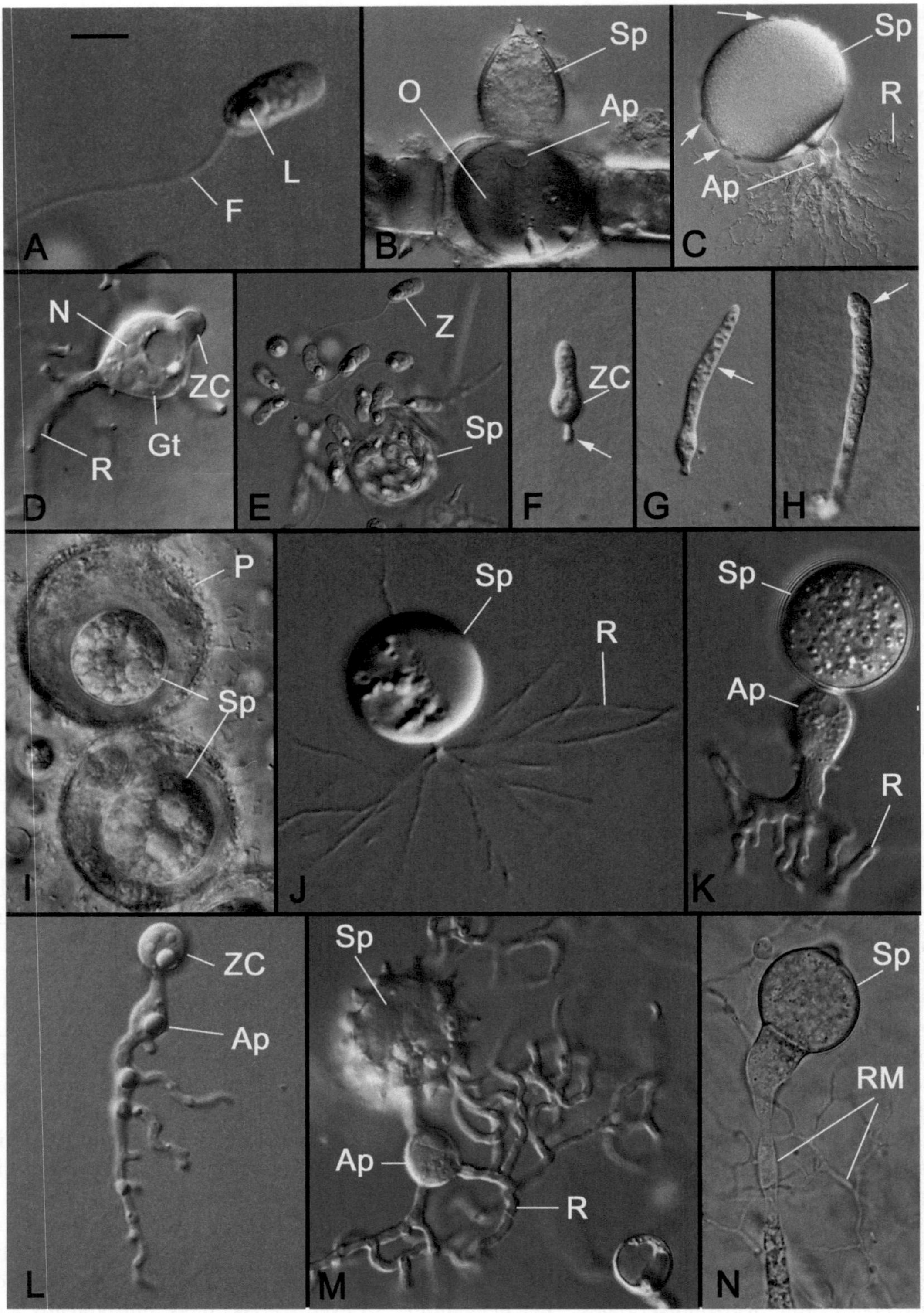

Fig. 6.2 Light microscopic morphological features of Chytridiomycota and Monoblepharidomycota. A. Oblong zoospore with prominent lipid globule (L) and posterior flagellum (F). B. *Chytridium olla* thallus growing on oospore in oogonium (O) of the green alga *Oedogonium*. The sporangium (Sp) bears an apical

reserves for energy (Powell 1976b). Their **pattern of swimming** is not one of a direct trajectory to a substrate but rather one of abrupt changes in direction, colliding into other chytrid zoospores and substrates before attachment, encystment, and germination. Chytrid and neocallimastigo zoospores are **attracted** to specific nutrient sources, and chytrid zoospores move toward blue wavelengths of light (Kazama 1972; Moss et al. 2008; Muehlstein et al. 1987, 1988; Orpin and Bountiff 1978; Strasburger 1878). In addition to dispersal, it seems that a primary role of the zoospore is the location of a suitable substrate or host on which to grow.

The fact that chytrids with identical ribosomal gene sequences can be isolated from soils in Australia and North America (Letcher et al. 2004) argues that they are not solely dependent upon zoospores for dispersal. Chytrid **resting spores** are thick-walled structures filled with glycogen, lipid, and protein reserves, and they may arise vegetatively or after sexual reproduction. Because they are commonly spherical, with or without wall ornamentation, and smaller than sporangia, they might easily be disseminated in soil, water, or air with a wider distribution than zoospores can provide. The ability of the two-celled resting spore of *Septosperma* to disarticulate from its substrate also argues for the importance of chytrid resting spores in wide dissemination (Powell and Blackwell 1991). Neocallimastigos seem to spread from mother to offspring through saliva during grooming and licking activities, and survival of resistant spores in dry dung may also enhance dispersal (Milne et al. 1989; Wubah et al. 1991).

Dung is widely recognized as an excellent substrate for a range of higher fungi (Webster 1970). As would be expected, neocallimastigos are found in dung; however, the occurrence of herbivore dung-inhabiting chytrids has only recently been discerned and is limited to Spizellomycetales and Lobulomycetales (Simmons et al. 2012; Wakefield et al. 2010). The presence of viable chytrids in freshly voided horse feces, as discovered in baiting experiments, indicates that chytrids can survive the digestive system of herbivores (Wakefield et al. 2010). Studies suggesting that birds and earthworms are **vectors** for chytrids are especially relevant in considerations of chytrids as dung fungi. Thornton (1970) demonstrated that earthworms could transport viable chytrids in two ways: among debris clinging to their mucilaginous surface or in casts they discharge after consuming soil. Supporting earthworms as dispersal agents for chytrids, Hampson and Coombes (1989) demonstrated that *Synchytrium endobioticum* dispersed greater distances when earthworms were present than when only zoospores were present. Birds have also been implicated in long-distance dispersal of chytrids, indicating that chytrids can survive the digestive system of birds and remain viable in bird dung (Thornton 1971). Whether birds acquire chytrids from

Fig. 6.2 (continued) apiculus. The chytrid penetrates the host with a haustorium-like apophysis (Ap). C. Thallus of *Gaertneriomyces semiglobifer* with multiple discharge pores (*arrows*) around sporangium (Sp) and finely branched rhizoidal system (R) bearing an apophysis (Ap). D. Germination of *Geranomyces variabilis* with exogenous development. Nuclei (N) have migrated from the zoospore cyst (ZC) into the germ tube (Gt), which expands and forms a sporangium with rhizoids (R). E. Release of zoospores (Z) from sporangium (Sp). F. Germination of zoospore cysts (ZC) of *Harpochytrium* sp. in a uniaxial thallus with basal holdfast (*arrow*). G. *Harpochytrium* sp. thallus with highly vacuolated, foamy appearing cytoplasm (*arrow*). H. *Harpochytrium* sp. cleaving zoospores beginning at apex (*arrow*) of thallus. I. *Olpidium* sp. monocentric holocarpic thalli endobiotic in pollen (P); the thallus is totally converted into a sporangium (Sp). J. *Alphamyces chaetifer* eucarpic thallus with spherical sporangium (Sp) and finely branched rhizoids (R) arising from a single axis. K. *Spizellomyces punctatus* eucarpic thallus with spherical sporangium (Sp) and coarsely branched rhizoids arising from a single apophysis (Ap); notice that the rhizoid tips are rounded and blunt. L. Germination of *Phlyctochytrium aureliae* with endogenous development. The nucleus remains in the zoospore cyst (ZC) and the germ tube branches into a rhizoidal system with an apophysis (Ap). M. *Phlyctochytrium aureliae* eucarpic thallus bearing sporangial (Sp) ornamentation and tubular rhizoids (R). A spherical apophysis (Ap) is far from the sporangium. N. Rhizomycelium (RM) of *Polychytrium aggregatum* is tubular with finely branched rhizoids and spherical sporangia (Sp). *Scale bar* shown in A = 3 μm in A; 6 μm in L; 10 μm in E–H; 15 μm in D, J, K, M, N; 20 μm in B, C, I

eating earthworms or plant debris harboring chytrids is not known. Aquatic birds carry thalli of the amphibian parasite *Batrachochytrium dendrobatidis* on the keratinous webbing of their feet and may also provide long-range dispersal of chytrids (Garmyn et al. 2012).

Plant pathogenic chytrids can be distributed by transport of contaminated plants and soil. *S. endobioticum*, the causal agent of potato wart disease, is readily introduced through infected seed potatoes and soils containing resting spores. There is no evidence that zoospores are effective at broad-scale dispersal of this disease (Hampson and Coombes 1989). *B. dendrobatidis* can be **transmitted between animals** (Rachowicz and Vredenburg 2004) and is believed to have been spread globally through movement of animals for food, medicine, research, and the global pet trade (Bai et al. 2010; Schloegel et al. 2012; Weldon et al. 2004).

III. Culture and Maintenance

Several informative references detail methods and techniques for isolation, culture, and growth of chytrids and monoblephs (Bills et al. 2004; Fuller and Jaworski 1987; Shearer et al. 2004). Most **chytrids** are extracted from a habitat by way of enrichment of environmental samples with heat-killed algae, chitin, cellulose, keratin, or pollen substrates and incubation at ambient temperatures for 2–3 days (Couch 1939; Barr and Désaulniers 1987). **Monoblephs** are frequently isolated from the surfaces of algae, fruits, and twigs in water (Emerson 1958, 1964; Emerson and Whisler 1968; Perrott 1955, 1958). **Neocallimastigos**, as obligate anaerobic fungi, must be grown under anaerobic conditions (Orpin 1975; Rezaeian et al. 2004). Most isolation studies extract neocallimastigos directly using a cannula collection system that aseptically penetrates into the rumen or other regions of the alimentary canal of herbivores (Orpin 1975).

A large number of zoosporic fungi are in culture, and the majority of these are in university-managed collections. Many zoosporic fungi survive on agar slants stored at 4 °C for up to 6 months. Advances have been made in **cryostorage** of chytrid cultures. Cultures grown in broth on cotton tips, transferred to a glycerin solution, and stored at −80 °C or in liquid nitrogen (Barr and Babcock 1994; Gleason et al. 2007) have been recovered after 15 years (C. E. Babcock, personal communication). Freezing techniques also facilitate the storage of plant-pathogenic chytrids. *Synchytrium solstitiale* stored in 0.5 M sucrose at −2 °C remained viable in host tissue for 3 months (Widmer 2006), and it may also remain viable as air-dried tissue for over 2 years (Bruckart et al. 2011).

IV. Phylogenetic Concepts of Zoosporic Fungi

The broadest ranging molecular phylogenetic analysis of zoosporic fungi was conducted by James et al. (2006b), and Fig. 6.1, which is based on that study, depicts our current phylogenetic hypothesis. Although sharing a common ancestor, the lineage including Neocallimastigomycota, Monoblepharidomycota, and Chytridiomycota diverges from the lineage that gave rise to the Blastocladiomycota and higher fungi (Fig. 6.1). Thus, it was unexpected that the Blastocladiomycota and the plant parasite *Olpidium brassicae* placed in a clade with non-zoosporic fungi. **Cellular characteristics** support the relationship of Blastocladiomycota with filamentous fungi, including sharing the loss of Golgi apparatus cisternal stacking (Powell and Letcher 2012). The phylogenetic placement of *Olpidium* spp. (James et al. 2006b; Sekimoto et al. 2011) based on molecular analyses is still perplexing when the mode of zoospore formation and zoospore structure are considered (Barr and Hartmann 1977; Lange and Olson 1979). *Rozella* spp., once classified in the Chytridiomycota (Barr 1980; Held 1975, 1981), are placed within the sister clade of all other fungi (James and Berbee 2011; James et al. 2006a, b).

Whereas the traditional classification of zoosporic fungi relied on **morphological features** of the thallus (Karling 1977; Sparrow 1960; Whiffen 1944), analyses of **zoospore ultrastructural features** and **molecular sequences** have revealed that many classically used morphological character states appear in multiple lineages and are convergent. For example, Sparrow (1960) used differences in zoospore discharge openings as a primary taxonomic characteristic for two series of chytrids (Operculatae and Inoperculatae), but we now know that operculate and inoperculate thalli may occur within a single evolutionary lineage. Only members of the Rhizophlyctidales and Spizellomycetales discharge zoospores exclusively through inoperculate openings (Powell 1976a). Five orders of Chytridiomycota (Chytridiales, Rhizophydiales, Cladochytriales, Lobulomycetales, and Polychytriales) include some members with operculate discharge and others with inoperculate discharge. Whether or not the underlying developmental mechanism for the production of an operculum in all orders is the same or different is not known, but differences have been described (Beakes et al. 1992; Powell et al. 2011; Taylor and Fuller 1981). As a second example, polycentric versus monocentric thallus complexity (Whiffen 1944) was used to distinguish families within Sparrow's (1960) two series of chytrids. However, recent phylogenetic analyses of the Chytridiales, Rhizophlyctidales, Cladochytriales, and Polychytriales have revealed members with polycentric and monocentric thalli within the same order. A third example is the so-called *Entophlyctis*-type of development, in which the germ tube rather than the zoospore cyst gives rise to the sporangium (Fig. 6.2D) (Blackwell et al. 2006). This exogenous type of development, along with endogenous development, is found in several lineages of Chytridiomycetes. Thus, it is clear that organisms within diverse evolutionary lineages, but with simple thalli growing in similar habitats and exposed to similar selective pressures, adapt with similar morphological phenotypes, resulting in a convergence of thallus features.

Contemporary taxa of chytrids are now delineated based on **molecular monophyly**. With this approach we look at a snapshot in time of the evolution of a species, with gene sequences serving as the primary taxonomic character. Because genes, zoospore ultrastructural characters, and thallus features evolve at different rates, we use a constellation of zoospore ultrastructural characters and thallus features to define taxa within monophyletic clades (Fig. 6.1). Zoosporic fungi are an ancient group of eukaryotes, and plesiomorphic character states shared with a common ancestor (=descent-based similarity; Hörandle and Stuessy 2010) may appear within diverging lineages only to be modified repeatedly or lost in multiple lineages. We have made great advances in delineating monophyletic orders, especially in circumscribing the limits of a monophyletic Chytridiales (Vélez et al. 2011) in the Chytridiomycota. Table 6.1 summarizes progress in the classification of zoosporic fungi with greater insights into their phylogenetic relationships.

V. Identification of Zoospore Ultrastructural Characters and Character States

Because of the stability of ultrastructural character states, ultrastructure is instrumental in understanding relationships among zoosporic fungi. Zoosporic fungi are notoriously apt at **thallus phenotypic diversity** (Powell and Koch 1977), which may adapt them well to changing environments but makes thallus-based identification challenging. With the added insight of molecular phylogenetic analyses, we have been able to identify and describe zoospore ultrastructural features useful for characterizing and delineating taxa. Koch (1961) first emphasized the "surprising diversity" of zoospore types in chytrids when he illustrated six major types, and from that beginning we now recognize a tremendous diversity in the architectural forms of chytrid zoospores. The two main regions of the zoospore that afford the richest supply of characters are the flagellar apparatus (Barr 1980, 2001; Barr and Désaulniers 1988) and the microbody–lipid globule complex (MLC) (Powell 1976b, 1978; Powell and Roychoudhury 1992).

The **flagellar apparatus** and auxiliary structures provide a range of characters and character states. Morphologies of **kinetosome-associated structures** (KASs) (Figs. 6.3F–H and 6.4F, G, I, J) are applicable in the determination of families (Letcher et al. 2008c; Vélez et al. 2011) and genera (Letcher and Powell 2005a; Longcore and Simmons 2012; Simmons 2011). In analyses of the precise configuration of the flagellar apparatus, the position of the KASs and microtubule roots has been described for several organisms and seems to be useful in generic delimitations (Barr and Désaulniers 1988; Roychoudhury and Powell 1992). The presence (Fig. 6.4D, E) or absence (Fig. 6.4C) of an electron-opaque core in the transition zone through which the axoneme doublets pass is a signature for several orders.

The **microbody–lipid globule complex** (MLC) is an assemblage of organelles consisting of lipid globules, a cisterna, mitochondria, and microbodies and is involved in the conversion of stored lipid into energy for zoospores (Powell 1976b, 1978). Because calcium may be sequestered in the MLC cisterna, which is positioned adjacent to the plasma membrane (Dorward and Powell 1982) and near the flagellar apparatus, it has been proposed to regulate flagellar beat and zoospore directionality (Powell 1983). How closely and the manner in which the organelles are linked in the MLC appear to be conserved indicators of phylogenetic relationships. Other features, such as the extensiveness or lobed nature of the microbody, are taxonomically informative (Barr and Désaulniers 1987; Letcher et al. 2008c). The MLC cisterna also provides character states for systematic comparisons. The MLC cisterna may be a simple cisterna (Fig. 6.3A) with no fenestrations, or it may contain a disk of honeycomb-patterned fenestrae (Fig. 6.3C) (Dorward and Powell 1982), termed the rumposome when first reported in the posterior portion of the zoospores of monoblephs (Fuller 1966; Fuller and Reichle 1968). Electron microscopic studies have eloquently demonstrated that the fenestrated disk of the cisterna is continuous with a nonfenestrated cisterna (Barr and Désaulniers 1987; McNitt 1974; Montecillo et al. 1980). The degree of fenestration may range from inconspicuous and minimal (Fig. 6.3B) to conspicuous and extensive (Fig. 6.3C), or it may be even more complex and multitiered (Barr and Désaulniers 1987; Fuller and Reichle 1968; Letcher et al. 2008c; Reichle 1972; Simmons et al. 2012).

VI. Characterization of Phyla

Molecular phylogenetic analyses have validated the application of zoospore ultrastructural characters in systematic considerations (James et al. 2006b; Letcher et al. 2008a, c; Longcore and Simmons 2012; Simmons 2011). We have repeatedly found that **molecular-based phylogenetic hypotheses predict zoospore ultrastructural types**. Using a constellation of character states, we can assign an organism to an order based on zoospore ultrastructural characters (Fig. 6.5). As we characterize a genetically more diverse sampling within orders, we are also uncovering more variation in zoospore architecture and can define more character states for each character (Letcher et al. 2008c, 2012a, b; Longcore and Simmons 2012; Picard et al. 2009; Simmons 2011; Simmons et al. 2012). Hence, **zoospore ultrastructural character states** can also be used to define families within orders and genera within families (Letcher et al. 2006, 2008a, c, 2012b; Longcore and Simmons 2012; Simmons 2011). There are zoospore types found in described species for which phylogenetic placement and classification into an order have not been resolved, and these species remain classified as *incertae sedis* (Beakes et al. 1988, 1993; Karpov et al. 2010; Nyvall et al. 1999; Powell 1981a, b).

Zoospores of each order are distinguished by a **suite of characters**, rather than a single defining feature. The constellation of ultrastructural states allows one to identify the order based on zoospore ultrastructural characters (Fig. 6.5). What complicates using zoospore ultrastructural characters alone to define orders is that, because of evolutionary descent, ancestral character states may be lost or transformed within multiple lineages with shared ancestry. For example, a MLC cisterna with fenestrae seems to be a character state shared with the last common ancestor of monoblephs and chytrids because it is present in both lineages. However, within diverging lineages,

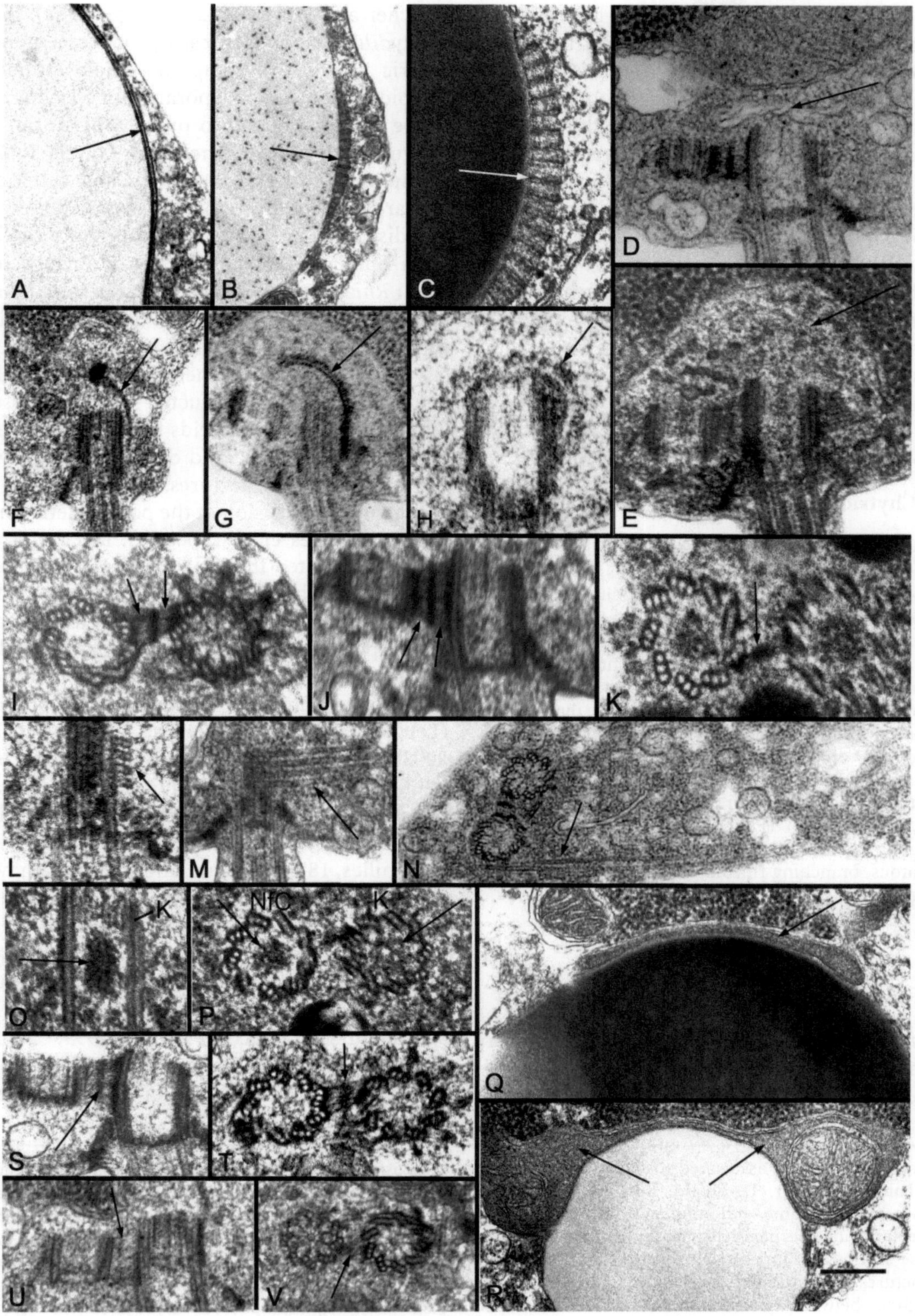

Fig. 6.3 Ultrastructural characters and character states in Rhizophydiales. *Arrows* indicate illustrated feature. A–C. Microbody–lipid globule complex cisterna. A. Simple. B. Inconspicuously fenestrated. C. Conspicuously fenestrated. D, E. Vesiculated region adjacent to kinetosome. D. Absent. E. Present. F–H. Kinetosome-

the fenestrae may be reduced or lost (Letcher et al. 2008c), and the cisterna may be absent (Longcore et al. 1999). When a fenestrated MLC cisterna is present, microtubule roots are typically also present (Barr and Désaulniers 1988; Dorward and Powell 1982). Conversely, when the MLC cisterna lacks fenestrae (=simple cisterna) or is absent, an organized microtubule root is typically absent (e.g., Picard et al. 2009; Powell et al. 2011). Thus, in using the concept of a characteristic zoospore type for each order, it is recognized that genes and morphology do not evolve at the same rate and molecular-based phylogenies allow tracking patterns of character state evolution.

A. Chytridiomycota

The Chytridiomycota is circumscribed as a monophyletic phylum containing a single class, Chytridiomycetes, with seven orders and two additional lineages. Doweld (2001) recognized the subclass Spizellomycetidae [=Spizomycetidae in Cavalier-Smith (1998)], but we do not use this subclass at this time (Table 6.1) because it would render subclass Chytridiomycetidae (Doweld 2001) polyphyletic (Fig. 6.1). The thalli of chytrids may grow endobiotically (Fig. 6.2I) or epibiotically (Fig. 6.2B) on a substrate or host, and the thallus may consist solely of a sporangium (holocarpic, monocentric) (Fig. 6.2I), a sporangium with rhizoids (eucarpic, monocentric) (Fig. 6.2C, J, K, M), or multiple sporangia (eucarpic, polycentric) growing along a filamentous, branching rhizoidal system (rhizomycelium) (Fig. 6.2N).

1. Rhizophydiales

Rhizophydium is among the larger and more complex genera of Chytridiomycetes and was traditionally classified in the Chytridiales (Letcher and Powell 2012; Sparrow 1960). *Rhizophydium* characteristically produces a monocentric thallus bearing a single tubular rhizoidal axis and a sporangium varying in shape from spherical, to oval, to pyriform, to irregularly lobed (Letcher and Powell 2012). Zoospores (Fig. 6.5A) are typically spherical and are released from one to several inoperculate discharge pores or tubes and, more rarely, from operculate openings. It was unexpected when *Rhizophydium* placed outside the Chytridiales clade in the James et al. (2000) molecular phylogenetic study. Thus, to explore the diversity in this genus, Letcher et al. (2004, 2006, 2008b, c, 2012b) conducted a broad-based global inventory of chytrids and revealed great molecular divergence and distinctive zoospore ultrastructural architectures. As the first step in the taxonomic revision of the polyphyletic Chytridiales (James et al. 2006a, b), Letcher et al. (2006) delineated the Rhizophydiales as a new order in the Chytridiomycota and designated a culture of *Rhizophydium globosum* as the epitype species of the genus. Rhizophydiales includes a large number of commonly collected and isolated chytrids as well as rare species (Letcher and Powell 2005a; Letcher et al. 2008b, c, 2012b; Longcore 2004; Longcore et al. 2011; Powell et al. 2011). Thus, what had once been a single genus with over 200 species (Letcher and Powell 2012) is now an order with 10 families, 18 genera, and lineages of unknown alliances. This clade also includes a wider range of thallus forms than previously realized. Thus far, all are monocentric except for *B. dendrobatidis*, which may be colonial. Several have multiple rhizoidal axes arising from the sporangium (Longcore et al. 1999, 2011). Although *Rhizophydium* species had been considered inoperculate, two operculate genera

Fig. 6.3 (continued) associated structure. F. Solid spur. G. Laminated spur. H. Shield. I–K. Fibrillar bridge between kinetosome and nonflagellated centriole. I. Fibrillar bridge perpendicular to two structures, transverse section. J. Fibrillar bridge perpendicular to two structures, longitudinal section. K. Fibrillar bridge diagonal between two structures, transverse section. L–N. Microtubular root. L. Oblique longitudinal section. M. Medial longitudinal section. N. Transverse section. O, P. Granular cylinder in kinetosome or nonflagellated centriole. O. Medial longitudinal section of kinetosome (K). P. Transverse section of nonflagellated centriole (NfC) and kinetosome (K). Q, R. Microbody. Q. Simple. R. Lobed. S–V. Zone of convergence of fibrils in fibrillar bridge between kinetosome and nonflagellated centriole. S. Wide (~0.075 μm), longitudinal section. T. Wide, transverse section U. Narrow (0.010–0.025 μm), longitudinal section. V. Narrow, transverse section. *Scale bar* shown in R = 0.15 μm in K; 0.16 μm in H, I; 0.20 μm in A–E, J, L, M, O–V; 0.25 μm in N; 0.33 μm in F, G

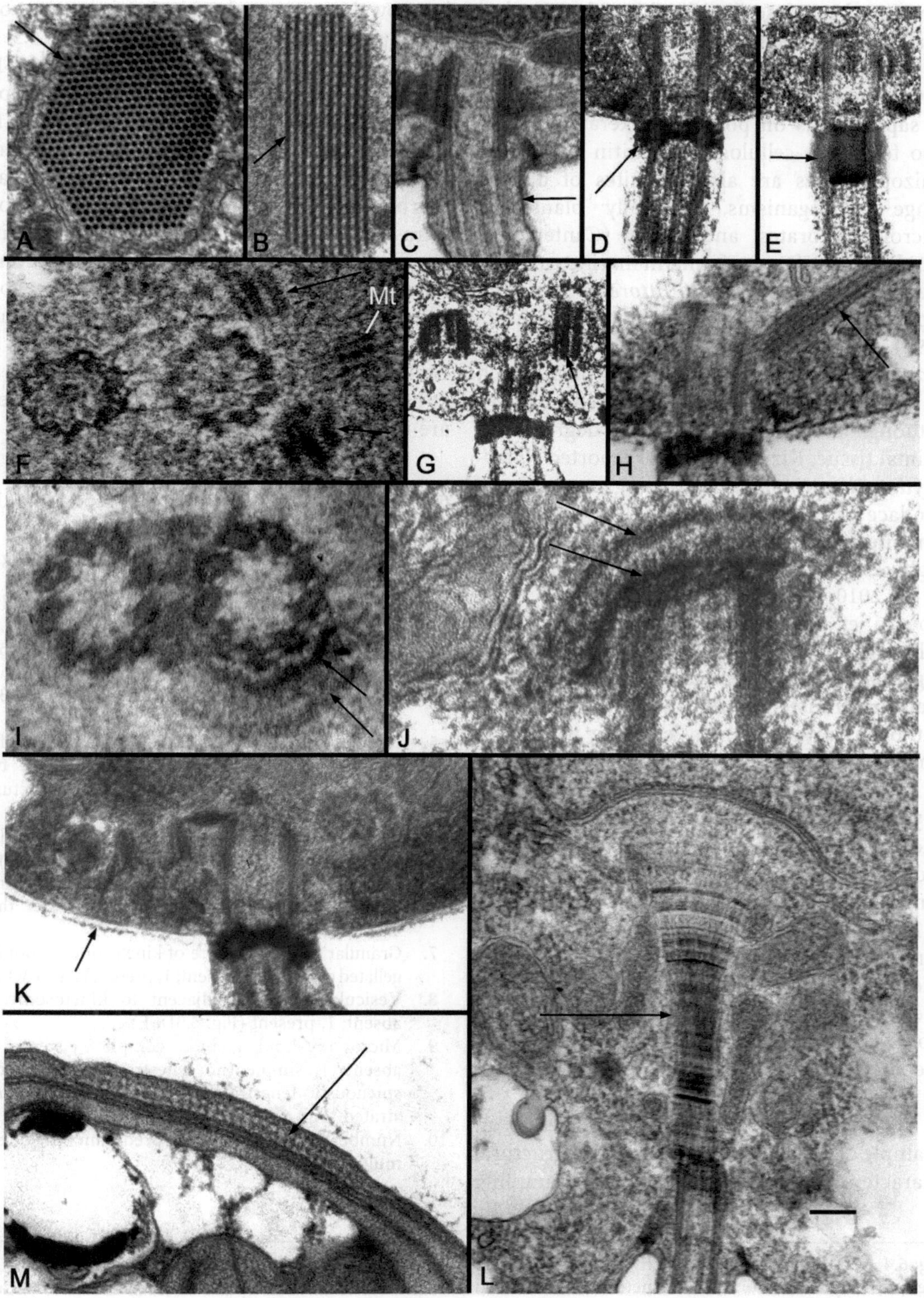

Fig. 6.4 Ultrastructural characters and character states in Chytridiales (A, B, D–K), Rhizophydiales (C), Rhizophlyctidales (L), and Monoblepharidomycota (M). *Arrows* indicate illustrated feature. A, B. Paracrystalline inclusion. A. Transverse section. B. Longitudinal section. C–E. Flagellar plug in base of flagellum. C. Absent.

have now been described for this order (Letcher et al. 2008c; Powell et al. 2011).

Members of the Rhizophydiales are environmentally diverse and commonly grow as **saprotrophs** on pollen and keratin but are also found on cellulose and chitin substrates. Rhizophydiales are also parasites of a wide range of organisms, especially planktonic microinvertebrates and algae (Canter and Lund 1951). A few are found in marine environments, and *Rhizophydium littoreum* has been reported as a **parasite** on crab eggs (Shields 1990) and algae (Kazama 1972). *Rhizophydium graminis* is a root parasite of higher plants, such as wheat, grasses, and a few dicots (Barr 1973). Although not generally considered degraders of animal tissue, Kiziewicz (2004) reported *Rhizophydium keratinophilum* growing on muscles of vendace fish in lakes. The only known **chytrid parasite of vertebrates** is *B. dendrobatidis*, the highly destructive pathogen of amphibians (Bai et al. 2010; Longcore et al. 1999, 2007; Piotrowski et al. 2004; Rosenblum et al. 2008; Schloegel et al. 2012; Voyles 2011). Evidence suggests that pathogenesis was acquired by **lateral gene transfers** from bacteria and oomycete pathogens rather than by evolving within the Rhizophydiales lineage (Sun et al. 2011).

Molecular-based ecological inventories of chytrids in lakes commonly detect novel clades and known species within the Rhizophydiales, indicating they may be a major component of fungal aquatic communities (Lefèvre et al. 2008, 2012; Monchy et al. 2011). It is possible that some of the novel phylotypes are chytrid parasites of plankton for which genes have not yet been sequenced and, hence, are not retrieved from public databases in BLAST searches (Lepère et al. 2008; Sønstebø and Rohrlack 2011).

The revision of the Rhizophydiales is an example of the value of zoospore ultrastructural characters and character states. Broad sampling has now demonstrated over 18 unique zoospore configurations in the order (e.g., Fig. 6.5 in Letcher et al. 2008c, 2012b; Powell et al. 2011), whereas in earlier studies *Rhizophydium* species were characterized as having a Group III-type zoospore (Barr and Hadland-Hartmann 1978). The key following the list of 14 characters below demonstrates how suites of zoospore ultrastructural character states distinguish families. Within the Rhizophydiales several lineages and subclades with distinctive zoospore types have now been described taxonomically and await greater sampling (Letcher et al. 2008b; Longcore et al. 1999, 2011; Powell et al. 2011; Powell and Roychoudhury 1992); these are not included in the key.

Characters and Character States of Zoospores in Rhizophydiales

1. Location of nucleus: 0, outside ribosomal aggregation; 1, embedded in ribosomal aggregation.
2. Endoplasmic reticulum ramifying through ribosomal aggregation: 0, absent; 1, present.
3. Kinetosome-associated structure: 0, absent; 1, solid spur; 2, laminated spur; 3, shield (Fig. 6.3E–H).
4. Microtubular root: 0, absent; 1, present (Fig. 6.3L–N)
5. Fibrillar bridge between kinetosome and nonflagellated centriole: 0, perpendicular to the two structures; 1, diagonal between the two structures (Fig. 6.3I, K).
6. Perpendicular zone of convergence in fibrillar bridge between kinetosome and nonflagellated centriole: 0, absent; 1, narrow (0.01–0.025 μm); 2, wide (approximately 0.075 μm; greater than 0.025 μm) (Fig. 6.3S–V).
7. Granular cylinder in core of kinetosome or nonflagellated centriole: 0, absent; 1, present (Fig. 6.3O, P).
8. Vesiculated region adjacent to kinetosome: 0, absent; 1, present (Fig. 6.3D, E).
9. Microbody–lipid globule complex cisterna: 0, absent; 1, simple (no fenestrations); 2, inconspicuously fenestrated; 3, conspicuously fenestrated (Fig. 6.3A–C).
10. Number of lipid globules: 0, predominantly one; 1, multiple.

Fig. 6.4 (continued) D. Present, Chytriomycetaceae. E. Present, Chytridiaceae. F, G. Kinetosome-associated structure a pair of stacked plates on either side of microtubular root. F. Transverse section, with microtubular root (Mt). G. Longitudinal section. H. Microtubular root, longitudinal section. I, J. Kinetosome-associated structure, a caplike body over kinetosome. I. Transverse section. J. Longitudinal section. K. Cell coat. L. Fibrillar rhizoplast between kinetosome and nucleus, longitudinal section. M. Rumposome (fenestrated cisterna) backed by microbody, longitudinal section. *Scale bar* shown in L = 0.08 μm in I, J; 0.10 μm in F; 0.12 μm in K, M; 0.13 μm in H; 0.15 μm in B–E; 0.18 μm in A, G; 0.20 μm in L

Key to Families in the Rhizophydiales using Ultrastructural Characters

A. Kinetosome associated, crescent-shaped structure a spur ....B

B. Spur solid (Fig. 6.3F) ....Terramycetaceae

BB. Spur layered (Fig. 6.3G) ....Rhizophydiaceae

AA. Kinetosome-associated-crescent-shaped spur absent (Fig. 6.3D)........................C

C. Fibrillar bridge between kinetosome and nonflagellated centriole diagonal and electron-opaque core in kinetosome (Fig. 6.3K) ....D

D. Electron-opaque core in kinetosome and in nonflagellated centriole (Fig. 6.3K) .... Kappamycetaceae

DD. Electron-opaque core in kinetosome and no electron-opaque core in nonflagellated centriole (Fig. 6.3O,P) ....Alphamycetaceae

CC. Fibrillar bridge between kinetosome and nonflagellated centriole perpendicular and no electron-opaque core in kinetosome or nonflagellated centriole (Fig. 6.3I, J) ....E

E. Nucleus embedded within ribosomal aggregation ....F

F. Perpendicular zone of convergence in fibrillar bridge between kinetosome and nonflagellated centriole wide, greater than 0.025 μm (Fig. 6.3S, T) ....Globomycetaceae

FF. Perpendicular zone of convergence in fibrillar bridge between kinetosome and nonflagellated centriole narrow, less than 0.025 μm (Fig. 6.3U, V) ....Gorgonomycetaceae

EE. Nucleus at surface of ribosomal aggregation ....G

G. Close association of a portion of the microbody with the kinetosome .... Angulomycetaceae

GG. No close association of a portion of the microbody with the kinetosome ....H

H. Perpendicular zone of convergence in fibrillar bridge between kinetosome and nonflagellated centriole wide, greater than 0.025 μm ....Aquamycetaceae

HH. Perpendicular zone of convergence in fibrillar bridge between kinetosome and nonflagellated centriole narrow, less than 0.025 μm I

I. Microbody not branching extensively away from lipid globule (Fig. 6.3Q) ....Pateramycetaceae

II. Microbody branching extensively away from lipid globule (Fig. 6.3R) ....Protrudomycetaceae

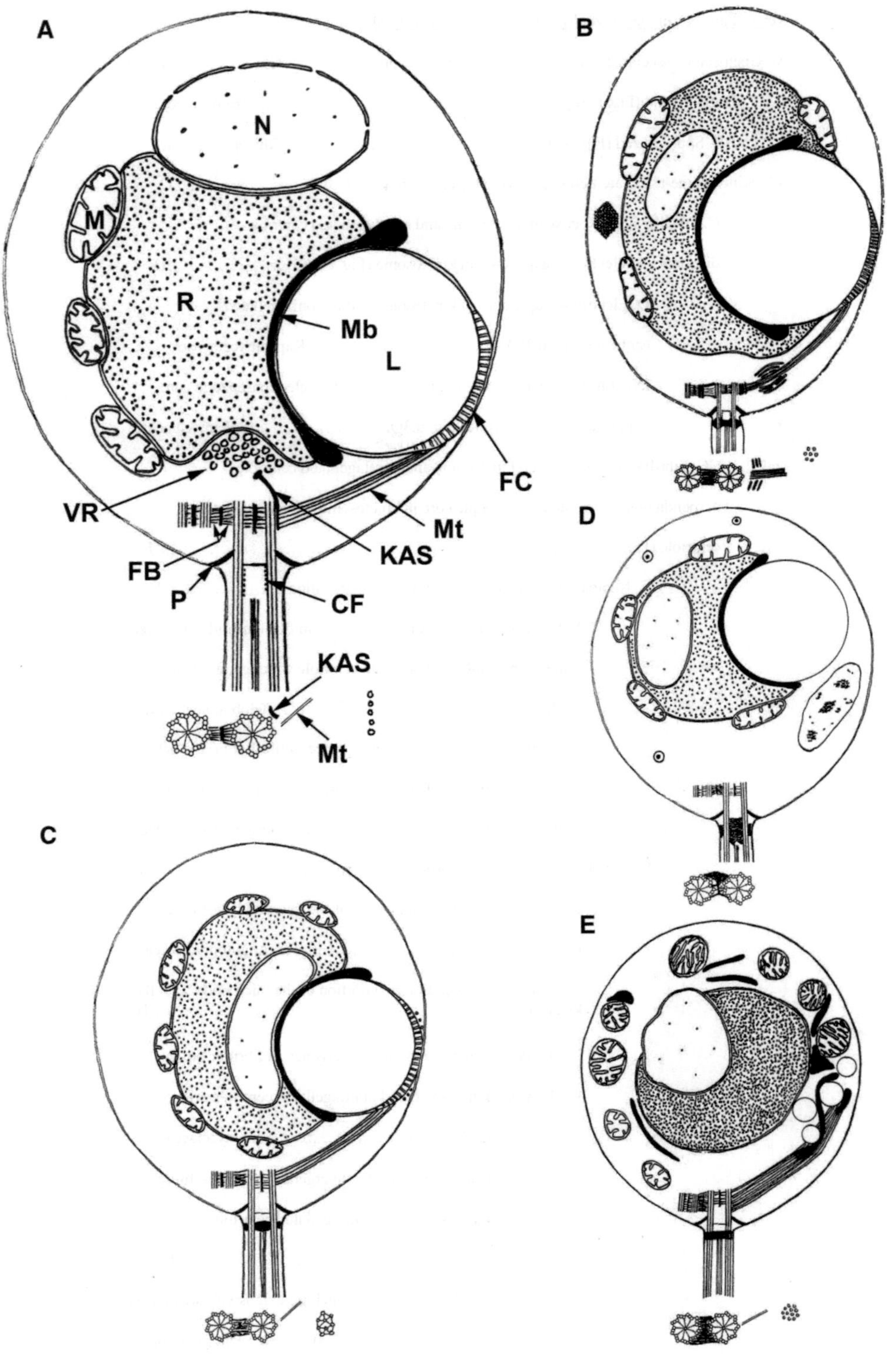

Fig. 6.5 (continued)

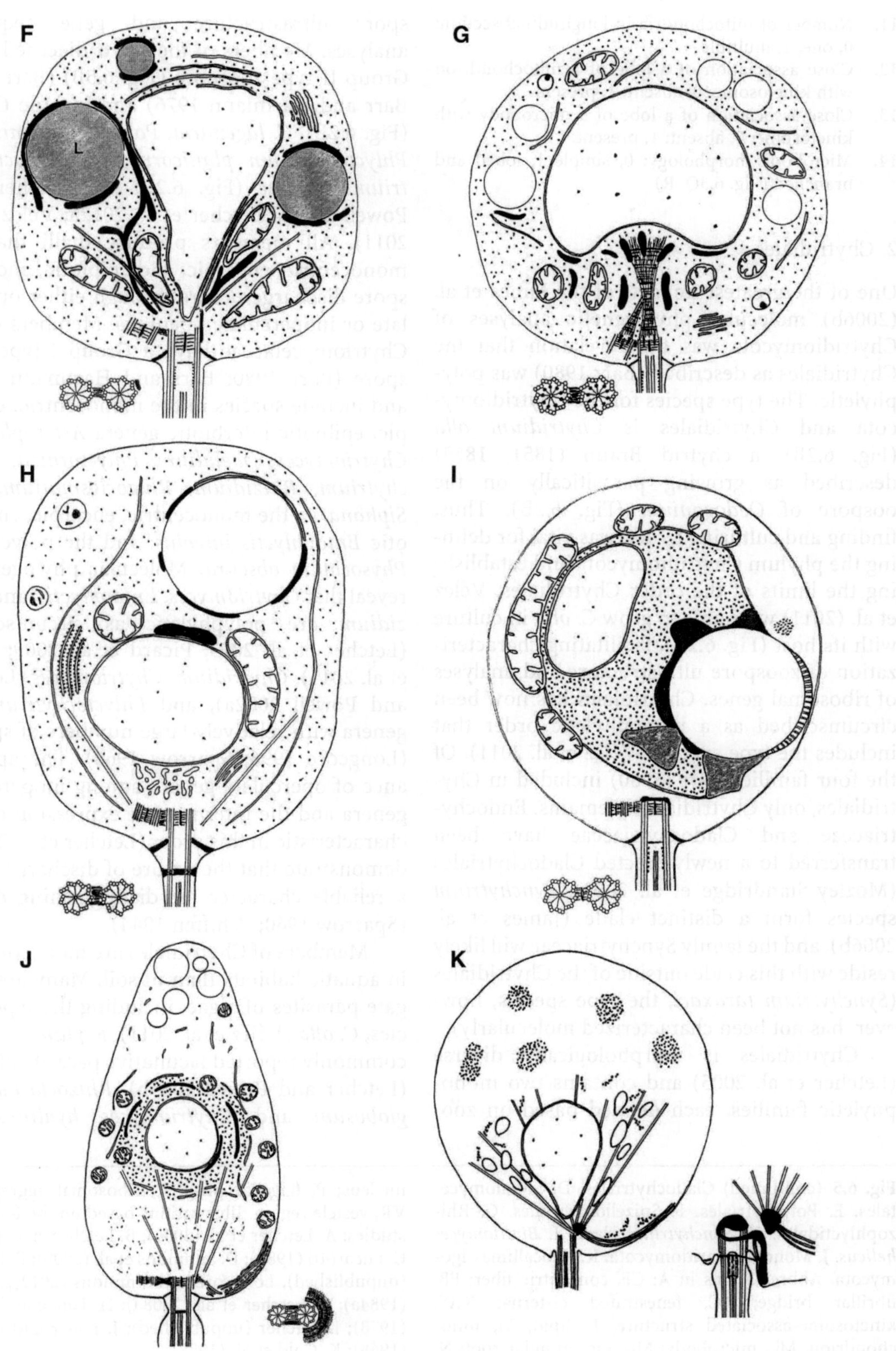

Fig. 6.5 Schematics of longitudinal sections through zoospores representative of 11 lineages of zoosporic fungi (A–K), with transverse sections through kinetosome, nonflagellated centriole, and microtubular root when present (A–J), and longitudinal section through kinetosome (K). A. Rhizophydiales. B. Chytridiales. C.

11. Number of mitochondria in longitudinal section: 0, one; 1, multiple.
12. Close association of a lobe of a mitochondrion with kinetosome: 0, absent; 1, present.
13. Close association of a lobe of a microbody with kinetosome: 0, absent; 1, present.
14. Microbody morphology: 0, simple 1, lobed and branched (Fig. 6.3Q, R).

2. Chytridiales

One of the greatest impacts of the James et al. (2006b) molecular phylogenetic analyses of Chytridiomycota was the revelation that the Chytridiales as described (Barr 1980) was polyphyletic. The type species for the Chytridiomycota and Chytridiales is *Chytridium olla* (Fig. 6.2B), a chytrid Braun (1851, 1855) described as growing parasitically on the oospore of *Oedogonium* (Fig. 6.2B). Thus, finding and culturing *C. olla* was vital for defining the phylum Chytridiomycota and establishing the limits of the order Chytridiales. Vélez et al. (2011) were able to grow *C. olla* in culture with its host (Fig. 6.2B), facilitating characterization of zoospore ultrastructure and analyses of ribosomal genes. Chytridiales has now been circumscribed as a monophyletic order that includes the type species (Vélez et al. 2011). Of the four families Barr (1980) included in Chytridiales, only Chytridiaceae remains. Endochytriaceae and Cladochytriaceae have been transferred to a newly erected Cladochytriales (Mozley-Standridge et al. 2009). *Synchytrium* species form a distinct clade (James et al. 2006b), and the family Synchytriaceae will likely reside with this clade outside of the Chytridiales (*Synchytrium taraxaci*, the type species, however, has not been characterized molecularly).

Chytridiales is morphologically diverse (Letcher et al. 2005) and contains two monophyletic families, each defined based on zoospore ultrastructure and gene sequence analyses. Members of the Chytridiaceae have a Group II-type zoospore (Fig. 6.5B) (Barr 1980; Barr and Hartmann 1976) and include *C. olla* (Fig. 6.2B), *C. lagenaria*, *Polyphlyctis unispina*, *Phlyctochytrium planicorne*, and *Phlyctochytrium aureliae* (Fig. 6.2L, M) (Letcher and Powell 2005b; Letcher et al. 2012a; Vélez et al. 2011). All members produce thalli that are monocentric, eucarpic, and epibiotic, and zoospore discharge occurs through either operculate or inoperculate openings. Members of the Chytriomycetaceae have a Group I-type zoospore (Barr 1980; Barr and Hartmann 1976) and include species in the monocentric, eucarpic, epibiotic/interbiotic genera *Asterophlyctis*, *Chytriomyces*, *Obelidium*, *Phlyctorhiza*, *Podochytrium*, *Rhizidium*, *Rhizoclosmatium*, and *Siphonaria*; the monocentric, eucarpic, endobiotic *Entophlyctis luteolus*; and the polycentric *Physocladia obscura*. Molecular phylogenetics reveal that *Chytriomyces*, *Entophlyctis*, and *Rhizidium* are polyphyletic as circumscribed (Letcher et al. 2005; Picard et al. 2009; Vélez et al. 2011). *Chytridium*, *Chytriomyces* (Letcher and Powell 2002a), and *Phlyctochytrium* are genera with relatively large numbers of species (Longcore 1996; Sparrow 1960). The appearance of operculate genera among inoperculate genera and the intermediate expression of this characteristic in this order (Letcher et al. 2012a) demonstrate that the nature of discharge is not a reliable character for distinguishing orders (Sparrow 1960; Whiffen 1944).

Members of Chytridiales are more common in aquatic habitats than in soil. Many are obligate parasites of algae, including the type species, *C. olla* (Vélez et al. 2011). *P. planicorne* is a commonly reported facultative parasite of algae (Letcher and Powell 2005b). *Rhizoclosmatium globosum* and *Chytriomyces hyalinus* are

Fig. 6.5 (continued) Cladochytriales. D. Lobulomycetales. E. Polychytriales. F. Spizellomycetales. G. Rhizophlyctidales. H. *Synchytrium* clade. I. *Blyttiomyces helicus*. J. Monoblepharidomycota. K. Neocallimastigomycota. Abbreviations in A: CF, concentric fiber; FB, fibrillar bridge; FC, fenestrated cisterna; KAS, kinetosome-associated structure; L, lipid; M, mitochondrion; Mb, microbody; Mt, microtubular root; N, nucleus; P, flagellar prop; R, ribosomal aggregation; VR, vesicle region. Illustrations based on the following studies: A. Letcher et al. (2006); B. Letcher et al. (2005); C. Lucarotti (1981); D. Simmons et al. (2009); E. Letcher (unpublished), Longcore and Simmons (2012); F. Barr (1984a); G. Letcher et al. (2008a); H. Lange and Olson (1978); I. Letcher (unpublished); J. Fuller and Reichle (1968); K. Gold et al. (1988)

among the most commonly reported and isolated chitinophilic chytrids from aquatic habitats. Isolated from soil, *Rhizidium phycophilum* grows in culture only in the company of a coccoid green alga, suggesting a symbiotic partnership (Picard et al. 2009).

Four features distinguish the zoospore of Chytridiales (Fig. 6.5B) from that of other orders: (1) the cordlike microtubule root is composed of approximately six to eight microtubules that are bundled together like a fist full of soda straws and extends laterally (Fig. 6.4F, H); (2) the kinetosome to nonflagellate centriole bridge is layered, with more electron-dense material at the anterior edge (Fig. 6.4K); (3) a paracrystalline structure composed of linear stacks of rods is present in the peripheral cytoplasm (Fig. 6.4A, B); and (4) a prominent cell coat (Dorward and Powell 1983) surrounds the zoospore body, but not the flagellar membrane (Fig. 6.4K). In their zoospores ribosomes aggregate at the center of the zoospore body, and organelles of the MLC are tightly packaged (Fig. 6.5B). When a microtubule root is present, it extends between the kinetosome and MLC cisterna, which is typically fenestrated. An axonemal basal plug is present in the transition region of the axoneme with axonemal microtubules passing through it, and the kinetosome and nonflagellated centriole are usually parallel (Fig. 6.5B) (Barr 1980; Barr and Désaulniers 1987, 1988; Barr and Hartmann 1976; Dorward and Powell 1982, 1983; Letcher and Powell 2005b; Letcher et al. 2005, 2012a; Longcore 1992b, 1995; Picard et al. 2009; Vélez et al. 2011).

KASs and the morphology of the electron-opaque plug in the transition region of the flagellum (FP) distinguish the Group I- and Group II-type zoospores (Barr 1980). In Chytridiaceae (Group II-type zoospore) the KASs are layered caplike structures that typically cover the anterior end and side of the kinetosome (Fig. 6.4I, J), and the FP is as long as it is wide (Fig. 6.4E). In Chytriomycetaceae (Group I-type zoospore) the KASs are stacked plates (Fig. 6.4F, G) between which the microtubule root extends from the kinetosome (Fig. 6.4F) to the MLC cisterna (Fig. 6.5B), and the FP is biconcave, shaped like a dog bone (Fig. 6.4D).

Investigations of genetically more diverse taxa within the two families of Chytridiales are revealing additional variations in each type of zoospore, with either modification or loss of a character. For example, *Phlyctochytrium aureliae* (Chytridiaceae) zoospores are patterned on the Group II-type zoospore, but in place of the caplike KAS there is an amorphous anvil-shaped KAS; and the fenestrations in the MLC cisternae are reduced in diameter (Letcher et al. 2012a). *R. phycophilum* (Chytriomycetaceae) zoospores are patterned on the Group I-type zoospore but have lost the stacked-plate KAS, microtubule root, and fenestrations in the MLC cisternae (Picard et al. 2009).

3. Cladochytriales

Cladochytriales was erected as a segregate from Chytridiales based on molecular monophyly and distinct zoospore ultrastructural characters (Mozley-Standridge et al. 2009). Molecular phylogenetic analyses (James et al. 2006b; Mozley-Standridge et al. 2009; Steiger et al. 2011) revealed that the order includes species of eight described genera, which are assigned to four families or are considered *incertae sedis*: *Catenochytridium (incertae sedis)*, *Cladochytrium* (Cladochytriaceae), *Cylindrochytridium (incertae sedis)*, *Nowakowskiella* (Nowakowskiellaceae), *Septochytrium* (Septochytriaceae), *Endochytrium* (Endochytriaceae), *Nephrochytrium* (*incertae sedis*), and *Allochytridium* (*incertae sedis*). However, in these analyses, *Allochytridium*, *Endochytrium*, and *Nephrochytrium* were polyphyletic (Mozley-Standridge et al. 2009). The order includes members with monocentric and polycentric thalli, epibiotic or endobiotic habits, apophysate and nonapophysate rhizoids, and operculate and inoperculate sporangia. The thallus structure may be variable as in *Septochytrium*, which is capable of producing either monocentric or polycentric thalli. The presence of **catenulate rhizoidal swellings** and intercalary swellings (=spindle organs, turbinate swellings) along the rhizomycelium appear to be a morphological feature characteristic of members of this order.

Members of this order are most commonly found on decaying plants and algae from aquatic habitats, suggesting they have a role in the initial degradation of cellulose-containing materials. With robust and extensively branched rhizoids or rhizomycelia, they are readily isolated from cellulosic baits and cultured on dilute soluble starch agar (Mozley-Standridge et al. 2009).

The distinguishing characteristic of the zoospore (Fig. 6.5C) is the structure of the lateral root, which consists of bundles of up to 25 microtubules with spaces between microtubules cross-linked with lateral fibrillar links (Barr and Désaulniers 1987, 1988; Lucarotti 1981; Mozley-Standridge et al. 2009). The basic zoospore design for this order is similar to that in the Chytridiales: a lateral root joins the fenestrated cisterna and kinetosome; an electron-opaque flagellar plug occupies the transition zone of the flagellar axoneme; ribosomes are aggregated in the core of the body of the zoospore; organelles of the MLC are tightly packaged; and the nonflagellated centriole is parallel to the kinetosome and joined by a dense fibrillar bridge. Variations in the states of some of these characters will be useful in distinguishing genera. For example, the MLC cisterna may have a thickened cisternal area containing the fenestrae (Barr 1986; Lucarotti 1981) or a narrow cisterna with a small fenestrated area (Barr et al. 1987), or it may contain two or three tiers of fenestrae in the cisterna (Barr and Désaulniers 1987). Structures associated with the kinetosome seem to distinguish genera and will be useful as the ultrastructure of more zoospores of this order is characterized. For example, in zoospores of *Allochytridium luteum* the microtubule root originates from a u-shaped structure connected to kinetosomal triplet 1, and in zoospores of *Catenochytridium hemicysti* rods are parallel and linked to kinetosomal triplets 9 and 2 with a bridge partially encircling the kinetosome and joining the two rods (Barr and Désaulniers 1988).

4. Lobulomycetales

In the James et al. (2006b) molecular analysis of Chytridiomycota, two species of *Chytriomyces*, *C. angularis* (Longcore 1992a) and *C. poculatus* (Willoughby and Townley 1961), placed outside of the clade that included the type of the genus, *Chytriomyces hyalinus* (Letcher and Powell 2002a). Comparative studies of *C. angularis* (Longcore 1992a) substantiated that the zoospore ultrastructure differed from that of chytridialian zoospores, and thallus features (fine, sparsely branched rhizoidal system and absence of a rhizoidal subsporangial swelling) were not characteristic of the type for *Chytriomyces*.

Additional collections and molecular environmental sequencing illuminated the diversity within this clade, leading to the establishment of a new order, Lobulomycetales, which includes four genera (*Alogomyces*, *Clydaea*, *Lobulomyces*, *Maunachytrium*) and six species (Simmons et al. 2009, 2012). Based on nuclear small subunit (SSU) ribosomal DNA sequence analysis (Müller et al. 1999), the marine algal parasite *Chytridium polysiphoniae* has been assigned to this order (Simmons et al. 2009). All members of the order are monocentric and include operculate or inoperculate organisms. They have been collected or their phylotypes detected from springs, *Sphagnum* in acidic lakes, ice-fed lakes, alpine barren soil, crop soils, acidic forest soils, tree-canopy detritus, and horse manure (Simmons et al. 2009, 2012). Environmental molecular sequencing studies often identify members of this order in lakes (Monchy et al. 2011) and deep-sea habitats (LeCalvez et al. 2009). Although the Lobulomycetales is a small group at this time, the extreme range in habitats in which its members are found suggests that this order is more diverse than presently described and is a common member of soil and aquatic microbial communities.

The most distinguishing zoospore ultrastructural characters (Fig. 6.5D) in this order are the anterior extensions on the electron-opaque plug in the transition region of the axoneme and dense amorphous material bridging the flagellum and nonflagellated centriole (Fig. 6.5D) (Longcore 1992a; Simmons et al. 2009). When Simmons et al. (2009) originally established the order, they reported the absence of MLC cisternae. However, a MLC cisterna with small fenestrae, morphologically quite distinct from the large honeycomb-patterned fenestrae in Chytridiales, was recently observed in zoospores of *Alogomyces* (Simmons et al. 2012). Zoospores contain a ribosomal aggregation around the nucleus and one to several lipid globules in the MLC. No organized microtubule root has been observed in any of the zoospores.

5. Polychytriales

The Polychytriales (Longcore and Simmons 2012) was erected based on the *Polychytrium* clade (James et al. 2006b). Its members are rhizophlyctoid chytrids (Dogma 1973) in which rhizoids emanate from multiple sites on the sporangium. All members grow on chitin, and all except *Karlingiomyces asterocystis* are able to grow on cellulose and keratin. The order consists of five genera (*Arkaya*, *Karlingiomyces*, *Lacustromyces*, *Neokarlingia*, *Polychytrium*), two of which are newly described and include new combinations with existing species: *Arkaya* with *Rhizophlyctis serpentina* and *Neokarlingia* with *Rhizophlyctis (Karlingia) chitinophila*. Three of the genera are monocentric, and two genera, *Polychytrium* and *Lacustromyces*, are polycentric with broadly tubular rhizomyceliums (Fig. 6.2N) **lacking the turbinate swellings** characteristic of polycentric members of the Cladochytriales. The only operculate genus is *Karlingiomyces* (Blackwell et al. 2004); the other genera release zoospores through inoperculate openings. Polychytriales is the sister group of Cladochytriales (James et al. 2006b), a clade that also includes monocentric and polycentric thalli but is characterized by growth on cellulose rather than chitin.

Each genus has a distinct suite of zoospore ultrastructural characters. The zoospore ultrastructure (Fig. 6.5E) is remarkably varied in this order (Longcore 1993; Longcore and Simmons 2012) and harkens diversity that is likely to be discovered. The zoospores are spherical and relatively large, typically greater than 4 μm in diameter. The zoospores are distinctive because the nonflagellated centriole is longer than that in other orders, with its length equal to or exceeding its diameter and with copious densely packaged fibrillar material joining the kinetosome the full length of the nonflagellated centriole. Microtubule roots range from three to none, and an electron-opaque plug in the transition region of the flagellum occurs in three of the five genera. *Lacustromyces* has the most extensive microtubule root system with three roots, one of which is massive and embedded in dense material (Longcore 1993). The MLC is varied and includes multiple lipid globules surrounded by or embedded in an extensive microbody in *Polychytrium aggregatum* and *Lacustromyces hiemalis*. The MLC cisterna is fenestrated only in zoospores of *Arkaya*, and the MLC cisterna is reported to be absent in the other genera.

6. Spizellomycetales

The earliest chytrids described were aquatic parasites of algae and were discovered by botanists observing algae (Braun 1851, 1855). Barr (1980) and Longcore et al. (1995) recognized that zoospores of more recently described soil-inhabiting species of two historic genera, *Phlyctochytrium* and *Entophlyctis*, were different from zoospores of the type species of these genera, which were algal parasites. Consequently, new genera were erected for soil-inhabiting species of *Phlyctochytrium* and *Entophlyctis* and were classified in a newly established order, Spizellomycetales (Barr 1980, 1984b; Longcore et al. 1995). Spizellomycetales was the first order separated from Chytridiales based on fundamental differences in zoospore ultrastructure (Barr 1980). Members of Spizellomycetales are distinct from other chytrids because they lack the translation elongation factor 1-alpha gene (EF-1 alpha) and instead possess the paralog, **elongation factor-like gene** (EFL) (James et al. 2006a; Keeling and Inagaki 2004; Simmons 2011; Simmons and Longcore 2012). Whether or not the paralog EFL is due to lateral gene transfer or to gene duplication and loss (Keeling and Inagaki 2004), its presence in all Spizellomycetales examined so far suggests a single evolutionary event corresponding to a major radiation of a chytrid lineage in soil.

Barr (1980) provisionally placed *Rhizophlyctis*, *Rozella*, *Olpidium*, and *Caulochytrium* in Spizellomycetales because ribosomes were dispersed in their zoospores and the nucleus was bridged to the kinetosome by either a striated rhizoplast or mitochondrion (Barr and Hadland 1977; Held 1975, 1981; Powell 1981b). Barr (2001) later questioned the relatedness of these taxa, emphasizing marked differences in nuclear features. *Rhizophlyctis*, *Rozella*, *Olpidium*, and *Caulochytrium* are now excluded from Spizellomycetales because phylogenetic placement in molecular analyses confirms Barr's (2001) doubts (James and Berbee 2011; James et al. 2006b; Karpov et al. 2010).

As Spizellomycetales is currently circumscribed (Simmons 2011; Wakefield et al. 2010), all are eucarpic, monocentric, and inoperculate. A great amount of genetic variation and diversity within Spizellomycetales is apparent, even for isolates collected within the same geographic location (Simmons 2011; Simmons and Longcore 2012; Wakefield et al. 2010). There are two monophyletic families, each corresponding to a specific mode of thallus development. Thalli in the Spizellomycetaceae grow epibiotically on substrates and exhibit endogenous development (the nucleus remains in the zoospore cyst, which develops into the sporangium, Fig. 6.2K) (Wakefield et al. 2010). Rhizoids often have a subsporangial swelling (apophysis), and the tips tend to be rounded or blunt (Fig. 6.2K). Spizellomycetaceae contains 4 genera (*Spizellomyces*, *Kochiomyces*, *Gaertneriomyces*, *Triparticalcar*) with 12 validly published species, but *Spizellomyces* and *Gaertneriomyces* are polyphyletic (Wakefield et al. 2010). Thalli in the Powellomycetaceae (Simmons 2011; Simmons and Longcore 2012) grow endobiotically and display exogenous development (the nucleus migrates from the zoospore cyst into the germ tube, and the germ tube grows into the sporangium with rhizoids) (Fig. 6.2D). Generally, the zoospore cyst persists attached to the sporangium and may function as the discharge tube (Powell and Koch 1977; Simmons 2011; Simmons and Longcore 2012). Powellomycetaceae contains four genera (*Fimicolochytrium*, *Geranomyces*, *Powellomyces*, *Thoreauomyces*) with eight species.

Spizellomycetalean chytrid are essentially ubiquitous in soils (Barr 1980; Wakefield et al. 2010). They are common saprobes of pollen and are found even in harsh and arid environments and in dung. Studies are beginning to explore the dynamics of Spizellomycetales in soil microbial communities and in nutrient dynamics and sustainability (Midgley et al. 2006). From studies focused on molecular detection of fungi from exposed soils at high elevations, spizellomycetalean chytrid phylotypes are prominent components of the fungal community (Freeman et al. 2009; Schmidt et al. 2012). As parasites of nematodes and oospores of downy mildews, they may have a beneficial impact on plants. On the other hand, as parasites of arbuscular mycorrhizae, they may be detrimental to plants [reviewed in Powell and Letcher (2012) and Wakefield et al. (2010)].

Zoospores of the Spizellomycetales can be recognized with a light microscope because they may become polymorphic even while swimming, shifting between elongate, round, or amoeboid. They sometimes swim with their flagellar insertion anterior, trailing the flagellum alongside the zoospore body (Fuller and Jaworski 1987). The constellation of their zoospore ultrastructural characters (Fig. 6.5F) is also distinctive because their ribosomes are dispersed, a portion of the nucleus is positioned adjacent to the kinetosome, an electron-opaque plug is absent from the flagellar transition zone, the nonflagellated centriole is at an acute to right angle with the kinetosome, organelles of the MLC are loosely packaged and the MLC cisterna is never fenestrated, and microtubule roots originate from kinetosome-associated structures and extend anteriorly but are not associated with the MLC. A constellation of zoospore ultrastructural character states distinguishes each genus in Spizellomycetales (Barr 1980, 1981, 1984a, b; Barr and Allan 1981; Longcore et al. 1995; Simmons 2011; Simmons and Longcore 2012). Where multiple types of kinetosome-associated structures were used within a single genus (*Spizellomyces* and *Gaertneriomyces*), molecular phylogenetics has demonstrated that the genus was polyphyletic (Barr 1980; Simmons 2011; Simmons and Longcore 2012; Wakefield et al. 2010). The variation in zoospore ultrastructure in the Spizellomycetales demonstrates well the intrinsic value of analyzing zoospore ultrastructure when describing new chytrid species.

7. Rhizophlyctidales

Rhizophlyctidales (Letcher et al. 2008a) was established as an order delineated from the Spizellomycetales (Barr 1980). In earlier electron microscopic analyses of zoospores of isolates putatively identified as *Rhizophlyctis rosea*, Barr and Désaulniers (1986) discovered four distinct zoospore subtypes. These observations presaged the great diversity Letcher et al. (2008a) later found in their analyses of morphology, zoospore ultrastructure, and nuclear large subunit (LSU) and internal transcribed spacer region (ITS) rRNA gene sequences of 49 isolates in the *R. rosea* complex from globally distributed soil samples, a study that included isolates previously studied ultrastructurally (Barr and Désaulniers 1986; Barr and Hartmann 1977). In molecular phylogenetic studies, Rhizophlyctidales places as the sister group of Spizellomycetales (Fig. 6.1) (James et al. 2006b), but the thalli, with multiple rhizoids emanating from the sporangial surface, and the distinctive zoospore ultrastructure distinguish its members from those in the Spizellomycetales (Letcher et al. 2008a). In addition to members with monocentric thalli, *Catenomyces persicinus* is a polycentric taxon in the order (James et al. 2006b). Rhizophlyctidales also differs from Spizellomycetales genetically, as evidenced by the possession of the translation elongation factor 1-alpha gene instead of the paralog elongation factor-like gene characteristic of Spizellomycetales (James et al. 2006a; Keeling and Inagaki 2004; Simmons 2011).

Rhizophlyctidales (Letcher et al. 2008a) includes four monophyletic families, each distinguishable morphologically and corresponding to one of the four zoospore subtypes (Barr and Désaulniers 1986). Each of the four families contains a single described genus (Rhizophlyctidaceae: *Rhizophlyctis*, Sonoraphlyctidaceae: *Sonoraphlyctis*, Arizonaphlyctidaceae: *Arizonaphlyctis*, and Borealophlyctidaceae: *Borealophlyctis*), but many isolates remain uncharacterized taxonomically. *R. rosea* is the type of *Rhizophlyctis* (Blackwell and Powell 1999), but the genus is not monophyletic because several of its species are known to reside in other orders (Letcher et al. 2006; Longcore and Simmons 2012). *R. rosea* is by far the most commonly collected and studied species within the order [reviewed in Letcher et al. (2008a)], and species of the other three genera are more rarely found (Letcher et al. 2008a). *R. rosea* can be considered a morphospecies because among 42 isolates in Rhizophlyctidaceae sequences were highly divergent, LSU > 91% and ITS > 60% (Letcher et al. 2008a). Despite the divergence of the morphospecies, phylotypes can be either cosmopolitan, with similar phylotypes found on different continents, or divergent, with dissimilar phylotypes found in the same location (Letcher et al. 2008a).

Members of Rhizophlyctidales are primarily terrestrial saprobes of cellulosic substrates. *R. rosea* is ubiquitous in agricultural and horticultural soils and vegetative debris matter and may survive periods of extended soil drying as desiccated sporangia (Gleason et al. 2004; Willoughby 1998; 2001) or in a resistive, resting stage (Johanson 1944). Thus, it was unexpected that molecular sequence analyses of environmental samples detected phylotypes of this clade in lakes (Monchy et al. 2011). It is possible that dispersion of propagules of these fungi may occur from soil perturbation via agricultural runoff or airborne desiccated sporangia or resting spores.

Although morphologically diverse, the four zoospore types in Rhizophlyctidales have in common a unique suite of ultrastructural features [see Fig. 6 in Letcher et al. (2008a)]. Their zoospores (Fig. 6.5G) lack a transition zone plug, organelles of the MLC are loosely arranged, the nonflagellated centriole is positioned at an acute angle to the kinetosome, and neither microtubule roots nor microtubules have been observed. In addition, the zoospore of *R. rosea* is characterized by dispersed ribosomes, multiple lipid globules, numerous mitochondria closely associated with the nucleus, a centrally located nucleus, and a fibrillar rhizoplast (Fig. 6.4L) extending from the kinetosome and nonflagellated centriole to the posterior end of the nucleus with closely clustered microbodies (Barr and Hartmann 1977; Letcher et al. 2008a).

8. *Synchytrium* Lineage

In sorting taxa formerly classified in the Chytridiales into monophyletic orders, several groups await official circumscriptions. One of these is the *Synchytrium* clade positioned as the sister group of the Lobulomycetales (James et al. 2006b). The family Synchytriaceae is excluded from Chytridiales because of a lack of monophyly with this order. The genus *Synchytrium* is composed of more than 250 described species of obligate plant and algal parasites (Karling 1964). At some stage of development the thallus is colonial, producing sporangia in sori, and sexual reproduction occurs with the fusion of motile gametes (Karling 1964). *S. endobioticum*, the causal agent of **potato wart disease**, has the potential to destroy crops, make soils unusable for potato crops for years, and result in the quarantine of a district's potatoes (Powell 1993). Although essentially eliminated from the USA (Putnam and Hampson 1989), *S. endobioticum* remains a threatening **pathogen of potatoes** in many regions of the world despite strict global quarantine regulations. *S. endobioticum* is on the US list of select agents and toxins that pose threats of economic damage to major agricultural crops if reintroduced (Rossman et al. 2006). On the other hand, species are being explored as potential **biocontrol agents** of invasive plants. Reports of *Synchytrium minutum* on kudzu *(Pueraria lobata)* vines in Korea demonstrate that this species is widespread in Asia and has generated interest as a biocontrol agent for kudzu in regions where it is invasive with no natural controls (Yun et al. 2011). Another species, *Synchytrium solstitiale*, is being evaluated as a biological control agent for the yellow starthistle, an invasive plant in the western USA (Bruckart et al. 2011).

The *Synchytrium* clade includes *S. decipiens*, *S. endobioticum*, *S. macrosporum*, *Synchytrium* sp. (James et al. 2006b), and *S. minutum* (Yun et al. 2011). Partial SSU sequences of these five isolates of *Synchytrium* are 85–95 % similar, while two strains of *S. endobioticum* (P-58 and AS-1) are only 90 % similar. *S. endobioticum* contains at least 20 pathotypes, and a cooperative global effort has attempted to unify the coding of pathotypes (Baayen et al. 2006; Stachewicz and De Boer 2002). Comparisons of sequence similarities indicate significant molecular divergence among *Synchytrium* taxa and potential issues with taxon identification based on morphology.

The zoospore ultrastructure is known only for *S. endobioticum* (Fig. 6.5H) (Lange and Olson 1978) and *S. macrosporum* (Montecillo et al. 1980), and the zoospores of these two species are remarkably different. Zoospores of these taxa have a suite of ultrastructural features that includes a single lipid globule, dispersed ribosomes, microbodies associated with both the lipid globule and the nucleus, an angled orientation of the nonflagellated centriole relative to the kinetosome, and microtubules that radiate from the kinetosome into the cytoplasm. The zoospore of *S. endobioticum* has an electron-opaque plug at the base of the flagellum and a simple MLC cisterna, while the zoospore of *S. macrosporum* has no plug in the flagellum base and has a fenestrated MLC cisterna.

9. *Blyttiomyces helicus* Lineage

Blyttiomyces helicus is the single taxon in a lineage sister of the group composed of Spizellomycetales and Rhizophlyctidales; and the inclusive grouping of *B. helicus*, Spizellomycetales, and Rhizophlyctidales is a sister group of Rhizophydiales (James et al. 2006b). *B. helicus* is morphologically stunning among chytrids, having a golden-brown sporangium ornamented with spiral bands (Sparrow and Barr 1955).

A saprobe of pollen, it has not been isolated into pure culture, and its SSU sequence was obtained from an enriched unifungal culture on pollen. From a different isolate on pollen, ultrastructural studies of the zoospore were conducted by sectioning multiple sporangia containing cleaved zoospores

The zoospore ultrastructure (Fig. 6.5I) of *B. helicus* is quite different from that of either Spizellomycetales or Rhizophlyctidales and more closely resembles that of Rhizophydiales. Ultrastructural features that are common to *B. helicus* and Rhizophydiales include aggregated ribosomes, a single lipid globule, a fenestrated MLC cisterna, shieldlike KASs, and the absence of an electron-opaque plug at the base of the flagellum. The zoospore of *B. helicus* is morphologically distinct, however, in having one or more prominent, often anvil-shaped, cisternae with granular matrices adjacent to the lipid globule, which seem to be continuous with the fenestrated MLC cisterna (Fig. 6.5I).

Blyttiomyces spinulosus is the type species of the genus and has not been isolated or characterized molecularly or ultrastructurally (Blackwell et al. 2011). Thus, whether *B. helicus* is molecularly and ultrastructurally representative of the genus is uncertain at this time.

B. Monoblepharidomycota

Among extensive nomenclatural revisions, Doweld (2001) elevated the order Monoblepharidales and formally described it as a phylum with a new class, order, and family (Table 6.1). Monoblepharidomycota is monophyletic and typically placed as the sister group of the Chytridiomycota (Bullerwell et al. 2003; James et al. 2006b) (Fig. 6.1), although the position of monoblephs in phylogenetic trees is not stable (Bullerwell and Lang 2005; Einax and Voigt 2003; Sekimoto et al. 2011). However, the phylum seems to be monophyletic in studies that include a broad range of zoosporic fungi (James et al. 2006b; Sekimoto et al. 2011). Determining whether proposed subphylum groupings are monophyletic awaits molecular analyses of a broader range of taxa.

Monoblepharidomycota is a distinctive group among zoosporic fungi because of their **oogamous sexual reproduction**, and this feature supports Doweld's (2001) recognition of monoblephs as a phylum. Monoblephs are saprotrophs, and no parasites are known (Emerson 1958, 1964; Emerson and Whisler 1968; Perrott 1955, 1958). Monoblephs and chytrids are also similar in a number of features, including **plasmodesmata** in septa (Powell 1974; Powell and Gillette 1987), **mitosis** with the nuclear envelope opened only at the poles at metaphase (Dolan and Fuller 1985; Powell 1975, 1980; Roychoudhury and Powell 1991; Whisler and Travland 1973), and initiation of **zoospore cleavage** before elongation of the flagellar axoneme (McNitt 1974). Monobleph thalli produce terminal sporangia and are filamentous with a basal holdfast or rhizoidal system, and vacuolated cytoplasm gives the thallus a foamy appearance (Fig. 6.2G). The filament may be short and mostly occupied by the sporangium as in *Harpochytrium* (Fig. 6.2F–H) or hypha-like as in *Monoblepharella*. The thallus of *Oedogoniomyces* attaches to a variety of substrates, including snail shells, seeds, and algae without penetration (Emerson and Whisler 1968). Molecular phylogenetic analysis unexpectedly revealed that the colorless green alga *Hyaloraphidium curvatum* placed as the sister group of all other taxa in the monoblephs (Forget et al. 2002; Ustinova et al. 2000). Its thallus is similar to *Harpochytrium*, but its sporangium releases **autospores** (Ustinova et al. 2000). Thus, it seems that loss of zoospore motility has occurred in monoblephs and chytrids (e.g., *Amoebochytrium*, *Sporophlyctis*).

As monobleph zoospores swim, they are elongate and tapered toward the anterior end. The zoospore ultrastructure (Fig. 6.5J) of all five genera that produce zoospores (*Gonapodya*, *Harpochytrium*, *Oedogoniomyces*, *Monoblepharis*, *Monoblepharella*) has been studied (Fuller 1966; Fuller and Reichle 1968; Gauriloff et al. 1980a, b; Mollicone and Longcore 1994, 1999; Reichle 1972; Travland and Whisler 1971). The most distinguishing features of the zoospore are their spherical mitochondria and the position of the MLC cisterna, which lies adjacent to the plasma membrane but backs the microbody, instead of lipid globules (Fig. 6.4M) (Dorward and Powell 1980) as in chytrids (Powell 1978). The MLC is fenestrated and was initially named the rumposome, but the complexity of the cisterna varies from a cisterna with shallow pores (Fuller and Reichle 1968) to deep pores (Reichle 1972). In zoospores of *Harpochytrium* the fenestrated cisterna connects to the striated rootlet (Travland and Whisler 1971). An electron-opaque plug is in the transition region of the flagellum but is lost in *Gonapodya polymorpha* (Mollicone and Longcore 1999), a lineage distinct from *Gonapodya prolifera* and characterized by a greater number of bases in the LSU C1_3 helix (Chambers 2003) and by the unique presence of a paraxonemal structure (Mollicone and Longcore 1999). Consistent with these observations, in initial molecular analyses, *G. polymorpha* and *G. prolifera* are not monophyletic (Chambers 2003). Zoospores of all genera are similar in that the endoplasmic reticulum both binds and penetrates the ribosomal aggregation surrounding the nucleus and a microtubule root radiates anteriorly from a striated disk that partially encircles the kinetosome. Lipids vary in their locations from predominantly posterior (Travland and Whisler 1971) to anterior (Fuller and Reichle 1968), but the reticulate microbody consistently extends between lipids and mitochondria and the area of the flagellar apparatus (Dorward and Powell 1980; Gauriloff et al. 1980a, b; Mollicone and Longcore 1999). Thus, despite the scattered nature of organelles of the MLC, they are interconnected, which is important for their functions (Powell 1976b, 1978).

C. Neocallimastigomycota

Neocallimastigomycota is comprised of obligate anaerobic zoosporic fungi and specialized commensals growing in the digestive system of herbivores, and their zoospores may bear from 1 to 20 posterior flagella. Flagella of polyflagellate neocallimastigos often adhere together, which might be adaptive to swimming through viscous digestive fluids (Gold et al. 1988). In most molecular phylogenetic analyses, the Neocallimastigomycota is placed as the sister group of the Monoblepharidomycota + Chytridiomycota (James et al. 2006b).

Searching for the true taxonomic affinities of these organisms, Orpin (1975, 1977) detected chitin in the cell walls of *Neocallimastix*, astutely ascribing their kinship to fungi. About a decade later, Heath et al. (1983) recognized that their posteriorly uniflagellate zoospores and microscopic thalli allied them to chytrid fungi. After another decade, Li et al. (1993) established the order Neocallimastigales within the Chytridiomycota. However, neocallimastigos are distinctive from all other groups of zoosporic fungi in the absence of flagellar props in zoospores (Gold et al. 1988). They also differ from monoblephs and chytrids in many developmental characteristics: the nuclear envelope is totally closed at metaphase [reviewed in Li et al. (1993)], plasmodesmata (Powell 1974) have not been observed in septa (Gold et al. 1988; Heath et al. 1983), during zoospore cleavage axonemes extend into flagellar vesicles prior to cleavage of zoospore bodies (Gold et al. 1988; Heath et al. 1983), the whole flagellum including the kinetosome is shed (Gold et al. 1988; Orpin 1975) instead of retracted during zoospore encystment (Koch 1968), no centrioles are associated with vegetative cell nuclei (Heath et al. 1986), and zoospores contain no nonflagellated centrioles (Heath et al. 1983; Li et al. 1991). Because of these differences and their position as a monophyletic group, neocallimastigos have recently been formally established as a phylum (Hibbett et al. 2007).

Neocallimastigos produce monocentric and polycentric thalli with extensive rhizoids or a bulbous haustorium-like structure (Gold et al. 1988; Ho and Barr 1995), and they discharge zoospores bearing one to several posterior flagella. Zoospores may be spherical, oval, or pyriform and are capable of amoeboid movement during which their form is irregular (Orpin 1975). Even within the same isolate, zoospore diameters vary, but those with single flagella are typically smaller than zoospores with multiple flagella. Evidence of sexual reproduction has never been observed among these organisms. Ho and Barr (1995) produced the most current monograph of the neocallimastigos, and only *Cyllamyces* has been added since then (Ozkose et al. 2001). The neocallimastigos constitute a relatively small group with 3 monocentric genera (*Caecomyces*, *Neocallimastix*, *Piromyces*), 3 polycentric genera (*Anaeromyces*, *Cyllamyces*, *Orpinomyces*), and 21 species (Eckart et al. 2010; Ho and Barr 1995; Ozkose et al. 2001). Most molecular phylogenetic studies of Neocallimastigomycota have utilized SSU and ITS1 ribosomal genes, where described genera have been supported but with genera showing varying degrees of divergence among species, as predicted with light and electron microscopic observations (Brookman et al. 2000; Ho and Barr 1995; Li et al. 1993). Molecular environmental studies have revealed that the diversity of this group is much greater than has been described, with more taxa to be characterized (Fliegerová et al. 2010; Liggenstoffer et al. 2010; Nicholson et al. 2010).

The neocallimastigos have attracted interest because of their **biotechnology** potentials in industrial applications (Chu et al. 2011), conversion of plant materials into **biofuels** [reviewed in Elshahed (2010)], and increased food efficiency when low-grade fibrous plant material is used as feed for herbivores (Ho and Barr 1995; Nagpal et al. 2011). As early colonizers of plant material in the rumen, neocallimastigos extensive rhizoidal system physically penetrates refractory, cellulose-containing fibrous plant materials and chemically degrades cellulose and other wall compounds with a whole battery of wall-degrading enzymes, allowing an increased surface area for additional degradation by cellulolytic bacteria and protozoa (Ho and Barr 1995; Montford and Orpin 1994; Orpin and Letcher 1979; Tachezy 2008). Because herbivorous mammals lack the enzymes to break down fibrous lignocellulosic-containing feed, neocallimastigos are vital to the **feed efficiency** of substrates that would otherwise be undigestible by host animals. Genome-sequencing projects, such as those for *Piromyces* and *Orpinomyces*, will greatly facilitate our understanding of genes that are potentially useful in biofuel production and the breakdown of cellulose (Griffith et al. 2010; Nagpal et al. 2011).

Zoospores of neocallimastigos (Fig. 6.5K) were once thought to be flagellated protozoa, but careful developmental studies link two life history stages, the zoospore stage and the monocentric or polycentric thallus stage found attached to fibrous feed (Orpin 1975, 1977; Orpin and Bountiff 1978). Like all flagellated opisthokonts, neocallimastigos have posteriorly directed flagella, and the possession of a transitional helix (=concentric fiber) (Barr 2001; Heath et al. 1983; Li et al. 1991) is a symplesiomorphic character shared with chytrids, monoblephs, and Blastocladiomycota. The ultrastructure of the zoospore differs in the cellular architecture and the range of flagella numbers from those of all other zoosporic fungi. Zoospores of neocallimastigos often have a protrusion opposite the flagellum, and the flagellum is inserted into a concave invagination at the posterior end of the zoospore (Gold et al. 1988). Unique to neocallimastigos zoospore, megatubules form a posterior dome. Instead of flagellar props characteristic of other flagellated fungi, they have a novel kinetosome/flagellar-associated complex with a circumflagellar ring lying just under the plasma membrane where the flagellum emerges from the zoospore body (Gold et al. 1988). A cup-shaped scoop covers the anterior end of the kinetosome, and several struts link the scoop to the circumflagellar ring (Gold et al. 1988). From a globular spur of electron-dense material near the anterior side of the kinetosome several microtubule roots arise, one root extending as a lateral sheet along the plasma membrane and another root flaring anteriorly toward the nucleus (Gold et al. 1988; Heath et al. 1983; Li et al. 1991). Hydrogenosomes cluster in the posterior end of the zoospore and along the side of the beaked extension of the nucleus (Fig. 6.5K). Ribosomes reportedly occur toward the anterior end of the cell as clusters and helices (Gold et al. 1988).

VII. Evolution

Rozella spp., once classified in Chytridiomycota, is placed as the sister group of all other fungi (Fig. 6.1). *Rozella* is an unwalled obligate endoparasite of other fungi and oomycetes (Held 1975, 1981). In some molecular studies (James and Berbee 2011; Karpov et al. 2013) *Rozella* spp. are affiliated with aphelids and the unwalled endoparasites of animals, Microsporidia. However, the phylogenetic placement of Microsporidia is still controversial (Corradi and Keeling 2009). The clade that includes *Rozella* spp. also includes a genetically diverse array of phylotypes detected from environmental samples of freshwater, marine sediments, and peat bogs and was informally named the Rozellida (Lara et al. 2010; Lepère et al. 2008; Nagahama et al. 2011) and formally circumscribed as the Cryptomycota (Jones et al. 2011). Cryptomycota is a highly derived and diverse group of organisms, and whether they are considered fungi is a matter of where the basal limits are drawn. Brown et al. (2009) proposed the Nucletmycea as a supergroup for the large clade that includes the filose-pseudopodia-forming nucleariids, the cellular slime mold *Fonticula alba*, the **Cryptomycota**, and fungi.

Neocallimastigos seem to share a common posteriorly uniflagellate aerobic ancestor with chytrids but diverged from chytrids in a lineage that adapted them to an anaerobic habitat (Fig. 6.1). Rather than respiration with mitochondria, they utilize hydrogenosomes that produce ATP by substrate-level phosphorylation (van der Giezen et al. 2003). **Hydrogenosomes** seem to be secondarily derived from mitochondria. Although these organelles have lost their mitochondrial genome (Bullerwell and Lang 2005), they retain two surrounding membranes and share protein-importing mechanisms (van der Giezen 2009; van der Giezen et al. 2003). From the close interaction between rumen bacteria and neocallimastigos in the gut of herbivores and the facility bacteria have in transferring genes, it seems that in the divergence of neocallimastigos from other zoosporic fungi they obtained numerous genes for enzymes important in the degradation of plant material by horizontal gene transfer from bacteria. This seems to be the case for glycosyl hydrolases (Garcia-Vallvé et al. 2000) and, perhaps, cellulase because of the presence of cellulosomes, multienzyme complexes that in neocallimastigos degrade crystal cellulose directly into glucose (Ljungdahl 2008).

Phylogenetic hypotheses based on the analyses of molecular sequences suggest that the ancestors of zoosporic fungi were unwalled nucleariid amoebae (Amaral Zettler et al. 2001; Brown et al. 2009; Bullerwell and Lang 2005; Liu et al. 2009; Steenkamp et al. 2006). Like fungi, nucleariid amoebae have flattened discoid mitochondrial cristae. With the production of a cyst wall and polarized growth, the filamentous (hyphalike) thallus form may have evolved leading to two basal and diverging zoosporic fungal groups, Blastocladiomycota and Monoblepharidomycota. Among the basal filamentous zoosporic fungi, a Spitzenkörper-like assemblage, which is characteristic of higher fungal hyphae, has only been found in *Allomyces* (Blastocladiomycota) (Vargas et al. 1993). The simpler eucarpic, monocentric thallus found among chytrids (Fig. 6.2C, J, K) may have been derived from ancestors with filamentous thalli consisting of short filaments bearing a basal holdfast and terminal sporangium, as in the monoblephs (Fig. 6.2 F–H). Thalli composed of anucleate rhizoids and a terminal sporangium occur in multiple lineages of chytrids. Much of the developmental variation among chytrid thalli depends upon the behavior of the nucleus following zoospore encystment and germination (Blackwell et al. 2006; Powell and Koch 1977). Whether the nucleus remains in the zoospore cyst (as in *Rhizophydium*), migrates out of the cyst into its germ tube (as in *Powellomyces*), moves with the protoplast into a host cell (as in *Synchytrium*), or continues to migrate along a rhizomycelium (as in *Nowakowskiella*) determines thallus morphology. Even within a clonal chytrid isolate, the pattern of nuclear migration can vary with differing environmental conditions (Powell and Koch 1977). The fact that monocentric and polycentric thallus forms occur within the same monophyletic order (such as Chytridiales, Cladochytriales, and Polychytriales) suggests that there is not great phylogenetic significance in the differences in thallus complexity within an order. Whether a chytrid is monocentric or polycentric may be a matter of how **nuclear migration genes** are regulated, and understanding the evolution and regulation of genes associated with nuclear positioning (Morris

2000) is key to understanding the evolutionary basis of chytrid development.

Substrate utilization may have been a selective factor in the evolution of saprobic chytrids. Rhizophlyctidales and Cladochytriales primarily use cellulose as a substrate, and Polychytriales and Chytridiales have lineages that use chitin exclusively. **Habitat** also seems to have exerted selective pressure in zoosporic evolution. Terminal groups (Fig. 6.1), such as Rhizophlyctidales and Spizellomycetales, are more often found in soil and dung than in bodies of water. No verification of sexual reproduction has been discovered in either of these orders, which means their adaptation to soil is primarily clonal.

Evidence suggests that basal fungi evolved in **freshwater aquatic systems** because no marine Blastocladiomycota or Monoblepharidomycota have been discovered, and most aquatic chytrids are found in freshwater. However, marine representatives are scattered among the orders of Chytridiomycota, having been described in genera now classified in Chytridiales, Rhizophydiales, Lobulomycetales, and Cladochytriales. Freshwater forms may have become secondarily adapted to halophytic soils, brackish waters, and marine habitats. There is a gradation of organisms that grow in estuarine areas (with great fluctuation in salinity) to authentic marine fungi growing on algal hosts or crab eggs (Amon 1976; Booth 1971; Johnson and Sparrow 1961; Karling 1977; Müller et al. 1999; Nyvall et al. 1999; Shields 1990; Sparrow 1960; reviewed in Gleason et al. 2011). The presence of chytrid phylotypes detected in deep-sea hydrothermal vents and cold seep sediments highlights the underexplored diversity of marine chytrids (LeCalvez et al. 2009; Nagahama et al. 2011).

With molecular-based phylogenetics we can begin to trace the pattern of inheritance of **zoospore ultrastructural features**, with character states being transformed or lost. In the phylogentic hypothesis with monoblephs as the sister group of chytrids (Fig. 6.1), the shared common ancestor would be aquatic with a filamentous/hypha-like thallus. Zoospores would have contained a fenestrated MLC cisterna, ribosomal aggregation, and electron-opaque plug in the flagellar transition region. Thus, in **terminal clades** such as Spizellomycetales, we find that in their divergence each of these characters has been lost and the organisms are well adapted to terrestrial habitats primarily as saprobes (Wakefield et al. 2010).

VIII. Conclusions

Exploration of new habitats and refinements in recognition of diversity among zoosporic fungi has revealed the untapped diversity of these organisms. Emerging molecular techniques for rapid sequencing of genes and total genomes will transform our understanding of zoosporic fungi in the next 10 years, as have applications of gene sequences and ultrastructural characters in the past decade. The genomes of only a few chytrids (*B. dendrobatidis* isolates JEL 423 and JAM 81; *Homolaphlyctis polyrhiza* JEL 142; *Spizellomyces punctatus* isolate BR 117), monoblephs (*Gonapodya* sp. isolate JEL 183), and neocallimastigos (*Piromyces* sp., *Orpinomyces* sp. isolate OUS 1) have been sequenced (Joneson et al. 2011; Stajich 2011). But even with the scant knowledge we have, access to sequenced genomes of zoosporic fungi is impacting our views of the evolution of genes in fungi. Rosenblum et al. (2008) have exploited the sequenced genome of *B. dendrobatidis* to use whole-genome arrays and track differential gene expression in zoospores and sporangia. Idnurm et al. (2010) used comparative genomics to identify putative light-receptive genes from the sequenced genome of *S. punctatus*. Zoosporic fungi have been considered intractable to genetic transformation because no procedures have been successfully developed with Chytridiomycetes, thwarting our ability to compare gene functions. However, a recently developed transformational system for Blastocladiomycota might prove useful for chytrids as well (Vieira and Camilo 2011). Recognizing the importance of zoosporic fungi in soil fertility (Midgley et al. 2006), in food webs (Kagami et al. 2007, 2011, 2012; Sime-Ngando et al. 2011), as pollution indicators (Dileo et al. 2010), and in biotechnology invites more extensive exploration of the diversity and ecological roles of these fascinating fungi.

Acknowledgements This review was supported by National Science Foundation Grants PEET DEB-0529694, REVSYS DEB-0949305, and AFTOL DEB 0732599.

References

Amaral Zettler LA, Nerad TA, O'Kelly CJ, Sogin ML (2001) The nucleariid amoebae: more protists at the animal-fungal boundary. J Eukaryot Microbiol 48:293–297

Amon JP (1976) An estuarine species of *Phlyctochytrium* (Chytridiales) having a transient requirement for sodium. Mycologia 68:470–480

Baayen RP, Cochius G, Hendriks H, Meffert JP, Bakker J, Bekker M, van den Boogert PHJF, Stachewicz H, van Leeuwen GCM (2006) History of potato wart disease in Europe—a proposal for harmonization in defining pathotypes. Eur J Plant Pathol 116:21–31

Bai C, Garner TWJ, Li Y (2010) First evidence of *Batrachochytrium dendrobatidis* in China: discovery of chytridiomycosis in introduced American bullfrogs and native amphibians in the Yunnan Province, China. EcoHealth 7:127–134

Bandoni RJ, Barr DJS (1976) On some zoosporic fungi from washed terrestrial litter. Trans Mycol Soc Japan 17:220–225

Barr DJS (1973) *Rhizophydium graminis* (Chytridiales): morphology, host range, and temperature effect. Can Plant Dis Surv 53:191–193

Barr DJS (1980) An outline for the reclassification of the Chytridiales, and for a new order, the Spizellomycetales. Can J Bot 58:2380–2394

Barr DJS (1981) Ultrastructure of the *Gaertneriomyces* zoospore (Spizellomycetales, Chytridiomycetes). Can J Bot 59:83–90

Barr DJS (1984a) Cytological variation in zoospores of *Spizellomyces* (Chytridiomycetes). Can J Bot 62:1202–1208

Barr DJS (1984b) The classification of *Spizellomyces*, *Gaertneriomyces*, *Triparticalcar*, and *Kochiomyces*. Can J Bot 62:1171–1201

Barr DJS (1986) *Allochytridium expandens* rediscovered: morphology, physiology and zoospore ultrastructure. Mycologia 78:439–448

Barr DJS (1990) Phylum Chytridiomycota. In: Margulis L, Corliss JO, Melkonian M, Chapman DJ (eds) Handbook of Protoctista. Jones and Bartlett, Boston, MA, pp 454–466

Barr DJS (2001) Chytridiomycota. In: Esser K, Lemke PA (eds) The mycota, systematics and evolution, vol VIIA. Springer, New York, pp 93–112

Barr DJS, Allan PME (1981) Ultrastructure of *Kochiomyces* and *Triparticalcar* zoospores (Spizellomycetales, Chytridiomycetes). Can J Bot 59:649–661

Barr DJS, Babcock CE (1994) Culture collection information: cryopreservation of unicellular zoosporic fungi. US Fed Cult Collections Newslett 24:6

Barr DJS, Désaulniers NL (1986) Four zoospore subtypes in the *Rhizophlyctis-Karlingia* complex (Chytridiomycetes). Can J Bot 64:561–572

Barr DJS, Désaulniers NL (1987) *Allochytridium luteum* n. sp.: morphology, physiology and zoospore ultrastructure. Mycologia 79:193–199

Barr DJS, Désaulniers NL (1988) Precise configuration of the chytrid zoospore. Can J Bot 66:869–876

Barr DJS, Hadland-Hartmann VE (1978) Zoospore ultrastructure in the genus *Rhizophydium* (Chytridiales). Can J Bot 56:2380–2404

Barr DJS, Hartmann VE (1976) Zoospore ultrastructure of three *Chytridium* species and *Rhizoclosmatium globosum*. Can J Bot 54:2000–2013

Barr DJS, Hartmann VE (1977) Zoospore ultrastructure of *Olpidium brassicae* and *Rhizophlyctis rosea*. Can J Bot 55:1221–1235

Barr DJS, Désaulniers NL, Knox JS (1987) *Catenochytridium hemicysti* n. sp.: morphology, physiology and zoospore ultrastructure. Mycologia 79:587–594

Beakes GW, Canter HM, Jaworski GHM (1988) Zoospore ultrastructure of *Zygorhizidium affluens* and *Z. planktonicum*, two chytrids parasitizing the diatom *Asterionella formosa*. Can J Bot 66:1054–1067

Beakes GW, Canter HM, Jaworski GHM (1992) Ultrastructural study of operculation (discharge apparatus) and zoospore discharge in zoosporangia of *Zygorhizidium affluens* and *Z. planktonicum*, chytrid parasites of *Asterionella formosa*. Mycol Res 96:1060–1067

Beakes GW, Canter HM, Jaworski GHM (1993) Sporangium differentiation and zoospore fine-structure of the chytrid *Rhizophydium planktonicum*, a fungal parasite of *Asterionella formosa*. Mycol Res 97:1059–1074

Bills GF, Christensen M, Powell MJ, Thorn G (2004) Saprobic soil fungi. In: Mueller GM, Bills GF, Foster MS (eds) Biodiversity of fungi: inventory and monitoring methods. Elsevier, Oxford, pp 271–302

Blackwell WH, Powell MJ (1999) The nomenclatural propriety of *Rhizophlyctis rosea*. Mycotaxon 70:213–217

Blackwell WH, Letcher PM, Powell MJ (2004) Synopsis and systematic reconsideration of *Karlingiomyces* (Chytridiomycota). Mycotaxon 89:259–276

Blackwell WH, Letcher PM, Powell MJ (2006) Thallus development and the systematics of Chytridiomycota: an additional developmental pattern represented by *Podochytrium*. Mycotaxon 97:91–109

Blackwell WH, Letcher PM, Powell MJ, Vélez CG (2011) The occurrence of *Blyttiomyces spinulosus* in Alabama and Argentina, and comments on the genus *Blyttiomyces* (Chytridiomycota). Phytologia 93:304–315

Booth T (1971) Ecotypic responses of chytrid and chytridiaceaous species to various salinity and temperature combinations. Can J Bot 49:1757–1767

Braun A (1851) Betrachtungen über die Erscheinung der Verjüngung in der Natur, insbesondere in der Lebens- und Bildungsgeschichte der Pflanze, Leipzig

Braun A (1855) Über *Chytridium*, eine Gattung einzelner Schmarotzergewächse auf Algen und Infusorien. Monatsber Berlin Akad 1855:378–384

Brookman JL, Mennim G, Trinci APJ, Theodorou MK, Tuckwell DS (2000) Identification and characterization of anerobic gut fungi using molecular methologies based on ribosomal ITS1 and 18S rRNA. Microbiology 146:393–403

Brown MW, Spiegel FW, Silberman JD (2009) Phylogeny of the "forgotten" cellular slime mold, *Fonticula alba*, reveals a key evolutionary branch within Opisthokonta. Mol Biol Evol 26:2699–2709

Bruckart WL III, Eskandari FM, Widmer TL (2011) *Synchytrium solstitiale*: reclassification based on the function and role of resting spores. Mycologia 103:775–778

Bruning K, Lingeman R, Ringelberg J (1992) Estimating the impact of fungal parasites on phytoplankton populations. Limnol Oceanogr 37:252–260

Bullerwell CE, Lang BF (2005) Fungal evolution: the case of the vanishing mitochondrion. Curr Opin Microbiol 8:362–369

Bullerwell CE, Forget L, Lang BF (2003) Evolution of monobleparidalean fungi based on complete mitochondrial genome sequences. Nucleic Acids Res 31:1614–1623

Canter HM, Lund JWG (1951) Studies on plankton parasites. III. Examples of the interaction between parasitism and other factors determining the growth of diatoms. Ann Bot (Oxford) 15:359–371

Cavalier-Smith T (1998) A revised six-kingdom system of life. Biol Rev 73:203–266

Chambers JG (2003) Ribosomal DNA, secondary structure, and phylogenetic relationships among the Chytridiomycota. Ph.D. dissertation, University of Alabama, Tuscaloosa, AL

Chen M, Chen F, Yu Y, Ji J, Kong F (2008) Genetic diversity of eukaryotic microorganisms in Lake Taihu, a large shallow subtropical lake in China. Microb Ecol 56:572–583

Chu C-Y, Tseng C-W, Yueh P-Y, Duan C-H, Liu J-R (2011) Molecular cloning and characterization of a β-glucanase from *Piromyces rhizinflatus*. J Biosci Bioeng 111:541–546

Corradi N, Keeling PJ (2009) Microsporidia: a journey through radical taxonomical revisions. Fungal Biol Rev 23:1–8

Couch JN (1939) Technic for collection, isolation and culture of chytrids. J Elisha Mitchell Sci Soc 55:208–214

Dileo K, Donat K, Min-Venditti A, Dighton J (2010) A correlation between chytrid abundance and ecological integrity in New Jersey pine barrens waters. Fungal Ecol 3:295–301

Dogma IJ (1973) Developmental and taxonomic studies on rhizophlyctoid fungi, Chytridiales I. Dehiscence mechanisms and generic concepts. Nova Hedwigia 24:393–411

Dolan TE, Fuller MS (1985) The ultrastructure of nuclear division in *Monoblepharella* sp. Mycologia 77:791–809

Dorward DW, Powell MJ (1980) Microbodies in *Monoblepharella* sp. Mycologia 72:549–557

Dorward DW, Powell MJ (1982) Cross-linking bridges associated with the microbody-lipid globule complex in *Chytriomyces aureus* and *Chytriomyces hyalinus*. Protoplasma 112:181–188

Dorward DW, Powell MJ (1983) Cytochemical detection of polysaccharides and the ultrastructure of the cell coat of zoospores of *Chytriomyces aureus* and *Chytriomyces hyalinus*. Mycologia 75:209–220

Doweld A (2001) Prosyllabus tracheophytorum: Tentamen systematis plantarum vascularium (Tracheophyta). Geos, Moscow

Ebersberger I, Simoes R, Kupczok A, Gube M, Kothe E, Voigt K, von Haeseler A (2012) A consistent phylogenetic backbone for the fungi. Mol Biol Evol 29:1319–1334

Eckart M, Fliegerová K, Hoffmann K, Voigt K (2010) Molecular identification of anaerobic rumen fungi. In: Gherbawy Y, Voigt K (eds) Molecular identification of fungi. Springer, Berlin, Heidelberg, New York, pp 297–313

Einax E, Voigt K (2003) Oligonucleotide primers for the universal amplification of b-tubulin genes facilitate phylogenetic analyses in the regnum Fungi. Org Divers Evol 3:185–194

Elshahed MS (2010) Microbiological aspects of biofuel production: current status and future directions. J Adv Res 1:103–111

Emerson R (1958) Mycological organization. Mycologia 5:589–621

Emerson R (1964) Performing fungi. Am Biol Teach 26:90–100

Emerson R, Whisler HC (1968) Cultural studies of *Oedogoniomyces* and *Harpochytrium*, and a proposal to place them in a new order of aquatic phycomycetes. Arch Mikrobiol 61:195–211

Fliegerová K, Mrázek F, Hoffmann K, Zábranská J, Voigt K (2010) Diversity of anerobic fungi within cow manure determined by ITS1 analysis. Folia Microbiol 55:319–325

Forget L, Ustinova J, Wang Z, Huss VAR, Lang BF (2002) *Hyaloraphidium curvatum*: a linear mitochondrial genome, tRNA editing, and an evolutionary link to lower fungi. Mol Biol Evol 19:310–319

Freeman KR, Martin AP, Karki D, Lynch RC, Mitter MS, Meyer AF, Longcore JE, Simmons DR, Schmidt SK (2009) Evidence that chytrids dominate fungal communities in high-elevation soils. Proc Natl Acad Sci USA 106:18315–18320

Fuller MS (1966) Structure of the uniflagellate zoospores of aquatic phycomycetes. In: Madelin MF (ed) The fungus spore, vol 18, Proceedings symposium of the Colston Research Society. Butterworth, London, pp 67–84

Fuller MS, Jaworski A (1987) Zoosporic fungi in teaching and research. Southeastern Publishing, Athens, GA

Fuller MS, Reichle RE (1968) The fine structure of *Monoblepharella* sp. zoospores. Can J Bot 46:279–283

Garcia-Vallvé S, Romeu A, Palau J (2000) Horizontal gene transfer of glycosyl hydrolases of the rumen fungi. Mol Biol Evol 17:352–361

Garmyn A, Van Rooij P, Pasmans F, Hellebuyck T, Van Den Broeck W, Haesebrouck F, Martel A (2012) Waterfowl: potential environmental reservoirs of the chytrid fungus *Batrachochytrium dendrobatidis*. PLoS One 7(4):e35038. doi:10.1371/journal.pone.0035038

Gauriloff LP, Delay RJ, Fuller MS (1980a) Comparative ultrastructure and biochemistry of chytridiomycetous fungi and the future of the Harpochytriales. Can J Bot 58:2098–2109

Gauriloff LP, Delay RJ, Fuller MS (1980b) The fine structure of zoospores of *Harpochytrium hedinii*. Can J Bot 58:2090–2097

Gleason FH, Letcher PM, McGee PA (2004) Some Chytridiomycota in soil recover from drying and high temperatures. Mycol Res 108:583–589

Gleason FH, Mozley-Standridge SE, Porter D, Boyle DG, Hyatt AD (2007) Preservation of Chytridiomycota in culture collections. Mycol Res 111:129–136

Gleason FH, Kagami M, Lefèvre E, Sime-Ngando T (2008) The ecology of chytrids in aquatic ecosystems: roles in food web dynamics. Fungal Biol Rev 22:17–25

Gleason FH, Küpper FC, Amon JP, Picard K, Gachon CMM, Marano AB, Sime-Ngando T, Lilje O (2011) Zoosporic true fungi in marine ecosystems: a review. Mar Freshw Res 62:383–393

Gold JJ, Heath IB, Bauchop T (1988) Ultrastructural description of a new chytrid genus of caecum anaerobe, *Caecomyces equi* gen. nov., sp. nov., assigned to the Neocallimasticaceae. BioSystems 21:403–415

Griffith GW, Baker S, Fliegerova K, Liggenstoffer A, van der Giezen M, Voigt K, Beakes G (2010) Anaerobic fungi: Neocallimastogomycota. IMA Fungus 1:181–185

Hampson MC, Coombes JW (1989) Pathogenesis of *Synchytrium endobioticum*. VII. Earthworms as vectors of wart disease of potato. Plant Soil 116:147–150

Heath IB, Bauchop T, Skipp RA (1983) Assignment of the rumen anerobe *Neocallimastix frontalis* to the Spizellomycetales (Chytridiomycetes) on the basis of its polyflagellate zoospore ultrastructure. Can J Bot 61:295–307

Heath IB, Kaminskyj SG, Bauchop T (1986) Basal body loss during fungal zoospore encystment: evidence against cenriole autonomy. J Cell Sci 83:135–140

Held AA (1975) The zoospore of *Rozella allomycis*: ultrastructure. Can J Bot 53:2212–2232

Held AA (1981) *Rozella* and *Rozellopsis*: naked endoparasitic fungi which dress-up as their hosts. Bot Rev 47:451–515

Hibbett DS, Binder M, Bischoff J, Blackwell M, Cannon P, Eriksson O, Huhndorf S, James T, Kirk P, Lucking R, Lumbsch T, Lutzoni F, Matheny P, McLaughlin D, Powell M, Redhead S, Schoch C, Spatafora J, Stalpers J, Vilgalys R et al (2007) A higher level phylogenetic classification of the fungi. Mycol Res 111:509–547

Ho YW, Barr DJS (1995) Classification of anaerobic gut fungi from herbivores with emphasis on rumen fungi from Malaysia. Mycologia 87:655–677

Holfeld H (2000) Relative abundance, rate of increase, and fungal infections of freshwater phytoplankton. J Plankton Res 22:987–995

Hörandle E, Stuessy TF (2010) Paraphyletic groups as natural units of biological classification. Taxon 59:1641–1653

Idnurm A, Verma S, Corrochano LM (2010) A glimpse into the basis of vision in the kingdom Mycota. Fungal Genet Biol 47:881–892

James TY, Berbee ML (2011) No jacket required—new fungal lineage defies dress code. Bioessays 34:94–102. doi:10.1002/bies.201100110

James TY, Porter D, Leander CA, Vilgalys R, Longcore JE (2000) Molecular phylogenetics of the Chytriomycota supports the utility of ultrastructural data in chytrid systematics. Can J Bot 78:336–350

James TY, Kauff F, Schoch CL, Matheny PB, Hofstetter V, Cox CJ, Celio G, Gueidan C, Fraker E, Miadlikowska J, Lumbsch HT, Rauhut A, Reeb V, Arnold AE, Amtoft A, Stajich JE, Hosaka K, Sung G, Johnson D, O'Rourke B, Crockett M, Binder M, Curtis JM, Slot JC, Wang Z, Wilson AW, Schüßler A, Longcore JE, O'Donnell K, Mozley-Standridge S, Porter D, Letcher PM, Powell MJ, Taylor JW, White MM, Griffith GW, Davies DR, Humber RA, Morton JB, Sugiyama J, Rossman AY, Rogers JD, Pfister DH, Hewitt D, Hansen K, Hambleton S, Shoemaker RA, Kohlmeyer J, Volkmann-Kohlmeyer B, Spotts RA, Serdani M, Crous PW, Hughes KW, Matsuura K, Langer E, Langer G, Untereiner WA, Lücking R, Büdel B, Geiser DM, Aptroot A, Diederich P, Schmitt I, Schultz M, Yahr R, Hibbett DS, Lutzoni F, McLaughlin DJ, Spatafora JW, Vilgalys R (2006a) Reconstructing the early evolution of Fungi using a six-gene phylogeny. Nature 443:818–822

James TY, Letcher PM, Longcore JE, Mozley-Standridge SE, Porter D, Powell MJ, Griffith GW, Vilgalys R (2006b) A molecular phylogeny of the flagellated Fungi (Chytridiomycota) and description of a new phylum (Blastocladiomycota). Mycologia 98:860–871

James TY, Porter TM, Martin WW (2014) Blastocladiomycota. In: McLaughlin DJ, Spatafora JW (eds) Systematics and evolution. Springer, Heidelberg

Johanson AE (1944) An endo-operculate chytridiaceous fungus: *Karlingia rosea* gen. nov. Am J Bot 31:397–404

Johnson TW, Sparrow FW (1961) Fungi in oceans and estuaries. J. Cramer, Weinheim, Germany

Jones MDM, Richards TA, Hawksworth DJ, Bass D (2011) Validation of the phylum name Cryptomycota phyl. nov. with notes on its recognition. IMA Fungus 2:173–175

Joneson S, Stajich JE, Shiu S-H, Rosenblum EB (2011) Genomic transition to pathogenicity in chytrid fungi. PLoS Pathog 7(11):e1002338. doi:10.1371/journal.ppat.1002338

Kagami M, deBruin A, Ibelings BW, van Donk E (2007) Parasitic chytrids: their effects on phytoplankton communities and food-web dynamics. Hydrobiologia 578:113–129

Kagami M, Helmsing NR, van Donk E (2011) Parasitic chytrids could promote copepod survival by mediating material transfer from inedible diatoms. Hydrobiologia 569:49–54

Kagami M, Amano Y, Ishii N (2012) Community structure of planktonic fungi and the impact of parasitic chytrids on phytoplankton in Lake Inba, Japan. Microb Ecol 63:358–368

Karling JS (1964) *Synchytrium*. Academic, New York

Karling JS (1977) Chytridiomycetarum Iconographia. Lubrecht and Cramer, Monticello, NY

Karpov SA, Letcher PM, Mamkaeva MA, Mamkaeva KA (2010) Phylogenetic position of the genus *Mesochytrium* (Chytridiomycota) based on zoospore ultrastructure and sequences from the 18S and 28S rRNA gene. Nova Hedwigia 90:81–94

Karpov SA, Mikhailov KV, Mirzaeva GS, Mirabdullaev IM, Mamkaeva KA, Titova NN, Aleoshin VV (2013) Obligately phagotrophic aphelids turned out to branch with the earliest-diverging fungi. Protist 164(2):195–205, http://dx.doi.org/10.1016/j.protis.2012.08.001

Kazama F (1972) Ultrastructure and phototaxis of the zoospores of *Phlyctochytrium* sp., an estuarine chytrid. J Gen Microbiol 71:555–566

Keeling PJ, Inagaki Y (2004) A class of eukaryotic GTPase with a punctate distribution suggesting multiple functional replacements of translation elongation factor 1alpha. Proc Natl Acad Sci USA 101:15380–15385

Kiziewicz B (2004) Aquatic fungi growing on the muscle of vendace (*Coregonus albula* L.), alpine bullhead (*Cottus poecilopus* H.) and lake trout (*Salmo trutta lacustris* L.) from Lake Hańcza (NE Poland). Zool Poloniae 49:85–95

Koch WJ (1961) Studies of the motile cells of chytrids. III. Major types. Am J Bot 48:786–788

Koch WJ (1968) Studies of the motile cells of chytrids. V. Flagellar retraction in posteriorly uniflagellate fungi. Am J Bot 55:841–856

Lange L, Olson LW (1978) The zoospore of *Synchytrium endobioticum*. Can J Bot 56:1229–1239

Lange L, Olson LW (1979) The uniflagellate Phycomycete zoospore. Dansk Bot Arkiv 33:1–95

Lara E, Moreira D, López-Garcia P (2010) The environmental clade LKM11 and *Rozella* form the deepest branching clade of fungi. Protist 161:116–121

LeCalvez T, Burgaud G, Mahé S, Barbier G, Vandenkoornhuyse P (2009) Fungal diversity in deep-sea hydrothermal ecosystems. Appl Environ Microbiol 75:6415–6421

Lefèvre E, Roussel B, Amblard C, Sime-Ngando T (2008) The molecular diversity of freshwater picoeukaryotes reveals high occurrence of putative parasitoids in the plankton. PLoS One 3(6):e2324. doi:10.1371/journal.pone.0002324

Lefèvre E, Letcher PM, Powell MJ (2012) Temporal variation of the small eukaryotic community in two freshwater lakes: emphasis on the zoosporic fungal community. Aquat Microb Ecol 67:91–105

Lepère C, Domaizon I, Debroas D (2008) Unexpected importance of potential parasites in the composition of the freshwater small-eukaryote community. Appl Environ Microbiol 74:2940–2949

Letcher PM, Powell MJ (2001) Distribution of zoosporic fungi in forest soils of the Blue Ridge and Appalachian Mountains of Virginia. Mycologia 93:1029–1041

Letcher PM, Powell MJ (2002a) A taxonomic summary of *Chytriomyces* (Chytridiomycota). Mycotaxon 84:447–487

Letcher PM, Powell MJ (2002b) Frequency and distribution patterns of zoosporic fungi from moss-covered and exposed soils. Mycologia 94:761–771

Letcher PM, Powell MJ (2005a) *Kappamyces*, a new genus in the Chytridiales (Chytridiomycota). Nova Hedwigia 80:115–133

Letcher PM, Powell MJ (2005b) Phylogenetic position of *Phlyctochytrium planicorne* (Chytridiales, Chytridiomycota) based on zoospore ultrastructure and partial nuclear LSU rRNA gene sequence analysis. Nova Hedwigia 80:134–146

Letcher PM, Powell MJ (2012) A taxonomic summary and revision of *Rhizophydium* (Rhizophydiales, Chytridiomycota). University Printing, Tuscaloosa, AL

Letcher PM, Powell MJ, Chambers JG, Holznagel WE (2004) Phylogenetic relationships among *Rhizophydium* isolates from North America and Australia. Mycologia 96:1339–1351

Letcher PM, Powell MJ, Chambers JG, Longcore JE, Churchill PF, Harris PM (2005) Ultrastructural

and molecular delineation of the Chytridiaceae (Chytridiales). Mycologia 83:1561–1573
Letcher PM, Powell MJ, Churchill PF, Chambers JG (2006) Ultrastructural and molecular phylogenetic delineation of a new order, the Rhizophydiales (Chytridiomycota). Mycol Res 110:898–915
Letcher PM, Powell MJ, Barr DJS, Churchill PF, Wakefield WS, Picard KT (2008a) Rhizophlyctidales—a new order in Chytridiomycota. Mycol Res 112:1031–1048
Letcher PM, Powell MJ, Viusent MC (2008b) Rediscovery of an unusual chytridiaceous fungus new to the order Rhizophydiales. Mycologia 100: 325–334
Letcher PM, Vélez CG, Barrantes ME, Powell MJ, Churchill PF, Wakefield WS (2008c) Ultrastructural and molecular analyses of Rhizophydiales (Chytridiomycota) isolates from North America and Argentina. Mycol Res 112:759–782
Letcher PM, Powell MJ, Picard KT (2012a) Zoospore ultrastructure and phylogenetic position of *Phlyctochytrium aureliae* Ajello is revealed (Chytridiaceae, Chytridiales, Chytridiomycota. Mycologia 104:410–418
Letcher PM, Vélez CG, Schultz S, Powell MJ (2012b) New taxa are delineated in Alphamycetaceae (Rhizophydiales, Chytridiomycota). Nova Hedwigia 94:9–29
Li J, Heath IB, Cheng K-J (1991) The development and zoospore ultrastructure of a polycentric chytridiomycete gut fungus, *Orphinomyces joyonii comb. nov.* Can J Bot 69:580–589
Li J, Heath IB, Packer L (1993) The phylogenetic relationships of the anaerobic chytridiomycetous gut fungi (Neocallimasticaceae) and the Chytridiomycota. II. Cladistic analysis of structural data and description of Neocallimasticales *ord. nov.* Can J Bot 71:393–407
Liggenstoffer AS, Youssef NH, Couger MB, Elshahed MS (2010) Phylogenetic diversity and community structure of anaerobic gut fungi (phylum Neocallimastigomycota) in ruminant and non-ruminant herbivores. ISME J 4:1225–1235
Liu Y, Steenkamp ET, Brinkmann H, Forget L, Philippe H, Lang BF (2009) Phylogenomic analyses predict sister group relationship of nucleariids and Fungi and paraphyly of zygomycetes with significant support. BMC Evol Biol 9:272. doi:10.1186/1471-2148-9-272
Ljungdahl LG (2008) The cellulose/hemicellulase system of the anaerobic fungus *Orpinomyces* PC-2 and aspects of its applied use. Ann N Y Acad Sci 1125:308–321
Lockhart RJ, Van Dyke MK, Beadle IR, Humphreys P, McCarthy AJ (2006) Molecular biological detection of anaerobic gut fungi (Neocallimastigales) from landfill sites. Appl Environ Microbiol 72:5659–5661
Longcore JE (1992a) Morphology and zoospore ultrastructure of *Chytriomyces angularis* sp. nov. (Chytridiales). Mycologia 84:442–451
Longcore JE (1992b) Morphology and zoospore ultrastructure of *Podochytrium dentatus* sp. nov. (Chytridiales). Mycologia 84:183–192
Longcore JE (1993) Morphology and zoospore ultrastructure of *Lacustromyces hiemalis* gen. et sp. nov. (Chytridiales). Can J Bot 71:414–425
Longcore JE (1995) Morphology and zoospore ultrastructure of *Entophlyctis luteolus* sp. nov. (Chytridiales): implications for chytrid taxonomy. Mycologia 87:25–33
Longcore JE (1996) Chytridiomycete taxonomy since 1960. Mycotaxon 60:149–174, http://www.umaine.edu/chytrids/Chytrid%20Bibliography/Bibliography-Index.html
Longcore JE (2004) *Rhizophydium brooksianum* sp. nov., a multipored chytrid from soil. Mycologia 96:162–171
Longcore JE (2005) Zoosporic fungi from Australian and New Zealand tree-canopy detritus. Aust J Bot 53:259–272
Longcore JE, Simmons DR (2012) The Polychytriales ord. nov. contains chitinophilic members of the rhizophlyctoid alliance. Mycologia 104:276–294
Longcore JE, Barr DJS, Désaulniers N (1995) *Powellomyces*, a new genus in the Spizellomycetales. Can J Bot 73:1385–1390
Longcore JE, Pessier AP, Nichols D (1999) *Batrachochytrium dendrobatidis* gen. et sp. nov., a chytrid pathogenic to amphibians. Mycologia 91:219–227
Longcore JR, Longcore JE, Pessier AP, Halteman WA (2007) Chytridiomycosis widespread in anurans of northeastern United States. J Wildl Manage 71:435–444
Longcore JE, Letcher PM, James TY (2011) *Homolaphlyctis polyrhiza* gen et sp. nov., a species in the Rhizophydiales (Chytridiomycetes) with multiple rhizoidal axes. Mycotaxon 118:433–440
Lucarotti CJ (1981) Zoospore ultrastructure of *Nowakowskiella elegans* and *Cladochytrium replicatum*. Can J Bot 59:137–148
McNitt R (1974) Zoosporogenesis in *Phlyctochytrium irregulare*. Cytobiologie 9:290–306
Midgley DJ, Letcher PM, McGee PA (2006) Access to organic and insoluble sources of phosphorous varies among soil Chytridiomycota. Arch Microbiol 186:211–217
Miki T, Takimoto G, Kagami M (2011) Role of parasitic fungi in aquatic food webs: a theoretical approach. Freshw Biol 56:173–1183
Milne A, Theodorou MK, Jordan MGC, King-Spooner C, Trinci APJ (1989) Survival of anaerobic fungi in feces, in saliva, and in pure culture. Exp Mycol 13:27–37
Mollicone MRN, Longcore JE (1994) Zoospore ultrastructure of *Monoblepharis polymorpha*. Mycologia 86:615–625

Mollicone MRN, Longcore JE (1999) Zoospore ultrastructure of *Gonapodya polymorpha*. Mycologia 91:727–734

Monchy S, Sanciu G, Jobard M, Rasconi S, Gerphagnon M, Chabé M, Cian A, Meloni D, Niquil N, Christaki U, Viscogliosi E, Sime-Ngando T (2011) Exploring and quantifying fungal diversity in freshwater lake ecosystems using rDNA cloning/sequencing and SSU tag pyrosequencing. Environ Microbiol 13:1433–1453

Montecillo CM, Bracker CE, Powell MJ (1980) Ultrastructure of *Synchytrium macrosporum* zoospores. Can J Bot 58:1885–1897

Montford DO, Orpin CG (1994) Anaerobic fungi: biology, ecology, and function. Dekker, New York

Morris NR (2000) Nuclear migration: from fungi to the mammalian brain. J Cell Biol 148:1097–1101

Moss AS, Reddy NS, Dortaj IM, San Francisco MJ (2008) Chemotaxis of the amphibian pathogen *Batrachochytrium dendrobatidis* and its response to a variety of attractants. Mycologia 100:1–5

Mozley-Standridge SE, Letcher PM, Longcore JE, Porter D, Simmons DR (2009) Cladochytriales—a new order in Chytridiomycota. Mycol Res 113:498–507

Muehlstein LK, Amon JP, Leffler DL (1987) Phototaxis in the marine fungus *Rhizophydium littoreum*. Appl Environ Microbiol 53:1819–1821

Muehlstein LK, Amon JP, Leffler DL (1988) Chemotaxis in the marine fungus *Rhizophydium littoreum*. Appl Environ Microbiol 54:1668–1672

Müller DG, Küpper FC, Küpper H (1999) Infection experiments reveal broad host ranges of *Eurychasma dicksonii* (Oomycota) and *Chytridium polysiphoniae* (Chytridiomycota), two eukaryotic parasites in marine brown algae (Phaeophyceae). Phycol Res 47:217–223

Nagahama T, Takahashi E, Nagano Y, Abdel-Wahab MA, Miyazaki M (2011) Molecular evidence that deep-branching fungi are major components in deep-sea methane cold-seep sediments. Environ Microbiol 13:2359–2370

Nagpal R, Puniya AK, Sehgal JP, Singh K (2011) In vitro fibrolytic potential of anaerobic rumen fungi from ruminants and non-ruminant herbivores. Mycoscience 52:31–38

Nicholson MJ, McSweeney CS, Mackie RI, Brookman JL (2010) Diversity of anaerobic gut fungal populations analysed using ribosomal ITS1 sequences in faeces of wild and domesticated herbivores. Anaerobe 16:66–73

Nikolcheva LG, Barlocher F (2004) Taxon specific fungal primers reveal unexpectedly high diversity during leaf decomposition in a stream. Mycol Prog 3:41–49

Nyvall P, Pedersén M, Longcore JE (1999) *Thalassochytrium gracilariopsidis* (Chytridiomycota), gen. et sp. nov., endosymbiotic in *Gracilariopsis* sp. (Rhodophyceae). J Phycol 35:176–185

Orpin CG (1975) Studies on the rumen flagellate *Neocallimastix frontalis*. J Gen Microbiol 91:249–262

Orpin CG (1977) Invasion of plant tissue in the rumen by the flagellate *Neocallimastix frontalis*. J Gen Microbiol 98:423–430

Orpin CB, Bountiff L (1978) Zoospore chemotaxis in the rumen phycomycete *Neocallimastix frontalis*. J Gen Microbiol 104:113–122

Orpin CG, Letcher AJ (1979) Utilization of cellulose, starch, xylan and other hemicelluloses for growth by the rumen phycomycete *Neocallimastix frontalis*. Curr Microbiol 3:121–124

Ozkose E, Thomas BJ, Davies DR, Griffith GW, Theodorou MK (2001) *Cyllamyces aberensis* gen. nov. sp. nov., a new anaerobic gut fungus with branched sporangiophores isolated from cattle. Can J Bot 79:666–673

Perrott E (1955) The genus *Monoblepharis*. Trans Br Mycol Soc 38:247–282

Perrott E (1958) Isolation and pure culture of *Monoblepharis*. Nature 182:1322–1324

Picard KT, Letcher PM, Powell MJ (2009) *Rhizidium phycophilum*, a new species in the Chytridiales. Mycologia 101:696–706

Piotrowski JS, Annis SL, Longcore JE (2004) Physiology of *Batrachochytrium dendrobatidis*, a chytrid pathogen of amphibians. Mycologia 96:9–15

Powell MJ (1974) Fine structure of plasmodesmata in a chytrid. Mycologia 66:606–614

Powell MJ (1975) Ultrastructural changes in nuclear membranes and organelle associations during mitosis of the aquatic fungus *Entophlyctis* sp. Can J Bot 53:627–646

Powell MJ (1976a) Development of the discharge apparatus in the fungus *Entophlyctis*. Arch Microbiol 111:59–71

Powell MJ (1976b) Ultrastructure and isolation of glyoxysomes (microbodies) in zoospores of the fungus *Entophlyctis* sp. Protoplasma 89:1–27

Powell MJ (1978) Phylogenetic implications of the microbody-lipid globule complex. BioSystems 10:167–180

Powell MJ (1980) Mitosis in the aquatic fungus *Rhizophydium sphaerotheca* (*Chytridiales*). Am J Bot 67:839–853

Powell MJ (1981a) Ultrastructure of *Polyphagus euglenae* zoospores. Can J Bot 59:2049–2061

Powell MJ (1981b) Zoospore structure of the mycoparasitic chytrid *Caulochytrium protostelioides* Olive. Am J Bot 68:1074–1089

Powell MJ (1983) Localization of antimonate-mediated precipitates of cations in zoospores of *Chytriomyces hyalinus*. Exp Mycol 7:266–277

Powell MJ (1993) Looking at mycology with a Janus face. A glimpse at Chytridiomycetes active in the environment. Mycologia 85:1–20

Powell MJ (1994) Production and modifications of extracellular structures during development of Chytridiomycetes. Protoplasma 181:123–141

Powell MJ, Blackwell WH (1991) A proposed dispersal mechanism for *Septosperma rhizophydii*. Mycologia 83:673–680

Powell MJ, Gillette L (1987) Septal structure of the chytrid *Rhizophlyctis harderi*. Mycologia 79:635–639

Powell MJ, Koch WJ (1977) Morphological variations in a new species of *Entophlyctis*. I. The species concept. Can J Bot 55:1668–1685

Powell MJ, Letcher PM (2012) From zoospores to molecules: the evolution and systematics of Chytridiomycota. In: Misra JK, Tewari JP, Deshmukh SK (eds) Systematics and evolution of fungi. CRC, Boca Raton, FL, pp 29–54

Powell MJ, Roychoudhury S (1992) Ultrastructural organization of *Rhizophlyctis harderi* zoospores and redefinition of the type I microbody-lipid globule complex. Can J Bot 70:750–761

Powell MJ, Letcher PM, Longcore JE (2011) *Operculomyces* is a new genus in the order Rhizophydiales. Mycologia 103:854–862

Putnam ML, Hampson MC (1989) Rediscovery of *Synchytrium endobioticum* in Maryland. Am Potato J 66:495–501

Rachowicz LJ, Vredenburg VT (2004) Transmission of *Batrachochytrium dendrobatidis* within and between amphibian life stages. Dis Aquat Org 61:75–83

Reichle RE (1972) Fine structure of *Oedogoniomyces* zoospores, with comparative observations on *Monoblepharella* zoospores. Can J Bot 50:819–824

Rezaeian M, Beakes G, Parker DS (2004) Methods for the isolation, culture and assessment of the status of anaerobic rumen chytrids in both *in vitro* and *in vivo* systems. Mycol Res 34:1215–1226

Rosenblum EB, Stajich JE, Maddox N, Eisen MB (2008) Global gene-expression profiles for life stages of the deadly amphibian pathogen *Batrachochytrium dendrobatidis*. Proc Natl Acad Sci USA 105:17034–17039

Rossman AY, Britton K, Luster D, Palm M, Royer M, Sherald J (2006) Evaluating the threat posed by fungi on the APHIS list of regulated plant pests. Plant Health Progress. doi:10.1094/PHP-2006-5050-01-PS. Available via Plant management network. http://www.plantmanagementnetwork.org/pub/php/perspective/2006/fungi/. Cited 5 May 2006

Roychoudhury S, Powell MJ (1991) Ultrastructure of mitosis in the algal parasitic fungus *Polyphagus euglenae*. Can J Bot 69:2201–2214

Roychoudhury S, Powell MJ (1992) Precise flagellar configuration of the *Rhizophlyctis harderii* zoospore. Can J Bot 70:762–771

Schloegel LM, Toledo LF, Longcore JE, Greenspan SE, Vieira CA, Lee M, Zhao S, Wangen C, Ferreira CM, Hipolito M, Davies AJ, Cuomo CA, Daszak P, James TY (2012) Novel, panzootic and hybrid genotypes of amphibian chytridiomycosis associated with the bullfrog trade. Mol Ecol 21:5162–5177. doi:10.1111/j.1365-294X.2012.05710.x

Schmidt SK, Naff CS, Lynch RC (2012) Fungal communities at the edge: ecological lessons from high alpine fungi. Fungal Ecol 5:443–452

Sekimoto S, Rochon D, Long JE, Dee JM, Berbee ML (2011) A multigene phylogeny of *Olpidium* and its implications for early fungal evolution. BMC Evol Biol 11:331. doi:10.1186/1471-2148-11-331

Shearer CA, Langsam DM, Longcore JE (2004) Fungi in freshwater habitats. In: Mueller GM, Bills GF, Foster MS (eds) Biodiversity of fungi: inventory and monitoring methods. Elsevier, Oxford, pp 513–531

Shields JD (1990) *Rhizophydium littoreum* on the eggs of *Cancer anthonyi*: parasite or saprobe? Biol Bull 179:201–206

Shields JP, Fuller MS (1996) Ultrastructure of chytridiomycete and oomycete zoospores using spray-freeze fixation. Protoplasma 191:84–95

Sime-Ngando T, Lefèvre E, Gleason FH (2011) Hidden diversity among aquatic heterotrophic flagellates: ecological potential of zoosporic fungi. Hydrobiologia 659:5–22

Simmons DR (2011) Phylogeny of Powellomycetaceae fam. nov. and description of *Geranomyces variabilis* gen. et comb. nov. Mycologia 103:1411–1420

Simmons DR, Longcore JE (2012) *Thoreauomyces* gen. nov., *Fimicolochytrium* gen. nov. and additional species in *Geranomyces*. Mycologia 104:1229–1243. doi:10.3852/12-015

Simmons DR, James TY, Meyer AF, Longcore JE (2009) Lobulomycetales, a new order in the Chytridiomycota. Mycol Res 113:450–460

Simmons DR, Letcher PM, Powell MJ, Longcore JE (2012) *Alogomyces tanneri* gen. et sp. nov., a chytrid in Lobulomycetales from horse manure. Mycologia 104:157–163

Sønstebø JH, Rohrlack T (2011) Possible implications of chytrid parasitism for population subdivision in freshwater cyanobacteria of the genus *Planktothrix*. Appl Environ Microbiol 77:1344–1351

Sparrow FK (1960) Aquatic phycomycetes, 2nd edn. University of Michigan Press, Ann Arbor

Sparrow FK, Barr ME (1955) Additions to the phycomycete flora of the Douglas Lake region I. New taxa and records. Mycology 47:546–556

Stachewicz H, De Boer SH (2002) Pathotype determination of potato wart from Prince Edward Island, Canada. Nachrichtebl Deut Pflanzenschutzd 54:269

Stajich JE (2011) Currently available fungal genome sequences. Available at http://fungalgenomes.org/wiki/Fungal_Genome_Links. Cited 4 July 2011

Stajich JE, Berbee ML, Blackwell M, Hibbett DS, James TY, Spatafora JW, Taylor JW (2009) The Fungi. Curr Biol 19:840–845

Steenkamp ET, Wright J, Baldauf SL (2006) The protistan origins of animals and fungi. Mol Biol Evol 23:93–106

Steiger RA, Simmons DR, Longcore JE (2011) *Cylindrochytridium johnstonii* is a member of the Cladochytriales. Mycotaxon 118:293–302

Stoeck T, Epstein S (2003) Novel eukaryotic lineages inferred from small-subunit rRNA analyses of oxygen-depleted marine environments. Appl Environ Microbiol 69:2657–2663

Stoeck T, Kasper J, Bunge J, Leslin C, Ilyin V, Epstein S (2007) Protistan diversity in the Arctic: a case of paleoclimate shaping modern biodiversity? PLoS One 2(8):e728. doi:10.1371/journal.pone.0000728

Strasburger E (1878) Wirkung des Lichtes und der Wärme auf Schwärmsporen. Jenaische Zeitschr f Naturwiss 2:551–625

Sun G, Yang Z, Kosch T, Summers K, Huang J (2011) Evidence for acquisition of virulence effectors in pathogenic chytrids. BMC Evol Biol 11:195. doi:10.1186/1471-2148/11/195

Tachezy J (2008) Hydrogenosomes and mitosomes: mitochondria of anaerobic eukaryotes. Springer, Berlin, Heidelberg, New York

Taylor JW, Fuller MS (1981) The Golgi apparatus, zoosporogenesis, and development of the zoospore discharge apparatus of *Chytridium confervae.* Exp Mycol 5:35–59

Thornton ML (1970) Transport of soil-dwelling aquatic phycomycetes by earthworms. Trans Br Mycol Soc 55:391–397

Thornton ML (1971) Potential for long-range dispersal of aquatic phycomycetes by internal transport in birds. Trans Br Mycol Soc 57:49–59

Travland LB, Whisler HC (1971) Ultrastructure of *Harpochytrium hedinii.* Mycologia 63:767–789

Ustinova I, Krienitz L, Huss VAR (2000) *Hyaloraphidium curvatum* is not a green alga, but a lower fungus; *Amoebidium parasiticum* is not a fungus, but a member of the DRIPs. Protist 151:253–262

van der Giezen M (2009) Hydrogenosomes and mitosomes: conservation and evolution of functions. J Eukaryot Microbiol 56:221–231

van der Giezen M, Birdsey GM, Horner DS, Lococq J, Dyal PL, Benchimol M, Danpure CJ, Embley TM (2003) Fungal hydrogenosomes contain mitochondria heat-shock proteins. Mol Biol Evol 20:1051–1061

Vargas M, Aronson JM, Roberson RW (1993) The cytoplasmic organization of hyphal tip cells in the fungus *Allomyces macrogynus.* Protoplasma 176:43–52

Vélez CG, Letcher PM, Schultz S, Powell MJ, Churchill PF (2011) Molecular phylogenetic and zoospore ultrastructural analyses of *Chytridum olla* establish the limits of a monophyletic Chytridiales. Mycologia 103:118–130

Vieira ALG, Camilo CM (2011) *Agrobacterium tumefasciens*-mediated transformation of the aquatic fungus *Blastocladiella emersonii.* Fungal Genet Biol 48:806–811

Voyles J (2011) Phenotypic profiling of *Batrachochytrium dendrobatidis*, a lethal fungal pathogen of amphibians. Fungal Ecol 4:196–200

Wakefield WS, Powell MJ, Letcher PM, Barr DJS, Churchill PF, Longcore JE, Chen S-F (2010) A molecular phylogenetic evaluation of the Spizellomycetales. Mycologia 102:596–604

Webster J (1970) Coprophilous fungi. Trans Br Mycol Soc 54:161–180

Weldon C, du Preez LH, Hyatt AD, Muller R, Speare R (2004) Origin of the amphibian chytrid fungus. Emerg Infect Dis 10:2100–2105

Whiffen J (1944) A discussion of taxonomic criteria in the Chytridiales. Farlowia 1:583–597

Whisler HC, Travland LB (1973) Mitosis in *Harpochytrium.* Arch Protistenk 115:69–74

Widmer TL (2006) Cryopreservation of *Synchytrium solstitiale* in plants. Plant Dis 90:429–432

Willoughby LG (1998) A quantitative ecological study on the monocentric soil chytrid, *Rhizophlyctis rosea*, in Provence. Mycol Res 102:1338–1342

Willoughby LG (2001) The activity of *Rhizophlyctis rosea* in soil: some deductions from laboratory observations. Mycologist 15:113–117

Willoughby LG, Townley PJ (1961) Two new saprophytic chytrids from the Lake District. Trans Br Mycol Soc 44:177–184

Wubah DA, Fuller MS, Akin DE (1991) Resistant body formation in *Neocallimastix* sp., an anaerobic fungus from the rumen of a cow. Mycologia 83:40–47

Yun HY, Kim YH, James TY (2011) First report of false rust caused by *Synchytrium minutum* on Kudzu in Korea. Dis Notes 95:358

7 Blastocladiomycota

Timothy Y. James[1], Teresita M. Porter[2], W. Wallace Martin[3]

CONTENTS

[1]Department of Ecology and Evolutionary Biology, University of Michigan, 830 N. University, Ann Arbor, MI 48109, USA; e-mail: tyjames@umich.edu
[2]Biology Department, Duke University, Durham, NC 27708, USA and Department of Biology, McMaster University, Hamilton, ON L8S 4K1, Canada; e-mail: terri@evol.mcmaster.ca
[3]Department of Biology, Randolph-Macon College, Ashland, VA 23005, USA; e-mail: wmartin@rmc.edu

I. Introduction

Blastocladiomycota are **zoosporic fungi** that comprise an early-diverging branch on the tree of true fungi possessing a number of distinguishing morphological and life history characteristics. These aquatic and soil fungi include genera typically considered so-called water molds, occurring as saprotrophs on decaying plants and animals. Other members of the clade are obligate parasites of invertebrates, plants, and algae. Beyond a basic understanding of global biodiversity, Blastocladiomycota present avenues for exciting research: they have served as models for fungal genetics and physiology (Olson 1984; Ribichich et al. 2005), they have potential as biocontrol agents of plant pests and disease vectors (Chapman 1985; Garcia 1983; Singh et al. 2007), and they interact with aquatic food webs and nutrient cycling through parasitism and consumption (Johnson et al. 2006). Convenient terms for the group include ***blastodad*** and ***blastoclad***, and we adopt the latter in reference to any fungus belonging to this phylum.

The blastoclads were historically considered to be closely related to other zoosporic true fungi, the Chytridiomycota (chytrids), because they reproduce with zoospores possessing a **single posteriorly directed flagellum** (Sparrow 1960). Blastocladiomycota, as well as the two zoosporic fungal phyla Chytridiomycota and Neocallimastigomycota, were shown to be members of the fungal kingdom in the earliest ribosomal DNA-based molecular phylogenies (Bowman et al. 1992; Bruns et al. 1992;

Systematics and Evolution, 2nd Edition
The Mycota VII Part A
D.J. McLaughlin and J.W. Spatafora (Eds.)

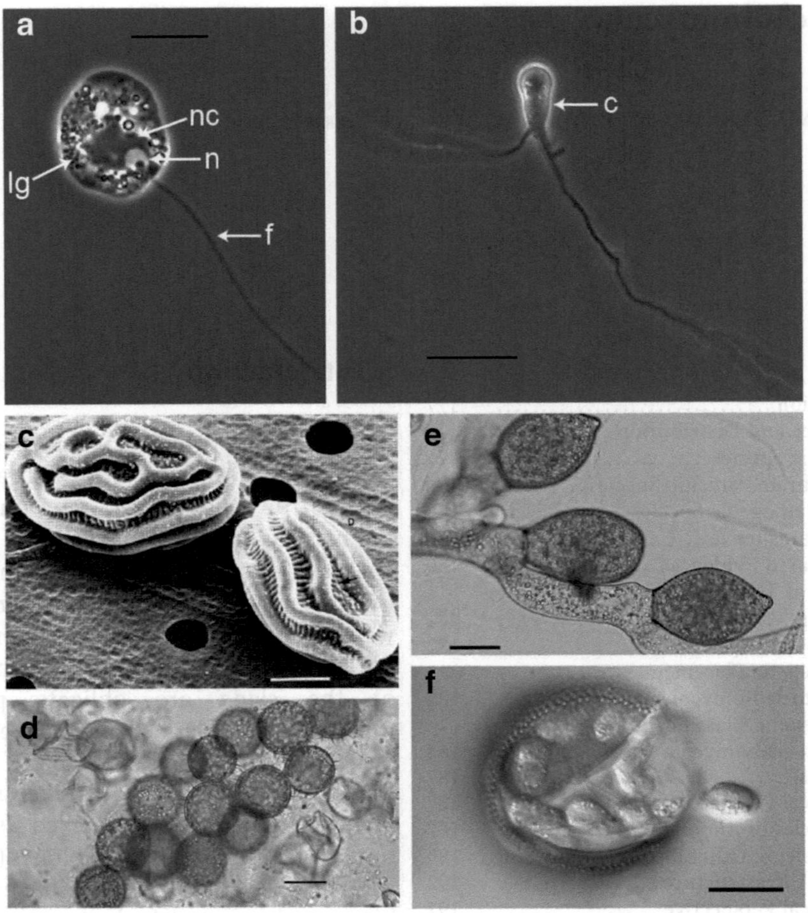

Fig. 7.1 Characteristic stages of blastoclads include zoospores and bipolar germination of encysted spores. (**a**) Uniflagellate zoospore of *Allomyces javanicus*. *f* flagellum, *n* nucleus, *nc* nuclear cap, *lg* lipoid granules. (**b**) Cyst (c) demonstrating bipolar germination with one end enlarging to form a hypha and the other end composed of thin rhizoids involved in substrate attachment and resource extraction. (**c**) Scanning electron micrograph of resting sporangia of *Coelomomyces iliensis* var. *iliensis*, a parasite of the mosquito *Culex antennatus* [photo from Couch and Bland (1985), used with permission from Elsevier]. (**d**) Spiny resting sporangia of *Catenaria spinosa*, a parasite of the midge *Chironomus decorus*. (**e**) Cluster of resting sporangia of *Allomyces moniliformis* demonstrating pitted outer wall. (**f**) Germinating resting sporangium of *A. javanicus* releasing meiospores. Scale bar in A–C = 10 μM; D, F = 30 μM; E = 20 μM

Förster et al. 1990). While it was obvious that posteriorly uniflagellate "phycomycetes" were allied with true fungi, molecular phylogenetic studies immediately detected a large phylogenetic distance between the blastoclads and the chytrids (Bruns et al. 1992; Nagahama et al. 1995). Despite an uncertain phylogenetic placement, the composition of blastoclads has remained largely consistent over time (with the exception of the genus *Physoderma*), and it has been accepted as monophyletic because of the presence of several distinctive characteristics. The most consistent characteristic of the group is reproduction by a zoospore with a prominent nuclear cap that in hyphal species often demonstrates bipolar germination (Fig. 7.1a, b). Another distinguishing feature is the thick-walled, darkly pigmented **resting sporangium** that is often ornamented with pits, ridges, or spines (Fig. 7.1c–f) and undergoes germination by the cracking of an outer wall through which an endosporangium protrudes. The resting sporangium is sometimes referred to as a resting spore; however, because the sporelike structure germinates to produce zoospores by internal cleavage, the term *sporangium* is more accurate. Also, in our discussion we refer to this structure as a *resting sporangium* rather than a *resistant sporangium* or *meiosporangium*.

Blastoclads are the only fungal group known to demonstrate an alternation of haploid and diploid generations. On the basis of this life cycle, ultrastructural characters, and overall phylogenetic distinctness from chytrids (James et al. 2000; Nagahama et al. 1995; Seif et al. 2005), the blastoclads have been considered a separate phylum, Blastocladiomycota, with a single class and order, Blastocladiomycetes and Blastocladiales, respectively (James et al. 2006b). A distinct life cycle, diverse ecological roles, and phylogenetics all suggest that the group is ancient. This ancient divergence is also corroborated by fossil evidence demonstrating an alternation of generations in the Devonian blastoclad ***Paleoblastocladia milleri*** (Remy et al. 1994) and *Allomyces*-like resting sporangia from the Devonian that are remarkably similar to those of extant species (Taylor et al. 1994).

Here, we provide a basic introduction to the blastoclads, covering their distribution and ecology, phylogenetic relationships, and morphological characteristics of their spores, and we present an overview of their unique life cycles and genetics.

II. Occurrence and Distribution

Blastoclads are globally distributed in numerous aquatic and terrestrial habitats. Within these habitats, blastoclads can be observed on decaying plant or insect material or as pathogens of aquatic organisms. However, other unique terrestrial habitats, such as in the photosynthetic tissues of vascular plants, and as parasites of terrestrial invertebrates, such as tardigrades and nematodes, are important habitats for blastoclads. Some genera, such as *Blastocladiella* and *Allomyces*, seem to be more common and speciose in tropical or subtropical habitats (Emerson 1941; Whisler 1987), a general trend in the diversity of zoosporic fungi and pseudo-fungi (Sparrow 1960). No marine or halophytic species are known.

The geographic distribution of certain species within well-studied genera suggests many widespread species of *Allomyces*, *Coelomomyces*, and *Blastocladia*. *Allomyces arbusculus* is known to occur on most continents, including temperate climates, while *A. moniliformis* and the hybrid species *Allomyces javanicus* are reported in both Old and New World tropical and subtropical habitats (Emerson 1941; Sparrow 1960; Wolf 1941). The species *Blastocladia pringsheimii* is thought to be widespread and ubiquitous and is known from all continents except Antarctica (Nascimento and Pires-Zottarelli 2010; Sparrow 1960). Numerous other species of *Blastocladia* are newly described and only known from India (Dasgupta and John 1988). It is important to note, however, that species concepts among blastoclads are largely untested, and current phylogenetic evidence suggests that species concepts in *Allomyces* are in need of revision (Porter et al. 2011). Other genera, for example, *Coelomomyces* spp., have species with rather distinct resting sporangium ornamentation, suggesting that they should be readily diagnosable taxa; these species also have been found on multiple continents (Couch and Bland 1985).

A. Saprobic Species

Most saprobic species are known from studies in which water or soil is baited with appropriate substrates such as hemp seeds, rosaceous fruits, pollen, or insect body parts. These species are thus likely to perform an active role in the decomposition of cellulosic, chitinous, or keratinic substrates within the ecosystem. Because the majority of saprobic blastoclads are known to produce a resistant sporangium, they are readily recovered from soils that are air-dried and then baited using an appropriate substrate (Whisler 1987). Soils that are periodically inundated represent good collecting locations for *Allomyces* (Sparrow 1960). The blastoclad-resistant sporangium may either provide an advantage in these habitats that undergo cycles of flooding and drying or merely explain the reason they are so readily recovered. Some *Blastocladia* spp. are **facultatively anaerobic** (Gleason and Gordon 1989) and are recovered from submerged twigs and fruits in stagnant water or by baiting under anaerobic conditions (Whisler 1987).

B. Invertebrate Parasitic Species

Both aquatic and terrestrial invertebrates are hosts for parasitic blastoclads. In cold streams, the larvae of black flies may be host to *Coelomycidium*. Midge eggs and mosquito larvae may be collected along the edges of ponds and streams and are frequently parasitized by *Catenaria* spp. and *Coelomomyces*, respectively (Martin 1987). Infected larvae typically die before metamorphosing into adults; however, fourth-instar female larvae infected by *Coelomomyces* may pupate into adults, become sterilized by the infection, and shed *Coelomomyces* resting sporangia when they attempt to oviposit (Lucarotti 1987). Additional hosts of parasitic blastoclads include nematodes, copepods, cladocerans, and trichopterans, where the parasites grow internally as walled or wall-less thalli, but they may be most readily spotted as gametangia or resting sporangia that may fill the entire body cavity. *Sorochytrium* is only known from the tardigrade *Milnesium* (Dewel et al. 1985) and has been collected only from the type locality in North Carolina. The genus *Myiophagus* is recorded from larvae and pupae from dipterans and scale insects (Karling 1977; Sparrow 1939). The wide distribution of blastoclad pathogens across Crustacea and Arthropoda suggests that many more genera and species remain to be detected. A recent review discussed all of the known genera and species of invertebrate blastoclad parasites (Gleason et al. 2010).

C. Plant-Pathogenic Species

The genera *Physoderma* and *Urophlyctis* are obligate parasites of vascular plants with a worldwide distribution. Symptoms include pustules and leaf streaks of darkly pigmented resting sporangia. The pathogen is observed primarily on the leaves and shoots of aquatic or terrestrial plants but may also appear in the root system (Sparrow 1965). Hosts are diverse and include water ferns, sedges, composites, and the crops *Zea mays* and *Medicago sativa*, but the economic importance of these pathogens seems minimal. Experimental inoculation studies have demonstrated that one unnamed *Physoderma* species from *Agropyron repens*, common quack grass, could infect all tested congeneric hosts but not nine other sympatric species from other families (Sparrow and Griffin 1964). Unlike many Chytridiomycota, only one monotypic genus is known as a parasite of algae, *Paraphysoderma* (Hoffman et al. 2008).

D. Mycoparasites

A single mycoparasitic species is known, *Catenaria allomycis*, which grows endobiotically in *Allomyces* spp. or *Blastocladiella simplex*. Experimental inoculations (Couch 1945) showed the parasite could infect all *Allomyces* spp. (to varying degrees), only one *Blastocladiella* sp., but not *Blastocladia parva*, *Catenaria anguillulae*, or any saprolegnialean hosts.

E. DNA-Based Evidence

Because most of what is known of the distribution and diversity of blastoclads is known from cultured studies using baiting rather than direct observation of samples, the investigation of **environmental DNA** has great potential to enlighten our understanding of the suitable habitats and diversity of blastoclads, as it has done in other fungal lineages in soil and plant roots (Freeman et al. 2009; Schadt et al. 2003; Vandenkoornhuyse et al. 2002). These studies involve obtaining an environmental sample, for example, soil or pond water, which is then subjected to DNA extraction, polymerase chain reaction using primers targeting ribosomal RNA genes, and analysis of DNA sequences to determine the organisms present in the sample. At least three published environmental DNA studies have detected blastoclads. One study

(Slapeta et al. 2005) recovered two sequence types of unknown blastoclads from the sediment of an anoxic pond that clustered with *Allomyces* and *Blastocladiella*. Later, when Porter et al. (2011) produced the first sequences of *Blastocladia*, these sequences were shown to be closely related to the sequences identified by Slapeta et al. (2005). These results are consistent with the role of *Blastocladia* as an obligately fermenting saprobic species. Other studies have revealed blastoclads from the oxycline (region of lower oxygen concentration) of a deep lake in France (Lefèvre et al. 2007) and from the surface water of a tropical lake (Chen et al. 2008). New sequence data from Porter et al. (2011) suggest that these two sequences are related to *Blastocladiella*.

III. Structure of Thallus and Reproductive Characters

Thalli of blastoclads vary greatly in size, extent, and position in relation to the substratum. Simple thalli may be **monocentric** (producing a single reproductive body) or **epibiotic** (with the reproductive body produced outside the substratum). A tube from a young epibiotic thallus penetrates the substratum and branches distally to form an **endobiotic** system of smaller branching tubes (**rhizoids**) that functions in the absorption of nutrients. The complex thalli of some genera are **mycelial** (filamentous with tubular hyphae) and **polycentric** (produce multiple reproductive bodies). Blastoclad thalli generally consist of a larger basal axis attached to the substratum by rhizoids, which may exhibit **determinate** (limited growth) or **indeterminate** (unlimited growth) apical branching and growth. The hyphae may be **septate** (with incomplete septa having central and lateral perforations) or **aseptate** and **coenocytic** (without septa). Polycentric thalli in some parasites take the form of slender rhizoidal elements alternating with spindle-shaped swellings (a **rhizomycelium**), while in other parasites the thalli may be reduced to unwalled coenocytic thalli (**hyphal bodies**) that lack rhizoids. Thalli lacking rhizoids that are completely converted into reproductive structures are **holocarpic**, whereas thalli bearing rhizoids or vegetative portions not contributing to reproductive structures are **eucarpic**.

Many blastoclads have been shown to share a life history involving the alternation of two developmentally associated phases: a haploid **gametophyte generation** (which produces gametes) and a diploid **sporophyte generation** (which produces asexual spores). Organisms in which the vegetative features of gametophyte and sporophyte are very similar have an **isomorphic alternation of generations**, whereas those in which the vegetative features of one generation differ substantially from those of the next have a **heteromorphic alternation of generations**. Sexual reproduction occurs when **gametes** produced from gametophytic thalli undergo **syngamy** (fusion) to form diploid **planozygotes** that give rise to sporophytic thalli. Sporophytic thalli carry out asexual reproduction through the formation of **thin-walled zoosporangia** (which produce zoospores that renew the sporophyte generation) and thick-walled **resistant sporangia** (which typically undergo **meiosis** and produce **meiospores** that renew the gametophyte generation). In some publications thin-walled zoosporangia and zoospores are referred to as mitosporangia and mitospores, respectively, whereas resistant sporangia are referred to as **resting spores**, **resting sporangia**, or **meiosporangia** and their products variously as **meiospores**, **RS zoospores**, or **RS planospores**. The resistant sporangium has a thick, pigmented outer wall that may be smooth or ornamented with pits, punctae, or a complex series of ridges (Fig. 7.1c). At germination the outer wall of the resistant sporangium cracks open at undetermined points in some species, while in others it opens along a preformed **germination slit** or a **circumcissile lid**. An elastic **inner wall** (also called the **endosporangium**) may swell and protrude through the outer wall at the time of germination. **Discharge papillae** are dome-shaped protrusions of gelatinous material that form temporary plugs in **discharge pores** or slits in the walls of zoosporangia, gametangia, and the inner wall of the resistant sporangium. Upon dissolution of the plug the spores pass through the pore or slit to the outside. Species producing gametangia on a single thallus whose nuclei are self-compatible are said to be **homothallic**, whereas those producing gametangia on different thalli whose nuclei are self-incompatible but cross-compatible are **heterothallic**. Morphologically distinct male and female **gametangia** or gametes occur in both homothallic and heterothallic species and may be distinguished by differences in size, color, and mating behavior. The alternative forms of physiologically distinguishable (but morphologically indistinguishable) heterothallic species are designated as **plus (+)** and **minus (-) mating types**. Sexual reproduction by fusion of flagellated gametes that are equal in size is referred to as **isogamy**, while fusion of flagellated gametes that are unequal in size is **anisogamy**. With reference to animal systems, the smaller gamete is designated as male while the larger is female.

IV. Phylogeny and Systematics

A. Phylogenetic Placement of Blastocladiomycota

The history of Blastocladiomycota began with the description of the first so-called *chytrid* genus, *Physoderma*, by Wallroth in 1833. In 1878, P. F. Reinsch described *Blastocladia pringheimii* as the single member of a new genus whose unusual features were puzzling to systematists who included the organism variously among the Saprolegniaceae (Fischer 1892) or the Leptomitaceae (Schroeter 1893). Petersen felt that these fungi differed substantially from the Saprolegniales and in 1909 established the order Blastocladiales containing the single family Blastocladiaceae to accommodate the genus *Blastocladia* (Petersen 1909), and a second genus, *Allomyces*, was soon added to the family (Butler 1911). Little additional knowledge of the group occurred until Kniep's researches on *Allomyces* (Kniep 1930), which resulted in the **discovery of a life cycle with sporic meiosis and a new type of sexuality (anisogamy) previously unknown in the fungi**. These findings stimulated further research and quickly led to a more refined concept of the Blastocladiales as a morphologically distinct group with greatest affinities to Scherffel's uniflagellate "Chytridineen" series of fungi (Scherffel 1925). This concept was followed by Sparrow in his monographic treatments in 1943 and 1960. Sparrow regarded the posteriorly uniflagellate Chytridiales, Blastocladiales, and Monoblepharidales as one of four lines of descent among the aquatic fungi and erected the class Chytridiomycetes into which they were transferred (Sparrow 1960).

Since the onset of molecular phylogenetic-based systematics, numerous studies have investigated the phylogenetic placement of the blastoclads in the fungal tree of life. Currently, the placement is disputed and ranges from being sister to the Chytridiomycota to being related to terrestrial zygomycete fungi (Fig. 7.2). The implications for the placement of the group within the fungal kingdom are important for understanding the traits of the most recent common ancestor of all fungi and of evolutionary trends in life cycles. Recently, Porter et al. (2011) produced the first comprehensive molecular phylogenetic study of genera and families of blastoclads. The results of these phylogenetic studies and additional ultrastructural studies led to the establishment of a new phylum, the Blastocladiomycota (James et al. 2006b), with the single class Blastocladiomycetes, into which were placed the genera included in Blastocladiales.

The tool of choice for fungal molecular phylogenetics has been nuclear-encoded ribosomal DNA (rDNA) because this multicopy region contains both conserved and divergent regions capable of providing multiple levels of phylogenetic resolution and is easy to amplify using conserved polymerase chain reaction primers (Bruns et al. 1991). Early phylogenetic studies of zoosporic fungi focused on ultrastructural characters of the zoospore that have been used to delimit the orders, families, and genera; these studies suggested a good correlation among ultrastructural characters and phylogenetic relationships based on rDNA (James et al. 2000; Letcher et al. 2006, 2008; Simmons et al. 2009). Subsequently, the rDNA operon (18S+5.8S+28S) also provided a framework for suggesting taxonomic revisions in the Blastocladiomycota (Porter et al. 2011).

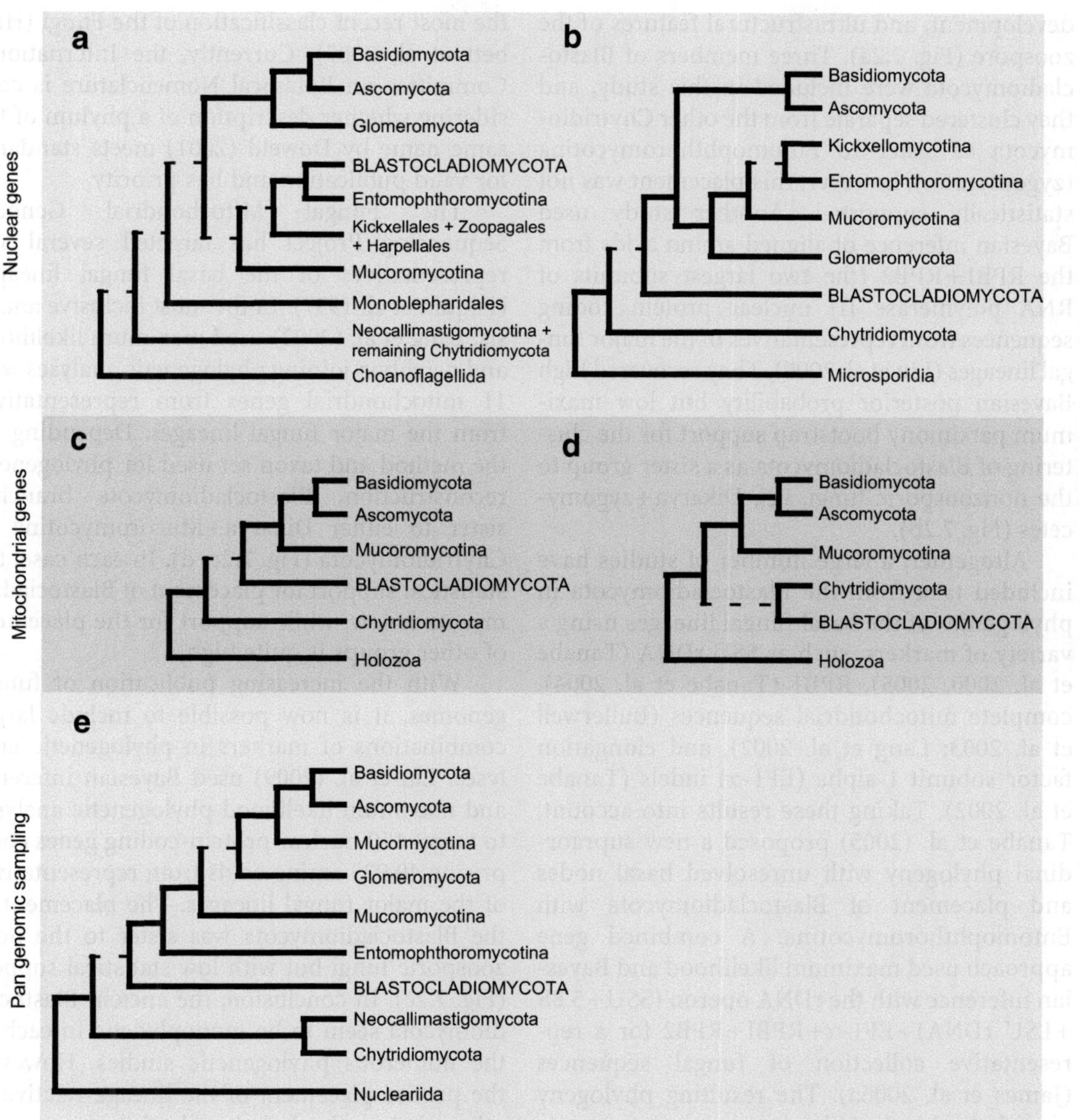

Fig. 7.2 Hypothesized phylogenetic placements of Blastocladiomycota among fungi. (**a**) James et al. (2000) with SSU rDNA analyzed using maximum parsimony. (**b**) Liu et al. (2006) with RPB1+RPB2 concatenated protein sequences analyzed using Bayesian inference and maximum parsimony. (**c, d**) Lang et al. (2002) with 11 concatenated mtDNA proteins analyzed using maximum likelihood (**c**) or a distance method (**d**). (**e**) Liu et al. (2009) with 150 concatenated proteins (40,925 amino acids) analyzed using Bayesian inference and maximum likelihood. Dashed lines: branches whose placement in phylogeny was not strongly supported by bootstrap analyses (<70 %)

In many of the following cited studies, species in the Blastocladiales are still classified with the Chytridiomycota; however, in Fig. 7.2 they are classified according to their present placement in Blastocladiomycota. Nuclear genes, such as rDNA and protein-coding genes, have been used with varying levels of success to classify the basal fungi. James et al. (2000) used small subunit (SSU) rDNA (18S rDNA) sequences to produce the first well-sampled phylogeny of zoosporic true fungi and addressed the phylogenetic consistency of zoospore discharge type, thallus development, and ultrastructural features of the zoospore (Fig. 7.2a). Three members of Blastocladiomycota were included in this study, and they clustered separate from the other Chytridiomycota as sister to Entomophthoromycotina (zygomycetes); however, this placement was not statistically supported. Another study used Bayesian inference of aligned amino acids from the RPB1+RPB2 (the two largest subunits of RNA polymerase II) nuclear protein coding sequences from representatives of the major fungal lineages (Liu et al. 2006). They recovered high Bayesian posterior probability but low maximum parsimony bootstrap support for the clustering of Blastocladiomycota as a sister group to the nonzoosporic fungi, i.e., Dikarya+zygomycetes (Fig. 7.2b).

Altogether, a large number of studies have included taxa from the Blastocladiomycota in phylogenies of the basal fungal lineages using a variety of markers, such as SSU rDNA (Tanabe et al. 2000, 2005), RPB1 (Tanabe et al. 2004), complete mitochondrial sequences (Bullerwell et al. 2003; Lang et al. 2002), and elongation factor subunit 1-alpha (EF1-α) indels (Tanabe et al. 2002). Taking these results into account, Tanabe et al. (2005) proposed a new supraordinal phylogeny with unresolved basal nodes and placement of Blastocladiomycota with Entomophthoromycotina. A combined gene approach used maximum likelihood and Bayesian inference with the rDNA operon (SSU+5.8S +LSU rDNA)+EF1-α+RPBI+RPB2 for a representative collection of fungal sequences (James et al. 2006a). The resulting phylogeny placed the blastoclads sister to nonzoosporic fungi. Altogether, studies using multiple combinations of phylogenetic markers have repeatedly found that the Blastocladiomycota are monophyletic and usually group separately from other lineages of zoosporic fungi.

Until 2006, the blastoclads were treated as an order (Blastocladiales) in Chytridiomycota. The distinctiveness of Blastocladiales from Chytridiomycota was formally addressed by James et al. (2006b), who described Blastocladiales as a separate phylum, Blastocladiomycota. This was based on rDNA phylogenetic analysis and ultrastructural characters. Blastocladiomycota was also recognized as a separate phylum in the most recent classification of the Fungi (Hibbett et al. 2007). Currently, the International Committee on Botanical Nomenclature is considering whether description of a phylum of the same name by Doweld (2001) meets standards for valid publication and has priority.

The Fungal Mitochondrial Genome Sequencing Project has targeted several key representatives of the basal fungal lineages (Paquin et al. 1997). In the most inclusive analysis, Lang et al. (2002) used maximum likelihood and neighbor-joining phylogenetic analyses with 11 mitochondrial genes from representatives from the major fungal lineages. Depending on the method and taxon set used for phylogenetic reconstruction, Blastocladiomycota branches sister to either Dikarya+Mucoromycotina or Chytridiomycota (Fig. 7.2c, d). In each case, the statistical support for placement of Blastocladiomycota is low, while support for the placement of other groups is quite high.

With the increasing publication of fungal genomes, it is now possible to include larger combinations of markers in phylogenetic analyses. Liu et al. (2009) used Bayesian inference and maximum likelihood phylogenetic analyses to target 150 nuclear protein-coding genes comprising 40,925 amino acids from representatives of the major fungal lineages. The placement of the Blastocladiomycota was sister to the nonzoosporic fungi but with low statistical support (Fig. 7.2e). In conclusion, the ancient Blastocladiomycota seem to be monophyletic in each of the numerous phylogenetic studies. However, the precise placement of the lineage relative to other phyla and subphyla of both zoosporic and nonzoosporic fungi is uncertain. In the future, additional genome and large-scale expressed sequence tag (EST) sequencing of many more representatives of the basal fungal lineages will be needed to resolve the question of when the blastoclads diverged among the fungi.

B. Phylogenetic Classification of Blastocladiomycota

Blastoclads comprise 5 families (Blastocladiaceae, Catenariaceae, Coelomomycetaceae, Physodermataceae and Sorochytriaceae), 14

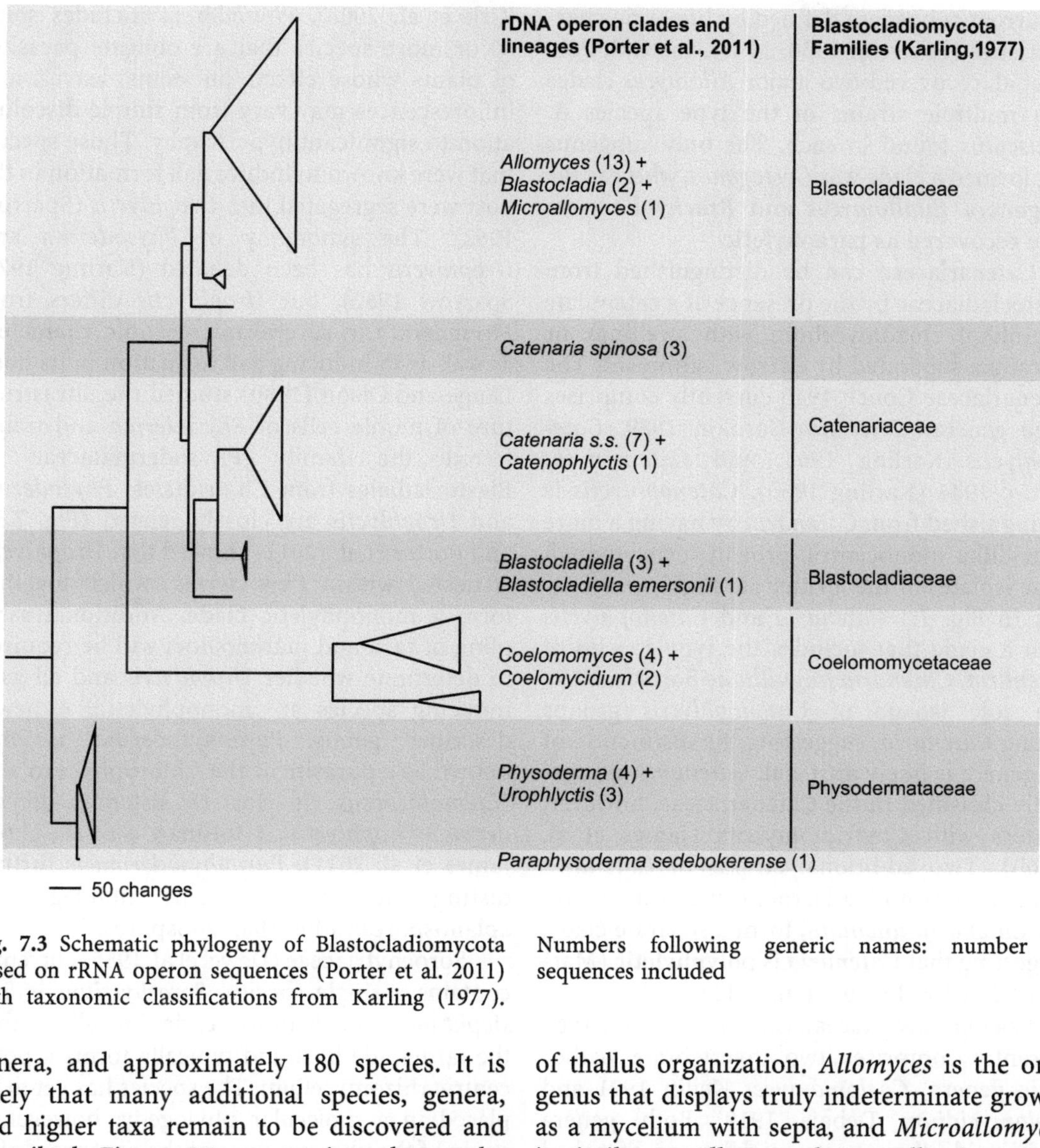

Fig. 7.3 Schematic phylogeny of Blastocladiomycota based on rRNA operon sequences (Porter et al. 2011) with taxonomic classifications from Karling (1977). Numbers following generic names: number of sequences included

genera, and approximately 180 species. It is likely that many additional species, genera, and higher taxa remain to be discovered and described. Figure 7.3 summarizes the results from a Bayesian rDNA (SSU+5.8S+LSU) phylogeny, including 45 gene sequences from the Blastocladiomycota representing 4 of the 5 families (Porter et al. 2011). Here we review the phylogenetic placement of the various families, the evidence of monophyly of genera, and the morphological characters that separate the various genera.

The Blastocladiaceae Petersen 1909 currently comprises five genera: *Allomyces* Butler 1911, *Blastocladia, Microallomyces* Emerson and Robertson 1974, *Blastocladiella* Matthews 1937, and *Blastocladiopsis* Sparrow 1950 (Karling 1977). Genera can be separated on the basis of thallus organization. *Allomyces* is the only genus that displays truly indeterminate growth as a mycelium with septa, and *Microallomyces* is similar to *Allomyces* but smaller in stature and lacking pseudosepta. Members of *Blastocladia* typically have a single trunklike basal cell, with septations only at reproductive organs. The poorly known *Blastocladiopsis* is distinguished from *Blastocladia* on the basis of an unpitted, loose resting sporangium. In Fig. 7.3, strains from *Allomyces, Blastocladia,* and *Microallomyces* form a clade, three strains of *Blastocladiella* and *Blastocladiella emersonii* Cantino and Hyatt 1953 cluster separately, and the genus *Blastocladiopsis* has yet to be placed in a molecular phylogeny. Though *Allomyces* forms a clade, Porter et al. (2011) showed that these taxa do not necessarily group according to current subgenera defined by life cycle characteristics (Emerson 1938, 1941). Instead, Porter et al. recovered two major *Allomyces* clades with multiple strains of the type species *A. arbusculus* found in each. The only subgenus that formed a clade was *Cystogenes*, whereas the **subgenera *Euallomyces* and *Brachyallomyces* were recovered as paraphyletic**.

Catenariaceae can be distinguished from Blastocladiaceae by the presence of a catenulate (chainlike) rhizomycelium with swellings or sporangia separated by narrow isthmuses. The Catenariaceae Couch 1945 currently comprises three genera: *Catenaria* Sorokin 1889, *Catenophlyctis* Karling 1965, and *Catenomyces* A.M. c 1944 (Karling 1965). *Catenophlyctis* is distinguished from *Catenaria* by having a more chytridlike monocentric growth form, though some isolates of the former are highly polycentric. In Fig. 7.3 *Catenaria* and *Catenophlyctis* form a clade that includes the type species of *Catenaria*, *Catenaria anguillulae* Sorokin 1876. The sole isolate of *Catenophlyctis* groups among *Catenaria*, suggesting the distinction of the genera is likely artificial. *Catenomyces*, currently classified in the Catenariaceae, however, clusters with Chytridiomycota (James et al. 2006b). Two additional strains of *Catenaria* isolated from midge larvae, *Catenaria spinosa* and *Catenaria uncinata*, form a separate clade, suggesting that ***Catenaria* is polyphyletic** (Martin 1975, 1978; Porter et al. 2011).

Coelomomycetaceae Couch ex Couch 1962 currently comprises two invertebrate pathogenic genera, *Coelomomyces* Keilin 1921 and *Coelomycidium* Debais 1919. Both genera grow inside their hosts in the form of naked protoplasts. *Coelomomyces* is distinguished from *Coelomycidium* on the basis of hosts, mosquitoes, and ostracods or copepods in the former and blackflies in the latter. In Fig. 7.3 ***Coelomomyces* and *Coelomycidium* are reciprocally monophyletic sister clades.**

Physodermataceae Sparrow 1952 comprises two plant pathogenic genera, *Physoderma* Wallr. 1833 and *Urophlyctis* J. Schrot. 1886. Both genera form an epibiotic, monocentric, sporangial stage and an endobiotic, polycentric phase. These two genera were synonymized (Karling 1950) and are generally still considered synonymous to this day (Karling 1977; Kirk et al. 2008). *Physoderma* includes some 80 or more species that are obligate parasites of plants whose effects on stems, leaves, and inflorescences may vary from simple discoloration to significant hypertrophy. Those species that were known to induce gall formation in the host were segregated into *Urophlyctis* (Sparrow 1962). The synonymy of *Physoderma* and *Urophlyctis* has been debated (Karling 1977; Sparrow 1962), but *Urophlyctis* differs from *Physoderma* in several microscopic characters as well as in inducing gall formation in its host. Lange and Olson (1980) studied the ultrastructure of motile cells of *Physoderma* and transferred the family Physodermataceae to Blastocladiales from Chytridiales. *Physoderma* and *Urophlyctis* are closely related (Fig. 7.3), and Porter et al. (2011) showed that *Urophlyctis* is nested within *Physoderma*, which together form a monophyletic clade. Additional sampling of taxa and markers/loci will be required to determine whether *Urophlyctis* and all gall-inducing species are monophyletic. A newly described genus, *Paraphysoderma*, is only known as a parasite of the Chlorophycean alga *Haematococcus*. It clusters sister to *Physoderma* + *Urophlyctis* (Hoffman et al. 2008; James et al. 2011). *Paraphysoderma* is further distinguished by producing nonflagellated aplanospores rather than zoospores.

Sorochytriaceae (Dewel et al. 1985) currently contains a single species, *Sorochytrium milnesiophthora*, which grows endobiotically within the tardigrade host and typically forms a polycentric rhizomycelium. The species has yet to be placed in a molecular phylogeny; however, a study of the ultrastructure of the zoospores of *S. milnesiophthora* clearly places the family with Blastocladiales (Dewel and Dewel 1990).

The time-consuming nature and specialist knowledge required to collect and identify new isolates means that many described members of Blastocladiomycota, particularly pathogenic species, have yet to be sequenced and placed in a molecular phylogeny, echoing a common pattern in science (Hibbett et al. 2007). Several additional organisms have been described in recent years whose affinities are clearly with the blastoclads but whose life cycles or development are incompletely known or understood. These include *Polycaryum laeve*, an endoparasite of *Daphnia* previously considered a haplosporidian. Phylogenetic evidence was used to affiliate *Polycaryum* with blastoclads, but only a partial sequence is available, precluding more precise placement (Johnson et al. 2006). An effort to collect and place type species in a molecular phylogeny would help to catalog the phylogenetic diversity in Blastocladiomycota and provide a framework for further ecological studies. Mention should also be made of *Callimastix cyclopis, a* parasite of the copepod *Cyclops* whose zoospore structure most resembles that of *Coelomomyces* (Vavra and Joyon 1966).

V. Life Cycles

A. Historical Perspective

By the early twentieth century the concept of an alternation of haploid and diploid generations in the life histories of lower plants and major algal groups was firmly established. However, the discovery by Kniep of an alternation of sporophyte and gametophyte generations in a new zoosporic fungus (*A. javanicus*) was unexpected and aroused great interest in the mycological community (Kniep 1930). Kniep established ploidy in the new fungus by reported **nuclear volume ratios of 1:2 between gametophyte and sporophyte**, and Emerson and Wilson (1949) and Wilson (1952) provided the cytological proof of "sporic" meiosis in the resistant sporangium. These studies, along with the later electron microscopic observations, confirmed that meiosis in *A. macrogynus* begins during early resistant sporangium formation with the appearance of a synaptonemal complex, is halted as **the resistant sporangium enters the resting phase in the late prophase (diplotene) of meiosis I, and is completed during germination** (Olson 1974).

Emerson's 1941 monograph was a seminal work detailing the results of a 6-year study involving the comparative development of 51 *Allomyces* isolates from around the world (Emerson 1941). In this work Emerson recognized three life cycle types that formed the basis of his classification of the genus into three subgenera: ***Euallomyces*** (to include so-called long-cycled species in which there is an isomorphic alternation of generations), ***Brachyallomyces*** (to include so-called short-cycled isolates in which there is no indication of sexuality), and ***Cystogenes*** (to include isolates in which the gametophyte thallus is reduced to a single-celled cyst). Since that time, new members of the Blastocladiomycota have been interpreted and described in relation to the life cycles of *Allomyces*, and this practice is reflected in much of the present classification. To understand the diversity of life histories now known for the blastoclads, it is necessary to review in some detail earlier research on the sexuality and life cycles of *Allomyces*.

B. Life Cycles of *Allomyces*

Euallomyces The subgenus *Euallomyces* includes long-cycled *Allomyces* species such as *A. javanicus* and *A. arbuscula* in which the vegetative thalli of both gametophyte and sporophyte generations are the same in appearance (an isomorphic alternation of generations). The life history of *A. arbuscula* (Emerson 1941; Hatch 1935) is shown in Fig. 7.4. A simplified diagram of the *Euallomyces* life cycle is presented in Fig. 7.6c. In these species the homothallic gametophyte generation bears gametangia that are typically paired and sexually dimorphic with smaller orange male and larger colorless female gametangia. Smaller, more active orange gametes released from male gametangia undergo anisogamous fusion with larger colorless gametes from female gametangia to produce biflagellate planozygotes that give rise to the diploid sporophyte generation. Sporophytes produce a mixture of thin-walled zoosporangia and thick-walled, brown, pitted, resistant sporangia. Zoospores from zoosporangia give rise asexually to additional sporophytic thalli, and resistant sporangia undergo meiosis and release haploid meiospores that produce the gametophyte generation. Deviations from these so-called normal patterns have been observed frequently, including the parthenogenetic development of both gametophyte and sporophyte thalli from single (nonfusing) female gametes and the development of sporophytic thalli from meiospores (Emerson 1941).

Cystogenes Emerson (1938, 1941) and Wilson (1952) described a very different type of life cycle in *A. moniliformis* and *A. neo-moniliformis* (=*A. cystogenus*) (Emerson 1941), a heteromorphic

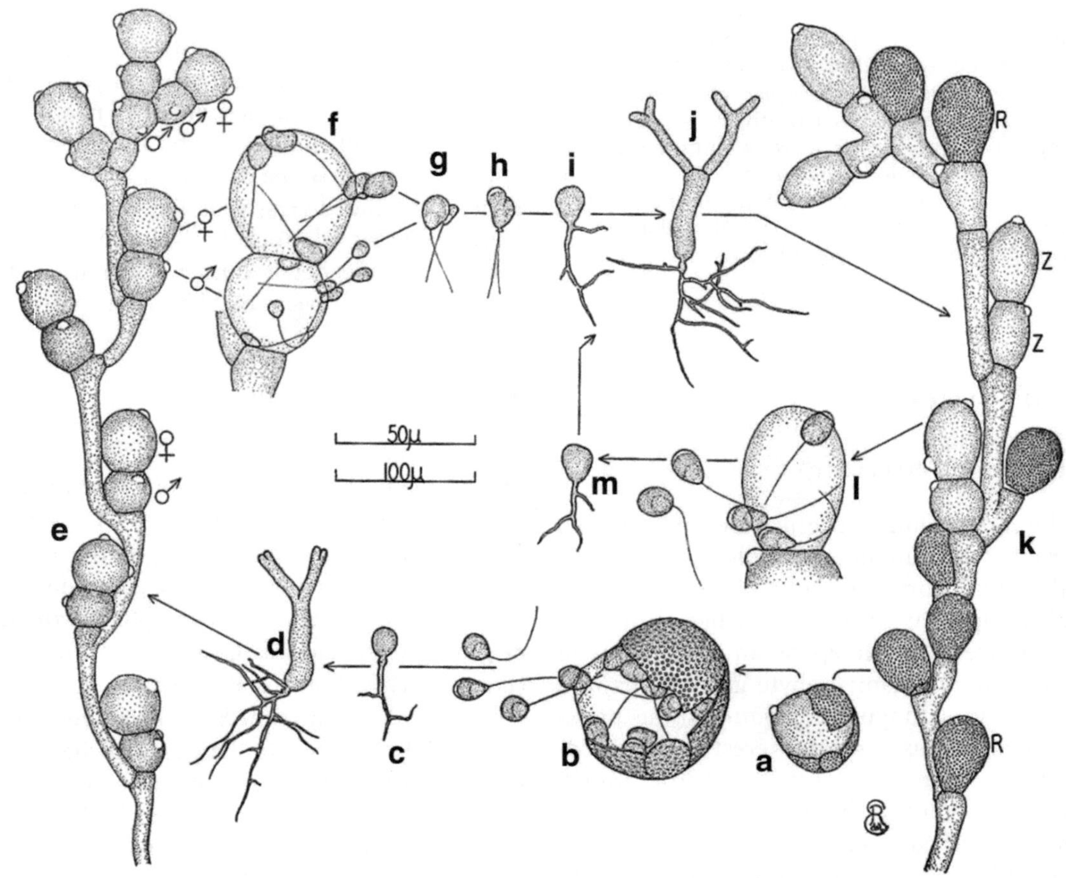

Fig. 7.4 (**a–m**) Life cycle of *Allomyces arbuscula*. (**a**) Germinating resistant sporangium with ruptured outer wall and swollen inner wall with discharge papilla. (**b**) Release of uniflagellate meiospores. (**c–e**) Stages in growth of young thallus (**c**, **d**) into mature gametophytic thallus bearing papillate male () and female () gametangia (**e**). (**f**) Release of male and female gametes from paired gametangia. (**g**) Syngamy of anisogametes. (**h**) Biflagellate planozygote. (**i**) Germination of encysted planozygote. (**j, k**) Young thallus (**j**) develops into mature sporophytic thallus (**k**) bearing resistant sporangia (R) and thin-walled zoosporangia (Z). (**l**) Release of zoospores from thin-walled zoosporangia. (**m**) Germination of encysted zoospore to form young thallus (**j**). Drawing from Emerson (1941), used with permission from Lloyd Library and Museum; American Society of Pharmacognosy

alternation of generations in which the gametophyte thallus is reduced to a single-celled cyst (Figs. 7.5 and 7.6d). The sporophytic thalli of *Cystogenes* species are like those of *Euallomyces* in structure and size, and meiosis takes place in the resistant sporangium (Emerson and Wilson 1949; Olson 1980; Wilson 1952). However, in *Cystogenes* isolates meiosis is followed by a pairing of haploid nuclei prior to cleavage into spores. Meiospores exit the resistant sporangium as binucleate cells that move sluggishly as biflagellate motile cells or as nonflagellated amoeboid cells before rapid encystment. The two cyst nuclei undergo a single mitotic division to produce four haploid cells (isogametes) that exit upon deliquescence of the single papilla. The fate of the quartet of cells emerging from the cyst was unknown to Emerson; however, the cells were shown to function as uniflagellate or aflagellate amoeboid isogametes that fuse to form zygotes (McCrainie 1942; Teter 1944).

Brachyallomyces Isolates in which the motile spores from the resistant sporangia regularly gave rise to asexual (sporophytic) rather than sexual (gametophytic) thalli were placed in a new species, *A. anomalus*, and included in the subgenus

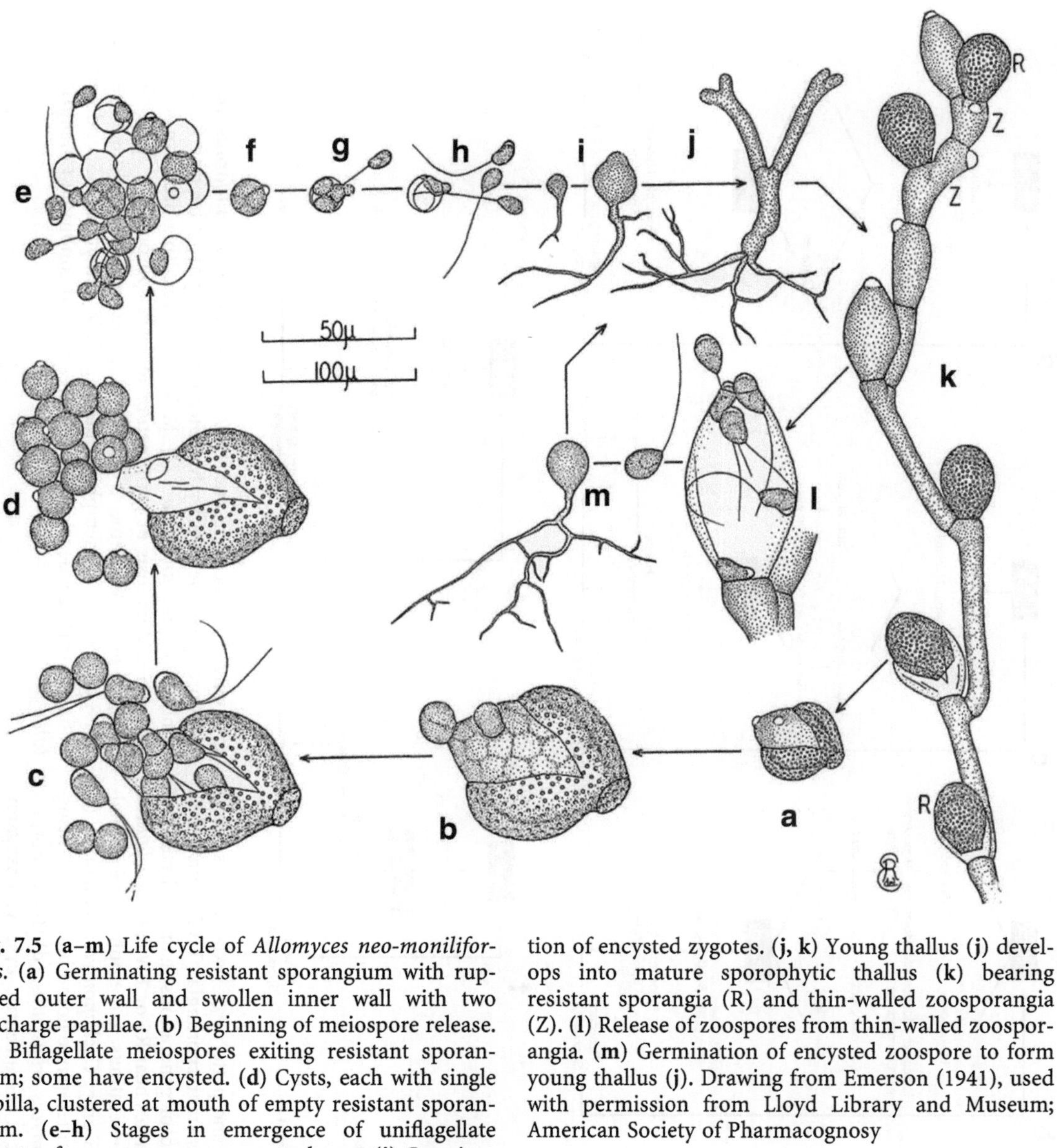

Fig. 7.5 (**a**–**m**) Life cycle of *Allomyces neo-moniliformis*. (**a**) Germinating resistant sporangium with ruptured outer wall and swollen inner wall with two discharge papillae. (**b**) Beginning of meiospore release. (**c**) Biflagellate meiospores exiting resistant sporangium; some have encysted. (**d**) Cysts, each with single papilla, clustered at mouth of empty resistant sporangium. (**e**–**h**) Stages in emergence of uniflagellate gametes from cysts; syngamy not shown. (**i**) Germination of encysted zygotes. (**j, k**) Young thallus (**j**) develops into mature sporophytic thallus (**k**) bearing resistant sporangia (R) and thin-walled zoosporangia (Z). (**l**) Release of zoospores from thin-walled zoosporangia. (**m**) Germination of encysted zoospore to form young thallus (**j**). Drawing from Emerson (1941), used with permission from Lloyd Library and Museum; American Society of Pharmacognosy

Brachyallomyces (Emerson 1941). Emerson found that by varying the substrate he could induce some putative *Brachyallomyces* isolates to form gametophytes, and he was careful to ascribe only those isolates that remained consistently asexual to the subgenus. It was later revealed that asexuality was maintained by mitosis in the resistant sporangia of some *A. anomalus* isolates (Fig. 7.6e), while in other isolates meiosis was presumably followed by endomitosis (nuclear replication without division) in germinating meiospores to reestablish the diploid condition (Fig. 7.6f) (Wilson and Flanagan 1968). The sporophytic thalli of *A. anomalus* isolates are entirely like those of *Euallomyces* species but lack a sexual stage.

C. Life Cycles of Other Blastocladiomycota

In the 70 years since the publication of Emerson's monograph on *Allomyces*, many new blastoclads have been discovered, and the diversity

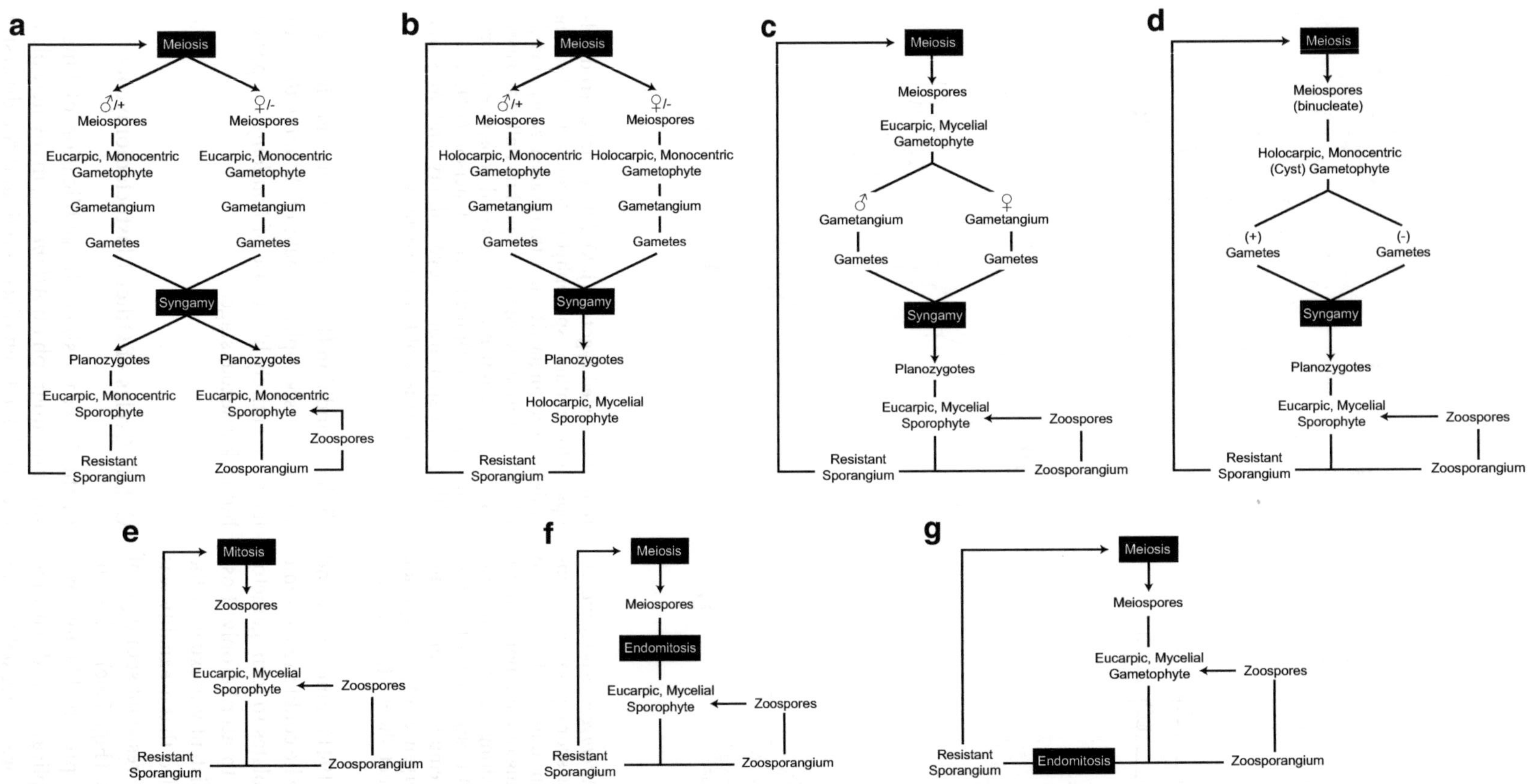

Fig. 7.6 Diagrams illustrating life cycles and patterns of sexuality in selected members of Blastocladiomycota. RS, resistant sporangia, TWZ, thin-walled zoosporangia. (**a**) Isomorphic alternation of generations, gametophytes heterothallic with isogamy, separate sporophytes bear RS or TWZ; *Blastocladiella variabilis*. (**b**) Isomorphic alternation of generations, gametophytes heterothallic with isogamy, sporophyte bearing RS only; *Coelomomyces punctatus*. (**c**) Isomorphic alternation of generations, gametophyte homothallic with anisogamy, sporophyte with RS and TWZ; *Allomyces arbusculus*. (**d**) Heteromorphic alternation of generations, gametophyte a monocentric cyst with isogamy, sporophyte produces RS and TWZ; *Allomyces neo-moniliformis*. (**e**) Asexual zoospores give rise directly by mitosis to sporophytes bearing RS and TWZ; some *Allomyces anomalus*. (**f**) Meiospores resulting from meiosis in RS undergo endomitosis at germination to form sporophytes; sporophyte produces RS and TWZ; some *A. anomalus*. (**g**) Meiospores resulting from meiosis in RS give rise to gametophytes bearing TWZ and young RS; endomitosis or selfing in early RS development restores diploid condition; *Catenaria anguillulae*

of life cycle types has increased. The *Euallomyces* life cycle of *Allomyces* (Fig. 7.6c) has an isomorphic alternation of generations with homothallic mating of morphologically distinct male and female gametes. **Anisogamy has been reported only in *Euallomyces* species.** Homothallism of the type displayed by *Allomyces* has not been reported in other isomorphic genera, but such a pattern may have existed in *Paleoblastocladia milleri* (Remy et al. 1994), a fossil blastoclad from the 400 million-year-old Rhynie chert. In addition to sporophytic thalli bearing zoosporangia and resistant sporangia there are similar (gametophytic?) thalli bearing paired cells that resemble rather remarkably the gametangia of *A. arbuscula*.

The *Cystogenes* life cycle of *Allomyces* (Fig. 7.6d) is a heteromorphic alternation of generations with homothallic mating of isogametes. This pattern has been reported in species of *Blastocladiella* (subgenus *Cystocladiella*) and in the mycoparasite *Catenaria allomycis*. However, the life cycle of these fungi differs slightly from that of *Cystogenes* in producing uninucleate and uniflagellate meiospores that undergo two mitotic divisions prior to gamete formation.

An asexual or *Brachyallomyces* life cycle seems to be widespread in many blastoclad genera, particularly those of the Blastocladiaceae. The types represented by mitosis in the resistant sporangium (Fig. 7.6e) and by meiosis followed by diploidization by endomitosis in germinating meiospores (Fig. 7.6f) have been reported only for *A. anomalus* isolates. A third asexual pattern (Fig. 7.6g) has been reported in *C. anguillulae* (Olson and Reichle 1978a), and a modification of this pattern is likely present in *B. emersonii* (Olson and Reichle 1978b). In both organisms synaptonemal complexes and meiotic divisions are present during resistant sporangium formation and germination; however, both meiospores and zoospores are reported to be haploid. Induction of diploidization and resistant sporangium formation in *C. anguillulae* occurred when the haploid zoosporangial thalli were transferred from a nutrient medium lacking starch to one containing it. A similar transformation of zoosporangia into resistant sporangia occurred in *B. emersonii* when bicarbonate and other salts were added to media (Cantino 1956) and in *B. brittanica* in response to darkness (Horenstein and Cantino 1962).

The life cycle of the *Coelomomyces* species that have been thoroughly studied is that of an isomorphic alternation of generations involving heterothallic mating of isogametes (Fig. 7.6b). While this heteroecious life cycle is classified as isomorphic, the gametophytic thalli in crustacean hosts are smaller and with fewer branches than sporophytic thalli in dipteran hosts. Some species of *Coelomomyces* have strikingly dimorphic gametangia, with the male mating type bright orange and the female amber or colorless. In other species the + and − gametangia have similar pigmentation or are colorless. The thin-walled zoosporangia produced in some *Coelomomyces* species do not seem to be homologous to the thin-walled zoosporangia of other blastoclads as they are structurally similar to resistant sporangia and, like them, produce meiospores rather than zoospores.

Physoderma species reported to be heterothallic have a life cycle similar to that of *Coelomomyces*, with notable exceptions. In *Physoderma* there is a heteromorphic alternation of generations owing to significant differences in the size and structure of the monocentric and epibiotic gametophyte thalli and the polycentric and endobiotic sporophyte thalli. In some species of *Physoderma* the epibiotic thalli are reported to mature into gametangia that produce isogametes or zoosporangia whose zoospores form additional epibiotic thalli. As in *Coelomomyces*, gametangia may be distinguished as orange or crimson males and colorless females in reportedly heterothallic species (Karling 1977).

Blastocladiella is similar to *Allomyces* but is monocentric and thus may produce only sporangia or gametangia at any given phase of the life cycle. Karling (1977) noted the similarity between the life cycles of *Blastocladiella* and *Allomyces* and erected subgenera of *Blastocladiella* corresponding to those of *Allomyces*: *Eucladiella* (corresponding to Emerson's *Euallomyces*), in which there is an isomorphic alternation of generations, *Cystocladiella* (corresponding to *Cystogenes*), in which the gametophyte generation is a single-celled cyst, and *Blastocladiella* (corresponding to *Brachyallomyces*), for short-cycled or asexual forms. The sporophyte generation of *Blastocladiella* generally consists of separate thalli bearing zoosporangia or resistant sporangia, although in some species zoosporangial thalli are lacking. **In species of *Eucladiella* the gametophyte generation consists of separate but equal-sized male and female thalli.** While the life cycles of *Cystocladiella* species such as *B. cystogena* are similar to those of *Cystogenes*, they differ in that they lack thin-walled zoosporangia (they produce thalli bearing resistant sporangia only) and produce cells from resistant sporangia that are uniflagellate and uninucleate rather than biflagellate and binucleate. The subgenus *Blastocladiella* includes short-cycled species such as *B. simplex* (Matthews 1937), *B. britannica* (Willoughby 1959), and *B. emersonii* (Cantino and Hyatt 1953), in which no sexual reproduction has been reported. Although much has been learned about the physiology of *B. emersonii*, questions remain about its sexuality and life cycle. Even though flagellated swarmers from orange and colorless (OC) cells were never observed to undergo karyogamy, they were reported to undergo plasmogamy and cytoplasmic exchange (Cantino and Horenstein 1954). Investigators have generally agreed that *B. emersonii* has a *Brachyallomyces* (or subgenus *Blastocladiella*) type of life cycle. Olson and Reichle (1978) found synaptonemal complexes and meiotic nuclear divisions in germinating *B. emersonii*-resistant sporangia, and Horgen et al. (1985) studied the fluorescence of mithramycin-stained nuclei and determined that meiospores contained half the DNA of zoospores of ordinary colorless cells (thin-walled zoosporangia). **The life cycle of *B. emersonii* is most similar to the *Brachyallomyces* pattern** in which meiosis is present (Fig. 7.6f).

The life cycle of *Blastocladiella variabilis* (Fig. 7.6a) is worthy of note because it perhaps has the greatest potential for development as a model genetic system among the flagellated fungi. Like *Coelomomyces* and *Physoderma*, its life cycle is an isomorphic alternation of generations with heterothallic mating. However, in *Blastocladiella* the thalli of both gametophyte and sporophyte generations are monocentric, epibiotic, and, thus, separate. Four distinct reproductive structures are produced on separate thalli in *B. variabilis*: two distinct heterothallic gametangia (one with reddish male gametangia and the other with colorless female gametangia) and two distinct sporophytic thalli (one bearing zoosporangia and the other bearing resistant sporangia) (Harder and Sörgel 1938).

VI. Zoospore Ultrastructure

A. Historical Perspective

In 1896 Thaxter examined the zoospores of *B. pringsheimii* and noted the posterior cilium and the broad and distinct mass of granular protoplasm in front of the large and subtriangular nucleus (Fig. 7.1a). Couch and Whiffen (1942) provided excellent illustrations of the zoospores and meiospores of *Blastocladiella cystogena* and rather prophetically stated, "This cap is undoubtedly of phylogenetic importance," and noted that such a structure had been reported previously in *Coelomycidium simulii* (Debaisieux 1920). Subsequently, the discovery of a new zoosporic fungus with posteriorly uniflagellate cells with a nuclear cap of the *Blastocladia* type became regarded as clear evidence of a relationship to the Blastocladiales.

Zoospores of the Chytridiales and Blastocladiales were among the earliest biological specimens to be examined with the electron microscope (Manton et al. 1952; Koch 1956). As new zoosporic fungi were discovered and the extent of variation in their vegetative and reproductive structures became known, the validity of morphology as revealed by light microscopy for taxonomic distinctions was questioned (Powell and Koch 1977a, b). Beginning in the 1960s detailed ultrastructural studies of the motile cells of various chytrid and blastoclad genera were conducted in search of additional characters of taxonomic and phylogenetic value. These studies revealed the identities of organelles previously observed with the light microscope and demonstrated new cytoplasmic similarities and differences among blastoclad genera. The nuclear cap was observed to be a membrane-bound cluster of ribosomes at the tip of a linear **axial assembly** found in all members of Blastocladiales. The side body (or Seitenkörper) of earlier studies was seen to be a mitochondrion and a component of a side body complex also found in all members of Blastocladiales.

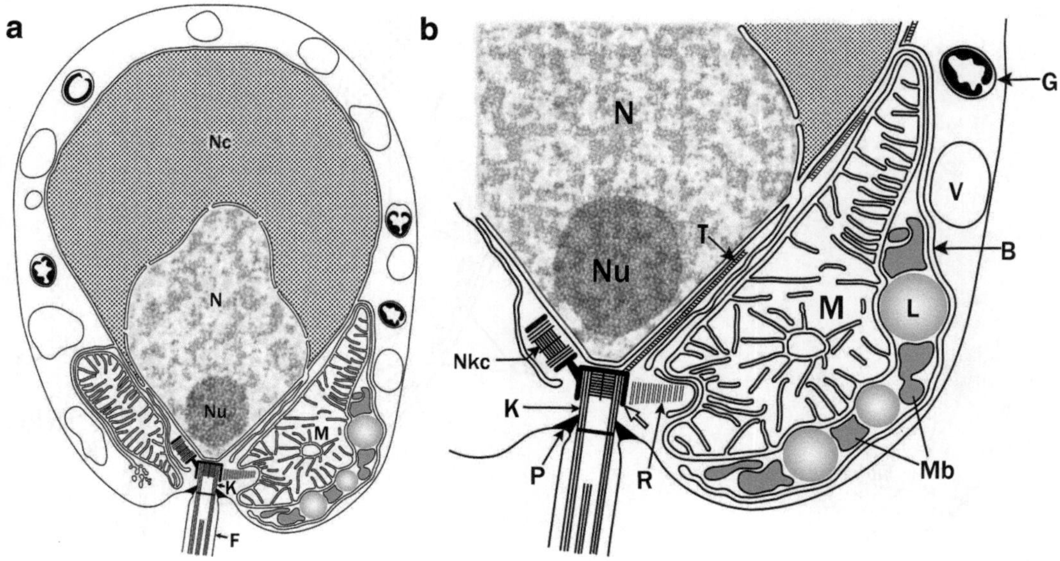

Fig. 7.7 (**a, b**) *Blastocladiella emersonii*. Interpretive drawings of zoospore showing ultrastructural details based primarily on several studies (Cantino and Truesdell 1970; Lessie and Lovett 1968; Reichle and Fuller 1967; Shaw and Cantino 1969). (**a**) Median longitudinal section through zoospore. (**b**) Detail of kinetosomal region and side body complex. *N* nucleus; *Nu* nucleolus; *Nc* nuclear cap; *F* flagellum; *K* functional kinetosome, *Nkc* nonkinetosomal centriole, *P* prop, *R* rhizoplast, *M* mitochondrion, *Mb* microbody, *L* lipid globule, *B* backing membrane, *G* gamma body, *T* cytoplasmic microtubule, *V* vacuole, *open arrow*, amorphous perikinetosomal material

B. Generalized Structure of Motile Cells of Blastocladiomycota

Surveys of ultrastructural characters in motile cells of the Chytridiomycetes have found four basic patterns of side body complexes or **microbody–lipid globule complexes (MLCs)** corresponding to the various orders of posteriorly uniflagellate fungi (Lange and Olson 1979; Powell 1978). The MLC of the Blastocladiales (type 4) was distinguished as an ordered arrangement of one or more mitochondria, microbodies, and lipid globules located along one side of an axially arranged nucleus and nuclear cap and enclosed by a backing membrane. Molecular phylogenetic studies provided most of the evidence by which Blastocladiales was raised to phylum status (James et al. 2006b); however, the formal description of phylum Blastocladiomycota was based largely on ultrastructural details. In the following account ultrastructural characters defining the phylum are discussed along with characters believed to be of systematic and phylogenetic importance in distinguishing major clades, subclades, and component genera.

1. Axial Assembly

The remarkable homogeneity of the nucleus, nuclear cap, and associated structures is revealed in drawings and photographs of motile cells of representatives of major clades of Blastocladiomycota, including a zoospore of *B. emersonii* (Fig. 7.7a, b), a zoospore of *Allomyces macrogynus* (Fig. 7.8a, b), a meiospore of *Coelomomyces punctatus* (Fig. 7.9), and a meiospore of *Physoderma maydis* (Fig. 7.10). The flagellated cells of all known members of Blastocladiomycota are characterized by a linear arrangement of organelles (axial assembly) consisting of a posterior flagellum with contained axoneme, functional **kinetosome**, nucleus with contained nucleolus, and anterior nuclear cap. The flagellum has the nine-plus-two arrangement typical of microtubules and narrows distally to a short whiplash portion. At the point where the flagellum joins the spore

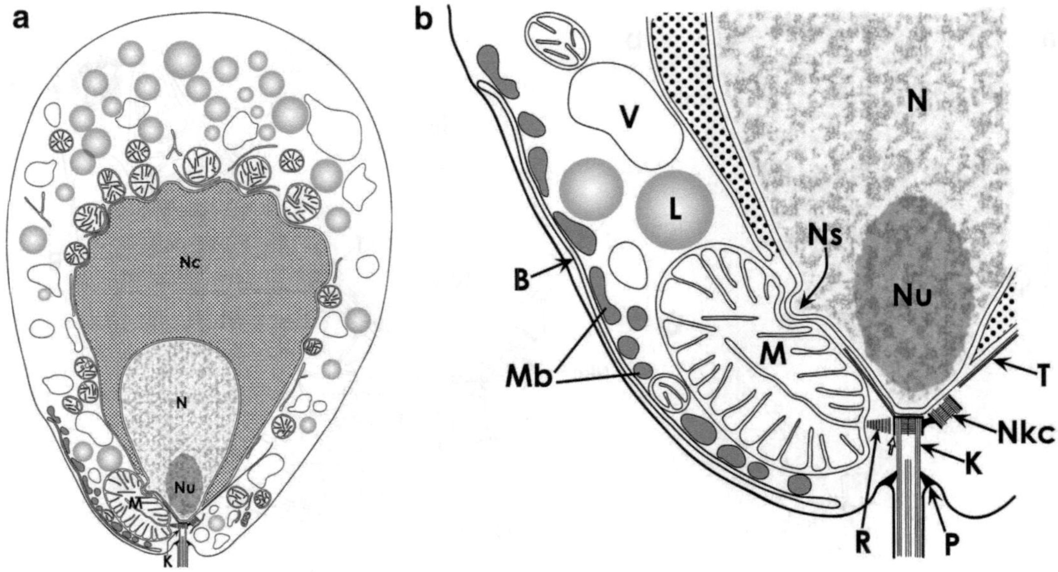

Fig. 7.8 (**a**, **b**) *Allomyces macrogynus*. Interpretive drawings of the zoospore showing ultrastructural details based primarily on two studies (Fuller and Olson 1971; Hill 1969). (**a**) Median longitudinal section through zoospore. (**b**) Detail of kinetosomal region and side body complex. *N* nucleus, *Nu* nucleolus, *Nc* nuclear cap, *Ns* nuclear spur, *K* functional kinetosome, *Nkc* nonkinetosomal centriole, *P* prop, *R* rhizoplast, *M* mitochondrion, *Mb* microbodies, *L* lipid globule, *B* backing membrane, *T* cytoplasmic microtubule, *V* vacuole, *open arrow* amorphous perikinetosomal material

body, nine electron-dense props extend inward to the axoneme or kinetosome. In all genera except *Coelomomyces* a **nonkinetosomal centriole** is present alongside the nucleus and is attached to the functional kinetosome by an electron-opaque bridge. A cap of electron-dense amorphous material surrounds the top of the functional kinetosome and extends down along the sides of the cartwheel region. In all genera cytoplasmic microtubules arise from the amorphous material at the extreme proximal end of the functional kinetosome and pass up and around the periphery of the nucleus and nuclear cap and into the cytoplasm (Figs. 7.7, 7.8, 7.9, and 7.10). It has been suggested that this area is a so-called organizing center for kinetosome-related microtubule formation (Dewel and Dewel 1990; Fuller and Calhoun 1968). In cross sections distal to the kinetosome the cytoplasmic microtubules typically occur in nine groups of three (triplets). The conical shape of the nucleus and nuclear cap is thought to be maintained by this basketlike arrangement of cytoplasmic microtubules. The ring or semicircle of amorphous material immediately below the organizing center is part of a flagellar apparatus that connects with the mitochondrion. The nuclear cap is composed of a dense cluster of dormant 80S ribosomes surrounded by double membranes, the outermost of which is continuous with the nuclear membrane (Jaworski and Stumhofer 1984).

2. Microbody–Lipid Globule Complex

Powell (1978) separated the type 4 MLC of Blastocladiomycota into two subtypes. Subtype 4A contains a single large mitochondrion that extends primarily along one side of the nucleus and nuclear cap and includes motile cells of *B. emersonii* (Fig. 7.7a, b), *C. punctatus* (Fig. 7.9), and *P. maydis* (Fig. 7.10). In *B. emersonii* and *C. punctatus* a cluster of lipid globules is located external to the long side of the mitochondrion and internal to a microbody, which is distended or completely penetrated by the rounded lipid globules. MLC subtype 4B occurs in motile cells of *A. macrogynus* (Fig. 7.8a, b) and *Allomyces*

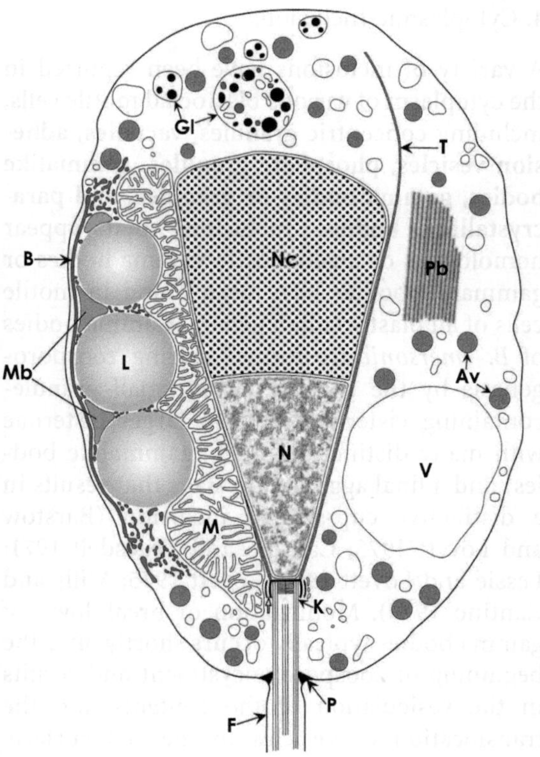

Fig. 7.9 *Coelomomyces punctatus*. Interpretive drawing of median longitudinal section through meiospore showing ultrastructural details based primarily on Martin (1971). *Av* adhesion vesicle, *B* backing membrane (contrasted with microbody to show tubular network), *Gl* gammalike body, *K* functional kinetosome, *L* lipid globule, *M* mitochondrion, *Mb* microbody, *N* nucleus, *Nc* nuclear cap, *P* prop, *F* flagellum, *Pb* paracrystalline body, *T* cytoplasmic microtubule, *V* vacuole, *open arrow* perikinetosomal striations

neo-moniliformis (Fuller and Olson 1971; Olson 1980) and is unlike other blastocladian genera in having components that are highly divided and less closely associated. Numerous rounded mitochondria are present in zoospores of *A. macrogynus*, and many are partially embedded in the nuclear cap. A larger cup-shaped mitochondrion partially surrounds the functional kinetosome and is the main component of a reduced and poorly organized MLC that contains several microbodies, a small number of lipid globules, and additional mitochondria. Motile cells of *Blastocladia ramosa*

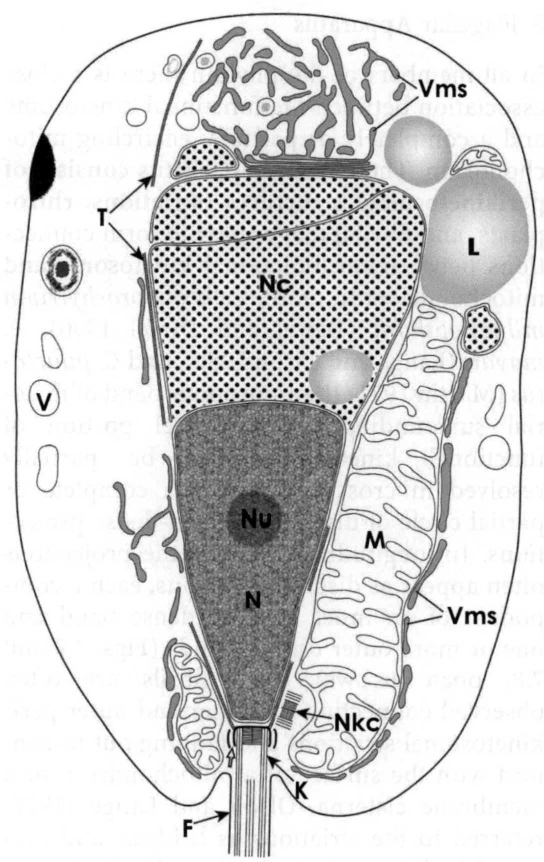

Fig. 7.10 *Physoderma maydis*. Interpretive drawing of median longitudinal section through meiospore showing ultrastructural details based primarily on Olson and Lange (1978) and Lange and Olson (1980). *F* flagellum, *K* functional kinetosome, *L* lipid globule, *M* mitochondrion, *N* nucleus, *Nc* nuclear cap, *Nkc* nonkinetosomal centriole, *Nu* nucleolus, *T* cytoplasmic microtubule, *V* vacuole, *Vms* vesicular–microbody system, *open arrow* perikinetosomal striations

contain multiple mitochondria but lack an organized MLC (Lingle and Barstow 1983). The outermost component of the MLC in subtypes 4A and 4B is a backing membrane that is usually continuous at several points with the outer membrane of the nuclear cap. The motile cells of *P. maydis* have a modified subtype 4A MLC in which the number of lipid globules may be reduced to one, and a **vesicular-microbody system** is present in place of a discrete backing membrane and microbody.

3. Flagellar Apparatus

In all members of the phylum there is a close association between the functional kinetosome and a completely or partially encircling mitochondrion. The **flagellar apparatus consists of perikinetosomal structures (striations, rhizoplasts, and fibrils)** that appear to form connections between the functional kinetosome and mitochondrion. In motile cells of *Sorochytrium milnesiopthora* (Dewel and Dewel 1990), *P. maydis* (Lange and Olson 1980), and *C. punctatus* (Martin 1971) the amorphous band of material surrounding the cartwheel portion of functional kinetosome may be partially resolved in cross sections as a complete or partial circle of discrete electron-dense projections. In longitudinal sections the projections often appear as discrete striations, each a component of an inner electron-dense band and one or more outer diffuse bands (Figs. 7.7 and 7.8, open arrows). Thin fibrils are often observed connecting the inner and outer perikinetosomal striations and fanning out to connect with the surface of a mitochondrion or a membrane cisterna. Olson and Lange (1978) referred to the striations as bridges, and presumably they are homologous with the transition fibers of *Coelomomyces dodgei* (Lucarotti and Federici 1984). The motile cells of *A. macrogynus* (Fig. 7.8), *A. neo-moniliformis* (Fuller and Olson 1971), *B. emersonii* (Fig. 7.7), *Blastocladiella brittanica* (Cantino and Truesdell 1971), and *C. anguillulae* (Olson et al. 1978) contain a rhizoplast (also called a striated or banded rootlet). The **rhizoplast** is a bar- or ribbon-shaped structure composed of a lateral series of equally spaced striae that is located laterally and in close proximity to the cartwheel portion of the functional kinetosome. Double membranes in the form of sheets or cisternae believed to originate from the outer nuclear or nuclear cap membrane are important components of the flagellar apparatus in many, if not all, blastoclad genera (Figs. 7.7b, 7.8b, 7.9). Powell (1983) has suggested that such cisternae may function in signal reception and transport between the cell surface and the flagellar apparatus.

4. Cytoplasmic Inclusions

A variety of inclusions have been reported in the cytoplasm of various blastoclad motile cells, including concentric granules, **vacuoles, adhesion vesicles, phosphate granules, gammalike bodies, gamma bodies or granules, and paracrystalline bodies.** Inclusions that appear homologous or analogous to gamma bodies or gammalike bodies have been found in motile cells of all blastocladian genera. Gamma bodies of *B. emersonii* are formed during zoosporogenesis by the coalescence of small granule-containing cisternae to form larger cisternae with many distinct granules (gammalike bodies) and a final aggregation stage that results in a distinctive cup-shaped inclusion (Barstow and Lovett 1975; Cantino and Truesdell 1971; Lessie and Lovett 1968; Lovett 1975; Mills and Cantino 1979). Mobilization or breakdown of gamma bodies typically occurs shortly after the beginning of zoospore encystment and results in the vesiculation of the contents and the translocation of vesicles to the cell surface. **Gamma bodies were once thought to function in the transport of chitin synthetase for cyst wall formation** (Barstow and Pommerville 1980; Mills and Cantino 1981); **however, later studies failed to support this hypothesis** (Dalley and Sonneborn 1982; Hohn et al. 1984). Olson and Lange (1983) interpreted the gamma bodies in motile spores of *Allomyces* as vesicle-generating structures that performed multiple functions upon mobilization or breakdown, including (1) the formation of water-expulsion vacuoles to maintain osmotic balance during zoospore motility, (2) production of vesicles that fuse to form axonemal and plasma membranes during sporogenesis, and (3) formation of vesicles that appear to be involved in cyst wall creation. The widespread occurrence of gamma bodies in the Blastocladiomycota prompted Dewel and Dewel (1990) to suggest that the gamma body should be considered a synapomorphy of the phylum. Adhesion vesicles are present in motile cells of *Coelomomyces* and are distinguished from gamma particles by their fine granular background and indistinct fibrous core (Fig. 7.9).

The contents of adhesion vesicles are released to form an adhesive plate that attaches meiospores to the intersegmental membranes of copepods and zygotes to the cuticle of mosquito larvae (Federici and Lucarotti 1986; Travland 1979).

A most unusual feature of both meiospores and gametes of *Coelomomyces* is the presence of one or more rod-shaped paracrystalline bodies that lie alongside the axial components opposite the MLC (Fig. 7.9) A similar structure has been reported in motile cells of *Callimastix cyclopis* (Manier and Loubes 1978; Vavra and Joyon 1966) and various members of the *Chytriomyces* clade of Chytridiomycota (Barr and Hartmann 1976; Picard et al. 2009; Taylor and Fuller 1981). The function of the paracrystalline body is unknown, but it has been speculated that it plays a role in infection (Madelin and Beckett 1972).

VII. Genetics and Physiology

A. Hybridization

The phenomenon of hybridization is not extensively documented in fungi, and verified instances of interspecific hybridization are rather rare (Brasier 2001; Schardl and Craven 2003). The classic work of Emerson and colleagues on hybridization between *A. arbusculus* and *A. macrogynus* provides convincing evidence for the existence of interspecific hybrids in the wild, but very little work has been conducted since the landmark paper by Emerson and Wilson (1954). Emerson and Wilson's work utilized clear differences between *A. arbusculus* and *A. macrogynus* in the arrangement of the pairs of male and female gametangia at hyphal tips. **Male gametangia (distinctively orange from gamma-carotene) are terminal or epigynous in *A. macrogynus*, while male gametangia are hypogynous in *A. arbusculus*.** Using controlled crosses between the two species, Emerson and Wilson found that F1 sporophytes were readily obtained, but the viability of meiospores produced by the sporophytes was greatly reduced. Among the viable meiospores, the F1 gametophyte generation displayed a range of gametangial arrangements, and putative hybrids typically showed a mixture of epigynous and hypogynous arrangements. This intermediate arrangement is also observed in **natural isolates of *A. javanicus*, which Emerson and Wilson hypothesized were hybrids of *A. arbusculus* x *A. macrogynus*.**

Emerson and Wilson further used cytology to verify the hybrid nature of the intermediate F1 gametophytes. Comparison of natural isolates suggested that *A. arbusculus* isolates were a polyploid series with a base chromosome number of 8, with the most common haploid (gametophyte) chromosome number of 16, implying that most *A. arbusculus* sporophytes are tetraploid. It was suggested that *A. macrogynus* had a base chromosome number of 14, but the common chromosomal types used by Emerson and Wilson possessed 28 chromosomes. Artificially produced *A. javanicus* would thus be expected to have 44 chromosomes before meiosis, and in crosses in which haploid *A. arbusculus* × *A macrogynus* fused, 44 chromosomes were observed, but only 1–5 bivalents were seen, indicating a lack of extensive homology between the chromosomes of the two species. This lack of pairing explains both the wide range of chromosome numbers in the artificial hybrid F1 gametophytes (improper segregation) and their low viability and agrees with the highly variable numbers of chromosomes seen in *A. javanicus* wild isolates. However, it was unclear why *A. javanicus* wild isolates had a variable but much lower (13–21) chromosome number than the artificial hybrids (20–44) (Emerson and Wilson 1954). It is also unclear whether the polyploid series within *A. arbusculus* and *A. macrogynus* are frequently generated by doubling or rarely generated and actually represent different species. Evidence that prolonged growth of *A. macrogynus* at 35 °C leads to a reduction in chromosome number that can be restored by growth at 23 °C (Borkhardt and Olson 1979; Olson and Borkhardt 1978) suggests that autopolyploidy by endomitosis could occur readily, but **the absence of many bivalents in F1 hybrid meiosis suggests, possibly, a more ancient origin.** Olson and Borkhardt additionally showed that when tetraploid resting sporangia are germinated and meiosis is blocked to induce the generation of sporophytic colonies with increased ploidy, they are usually unstable (Olson and Borkhardt 1978).

The highly selfing nature of *Allomyces* allows for the rapid stabilization and fertility of hybrids in the F2 and F3 generations as each haploid chromosome of viable F1 hybrids would find an identical homologous chromosome to pair with following the fusion of genetically identical selfed male and female gametes. Thus, selfing may have facilitated the recovery of natural hybrids in *Allomyces*. Phylogenetic analyses have now begun to shed light on the relationships between these hybridizing species of section *Euallomyces*. Neither *A. arbusculus* nor *A. macrogynus* seems to be monophyletic (Porter et al. 2011). Thus, the simple designations used to designate the species based on gametangial arrangements seem to be artificial, and the extensive morphological variation and polyploid series must be reevaluated by a combined study of chromosomes, phylogenies, and crosses.

Experimental hybridization has also been conducted in *Coelomomyces*, between *C. dodgei* and *C. punctatus* (Federici 1979, 1982). Utilizing the orange pigment of male gametangia and a common gametophyte host, the copepod *Cyclops vernalis*, Federici fused isogamous gametes of opposite mating type between the two species and then demonstrated that the hybrids could infect a common mosquito host (either *Anopheles freeborni* or *Anopheles quadrimaculatus*), proliferate as a sporophyte, and produce meiosporangia. The resting sporangia produced by the hybrid sporophytes displayed a wide range of characteristics but were mostly similar to one or the other parental species. The resting sporangia dehisced and released meiospores that encysted on the copepod host; however, **no gametophytes were ever produced**. These results demonstrate that the germination and growth of the haploid gametophyte is the most disrupted phase among hybrids of both *Coelomomyces* and *Allomyces*, as predicted by genetics. These studies also produce a working model for testing biological species; however, most of the crossing manipulations are extremely laborious.

B. Mitosis

Mitosis in the Blastocladiomycota has been well characterized using a combination of light and electron microscopy, and several innovations have been developed to study the process in the group. The process of nuclear division was often described as a part of a larger description of the development of hyphal, zoosporangial, or gametangial development from germinating zoospores at a time when *Allomyces* and *Blastocladiella* were still considered model organisms in genetics. Kniep (1930) observed that, although hyphal nuclei of *Allomyces* are large compared with other fungi, they were difficult subjects for the study of mitosis. Another early light-microscopy study described nuclear behavior in detail for *A. arbusculus* (Hatch 1935). Hatch described the development of germinating spores into coenocytic multinucleate hyphae that in turn develop into either a gametophyte bearing gametangia or into a sporophyte bearing zoosporangia. The nuclear count at the start of septation, when the first septum forms on a hypha, delimiting the apical female gametangium, and the second septum that forms further behind on the hypha, delimiting the male gametangium, showed roughly equal numbers of nuclei in each gametangium. At the end of gametangial differentiation a two-fold increase in the number of nuclei in the male gametangium was observed as a result of repeated nuclear divisions. It was also observed that mitotic divisions were not synchronous, and drawings of actively dividing nuclei with a spindle as well as anaphase and telophase chromosome configurations were provided. Though nuclei in the female gametangium were about twice the size of nuclei in the male gametangium, both gametes contain only six chromosomes. Hatch (1935) suggested that the size difference between male and female nuclei may be related to maintaining a particular nuclear-plasma ratio, though how this was related to an increased number of mitotic divisions in the male gametangium could not be explained. Another light-microscopy study by Wilson and Flanagan (1968) followed mitosis in resistant sporangia and hyphae in *Brachyallomyces* strains and noted that somatic nuclei were smaller than those in resistant sporangia.

Either closed mitosis or partially open mitosis has been shown in fungi. So far, **only closed mitosis, or intranuclear division, has been described in blastoclads.** In closed mitosis the nuclear membrane remains intact or largely intact and the spindle forms inside the nucleus (DeSouza and Osmani 2007; Heath 1980). A persistent nuclear membrane during somatic mitosis has been demonstrated in *Allomyces* spp. (Olson 1984), *C. anguillulae* (Ichida and Fuller 1968), and *Coelomomyces indicus* (Madelin and Beckett 1972). Lessie and Lovett (1968) reported intranuclear mitosis in *B. emersonii* comprised of a typical microtubular spindle apparatus and paired but unequal extranuclear centrioles at each pole.

C. Taxis

Taxis refers to the ability of motile cells or organisms to move across a gradient in a directed manner. Such behavior is clearly advantageous for zoospores and gametes of blastoclads as they disperse to find a new food source or mate. Both phototaxis and chemotaxis have been well documented in blastoclads. Positive phototaxis toward light was demonstrated in both *Allomyces* spp. (Olson 1984; Robertson 1972) and *Coelomomyces* (Martin 1970). Positive phototaxis may provide a mechanism by which gametes or zoospores may emerge from sediments. The attraction of zoospores of *Allomyces* to cellulose and chitin irrespective of light has been shown (Mitchell and Deacon 1986), as has positive chemotaxis toward several amino acids (Machlis 1969; Stumm et al. 1976).

Chemotaxis during mating should facilitate motile gametes seeking a compatible partner. The diffusible hormone **sirenin** is produced by female gametes of *Allomyces* (Machlis 1958a, b) and **has activity at very low concentrations** (10^{-10} to 10^{-5} M) (Carlile and Machlis 1965). Sirenin was the first fungal hormone to be chemically characterized and shown to be a **bicyclic sequiterpenediol** (Machlis 1968). Pommerville has also provided evidence that male gametes produce a hormone, though the swimming ability of female gametes is much reduced compared to those of males (Pommerville 1977, 1978).

D. Substrate Utilization and Respiration

Completely defined media have been constructed for studying nutrition in *Allomyces* (Ingraham and Emerson 1954). Growth on glucose, maltose, and starch as a sole carbon source has been shown for *Allomyces* (Ingraham and Emerson 1954), *C. anguillulae* (Nolan 1970), and *B. pringsheimii* (Crasemann 1957; Emerson and Cantino 1948; Gleason and Gordon 1989). Nitrogen utilization varies among taxa, with ***Allomyces* capable of using inorganic nitrogen and *Blastocladiella* and *Catenaria* using only organic sources** (Barner and Cantino 1952). An absolute requirement for an organic source of sulfur in the medium has also been demonstrated for the saprotrophic genera (Cantino and Turian 1959; Nolan 1969). Nutritional studies have facilitated the isolation of auxotrophic mutants; however, many mutants reported in the literature have apparently been unstable or displayed non-Mendelian inheritance due to ploidy (Olson 1984).

The obligately biotrophic parasites have been nearly impossible to isolate into pure culture. Numerous methods employing a "shotgun" approach have been tried for the growth of *Coelomomyces*, which would have clear benefits for biocontrol (Bland 1985; Nolan 1985). Several media, such as BHM (comprised of brain-heart infusion, mosquito larval extract, fetal bovine serum, and corn stunt spiroplasma media, to name a few ingredients!), have supported the growth of *Coelomomyces*, including the production of inviable sporangia (Bland 1985; Castillo and Roberts 1980). Key to the successful deployment of *Coelomomyces* inoculum as a biocontrol agent will be the in vitro culture of the gametophyte stage from copepods, needed to produce infective zygotes. Alternatively, the development of species such as *Coelomomyces iliensis* var. *iliensis* whose sporophyte stage makes asexual diploid zoospores that can reinfect the mosquito host should be pursued. Among the plant parasitic genera, the algal parasite *Paraphysoderma* is the only one that can be grown in vitro (Hoffman et al. 2008). This observation suggests inroads to cultivating *Physoderma* may be found by careful study of the nutritional requirements of the former.

The genus *Blastocladia* has been shown to be obligately fermentative and facultatively anaerobic using with at least one species, *B. ramosa,* conforming to its observed niche of stagnant waters (Held et al. 1969). *Blastocladia* cultures responded positively to the addition of CO_2 to 20 %, suggesting they may be able to convert CO_2 into organic acids (Tabak and Cooke 1968). Interestingly, electron micrographs of *B. ramosa* showed double-membrane structures like mitochondria lacking any cristae in germlings (Held et al. 1969); however, with improved fixation techniques the single mitochondrion of the zoospores of ***B. ramosa* did indeed have cristae but not as many as the obligately aerobic genera** (Lingle and Barstow 1983). Microaerophily, or improved development under low oxygen conditions, has been suggested for other members of Blastocladiaceae: *Allomyces reticulatus* and *Microallomyces dendroideus* (Emerson and Robertson 1974).

E. Genomics

Our knowledge of fungal genomes is biased toward mainly Ascomycota species, particularly model organisms and pathogenic species. The only Blastocladiomycota genome in progress is for ***A. macrogynus* ATCC38327,** sequenced by the Broad Institute's Origins of Multicellularity project (Ruiz-Trillo et al. 2007). Data on the expressed portion of Blastocladiomycota genomes are available from EST projects such as through the Taxonomically Broad EST Database (TBestDB) for *A. macrogynus* (submitted by B.F. Lang, University of Montreal) and the National Center for Bioinformatics Information (NCBI) for *B. emersonii* (Ribichich et al. 2005). Our knowledge of fungal mitochondrial genomes is similarly biased, but mitochondrial genomes of the basal fungal lineages are better represented. Two Blastocladiomycota mitochondrial genomes have been completed for *A. macrogynus* and *B. emersonii* (Paquin and Lang 1996; Tambor et al. 2008).

Mitochondrial chromosomes (mtDNA) usually encode proteins involved in the electron transport chain, adenosine triphosphate (ATP) synthesis, structural proteins, and proteins of unknown function that may be found as open reading frames in introns (Griffiths 1996). Though mitochondrial function is basically the same in all organisms, fungal mitochondrial genomes may show great differences in size and gene organization due to the presence of introns and size variation in intergenic spacer regions (Lang et al. 2007). The mitochondrial genomes examined so far in Blastocladiomycota have been shown by electron microscopy, restriction enzyme analysis, and sequencing to be circular, although more recently it has been shown that linear forms may also be present in vivo for many fungi (Bendich 1993, 1996, 2010; Burger et al. 2003).

Allomyces macrogynus The first mtDNA physical maps for the aquatic fungi were for *A. macrogynus* (Borkhardt and Delius 1983; Borkhardt et al. 1988). The mitochondrial genome sequence for *A. macrogynus* confirmed its circular nature, total size of 57,473 bp, and slightly enriched A+T base content of 60.5 % (Paquin and Lang 1996). All mitochondrial genes seemed to be transcribed from the same DNA strand, similar to many other fungi (Paquin and Lang 1996).

It has been hypothesized that the universal mitochondrial code is an ancestral trait in fungal mitochondria (Paquin et al. 1997). Similar to plants and protists, *A. macrogynus* only uses the UGG codon for tryptophan (Paquin and Lang 1996). Other fungi that share this trait include the blastoclad *B. emersonii*, the zygomycete *Rhizopus stolonifer*, the chytrids *Spizellomyces*, *Monoblepharella*, and *Harpochytrium*, and the basal Ascomycete fission yeast *Schizosaccharomyces pombe* (Massey and Garey 2007; Paquin et al. 1997; Tambor et al. 2008). In most other fungi and animals, the UGA stop codon has been reassigned to code for tryptophan. Additionally, *A. macrogynus* uses the UAG and UAA stop codons equally to signal termination (*A. macrogynus* does not use UGA), whereas most other fungi uniquely or preferentially use UAA (Paquin and Lang 1996). The only other organism known to use both stop codons equally is *Paramecium* (Pritchard et al. 1990).

Structural RNAs included the large and small mitochondrial-encoded rRNA genes (*rnl* and *rns*) and a set of 25 transfer RNAs (tRNAs). Protein-coding genes include three ATPase subunits (*atp6*, *apt8*, and *atp9*), apocytochrome b (*cob*), three subunits of the cytochrome oxidase complex (*cox1*, *cox2*, and *cox3*), and seven subunits of the NADH dehydrogenase complex (*nad1*, *nad2*, *nad3*, *nad4*, *nad4L*, *nad5*, and *nad6*). Generally, the observed mitochondrial gene content is typical of that found in other fungi (Paquin and Lang 1996). The presence of introns in mitochondrial genes was observed in the first mtDNA genome sequences of fungi, including *A. macrogynus*, and has important consequences for the development of mtDNA genes for DNA barcoding efforts. For example, hybridization studies of the *A. macrogynus cox1* gene suggest the presence of intronic sequences (Borkhardt et al. 1988). The *cox1* gene is now known to contain 12 introns and has a total size of more than 11 kbp (Paquin and Lang 1996).

Blastocladiella emersonii The physical map of the *B. emersonii* mtDNA genome showed this to be the smallest mitochondrial genome of the zoosporic fungi sequenced so far, with a circular structure and a size of 36,503 bp and an enriched A+T base content of 64.9 % (Borkhardt and Olson 1986; Tambor et al. 2008). All mtDNA genes use the universal translation code and are found in the same orientation (Tambor et al. 2008). However, *B. emersonii* mtDNA differs from that of *A. macrogynus* in several ways. First, *B. emersonii* has a smaller mtDNA genome relative to *A. macrogynus*, caused by differences in the number and size of introns, intergenic spacer regions, and double-hairpin DNA elements (DHEs) (Paquin and Lang 1996; Tambor et al. 2008). Specifically, the mitochondrial genome of *B. emersonii* contains only 2 introns (both in the *cox1* gene), compared with 28 introns across the mitochondrial genome of *A. macrogynus*. The intergenic spacer regions comprise 47.9 % of the mitochondrial genome in *B. emersonii*, compared with 22 % in *A. macrogynus*.

VIII. Conclusions and Future Directions

What the blastoclads lack in species number they make up for in phylogenetic and ecological diversity. They are the only group of fungi known to possess an alternation of haploid and diploid generations. This trait, unlike anisogamy, which evolved in the ancestor of subgenus *Euallomyces*, seems to be ancestral to blastoclads. Does a life cycle alternating between haploid and diploid generations suggest that blastoclads might be the first diverging branch in fungi and that they inherited this trait from the most recent common ancestor of all fungi? The answer is, unfortunately, unclear because the phylogenetic placement of the group has yet to be definitively resolved, and the life cycles of the putative outgroups of fungi are not completely known (Brown et al. 2009; Jones et al. 2011). On the other hand, closed mitosis, Golgi equivalents rather than a Golgi apparatus, and true mycelial growth are characteristics of blastoclads that are more similar to the more derived nonzoosporic fungi.

Variations in life cycles are common throughout the group, and they have been used to define subgenera and species in *Allomyces* and *Blastocladiella*. Yet all indications are that life cycle variants can occur within species and that species may hybridize readily and differ greatly in ploidy. Thus, it is no surprise that phylogenetic analysis of *Allomyces* reveals serious problems with traditional species concepts (Porter et al. 2011). All five of the species of *Allomyces* in the phylogeny represented by more than one strain were shown to be nonmonophyletic. In the phylogeny, roughly 12 terminal clades that are suggestive of species were observed, whereas only 9 names are currently valid. Thus, in the minimal sampling employed, additional species must be described. A major unanswered question is how many good blastoclad species have been proposed and how many are in need of revision. Modern mycology emphasizes a holistic view of fungal species that includes morphological, physiological, phylogenetic, and compatibility data (Cai et al. 2011; Taylor et al. 2000).

While species concepts in blastoclads must be redefined, a concerted effort to enumerate what are likely to be many undescribed species should also be undertaken. The mere 180 species named is obviously a gross underestimate of the true diversity, and future taxonomists of the group will be required to document and describe new species as they will inevitably be encountered. Discovery of these species will be facilitated by DNA-based evidence emerging from sampling of environments likely to be rich in blastoclads, such as suboxic sediments, periodically inundated soils, and invertebrate hosts. How we apply species concepts and delimit taxa in early-diverging fungi is a question that has largely been avoided, and there are very little data to address the question. Several lines of evidence suggest major revisions are warranted. Firstly, hybridization and horizontal gene transfer are poorly documented phenomena in fungi, but recent studies suggest that they may be as common in fungi as in other eukaryotic groups (Brasier 2001; Schardl and Craven 2003). Perhaps the best example of hybridization in fungi comes from experimental and natural hybrids in the species *A. javanicus*, the presumed hybrid of *A. macrogynus* and *A. arbusculus*. However, experiments to characterize *A. javanicus* genetically have never been done, and the role of hybridization in speciation has not been addressed in the blastoclads.

The coevolution of host and pathogen has likely driven diversification in blastoclads. In the future, studies of host specificity must be integrated into studies of taxonomy and systematics so that the simple assumption of one pathogen species per host species does not lead to erroneous classifications. Because most of the *Physoderma* species were named under the assumption that each host species had distinct parasites, and there is evidence to suggest that the species of *Physoderma* are less than host-species-specific in inoculation studies (Sparrow and Griffin 1964), it may be that the species diversity in this group is much lower than the list of taxonomically accepted names. Similar difficulties may arise in other parasitic genera, for example, *Coelomomyces* and *Catenaria*; however, the phylogenetic diversity of *Physoderma* (as measured by branch lengths in a phylogeny) is much shallower than observed in the *Coelomomyces* clade.

References

Barner HD, Cantino EC (1952) Nutritional relationships in a new species of *Blastocladiella*. Am J Bot 39:746–751

Barr DJS, Hartmann VE (1976) Zoospore ultrastructure of three *Chytridium* species and *Rhizoclosmatium globosum*. Can J Bot 54:2000–2013

Barstow WE, Lovett JS (1975) Formation of gamma-particles during zoosporogenesis in *Blastocladiella emersonii*. Mycologia 67:518–529

Barstow WE, Pommerville J (1980) The ultrastructure of cell wall formation and of gamma-particles during encystment of *Allomyces macrogynus* zoospores. Arch Microbiol 128:179–189

Bendich AJ (1993) Reaching for the ring: the study of mitochondrial genome structure. Curr Genet 24:279–290

Bendich AJ (1996) Structural analysis of mitochondrial DNA molecules from Fungi and Plants using moving pictures and pulsed-field gel elctrophoresis. J Mol Biol 255:564–588

Bendich AJ (2010) The end of the circle for yeast mitochondrial DNA. Mol Cell 39:831–832

Bland CE (1985) Culture. In: Couch JN, Bland CE (eds) The genus *Coelomomyces*. Academic, Orlando, FL, pp 349–359

Borkhardt B, Delius H (1983) Physical map of the mitochondrial DNA from the phycomycete *Allomyces macrogynus* including the position of the ribosomal RNA genes and of an intervening sequence in the large rRNA gene. Curr Genet 7:327–333

Borkhardt B, Olson LW (1979) Meiotic prophase in diploid and tetraploid strains of *Allomyces macrogynus*. Protoplasma 100:323–343

Borkhardt B, Olson LW (1986) The mitochondrial genome of the aquatic phycomycete *Blastocladiella emersonii*. Curr Genet 11:139–143

Borkhardt B, Brown TA, Thim P, Olson LW (1988) The mitochondrial genome of the aquatic phycomycete *Allomyces macrogynus*. Physical mapping and mitochondrial DNA instability. Curr Genet 13:41–47

Bowman BH, Taylor JW, Brownlee AG, Lee J, Lu SD, White TJ (1992) Molecular evolution of the fungi-relationship of the Basidiomycetes, Ascomycetes, and Chytridiomycetes. Mol Biol Evol 9:285–296

Brasier CM (2001) Rapid evolution of introduced plant pathogens via interspecific hybridization. Bioscience 51:123–133

Brown MW, Spiegel FW, Silberman JD (2009) Phylogeny of the "forgotten" cellular slime mold, *Fonticula alba*, reveals a key evolutionary branch within Opisthokonta. Mol Biol Evol 26:2699–2709

Bruns TD, White TJ, Taylor JW (1991) Fungal molecular systematics. Annu Rev Ecol Syst 22:525–564

Bruns TD, Vilgalys R, Barns SM, Gonzalez D, Hibbett DS, Lane DJ, Simon L, Stickel S, Szaro TM, Weisburg WG, Sogin ML (1992) Evolutionary relationships within the Fungi: analyses of nuclear small subunit rRNA sequences. Mol Phylogenet Evol 1:231–241

Bullerwell CE, Forget L, Lang BF (2003) Evolution of monoblepharidalean fungi based on complete mitochondrial genome sequences. Nucleic Acids Res 31:1614–1623

Burger G, Forget L, Zhu Y, Gray MW, Lang BF (2003) Unique mitochondrial genome architecture in unicellular relatives of animals. Proc Natl Acad Sci USA 100:892–897

Butler EJ (1911) On *Allomyces*, a new aquatic fungus. Ann Bot 25:1023–1035

Cai L, Giraud T, Zhang N, Begerow D, Cai GH, Shivas RG (2011) The evolution of species concepts and species recognition criteria in plant pathogenic fungi. Fungal Divers 50:121–133

Cantino EC (1956) The relation between cellular metabolism and morphogenesis in *Blastocladiella*. Mycologia 48:225–240

Cantino EC, Horenstein EA (1954) Cytoplasmic exchange without gametic copulation in the water mold *Blastocladiella emersonii*. Am Nat 88:143–154

Cantino EC, Hyatt MT (1953) Phenotypic "sex" determination in the life history of a new species of *Blastocladiella*, B emersonii. Antonie Van Leeuwenhoek 19:25–70

Cantino EC, Truesdell LC (1970) Organization and fine structure of side body and its lipid sac in zoospore of *Blastocladiella emersonii*. Mycologia 62:548–567

Cantino EC, Truesdell LC (1971) Cytoplasmic gamma-like particles and other ultrastructural aspects of zoospores of *Blastocladiella britannica*. Trans Br Mycol Soc 56:169–179

Cantino EC, Turian GF (1959) Physiology and development of lower fungi (Phycomycetes). Annu Rev Microbiol 13:97–124

Carlile MJ, Machlis L (1965) Response of male gametes of *Allomyces* to sexual hormone sirenin. Am J Bot 52:478–483

Castillo JM, Roberts DW (1980) *In vitro* studies of *Coelomomyces punctatus* from *Anopheles quadrimaculatus* and *Cyclops vernalis*. J Invertebr Pathol 35:144–157

Chapman HC (1985) Ecology and use of *Coelomomyces* species in biological control: a review. In: Couch JN, Bland CE (eds) The genus *Coelomomyces*. Academic, Orlando, FL, pp 361–368

Chen MJ, Chen FZ, Yu Y, Ji J, Kong FX (2008) Genetic diversity of eukaryotic microorganisms in Lake Taihu, a large shallow subtropical lake in China. Microb Ecol 56:572–583

Couch JN (1945) Observations on the genus *Catenaria*. Mycologia 37:163–191

Couch JN, Bland CE (1985) The genus *Coelomomyces*. Academic, New York

Couch JN, Whiffen AJ (1942) Observations on the genus *Blastocladiella*. Am J Bot 29:582–591

Crasemann JM (1957) Comparative nutrition of two species of *Blastocladia*. Am J Bot 44:218–224

Dalley NE, Sonneborn DR (1982) Evidence that *Blastocladiella emersonii* zoospore chitin synthetase is located at the plasma membrane. Biochim Biophys Acta 686:65–76

Dasgupta SN, John R (1988) A contribution to our knowledge of the genus *Blastocladia*. Indian Phytopathol 41:521–547

Debaisieux P (1920) *Coelomycidium simulii* nov. gen., nov. sp. et remarques sur l'*Amoebidium* des larves de *Simulium*. Cellule 30:249–271

DeSouza CPC, Osmani SA (2007) Mitosis, not just open or closed. Eukaryot Cell 6:1521–1527

Dewel RA, Dewel WC (1990) The fine structure of the zoospore of *Sorochytrium milnesiophthora*. Can J Bot 68:1968–1977

Dewel RA, Joines JD, Bond JJ (1985) A new chytridiomycete parasitizing the tardigrade *Milnesium tardigradum*. Can J Bot 63:1525–1534

Doweld AB (2001) Prosyllabus tracheophytorum. Tentamen systematis plantarum vascularium (Tracheophyta). Geos, Moscow

Emerson R (1938) A new life cycle involving cyst-formation in *Allomyces*. Mycologia 30:120–132

Emerson R (1941) An experimental study of the life cycles and taxonomy of *Allomyces*. Lloydia 4:77–144

Emerson R, Cantino EC (1948) The isolation, growth, and metabolism of *Blastocladia* in pure culture. Am J Bot 35:157–171

Emerson R, Robertson JA (1974) Two new members of Blastocladiaceae. I. Taxonomy, with an evaluation of genera and interrelationships in family. Am J Bot 61:303–317

Emerson R, Wilson CM (1949) The significance of meiosis in *Allomyces*. Science 110:86–88

Emerson R, Wilson CM (1954) Interspecific hybrids and the cytogenetics and cytotaxonomy of Euallomyces. Mycologia 46:393–434

Federici BA (1979) Experimental hybridization of *Coelomomyces dodgei* and *Coelomomyces punctatus*. Proc Natl Acad Sci USA 76:4425–4428

Federici BA (1982) Inviability of interspecific hybrids in the *Coelomomyces dodgei* complex. Mycologia 74:555–562
Federici BA, Lucarotti CJ (1986) Structure and behavior of the meiospore of *Coelomomyces dodgei* during encystment on the copepod host, *Acanthocyclops vernalis*. J Invertebr Pathol 48:259–268
Fischer A (1892) Phycomycetes. Die Pilze Deutschlands, Oesterreichs und der Schweiz. Kryptogamen-Fl 1:1–490
Förster H, Coffey MD, Elwood H, Sogin ML (1990) Sequence analysis of the small subunit ribosomal RNAs of three zoosporic fungi and implications for fungal evolution. Mycologia 82:306–312
Freeman KR, Martin AP, Karki D, Lynch RC, Mitter MS, Meyer AF, Longcore JE, Simmons DR, Schmidt SK (2009) Evidence that chytrids dominate fungal communities in high-elevation soils. Proc Natl Acad Sci USA 106:18315–18320
Fuller MS, Calhoun SA (1968) Microtubule-kinetosome relationships in motile cells of Blastocladiales. Z Zellforsch Mikrosk Anat 87:526–533
Fuller MS, Olson LW (1971) Zoospore of *Allomyces*. J Gen Microbiol 66:171–183
Garcia R (1983) Mosquito management - ecological approaches. Environ Manage 7:73–78
Gleason FH, Gordon GLR (1989) Anaerobic growth and fermentation in *Blastocladia*. Mycologia 81:811–815
Gleason FH, Marano AV, Johnson P, Martin WW (2010) Blastocladian parasites of invertebrates. Fungal Biol Rev 24:56–67
Griffiths AJF (1996) Mitochondrial inheritance in filamentous fungi. J Genet 75:403–414
Harder R, Sörgel G (1938) Über einen neuen planoisogamen Phycomyceten mit Generationswechel und seine phylogenetische Bedeutung. Nachrichten Gesell Wiss Göttingen Fachgruppe VI (Biol) 3:119–127
Hatch WR (1935) Gametogenesis in *Allomyces arbuscula*. Ann Bot 49:623–649
Heath IB (1980) Variant mitoses in lower eukaryotes: indicators of the evolution of mitosis? Int Rev Cytol 64:1–80
Held AA, Emerson R, Fuller MS, Gleason FH (1969) *Blastocladia* and *Aqualinderella* - fermentative water molds with high carbon dioxide optima. Science 165:706–709
Hibbett DS, Binder M, Bischoff JF, Blackwell M, Cannon PF, Eriksson OE, Huhndorf S, James T, Kirk PM, Lucking R, Lumbsch HT, Lutzoni F, Matheny PB, McLaughlin DJ, Powell MJ, Redhead S, Schoch CL, Spatafora JW, Stalpers JA, Vilgalys R, Aime MC, Aptroot A, Bauer R, Begerow D, Benny GL, Castlebury LA, Crous PW, Dai YC, Gams W, Geiser DM, Griffith GW, Gueidan C, Hawksworth DL, Hestmark G, Hosaka K, Humber RA, Hyde KD, Ironside JE, Koljalg U, Kurtzman CP, Larsson KH, Lichtwardt R, Longcore J, Miadlikowska J, Miller A, Moncalvo JM, Mozley-Standridge S, Oberwinkler F, Parmasto E, Reeb V, Rogers JD, Roux C, Ryvarden L, Sampaio JP, Schussler A, Sugiyama J, Thorn RG, Tibell L, Untereiner WA, Walker C, Wang Z, Weir A, Weiss M, White MM, Winka K, Yao YJ, Zhang N (2007) A higher-level phylogenetic classification of the Fungi. Mycol Res 111:509–547
Hill EP (1969) Fine structure of zoospores and cysts of *Allomyces macrogynus*. J Gen Microbiol 56:125–130
Hoffman Y, Aflalo C, Zarka A, Gutman J, James TY, Boussiba S (2008) Isolation and characterization of a novel chytrid species (phylum Blastocladiomycota), parasitic on the green alga *Haematococcus*. Mycol Res 112:70–81
Hohn TM, Lovett JS, Bracker CE (1984) Characterization of the major proteins in gamma particles, cytoplasmic organelles in *Blastocladiella emersonii* zoospores. J Bacteriol 158:253–263
Horenstein EA, Cantino EC (1962) Dark-induced morphogenesis in synchronized cultures of *Blastocladiella britannica*. J Bacteriol 84:37–45
Horgen PA, Meyer RJ, Franklin AL, Anderson JB, Filion WG (1985) Motile spores from resistant sporangia of *Blastocladiella emersonii* possess one-half the DNA of spores from ordinary colorless sporangia. Exp Mycol 9:70–73
Ichida AA, Fuller MS (1968) Ultrastructure of mitosis in the aquatic fungus *Catenaria anguillulae*. Mycologia 60:141–155
Ingraham JL, Emerson R (1954) Studies of the nutrition and metabolism of the aquatic phycomycete, *Allomyces*. Am J Bot 41:146–152
James TY, Porter D, Leander CA, Vilgalys R, Longcore JE (2000) Molecular phylogenetics of the Chytridiomycota supports the utility of ultrastructural data in chytrid systematics. Can J Bot 78:336–350
James TY, Kauff F, Schoch C, Matheny PB, Hofstetter V, Cox C, Celio G, Gueidan C, Fraker E, Miadlikowska J, Lumbsch HT, Rauhut A, Reeb V, Arnold AE, Amtoft A, Stajich JE, Hosaka K, Sung G-H, Johnson D, O'Rourke B, Crockett M, Binder M, Curtis JM, Slot JC, Wang Z, Wilson AW, Schüßler A, Longcore JE, O'Donnell K, Mozley-Standridge S, Porter D, Letcher PM, Powell MJ, Taylor JW, White MM, Griffith GW, Davies DR, Humber RA, Morton JB, Sugiyama J, Rossman AY, Rogers JD, Pfister DH, Hewitt D, Hansen K, Hambleton S, Shoemaker RA, Kohlmeyer J, Volkmann-Kohlmeyer B, Spotts RA, Serdani M, Crous PW, Hughes KW, Matsuura K, Langer E, Langer G, Untereiner WA, Lücking R, Büdel B, Geiser DM, Aptroot A, Diederich P, Schmitt I, Schultz M, Yahr R, Hibbett D, Lutzoni F, McLaughlin D, Spatafora J, Vilgalys R (2006a) Reconstructing the early evolution of the fungi using a six gene phylogeny. Nature 443:818–822
James TY, Letcher PM, Longcore JE, Mozley-Standridge SE, Porter D, Powell MJ, Griffith GW, Vilgalys R (2006b) A molecular phylogeny of the flagellated fungi (Chytridiomycota) and description of a new phylum (Blastocladiomycota). Mycologia 98:860–871

James TY, Hoffman Y, Zarka A, Boussiba S (2011) *Paraphysoderma sedebokerense*, gen. et sp. nov., an aplanosporic relative of *Physoderma* (*Blastocladiomycota*). MycoTaxon 118:177–180

Jaworski AJ, Stumhofer P (1984) Dormant ribosomes in *Blastocladiella emersonii* zoospores are arrested at elongation. Exp Mycol 8:13–24

Johnson PTJ, Longcore JE, Stanton DE, Carnegie RB, Shields JD, Preu ER (2006) Chytrid infections of *Daphnia pulicaria*: development, ecology, pathology and phylogeny of *Polycaryum laeve*. Freshw Biol 51:634–648

Jones MDM, Forn I, Gadelha C, Egan MJ, Bass D, Massana R, Richards TA (2011) Discovery of novel intermediate forms redefines the fungal tree of life. Nature 474:200–234

Karling JS (1950) The genus *Physoderma*. Lloydia 13:29–71

Karling JS (1965) *Catenophlyctis*, a new genus of Catenariaceae. Am J Bot 52:133–138

Karling JS (1977) Chytridiomycetarum Iconographia. Lubrecht and Cramer, Monticello, NY

Kirk PM, Cannon PF, Minter DW, Stalpers JA (2008) Dictionary of the Fungi, 10th edn. CABI Publishing, The Netherlands

Kniep H (1930) Über den Generationswechsel von *Allomyces*. Zeitschrift für Botanik 22:433–441

Koch WJ (1956) Studies of the motile cells of chytrids. I. Electron microscope observations of the flagellum, blepharoplast and rhizoplast. Am J Bot 43:811–819

Lang BF, O'Kelly C, Nerad T, Gray MW, Burger G (2002) The closest unicellular relatives of animals. Curr Biol 12:1773–1778

Lang BF, Laforest M-J, Burger G (2007) Mitochondrial introns: a critical view. Trends Genet 23:119–125

Lange L, Olson LW (1979) Uniflagellate phycomycete zoospore. Dansk Botanisk Arkiv 33:7–95

Lange L, Olson LW (1980) Transfer of the Physodermataceae from the Chytridiales to the Blastocladiales. Trans Br Mycol Soc 74:449–457

Lefèvre E, Bardot C, Noel C, Carrias JF, Viscogliosi E, Amblard C, Sime-Ngando T (2007) Unveiling fungal zooflagellates as members of freshwater picoeukaryotes: evidence from a molecular diversity study in a deep meromictic lake. Environ Microbiol 9:61–71

Lessie PE, Lovett JS (1968) Ultrastructural changes during sporangium formation and zoospore differentiation in *Blastocladiella emersonii*. Am J Bot 55:220–236

Letcher PM, Powell MJ, Churchill PF, Chambers JG (2006) Ultrastructural and molecular phylogenetic delineation of a new order, the Rhizophydiales (Chytridiomycota). Mycol Res 110:898–915

Letcher PM, Velez CG, Barrantes ME, Powell MJ, Churchill PF, Wakefield WS (2008) Ultrastructural and molecular analyses of Rhizophydiales (Chytridiomycota) isolates from North America and Argentina. Mycol Res 112:759–782

Lingle WL, Barstow WE (1983) Ultrastructure of the zoospore of *Blastocladia ramosa* (Blastocladiales). Can J Bot 61:3502–3513

Liu YJ, Hodson MC, Hall BD (2006) Loss of the flagellum happened only once in the fungal lineage: phylogenetic structure of Kingdom Fungi inferred from RNA polymerase II subunit genes. BMC Evol Biol 6:13

Liu Y, Steenkamp ET, Brinkmann H, Forget L, Philippe H, Lang BF (2009) Phylogenomic analyses predict sistergroup relationship of nucleariids and Fungi and paraphyly of zygomycetes with significant support. BMC Evol Biol 9:11

Lovett JS (1975) Growth and differentiation of water mold *Blastocladiella emersonii* - cytodifferentiation and role of ribonucleic acid and protein synthesis. Bacteriol Rev 39:345–404

Lucarotti CJ (1987) *Coelomomyces stegomyiae* infection in adult *Aedes aegypti*. Mycologia 79:362–369

Lucarotti CJ, Federici BA (1984) Ultrastructure of the gametes of *Coelomomyces dodgei* Couch (Blastocladiales, Chytridiomycetes). Protoplasma 121:77–86

Machlis L (1958a) A study of sirenin, the chemotactic sexual hormone from the watermold *Allomyces*. Physiol Plant 11:845–854

Machlis L (1958b) Evidence for a sexual hormone in *Allomyces*. Physiol Plant 11:181–192

Machlis L (1968) Response of male gametes of *Allomyces* to sexual hormone sirenin. Plant Physiol 43:1319–1320

Machlis L (1969) Zoospore chemotaxis in the water mold *Allomyces*. Physiol Plant 22:126–139

Madelin MF, Beckett A (1972) Production of planonts by thin-walled sporangia of fungus *Coelomomyces indicus* - parasite of mosquitos. J Gen Microbiol 72:185–200

Manier JF, Loubes C (1978) *Callimastix cyclopis* Weissenberg, 1912 (Phycomycetes, Blastocladiales) a parasite of *Microcyclops claus*, 1893 (Copepoda, Cyclopoida) from Tchad - ultrastructural features. Protistologica 14:493–501

Manton I, Clarke B, Greenwood AD, Flint EA (1952) Further observations on the structure of plant cilia by combination of visual and electron microscopy. J Exp Bot 3:204–215

Martin WW (1970) A morphological and cytological study of *Coelomomyces punctatus*. M.Sc., University of North Carolina, Chapel Hill

Martin WW (1971) The ultrastructure of *Coelomomyces punctatus* zoospores. J Elisha Mitchell Sci Soc 87:209–221

Martin WW (1975) A new species of *Catenaria* parasitic in midge eggs. Mycologia 67:264–272

Martin WW (1978) Two additional species of *Catenaria* (Chytridiomycetes, Blastocladiales) parasitic in midge eggs. Mycologia 70:461–467

Martin WW (1987) Zoosporic parasites of aquatic insects: collection, identification, and culture. In: Fuller MS, Jaworski A (eds) Zoosporic fungi in teaching and research. Southeastern, Athens, GA, pp 137–142

Massey SE, Garey JR (2007) A comparative genomics analysis of codon reassignments reveals a link with mitochondrial proteome size and a mechanism of genetic code change via suppressor tRNAs. J Mol Evol 64:399–410

Matthews VD (1937) A new genus of the Blastocladiaceae. J Elisha Mitchell Sci Soc 53:191–195

McCrainie J (1942) Sexuality in *Allomyces cystogenus*. Mycologia 34:209–213

Mills GL, Cantino EC (1979) Trimodal formation of microbodies and associated biochemical and cytochemical changes during development in *Blastocladiella emersonii*. Exp Mycol 3:53–69

Mills GL, Cantino EC (1981) Chitosome-like vesicles from gamma-particles of *Blastocladiella emersonii* synthesize chitin. Arch Microbiol 130:72–77

Mitchell RT, Deacon JW (1986) Selective accumulation of zoospores of chytridiomycetes and oomycetes on cellulose and chitin. Trans Br Mycol Soc 86:219–223

Nagahama T, Sato H, Shimazu M, Sugiyama J (1995) Phylogenetic divergence of the entomophthoralean fungi: evidence from nuclear 18S ribosomal RNA gene sequences. Mycologia 87:203–209

Nascimento CDA, Pires-Zottarelli CLA (2010) Blastocladiales e Spizellomycetales do Parque Estadual da Serra da Cantareira, São Paulo, Brasil. Revista Brasileira de Botânica 33:693–704

Nolan RA (1969) Nutritional requirements for species of *Allomyces*. Mycologia 61:641–644

Nolan RA (1970) Sulfur source and vitamin requirements of aquatic phycomycete, *Catenaria anguillulae*. Mycologia 62:568–577

Nolan RA (1985) Physiology and biochemistry. In: Couch JN, Bland CE (eds) The genus *Coelomomyces*. Academic, Orlando, FL, pp 321–348

Olson LW (1974) Meiosis in the aquatic Phycomycete *Allomyces macrogynus*. Comptes rendus des travaux du laboratoire Carlsberg 40:113–124

Olson LW (1980) *Allomyces neo-moniliformis* gametogenesis. The Cystogenes life cycle. Protoplasma 105:87–106

Olson LW (1984) *Allomyces* – a different fungus. Opera Bot 73:5–96

Olson LW, Borkhardt B (1978) Polyploidy and its control in *Allomyces macrogynus*. Trans Br Mycol Soc 71:65–76

Olson LW, Lange L (1978) Meiospore of *Physoderma maydis* – causal agent of physoderma disease of maize. Protoplasma 97:275–290

Olson LW, Lange L (1983) The gamma body- a vesicle generating structure. Nord J Bot 3:673–680

Olson LW, Reichle R (1978a) Meiosis and diploidization in the aquatic Phycomycete *Catenaria anguillulae*. Trans Br Mycol Soc 70:423–437

Olson LW, Reichle R (1978b) Synaptonemal complex formation and meiosis in the resting sporangium of *Blastocladiella emersonii*. Protoplasma 97:261–273

Olson LW, Lange L, Reichle R (1978) Zoospore and meiospore of aquatic phycomycete *Catenaria anguillulae*. Protoplasma 94:53–71

Paquin B, Lang BF (1996) The mitochondrial DNA of *Allomyces macrogynus*: the complete genomic sequence from an ancestral fungus. J Mol Biol 255:688–701

Paquin B, Laforest MJ, Forget L, Roewer I, Wang Z, Longcore J, Lang BF (1997) The fungal mitochondrial genome project: evolution of fungal mitochondrial genomes and their gene expression. Curr Genet 31:380–395

Petersen HE (1909) Studier over Ferskvands-Phycomyceter. Bidrag til Kundskaben om de submerse Phykomyceters Biologi og Systematik, samt om deres Udbredelse i Danmark. Bot Tidskrift 29:345–440

Picard KT, Letcher PM, Powell MJ (2009) *Rhizidium phycophilum*, a new species in Chytridiales. Mycologia 101:696–706

Pommerville J (1977) Chemotaxis of *Allomyces* gametes. Exp Cell Res 109:43–51

Pommerville J (1978) Analysis of gamete and zygote motility in *Allomyces*. Exp Cell Res 113:161–172

Porter TM, Martin WM, James TY, Longcore JE, Gleason F, Adler PH, Letcher PM, Vilgalys R (2011) Molecular phylogeny of the Blastocladiomycota (Fungi) based on nuclear ribosomal data. Fungal Biol 115:381–392

Powell MJ (1978) Phylogenetic implications of microbody-lipid globule complex in zoosporic fungi. BioSyst 10:167–180

Powell MJ (1983) Localization of antimonate-mediated precipitates of cations in zoospores of *Chytriomyces hyalinus*. Exp Mycol 7:266–277

Powell MJ, Koch WJ (1977a) Morphological variations in a new species of *Entophlyctis*. I. Species concept. Can J Bot 55:1668–1685

Powell MJ, Koch WJ (1977b) Morphological variations in a new species of *Entophlyctis*. II. Influence of growth-conditions on morphology. Can J Bot 55:1686–1695

Pritchard AE, Seilhammer JJ, Mahalingam R, Sable CL, Venuti SE, Cummings DJ (1990) Nucleotide sequence of the mitochondrial genome of *Paramecium*. Nucleic Acids Res 18:173–180

Reichle RE, Fuller MS (1967) Fine structure of *Blastocladiella emersonii* zoospores. Am J Bot 54:81–92

Remy W, Taylor TN, Hass H (1994) Early Devonian fungi – a Blastocladalean fungus with sexual reproduction. Am J Bot 81:690–702

Ribichich KF, Salem-Izacc SM, Georg RC, Vencio RZN, Navarro LD, Gomes SL (2005) Gene discovery and expression profile analysis through sequencing of expressed sequence tags from different develop-

mental stages of the chytridiomycete *Blastocladiella emersonii*. Eukaryot Cell 4:455–464
Robertson JA (1972) Phototaxis in a new *Allomyces*. Arch f Mikrobiol 85:259–266
Ruiz-Trillo I, Burger G, Holland PWH, King N, Lang BF, Roger AJ, Gray MW (2007) The origins of multicellularity: a multi-taxon genome initiative. Trends Genet 23:113–118
Schadt CW, Martin AP, Lipson DA, Schmidt SK (2003) Seasonal dynamics of previously unknown fungal lineages in tundra soils. Science 301:1359–1361
Schardl CL, Craven KD (2003) Interspecific hybridization in plant-associated fungi and oomycetes: a review. Mol Ecol 12:2861–2873
Scherffel A (1925) Endophytische Phycomyceten-Parasiten der Bacillariaceen und einege neue Monadinen. Ein Beitrag zur Phylogenie der Oomyceten (Schröter). Archiv Protistenk 52:1–141
Schroeter J (1893) Phycomycetes. Natürlichen Pflanzenfam 1:63–141
Seif E, Leigh J, Liu Y, Roewer I, Forget L, Lang BF (2005) Comparative mitochondrial genomics in zygomycetes: bacteria-like RNase P RNAs, mobile elements and a close source of the group I intron invasion in angiosperms. Nucleic Acids Res 33:734–744
Shaw DS, Cantino EC (1969) An albino mutant of *Blastocladiella emersonii* - comparative studies of zoospore behaviour and fine structure. J Gen Microbiol 59:369–382
Simmons DR, James TY, Meyer AF, Longcore JE (2009) Lobulomycetales, a new order in the Chytridiomycota. Mycol Res 113:450–460
Singh KP, Jaiswal RK, Kumar N (2007) *Catenaria anguillulae* Sorokin: a natural biocontrol agent of *Meloidogyne graminicola* causing root knot disease of rice (Oryza sativa L.). World J Microbiol Biotechnol 23:291–294
Slapeta J, Moreira D, Lopez-Garcia P (2005) The extent of protist diversity: insights from molecular ecology of freshwater eukaryotes. Proc R Soc B Biol Sci 272:2073–2081
Sparrow FK (1939) The entomogenous chytrid *Myrophagus* Thaxter. Mycologia 304:113–116
Sparrow FK (1960) Aquatic phycomycetes. University of Michigan Press, Ann Arbor, MI
Sparrow FK (1962) *Urophlyctis* and *Physoderma*. Trans Mycol Soc Jpn 3:16–18
Sparrow FK (1965) Concerning *Physoderma graminis*. Mycologia 57:624–627
Sparrow FK, Griffin JE (1964) Observation on chytridiaceous parasites of phanerograms. XV. Host range and species concepts studies in Physoderma. Arch f Mikrobiol 40:275–282
Stumm C, Hermans JMH, Croes AF, Bucks JH (1976) Chemotaxis and transport of amino acids in *Allomyces arbuscula*. Ant v Leeuwenhoek 42:203–209
Tabak HH, Cooke WB (1968) Effects of gaseous environments on growth and metabolism of fungi. Bot Rev 34:126–252
Tambor JHM, Ribichich KF, Gomes SL (2008) The mitochondrial view of *Blastocladiella emersonii*. Gene 424:33–39
Tanabe Y, O'Donnell K, Saikawa M, Sugiyama J (2000) Molecular phylogeny of parasitic Zygomycota (Dimargaritales, Zoopagales) based on nuclear small subunit ribosomal DNA sequences. Mol Phylogenet Evol 16:253–262
Tanabe Y, Watanabe MM, Sugiyama J (2002) Are Microsporidia really related to Fungi?: a reappraisal based on additional gene sequences from basal fungi. Mycol Res 106:1380–1391
Tanabe Y, Saikawa M, Watanabe MM, Sugiyama J (2004) Molecular phylogeny of Zygomycota based on EF-1 and RPB1 sequences: limitations and utility of alternative markers to rDNA. Mol Phylogenet Evol 30:438–449
Tanabe Y, Watanabe MM, Sugiyama J (2005) Evolutionary relationships among basal fungi (Chytridiomycota and Zygomycota): insights from molecular phylogenetics. J Gen Appl Microbiol 51:267–276
Taylor JW, Fuller MS (1981) The Golgi apparatus, zoosporogenesis, and development of the zoospore discharge apparatus of *Chytridium confervae*. Exp Mycol 5:35–59
Taylor TN, Remy W, Hass H (1994) *Allomyces* in the Devonian. Nature 367:601–601
Taylor JW, Jacobson DJ, Kroken S, Kasuga T, Geiser DM, Hibbett DS, Fisher MC (2000) Phylogenetic species recognition and species concepts in fungi. Fungal Genet Biol 31:21–32
Teter HE (1944) Isogamous sexuality in a new strain of *Allomyces*. Mycologia 36:194–210
Travland LB (1979) Structures of the motile cells of *Coelomomyces psorophorae* and function of the zygote in encystment on a host. Can J Bot 57:1021–1035
Vandenkoornhuyse P, Baldauf SL, Leyval C, Straczek J, Young JPW (2002) Evolution - Extensive fungal diversity in plant roots. Science 295:2051–2051
Vavra J, Joyon L (1966) Etude sur la morphologie le cycle evolutif et la position systematique de *Callimastix cyclopis* Weissenberg. Protistologica 2:5–15
Whisler HC (1987) On the isolation and culture of water molds: the Blastocladiales and Monoblepharidales. In: Fuller MS, Jaworski A (eds) Zoosporic fungi in teaching and research. Southeastern, Athens, GA, pp 121–124
Willoughby LG (1959) A new species of *Blastocladiella* from Great Britain. Trans Br Mycol Soc 42:287–291
Wilson CM (1952) Meiosis in *Allomyces*. Bull Torr Bot Club 79:139–160
Wilson CM, Flanagan PW (1968) The life cycle and cytology of *Brachyallomyces*. Can J Bot 46:1361–1367
Wolf FT (1941) A contribution to the life history and geographic distribution of the genus *Allomyces*. Mycologia 33:158–173

8 Zygomycetous Fungi: Phylum Entomophthoromycota and Subphyla Kickxellomycotina, Mortierellomycotina, Mucoromycotina, and Zoopagomycotina

Gerald L. Benny[1], Richard A. Humber[2], Kerstin Voigt[3]

CONTENTS

[1]Department of Plant Pathology, University of Florida, 1453 Fifield Hall, Gainesville, FL 32611-0680, USA; e-mail: gbenny@ufl.edu

[2]USDA-ARS Biological Integrated Pest Management Unit, R.W. Holley Center for Agriculture and Health, 538 Tower Road, Ithaca, NY 10453, USA; e-mail: richard.humber@ars.usda.gov

[3]Jena Microbial Resource Collection (JMRC), Institute of Microbiology, University of Jena, Neugasse 07743, Jena, Germany; e-mail: Kerstin.Voigt@uni-Jena.de

I. Introduction

Members of the phylum Zygomycota were discussed previously as Trichomycetes and Zygomycetes (Benny 2001; Benny et al. 2001). The phylum was never validly published according to the International Code of Botanical Nomenclature (McNeill et al. 2012). Hibbett et al. (2007) recommended that the phylum name not be validated and that the taxon not be used formally. The trichomycetes and zygomycetes are no longer used here as classes, but the included organisms share a similar habit and habitat, and therefore the names are used in lowercase to demonstrate these affinities.

Zygomycotan fungi are heterotrophs that reproduce sexually, where known, by the formation of zygospores. Asexual reproduction is by the formation of aplanospores (sporangiospores, trichospores, conidia, yeast cells, arthrospores, chlamydospores). Nutrient uptake is by absorption. The cell wall is composed of chitosan in the order Mucorales or chitin in the other orders, as far as is known.

Zygomycetous fungi are symbionts in the orders Asellariales and Harpellales) in the gut of arthopods, including insects and their larvae, that attach to the host via a cellular or

Systematics and Evolution, 2nd Edition
The Mycota VII Part A
D.J. McLaughlin and J.W. Spatafora (Eds.)

noncellular holdfast, whereas the remaining orders are saprobes, ectomycorrhizal, or haustorial or nonhaustorial parasites of animals or plants (Benny 2009, 2012; Benny et al. 2001; Lichtwardt 1986; Lichtwardt et al. 2001). Taxa of the Asellariales and Harpellales can be reviewed on the trichomycete Web site (http://www.nhm.ku.edu/~fungi/), and many of the remaining orders are presented on the zygomycete Web site (http://www.zygomycetes.org/).

Benjamin (1979) considered the mode of nutrition an important taxonomic criterion at the ordinal level and defined the following specialties: (1) saprobic or nonhaustorial (facultative) parasites (Basidiobolales, Mortierellales, Mucorales, Kickxellales, some Entomophthorales, and Endogonales); (2) haustorial (Dimargaritales, some Zoopagales) or nonhaustorial parasites forming hyphal bodies (most Entomophthorales) or coiled thalli (some Zoopagales); and (3) ectomycorrhizae (some Endogonales).

The classification of zygomycotan fungi presented in the first edition (Benny 2001; Benny et al. 2001) is based mainly on the morphology of reproductive structures and the presence of septa (Benjamin 1959, 1979). The orders discussed by Benny (2001) and Benny et al. (2001) in the trichomycetes and zygomycetes are revised here. Four of the orders presented in the aforementioned chapters are now known not to be zygomycotan fungi. These include protozoans (Amoebidales, Eccrinales) (Benny and O'Donnell 2000; Cafaro 2005) or members of the Glomeromycota (Geosiphonales, Glomerales—as Glomales) (Schüssler et al. 2001; see Redecker and Schüssler 2014). These deletions resulted in the retention of nine orders (Asellariales, Basidiobolales, Dimargaritales, Endogonales, Entomophthorales, Kickxellales, Mortierellales, Mucorales, Zoopagales), and a tenth order was described recently (Neozygitales; Humber 2012b). Phylogenetic studies resulted in the description of one phylum, Entomophthoromycota, and four subphyla, Kickxellomycotina, Mortierellomycotina, Mucoromycotina, and Zoopagomycotina, for these orders (Hibbett et al. 2007; Hoffmann et al. 2011; Humber 2012b) as discussed subsequently.

II. Occurrence and Distribution

The members of Zygomycota are more or less cosmopolitan, subsist on soluble nutrients, and utilize a remarkable spectrum of substrates. The most common sources of Mucorales, Dimargaritales, Kickxellales, some Zoopagales (Helicocephalidaceae, Piptocephalidaceae, Sigmoideomycetaceae), and a few genera of Entomophthorales (*Basidiobolus*, *Conidiobolus*) are dung and soil. Many zygomycetes can be isolated from soil, but the same species can often be found on dung or other organic material. A few taxa (e.g., *Hesseltinella*, *Zygorhynchus*) are found only in soil.

Pilobolus species (Grove 1934) are obligate coprophiles. The other species found on dung are not obligate coprophiles, but many of these fungi usually are not isolated from other substrates. Herbivore dung, especially from horses or cows, is the most reliable source for *Pilobolus* spp. The dung of small, omnivorous rats and mice is often the best source of the merosporangiferous Mucorales [fungi now in Mucorales (Syncephalastraceae), Kickxellales, Dimargaritales, and Zoopagales (Piptocephaladiaceae)] (Benjamin 1959) and many other taxa in Mucorales.

Fungi in Endogonales, as currently circumscribed (Morton and Benny 1990), produce zygospores with apposed suspensors in yellow to orange sporocarps that are formed on or in the soil or on organic debris such as leaves or wood. Some species are ectomycorrhizal, but others are saprobes. Members of Endogonales are infrequently collected but may occur worldwide.

Entomophthoralean fungi can be found in virtually all temperate and tropical parts of the world and are most commonly encountered as pathogens of a very wide range of insects and some phytophagous mites. The insects most commonly affected by entomophthoralean pathogens worldwide are aphids (Homoptera: Aphididae) and a variety of lepidopterans (mostly moths from many families), adult flies, and grasshoppers and locusts (Orthoptera: Acrididae). A much smaller set of genera and species attacks soil invertebrates such as nematodes.

Species of *Conidiobolus* and *Basidiobolus*, whose natural habit appears to be saprobic in plant detritus and soil, also have a worldwide distribution. A small number of *Conidiobolus* spp. are known only as entomopathogens, but most are saprobes. *Conidiobolus thromboides* Drechsler may be found in China and India as a soil saprobe, but the species is a significant pathogen of a fairly broad range of insects (Latgé et al. 1980), and *Conidiobolus coronatus* (Costantin) Batko is a common saprobe, a weak but widely known entomopathogen, and, especially in the tropics, may cause mycoses of humans or other mammals (Sect. IIIC).

Thermophilic or thermotolerant zygomycetes are mostly restricted to Mucorales in *Rhizomucor* sensu Schipper (1978), *Thermomucor* (Schipper 1979), *Lichtheimia* (Hoffmann 2010), and a few species of several other genera. *Calcarisporiella*, a genus of unknown affinities in Mucoromycotina, is thermotolerant (Hirose et al. 2012). Psychrophilic or psychrotolerant zygomycetes can be found in the genera *Chaetocladium*, *Dicranophora*, *Dissophora*, *Helicostylum*, and *Spinellus* and other taxa including species of *Mortierella*, *Mucor*, *Thamnidium*, and *Zygorhynchus*. Schmidt et al. (2008) isolated *Mucor mucedo* Fresen., *Helicostylum elegans* Corda, *Mortierella* spp., and several unidentified members of Mortierellales and Mucorales from a subalpine forest in Colorado, USA, and *Pirella circinans* Bainier was reported on a beetle collected on a subantarctic island (Bridge et al. 2008).

III. Economic Importance and Biology

A. Food

Members of Mucorales are used to make food in Asia (Benny 2012) and can cause food spoilage. *Thamnidium elegans* Link and *Helicostylum pulchrum* (Preuss) Pidopl. & Milko can grow on cut meat stored in walk-in refrigerators. This so-called whiskery beef imparts a desirable flavor, and these fungi are intentionally inoculated (Benny 2012).

B. Plant Pathogens (*Choanephora* Fruit Rot) and Storage Rots

Fruit storage rots are caused by a few species of Mucorales, including *Rhizopus stolonifer* (Ehrenb.: Fr.) Vuill. (e.g., strawberries, sweet potatoes, peanut seedlings), which may be more or less cosmopolitan. *Gilbertella persicaria* (E.D. Eddy) Hesselt. can cause storage rots of nectarines, peaches, and tomatoes in California, the southeastern USA, India, and China; storage rot of dragon fruit [*Hylocereus costaricensis* (F.A.C. Weber) Britt. & Rose; Cactaceae] in China; and *Gilbertella* stem rot of pithya [*Hylocereus undatus* (Haw.) Britt. & Rose; Cactaceae] in Okinawa, Japan (Guo et al. 2012; Taba et al. 2011). *Mucor piriformis* A. Fisch. is the agent of pear rots, even in orchards, in the USA (California, Oregon, Washington) and Chile (Michailides 1991; Michaelides and Spotts 1990). Cold storage retards the growth of *R. stolonifer* and *G. persicaria* and is the best method to inhibit the growth of these fungi. *M. piriformis*, however, grows not only at 20 °C but also at refrigerator temperature (4–7 °C) and as a result can still damage fruit during long-term cold storage (Michaelides and Spotts 1990; Ogawa et al. 1992). *Rhizopus oryzae* Went. & Prinsen-Geerlings also can cause a soft rot of fruits and vegetables, especially in combination with other fungi (Holliday 1980). Both *R. oryzae* and *R. stolonifer* can cause *Rhizopus* head rot of sunflower (Shtienberg 1997).

***Choanephora cucurbitarum* (Berk. & Ravenel) Thaxt. is a pathogen (so-called wet rot or blossum end rot) of many crop plants,** including green beans and okra, but commonly affects Cucurbitaceae, especially yellow summer squash (Wolf 1917). Plant diseases caused by *C. cucurbitarum* include *Choanephora* rots or blights of chili peppers, Egyptian henbane, ice plant, and other plants (Abdel-Motaal et al. 2010; Kagiwada et al. 2010; Sinha 1940b). *C. cucurbitarum* can be isolated from soil and is usually found when it is warm and humid; this species is common during rainy summers in the southeastern USA and globally in other regions with a similar climate. An unidentified species of *Choanephora* was found on the male inflorescence of *Artocarpus integer* (Thunb.) Merr. and is eaten by the pollinator, a species of *Contarinia* Rodani (a gall midge). These insects lay eggs in inflorescences, and the larvae feed on the hyphae of *Choanephora*; this mutualism was observed in Borneo (Sakai et al. 2000).

A related fungus, *Blakeslea trispora* Thaxt., can also be a plant pathogen in India. This includes a wet rot of *Colocasia antiquorum* Schott [syn.–*C. esculenta* (L.) Schott] leaves by both *B. trispora* and *C. cucurbitarum* (Sinha 1940a) and a seedling blight of brinjal, cauliflower, chilli peppers, and tomato (Saksena and Narain 1983).

C. Zygomycosis

Zygomycosis is caused by zygomycetous fungi that can grow at 37 °C and includes the causative agents of Basidiobolomycosis (*Basidiobolus*), Entomophthoromycosis (*Conidiobolus*), and mucormycosis (*Apophysomyces, Cokeromyces, Cunninghamella, Lichtheimia, Mucor, Rhizomucor, Rhizopus,* and *Saksenaea*—all members of Mucorales) as reviewed by Ribes et al. (2000) and Gomes et al. (2011). One zygomycotic species that was recently renamed is *Rhizomucor variabilis* var. *variabilis* R.Y. Zheng & G.Q. Chen, which is now *Mucor irregularis* Stchigel, Cano, Guarro & Ed. Álvarez (Lu et al. 2013).

A small number of species of *Conidiobolus* [*C. coronatus* (Costantin) A. Batko, *C. incongruus* Drechsler, *C. lamprauges* Drechsler; Kimura et al. 2011; Prabhu and Patel 2004] and *Basidiobolus ranarum* Eidam may cause facultative mycoses of humans, equines, or other mammals (Humber et al. 1988; Kwon-Chung and Bennett 1992). Conidiobolomycosis is usually confined to the nasal tract and adjacent superficial facial tissues, although the resulting mycoses can be fatal in some animals and may be severely disfiguring in humans (Kwon-Chung and Bennett 1992). Basidiobolomycosis is often a subcutaneous disease affecting the limbs but may become systemic in debilitated hosts; gastrointestinal mycosis is becoming more prevalent (Liu 2011).

Many zygomycetes that are potential or known human or animal pathogens include species of *Apophysomyces, Cokeromyces, Cunninghamella, Lichtheimia, Mucor, Rhizomucor, Rhizopus,* and *Saksenaea.* Some of the fungi that cause mucormycosis are geographically widespread (*Lichtheimia, Mucor, Rhizomucor, Rhizopus*) or localized (*Apophysomyces, Cunninghamella* in the USA or *Saksenaea* in the USA and Asia). **Predisposing factors for mucormycosis include burns, acidosis, hyperglycemia, and leucopenia.** A patient's underlying disease may affect the site of initial infection. For example, paranasal sinus infection by *Rhizopus* sp. is often associated with diabetes mellitis, or a local trauma infection by *Apophysomyces, Rhizopus,* or *Saksenaea* may be associated with immunosuppression or diabetes. A form of chronic cystitis is caused by the yeast phase of *Cokeromyces recurvatus* Poitras. Treatment of mucormycosis is by the surgical removal of infected tissue and the appropriate antifungal agent, along with control of the predisposing ailment (Kwon-Chung and Bennett 1992).

Other fungi of interest as the causative agents of zygomycosis are *Mucor amphibiorum* Schpper (platypus in Tasmania, toads on the Australian mainland) and *Mortierella wolfii* B.S. Mehrotra & Baijal, which can cause mycotic abortion and mycoses in cattle (Papp et al. 2011). *M. wolfii* is not known to infect humans.

The detection and identification of the organisms that cause zygomycosis traditionally have been by means of making cultures and at maturity microscopically examining the fungus. These classical methods are time consuming, and recently developed molecular methods are increasingly being used for the rapid detection and identification of these fungi (Liu 2011).

D. Industrial Uses of Zygomycotan Fungi: Biotechnology, Biodegradation, Biosorption, Bioremediation, Biotransformation

Lycopene production has been optimized by the use of mated cultures of *B. trispora* Thaxt. (Wang et al. 2011). **Arachidonic acid (AA) is**

produced by *Mortierella alpina* Peyronel and *M. alliacea* Linnem. using glucose or waste glycerol, a byproduct of biodiesel production, as the carbon source (Jermsuntiea et al. 2011; Khanna et al. 2011).

Biodiesel can be made from several plant oils (rapeseed, soybean, sunflower) that are high in fatty acid methyl esters. A comparison was made of the biodiesel produced from rapeseed oil and *Cunninghamella echinulata* (Thaxt.) Thaxt. ex Blakeslee, and they both conformed to European standards (Sergeeva et al. 2011). ***M. alpina* and species of *Rhizopus*, *Umbelopsis*, and *Zygorhynchus* have been used to produce lipids that are precursors for biodiesel production** (Kosa and Ragauskas 2011). Micro-Raman spectroscopy has been used to determine the composition and degree of saturation of hyphal oil in *M. alpina* and *M. elongata* Linnem. (Münchberg et al. 2012); this procedure may be useful for screening other fungi for fatty acid production.

Members of Mucorales, including species of *Cunninghamella* and *Rhizopus*, have been used for the **biosorption** of polycyclic aromatic hydrocarbons and treatment of textile waste water (Tigini et al. 2011). **Species of *Cunninghamella* have been used for biotransformation** (Amadio et al. (2010) and to simulate mammalian physiology in studies of drug metabolism (Asha and Vidyavathi 2009). Endosulfan, an insecticide with a long half-life in soil, has been degraded by *Mortierella* sp. (Kataoka et al. 2011).

E. Mucorales and Endobacteria

Early studies of *Rhizopus* rice seedling blight (Mew and Gonzalez 2002) initially concentrated on the fungus because an endobacterium was not known to be present. Partida-Martinez and Hertweck (2006) discovered that ***Rhizopus microsporus* Tiegh. also contained a toxin (rhizonin) producing the endosymbiotic bacterium** *Berkholderia rhizoxinica* Partida-Martinez, I. Groth, I. Schmitt, W. Richer, M. Roth, & C. Hertweck (Partida-Martinez et al. 2007a, b). Later, Chamilos et al. (2007) hypothesized that the emergence of zygomycosis might be due to the presence of endosymbiotic bacteria that conferred multidrug resistance, but endosymbiots were not present in the clinical isolates of *Rhizopus*; the fungi were the pathogens (Partida-Martinez et al. 2008). A strain of *R. microsporus* used in Sufu production contained a toxin-forming endosymbiotic bacterium, and the hyphae of four stains of *Mortierella elongata* contained endotoxin-forming betaproteobacteria (Rohm et al. 2010; Sato et al. 2010).

F. Light

The photobiology of many fungi, including *Phycomyces blakesleeanus* Burgeff, has been studied (Corrochano and Galland 2006; Corrochano and Garre 2010). The sporangiophores of *P. blakesleeanus* are influenced by air movement, gravity, light, touch, and the presence of close structures. Light induces the production of both macro- and microsporangiophores; blue light stimulates only macrosporangiophore formation. Light is also important in the production of mycelial β-carotene, which is the precursor of the sexual pheromone trisporic acid (Schimek and Wöstemeyer 2006).

All species of *Pilobolus* periodically form sporangia in alternating light and dark. The various taxa have differing light requirements for the production of trophocysts, sporangiophores, and sporangia; trophocysts always precede the formation of sporangiophores in both *Pilobolus* and *Utharomyces* (Kirk and Benny 1980; Page 1962). In one study, light did not have an effect on sporulation, whereas the type of medium used was important in *B. trispora* sporangia and sporangiola formation (Goldring 1936). *C. cucurbitarum*, however, forms more sporangiola after 8 h of continuous light, and a temperature of 31 °C stimulated formation of sporangia, but at 25 °C fewer of these structures were produced (Barnett and Lilly 1950).

G. Mating

Trisporic acid formation during sexual reproduction has been observed in members of

Mortierellales and Mucorales. Several species of *Mortierella* were screened and compared with members of Mucorales for the formation of trisporic acid and its precursors. The gene (*TSP*1) for 4-dihydromethyltrisporate dehydrogenase was found in *Mortierella* and selected members of Mucorales and may occur in other zygomycotan fungi (Schimek et al. 2003).

Werkman (1976) studied mating in *Zygorhynchus moelleri* Vuill., a homothallic member of Mucorales. The minus (−) gametangium is a cell on the main sporangiophore and becomes the smaller of the two suspensors, whereas the other gametangium is plus (+); it is borne on a side branch and at maturity is the larger of the two suspensors.

Schimek and Wöstemeyer (2006, 2009) and Wöstemeyer and Schimek (2007) reviewed the chemical pathway for the production of trisporic acid and its precursors in three heterothallic members of Mucorales, *B. trispora*, *Mucor mucedo* Fresen., and *Phycomyces blakesleeanus*, in which the process has been studied. Feofilova (2006) discussed heterothallism and its application in industry for the increased production of β-carotene, lycopene, sterols, and other isoprenoids by mated cultures of *B. trispora*.

Parasitella parasitica (Bainier) Syd. is a biotrophic, gall-forming parasite of other Mucorales members. This species induces hosts to produce galls (Thaxter 1895, Plate XXXIV, Figs. 10–13) or sikyotic cells that mature to form sikyospores. *P. parasitica* is heterothallic, and parasitism occurs only on a host of the opposite mating type. *Absidia glauca* Hagem is the host used in mating studies. Zygospores are formed by *P. parasitica* only if one of the mating types is parasitizing a host. Trisporic acid is involved in host–parasite recognition (Wöstemeyer et al. 1995).

IV. Development of Taxonomic Theory

Many taxa of Mucorales s.l. were described by the mid-nineteenth century. Van Tieghem (1878) published the first classification of Mucorales. Thaxter's (1888) monograph of Entomophthoreae (Entomophthorales) formed the basis for many, especially early, treatments of the group (Waterhouse 1973). A very distinct and nearly equally influential approach to entomophthoralean taxonomy was developed even earlier in Europe (Brefeld 1870; Nowakowski 1883). The earliest molecular studies on the phylogeny of Entomophthorales (Jensen et al. 1998; Nagahama et al. 1995) suggested that these fungi were readily distinguished from other zygomycetes.

Fungi in Entomophthorales were not included in van Tieghem's (1878) concept of Mucorales, a scheme followed by most students of zygomycetes (Benjamin 1979). Many early students of Mucorales placed considerable emphasis on the production of unispored sporangiola, or so-called conidial fungi in their opinion (Benny and Benjamin 1975). Thaxter (1922) monographed the sporocarpic Endogonaeae; many of these taxa are now in the phylum Glomeromycota (see Redecker and Schüssler 2014; Schüssler et al. 2001).

Several taxonomic treatments of Mucorales s.l. were published in the first part of the twentieth century (Benjamin 1979; Benny and Benjamin 1975). Hesseltine's (1955) classification of Mucorales was later expanded (Hesseltine and Ellis 1973) to include Benjamin's (1959) merosporangiferous Mucorales. The classification proposed by Benjamin (1979) is the basis for the ordinal arrangement of many taxonomic schemes (Alexopoulos et al. 1996; Kirk et al. 2008). Benjamin (1979) recognized the orders Mucorales, Entomophthorales, and Zoopagales, validated Endogonales, Kickxellales, and Zoopagales, and described Dimargaritales. His inclusion of Harpellales (trichomycetes) (Lichtwardt 1986) in zygomycetes was less widely accepted (Alexopoulos et al. 1995; Hawksworth et al. 1995). Much of the research in the two decades between Benjamin's (1979) review and that in the first edition (Benny 2001; Benny et al. 2001) of this book was an attempt to create a more natural morphology-based classification.

Batko (1964a, b, c, d) combined the divergent taxonomic approaches of Nowakowski (1883) and Thaxter (1888) to provide the basis for a more natural entomophthoralean classification that resulted in the recognition of six families (Humber 1989). Balazy's (1993) five-family classification of the Entomophthorales follows the Batkoan classification more closely than those developed later (Ben-Ze'ev and Kenneth 1982; Humber 1981, 1989; Remaudière and Keller 1980). The majority of the research on Entomophthoromycota is on insect pathogens because these taxa are the most common and economically important members of the order. Fewer mycologists have studied the saprobic Entomophthoromycota, but they are also taxonomically diverse. The genus *Ancylistes* was long treated as the type genus of an order of zoosporic fungi until it was transferred to Entomophthorales (Berdan 1938).

Partial zygomycete phylogenies based on 18S rDNA for Entomophthorales and Kickxellales are available (Jensen et al. 1998; Nagahama et al. 1995; O'Donnell et al. 1998). There is a phylogenetic analysis using 18S and 28S rDNA sequences from the medically important species of Basidiobolales, Entomophthorales, Mortierellales, and Mucorales (Voigt et al. 1999). Two phylogenetic analyses, based on three genes, include at least one member of every recognized genus of Mucorales in culture (O'Donnell et al. 2001: 18S rRNA, 28S rRNA, Ef-1α, morphology; Voigt and Wöstemeyer 2001: 18S rRNA, actin, Ef-1α). O'Donnell et al. (2001) revealed that only a few of the morphological characteristics, such as spore appendages and trophocysts, were phylogenetically informative.

A multigene phylogenetic study of all fungal phyla, including zygomycotan fungi, was published by Lutzoni et al. (2004); this data set was composed of nucLSU and nucSSU for Chytridiomycota, Glomeromycota, and zygomycotan fungi, whereas additional sequences were included for the remaining phyla (Ascomycota, Basidiomycota).

Multigene phylogenies of the fungi, including zygomycota, were subsequently published (James et al. 2006; Liu et al. 2006). A cladogram of the zygomycotan fungi based on three genes (5.8S rRNA, 18S rRNA, 28S rRNA) was published by White et al. (2006a). White (2006) presented a multigene phylogeny based on rRNA from the available members of the Harpellales. Hibbett et al. (2007) published a new classification of the fungi down to order based on molecular phylogenetic studies of all fungal groups, and four subphyla (Entomophthoromycotina, Kickxellomycotina, Mucoromycotina, Zoopagomycotina) with unknown affinities were reported. Later, a fifth subphylum, Mortierellomycotina, was described, also with unknown affinities (Hoffmann et al. 2011).

Humber (2012b) described a new phylum, Entomophthoromycota, and several higher taxa for the entomophthoralean fungi. This classification was based on a multigene phylogeny presented by Gryganskyi et al. (2012, 2013). Two recent papers summarize the classification of the zygomycotan fungi but differ in Mucorales family recognition (Benny 2012; Voigt 2012).

V. Reproduction and Dispersal

A. Growth

Zygomyceous fungi can grow quite rapidly, producing sporangiospores (or conidia) and, under optimal conditions, even zygospores in 1–3 weeks. Many members of Mucorales and Kickxellales are saprobes that can be encountered on diverse organic substrates. Many or all taxa in Zoopagales, Dimargaritales, and Entomophthoromycota are obligate parasites; a few are nonhaustorial. Many members of Zoopagales are mycoparasites, but the remaining ones, and some members of Entomophthoromycota, are parasites of nematodes or their eggs, amoebae, rotifers, and tardigrades (Barron 2004; Tucker 1981). Other members of Entomophthoromycota are obligate parasites of algae (*Ancylistes*), fern prothallia (*Completoria*), or insects or mites (most taxa); most *Basidiobolus* and *Conidiobolus* spp. are saprobes (Humber 1989).

B. Dispersal

Spore dispersal may be active (forcible discharge), as in *Pilobolus* and Entomophthoromycota (except *Massospora*), or passive (contact, water droplets, or air currents), in the remaining zygomycete taxa (Ingold 1978; Tucker 1981). In the latter organisms the mature asexual reproductive structures (sporangia, sporangiola, merosporangia, conidia) are either wet (forming spore drops) or dry and are dispersed by contact (e.g., small insects, mites), air (wind currents), or water (rain). Asexual spores of Entomophthorales are not produced in any sort of distinct slime but usually adhere firmly to any substrate they contact.

Taxa in Choanephoraceae and Pilobolaceae (Mucorales) have highly specialized dispersal methods. Species of *Blakeslea*, *Choanephora*, *Gilbertella*, and *Poitrasia* (Choanephoraceae; Voigt and Olsson 2008) possess sporangia with persistent, sutured walls and appendaged sporangiospores (O'Donnell 1979; Voigt and Olsson 2008). Many of these fungi are plant parasites, and the appendages may aid insects in dispersing spores from flower to flower or by other methods such as water droplets. Species of *Pilobolus* and *Utharomyces* are coprophilous (Grove 1934; Kirk and Benny 1980), and all have developed mechanisms for efficiently dispersing spores onto vegetative material. The sporangia of *Pilobolus* are directed toward a light source and are actively discharged by a pressurized string of cytoplasm (Page 1964; Yafetto et al. 2008) up to a distance of 2 m or more, where they attach to any substrate they contact (Page 1962). The larval stage of cattle lungworm (*Dictyocaulus viviparus* Bloch) can be dispersed by *Pilobolus* sporangia (Eysker 1991).

Utharomyces (Pilobolaceae) has phototropic sporangiophores that rapidly elongate until contacting a surface. The subsporangial vesicle of *Utharomyces* ruptures on contact with a solid substrate, releasing the intact, stalked sporangium (Kirk and Benny 1980).

Many sporangia and other aerial structures are adorned with calcium oxalate crystals or crystal-bearing spines. The sporangiola of *Cunninghamella* and *Hesseltinella* bear long acicular spines (O'Donnell 1979). The distribution and type of calcium oxalate deposits may aid in propagule dispersal (Birkby and Preece 1988).

VI. Classification

A. Phylum, Subphyla, Classes, and Orders

The zygomycetous fungi, as treated by Benjamin (1979), are defined as having a thallus consisting of aseptate or regularly septate hyphae, yeast cells, hyphal bodies, or protoplasts. Asexual reproduction is by the production of sporangiospores in sporangia, sporangiola, or merosporangia, or chlamydospores, arthrospores, or conidia. Sexual reproduction, where known, is by zygosporogenesis, although some taxa form azygospores.

Zygospores with opposed suspensors usually form at or above the substrate surface, whereas those with apposed or undifferentiated (hyphoid) suspensors occur at or below the substrate surface. Taxa where zygospores have never been observed were classified by their vegetative and asexual reproductive characters (O'Donnell 1979).

Zygospores form after the hormonally mediated fusion of two gametangia that arise, in turn, from undifferentiated vegetative cells, hyphae, or differentiated hyphae called zygophores. The majority of Mucorales are heterothallic (Schipper and Stalpers 1980) and form zygospores only after two separate thalli (designated + and −) are crossed. Most other orders have a high percentage of homothallic species showing no discernible mating types and, therefore, do not require outcrossing. All zygosporogenesis in Entomophthorales members is homothallic. Azygospores (sometimes referred to as parthenospores) are zygosporelike spores formed without a prior gametangial conjugation. Azygospores have been reported in Mucorales (Benjamin and Mehrotra 1963), and development has been studied ultrastructurally in two species of *Mucor* (Ginman and Young 1989).

B. Ordinal Distribution

Based on differences in nutrition and vegetative and reproductive morphology, ten orders of the zygomycotan fungi are recognized and discussed here (Asellariales, Basidiobolales, Dimargaritales, Endogonales, Entomophthorales, Kickxellales, Mortierellales, Mucorales, Neozygitales, and Zoopagales). Based on phylogenetic analysis, these orders are distributed in one phylum, Entomophthoromycota, and four subphyla, Kickxellomycotina, Mortierellomycotina, Mucoromycotina, and Zoopagomycotina (Hibbett et al. 2007; Hoffmann et al. 2011; Humber 2012b) (Table 8.1).

C. Phylum Entomophthoromycota (Fig. 8.1)

The members of Entomophthoromycota are saprobes or arthopod pathogens that form simple or branched conidiophores, each of which apically produces a single primary conidium. Primary conidia are composed of wall layers that are continuous with those of the conidiophore and are usually forcibly discharged; if they land on an unfavorable substrate, then a secondary conidium can be produced. Secondary conidia can be actively or passively discharged and may or may not have the same shape as the primary conidium. Resting spores have a two-layered, relatively thick wall, and each contains two to many nuclei. Resting spores can be formed as a result of conjugation (zygospore) or without conjugation (azygospore). Arthropod pathogens may produce rhizoids that may or may not possess holdfasts to attach the infected host to the substrate and cystidia to break the host cuticle to facilitate conidiophore emergence.

1. Morphology

a) Vegetative Structures

The vegetative structures of most pathogenic taxa in Entomophthoromycota are short segments (hyphal bodies), whereas coenocytic to sparingly septate hyphae are more characteristic of saprobic taxa in this phylum. Multinucleate hyphal bodies may be walled or, in many entomopathogens, wall-less protoplasts whose morphology may rapidly change in shape but are generally fusoid or in highly irregular (or even beadlike) filaments with prominent filopodia (Tyrrell 1977; Macleod et al. 1980) that are immobile and hyphoid (Butt et al. 1981). Septate hyphae with uninucleate cells occur only in *Basidiobolus*.

b) Asexual Spores (Conidia)

Entomophthoralean conidia [Fig. 8.1(5–8)] are covered by the same outer electron-dense and inner electron-lucent wall layers covering the conidiogenous cell [Fig. 8.1(4)]. The spores are cut off by a centripetal infolding of the inner (electron-lucent) wall layer, resulting in a bilayered septum composed of two electron-lucent layers; no new internal (sporangiospore) wall is formed at any time during spore formation (Benny et al. 2001). Conidial discharge occurs when the outer (electron-dense) layer breaks at the septum, and in most taxa, hydrostatic pressure in the conidium forces the sudden eversion of the electron-lucent septal wall layer. The often-visible line of demarcation between the conidial body and the papilla (the everted septum) is the margin of the outer wall layer. In the bitunicate conidia [Fig. 8.1(8)] of species of *Erynia*, *Furia*, *Pandora*, *Strongwellsea*, and *Zoophthora*, the outer wall layer may separate from the spore surface in liquid and give a false impression that the spores are monosporic sporangioles; conidial wall layers of the unitunicate conidia [Fig. 8.1(5, 6)] in all other genera do not separate (Remaudière and Hennebert 1980).

Secondary conidia may be formed by species in all genera, except *Massospora*, when conidia land on substrates unsuitable for initiating germ tubes. The morphological types of secondary conidia are discussed in detail by Ben-Ze'ev and Kenneth (1982). Forcibly discharged secondary conidia are uniformly shot off by papillar eversion regardless of the mode of discharge for the primary conidia. Forcibly discharged secondary conidia in genera that may produce passively discharged capilliconidia [Fig. 8.1(9), *Zoophthora radicans*] may form forcibly discharged tertiary conidia or passively dispersed tertiary capilliconidia. Resporulation by passively dispersed capilliconidia is always by production of another capillary conidiophore and capilliconidium. The genus *Orthomyces* (Steinkraus et al. 1998) was described as differing from *Zoophthora* in part because the secondary capilliconidia of *Orthomyces aleyrodisone* are

Table 8.1 Synopsis of the classification of Zygomycotan fungi[a]

ENTOMOPHTHOROMYCOTA Humber
Basidiobolomycetes Humber
Basidiobolales Caval.-Sm
Basidiobolaceae Engl. & E. Gilg
Basidiobolus Eidam, Schizangiella and Drechslerosporium (not formally described)
Entomophthoromycetes Humber
Entomophthorales G. Winter
= Ancylistales J. Schröter [as Ancylistineae]
Ancylistaceae J. Schröt.
Ancylistes Pfitzer, *Conidiobolus* Bref.
Macrobiophthora Reukauf emend. B.E. Tucker
Completoriaceae Humber
Completoria Lohde
Entomophthoraceae Nowakowski
Entomophthoroideae S. Keller, = Massosporoideae S. Keller
Batkoa Humber
Entomophaga Batko emend. Humber
Entomophthora Fresenius
Genus incertae sedis *Eryniopsis* Humber
Erynioideae S. Keller
Erynia (Nowak. ex A. Batko) Remaud. & Hennebert emend. Humber
Furia (A. Batko) Humber
Massospora Peck emend. R.S. Soper
Orthomyces Steinkr., Humber & J.B. Oliv.
Pandora Humber
Strongwellsea A. Batko & J. Weiser emend. Humber
Zoophthora A. Batko emend. Ben-Ze'ev & Kenneth
Meristacraceae Humber
Meristacrum Drechsler emend. B.E. Tucker & Humber
Tabanomyces Couch, R.V. Andrejeva, Laird & Nolan
Genus incertae sedis—*Tarichium* Cohn
Ballocephala Drechsler and *Zygnemomyces* Miura transferred to the Kickxellomycotina
Neozygitomycetes Humber
Neozygitales Humber
Neozygitaceae I. Ben-Ze'ev, R.G Kenneth & Uziel
Apterivorax S. Keller
Neozygites Witlaczil, T haxterosporium Ben-Ze'ev & R.G. Kenneth

Classification of remaining zygomycotan fungi to subphylum, order, family and genus

KICKXELLOMYCOTINA Benny
Asellariales Manier ex Manier & Lichtw
Asellariaceae Manier ex Manier & Lichtw
Asellaria R.A. Poiss., *Orchesellaria* Manier ex Manier & Lichtw
Genus of unknown affinity
Baltomyces Cafaro emend. Oman & M.M. White
Dimargaritales R.K. Benj
Dimargaritaceae R.K. Benj
Dimargaris Tiegh., *Dispira* Tiegh
Tieghemiomyces R.K. Benj
Genus of unknown affinity
Spinalia Vuil
Harpellales Lichtw. & Manier
Harpellaceae L. Lger & Duboscq ex P.M. Kirk & P.F. Cannon
Carouxella Manier, J.-A. Rioux & Whisler ex Manier, J.-A. Roux & Lichtw.
Harpella L. Léger & O. Duboscq
Harpellomyces Lichtw. & S.T. Moss emend Lichtw., M.M. White and Colbo
Stachylina L. Léger & M. Gauthier
Stachylinoides Lichw. & López-Lastra
Legeriomycetaceae Pouzar
= Genestellaceae L. Léger & M. Gauthier
Allantomyces M.C. Williams & Lichw.
Legeriomyces Pouzar, = *Genistella* L. Léger & M. Gauthier
Austrosmittium Lichtw. & M.C. Williams
Bactromyces R.T. William & Strongman
Baetimyces L.G. Valle & Santam
Barbatospora M.M. White, Siri & Lichtw
Bojamyces Longcore emend. L.G. Valle & Santam., *Capniomyces* S.W. Peterson & Lichtw
Caudomyces Lichtw., Kobayasi & Inhoh
Coleopteromyces Ferrington, Lichw. & López-Lastra
Dacrodiomyces Lichtw.
Ejectosporus S.W. Peterson, Lichtw. & M.C. Williams emend. Strongman
Ephemerellomyces M.M. White & Lichtw
Furculomyces Lichtw. & M.C. Williams
Gauthieromyces Lichtw
Genistelloides S.W. Peterson, Lichtw. & B.W. Horn
Genistellospora Lichtw
Glotzia M. Gauthier ex Manier & Lichtw.
Graminella L. Lger & M. Gauthier ex Manier
Graminelloides Lichtw.
Klastostachys Lichtw., M.C. Williams & M.M. White
Laculus R.T. William & Strongman
Lancisporomyces Santam., *Legeriodes* M.M. White
Endogonaceae Paoletti emend. J.B. Morton & Benny

(continued)

Table 8.1 (continued)

Legeriosimilis M.C. Williams, Lichtw., M.M. White & J.K. Misra,
Orphella L. Léger & M. Gauthier emend. Santam. & Girbal
Pennella Manier ex Manier
Plecopteromyces Lichtw., Ferrington & López-Lastra
Pseudoharpella Ferrington, M.M. White & Lichtw.
Pteromaktron Whisler
Simuliomyces Lichtw.
Sinotrichium Juan Wang, S.Q. Xu & Strongman
Smittium R.A. Poiss.
Spartiella Tuzet & Manier ex Manier
Stipella L. Léger & M. Gauthier
Tectimyces L.G. Valle & Santam
Trichozygospora Lichtw.
Trifoliellum Strongman & M.M. White
Zancudomyces Yan Wang, Tretter, Lichtw. & M.M. White
Zygopolaris S.T. Moss, Lichtw. & Manier
Kickxellales Kreisel ex R.K. Benj.
Kickxellaceae Linder
Coemansia Tiegh. & G. Le Monn.
Dipsacomyces R.K. Benj.
Kickxella Coem
Linderina Raper & Fennell
Martensella Coem
Martensiomyces Meyer
Mycoëmilia Kurihara, Degawa & Tokum.
Myconymphaea Kurihara, Degawa & Tokum.
Pinnaticoemansia Kurihara & Degawa
Ramicandelaber Y. Ogawa, S. Hayashi, Degawa & Yaguchi
Spirodactylon R.K. Benj., *Spiromyces* R.K. Benj
Possible members of the Kickxellomycotina
Ballocephala Drechsler
Zygnemomyces Miura
MORTIERELLOMYCOTINA Kersten Hoffm., K. Voigt & P.M. Kirk
Mortierellales Caval.-Sm.
Mortierellaceae A. Fischer
Aquamortierella Embree & Indoh, *Dissophora* Thaxt., *Echinochlamydosporium* X.Z. Jiang X.Y. Liu, Xing Z. Liu, *Gamsiella* (R.K. Benj.) Benny & M. Blackw., *Lobosporangium* M. Blackw. & Benny, = *Echinosporangium* Malloch, non *Echinosporangium* Kylin
Modicella Kanouse, *Mortierella* Coemans, = *Haplosporangium* Thaxt., = *Azygozygum* Chesters, = *Actinomortierella* Chalab.
Genus of unknown affinity *Nothadelphia* Degawa & W. Gams
MUCOROMYCOTINA Benny
Endogonales Moreau ex R.K. Benj. emend. Morton & Benny
Endogone Link,
Peridiospora C.G. Wu & J. Lin,
Sclerogone Warcup,
Youngiomyces Y.J. Yao.
Mucorales Fr.
Backusellaceae K. Voigt & P.M. Kirk
Backusella Hesselt. & J.J. Ellis
Choanephoraceae J. Schröt., ≡ Gilbertellaceae Benny
Choanephoroideae K. Voigt & P.M. Kirk
Blakeslea Thaxt.
Choanephora Curr.
Poitrasia P.M. Kirk = *Abradeosporangium* Subrahm. & Swathi Sri
Gilbertelloideae K. Voigt & P.M. Kirk
Gilbertella Hesselt.
Cunninghamellaceae R.K. Benj. emend. Benny, R.K. Benj. & P.M. Kirk, ≡ Absidiaceae Arx
Cunninghamelloideae K. Voigt & P.M. Kirk
Cunninghamella Matr.
Absidioideae K. Voigt & P.M. Kirk
Absidia Tiegh. s.s., = *Tieghemella* Berl. & De Toni, = *Proabsidia* Vuill.,
Chlamydoabsidia Hesselt. & J.J. Ellis,
Gongronella Ribaldi,
Halteromyces Shipton & Schipper, *Hesseltinella* H.P. Upadhyay,
Lentamycetaceae K. Voigt & P.M. Kirk,
Lentamyces Kerst. Hoffm. & K. Voigt,
Lichtheimiaceae Kerst. Hoffm., G. Walter & K. Voigt
Dichotomocladioideae K. Voigt & P.M. Kirk
Dichotomocladium Benny & R.K. Benj.
Lichtheimioideae K. Voigt & P.M. Kirk
Lichtheimia Vuill.
Rhizomucoroideae K. Voigt & P.M. Kirk
Rhizomucor Lucet & Costanin,
Thermomucor Subrahm., B.S.Mehrotra & Thirum.
Mucoraceae Dumort., ≡ Chaetocladiaceae A. Fisch., ≡ Thamnidiaceae Fitzp., ≡ Dicranophoraceae J.H. Mirza
Dicranophoroideae K. Voigt & P.M. Kirk
Dicranophora J. Schröt.
Chaetocladioideae K. Voigt & P.M. Kirk
Chaetocladium Fresen.
Mucoroideae K. Voigt & P.M. Kirk
Actinomucor Schostak., = *Glomerula* Bainier
Circinomucor Arx
Ellisomyces Benny & R.K. Benj.
Helicostylum Corda emend. Benny
Hyphomucor Schipper & Lunn
?*Isomucor* J.I. Souza, Pires-Zottar. & Harakava
Mucor Fresen., ≡ ?*Zygorhynchus* Vuill.
Parasitella Bainier
Pilaira Tiegh.
Pirella Bainier

(continued)

Table 8.1 (continued)

Mycocladaceae K. Hoffmann, S. Discher & K. Voigt (family invalid if genus invalid)
Mycocladus Beauverie (genus may be invalid)
Mycotyphaceae Benny & R.K. Benj
Cokeromycetoideae K. Voigt & P.M. Kirk
Benjaminiella Arx
Cokeromyces Shanor
Kirkomycetoideae K. Voigt & P.M. Kirk
Kirkomyces Benny, = *Kirkia* Benny, non *Kirkia* Oliv.
Mycotyphoideae K. Voigt & P.M. Kirk
Mycotypha Fenner
Phycomycetaceae Arx
Phycomyces Kunze
Spinellus Tiegh.
Pilobolaceae Corda
Pilobolus Tode,
Utharomyces Boedijn
Radiomycetaceae Hesseltine & J.J. Ellis
Radiomyces Embree
Rhizopodaceae K. Voigt & P.M. Kirk
Amylomyces Calmette,
Rhizopus Ehrenb.,
Sporodiniella Boedijn,
Syzygites Ehrenb.
Saksenaeaceae Hesselt. & J.J Ellis
Apophysomyces P.C. Misra, *Saksenaea* S.B. Saksena
Syncephalastraceae Naumov ex R.K. Benj.
Circinella Tiegh. & G. Le Monn.,
Fennellomyces Benny & R.K. Benj.,
Phascolomyces Boedijn,
Protomycocladus Schipper & Samson,
Syncephalastrum J. Schröt.,
Thamnostylum Arx & H.P. Upadhyay,
Zychaea Benny & R.K. Benj.,
Umbelopsidaceae W Gams & W. Meyer
Umbelopsis R.E. Amos & H.L. Barnett, = *Micromucor* (W. Gams) Arx
Thamnidium Link, non *Thamnidium* Tuck. ex Schwend
?*Zygorhynchus* Vuill
Genera of unknown family affiliations
Calcarisporiella de Hoog
?*Isomucor* J.I. Souza, Pires-Zottar. & Harakava
Rhizopodopsis Boedijn
Siepmannia Kwaśna & Nirenberg ex Nirenberg & Kwaśna
ZOOPAGOMYCOTINA Benny
Zoopagales Bessey ex R.K. Benj.
Cochlonemataceae Dudd.
Amoebophilus P.A. Dang.
Aplectosoma Drechsler
Bdellospora Drechsler,
Cochlonema Drechsler,
Endocochlus Drechsler,
Euryancale Drechsler,
Possible member of the family *Aenigmatomyces* R.F. Castañeda & W.B. Kendr.
Helicocephalidaceae Boedijn
Brachymyces G.L. Barron,
Helicocephalum Thaxt.,
Rhopalomyces Corda
Piptocephalidaceae J Schröt.
Kuzuhaea R.K. Benj.,
Piptocephalis de Bary,
Syncephalis Tiegh. & G. Le Monn.
Sigmoideomycetaceae Benny, R.K. Benj. & P.M. Kirk
Reticulocephalis Benny, R.K. Benj. & P.M. Kirk,
Sigmoideomyces Thaxt.,
Thamnocephalis Blakeslee
Zoopagaceae Drechsler emend. Dudd.
Acaulopage Drechsler,
Cystopage Drechsler,
Stylopage Drechsler,
Zoopage Drechsler,
Zoophagus Sommerst. emend. M.W. Dick

[a]Kirk et al. (28),synopsis of taxa; Benny (2012),synopsis of taxa annotated with citations of important literature; Voigt (212), classification of the zygomycota; Mucorales revised to family and subfamily (See Mycobank and the CABI Bioscience Databases (http://www.indexfungorum.org) Index Fungorum for a list of species, with literature citations, for each accepted genus)

globose rather than elongate and have a distinct papilla that may be large enough to cause at least the active dislodgement of the capilliconidia, if not their forcible discharge.

c) Auxiliary Hyphal Structures

Many entomopathogenic species produce cystidia [Fig. 8.1(1)] or rhizoids [Fig. 8.1(3)]. These auxiliary cells are among the first fungal cells to emerge from a diseased host and serve, respectively, to perforate the exoskeleton (thereby aiding the emergence of growing conidiophores) or to anchor the host to the substrate. Cystidia usually project well above the conidiogenous hymenium on the cuticle of the diseased host; these cells may collapse and become indistinguishable during the later stages of conidiogenesis and discharge. Rhizoids may or may not have differentiated terminal holdfasts [Fig. 8.1(2)]. Rhizoids and

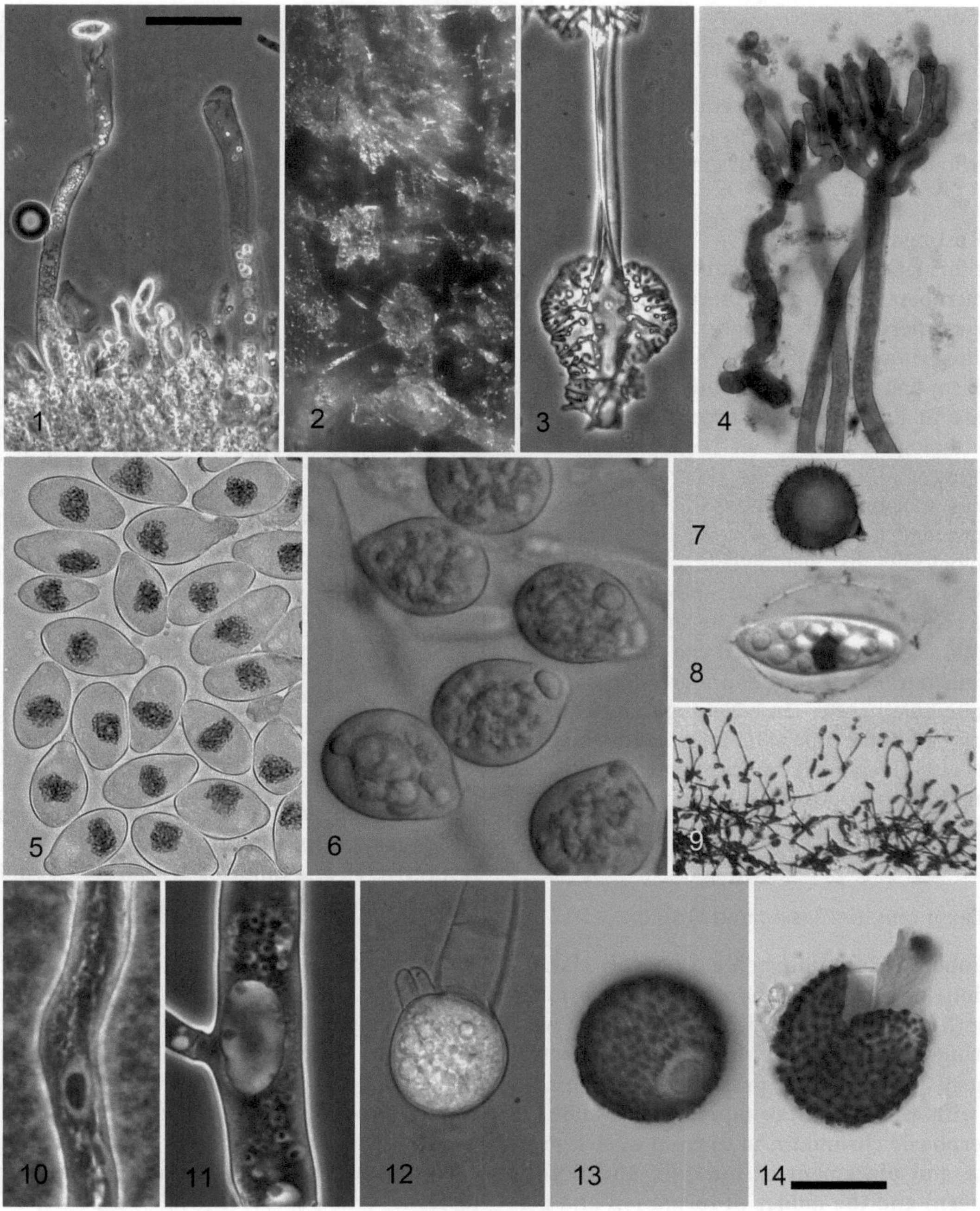

Fig. 8.1 Representatives of two classes of Entomophthoromycota showing some selected morphological features. **10, 12.** Basidiobolomycetes. **1–9, 11, 13, 14.** Entomophthoromycetes. **1.** Cystidia of *Erynia sepulchralis* projecting aforementioned conidiogenous hymenium. **2.** Discoid holdfasts of *Pandora neoaphidis* attach aphids to plant surfaces. **3.** Rhizoid and terminal holdfast of *Pandora ithacensis*. **4.** Digitately branched conidiophores of *Zoophthora radicans*. **5.** Primary conidia of *Furia gastropachae* with highly granular nuclear contents stained by aceto-orcein. **6.** Primary conidia of *Conidiobolus obscurus*. **7.** Villose conidium of *Conidiobolus coronatus* seems to function like more typical resting spore. **8.** Bitunicate conidium of *Erynia sepulchralis* has a separable outer wall layer. **9.** Secondary capilliconidia on capillary secondary conidiophores of *Zoophthora radicans*. **10.** Nucleus of *Basidiobolus* sp. with a large central nucleolus. **11.** Multiple small nuclei of *Conidiobolus coronatus* flanking a vacuole resemble those of *Basidiobolus* in having clear nucleoplasm and a

cystidia are morphologically and developmentally distinct cell types (Brobyn and Wilding 1977) that may occur together or independently in a species. Rhizoids or cystidia are not formed by many entomopathogens or, apparently, any saprobic Entomophthorales.

d) Resting Spores

The formation of resting spores [Fig. 8.1(12–14)] in members of Entomophthoromycota, whether as zygospores or azygospores, is strictly homothallic (Humber 1989); no evidence for heterothallism has been found in this order. Research needs to be done to distinguish between the morphological [zygospores (Fig. 8.1(12)) versus azygospores] and genetic (karyogamic/meiotic versus apogamic) definitions of sex. While karyogamy and meiosis have never been confirmed cytologically in zygomycotan fungi, resting spores in most families of Entomophthoromycota are initially multinucleate and either remain multinucleate or are reduced to a binucleate state in fully dormant mature spores; resting spores in Basidiobolomycetes and Neozygitomycetes are binucleate at the time of their formation. McCabe et al. (1984) regarded binucleate resting spores as being likely to undergo karyogamy and meiosis during germination.

2. Taxonomy to Classes and Orders

Humber (1989) published a six-family classification of Entomophthoromycota (as Entomopthorales), in which the taxa were distinguished primarily by nuclear characters [Fig. 8.1(10, 11)] (size, presence, and stainability with aceto-orcein or other nuclear stains of interphasic chromatin in unfixed nuclei, presence and placement of nucleoli, and mitotic pattern) and the modes of formation and germination of resting spores (Humber 1989). Balazy (1993) only recognized five families and a generic taxonomy closer to that originally outlined by Batko (1964a, b, c, d).

All genera in Entomophthoraceae (Entomophthoromycetes) **and Neozygitaceae** (Neozygitomycetes) **are obligate pathogens of insects and allied arthropods**; the latter taxa are also distinguished by pigmentation. The monotypic family **Completoriaceae** (***Completoria complens*** **Lohde) parasitizes fern prothallia.** Members of **Meristacraceae, distinguished by erect conidiophores bearing several lateral conidia, parasitize nematodes, tardigrades, and, in one reported instance, insects** (Couch et al. 1979). Two families, Ancylistaceae (Entomophthoromycetes) and Basidiobolaceae (Basidiobolomycetes), are primarily saprobic in plant detritus, organic litter, and the dung of cold-blooded vertebrates. ***Ancylistes*** **is obligately parasitic on species of the desmid alga** ***Closterium*** **Nitzsch ex Ralfs.** ***Basidiobolus*** **or** ***Conidiobolus*** **species may be obligately or facultatively zoophilic (and may cause mycoses in vertebrates).**

Generic characters distinguishing the genera of Entomophthoromycota include (1) conidial characters such as shape, nuclear number, and bi- or unitunicate status (with a separable or inseparable outer wall layer, respectively), and mode of discharge; (2) conidiophore characters such as branching (simple, branched, or an erect multisporic axis); and (3) general habit and host range (Humber 1981, 1989, 1997a). These primary generic characters may be augmented with secondary characters, such as the absence or presence (and morphology) of rhizoids and cystidia, types of secondary conidia produced, morphological characters of the resting spores, vegetative characters, especially whether vegetative cells are walled or protoplastic, and pathobiological characters, such

Fig. 8.1 (continued) central nucleolus. **12.** Immature zygospore of *Basidiobolus ranarum* showing beaklike projections with remnants of prezygotic mitotic nuclei. **13.** Mature resting spore of *Tarichium* sp. from tipulid fly with a dark, roughened outer wall showing clear fenestra where spore was attached to parental cell during development. **14.** Mature resting spore, as in **13**, cracked open to show smooth, uncolored inner wall layer and projecting cytoplasm, including a stained nucleus. Scale bars in **1** and **14** are the same size—bars give magnification for each figure. **1, 9.** *Bar* 200 μm. **2.** *Bars* 150 μm. **3.** *Bars* 75 μm. **4.** *Bars* 85 μm. **5, 6, 8, 10, 11.** *Bars* 50 μm. **7.** *Bars* 30 μm. **12, 13.** *Bars* 70 μm. **14.** *Bars* 80 μm

as host range and nature of fungal interaction with the host (Humber 1989).

a) Basidiobolomycetes and Basidiobolales

The members of Basidiobolales are either saprobes or they can cause basidiomycoses in humans. The vegetative mycelium is septate and uninucleate [Fig. 8.1(10)], or cells can be produced that are yeastlike. Each nucleus is at least 10 μm long, but it can be longer, and the nucleolus is prominent. The conidiophores are not branched, and each bears a subsporangial vesicle. Each conidium is uninucleate and produces a two-layered wall, but the layers are not separable. The conidia are actively discharged by a rocketlike mechanism upon the rupture of a circumscissile weakening in the wall of a turgid subconidial swelling. The mature resting spores [Fig. 8.1(12), zygospores] have two nuclei, one from each of the conjugating cells.

Basidiobolales was proposed (Cavalier-Smith 1998) for members of the genus *Basidiobolus*. These fungi form a spindle-pole body (SPB) (see McLaughlin et al. 2014) or a nucleus-associated organelle according to McKerracher and Heath (1985). Each SPB is composed of 11–12 singlet microtubules and is a centriole, according to Cavalier-Smith (1998). The centrioles and kinetosomes of all eukaryotes with flagella have microtubules arranged in a 9 × 3 pattern. The SPB is not part of a flagellum and, therefore, is not a centriole, according to Humber (2012b).

These fungi form forcibly released conidia and protuberances with terminal cells (Benjamin 1962; Drechsler 1956). Early phylogenetic studies (Jensen et al. 1998; Nagahama et al. 1995) indicated that *B. ranarum* Eidam is the sister species to *Olpidium brassicae* (Woronin) P.A. Dang. *B. ranarum* and other taxa in Basidiobolales are members of Entomophthoromycota based on biology and the latest phylogenetic studies (Gryganskyi et al. 2012, 2013; Humber 2012b).

b) Entomophthoromycetes and Entomophthorales s.s.

Vegetatively the members of Entomophthorales sensu Humber (2012b) can produce hyphal bodies (arthropod parasites) or coenocytic mycelium (saprobic species of *Conidiobolus*) [Fig. 8.1(1)]. Simple or branched conidiophores [Fig. 8.1(4)] are produced that bear an apical conidiogenous cell (Ancylistaceae, Completoriaceae, Entomophthoraceae), or the fertile branch may be erect and septate, and one propagule is produced by each cell (Meristacraceae). The primary conidia are either uni- or bitunicate [Fig. 8.1(5–8)] and are usually actively discharged by papillar eversion. The secondary conidia, if formed, can be either actively or passively discharged. The interphase nuclear appearance varies depending on the specific taxon examined. The resting spores [Fig. 8.1(13, 14)] can be either zygospores or azygospores.

i. Entomophthorales Sensu Humber (2012a) and Four Families (Ancylistaceae, Completoriaceae, Entomophthoraceae, Meristacraceae)

Many members of Entomophthoromycetes (Entomophthorales sensu Humber 2012b), or Entomophthorales without the Basidiobolaceae and Neozygitaceae of Humber (1989), are obligate parasites of various animals (e.g., insects, nematodes, mites). *Conidiobolus* is mainly saprobic in soil, detritus, and dung, although species in this genus may be pathogens of insects or cause mycoses in humans or other mammals. *Ancylistes* and *Completoria* are parasites of desmid algae and fern prothallia, respectively. Species of *Macrobiophthora* and *Meristacrum* are parasites of nonarthropodous invertebrates such as nematodes and tardigrades (Tucker 1981). *Tabanomyces* (Couch et al. 1979) is a member of Meristacraceae (Humber 2012b).

Entomophthoralean pathogens (especially *Erynia* spp.) affecting insects in very wet environments, for example, ovipositing blackflies (Diptera: Simuliidae), may become wetted during the development of primary or secondary conidia and then may produce "tetraradiate" spores bearing several blunt branches. Such spores are passively rather than actively dispersed (Descals and Webster 1984).

The members of Entomophthorales are best known from temperate climates to the semiarid tropics; these fungi are relatively little collected in forests of the humid tropics (Evans 1982), although this paucity of collections does not mean that they are rare in such habitats. Entomopathogenic entomophthoraleans are easily seen and collected, and they are much more widely collected and described worldwide than the "nonentomogenous taxa (Tucker 1981) whose microscopic hosts occur in litter or soil habitats. . . ."

Entomophthoralean asexual spores differ markedly from those of other zygomycotan fungi in that they are true conidia (not sporangiospores!) formed by blastosporogenesis, are forcibly discharged in all genera, except *Massospora*, and tend strongly to form a variety of secondary conidia. Ultrastructural confirmation of the conidial nature of these spores is available for taxa in several families of this order (Dykstra 1994; Eilenberg et al. 1986, 1995; Latgé et al. 1989). Ultrastructural studies of some taxa in Zoopagales demonstrate that their spores also are conidia (Saikawa 2011b; Saikawa and Katsurashima 1993; Saikawa and Sato 1991). Entomophthoralean fungi actively discharge their conidia [Fig. 8.1(5–8)] using several diverse mechanisms (Humber 1989). Turgid cells round off in most genera to cause the sudden eversion of a papilla (Couch 1939). The conidia of *Entomophthora* spp. discharge in a cannonlike manner when the conidiogenous cell apex ruptures, although Eilenberg et al. (1986) proposed an alternative explanation of this discharge mechanism. *Massospora* is the only genus that forms conidia that are passively rather than actively dispersed.

Nearly all genera of the order Entomophthorales s.s. sensu Humber (2012b), Entomophthoromycetes, form one or more types of secondary conidia when the primary conidia land on nonnutritive or other unsuitable substrates. Secondary conidia can be the major infective units for some of the entomopathogenic taxa (e.g., *Entomophthora* spp.). Humber (1981), King and Humber (1981), and Ben-Ze'ev and Kenneth (1982) discuss the types of secondary conidia formed; these include one or more forcibly discharged secondary conidia or passively dispersed conidia at the apex of long, capillarylike secondary conidiophores.

Resting spores [Fig. 8.1(13, 14)], zygospores, and azygospores are thick-walled and form between conjugating cells (gametangia) of different or the same hyphae or hyphal bodies; they are considered zygospores. Spores formed without prior gametangial conjugations of hyphae or hyphal bodies are azygospores. Evidence suggests that karyogamy and meiosis, if they occur in a species, take place in the resting spores regardless of their zygosporic or azygosporic origins (McCabe et al. 1984). Illustrations of the characters used in the identification of the entomophthoralean fungi can be found in Thaxter (1888), Balazy (1993), Humber (1997a), Keller and Petrini (2005), and Gryganskyi et al. (2012, 2013).

c) Neozygitomycetes and Neozygitales

The members of Neozygitomycetes (Humber 2012b) **produce hyphal bodies that contain 3–5, usually four, nuclei in the vegetative cells. The conidiophore is not branched, a single conidium is produced, and the primary conidia are subglobose to broadly ovoid and have a short basal papillum. Conidial discharge is active by means of papillar eversion, and secondary conidia are rapidly produced. Each resting spore is produced on a relatively short conjugation bridge that arises from the gametangia that are rounded hyphal bodies. All portions of Neozygitomycetes possess melanin pigments; the conidia are smoky gray, and the resting spores are dark gray to black. The formation of melanized structures is a cardinal feature of Neozygitomycetes.** The members of Neozygitomycetes were not proposed based on molecular evidence but because of the production of pigmentation.

D. Nonentomophthoralean Subphyla (Kickxellomycotina, Mortierellomycotina, Mucoromycotina, Zoopagomycotina)

Asexual reproduction in all zygomycete orders is by nonmotile, single-celled endospores formed as conidia or in sporangia, sporangiola, or merosporangia, and by the formation of chlamydospores, arthrospores, or yeast cells (Benjamin 1979) (Figs. 8.2, 8.3, 8.4, and 8.5).

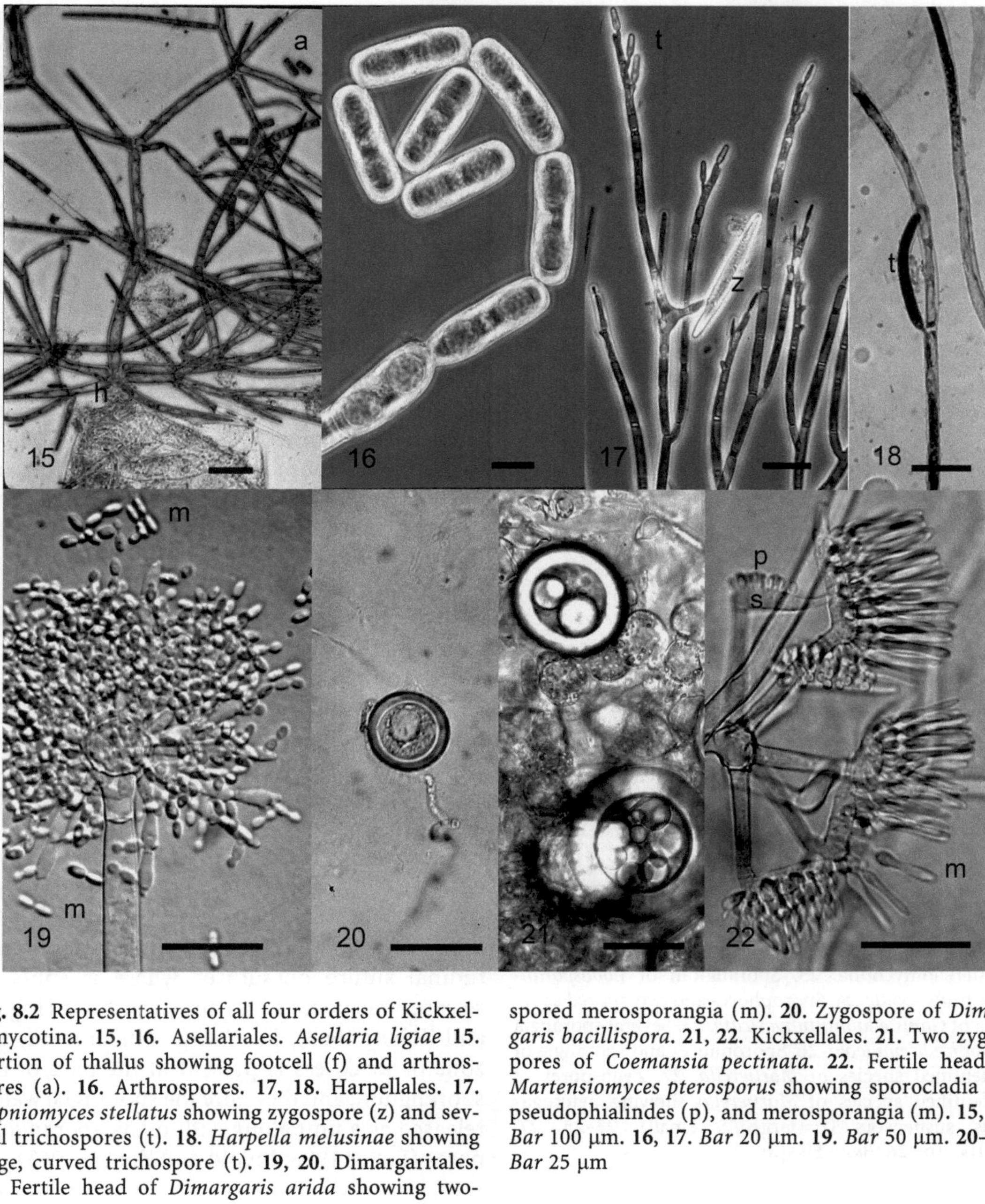

Fig. 8.2 Representatives of all four orders of Kickxellomycotina. **15, 16.** Asellariales. *Asellaria ligiae* **15.** Portion of thallus showing footcell (f) and arthrospores (a). **16.** Arthrospores. **17, 18.** Harpellales. **17.** *Capniomyces stellatus* showing zygospore (z) and several trichospores (t). **18.** *Harpella melusinae* showing large, curved trichospore (t). **19, 20.** Dimargaritales. **19.** Fertile head of *Dimargaris arida* showing two-spored merosporangia (m). **20.** Zygospore of *Dimargaris bacillispora*. **21, 22.** Kickxellales. **21.** Two zygospores of *Coemansia pectinata*. **22.** Fertile head of *Martensiomyces pterosporus* showing sporocladia (s), pseudophialindes (p), and merosporangia (m). **15, 18.** *Bar* 100 μm. **16, 17.** *Bar* 20 μm. **19.** *Bar* 50 μm. **20–22.** *Bar* 25 μm

1. Mitospores

a) Sporangiospores

The primary mode of asexual reproduction in zygomycotan fungi is the unicelled sporangiospore. In multispored sporangia, sporangiola, and some merosporangia (Mucorales), spores are delimited from the cytoplasm by cleavage vesicle fusion. In other merosporangia (Zoopagales), spores are formed as a result of the fusion of the invaginating plasmalemma.

Sporangiospores are unicelled, uni-, bi-, or multinucleate, and often are smooth-walled, although some taxa have external spines, warts, or striations, and many members of the Kickxellales have spines embedded in their walls. The majority of species produce a hyaline spore, but a few fungi have pigmented spores.

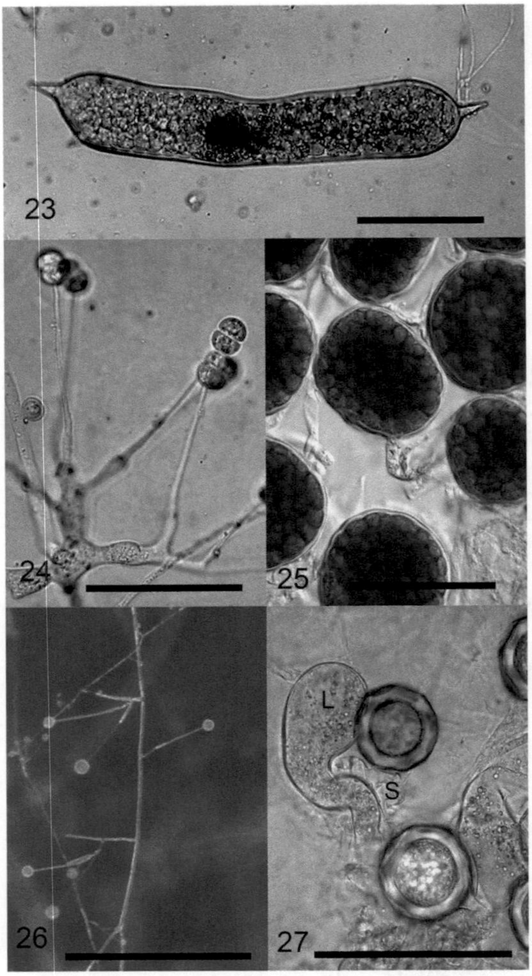

Fig. 8.3 Representatives of only order (Mortierellales) of Mortierellomycotina. **23.** Sporangium of *Lobosporangium transversale*. **24.** Sporophores and two-spored sporangia of *Gamsiella multidivericata*. **25.** Several sporangia of *Modicella malleola*. **26.** Sporophores and sporangia of unidentified species of *Mortierella*. **27.** Zygospores of unidentified species of *Mortierella* showing heterogamous suspensors (L=large, S=small). **23–25, 27.** *Bar* 100 μm. **26.** *Bar*=500 μm

Spores of members of Choanephoraceae bear hyaline appendages; the ultrastructure of sporangiospore ontogeny was published for *Zygorhynchus heterogamus* (Vuill.) Vuill. (Edelmann and Komparens 1994).

b) "Conidia" Versus True Conidia

The term *conidium* has often been applied to the unispored propagules of several taxa in the Mucorales, including *Chaetocladium*, *Choanephora*, and *Cunninghamella*. **These so-called conidia, however, are unispored sporangiola in which a sporangiospore wall is deposited inside the sporangiolar wall. True conidia lack the sporangiospore wall and are only formed by members of Zoopagomycotina (in these subphyla).**

2. Endospore-Forming Structures

a) Sporangia

The sporangium is a reproductive structure characteristic of Mortierellales and Mucorales. Sporangia contain between 100 and 100,000 endogenous sporangiospores, depending on the species. The majority of mucoralean fungi produce sporangia that are globose to obpiriform. Sporangia [Fig. 8.4(30, 31, 33)] are formed on the apex of the main sporophore or its branches; a columella is usually present. The region of the sporophore immediately below the sporangium can be nonapophysate (constricted), apophysate (expanded or hemiconical), or vesiculate (expanded and then constricted); these sporangial types are illustrated by Zycha et al. (1969), Benjamin (1979), and O'Donnell (1979). The sporangial wall can be persistent [Fig. 8.4(33)], deliquescent, or evanescent (fugaceous), and it can be variously colored. In Choanephoraceae [Fig. 8.4(33, 34)] the sporangial wall is persistent and has a longitudinal suture or sutures where it splits to release the spores in a droplet of fluid (Ingold and Zoberi 1963). In *Pilaira* (Mucoraceae) the zone of weakness is present at the base of the sporangium, and therefore the sporangial wall is released as a unit along with many sporangiospores (Fuller 1978; Ingold and Zoberi 1963).

b) Sporangiola

The sporangiolum is a small sporangium; it has a persistent wall and is borne on a pedicel or denticle. A pedicel arises from a sporophore or fertile vesicle, whereas a denticle is produced only from a fertile vesicle (Benny 1995; Benny and Benjamin 1975, 1976). Sporangiola are globose to obpiriform, apophysate or nonapophysate, and can contain one [Fig. 8.4(37)]

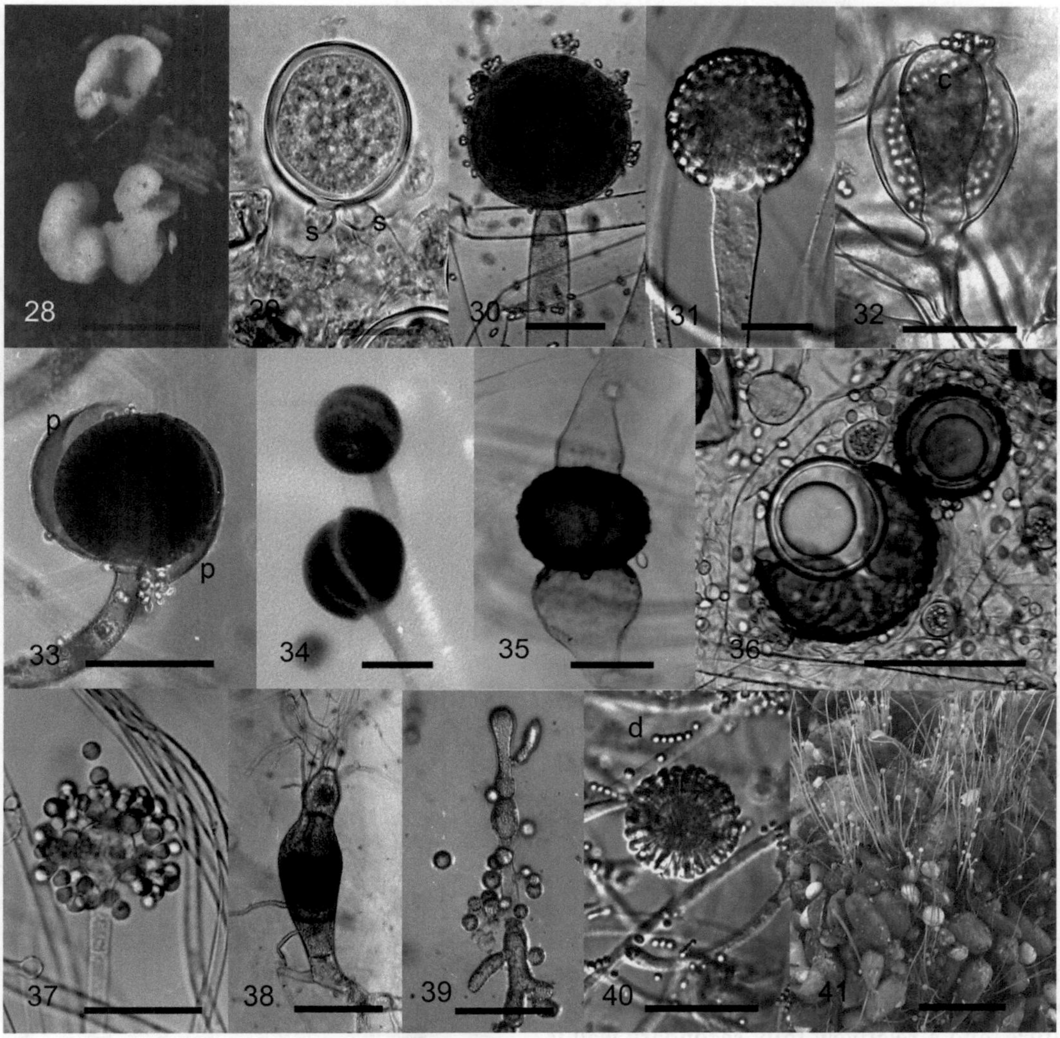

Fig. 8.4 Representatives of two orders of Mucoromycotina. **28, 29.** Endogonales. *Endogone pisiformis*. **28.** Two sporocarps. **29.** Young zygospore with two suspensors (s). **30–41.** Mucorales. **30.** Sporangium of *Pirella naumovii*. **31.** Sporangium with subsporangial sporophore vesicle of *Fennellomyces linderi*. **32.** Sporangiolum of *Pirella circinans* showing columella (c). **33–36.** *Gilbertella persicaria*. **33.** Sporangium in wet mount showing persistent sporangial wall (p), dark spore mass, and subtending sporophore. **34.** Two dry sporangia showing dark wall split open revealing lighter spore mass. **35.** Typical zygospore showing dark zygosporangium and two suspensors. **36.** Two dark zygosporangia with one broken open to reveal zygospore containing a single eccentric globule. **37, 38.** *Phascolomyces articulosus*. **37.** Fertile head showing unispored sporangiola borne on long, straight pedicels. **38.** A typical chlamydospore. **39.** Irregular hypha and globose yeast cells produced in nutrient-rich agar culture media by *Mycotypha africana*. **40.** Fertile head of *Syncephalastrum racemosum* covered with multispored merosporangia and some dehisced (d). **41.** Colony of immature sporophores and sporangia of *Phycomyces blakesleeanus* growing from cake of pressed birdseed. **28.** *Bar*=5 mm. **29.** *Bar* 25 µm. **30, 33, 34, 36, 38, 39.** *Bar* 100 µm. **31, 32, 35, 37, 40.** *Bar* 50 µm. **41.** *Bar* 20 mm

to several hundred spores. *Pirella* spp. (Benny and Schipper 1992) produce sporangiola [Fig. 8.4(32)] that can be the same size as the sporangia.

The columella [Fig. 8.4(32)] of multispored sporangiola can be large and nearly fill the sporangiolum, as in *Pirella circinans* Bain. (Benny and Schipper 1992). The columella in other taxa can be relatively small and hemispherical (often compressed) in *Backusella* and several other thamnidiaceous Mucorales, discoid, or even lacking in the unispored sporangiola of species of *Benjaminiella*, *Chaetocladium*, *Dichotomocladium*, and *Mycotypha* (Benny and Benjamin 1975, 1976; Benny et al. 1985).

c) Merosporangia

Merosporangia are simple or branched endospore-forming structures that contain 1–20 or more spores. The multispored forms are usually more or less cylindrical, whereas unispored merosporangia can be variously shaped. In multispored taxa the spores are **usually borne uniseriately**, except in *Syncephalastrum racemosum* Schroeter, where the occasional spore may be biseriate. The merosporangial wall can be evanescent or persistent. Merosporangia are formed in one family (Syncephalastraceae) of Mucorales [Fig. 8.4(40)] and in Dimargaritales, Kickxellales [Fig. 8.2(19, 22)] (Benjamin 1959), and some members of Zoopagales [Fig. 8.5(44, 47)] (Piptocephalidaceae).

3. Thallospores

a) Arthrospores

Arthrospores are formed singly or in chains from vegetative hyphae with a primary function of dissemination. The cells swell as they mature, and a relatively thin, secondary wall is deposited inside the primary wall. The arthrospores of *Ellisomyces* (Mucorales) are dehisced either rhexolytically or schizolytically (Beakes et al. 1984). The asexual propagules of *Helicocephalum* (Zoopagales) are arthrospores (Barron 1975).

b) Chlamydospores

Chlamydospores are formed in swellings that arise in coenocytic, young hyphae. These spores are borne singly or in chains, intercalary or terminal, often thick-walled cells with perennation as the primary function [Fig. 8.4 (38)] (Griffiths 1974). The ontogeny of chlamydospores has been observed, using light and transmission electron microscopy, in *Gilber-*

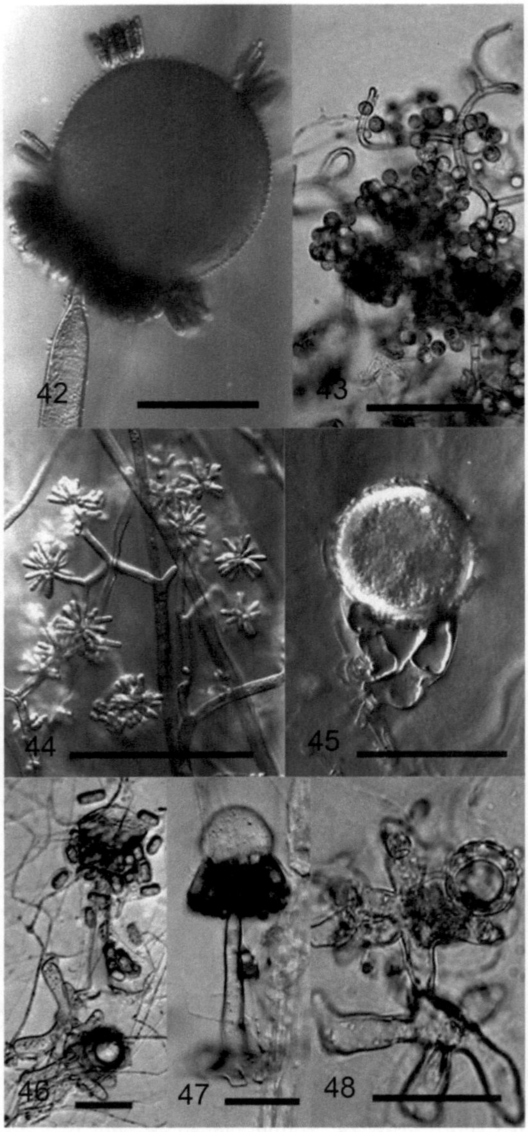

Fig. 8.5 Representatives of only order (Zoopagales) of Zoopagomycotina. **42.** Sporophore and fertile vesicle bearing some pigmented, unispored sporangiola of *Rhopalomyces strangulatus*. **43.** Portion of fertile head of *Thamnocephalis sphaerospora* showing septate fertile hyphae, and fertile vesicles bearing unispored sporangiola. **44.** A few terminal branches of *Piptocephalis xenophila* bearing numerous multispored merosporangia. **45.** Young zygospore of Piptocephalis lepidula showing suspensors that lack vesicles. **46–48.** *Syncephalis hypogena*. **46.** Typical sporophore and zygospore. **47.** Asexual reproductive structure (rhizoids, sporophore, globose vesicle, merosporangia) showing merosporangia borne on bottom of vesicle. **48.** Typical zygospore (*top right*) borne on vesiculate suspensors. **42.** *Bar* 100 μm. **43–48.** *Bar* 50 μm

tella persicaria (E.D. Eddy) Hesselt. (Powell et al. 1981).

c) Yeast Cells

Anaerobic formation of yeast cells has been described in *Mucor rouxii* by Bartnicki-Garcia (1978). **In *C. recurvatus* and species of *Benjaminiella* and *Mycotypha*, yeast cells are produced aerobically when the spores germinate on the surface of a nutrient-rich culture medium** [Fig. 8.4(39)] (Benny and Benjamin 1976; Benny et al. 1985). Changes in cell wall chemistry (carbohydrates, hexosamine, protein, lipids, fatty acids, ash) were noted as the yeast cells passed through an intermediate phase and, finally, to vegetative hyphae in *Benjaminiella poitrasii* (R.K. Benj.) von Arx (Cole et al. 1980).

4. Meiospores

a) Zygospores

Zygospores are formed as a result of the copulation between two or, rarely, three, usually differentiated, gametangia. The stages in zygospore formation are illustrated for *Syzygites megalocarpus* by Fuller (1978) and O'Donnell (1979) and as part of a life cycle of *R. stolonifer* by Alexopoulos et al. (1995). Ultrastructural studies of zygospore formation are demonstrated using *R. stolonifer* (isogamous, heterothallic) and *Zygohynchus heterogamus* (Vuill.) Vuill. (heterogamous, homothallic) (Edelmann and Komparens 1995; Ho and Chen 1998).

The zygospore is thick-walled and hyaline and usually contains a single globule, and only one is formed in each zygosporangium. The characteristics of the "zygospore" noted in species descriptions are really the color and ornamentation of the zygosporangial wall (Schipper et al. 1975). The zygosporangium formed by species of Mucorales can be smooth to variously ornamentated and pigmented [Fig. 8.4(35)]. The zygosporangia can be dark brown or black and opaque, lighter brown and translucent, or light brown and transparent; these light brown spores can have either an undulate or smooth wall. These zygospores, all found in Mucorales, have opposed suspensors [Fig. 8.4(35)] and are formed aerially. The zygospore when released is hyaline and thick-walled and usually contains a single yellow globule when mature [Fig. 8.4(36)]. The "zygospore" conforms to the internal shape of the zygosporangium and the remnants of the suspensors (Benny and Schipper 1992).

Other zygospores produce apposed suspensors that are often entwined or bear protuberances. These zygospores are usually produced at or below the surface of the substrate. Members of Endogonales [Fig. 8.4(29)], Mortierellales [Fig. 8.3(27)], Zoopagales [Fig. 8.5(45, 46, 48)], and some families of Mucorales (Choanephoraceae, Mucoraceae, Phycomycetaceae, Pilobolaceae) have zygospores with apposed suspensors (Benjamin 1959, 1966, 1979, 1985b; Degawa and Tokumasu 1998; Fuller 1978; Kuhlman 1972; Thaxter 1922). Dimargaritales and Kickxellales produce hyaline, globose zygospores in the substrate and have two or even three undifferentiated suspensors [Fig. 8.2(20, 21)] (Benjamin 1959; Kurihara et al. 2004).

Meiospore formation requires the presence of two compatible hyphae (denoted by + and −) in heterothallic species, whereas only a single hyphal type is required for homothallic taxa. The majority of Mucorales members are heterothallic (Feofilova 2006; Schipper and Stalpers 1980), but in the remaining orders, where sexual reproduction is known, the species are homothallic. **In Mucorales, zygospore formation is by light-induced biosynthesis of β-carotene to the hormone trisporic acid, which initiates the conversion of vegetative hyphae to zygophores; intermediate molecules act like pheromones** (Corrochano and Garre 2010; Schimek and Wöstemeyer 2006, 2009).

b) Azygospores

Azygospores are parthenogenetically formed zygospores. These structures look like zygospores, but they are formed on only a single suspensor. Azygospores have been reported in many of the orders in which zygospores are produced (Benjamin and Mehrotra 1963; Ginman and Young 1989).

5. Kickxellomycotina

The members of Kickellomycotina (Fig. 8.2) are characterized by the formation of regularly septate hyphae. The septa produce a lenticular cavity that contains a single plug that is more or less lens-shaped. The members of Dimargaritales form plugs with two protuberances, one on each side, that protrude into the cytoplasm of the adjoining hyphal cells. The septal plugs formed by the remaining orders lack protuberances. Asexual reproduction is by arthrospores [Fig. 8.2(16)], one- or two-spored merosporangia [Fig. 8.2(19, 22)], or trichospores [Fig. 8.2 (17, 18)]. The zygospores are globose, broadly fusiform, hemifusiform, or long-cylindrical and coiled. These fungi are terrestrial parasites or saprobes, or symbionts in the mid- or hindgut of insects or isopods. Taxa of Asellariales and Harpellales can be found on the trichomycetes Web site (http://www.nhm.ku.edu/~fungi/).

Humber (2012b) transferred *Ballocephala* and *Zynemomyces* from Entomophthoromycota to Kickxellomycotina as taxa of unknown affinities because a member of both taxa produces a septum and septal plug characteristic of this subphylum (Saikawa 1989; Saikawa et al. 1997). Tretter et al. (2013) report that *Barbatospora*, *Orphella* (Harpellales), *Ramicandelaber*, and *Spiromyces* (Kickxellales) may represent unique clades in Kickxellomycotina.

a) Asellariales

The members of Asellariales [Fig. 8.2(15, 16)] produce branched, multicelled thalli attached to the hindgut of their host by a cellular or noncellular holdfast [Fig. 8.2(15)]. Asexual reproduction is by arthrospore [Fig. 8.2(16)] formation. Sexual reproduction has been reported in *Asellaria jatibonicua* L.G. Valle & Cafaro (Valle and Cafaro 2008). In *Asellaria ligiae* Tuzet & Manier ex Manier, arthrospore development is similar to trichospore formation in some members of Harpellales, especially *Carouxella* (Lichtwardt et al. 2001). Lichtwardt (1986) recognized two genera, Asellaria and Orchesellaria (Degawa 2009; Valle and Cafaro 2008). One monotypic genus, *Baltomyces* (Oman and White 2011), has been described that may belong in Asellariales, but its final disposition will depend on further study. Several genes have been sequenced for *A. ligiae* (E. Tretter, pers. comm.); it appears to be a member of Kickxellomycotina.

b) Dimargaritales

All members of Dimargaritales [Fig. 8.2(19, 20)] are haustorial parasites of other fungi, especially Mucorales, but some *Dispira* spp. are parasitic on *Chaetomium* (Ascomycota; Benjamin 1959, 1965, 1966, 1979). Subaerial hyphae and sporophores are regularly septate and simple or branched. **The septa are incomplete and have a lenticular cavity containing a single plug with polar protuberances. The plugs dissolve in 2–3 % KOH** (Benjamin 1959, 1966, 1979). **Asexual reproduction is by two-spored merosporangia [Fig. 8.2(19)]** formed directly on a fertile vesicle (*Spinalia*) (Benjamin 1959) or on simple or branched sporiferous branchlets that arise from inflated or unmodified sporophore apices (*Dimargaris*, *Dispira*, *Tieghemiomyces*) (Benjamin 1959, 1965). All species of *Dispira* and *Tieghemiomyces* and two species of *Dimargaris* are dry-spored at maturity; all other *Dimargaris* species are wet-spored (Benjamin 1959, 1965). Merosporangiosporogenesis is successive in most Dimargaritales but simultaneous in two *Dimargaris* spp. (Benjamin 1959, 1965). **Sexual reproduction is by hyaline, smooth, or ornamentated zygospores formed in the substrate on undifferentiated sexual hyphae [Fig. 8.2(20)]** (Benjamin 1966). Zygospores are similar in all Dimargaritales members, and their formation is enhanced when a culture is grown on non-slanted YpSs agar (Benny 2008) medium in a test tube.

The haustorial parasites in Dimargaritales, *Dimargaris*, *Dispira*, and *Tieghemiomyces* can grown on a medium containing glycerol as a carbon source (Barnett 1970), but growth is extremely slow and most taxa do not sporulate. Benjamin (1959) reported that *Tieghemiomyces californicus* R.K. Benj. grows and sporulates normally without a host on malt extract–yeast extract agar (Benny 2008). A few *Dimargaris* species have been isolated from soil, but most members of Dimargaritales are found on dung, especially of small rodents (Benjamin 1959, 1965). Several species may have a global distribution but are seldom recognized or cultured. *Spinalia radians* Vuill. is a member of Dimargaritales but not in culture (Benjamin 1959; Vuillemin 1904).

c) Harpellales

The members of Harpellales [Fig. 8.2(17, 18)] are found in the mid- or hindgut of immature aquatic insects usually collected from fast-flowing streams, but they may also be found in ponds and other still waters. The thalli are septate and are attached to the lining of the host's gut by a noncellular holdfast. Asexual reproduction is by unispored sporangia [trichospores (Fig. 8.2(17, 18))] that usually bear one or more long, basal appendages. Zygospores [Fig. 8.2 (17)], where known, are more or less conical, biconical, or long-cylindrical and coiled.

Two families are recognized: Harpellaceae is reserved for those taxa with unbranched thalli, whereas Legeriomycetaceae contains only fungi that are branched. The shape of the trichospore and whether or not it bears a basal collar, the number of appendages, and the nature of the holdfast are all important characters in identifying a genus in Harpellales. The genera of Harpellales are listed in Table 8.1 [see Benny (2012) for the literature on the majority of the genera].

Zygospores are variously shaped (types according to Moss et al. 1975): the biconical zygospore types (I–III) are in the following position in relation to the zygosporophore: (1) perpendicular or type I (*Allantomyces*, *Bojamyces*, *Harpella*, *Genistelloides*, *Klastostachys*, *Simuliomyces*, *Spartiella*, *Stipella*); (2) oblique or type II (*Austrosmittium*, *Capniomyces* [Fig. 8.2(17)], *Legeriomyces*, *Glotzia*, *Graminella*, *Harpellomyces*, *Laculus*, *Legerioides*, *Legeriomyces*, *Legeriosimilis*, *Sinotrichium*, *Smittium*, *Tectimyces*, *Trichozygospora*, *Zancudomyces*); (3) parallel or type III (*Genistellospora*, *Pennella*); or (4) pointed with a round base and attached to the zygosporophore at the base or type IV (*Carouxella*, *Dacryodiomyces*, *Lancisporomyces*, *Plecopteromyces*, *Zygopolaris*). The ultrastructure of selected harpellalean zygospores is illustrated by Moss and Lichtwardt (1977). A fifth type of zygospore formation has been reported in *Orphella*; these zygospores are elongate-cylindrical and coiled or basally coiled and then straight medially to distally (Valle and Santamaria 2005). Zygospores are unknown in the remaining genera of Harpellales.

In many taxa the upper portion of the wall of the generative cell may remain with the trichospore as a collar. The trichospore and generative cell remain intact in *Carouxella*, and dehiscence is the result of thallus separation. In *Orphella* (Lichtwardt et al. 1991) the trichospore, generative cell, and a sterile terminal cell remain attached and form a dehiscent complex reproductive unit. In a few taxa, trichospores are produced from germinating blackfly ovarian cysts that may infect blackfly larvae, potentially adding infectious propagules to the life cycle (Labeyrie et al. 1996); zygospores were observed later (Rizzo and Pang 2005). White et al. (2006a) used sequence data on blackfly larvae to identify the cyst inducing Harpellales.

Horn (1989a, 1990) showed that extrusion (germination in some authors) of *Smittium* trichospores occurs only after passage through the gut of a mosquito larva. This process occurs in two parts: in phase I the concentration of potassium is relatively high (40 mM) and the pH is 10 (environment of larval midgut), and in phase II the pH is ca. 7 (6–8) (condition in the larval hindgut). Extrusion of trichospores occur after exposure of 15 min or less to each phase. Holdfast formation of *Zancudomyces culisetae* (Lichtw.) Y. Wang, Tretter, Lichtw. & M.M. White (as *Smittium culisetae* Lichtw.) and *S. culicis* Manier occur after phase II, but spore extrusion takes place after phase I (*S. culicis*) or phase II (*Z. culisetae*, as *S. culisetae*). An ultrastructural study of trichospore extrusion and holdfast formation in the aforementioned *Smittium* spp. showed canals and an interwall layer at both ends of the spore. Spore bodies, structures that hold the initial holdfast material, accumulate at the apical end of the trichospore after phase I. Holdfast formation occurs during phase II in 10 s or less. Spore bodies are randomly dispersed in the trichospore cytoplasm before it is exposed to the phase I environment (Horn 1989b). Additional papers on trichospore ultrastructure have been published (Moss and Lichtwardt 1976; Sato 2002).

d) Kickxellales

The 12 genera (Benjamin 1966; Kurihara and Degawa 2006) in the order **Kickxellales [Fig. 8.2 (21, 22)] produce septa and zygospores [Fig. 8.2 (21)] similar to those of Dimargaritales except that the septal plugs lack protuberances and do not dissolve in KOH** (Benjamin 1959, 1966, 1979). Members of **Kickxellales usually form unispored merosporangia [Fig. 8.2(22m)] from a phialidelike cell, the pseudophialide [Fig. 8.2(22p)], borne on a specialized fertile structure, the sporocladium [Fig. 8.2(22s)]** (Benjamin 1959, 1966). **Pseudophilalides [Fig. 8.2 (22p)] are not produced by species of *Mycoëmilia* and *Spiromyces*** (Benjamin 1966; Kurihara et al. 2004; O'Donnell et al. 1998), whereas the other relatively recently described genera (*Myconymphaea*, *Ramicandelaber*, *Pinnaticoemansia*) produce pseudophialides (Kurihara and Degawa 2006; Kurihara et al. 2001, 2004; Ogawa et al. 2001). Four sporocladial types are known in Kickxellales (Benjamin 1966): two are aseptate [*Linderina*, *Mycoëmilia*, *Myconymphaea* (occasionally two-celled), *Ramicandelaber*, *Spiromyces*], and the remaining types have two (*Kickxella*) or many septa (*Coemansia*, *Dipsacomyces*, *Martensiomyces*, *Martensella*, *Pinnaticoemansia*, *Spirodactylon*). *Martensella corticii* Thaxt. (Benjamin 1959) is a rare parasite of the basidiomycete *Corticium radiosum* (Fr.) Fr.; it is widespread and only occurs in Canada and the adjacent USA (Jackson and Dearden 1948). *Spiromyces minutus* R.K. Benj. and *Spirodactylon* grow slowly in pure culture. *Spiromyces aspiralis* R.K. Benj. & Benny grows and sporulates axenically on malt extract–yeast extract agar (O'Donnell et al. 1998). All other taxa of Kickxellales are saprobic.

Most members of Kickxellales (*Coemansia*, *Dipsacomyces*, *Kickxella*, *Linderina*, *Martensiomyces*, *Martensella*, *Mycoëmilia*, *Myconymphaea*, *Pinnaticoemansia*, *Ramicandelaber*) release spores in a droplet of fluid, but species of *Spiromyces* and *Spirodactylon* are dry-spored at maturity (Ingold 1978).

A transmision electron microscope reveals the presence of spines embedded in the spore walls of all wet-spored Kickxellales except *Coemansia reversa* Tiegh. & Le Monn., and the spine apices produce a characteristic pattern on the merosporangial wall (Benny and Aldrich 1975; Young 1968; Zain et al. 2012). A saclike organelle is attached to the base of the septum between the merosporangium and the pseudophialide and occupies the upper portion of the latter structure. This saccate structure has been observed in *Kickxella alabastrina* Coem. [labyrinuliform organelle of Young (1974)] and *Linderina pennispora* Raper & Fennell [the abscission vacuole of Benny and Aldrich (1975)]. The dry-spored members of Kickxellales have spines or warts on the merosporangial surface (O'Donnell et al. 1998; Young 1968). Ultrastructural observations show that the septal plug lacks protuberances and that the aerial hyphae are adorned with noncrystalline spines (Benny and Aldrich 1975; O'Donnell 1979; Young 1974).

Members of Kickxellales can be isolated from soil, but they also occur on dung (Benjamin 1959; 1966). *Coemansia* is the only common member of the order; many species can be isolated from both dung and soil. Two coprophilous genera, *Kickxella* and *Spiromyces*, are rarely encountered in nature on dung (Benjamin 1966; O'Donnell et al. 1998). All remaining taxa of Kickxellales are known from only one or two collections, usually from soil. Soil baited with dead, sterile arthropods added three genera (*Mycoëmilia*, *Myconymphaea*, *Pinnaticoemansia*). The fourth new genus (*Ramicandelaber*) can be isolated from unbaited soil.

Kickxella is coprophilous and psychrotolerant and is thus found only in localities where its temperature requirements are met. *Linderina* is a genus with two known species, and one or both taxa can be isolated from soil collected in subtropical or tropical parts of the world (Chuang and Ho 2009; Ho et al. 2007; Kurihara et al. 2008). Four species have been described in *Ramicandelaber*, all isolated from soil (Chuang et al. 2013).

6. Mortierellomycotina and Mortierellales

This order was described by Cavalier-Smith (1998). Members of **Mortierellales (Fig. 8.3) have sporangia [Fig. 8.3(23)] and sporangiola [Fig. 8.3(24)] with columellae that are absent**

or small and produce hyaline and smooth or angular zygospores with apposed suspensors [Fig. 8.3(27)] that are surrounded by hyphae in a few species (Kirk et al. 2008). **The bases of the sporangiophores are often inflated [Fig. 8.3 (24)], and the colony may produce a garlic- or onionlike odor** (Hoffmann et al. 2011; Petkovits et al. 2011). These fungi are discussed and illustrated by Gams (1976, 1977a, b) and Zycha et al. (1969).

Mortierellomycotina is based on the phylogenetic analysis of a multigene data set composed of 18S and 28S rRNA, actin, α- and β-tubulin, and RPB1 and RPB2 genes (Hoffmann et al. 2011), a separation supported by an analysis of fungal sterols (Weete et al. 2010). **The cardinal morphological traits for this subphylum are the lack of a columella, basally inflated sporangiophores, and the morphology of the zygospore** (Degawa and Tokumasu 1998; Kuhlman 1972). Some taxa produce a characteristic odor and **colonies with a more or less regular, flattened, wavy or undulate growth pattern** (Petkovits et al. 2011).

Petkovits et al. (2011) and Wagner et al. (2013) analyzed a phylogeny of Mortierellaceae and observed that the clades did not correlate with the sections of *Mortierella* [Gams 1977a; illustrated in Gams (1977b)]. The data set was composed of 90 reference and type strains, representing 53 different species of *Mortierella* and four taxa from *Dissophora*, *Gamsiella* [Fig. 8.3(24)], and *Lobosporangium* [Fig. 8.3 (23)], from Mortierellales. Additional isolates (more than 400) were included in a data set of Mortierellales that demonstrated that the order includes unresolved taxa, including species complexes. Nagy et al. (2011) found 55 % of the described taxa of *Mortierella* [Fig. 8.3(26)], but there may eventually be 127 species worldwide by the time all members of the genus are described. Kirk et al. (2008) estimate that approximately 85 *Mortierella* spp. have been described. Hibbett and Glotzer (2011) revealed that of all the species of the Mortierellales described in the last 140 years by taxonomists, approximately 50 % of these taxa were found in 6 years by molecular ecologists.

One taxon, *Aquamortierella* (Embree and Indoh 1967), has appendaged sporangiospores and is known only in aquatic habitats. Other taxa in Mortierellales are *Dissophora*, a psychrophile found on dung and in soil (Benny 1995; Thaxter 1914), *Echinochlamydosporium*, which colonizes soybean cyst nematode juveniles (Jiang et al. 2011), and *Gamsiella* [Fig. 8.3(24)] and *Lobosporangium* [Fig. 8.3 (23)], both isolated from soil or humus (Benny and Blackwell 2004). The aforementioned taxa are all known from one or two reports in the literature. The majority of the species, however, are members of the genus *Mortierella* [Fig. 8.3(26)] (Gams 1977a). Sporocarps of *Modicella* [Fig. 8.3(25)] have been collected numerous times in the field and several reports have been published on the genus. *Modicella* has two known species that have never been cultured (Gerdemann and Trappe 1974). *Nothadelphia* is a genus that is parasitic on a species of *Mortierella*; it may be a member of Mortierellales, but its true affinities are unknown (Degawa and Gams 2004).

7. Mucoromycotina

a) Endogonales

This order was validated by Benjamin (1979). **The members of Endogonales [Fig. 8.4(28, 29)] are either saprobes or ectomycorrhizal** (Warcup 1990) **and reproduce by forming zygospores with apposed suspensors [Fig. 8.4 (29)] in uni- or multizygosporic sporocarps [Fig. 8.4(28)]**. Historically the members of Endogonales have been associated with the endomycorrhizal fungi (Morton and Benny 1990). Initially, members of Endogonales (as the family Endogonaceae) were in the order Mucorales (Hesseltine and Ellis 1973). The members of Endogonales consisted only of sporocarpic taxa in *Endogone* (Thaxter 1922), including species now in Glomeromycota (Schüssler et al. 2001).

The zygospores of *Endogone pisiformis* Link have been studied ultrastructurally from sporocarps collected in the field [Fig. 8.4(28, 29)] (Gibson et al. 1986). Cultures made from germinating zygospores and grown on ordinary laboratory culture media only form mycelium (Dalpé 1990). Sporocarps of *E. pisiformis* have been produced in axenic culture with Scots pine (*Pinus sylvestris* L.) seedlings (Berch and Castellano 1986).

Four genera are currently known: *Endogone*, *Peridiospora*, *Sclerogone*, and *Youngiomyces* (Gerdemann and Trappe 1974; Wu and Lin 1997; Yao et al. 1996). Species of *Endogone*, *Sclerogone*, and *Youngiomyces* were described and illustrated by Yao et al. (1996).

A fifth genus, *Densospora* P.A. McGee (McGee 1996), with *Glomus*-like spores, is included in Endogonales because it produces ectomycorrhizae. *Densospora*, however, does not produce zygospores or arbuscules and cannot be comfortably included in either Endogonales or Glomeromycota.

Dibartono et al. (2011) presented evidence that fungi similar to *Endogone* may have been mycorrhizal associates of early land plants. Several extant species of *Endogone* are mycorrhizal with the earliest branch of the liverworts. These mycorrhizal species of *Endogone* resemble the fungi found in the Rhynie Chert fossil plant *Nothia aphylla* Lyon ex El-Saadawy & Lacy (Krings et al. 2007). Members of Glomeromycota (Schüssler et al. 2001) were the endomycorrhizal associates of the land plants that evolved later, including the thalloid liverworts. Members of Glomeromycota are present in the roots of plants in early Devonian Rhynie Chert (Krings et al. 2007; Taylor et al. 2003, 2014) [cf. zygomyceteous fossils reviewed by Krings et al. (2013)].

b) Mucorales

The members of Mucorales [Fig. 8.4(30–41)] are saprobes or facultative (nonhaustorial) parasites in nature that can usually be grown axenically in the laboratory. The thallus generally consists of coenocytic hyphae, and spores are formed in sporangia [Fig. 8.4(30, 31, 33, 34, 41)], sporangiola [Fig. 8.4(32, 37)], or merosporangia [Fig. 8.4(40)]. The typical zygospore (O'Donnell 1979) has a rough, carbonaceous zygosporangium and differentiated, nonappendaged, opposed suspensors [Fig. 8.4(35)]. The sexual spore contains a single endospore with a hyaline wall and an eccentric globule [Fig. 8.4(36)]. Appendaged suspensors are produced by some taxa (*Absidia*, *Phycomyces*, *Radiomyces*). *Pilaira*, *Pilobolus*, *Blakeslea*, *Choanephora*, and *Poitrasia* produce zygospores with apposed suspensors. *Phycomyces* has large rough-walled zygospores with tongslike appendaged suspensors (O'Donnell et al. 1978). The members of Mucorales can form chlamydospores but rarely produce yeast cells [Fig. 8.4(38, 39)].

Thirteen Mucorales families have been recognized (Benny et al. 2001) based on morphological characters (Alexopoulos et al. 1995; Hawksworth et al. 1995). The phylogenetic analysis of O'Donnell et al. (2001) and Voigt and Wöstemeyer (2001), however, demonstrated that most of the morphologically based Mucorales families that contain two or more genera are not concordant with clades based on molecular data. A critical examination of the morphological characters did reveal which were phylogenetically informative (O'Donnell et al. 2001), such as trophocyst formation by *Pilobolus* and *Utharomyces* (Kirk and Benny 1980; Page 1962).

Fourteen families currently are recognized (Voigt 2012) in Mucorales based on the results of a sequence-based data set; they are discussed by Voigt (2012) and Hoffmann et al. (2013). The new families and subfamilies (Table 8.1) mentioned by Voigt (2012) have been validated (Kirk 2012; Kirk and Voigt 2012).

Important new taxa, all supported by phylogenetic analyses, include the revised *Absidia* s.l. (*Absidia* s.s., *Lentamyces*, *Lichtheimia*, *Siepmannia*), new species in *Apophysomyces* and *Saksenaea*, and some mesophilic species of *Rhizomucor* that have been transferred to *Mucor*; species in the latter three genera can cause mucormycosis (Alvarez et al. 2010a, b, 2011; Hoffmann 2010; Liu 2011; Lu et al. 2013). A new *Mucor*-like genus, *Isomucor*, has been described, and *Calcarisporiella*, formerly an anamorphic ascomycete, is now recognized as a member of Mucoromycotina, possibly Mucorales (Hirose et al. 2012; de Souza et al. 2012).

Cunninghamella, *Pilaira*, and *Rhizopus* were monographed or revised (Zheng and Chen 2001; Zheng and Liu 2009; Zheng et al. 2007). These classifications are supported by phylogenetic studies (Liu et al. 2001, 2007, 2012).

The genera *Dicranophora*, *Spinellus*, and *Syzygites* (Mucorales) are found in nature only as mushroom parasites. *Dicranophora* and *Spinellus* are psychrophilic and only grow at temperatures below 20 °C; 15 °C is optimal (Voglmayr and Krisai-Greilhuber 1996; Watson 1964). *Syzygites* can grow at 25 °C but not 30 °C (Wenger and Lilly 1966). All three genera grow in pure culture on nutrient-rich media. *Syzygites megalocarpus* Ehrenb.: Fr. has the most extensive host range (98 species in 22 basidiomycete families and 4 ascomycete species) (Kovacs and Sundberg 1996).

The facultative parasites of mushrooms, arthropods (*Sporodiniella umbellata*), and other members of Mucorales (*Lentamyces parricida*, *Chaetocladium* spp., *Parasitella simplex*) may only be found on a few, or many, hosts in nature and can all be grown on ordinary culture media. Those parasitic taxa that grow axenically on culture are usually not altered morphologically. The mucoralean mycoparasites *Chaetocladium*, *Lentamyces*, and *Parasitella* are nonhaustorial biotrophic fusion parasites that induce gall formation in a host (Hoffmann and Voigt 2009). *Sporodiniella* only is found in nature on insects and spiders in the subtropics and tropics (Chien and Hwang 1997; Evans and Samson 1977).

8. Zoopagomycotina and Zoopagales

All members of Zoopagales sensu Benjamin (1979) (Fig. 8.5) are obligate parasites (predaceous, ectoparasites, or endoparasites) of other fungi or small animals (amoebae, rotifers and nematodes and their eggs), with predaceous species and ectoparasites forming haustoria in the host. The hyphae are coenocytic or septate. The thallus is a branched or unbranched, inflated hyphal coil in the host, or it consists of branched external hyphae. Asexual reproduction is by conidia [Fig. 8.5(42)] or, possibly, unispored [Fig. 8.5(43)] (G.L. Benny, unpublished data) or multispored merosporangia [Fig. 8.5(44, 47)]. The conidia are formed as single spores or in simple or branched chains of many merosporangia. Other means of asexual reproduction are by arthrospores and chlamydospores. Sexual reproduction is by the formation of zygospores [Fig. 8.5(45, 46, 48)]. Zygospores, when formed, are more or less globose, have opposed suspensors, and either do not [Fig. 8.5(45)] or may bear suspensor outgrowths [Fig. 8.5(46, 48)].

Zoopagales contained only the type family until Duddington (1973) removed the ecto- and endoparasites and transferred them to Cochlonemataceae, leaving only the predaceous forms in the Zoopagaceae. Benjamin (1979) transferred the haustorial parasitic members (Helicocephalidaceae, Piptocephalidaceae) of Mucorales sensu Hesseltine and Ellis (1973) to Zoopagales. Helicocephalidaceae, including *Rhopalomyces*, are parasites of nematodes and their eggs and of rotifer eggs, whereas Piptocephalidaceae are mycoparasites, especially of Mucorales. The mycoparasitic members of Zoopagales are restricted to the Piptocephalidaceae and Sigmoideomycetaceae. *Piptocephalis* and *Syncephalis* (Piptocephalidaceae) (Benjamin 1959) contain numerous species and may be found anywhere the host (usually a member of the Mucorales) occurs. These genera are more or less cosmopolitan, whereas the third genus, *Kuzuhaea* (Benjamin 1985a), is known only from the original isolation in Japan.

Ultrastructural evidence on the number and appearance of wall layers shows that the asexual spores of *Stylopage rhabdospora* Drechsler (Saikawa 1986) are catenate conidia with a wall structure like that of Entomophthorales (Benny et al. 2001) [Fig. 8.4(32–35)] rather than merosporangiospores, as suggested by Benjamin (1979). The other members of Cochlonemataceae and Zoopagaceae that have been examined ultrastructurally also produce conidia (Saikawa 2011b).

Saikawa and coworkers made several observations on selected taxa of two families of Zoopagales (Cochlonemataceae, Zoopagaceae) using both light and electron microscopy. Saikawa et al. (2011) described the ontogeny and germination of *Acaulopage pectospora* zygospores. Observations made by Saikawa and Morikawa (1985) and Saikawa et al. (1988) indicated that *A. pectospora* was similar to *Zoophagus insidians* Sommerst. Dick (1990) observed the zygospores of *A. pectospora*, transferred the species to *Zoophagus*, and revised the genus description. A phylogenetic analysis determined that *Z. insidians* is a member of Zoopagales (Tanabe et al. 2000).

A new species of *Euryancale*, *E. phallospora* Saikawa & Katsurashima (Saikawa and Katsurashima 1993), was described that forms phalloid conidia. Saikawa and Katsurashima (1993) mentioned that *E. phallospora* produced a pouchlike appendage resembling those of *Euryancale marsipospora* Drechsler and *Euryancale sacciospora* Drechsler (Saikawa and Aoki 1991). Another new species, *Euryancale marsipioides* Aoki, has been confused with *E. marsipospora* (Saikawa and Aoki 1995). Saikawa and Sato (1986) published a report of zygospore formation in *E. marsipospora* and *E. sacciospora* Drechsler; both fungi were found on water agar plates next to dead nematodes.

Saikawa and Kadowaki (2002) documented the capture of amoebae in water and spore formation on the water surface by two species of *Acaulopage*. When the mycelium was transferred to a fresh water plate, there was new mycelial growth, and amoebae were captured along the length of the freshly grown hyphae. Conidial germination is always from the basal end 3 h after transfer to a fresh plate. Saikawa and Kadowaki reported that aging cultures produced zygospores, and they were observed for the first time in *Acaulopage dichotoma* Drechsler. Shimada and Saikawa (2006) reported on nematode capture and chlamydospore germination by *Cystopage cladospora* Drechsler.

The morphology and subsequent germination of the zygospores formed by two species of *Cochlonema* and an *Acaulopage* parasitizing amoebae was reported by Hirotane-Akabane and Saikawa (2010). The zygospore of *Acaulopage lophospora* Drechsler germinated to produce a hypha, whereas the sexual spore of both species of *Cochlonema* (*C. cerasphorum* Drechsler, *C. megalosomum* Drechsler) formed a conidiophore bearing a conidial chain. Keys to some taxa of Zoopagales are available (Dayal 1973/1974; Saikawa 2011a). Saikawa (2012) published a paper that included color light micrographs of selected species in many genera in the Cochlonemataceae and Zoopagaceae and keys to the genera and species.

Koehsler et al. (2007) sequenced the 18S rDNA of *Cochlonema euryblastum* Drechsler isolated in Germany from sediment collected in a rain gutter; it is a parasite of *Thecamoeba quadrilineata* Carter. Light and transmission electron micrographs of both the host and parasite were included.

Amoebophilus simplex Barron (1983) has been the subject of several papers and Internet reports. The *Amoebophilus* conidium attaches to the host amoeba and then produces a knobby haustorium; the conidium becomes the thallus that gives rise to either a short chain of conidia or zygospores (Barron 1983; Laber 2009). Brief (2005) reported that *Ouramoeba botulacauda* Leidy (1879, Plate IX, Figs. 13–18) is a species of the amoeba *Mayorella* Schaeffer parasitized by *A. simplex*. Mrva (2008, 2011) identified two species of *Mayorella* (*Mayorella penardi* Page, *M. vespertilioides* Page) as the hosts of *A. simplex*. Siemensma (2012) believed that *M. penardi* may be the only host for *A. simplex*. Infected amoebae are usually rare but were abundant on one occasion (Siemensma 2012). Another fungal parasite similar to *Ouramoeba vorax* Leidy (1879, Plate IX, Figs. 1–12) lacked the infective conidium as the thallus; it has been reported at least twice (Hippe 2007; Kreutz 2010). The vegetative hyphae often branch several times before the conidial chains are produced; the host is *Amoeba proteus* (Pal.). The latter taxon may be a species of *Zoopage*.

Members of Sigmoideomycetaceae (Benny et al. 1992) are also mycoparasites (Chien 2000). *Thamnocephalis quadrupedata* Blakeslee is a parasite of Basidiobolus sp. in nature and T. sphaerospora Benny, R.K. Benj. & P.M. Kirk grows on *C. recurvatus* in the laboratory. Members of Zoopagales occur in soil, although Members of Zoopagales occur in soil, although the latter three species are also found on dung. Members of Sigmoideomycetaceae (Benny et al. 1992) are haustorial parasites (Chien 2000) that are rarely collected. The host ranges for members (*Reticulocephalis*, *Sigmoideomyces*, *Thamnocephalis*) of this family are also unknown.

Piptocephalidaceae (*Piptocephalis*, *Syncephalis*) are the most commonly encountered members of the order because both genera contain many species, both the host and parasite are relatively large, and the host fungi are usually readily grown in the laboratory. *Piptoce-*

phalis should be isolated and grown below 20 °C (18 °C works well). Members of Kickxellales and some members of the genus *Mortierella* (Mortierellales) can be hosts for zygomycete mycoparasites, especially species of *Piptocephalis* and *Syncephalis* (Cuthbert and Jeffries 1984; Jeffries and Young 1978). Benjamin (1985b) discussed and illustrated the ontogeny of both the anamorph and teleomorph of *Syncephalis hypogena* R.K. Benj. Ultrastructural studies on merosporangiospore ontogeny in *Piptocephalis indica* B.S. Mehrotra & Baijal and *Syncephalis sphaerica* Tiegh. (Baker 1979; Baker et al. 1977) indicate that they are similar (invagination of plasmalemma) but differ from that of *Syncephalastrum racemosum* Cohn ex J. Schröt. (cytoplasmic cleavage) (Fletcher 1972).

Syncephalis often grows best on the original host, but some taxa may not survive more than one or two transfers. Many species of *Syncephalis*, however, can be transferred to another host, such as *Zygorhynchus* or *Cokeromyces*. Other taxa of *Syncephalis* are restricted to a species of *Mortierella*. Some members of Zoopagales can be found in an aquatic environment including *Zoophagus* and some species of the Cochlonemataceae.

Aenigmatomyces R.F. Castañeda & W.B. Kendr. is an ascomycete according to Mycobank (http://www.mycobank.org/) and Species Fungorum (http://www.speciesfungorum.org/). *Aenigmatomyces*, however, is probably a member of Zoopagomycotina (Cochlonemataceae, Zoopagales) according to Degawa (2002a) and Seifert et al. (2011). *Aenigmatomyces* zygospores were illustrated by Degawa (2002b). *Basidiolum fimbriatum* Cienk. (White 2003), a possible member of Zoopagales, resembles a small, aquatic *Syncephalis*.

Lecophagus and *Zoophagus* (Dick 1990) were both treated as members of Zoopagales. Recent morphological (light microscopy and transmission electron microscopy) and molecular (small subunit sequences) studies demonstrate that members of *Lecophagus* are ascomycete anamorphs (Morikawa et al. 1993; Tanabe et al. 1999). *Zoophagus* was retained in the zygomycota based on the comparative analysis of 18S rDNA sequences (Tanabe et al. 1999).

VII. Maintenance and Culture

Accounts of the collection and incubation of substrates and the isolation and cultivation of most zygomycetes are discussed in detail by Benjamin (1959), O'Donnell (1979), Krug et al. (2004), and Benny (2008). Similar information can be found for Entomophthoromycota (Entomophthorales) (Fuller 1978). Ellis (1963, 1966) described methods for the culture of *Rhopalomyces elegans* Corda and species of *Syncephalis* without a host.

Tansey (1984) recommended incubating samples at 45 °C on media emended with antibiotics and benomyl when isolating thermophilic and thermotolerant Mucorales. Isolation of both psychrophilic and psychrotolerant zygomycetes is best done on media incubated at 5 °C (Botha et al. 1999). The addition of antibiotics and benomyl to the media might inhance isolation, but it could eliminate some *Mortierella* species (Strauss et al. 2000).

Barron (2004) described techniques useful for studying nematode parasites. Benjamin (1959, 1985b), Jeffries and Kirk (1976), and Benny (2008) include several techniques that are valuable for making microscope slide preparations and for isolating, growing, and culturing zygomycetous fungi.

Fungi in Mucorales, Dimargaritales, Kickxellales, some Zoopagales (Helicocephalidaceae, Piptocephalidaceae), and many species of Entomophthorales can be grown on ordinary laboratory culture media (Benny 2008; Humber 2008). Potentially, these fungi may be isolated from any organic substrate (e.g., dung, humus, leaf mold, bark).

The addition of the antibiotics streptomycin sulfate, 0.1 g^{-1} l (filter sterilized), and chlortetracycline HCl, 0.05 g^{-1} l (in 95 % ethanol), and the fungicide benomyl, 10–20 ppm (aqueous), to the isolation medium increases the chances of isolating *Coemansia* spp. and the recovery of other zygomycotan taxa from soil; see Strauss et al. (2000) for other formulations.

The majority of members of Mucorales and Kickxellales grow best at pH 6 and require an exogenous thiamine source. *Pilobolus* is an obligate coprophile but will grow and sporulate only on slightly alkaline culture media containing hemin or dung extract (Levetin and Caroselli 1976).

Members of Dimargaritales and Piptocephalidaceae (Zoopagales) are haustorial parasites of other fungi, especially Mucorales; *C. recurvatus* Poitras is the host of choice (Benjamin 1959). Species of *Mortierella*, *Umbelopsis*, *Penicillium*, and *Chaetomium* also may be hosts for a few mycoparasites (Benjamin 1959, 1979) in the laboratory. The optimal isolation and growth temperature for species of *Piptocephalis* is 18 °C; reproduction is normal, but growth is slightly slower.

Cryogenic storage at −196 to −80 °C in ultracold mechanical freezers or in liquid nitrogen in special Dewar flasks is possible for nearly all zygomycotan fungi and may be required for some groups (especially Entomophthorales). Fungi can also be maintained in pure axenic (two-membered for parasites) culture, refrigerated after sporulation, and transferred periodically (every 6–12 months for most taxa but as often as every 2–3 weeks for some Entomophthorales, e.g., *Basidiobolus* and *Conidiobolus*). Many of the remaining zygomycetes can be lyophilized or stored under liquid nitrogen, sterile distilled water, or sterile mineral oil for long-term preservation.

Most entomophthoraleans can be isolated on pure culture on relatively simple media, such as Sabouraud dextrose agar +1 % yeast extract, or on a variety of media incorporating coagulated egg yolk (Papierok and Hajek 1997). Cultures of members of Entomophthorales cannot be preserved by lyophilization but are generally readily storable in cryogenic freezers, preferably at −196 °C (immersed in liquid nitrogen) using 10 % glycerol as cryoprotectant. Humber (1994, 1997b) discusses several of the problems and techniques of culturing and preserving entomophthoraleans.

Commercial-scale growth of entomophthoraleans for use in biological control may not be possible for taxa whose vegetative stages are naturally protoplastic, such as *Entomophaga* spp. that are pathogenic for caterpillars, grasshoppers, and locusts. Taxa whose vegetative growth is walled, such as *Zoophthora* spp. affecting a very wide range of hosts, are more tractable subjects for large-scale aerobic fermentation. The best strategy for the mass production and use of entomophthoraleans for insect biocontrol is by the controlled drying of the mycelium (McCabe and Soper 1985), which when ground into flakes, rehydrated, and applied produces infective spores in the midst of the target population. There is no adequate or practical strategy for the production and use of entomophthoralean conidia or resting spores for biological control.

Entomophthoralean saprobes in soil and plant detritus, mainly species of *Conidiobolus* and *Basidiobolus*, are usually detected only by using Drechsler's (1952) canopy plating isolation technique. Independent studies on terricolous, saprobic entomophthoraleans in the USA by C. Drechsler and D. King, and in India by Srinivasan, Thirumalachar, and Narasimhan (cited in King 1977) suggest that these fungi may be taxonomically diverse and relatively abundant wherever they are specifically sought.

The species that have been cultured are primarily the entomopathogens and species of *Conidiobolus* and *Basidiobolus*; no species of the Meristacraceae (*Meristacrum*, *Tabanomyces*) or the phytopathogenic genera *Ancylistes* and *Completoria* have been isolated in axenic cultures.

Members of Harpellales currently in axenic culture only consist of a number of taxa from the hindguts of many types of aquatic insect larvae: many *Smittium* spp. from six families of dipteran larvae, *Austrosmittium* sp. and *Trichozygospora chironomidarum* Lichtw. from midge larvae, and *Capniomyces stellatus* S.W. Peterson & Lichtw., *Genistelloides hibernus*, and *Simuliomyces spica* S.W. Peterson & Lichtw. from winter-emerging stonefly nymphs.

Ordinary transfer procedures will not work with members of Harpellales because they are aquatic fungi. Details of the culture methods and physiology of Harpellales can be found in Lichtwardt (1986) and Lichtwardt et al. (2001).

One of the best culture media is dilute brain-heart infusion (one-tenth the normal concentration) agar with vitamins (one-tenth BHIv) (Lichtwardt 1986). For routine culture maintenance, one-tenth BHIv can be prepared as slants in screw-capped test tubes. Then add a small amount of sterile, distilled water (1.5 cm) at the bottom of the tube when it is upright. WATER IS

ESSENTIAL! To study thalli, trichospores, etc., Harpellales can be grown in small (60 × 15 mm) petri dishes. After the one-tenth BHIv agar solidifies, pour a very thin layer of sterile, distilled water over the medium sufficient to cover the surface. Small petri dishes are recommended because they are easier to manipulate than larger ones.

All cultures should be stored in the refrigerator soon after good growth has occurred, preferably in screw-capped test tubes. Most isolates must be transferred every 4 months and slow-growing cultures every 2–3 months. Storage in liquid nitrogen is the best method for long-term preservation using conventional techniques for freezing and thawing. USE YOUNG, ACTIVELY GROWING THALLI (NOT SPORES). Do NOT lyophilize cultures of Harpellales!

With few exceptions, trichospores of Harpellales do not germinate in vitro. This necessitates the transfer of portions of thalli that are broken into smaller pieces in the new culture media. This is best done with a standard bacteriological transfer loop.

In slants, the loop should be rapidly and vigorously agitated sideways in the water laver at the bottom of the tube until the thalli are broken up sufficiently. In petri dishes, the loop can be agitated sideways in the water layer or vibrated by dragging the wire of the loop back and forth over the upper edge of the plastic petri dish bottom. For more precise handling of fungal material in petri dishes, a small loop (4–5 mm in diameter) is preferable to the more standard size.

Thalli developing in petri dishes can be left undisturbed when new growth is being produced. In test tubes, however, the slants should be tilted daily to allow the water to flow over the surface of the culture medium. This should be done for 3–7 days and stopped when good growth of new thalli is observed on the slant both above and below the surface of the water. Some taxa grow best at 24 °C, but others have their optimum growth temperature near 18 °C.

Cultures of many zygomycetes are available from the American Type Culture Collection (ATCC: Manassas, Virginia, USA), Centraalbureau voor Schimmelcultures (CBS: Utrecht, the Netherlands), International Mycological Institute (IMI: Egham, UK), and the USDA-ARS Culture Collection (NRRL: National Center for Agricultural Utilization Research, Peoria, Illinois, USA). Entomopathogenic Entomophthorales are available from the USDA-ARS Collection of Entomopathogenic Fungal Cultures (ARSEF: Ithaca, New York, USA; Humber 2012b).

VIII. Conclusions

The conidiogenesis of the Entomophthoromycota and some members of the Zoopagomycotina should be regarded as fundamentally distinct from the sporangiogenesis in Mucorales and related orders, from the production of cylindrical sporangia in Kickxellales, Dimargaritales, and several taxa of Zoopagales, and from the highly modified monosporic sporangiola of Asellariales and Harpellales. Similarly, the relatively undifferentiated state of the conidiogenesis of Entomophthoromycota and Zoopagomycotina is markedly different from the obviously sporangiate zygomycetes whose asexual reproductive systems are usually well differentiated (often elaborately so) from the vegetative mycelium. Members of Entomophthoromycota are further distinguished from the remaining zygomycotan fungi by strictly homothallic zygosporogenesis and a marked (secondary?) tendency for azygosporogenesis as opposed to the well-differentiated and morphologically complex heterothallic zygosporogenesis in the sporangiate zygomycetes and most members of Zoopagales.

The phylogeny of the zygomycotan fungi does not conform to the simpler unitary schemes of past decades based on morphology, development, and biology. The initial phylogenetic studies suggested that members of Entomophthorales were paraphyletic, with *Basidiobolus* being notably distinct from the remainder of the phylum (Jensen et al. 1998; Nagahama et al. 1995). The recent papers by Gryganskyi et al. (2012, 2013) clearly demonstrate that *Basidiobolus* is a member of the Entomophthoromycota.

Initially, only a few zygomycetes were ever included in a phylogenetic analysis. These treatments have usually included small subunit rDNA (Sugiyama 1998), but there has never been enough sequence information available for these analyses to provide an accurate representation of the phylogeny of the zygomycetes. Recent studies, however, have included many more genes in the data sets used for phylogenetic analysis (Gryganskyi et al. 2012, 2013; James et al. 2006; White et al. 2006a).

Identification of Mucorales, and other fungi discussed here, will be facilitated by barcoding (ITS1-5.8S-ITS2) of type or authentic cultures (Schoch et al. 2012). Barcoding was discussed for members of Mucorales (Walther et al. 2013), and the procedure will aid in species determination for biotechnology and classification (Voigt and Kirk 2011).

References

Abdel-Motaal FF, El-Sayed MA, El-Zayat SA, Nassar MSM, Ito SI (2010) *Choanephora* rot of floral tops of *Hyoscyamus muticus* caused by *Choanephora cucurbitarum*. J Gen Plant Pathol 76:358–361

Alexopoulos CJ, Mims CW, Blackwell MM (1996) Introductory mycology, 4th edn. Wiley, New York, NY

Alvarez E, Garcia-Hermoso D, Sutton DA, Cano JF, Stchigal AM, Hoinard D, Fothergill AW, Rinaldi MG, Dromer F, Guarro J (2010a) Molecular phylogeny and proposal of two new species of the emerging pathogenic fungus *Saksenaea*. J Clin Microbiol 48:4410–4416

Alvarez E, Stchigel AM, Cano J, Sutton DA, Fothergill AW, Chander J, Salas V, Rinaldi MG, Guarro J (2010b) Molecular phylogenetic diversity of the emerging mucoralean fungus *Apophysomyces*: proposal of three new species. Rev Iberoam Micol 27:80–89

Álvarez E, Cano J, Stchigel AM, Sutton DA, Fothergill AW, Salas V, Rinaldi MG, Guarro J (2011) Two new species of *Mucor* from clinical samples. Med Mycol 49:62–72

Amadio J, Gordon K, Murphy CD (2010) Biotransformation of flurbiprofen by *Cunninghamella* species. Appl Environ Microbiol 76:6299–6303

Asha S, Vidyavathi M (2009) *Cunninghamella* – a microbial model from drug metabolism studies – a review. Biotechnol Adv 27:16–29

Baker K (1979) Ultrastructural development of merosporangia in the mycoparasite *Piptocephalis indica*. Can J Bot 57:2460–2465

Baker KL, Hooper GR, Beneke ES (1977) Ultrastructural development of rnerosporangia in the mycoparasite *Syncephalis sphaerica* (Mucorales). Can J Bot 55:2207–2215

Balazy S (1993) Flora of Poland. Fungi (Mycota), vol XXIV, Entomophthorales. Polish Academy of Sciences, Krakow, Poland

Barnett HL (1970) Nutritional requirements for the axenic growth of some haustorial mycoparasites. Mycologia 62:750–760

Barnett HL, Lilly VG (1950) Influence of nutritional and environmental factors upon asexual reproduction of *Choanephora cucurbitarum* in culture. Phytopathology 40:80–89

Barron GL (1975) Nematophagus fungi: *Helicocephalum*. Trans Br Mycol Soc 65:309–310

Barron GL (1983) A new *Amoebophilus* (Zygomycetes) parasitic on amoebae. Can J Bot 61:3091–3094

Barron GL (2004) Fungal parasites and predators of rotifers, nematodes and other invertebrates. In: Mueller GM, Bills GF, Foster MS (eds) Biodiversity of fungi. Inventory and monitoring methods. Elsevier, Burlington, MA, pp 435–450

Bartnicki-Garcia S (1978) *Mucor rouxii*: mold-yeast dimorphism. In: Fuller MS (ed) Lower fungi in the laboratory. Palfrey Contributions in Botany No. 1, Department of Botany, University of Georgia, Athens, GA, pp 146–148

Batko A (1964a) On the genera: *Zoophthora* gen. nov., *Triplosporium* (Thaxter) gen. nov. and *Entomophaga* gen. nov. (Phycomycetes: Entomophthoraceae). Bull Acad Pol Sci Club, Sér II 12:323–326

Batko A (1964b) Remarks on the genus *Entomophthora* Fresenius 1856 non Nowakowski 1883 (Phycomycetes: Entomophthoraceae). Bull Acad Pol Sci Club, Sér II 12:319–321

Batko A (1964c) Remarks on the genus *Lamia* Nowakowski 1883 vs. *Culicola* Nieuland 1916 (Phycomycetes: Entomophthoraceae). Bull Acad Pol Sci Club, Sr II 12:399–402

Batko A (1964d) Some new combinations in the fungus family Entomophthoraceae (Phycomycetes). Bull Acad Pol Sci Club, Sér II 12:403–406

Beakes GW, Campos-Takaki GM, Takaki M (1984) The effect of glucose concentration and light on thallospore structure in *Ellisomyces anomalus* (Thamnidiaceae, Mucorales). Can J Bot 62:2677–2687

Benjamin RK (1959) The merosporangiferous Mucorales. Aliso 4:321–433

Benjamin RK (1965) Addenda to "The merosporangiferous Mucorales" III. Dimargaris. Aliso 6:1–10

Benjamin RK (1966) A new *Basidiobolus* that forms microspores. Aliso 5:223–233

Benjamin RK (1966) The merosporangium. Mycologia 58:1–42

Benjamin RK (1979) Zygomycetes and their spores. In: Kendrick B (ed) The whole fungus. National Museum of Natural Sciences and the Kanaskis Foundation, Ottawa, pp 573–616

Benjamin RK (1985a) A new genus of the Piptocephalidaceae (Zoopagales) from Japan. Bot J Linn Soc 91:117–133

Benjamin RK (1985b) A novel new species of *Syncephalis* (Zoopagales: Piptocephalidaceae) from California that forms hypogenous merosporangia. Aliso 11:1–15

Benjamin RK, Mehrotra BS (1963) Obligate azygospore formation in two species of *Mucor*. Aliso 5:235–245

Benny GL (1995) Classical morphology in zygomycete taxonomy. Can J Bot 73(Suppl 1):S725–S730

Benny GL (2001) Zygomycota: Trichomycetes. In: McLaughlin DJ, McLaughlin EG, Lemke PA (eds) The Mycota, vol VIIA. Springer, Berlin, pp 147–160

Benny GL (2008) The methods used by Dr. R.K. Benjamin, and other mycologists, to isolate Zygomycetes. Aliso 26:37–61

Benny GL (2009) Zygomycetes. http://www.zygomycetes.org. Accessed 17 July 2010

Benny GL (2012) Current systematics of the zygomycotan fungi with a brief discussion of their biology. In: Misra JK, Tewari JP, Deshmukh SK (eds) Systematics and evolution of fungi. Progress in mycological research. Science Publishers, Enfield, NH, pp 55–105

Benny GL, Aldrich HA (1975) Ultrastructural observations on septal and merosporangial ontogeny in *Linderina pennispora* (Kickxellales; Zygomycetes). Can J Bot 53:2325–2335

Benny GL, Benjamin RK (1975) Observations on Thamnidiaceae (Mucorales). New taxa, new combinations, and notes on selected species. Aliso 8:301–351

Benny GL, Benjamin RK (1976) Observations on Thamnidiaceae (Mucorales). II. *Chaetocladium*, *Cokeromyces*, *Mycotypha*, and *Phascolomyces*. Aliso 8:391–424

Benny GL, Blackwell M (2004) *Lobosporangium*, a new name for *Echinosporangium* Malloch, and *Gamsiella*, a new genus for *Mortierella multidivaricata*. Mycologia 96:143–149

Benny GL, O'Donnell K (2000) *Amoebidium parasiticum* is a protozoan, not a trichomycete. Mycologia 92:1133–1137

Benny GL, Schipper MAA (1992) Observations on Thamnidiaceae (Mucorales). IV. Pirella. Mycologia 84:52–63

Benny GL, Kirk PM, Benjamin RK (1985) Observations on Thamnidiaceae (Mucorales). III. Mycotyphaceae fam. nov. and a re-evaluation of Mycotypha sensu Benny & Benjamin illustrated by two new species. Mycotaxon 22:119–148

Benny GL, Benjamin RK, Kirk PM (1992) A reevaluation of Cunninghamellaceae (Mucorales). Sigmoideomycetaceae fam. nov. and *Reticulocephalis* gen. nov.; cladistic analysis and description of two new species. Mycologia 84:615–641

Benny GL, Humber RA, Morton JB (2001) Zygomycota: Zygomycetes. In: McLaughlin DJ, McLaughlin EG, Lemke PA (eds) The Mycota, vol VIIA. Springer, Berlin, pp 113–146

Ben-Ze'ev I, Kenneth RG (1982) Features-criteria of taxonomic value in the Entomophthorales: I. A revison of the Batkoan classification. Mycotaxon 14:393–455

Berch SM, Castellano MA (1986) Sporulation of *Endogone pisiformis* in axenic and monoaxenic culture. Mycologia 78:292–295

Berdan H (1938) Revision of the genus *Ancylistes*. Mycologia 30:396–415

Birkby KM, Preece TF (1988) Calcium oxalate crystals on the sporangium of *Pilobolus*. Mycologist 2(2):68–69

Botha A, Paul I, Roux C, Kock JFL, Coetzee DJ, Strauss MC (1999) An isolation procedure for arachidonic acid prodicing *Mortierella* species. Antonie van Leeuwenhoek 76:253–256

Brefeld O (1870) Entwicklungsgeschichte der *Empusa muscae* und *E. radicans*. Bot Zeit 28:161–166, 177–186

Bridge PD, Hughes KA, Denton JO (2008) Association of the coprophilous fungus *Pirella circinans* with an indigenous beetle on the sub-Antarctic Bird Island. Polar Biol 31:657–661

Brief J (2005) *Ouramoeba botulicauda*, Leidys Amöebe mit Appendix—Ein von *Amoebophilus simplex* parasitierte *Mayorella*. Mikrokosmos 94:16–21

Brobyn PJ, Wilding N (1977) Invasive and developmental processes of *Entomophthora* species infecting aphids. Trans Br Mycol Soc 69:349–366

Butt TM, Beckett A, Wilding N (1981) Protoplasts in the in vivo life cycle of *Erynia neoaphidis* and *Erynia neoaphidis*. J Gen Microbiol 127:417–421

Cafaro MJ (2005) Eccrinales (Trichomycetes) are not fungi, but a clade of protests at the early divergence of animals and fungi. Mol Phylogenet Evol 35:21–34

Cavalier-Smith T (1998) A revised six-kingdom system of life. Biol Rev 73:203–266

Chamilos G, Lewis RE, Kontoyiannis DP (2007) Multidrug-resistant endosymbiotic bacteria account for the emergence of zygomycosis: a hypothesis. Fungal Genet Biol 44:88–92

Chien CY (2000) *Thamnocephalis quadrupedata* (Mucorales) as a mycoparasite of the entomophthoraceous fungus *Basidiobolus ranarum*. Cytobios 103:71–78

Chien CY, Hwang BC (1997) First record of the occurrence of *Sporodiniella umbellata* (Mucorales) in Taiwan. Mycoscience 38:343–346

Chuang SC, Ho HM (2009) Notes on Zygomycetes of Taiwan (VII): Two kickxellean species, *Linderina macrospora* and *Ramicandelaber brevisporus* new to Taiwan. Fungal Sci 24:23–28

Chuang SC, Ho HM, Benny GL, Lee CF (2013) Two new *Ramicandelaber* species from Taiwan. Mycologia 105:320–334

Cole GT, Sekiya R, Kasai R, Nozawa Y (1980) Morphogenesis and wall chemistry of the yeast, "intermediate," and hyphal phases of the dimorphic fungus, *Mycotypha poitrasii*. Can J Microbiol 26:36–49

Corrochano LM, Galland P (2006) Photomorphologenesis and gravitropism in fungi. In: Kües U, Fischer R (eds) Growth, differentiation and sexuality, vol 1, 2nd edn, The Mycota. Springer, Berlin, pp 233–259

Corrochano LM, Garre V (2010) Photobiology in the Zygomycota: multiple photoreceptor genes for complex responses to light. Fungal Genet Biol 47:893–899

Couch JN (1939) A new *Conidiobolus* with sexual reproduction. Am J Bot 26:119–130

Couch JN, Andreeva RB, Laird M, Nolan R (1979) *Tabanomyces milkoi* (Dudka and Koval) emended, genus novum, a fungal pathogen of horseflies. Proc Natl Acad Sci USA 76:2299–2302

Cuthbert A, Jeffries P (1984) Mycoparasitism of *Piptocephalis unispora* within the Mortierellaceae. Trans Br Mycol Soc 83:700–702

Dalpé Y (1990) *Endogone pisiformis*: croissance, morphologie et conservation des cultures. Can J Bot 68:910–915

Dayal R (1973/1974) Key to Phycomycetes predaceous or parasitic in nematodes or amoebae. I. Zoopagales. Sydowia 27:293–301

De Souza JI, Pires-Zottarelli CIA, dos Santos JF, Costa JP, Harakava R (2012) *Isomucor* (Mucoromycotina): a new genus from a cerrado reserve in state of São Paulo, Brazil. Mycologia 104:232–241

Degawa Y (2002a) 655 - Elucidating the 'enigma' of *Aenigmatomyces ampullisporus* which is now found as a Zygomycetous fungus parasitizes a spermatophore of collembolan insect. Organizing Committee (eds) Oslo, Norway, IMC7 Book of Abstracts, p 198

Degawa Y (2002b) On *Aenigmatomyces ampullisporus*. Newslett. Kanagawa Prefect. Mus Nat Hist 8(4):1–4. [http://nh.kanagawa-museum.jp/tobira/8-4/degawa.html. Accessed 14 Mar 2012; in Japanese—Google translation]

Degawa Y (2009) Secondary spore formation in *Orchesellaria mauguioi* (Asellariales, Trichomycetes) and its taxonomic and ecological implications. Mycoscience 50:247–252

Degawa Y, Gams W (2004) A new species of *Mortierella*, and an associated sporangiiferous mycoparasite in a new genus, *Nothadelphia*. Stud Mycol 50:567–572

Degawa Y, Tokumasu S (1998) Two new homothallic species of *Mortierella*, *M. cogitans*, and *M. microzygospora*, and their zygospore formation. Mycologia 90:1040–1046

Descals E, Webster J (1984) Branched aquatic conidia in *Erynia* and *Entomophthora sensu* lato. Trans Br Mycol Soc 83:669–682

Dibartono MI, Read DJ, Trappe JM, Merckx V, Ligrone R, Duckett JG (2011) The dawn of symbiosis between plants and fungi. Biol Lett 7:574–577

Dick MW (1990) The systematic position *of Zoophagus insidans*. Mycol Res 94:347–354

Drechsler C (1952) Widespread distribution of *Delacroixia coronata* and other saprophytic Entomophthoraceae in plant detritus. Science 115:575–576

Drechsler C (1956) Supplementary developmental stages of *Basidiobolus ranarum* and *Basidiobolus haptosporus*. Mycologia 48:655–676

Duddington CL (1973) Zoopagales. In: Ainsworth GC, Sparrow FK, Sussman AS (eds) The fungi, vol 4B. Academic, New York, NY, pp 231–234

Dykstra MJ (1994) Ballistosporic conidia in *Basidiobolus ranarum*: the influence of light and nutrition on the production of conidia and endospores (sporangiospores). Mycologia 86:494–501

Edelmann RE, Komparens KL (1994) The ultrastructural development of sporangiospores in multispored sporangia of *Zygorhynchus* heterogamous with a hypothesis for sporangial wall dissolution. Mycologia 86:57–71

Edelmann RE, Komparens KL (1995) Zygosporogenesis in *Zygorhynchus heterogamous*, with a proposal for standardization for structural nomenclature. Mycologia 87:304–318

Eilenberg J, Bresciani J, Latgé J-P (1986) Ultrastructural studies of primary spore formation and discharge in the genus *Entomophthora*. J Invertebr Pathol 48:318–324

Eilenberg J, Bresciani J, Olesen U, Olson L (1995) Ultrastructural studies of secondary spore formation and discharge in the genus *Entomophthora*. J Invertebr Pathol 65:179–185

Ellis JJ (1963) A study of *Rhopalomyces elegans* in pure culture. Mycologia 55:183–193

Ellis JJ (1966) On growing *Syncephalis* in pure culture. Mycologia 58:465–469

Embree RW, Indoh H (1967) *Aquamortierella*, a new genus in the Mucorales. Bull Torrey Bot Club 94:464–467

Evans HC (1982) Entomogenous fungi in tropical forest ecosystems: an appraisal. Ecol Entomol 7:47–60

Evans HC, Samson RA (1977) *Sporodiniella umbellata*, an entomogenous fungus of the Mucorales from cocoa farms in Ecuador. Can J Bot 55:2981–2984

Eysker M (1991) Direct measurement of dispersal of *Dictyocaulus viviparus* in sporangia of *Pilobolus* species. Res Vet Sci 50:29–32

Feofilova EP (2006) Heterothallism of mucoraceous fungi: a review of biological implications and uses in biotechnology. Microbiology 42:501–519

Fletcher J (1972) Fine structure of developing merosporangia and sporangiospores of *Syncephalastrum racemosum*. Arch Mikrobiol 87:269–284

Fuller MS (ed) (1978) Lower fungi in the laboratory. Department of Botany, University of Georgia, Athens, GA

Gams W (1976) Some new or noteworthy species of *Mortierella*. Persoonia 9:111–140

Gams W (1977a) A key to the species of *Mortierella*. Persoonia 9:381–391

Gams W (1977b) Taxonomic problems in *Mortierella*. In: Bigelow HE, Simmons EG (eds) Second international mycological congress, Abstracts Volume A–L, Tampa, FL, p 216

Gerdemann JH, Trappe JM (1974) The Endogonaceae in the Pacific northwest. Mycol Mem 5:1–76

Gibson JL, Kimbrough JW, Benny GL (1986) Ultrastructural observations on Endogonaceae: *Endogone pisiformis*. Mycologia 78:543–553

Ginman A, Young TWK (1989) Azygospore morphology in *Mucor azygospora* and *M. bainieri*. Mycol Res 93:314–320

Goldring D (1936) The effect of environment upon the production of sporangia and sporangiola in *Blakeslea trispora* Thaxter. Ann Missouri Bot Garden 23:527–542

Gomes MZ, Lewis RE, Kontoyiannis DP (2011) Mucormycosis caused by unusual Mucormycetes, non-*Rhizopus*, -*Mucor* and -*Lichtheimia* species. Clin Microbiol Rev 24:411–445

Griffiths DA (1974) The origin, structure and function of chlamydospores in fungi. Nova Hedwigia 25:503–547

Grove WB (1934) A systematic account and arrangement of the Pilobolidae. In: Buller AHR (ed) Researches on fungi, vol VI. Longmans, Green and Company, London, pp 190–224

Gryganskyi AP, Humber RA, Smith ME, Miadlikovska J, Wu S, Voigt K, Walther G, Anishchenko IM, Vilgalys R (2012) Molecular phylogeny of the Entomophthoromycota. Mol Phylogenet Evol 65:682–694

Gryganskyi AP, Humber RA, Smith ME, Hodge K, Huang B, Voigt K, Vilgalys R (2013) Phylogenetic lineages in Entomophthoromycota. Persoonia 30:94–105

Guo LW, Wu YX, Mao ZC, Ho HH, He YQ (2012) Storage rot of dragon fruit caused by *Gilbertella persicaria*. Plant Dis 96:1826

Hawksworth DL, Kirk PM, Sutton BC, Pegler DN (1995) Ainsworth & Bisby's dictionary of the fungi, 8th edn. CAB International, Wallingford

Hesseltine CW (1955) Genera of Mucorales with notes on their synonomy. Mycologia 47:344–363

Hesseltine CW, Ellis JJ (1973) Mucorales. In: Ainsworth GC, Sparrow FK, Sussman AS (eds) The Fungi, vol 4b. Academic, New York, NY, pp 187–217

Hibbett D, Glotzer D (2011) Where are all the undocumented fungal species? A study of *Mortierella* demonstrates the need for sequence- based classification. New Phytol 191:592–596

Hibbett DS, Binder M, Bischoff JF, Blackwell M, Cannon PF, Eriksson OE, Huhndorf S, James T, Kirk PM, Lücking R, Lumbsch H, Lutzoni F, Matheny PB, McLaughlin DJ, Powell MJ, Redhead S, Schoch CL, Spatafora JW, Stalpers JA, Vilgalys R, Aime MC, Aptroot A, Bauer R, Begerow D, Benny GL, Castlebury LA, Crous PW, Dai YC, Gams W, Geiser DM, Griffith GW, Gueidan C, Hawksworth DL, Hestmark G, Hosaka K, Humber RA, Hyde KD, Ironside JE, Kõljalg U, Kurtzman CP, Larsson KH, Lichtwardt R, Longcore J, Miądikowska J, Miller A, Moncalvo JM, Mozley-Standridge S, Oberwinkler F, Parmasto E, Reeb V, Rogers JD, Roux C, Ryvarden L, Sampaio JP, Schüssler A, Sugiyama J, Thorn RG, Tibell L, Untereiner WA, Walker C, Wang Z, Weir A, Weiss M, White MM, Winka K, Yao YJ, Zhang N (2007) A higher-level phylogenetic classification of the Fungi. Mycol Res 111:509–547

Hippe E (2007) Die Borstenamöebe—Leidys *Ouramoeba vorax* mit ihr Pilze. Mikrokosmos 96:204–206

Hirose D, Degawa Y, Inaba Tokumasu S (2012) The anamorphic genus *Calcarisporiella* is a new member of the Mucoromycotina. Mycoscience 53:256–260

Hirotane-Akabane W, Saikawa M (2010) Germination and morphology of zygospores in two *Cochlonema* and *Acaulopage* species. Mycologia 102:39–43

Ho HM, Chen ZC (1998) Ultrastructural study of wall ontogeny during zygosporogenesis in *Rhizopus stolonifer* (Mucoraceae), an amended model. Bot Bull Acad Sin 39:269–277

Ho HM, Chien CY, Chuang SC (2007) Notes on Zygomycetes of Taiwan (V): *Linderina pennispora* new to Taiwan. Fungal Sci 22:35–38

Hoffmann K (2010) Identification of the genus *Absidia* (Mucorales, Zygomycetes): a comprehensive taxonomic revision. In: Gherbawy Y, Voigt K (eds) Molecular identification of fungi. Springer, Berlin, pp 439–460

Hoffmann K, Voigt K (2009) *Absidia parricida* plays a dominant role in biotrophic fusion parasitism among mucoralean fungi (Zygomycetes): *Lentamyces*, a new genus for *A. parricida* and *A. zychae*. Plant Biol 11:537–554

Hoffmann K, Voigt K, Kirk PM (2011) Mortierellomycotina subphyl. nov., based on multi-gene genealogies. Mycotaxon 115:353–363

Hoffmann K, Pawłowska J, Walther G, Wrzosek M, de Hoog GS, Benny GL, Kirk PM, Voigt K (2013) The family structure of the Mucorales: a synoptic revision based on comprehensive multigene-genealogies. Persoonia 30:57–76

Holliday P (1980) Fungus disease of tropical plants. Cambridge University Press, Cambridge

Horn BW (1989a) Requirement for potassium and pH shift in host-mediated sporangiospore extrusion from trichospores of *Smittium culisetae* and other *Smittium* species. Mycol Res 93:303–313

Horn BW (1989b) Ultrastructural changes in trichospores of *Smittium culisetae* and *S. culicis* during in vitro sporangiospore extrusion and holdfast formation. Mycologia 81:742–753

Horn BW (1990) Physiological changes associated with sporangiospore extrusion from trichospores of *Smittium culisetae*. Exp Mycol 14:113–123

Humber RA (1981) An alternative view of certain taxonomic criteria used in the Entomophthorales (Zygomycetes). Mycotaxon 13:191–240

Humber RA (1989) Synopsis of a revised classification for the Entomophthorales (Zygomycotina). Mycotaxon 34:441–460

Humber RA (1994) Special considerations for operating a culture collection of fastidious fungal pathogens. J Ind Microbiol 13:195–196

Humber RA (1997a) Fungi: indentification. In: Lacey L (ed) Manual of techniques in insect pathology. Academic, San Diego, CA, pp 153–185

Humber RA (1997b) Fungi: preservation of cultures. In: Lacey L (ed) Manual of techniques in insect pathology. Academic, San Diego, CA, pp 269–279

Humber RA (2008) Recipes for commonly used culture media. http://www.ars.usda.gov/SP2UserFiles/Place/19070510/media_recipes.pdf. Accessed 19 Mar 2012

Humber RA (2012a) ARESF collection of entomopathogenic fungal cultures: Entomophthoroimycota: Basidiobolomycetes, Neozygitomycetes and Entomophthoromycetes. http://www.ars.usda.gov/SP2UserFiles/Place/19070510/ARSEFpdfs/Entomophthoromycota.pdf. Accessed 25 Jan 2012

Humber RA (2012b) Entomophthoromycota: a new phylum and reclassification for entomophthoroid fungi. Mycotaxon 120:477–492

Humber RA, Brown CC, Kornegay RW (1988) Equine zygomycosis caused by *Conidiobolus lamprauges*. J Clin Microbiol 27:573–576

Ingold CT (1978) Biology of *Mucor* and its allies. Edward Arnold, London

Ingold CT, Zoberi MH (1963) The asexual apparatus of Mucorales in relation to spore liberation. Trans Br Mycol Soc 46:115–134

Jackson HS, Dearden ER (1948) *Martensella corticii* Thaxter and its distribution. Mycologia 40:168–176

James TY, Kauff F, Schoch CL, Matheny PB, Hofstetter V, Cox C, Celio G, Gueidan C, Fraker E, Miądikowska J, Lumbsch HT, Rauhut A, Reeb V, Arnold EA, Amtoft A, Stajich JE, Hosaka K, Sung G-H, Johnson D, O'Rourke B, Crockett M, Binder M, Curtis JM, Slot JC, Wang Z, Wilson AW, Schüßler A, Longcore JE, O'Donnell K, Mozley-Standridge S, Porter D, Letcher PM, Powell MJ, Taylor JW, White MM, Griffith GW, Davies DR, Humber RA, Morton J, Sugiyama J, Rossman AY, Rogers JD, Pfister DH, Hewitt D, Hansen K, Hambleton S, Shoemaker RA, Kohlmeyer J, Volkmann-Kohlmeyer B, Spotts RA, Serdani M, Crous PW, Hughes KW, Matsuura K, Langer E, Langer G, Untereiner WA, Lacking R, Badel B, Geiser DM, Aptroot A, Diederich P, Schmitt I, Schultz M, Yahr R, Hibbett DS, Lutzoni F, McLaughlin D, Spatafora J, Vilgalys R (2006) Reconstructing the early evolution of the fungi using a six gene phylogeny. Nature 443:818–822

Jeffries P, Kirk PM (1976) New technique for the isolation of mycoparasitic Mucorales. Trans Br Mycol Soc 66:541–543

Jeffries P, Young TWK (1978) Mycoparasitism by *Piptocephalis unispora* (Mucorales): host range and reaction with *Phascolomyces articulosus*. Can J Bot 56:2449–2459

Jensen AB, Gargas A, Eilenberg J, Rosendahl S (1998) Relationships of the insect-pathogenic order Entomophthorales (Zygomycota, Fungi) based on phylogenetic analysis of nuclear small subunitribosomal DNA sequences (SSU rDNA). Fungal Genet Biol 24:325–334

Jermsuntiea W, Aki T, Kawamoto S, Ono K (2011) Matabolism and synthesis of lipids in the polyunsaturated fatty acid-producing fungus *Mortierella alliacea*. J Oleo Sci 60:11–17

Jiang X, Yu H, Xiang M, Liu X, Liu X (2011) *Echinochlamydosporium variable*, a new genus and species of Zygomycota from soil nematodes. Fungal Divers 46:43–51

Kagiwada S, Kayano Y, Hoshi H, Kawanishi T, Oshima K, Hamamoto H, Horie H, Namba S (2010) First report of *Choanephora* rot of ice plant (*Mesembryanthemum crystallinum*) caused by *Choanephora cucurbitarum* in Japan. J Gen Plant Pathol 76:345–347

Kataoka R, Takgi K, Sakakibara F (2011) Biodegregation of endosulfan by *Mortierella* sp. strain W8 in soil: influence of different substrates on biodegregation. Chemosphere 85:548–552

Keller S, Petrini O (2005) Keys to the identification of the arthropod pathogenic genera of the families Entomophthoraceae and Neozygitaceae (Zygomycetes), with descriptions of three new subfamilies and a new genus. Sydowia 57:23–53

Khanna S, Goyal A, Moholkar VS (2011) Microbial conversion of glycerol: present status and future prospects. Crit Rev Biotechnol 2011:1–28

Kimura M, Yaguchi T, Sutton DA, Fothergill AW, Thompson EH, Wickes BL (2011) Disseminated human conidiobolomycosis due to *Conidiobolus lamprauges*. J Clin Microbiol 49:752–756

King DS (1977) Systematics of *Conidiobolus* (Entomophthorales) using numerical taxonomy. III. Descriptions of recognized species. Can J Bot 55:718–729

King DS, Humber RA (1981) Identification of the Entomophthorales. In: Burges HD (ed) Microbial control of pests and plant diseases 1970–1980. Academic, London, pp 107–127

Kirk PM (2012) Nomenclatural novelties. Index Fungorum no. 11:1

Kirk PM, Benny GL (1980) The genus *Utharomyces* Boedijn (Pilobolaceae: Zygomycetes). Trans Br Mycol Soc 75:123–131

Kirk PM, Voigt K (2012) Nomenclatural novelties. Index Fungorum no. 13:1

Kirk PM, Cannon PF, Minter DW, Stalpers JA (2008) Ainsworth & Bisby's dictionary of the fungi, 10th edn. CAB International, Wallingford

Koehsler M, Michel R, Wylezich C, Lugauer J, Walochnik J (2007) Molecular identification and classification of *Cochlonema euryblastum*, a zoopagalean parasite of *Thecamoeba quadrilineata*. Mycologia 99:215–221

Kosa M, Ragauskas AJ (2011) Lipids from heterotrophic microbes: advances in metabolism research. Trends Biotechnol 29:53–61

Kovacs RL, Sundberg WJ (1996) An updated report on Syzygites megalocarpus (Zygomycetes, Mucorales) in the contiguous United States. 1996 MSA Abstracts. Inoculum 47(3):17

Kreutz M (2010) Leidys *Ouramoeba vorax* - mit video. http://www.mikroskopie-forum.de/index.php?topic=5762.msg39015#msg39015. Accessed 7 Nov 2011

Krings M, Taylor TN, Hass H, Kerp H, Dotzler N, Hermsen EJ (2007) Fungal endophytes in a 400-million-yr-old plant: infection pathways, spatial distribution, and host responses. New Phytol 174:648–657

Krings M, Taylor TN, Dotzler N (2013) Fossil evidence of the zygomycetous fungi. Persoonia 30:1–10

Krug JC, Benny GL, Keller HW (2004) Coprophilous fungi. In: Mueller GM, Bills GF, Foster MS (eds) Biodiversity of fungi. Inventory and monitoring methods. Elsevier, Burlington, MA, pp 467–499

Kuhlman EG (1972) Variation in zygospore formation among species of *Mortierella*. Mycologia 64:325–341

Kurihara Y, Degawa Y (2006) *Pinnaticoemansia*, a new genus of Kickxellales, with a revised key to the genera of Kickxellales. Mycoscience 47:205–211

Kurihara Y, Degawa Y, Tokumasu Y (2001) A new genus *Myconymphaea* (Kickxellales), with peculiar septal plugs. Mycol Res 105:1397–1402

Kurihara Y, Degawa Y, Tokumasu S (2004) Two novel kickxellalean fungi, *Mycoëmilia scoparia* gen. sp. nov. and *Ramicandelaber brevisporus* sp. nov. Mycol Res 108:1143–1152

Kurihara Y, Sukarno N, Ilyas M, Yuniarti E, Mangunwardoyo W, Park J-Y, Saraswati R, Widyastuti Y, Ando K (2008) Indonesian Kickxellales: two species of *Coemansia* and *Linderina*. Mycoscience 49:250–257

Kwon-Chung KJ, Bennett JE (1992) Medical mycology. Lea & Febiger, Philadelphia, PA

Laber B (2009) Amoeba attack by fungal parasite. (http://www.photomacrography.net/forum/viewtopic.php?t=8076&highlight=amoebophilus/. Accessed 8 Oct 2011

Labeyrie ES, Molloy DP, Lichtwardt RW (1996) An investigation of Harpellales (Trichomycetes) in New York state blackflies (Diptera: Simuliidae). J Invertebr Pathol 68:293–298

Latgé J-P, King DS, Papierok B (1980) Synonymie de *Entomophthora virulenta* Hall et Dunn et de *Conidiobolus thromboides* Drechsler. Mycotaxon 11:255–268

Latgé JP, Perry DF, Prévost MC, Samson RA (1989) Ultrastuctural studies of primary spores of *Conidiobolus*, *Erynia*, and related Entomophthorales. Can J Bot 67:2576–2589

Leidy J (1879) Freshwater rhizopods of North America. In: U.S. geological survey of the territories, vol XII. U.S. Government Printing Office, Washington, DC

Levetin E, Caroselli NE (1976) A simplified medium for growth and sporulation of *Pilobolus* species. Mycologia 68:1254–1258

Lichtwardt RW (1986) The Trichomycetes. Fungal associates of arthropods. Springer, New York, NY

Lichtwardt RW, Peterson SW, Huss MJ (1991) *Orphella hiemalis*: a new and rare trichomycete occurring in winter-emerging stoneflies (Plectoptera, Capniidae). Mycologia 83:214–219

Lichtwardt RW, Cafaro MJ, White MM (2001) The Trichomycetes. Fungal associates of arthropods. Revised Edition. http://www.nhm.ku.edu/fungi/monograph/mono.htm/. Accessed 2 Apr 2009

Liu D (ed) (2011) Molecular detection of human fungal pathogens. CRC, Boca Raton, FL

Liu X-Y, Huang H, Zheng R-Y (2001) Relationships within *Cunninghamella* based on sequence analysis of ITS rDNA. Mycotaxon 80:77–95

Liu YJ, Hodson MC, Hall BD (2006) Loss of the flagellum happened only once in the fungal lineage: phylogenetic structure of kingdom Fungi inferred from RNA polymerase II subunit genes. BMC Mol Biol 6:74. doi:10.1186/1471-2148-6-74

Liu X-Y, Huang H, Zheng R-Y (2007) Molecular phylogenetic relationships within *Rhizopus* based on combined analysis of ITS rDNA and pyrG gene sequences. Sydowia 59:235–253

Liu X-Y, Wang Y-N, Zheng R-Y (2012) Molecular phylogeny of *Pilaira* (Mucorales, Zygomycetes) inferred from ITS rDNA and *pyrG* sequences. Sydowia 64:55–66

Lu XL, Najafzedeh MJ, Dolatabadi S, Ran YP, Gerrits van den Ende AGH, Shen YN, Li CY, Xi LY, Hao YF, Zhang QQ, Li RY, Hu ZM, Lu GX, Wang JJ, Drogari-Apiranthitou M, Klaassen C, Meis JF, Hagen F, Liu WD, de Hoog GS (2013) Taxonomy and epidemiology of *Mucor irregularis*, agent of chronic cutaneous mucormycosis. Persoonia 30:48–56

Lutzoni F, Kauff F, Cox CJ, McLaughlin D, Celio G, Dentinger B, Padamsee M, Hibbett D, James TY, Baloch E, Grube M, Reeb V, Hofstetter V, Schoch C, Arnold AE, Miadlikowska J, Spatafora J, Johnson D, Hambleton S, Crockett M, Shoemaker R, Sung GH, Lücking R, Lumbsch T, O'Donnell K, Binder M, Diederich P, Ertz D, Gueidan C, Hansen K, Harris RC, Hosaka K, Lim YW, Matheny B, Nishida H, Pfister D, Rogers J, Rossman A, Schmitt I, Sipman H, Stone J, Sugiyama J, Yahr R, Vilgalys R (2004) Assembling the fungal tree of life progress, classification, and evolution of subcellular traits. Am J Bot 91:1446–1480

McLaughlin DJ, Kumar TKA, Blackwell M, Letcher PM, Roberson RW (2014) Subcelluar structure and biochemical characters in fungal phylogeny. In: McLaughlin DJ, Spatafora JW (eds) The Mycota, vol VIIA, 2nd edn. Springer, Berlin, In press

MacLeod DM, Tyrrell D, Welton MA (1980) Isolation and growth of the grasshopper pathogen, *Entomophthora grylli*. J Invertebr Pathol 36:85–89

McCabe DE, Soper RS (1985) Preparation of an entomopathogenic fungal insect control agent. US Patent 4,530,834, 23 July 1985

McCabe DE, Humber RA, Soper RS (1984) Observation and interpretation of nuclear reductions during maturation and germination of entomophthoralean resting spores. Mycologia 76:1104–1107

McGee PA (1996) The Australian zygomycetous mycorrhizal fungi: the genus *Densospora* gen. nov. Aust Syst Bot 9:329–336

McKerracher LJ, Heath IB (1985) The structure and cycle of the nucleus-associated organelle in two species of *Basidiobolus*. Mycologia 77:412–417

McNeill J, Barrie FR, Buck WR, Demoulin V, Greuter W, Hawksworth DL, Herendeen PS, Knapp S, Marhold K, Prado J, Prud'homme van Reine WF, Smith GF, Wiersema JH, Turland N (eds & comps) (2012) International Code of Nomenclature for algae, fungi, and plants (Melbourne Code), adopted by the Eighteenth International Botanical Congress, Melbourne, Australia, July 2011. Koeltz Scientific Books, Königstein, Germany (Regnum Vegetabile vol 154, and http://www.iapt-taxon.org/nomen/main.php?page=title. Cited 1 May 2013)

Mew TW, Gonzalez P (2002) A handbook of rice seedborne fungi. International Rice Research Institute, Los Banós, Philippines

Michaelides TJ, Spotts RA (1990) Postharvest diseases of pome and stone fruits caused by *Mucor piriformis* in the Pacific northwest and California. Plant Dis 74:537–543

Michailides TJ (1991) Characterization and comparative studies of *Mucor* isolates from stone fruits from California and Chile. Plant Dis 75:373–380

Morikawa C, Saikawa M, Barron GL (1993) Fungal predators of rotifers–a comparative study of Zoophagus, Lecophagus and Cephaliophora. Mycol Res 97:421–428

Morton JB, Benny GL (1990) Revised classification of arbuscular mycorrhizal fungi (Zygomycetes): a new order, Glomales, two new suborders, Glomineae and Gigasporineae and two new families, Acaulosporaceae and Gigasporaceae, with an emendation of Glomaceae. Mycotaxon 37:471–491

Moss ST, Lichtwardt R (1976) Development of trichospores and their appendages in *Genistellospora homothallica* and other Harpellales and fine-structural evidence for the sporangial nature of trichospores. Can J Bot 54:2346–2364

Moss ST, Lichtwardt RW (1977) Zygospores of the Harpellales: an ultrastructural study. Can J Bot 55:3099–3110

Moss ST, Young TWK (1978) Phyletic considerations of the Harpellales and Asellariales (Trichomycetes, Zygomycotina) and Kickxellales (Zygomycetes, Zygomycotina). Mycologia 70:944–963

Moss ST, Lichtwardt RW, Manier J-F (1975) *Zygopolaris*, a new genus of Trichomycetes producing zygospores with polar attachment. Mycologia 67:120–127

Mrva M (2008) Infection of *Mayorella penardi* (Gymnamoebia) in oak-hornbeam forest soil by the ectoparasitic fungus *Amoebophilus simplex* (Zygomycota). Eur J Soil Biol 44:80–83

Mrva M (2011) *Mayorella vespertioides* Page, 1983 (Amoebozoa) - a new host for the ectoparasitic fungus *Amoebophilus simplex* (Zygomycota). Biologia 66:645–647

Münchberg U, Wagner L, Spielberg ET, Voigt K, Rösch P, Popp J (2012) Spatially resolved investigation of the oil composition in single intact hyphae of *Mortierella* spp. with micro-Raman spectroscopy. Biochim Biophys Acta 1831:341–349

Nagahama T, Sato H, Shimazu M, Sugiyama J (1995) Phylogenetic divergence of the entomophthoralean fungi: evidence from nuclear 18S ribosomal RNA gene sequences. Mycologia 87:203–209

Nagy LG, Petkovits T, Kovács GM, Voigt K, Vágvölgyi C, Papp T (2011) Where is the unseen fungal diversity hidden? A study of *Mortierella* reveals a large contribution of reference collections to the identification of fungal environmental sequences. New Phytol 191:789–794

Nowakowski L (1883) Entomophthoreae. Przycrynek do znajomósci pasorzytnych grzybków sprawiajacych pomór owadów [Entomophthoraceae. Contribution to knowledge of parasitic fungi causing insect deaths]. Pamietnik Wydz Akad Umiej Krakow 8:153–183

O'Donnell KL (1979) Zygomycetes in culture. Department of Botany, University of Georgia, Athens, GA

O'Donnell KL, Flegler SL, Hooper GR (1978) Zygosporangium and zygospore formation in *Phycomyces nitens*. Can J Bot 56:96–100

O'Donnell K, Cigelnik E, Benny GL (1998) Phylogenetic relationships among the Harpellales and Kickxellales. Mycologia 90:624–639

O'Donnell K, Lutzoni FM, Ward TJ, Benny GL (2001) Evolutionary relationships among mucoralean fungi (Zygomycota): evidence for family polyphyly on a large scale. Mycologia 93:286–296

Ogawa JM, Sonada RM, English H (1992) Postharvest diseases of tree fruit. In: Kumar J, Chambe HS, Singh US, Mukhopadhyay AN (eds) Plant diseases of international importance. III. Diseases of fruit crops. Prentice Hall, Englewood Cliffs, NJ, pp 405–422

Ogawa Y, Hayashi S, Degawa Y, Yaguchi Y (2001) *Ramicandelaber*, a new genus of the Kickxellales, Zygomycetes. Mycoscience 42:193–199

Oman SJ, White MM (2011) Extended studies of *Baltomyces styrax* in Idaho and expanded distribution of this isopod gut fungus in USA. Mycologia 104:313–320

Page RM (1962) Light and asexual reproduction of *Pilobolus*. Science 138:1238–1245
Page RM (1964) Sporangium discharge in *Pilobolus*: a photographic study. Science 146:925–927
Papp T, Hoffmann K, Nyilasi I, Petkovits T, Wagner L, Vágvölgyi C, Voigt K (2011) *Mortierella*. In: Liu D (ed) Molecular detection of human fungal pathogens. CRC, Boca Raton, FL, pp 749–757
Papierok B, Hajek AE (1997) Fungi: Entomophthorales. In: Lacey L (ed) Manual of techniques in insect pathology. Academic, San Diego, CA, pp 187–212
Partida-Martinez LP, Hertweck C (2006) Pathogenic fungus harbours endosymbiotic bacteria for toxin production. Nature 437:884–888
Partida-Martinez LP, de Looß CF, Ishida K, Ishida M, Roth M, Buder K, Hertweck C (2007a) Rhizonin, the first mycotoxin isolated from the Zygomycota, is not a fungal metabolite but is produced by bacterial endosymbionts. Appl Environ Microbiol 73:793–797
Partida-Martinez LP, Groth I, Schmitt I, Richer W, Roth M, Hertweck C (2007b) *Burkholderia rhizoxinica* sp. nov. and *Burkholderia endofungorum* sp. nov., bacterial endosymbionts of the plant-pathogenic fungus *Rhizopus microsporus*. Int J Syst Evol Microbiol 57:2583–2590
Partida-Martinez LP, Bandemer S, Rüchel R, Dannaoui E, Hertweck C (2008) Lack of evidence of endosymbiotic toxin-producing bacteria in clinical *Rhizopus* isolates. Mycoses 51:266–269
Petkovits T, Nagy LG, Hoffmann K, Wagner L, Nyilasi I, Griebel T, Schnabelrauch D, Vogel H, Voigt K, Vágvölgyi C, Papp T (2011) Data partitions, Bayesian analysis and phylogeny of the zygomycetous fungal family Mortierellaceae, inferred from nuclear ribosomal DNA sequences. PLoS One 6 (11):e27057
Powell MJ, Bracker CE, Sternshein DJ (1981) Formation of chlamydospores in *Gilbertella persicaria*. Can J Bot 59:908–928
Prabhu RM, Patel R (2004) Mucormycosis and entomophthoromycosis: a review of the clinical manifestations, diagnosis and treatment. Clin Microbiol Infect 10(Suppl 1):31–47
Redecker D, Schüssler A (2014) Glomeromycota. In: McLaughlin DJ, Spatafora JW (eds) Systematics and evolution: Part A. Springer, Heidelberg
Remaudière G, Hennebert GL (1980) Révision systématique de *Entomophthora aphidis* Hoffm. in Fres. Description de deux nouveaux pathogènes d'aphides. Mycotaxon 11:269–321
Remaudière G, Keller S (1980) Revision systematique des genres d'Entomophthoraceae a potentiale entomopàthogene. Mycotaxon 11:323–338
Ribes JA, Vanover-Sams CL, Baker DL (2000) Zygomycetes in human disease. Clin Microbiol Rev 13:236–301
Rizzo AM, Pang KL (2005) A zygosporic stage of the Harpellales (Trichomycetes, Zygomycota) associated with simuliid egg masses. Mycol Res 109:266–270
Rohm B, Scherlach K, Möbius N, Partida-Martinez LP, Hertweck C (2010) Toxin production by bacterial endosymbionts of a *Rhizopus microsporus* strain used for tempe/sufu processing. Int J Food Microbiol 136:368–371
Saikawa M (1986) Ultrastructure of *Stylopage rhabdospora* an amoeba capturing zoopagaceous fungus. Mycologia 78:309–312
Saikawa M (1989) Ultrastructure of the septum in *Ballocephala verrucospora* (Entomophthorales, Zygomycetes). Can J Bot 67:2484–2488
Saikawa M (2011a) On fungi in the Zoopagaceae and Cochlonemataceae. Nihon Kingakkai Kaiho (Jap J Mycol) 52:19–27
Saikawa M (2011b) Ultrastructural studies on zygomycetan fungi in the Zoopagaceae and Cochlonemataceae. Mycoscience 52:83–90
Saikawa M (2012) Morphological studies on fungi in the Zoopagaceae and Cochlonemataceae. Bull Tokyo Gakugei Univ Div Nat Sci 64:55–76
Saikawa M, Aoki Y (1991) Pouchy appendages on the conidia of *Euryancale sacciospora* and *Euryancale maripospora* and the rediscovery of *Euryancale obliqua* from Japan. Trans Mycol Soc Jpn 32:509–515
Saikawa M, Aoki Y (1995) Taxonomic studies on *Euryancale marsipospora* and *E. sacciospora*. Nova Hedwigia 60:571–581
Saikawa M, Kadowaki T (2002) Studies on *Acaulopage dichotoma* and *A. tetraceros* (Zoopagales, Zygomycota) capturing soil amoebae. Nova Hedwigia 74:365–371
Saikawa M, Katsurashima E (1993) Light and electron microscopy of a new species of *Euryancale* producing phallus-shaped conidia. Mycologia 85:24–29
Saikawa M, Morikawa C (1985) Electron microscopy on a nematode-trapping fungus, *Acaulopage pectospora*. Can J Bot 63:1386–1390
Saikawa M, Sato H (1986) Zygosporic structures in two species of *Euryancale* (Zoopagales). Trans Br Mycol Soc 87:337–340
Saikawa M, Sato H (1991) Ultrastructure of *Cochlonema odontosperma* an endoparasite in amoebae. Mycologia 83:403–408
Saikawa M, Yamaguchi K, Morikawa C (1988) Capture of rotifers by *Acaulopage pectospora*, and further evidence of its similarity to *Zoophagus insidians*. Mycologia 80:880–884
Saikawa M, Oguchi M, Castañeda Ruiz RF (1997) Electron microscopy of two nematode destroying fungi, *Meristacrum asterospermum* and *Zygnemomyces echinulatus* (Meristacraceae, Entomophthorales). Can J Bot 75:762–768
Saikawa M, Wakai Y, Katsushima N (2011) Electron microscope studies on the development and germination of zygospores in *Zoophagus pectos-*

porus, a zoopagaceous fungus capturing nematodes. Bull Tokyo Gakugei Univ Div Nat Sci 63:91–100

Sakai S, Kato M, Nagamasu H (2000) *Artocarpus* (Moraceae)—gall midge flower pollination mutualism merdiated by a male-flower parasitic fungus. Am J Bot 87:440–445

Saksena HK, Narain U (1983) Blight of vegetabile seedlings. Indian J Hort 40:129–131

Sato H (2002) Two ultrastructural aspects of the trichospore of *Pennella angustispora* (Harpellales): canals in the sporangiospore cell wall and appendage formation. Mycoscience 43:33–36

Sato Y, Narisawa K, Tsuruta K, Umezu M, Nishizawa T, Tanaka K, Yamiguchi M, Komatsuzaki M, Ohta H (2010) Detection of Betaproteobacteria inside the Mycelium of the Fungus *Mortierella elongata*. Microbes Environ 25:321–324

Schimek C, Wöstemeyer J (2006) Pheromone action in the fungal groups Chytridiomycota and Zygomycota, and in the Oomycota. In: Kües U, Fischer R (eds) Growth, differentiation and sexuality, vol 1, 2nd edn, The Mycota. Springer, Berlin, pp 216–231

Schimek C, Wöstemeyer J (2009) Carotene derivatives in sexual communication of zygomycete fungi. Phytochemistry 70:1867–1875

Schimek C, Kleppe K, Saleem AH, Voigt K, Burmester A, Wöstemeyer J (2003) Sexual reactions in Mortierellales are mediated by the trisporic acid system. Mycol Res 107:736–747

Schipper MAA (1978) On the genera *Rhizomucor* and *Parasitella*. Stud Mycol 17:53–71

Schipper MAA (1979) *Thermomucor* (Mucorales). Antonie van Leeuwenhoek 45:275–280

Schipper MAA, Stalpers JA (1980) Various aspects of the mating system in Mucorales. Persoonia 11:53–63

Schipper MAA, Samson RA, Stalpers JA (1975) Zygospore ornamentation in the genera *Mucor* and *Zygorhynchus*. Persoonia 8:321–328

Schmidt SK, Wilson KLL, Meyer AF, Gebauer MM, King AJ (2008) Phylogeny and ecophysiology of opptunistic "snow molds" from a subalpine forest ecosystem. Microbiol Ecol 56:681–687

Schoch CL, Seifert KA, Huhndorf S, Robert V, Sprouge JL, Levesque CA, Chen W, Fungal Barcode Consortium (2012) Nuclear ribosomal internal transcribed spacer (ITS) region as a universal DNA barcode marker for Fungi. Proc Natl Acad Sci USA 109:6241–6246

Schüssler A, Schwarzott D, Walker C (2001) A new fungal phylum, the Glomeromycota: phylogeny and evolution. Mycol Res 105:1413–1421

Seifert K, Morgan-Jones G, Gams W, Kendrick B (2011) The genera of Hyphomycetes. CBS Biodiversity Series 9. CBS-KNAW Fungal Biodiversity Centre, Utrecht, The Netherlands

Sergeeva YE, Galanina LA, Lunin VV, Feofilova EP (2011) A biotechnology for the production of new biodiesel fuel. Catal Indus 3:53–56

Shimada H, Saikawa M (2006) Observations on germination of chlamydospores and capture of nematodes in *Cystopage cladospora* (Zoopagales, Zygomycota). Nippon Kingakukai Kaiho 47:1–9

Shtienberg D (1997) *Rhizopus* Head rot of confectionery sunflower: effects on yield quantity and quality and implications for disease management. Phytopathology 87:1226–1232

Siemensma F (2012) *Mayorella* infected by the fungus *Amoebophilus simplex*. http://www.arcella.nl/fungus-Mayorella?. Accessed 21 Oct 2012

Sinha S (1940a) A wet rot of leaves of *Colocasia antiquorum* due to secondary infection by *Choanephora cucurbitarum* Thaxter and *Choanephora trispora* Thaxter sp. (=*Blakeslea trispora* Thaxter). Proc Indian Acad Sci Sect B 11:167–176

Sinha S (1940b) On the characters of *Choanephora cucurbitarum* Thaxter on chillies (*Capsicum* spp.). Proc Indian Acad Sci Sect B 11:162–166

Steinkraus DC, Oliver JB, Humber RA, Gaylor MJ (1998) Mycosis of bandedwinged whitefly (*Trialeurodes abutilonea*) (Homoptera: Aleyrodidae) caused by *Orthomyces aleyrodis* gen. et. sp. nov. (Entomophthorales: Entomophthoraceae). J Invertebr Pathol 72:1–8

Strauss T, Botha A, Greyling D, Mostert T, du Preez PJ, Kock JLF (2000) Development and testing of selective media for mucoralean fungi. S Afr J Sci 96:597–602

Sugiyama J (1998) Relatedness, phylogeny, and evolution of the fungi. Mycoscience 38:487–511

Taba S, Nakazato M, Nasu K, Takushi T, Moromitzato Z (2011) *Gilbertella* stem rot of pitaya (*Hylocereus undatus*), a new disease caused by *Gilbertella persicaria*. Jpn J Phytopathol 77:291–294 (in Japanese)

Tanabe Y, Nagahama T, Saikawa M, Sugiyama J (1999) Phylogenetic relationship of *Cephaliophora* to nematophagus hyphomycetes iincluding taxonomic and nomenclatural emendations of the genus *Lecophagus*. Mycologia 91:830–835

Tanabe Y, O'Donnell K, Saikawa M, Sugiyama J (2000) Molecular phylogeny of parasitic Zygomycota (Dimargaritales, Zoopagales) based on nuclear small subunit ribosomal DNA sequences. Mol Phylogenet Evol 16:253–262

Tansey MR (1984) Efficient isolation of thermophilic and thermotolerent mucoralean fungi. Mycopathologia 85:31–42

Taylor TN, Klavins KD, Krings M, Taylor EL, Kerp H, Hass H (2003) Fungi from the Rhynie chert: a view from the dark side. Trans R Soc Edinb Earth Sci 94:457–473

Taylor TN, Krings M, Taylor EL (2014a) Fungal diversity in the fungal record. In: McLaughlin DJ, Spatafora JW (eds) The Mycota, vol VIIA, 2nd edn. Springer, Berlin, In press

Thaxter R (1888) The Entomophthoreae of the United States. Mem Boston Soc Nat Hist 4:133–201

Thaxter R (1895) New or peculiar American Zygomycetes. I. *Dispira*. Bot Gaz 20:513–518

Thaxter R (1914) Nw or peculiar Zygomycetes: 3: *Blakeslea*, *Dissophora*, and *Haplosporangium*, nova genera. Bot Gaz 58:353–366

Thaxter R (1922) A revision of the Endogoneae. Proc Am Acad Arts Sci 57:291–351

Tigini V, Prigione V, Donelli I, Anastasi A, Freddi G, Giansanti P, Mangiavillano A, Varese GC (2011) *Cunninghamella elegans* biomass optimisation for textile wastewater biosorption treatment: an analytical and ecotoxicological approach. Appl Microbiol Biotechnol 90:343–352

Tretter ED, Johnson EM, Wang Y, Kandel P, White MM (2013) Examining new phylogenetic markers to uncover the evolutionary history of early diverging fungi: comparing MCM7, TSR1, and rRNA genes for single- and multi-gene analysis of the Kickxellomycotina. Persoonia 30:106–125

Tucker BE (1981) A review of the nonentomogenous Entomophthorales. Mycotaxon 13:481–505

Tyrrell D (1977) Occurrence of protoplasts in the natural life cycle of *Entomophthora egressa*. Exp Mycol 1:259–263

Valle LG, Cafaro MJ (2008) First report of zygospores in Asellariales and new species from the Caribbean. Mycologia 100:122–131

Valle LG, Santamaria S (2005) Zygospores as evidence of sexual reproduction in the genus *Orphella*. Mycologia 97:1335–1347

van Tieghem P (1878) Troisième mémoire sur les Mucorinées. Ann Sci Nat Bot, Sér 6, 4:312–398

Voglmayr H, Krisai-Greilhuber I (1996) *Dicranophora fulva*, a rare mucoraceous fungus growing on boletes. Mycol Res 100:583–590

Voigt K (2012) Zygomycota Moreau. In: Frey W (ed) Syllabus of plant families. A. Engler's Syllabus der Pflanzenfamilien, 13th edn. Part 1/1. Blue-green algae, Myxomycetes and Myxomycete-like organisms, Phytoparasitic protists, heterotrophic Heterokontobiota and Fungi p.p. Borntraeger Science Publishers, Berlin, pp 130–162

Voigt K, Kirk PM (2011) Recent developments in the taxonomic affiliation and phylogenetic positioning of fungi: impact in applied microbiology and environmental biotechnology. Appl Microbiol Biotechnol 90:41–57

Voigt K, Olsson I (2008) Molecular phylogenetic and scanning electron microscopical analysis places the Choanephoraceae and the Gilbertellaceae in a monophyletic group within the Mucorales (Zygomycetes, Fungi). Acta Biol Hung 59:365–383

Voigt K, Wöstemeyer J (2001) Phylogeny and origin of 82 Zygomycetes from all 54 genera of the Mucorales and Mortierellales based on combined analysis of actin and translation elongation factor EF-1α genes. Gene 270:113–120

Voigt K, Cigelnik E, O'Donnell K (1999) Phylogeny and PCR identification of clinically important Zygomycetes based nuclear ribosomal-DNA sequence data. J Clin Microbiol 37:3957–3964

Vuillemin P (1904) Le *Spinalia radians* g. et sp. nov. et la série des Dispirées. Bull Soc Mycol France 20:26–33

Wagner L, Stielow B, Hoffmann K, Petkovits T, Papp T, Vágvölgyi C, de Hoog GS, Verkley G, Voigt K (2013) A comprehensive molecular phylogeny of the Mortierellales (Mortierellomycotina) based on nuclear ribosomal DNA. Persoonia 30:77–93

Walther G, Pawłowska P, Alastruey-Izquierdo A, Wrzosek M, Rodriguez-Tudela JL, Dolatabadi S, Chakrabarti A, de Hoog GS (2013) DNA barcoding in Mucorales: an inventory of biodiversity. Persoonia 30:11–47

Wang J-F, Liu X-J, Liu R-S, H-Mei L, Tang Y-J (2011) Optimization of the mated fermentation process for the production of lycopene by *Blakeslea trispora* NRRL 2895 (+) and NRRL 2896 (−). Bioprocess Biosyst Eng 35:553–564

Warcup JH (1990) Taxonomy, culture and mycorrhizal associations of some zygosporic Endogonaceae. Mycol Res 94:173–178

Waterhouse GM (1973) Entomophthorales. In: Ainsworth GC, Sparrow FK, Sussman AS (eds) The Fungi, vol 4b. Academic, New York, NY, pp 219–229

Watson P (1964) Spore germination in *Spinellus macrocarpus*. Trans Br Mycol Soc 47:239–245

Weete JD, Abril M, Blackwell M (2010) Phylogenetic distribution of fungal sterols. PLoS One 5(5): e10899. doi:10.1371/journal.pone0010899

Wenger CJ, Lilly VG (1966) The effects of light on carotenogensis, growth, and sporulation of *Syzygites megalocarpus*. Mycologia 58:671–680

Werkman BA (1976) Localization and partial characterization of a sex-specific enzyme in homothallic and heterothallic Mucorales. Arch Microbiol 109:209–213

White MM (2003) First report of *Basidiolum fimbriatum* since 1861, with comments on its development, occurrence, distribution and relationship with other fungi. Mycol Res 107:243–250

White MM (2006) Evolutionary implications of a rRNA-based phylogeny of Harpellales. Mycol Res 110:1011–1024

White MM, Lichtwardt RW, Colbo MH (2006a) Confirmation and identification of parasitic stages of obligate endobionts (Harpellales) in blackflies (Simuliidae) by means of rRNA sequence data. Mycol Res 110:1070–1079

White MM, James TY, O'Donnell K, Cafaro MJ, Tanabe Y, Sugiyama J (2006b) Phylogeny of the Zygomycota based on nuclear ribosomal sequence data. Mycologia 98:872–884

Wolf FA (1917) A squash disease caused by *Choanephora cucurbitarum*. J Agric Res 8:319–327

Wöstemeyer J, Schimek C (2007) Trisporic acid and mating in Zygomycetes. In: Heitman J, Kronstad JW, Taylor JW, Casselton LA (eds) Sex in fungi. Molecular determination and evolutionary implications. ASM, Washington, DC, pp 433–443

Wöstemeyer J, Wöstemeyer A, Burmester A, Czempinski K (1995) Relationships between sexual process and parasitic interactions in the host—pathogen system *Absidia glauca—Parasitella parasitica*. Can J Bot 73(Suppl 1):S243–S250

Wu G-G, Lin S-J (1997) Endogonales in Taiwan: a new genus with unizygosporic sporocarps and a hyphal mantle. Mycotaxon 64:179–188

Yafetto L, Carroll L, Cui Y, Davis DJ, Fischer MWF, Henterly AC, Kessler JD, Kilroy HA, Shidler JB, Stolze-Rybczynski JL, Sugawara Z, Money NP (2008) The fastest flights in nature: high-speed spore discharge mechanisms among fungi. PLoS One 3(9):e3237 (also with Video S4 of *Pilobolus kleinii* sporangial discharge, 50,000 fps. doi:10.1371/journal.pone.0003237.s003)

Yao Y-J, Pegler DN, Young TWK (1996) Genera of endogonales. Royal Botanic Gardens, Kew

Young TWK (1968) Electron microscopic study of asexual spores in Kickxellaceae. New Phytol 67:823–836

Young TWK (1974) Ultrastructure of the sporangiospore of *Kickxella alabastrina* (Mucorales). Ann Bot 38:873–876

Zain ME, Moss ST, El-Sheikh HH (2012) Development of merosporangia in *Linderina pennispora* (Kickxellales, Kickxellaceae). IMA Fungus 3:103–108

Zheng R-Y, Chen G-Q (2001) A monograph of *Cunninghamella*. Mycotaxon 80:1–75

Zheng R-Y, Liu X-Y (2009) Taxa of *Pilaira* (Mucorales, Zygomycota) from China. Nova Hedwigia 88:255–267

Zheng R-Y, Chen G-Q, Liu X-Y (2007) A monograph of *Rhizopus*. Sydowia 59:273–372

Zycha H, Siepmann R, Linnemann G (1969) Mucorales eine Beschreibung aller Gattungen und Arten dieser Pilzgruppe. J. Cramer, Lehre, Germany, 355 p

9 Glomeromycota

D. Redecker[1], A. Schüßler[2]

CONTENTS

[1]Université de Bourgogne, UMR1347 Agroécologie, 17 rue Sully, BP 86510, 21000 Dijon, France; e-mail: dirk.redecker@dijon.inra.fr

[2]Department of Biology I, Genetics, Ludwig-Maximilians-University Munich, Großhaderner Str. 2-4, 82152 Martinsried, Germany; e-mail: arthur.schuessler@lmu.de

I. Introduction

The Glomeromycota are a monophyletic group of fungi living as obligate biotrophs forming arbuscular mycorrhiza (AM) or (in one instance) an endosymbiosis with cyanobacteria (Schüßler et al. 2001b). Being one of the smallest of the fungal phyla, the Glomeromycota presently include only approximately 230 described species (Schüßler and Walker 2010). Taxa have been traditionally described based on the morphology of the large, multinucleate spores, which are sometimes organized in spore aggregates or sporocarps. However, due to the paucity of morphological characters, molecular data have been increasingly used for taxon description from the phylum down to species.

Within the true fungi, the Glomeromycota have been placed as a sister group to Ascomycota and Basidiomycota (Dikarya) in rDNA-based phylogenies, but they group among lineages of the paraphyletic zygomycetous fungi when protein-coding genes are used (Lee and Young 2009; Liu et al. 2009; Redecker and Raab 2006). In the case of a sister group relationship to the Dikarya, the clade uniting the two would be characterized by the ability to form mutualistic symbioses with plants or algae, which is rarely found in other clades. Zygomycetous fungi and Glomeromycota both have coenocytic (non-septate) mycelium and a certain similarity of the spores and sporocarps, but both could be shared ancestral traits.

Molecular-marker-based field studies have recently revealed a considerable diversity of AM fungi (AMF) that could not be assigned to formally described species, possibly due to a

Systematics and Evolution, 2nd Edition
The Mycota VII Part A
D.J. McLaughlin and J.W. Spatafora (Eds.)

high proportion of species rarely or never producing spores.

Here, we review the recent state of glomeromycotan systematics. The numerous changes and revisions in AMF systematics over the last decade are presented in the context of the historical background and also of their implications for ecology and evolution.

II. Arbuscular Mycorrhizal Symbiosis

AM is the most widespread type of mycorrhizal symbiosis, a mutualistic association between plants and fungi. The great majority of land plants, among them vascular plants, up to around 80 % of investigated species, are known to form AM (Brundrett 2009). The remainder either is nonmycorrhizal or forms one of the other types of mycorrhiza, i.e., ectomycorrhiza, ericoid, or orchid mycorrhiza. It should be noted, however, that many gymnosperms, pteridophytes, and even nonvascular plants, like liverworts and hornworts, form AM or AM-like associations (Smith and Read 2008). Thus, AMF are found ubiquitously in soils wherever their plant hosts are available.

As obligate symbionts, the fungi depend entirely on the support of reduced carbon compounds by plants. The dependence of plants on their mycosymbionts varies according to plant species and environmental conditions, but in any case AM is one of the major factors in plant nutrient uptake and nutrient cycling in the soil. Plants benefit in particular from the transport of immobile ions, such as phosphate, which are difficult for the root to reach (Smith and Read 2008). Significant transport of nitrogen has also been reported (Jin et al. 2005). The hyphae of extraradical mycelia are by an order of magnitude finer than the root hairs and therefore much more efficient in taking up ions from small soil pores and extending the volume of exploration for immobile nutrients well beyond the depletion zone found around the root. Among other benefits to plants, which may be due in part to better mineral nutrition but also to less-investigated, more specific effects, improved resistance against root and other pathogens has been reported (Azcon-Aguilar and Barea 1996).

As most crops form AM, this symbiosis also has considerable economic importance (Gianinazzi et al. 2010). However, fungal diversity in agricultural settings seems to be strongly diminished by management practices such as plowing or by fungicide application (Helgason et al. 1998; Oehl et al. 2003).

Typically, AMF form finely branched tree-shaped structures within root cells, the eponymous arbuscules. Plant and fungal cytoplasm are only separated by plasma membranes and a very thin layer of amorphous wall polymers, facilitating the exchange of nutrients between symbionts (Bonfante-Fasolo and Grippiolo 1982). In fact, an exchange of phosphate from fungus to plant across the arbuscules has been demonstrated, whereas hexoses apparently are also transferred elsewhere from the plant to the fungus (Smith et al. 2001). Some glomeromycotan families also form storage organs inside roots, the vesicles, which usually appear at later stages of the association.

The morphology of intraradical (within root) symbiotic structures in the AM has been classified into two types, the *Paris* and the *Arum* types, according to the two host plants where they were first described. In *Arum*-type colonization, the fungus proliferates along the root in the intercellular spaces and arbuscules enter into the cells from the resulting axes. In the *Paris* type, the fungi spread from cell to cell, and in many cases intracellular hyphal coils are formed instead of or together with arbuscules. Thus, in many cases plants forming AM do not necessarily show arbuscule formation. The two types, however, just represent two ends of a continuum of structures that are determined by the plant host, the identity of the fungus, or the interaction of the two (Dickson et al. 2007). Thus, they may even be present in the same root.

Arbuscular mycorrhizal fungi are found everywhere where hosts to this symbiosis occur. Non-AM plants may have other kinds of mycorrhiza, e.g., many woody species, in particular the Pinaceae, which have ectomycorrhiza, orchids and ericoid plants with their own associations, and some families typically regarded as nonmycorrhizal, such as the Brassicaceae, Chenopodiaceae, and Cyperaceae, may still have members that form these associations (Smith and Read 2008). Therefore, the habitats of these fungi include most plant ecosystems, even submerged plants (Sondergaard and Laegaard 1977), plants in geothermal soils (Appoloni et al. 2008; Bunn and Zabinski 2003), and deserts (Stutz and Morton 1996).

III. Morphology and Reproduction

The Glomeromycota form a coenocytic mycelium of narrow to broad (2–10 μm, sometimes up to 20 μm), often knobby hyphae. Anastomoses, resulting in an interconnected hyphal network, have been reported frequently from the Glomeraceae but do not seem to occur or are rare in the Gigasporaceae, although the latter possess the ability to form end-to-end anastomoses to bridge interrupted hyphal connections (de la Providencia et al. 2005; Gerdemann 1955a; Purin and Morton 2011). Septa are formed in senescent parts of the mycelium, when the fungus retracts the cytoplasm, or after spore formation.

Germ tubes emerge from spores in different ways, according to the taxon: through the attachment of the subtending hypha or through the spore wall (in some taxa both modes exist) and with or without the involvement of a membranous germination structure (germination shield, germination coil; see following sections for details). Spore germination may be enhanced by plant-produced factors (Bécard et al. 1995). Strigolactones have been identified as compounds inducing spore germination or hyphal branching near a prospective host, thereby maximizing the chance to colonize it (Akiyama et al. 2005; Besserer et al. 2006). On the root surface, appressoria (hyphopodia) are formed that allow the fungus to enter the epidermal cells. The formation by the plant of a prepenetration apparatus facilitates and directs the entrance and the transit of hyphae across the epidermal and cortical root cells (Genre et al. 2008).

Inside the root the fungus may form arbuscules, hyphal coils, or vesicles. Depending on physiological factors, spore formation may be triggered after some time. These spores are always multinucleate and, depending on the size, may contain between fewer than 50 and several thousand nuclei (Bécard and Pfeffer 1993; Marleau et al. 2011). The question of whether these nuclei are genetically homogeneous or constitute a mixed "population" of genotypes has been the subject of a long-standing debate [for overviews see Rosendahl (2008) and Young (2008)]. New roots may be colonized from spores after germination or in many taxa also directly by mycelia emanating from a colonized root. Exceptions to the latter again are members of the Gigasporaceae, which apparently always colonize roots starting from spores. Hyphal fragments in the soil may also act as infective propagules.

No morphological evidence for sexual reproduction has been confirmed in the Glomeromycota. Therefore, their spores, despite a certain resemblance to *Endogone* zygospores, are assumed to be formed asexually. Close examination of nuclear migration during spore formation provided no hint of sexual processes (Jany and Pawlowska 2010). However, studies combining microscopic examination and molecular genetics have provided evidence for an exchange of genetic markers between different strains and, thus, for genetic recombination (Sanders and Croll 2010), at least in the model AMF *Rhizophagus irregularis* (formerly known as *Glomus intraradices* or *Glomus irregulare*).

IV. Dispersal and Host Relations

A. Geographical Distribution

Due to the cryptic nature of their association with plants, data about the geographical distribution of glomeromycotan taxa are scarce. Large regions of the world have not been surveyed, even for AMF spores, which would allow at least limited insight into local glomeromycotan diversity. A number of species have been found in only a single location and could be endemic, while others are surprisingly widespread globally. It is indeed puzzling when approximately 20 % of all described morphospecies are found in one region and approximately 12 % in a single field site (Oehl et al. 2003). Molecular field surveys confirmed the pattern of widespread (bona fide) endemism on one hand but extremely widespread dispersal of other taxa on the other (Öpik et al. 2006). However, certain species that were proposed as possibly specific to a certain environment or altitude were later detected, on the basis of molecular markers, in very different habitats (Krüger et al. 2011). Thus, it must be concluded that much remains to be discovered in this respect and that it is too early to make concise statements about the biogeography of most AMF taxa. The well-studied species *R. irregularis* has been detected in a multitude of habitats and regions, often as the dominant molecular taxon (e.g., Appoloni et al. 2008; Sýkorová et al. 2007), and disturbance-adapted species such as *Glomus mosseae* (recently renamed *Funneliformis mosseae*) are also extremely widespread, especially in agricultural soils (e.g., Daniell et al. 2001; Helgason et al. 1998, Hijri et al. 2006). Interestingly, genotypes of this species seem to be rather uniform worldwide, with no geographic structure detectable. Based on these data, Rosendahl (2008) concluded that the species probably has been relatively recently spread by agricultural practice around the world. The more thorough and defined use of molecular operational taxonomic units (MOTUs) (Hibbett et al. 2011) might facilitate a better understanding of AMF biogeography in the future, providing their clear definition (Hawksworth et al. 2011).

Dispersal has not been well studied in the Glomeromycota. Hyphal spread from colonized plants and spores transported with soil particles may be the predominant nonhuman-mediated means of dispersal, but translocation of spores by earthworms or mammals has also been reported (Gange 1993). Some sporocarpic species might also be spread through the feces of rodents (Mangan and Adler 2002).

B. Host Specificity

Considering the relation between glomeromycotan species number and the richness of potential host plants there does not seem to be much room for host specificity in AM. Indeed, greenhouse experiments, combining single species of plant host and mycobiont, indicated almost universal compatibility (Klironomos et al. 2000). It is clear, however, that species cultivatable in the greenhouse are most likely not representative of what occurs in the field, and the diversity of cultured AMF may be strongly biased toward generalists. Molecular approaches allowed this question to be addressed in the field, and the results generally showed the absence of strict specificity. Most plants associate with several glomeromycotan species at the same time, and most glomeromycotan species are linked to different species of plants. However, a certain degree of host preferences (Helgason et al. 2002; Sýkorová et al. 2007) was demonstrated in some studies. Strict host specificity in the sense of a limited spectrum of fungal associates of a host plant was found only in mycoheterotrophic plants that parasitize the mycorrhizal association (Bidartondo et al. 2002).

V. Development of Taxonomic Theory

The history of AM research and glomeromycotan taxonomy has been reviewed by Koide and Mosse (2004) and was described as comprising four major periods (Stürmer 2012): the discovery period (1845–1974), the alpha taxonomy period (1975–1989), the cladistics period (1990–2000), and the phylogenetic synthesis period (since 2001). Spores and sporocarps of glomeromycotan fungi had in fact been collected and described long before it became clear that these fungi formed a mycorrhizal association. Initially, nearly exclusively sporocarp-forming species were the focus, starting with the first *Glomus* species described by Charles and Edmond Tulasne (Tulasne and Tulasne 1844), other species initially placed in the genus *Endogone*, previously erected by

Link (1809), and species of *Sclerocystis* (Berkeley and Broome 1873), all of these in the family Endogonaceae.

The first observation of what may constitute an AM was reported by Nägeli (1842), who found "fungi within cells" in *Iris* roots, but by the end of the nineteenth century several researchers had published descriptions that definitely showed this type of mycorrhiza (e.g., Janse 1897; Schlicht 1889). In 1885, the term mycorrhiza was used by Frank; however, it was ectomycorrhiza that was first recognized as a mutualistic symbiosis between plants and fungi (Frank 1885). Later "endotrophic mycorrhiza" or "vesicular-arbuscular mycorrhiza" (VAM) began to receive attention (Gallaud 1905; Janse 1897; Peyronel 1923). The term vesicular eventually was dropped because it became clear that some major groups in the Glomeromycota do not form vesicles.

Hyphal connections between mycorrhizal roots and sporocarps were noticed (Peyronel 1923). To establish a causal link between sporocarps and mycorrhizal infection, i.e., to fulfill Koch's postulates, took another three decades until the work of Mosse (1953) and Gerdemann (1955b). Now it was also possible to set up cultures of a defined mycorrhizal fungus together with a host plant to propagate it separately from other species and study its biology.

After pioneering studies, such as that by Nicolson and Gerdemann (1968), describing AM fungal species within the concept of the genus *Endogone*, the monograph by Gerdemann and Trappe (1974) constituted the birth date of the taxonomy of known AMF. For the first time, these authors placed all taxa of AMF known at the time in a stringent Linnaean context. They removed all AM-forming, nonzygosporic species from *Endogone* and placed them in the genera *Glomus*, *Sclerocystis*, and (newly described) *Acaulospora* and *Gigaspora*. For the first time, the mode of spore formation, that is, the way spores are formed on hyphae (see below for details), was recognized as a taxonomically useful character. Still, sporocarpic species accounted for a large proportion of the species listed in this account, reflecting the searching strategies of early mycorrhizologists, which very much resembled truffle hunting. However, the wet-sieving and decanting method of isolating glomeromycotan spores formed singly or in small clusters in the soil had already been reported by Gerdemann and Nicolson (1963), and in the ensuing 20 years, the sporocarpic species were destined to become a relatively marginal phenomenon, so that in 1990 they only accounted for approximately 42 % of *Glomus* (including *Sclerocystis*) species, compared to 95 % in 1974.

The growing interest in AM as a potential resource for agriculture and its recognition as an ecologically important factor also raised interest in the species diversity of these fungi, resulting in numerous descriptions in the 1970s and 1980s. It must be noted, however, that up to the present the mycorrhiza formation of many species is implied by analogy and has been proven only for a subset of species by pure culture on a host plant. The spore wall structure of the glomeromycetes was recognized as a crucial character for distinguishing species. The method of visualizing its components by crushing the spores gently on a microscope slide under a cover slip in a mountant, such as polyvinyl alcohol lactoglycerol (PVLG), became common. To better describe the multitude of wall structures, Walker (1983) created a standardized system of "walls" (discernible substructures of the spore wall) and "wall groups" (arrangements of walls staying attached to each other during this treatment). This standardization was an important step forward to compare different species more efficiently.

In 1990, Morton and Benny placed the genera known by then in a hierarchical taxonomic structure, removing AMF from the Endogonacae and placing them in their own order, Glomales (the orthography of which was later corrected to Glomerales). Cladistic analyses of the characters of spore morphology were used to provide the first putatively phylogenetic framework for the Glomerales, separating two major clades, the suborders Gigasporineae and Glomineae, and the families Glomeraceae (as Glomaceae), Acaulosporaceae, and Gigasporacaceae. Another advance was the inclusion of the spore ontogeny to group the spore wall structure hierarchically in contrast to the strictly phenetic system of Walker. This was based on the observation that there is a predictable sequence of the formation of the respective "walls" or "wall layers" in the different taxa. It had already been noticed that certain taxa possess flexible inner walls, which sometimes are involved in spore germination and bear specialized structures (germination shields, germination orbs) playing a role in this process (Morton 1995; Walker and Sanders 1986).

Despite attempts to classify them using fatty acid profiles (Bentivenga and Morton 1996), isozymes, or monoclonal antibodies, the phylogenetic position of the Glomeromycota remained the subject of much speculation. After Simon et al. (1993) provided the first DNA sequences of the nuclear small subunit (SSU) ribosomal RNA gene from three AM fungal species, it was clear that they were a lineage of the true fungi, but, due to the limited taxon sampling and the absence of DNA sequences for many other basal fungal lineages, their exact placement could not be determined. Nevertheless, these data led to the first attempts to detect AMF by molecular methods in the environment (Clapp et al. 1995). At the time, methods to study the diversity of ectomycorrhizal fungi were far ahead of those for AMF because they were easier to study and had already been used to show the discrepancy between the diversity of mycorrhizal symbionts analyzed directly from roots and the diversity of their fruiting structures (Gardes and Bruns 1993). These findings stimulated the design of molecular tools to also analyze AM fungal species' richness in nature.

Molecular DNA data then became more common in elucidating the phylogenetic relationships among glomeromycotan fungi. The fact that *Geosiphon pyriformis*, a fungus forming an endosymbiosis with cyanobacteria, belongs in a basal glomeromycotan lineage was elucidated by SSU sequences (Gehrig et al. 1996) following the recognition of the similarity of its spores with those of some AMF (Fig. 9.1i) (Schüßler et al. 1994). Nowadays this makes *G. pyriformis* an interesting model for molecular biological studies of symbiosis-related genes (Schüßler 2012; Schüßler et al. 2006).

At the same time, molecular data demonstrated that morphological characteristics previously used to distinguish higher-level taxa, such as genera and families, were poor predictors of phylogenetic relationships. As an example, the genus *Paraglomus*, a deeply diverging lineage in the Glomeromycota, has spores that are morphologically indistinguishable at the genus level from those of *Glomus*, but the two genera are separated by hundreds of million years of evolutionary history (Morton and Redecker 2001; Redecker et al. 2000b).

In the Archaeosporaceae, some taxa even produced on the same fungal thallus glomoid spore types thought to be indicative of the genus *Glomus* and spores typical for the genus *Acaulospora* (Fig. 9.1h) (Morton and Redecker 2001; Morton et al. 1997). Molecular data revealed that they belonged to neither genus but rather constituted another deeply divergent lineage. This was the case for *Archaeospora leptoticha*, later placed in the genus *Ambispora* (Redecker et al. 2000b). Similarly, acaulosporoid and entrophosporoid spore formation in *Archaeospora trappei* and *Archaeospora schenckii* does not imply a close phylogenetic relation with *Acaulospora* or *Entrophospora*. The phylogenetic position of *Entrophospora infrequens*, however, has been impossible to determine because DNA analyses from different laboratories yielded a variety of sequences, often related to *Claroideoglomus* (Rodriguez et al. 2001), a fact that has been impossible to explain up to now.

With increasing knowledge of the phylogeny and biology of AMF, similarities to zygomycetes, such as *Endogone*, appeared to be more and more superficial, and the zygomycetous fungi began to emerge as an ill-defined, paraphyletic assortment of fungal lineages. Consequently, the monophyletic Glomeromycota were separated in their own monophyletic phylum (Schüßler et al. 2001b).

The species in two recently described genera, *Diversispora* and *Redeckera*, were previously placed in *Glomus* but shown to be phylogenetically very distant (Redecker et al. 2007; Schüßler and Walker 2010; Walker and

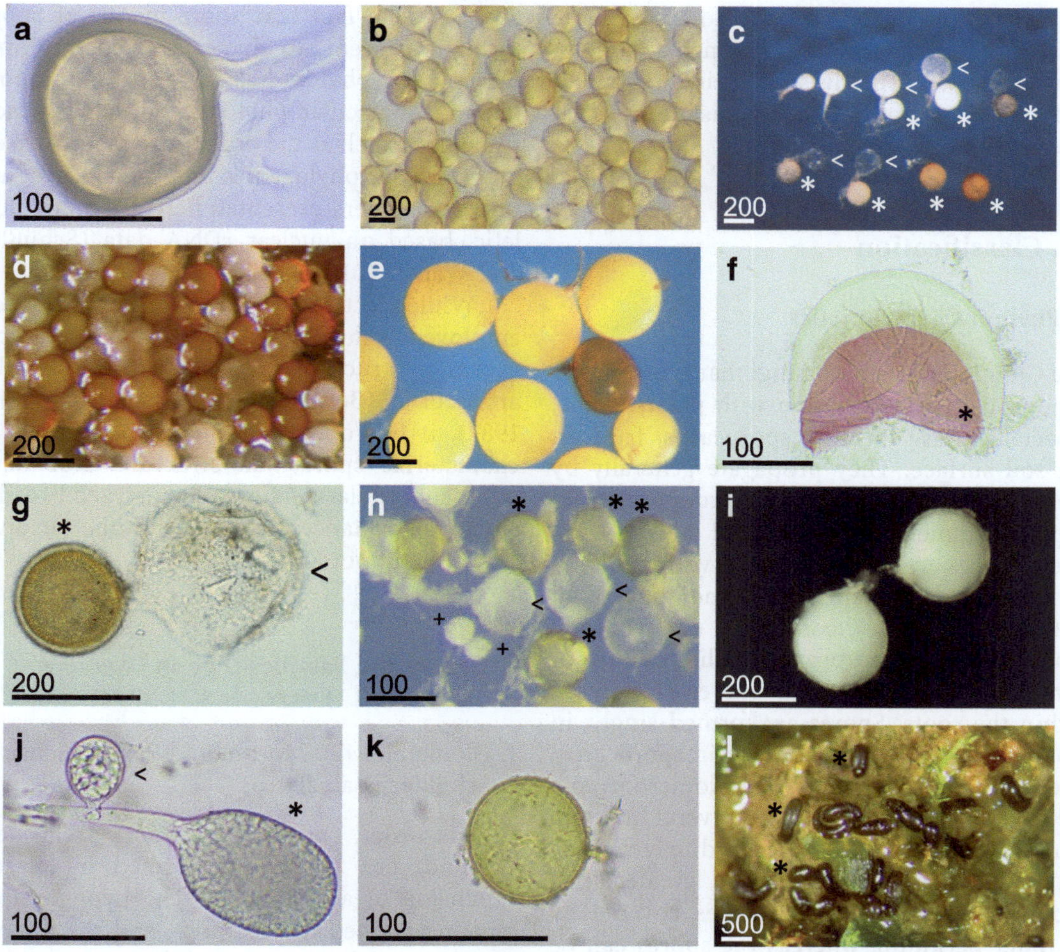

Fig. 9.1 From *Top left* to *bottom right* (**a**) *Glomus macrocarpum* (Glomeraceae; from >150 year-old type material); (**b**) *Claroideoglomus claroideum* (Claroideoglomeraceae); (**c**) *Acaulospora spinosa* (Acaulosporaceae; *asterisk* sporiferous saccules *open angular bracket* spores); (**d**) *Diversispora epigaea* (Diversisporaceae; BEG47, a culture frequently used in AM research); (**e**) *Gigaspora gigantea* (Gigasporaceae); (**f**) *Pacispora franciscana* (Pacisporaceae; *asterisk* germinal wall stained with Melzer's); (**g**) *Entrophospora infrequens* (Entrophosporaceae; *asterisk* sporiferous saccule *open angular bracket* spore); (**h**) *Ambispora fennica* (Ambisporaceae; *asterisk* sporiferous saccules *open angular bracket* acaulosporoid spores *plus symbol* glomoid spores); (**i**) *Geosiphon pyriformis* spores (Geosiphonaceae); (**j**) *Archaeospora trappei* (Archaeosporaceae; *asterisk* sporiferous saccule *open angular bracket* acaulosporoid spore); (**k**) *Paraglomus occultum* (Paraglomeraceae); (**l**) *Geosiphon pyriformis* symbiotic bladders (*asterisk* dark vesicles, harboring cyanobacteria). *Scale bars* in micrometers

Schüßler 2004). Another example of where molecular data guided morphological analyses in defining new taxa was *Pacispora*, uniting characteristics like the glomoid spore formation and germinal walls similar to *Scutellospora* (Oehl and Sieverding 2004; Walker and Schüßler 2004; Walker et al. 2004).

Based on nuclear ribosomal DNA (rDNA) sequences, Schüßler and Walker (2010) redefined more genera in the Glomeromycota and created the new family Claroideoglomeracae for other species previously in *Glomus*. More and more species descriptions have been accompanied by molecular data, illustrating the usefulness of such data in separating taxa with few morphological characters. The need to be able to assign these taxa to sequences from environmental studies, allowing new perspectives on their distribution and phylogeography, is also satisfied by such studies.

VI. Classification

A. Phylum Characteristics

The Glomeromycota are fungi that grow mostly hypogeously in association with plants; some, especially those forming sporocarps, fruit on the soil surface. They propagate generally by asexual spores, but in some groups also by hyphal fragments or colonized root pieces. Their spores are relatively large, with a diameter between less than 40 μm and more than 1,000 μm (Fig. 9.1e), containing up to several thousand nuclei and prominent lipid and protein globules. In some taxa, spores are formed within the roots. Spores are formed singly, in loose clusters, dense masses, or in sporocarps. The sporocarps formed by the Glomeromycota are agglomerations of a few to several hundred thousand spores, their size accordingly varying between less than 500 μm and greater than 4 cm. Sporocarps are sometimes covered by an outer peridium, whereas the spores can be embedded in mycelium or in some cases be radially arranged around a hyphal plexus.

B. Morphological Criteria Used for Classification

The color, size, and shape of spores and the characteristics of hyphal attachment of the spores are important morphological criteria for determining taxa. The color, number, thickness, and consistency of wall layers have been used to distinguish species, whereas the presence or absence of flexible "germinal walls" and the morphology of the hyphal attachment (the so-called mode of spore formation) traditionally were used to determine the genus or family (Morton 1988). The staining behavior of the intraradical structures was also used to distinguish taxa, but it is variable within some families. The fact that some modes of spore formation seem to have evolved multiple times in the phylum has increased the importance of molecular phylogenetic data; in fact, some taxa, such as the orders, are mainly based on molecular phylogenies and sequence signatures. All orders presented here are monophyletic, based on nuclear rDNA data (Schüßler and Walker 2010; Schüßler et al. 2001b).

While electron microscopy (EM) has been employed widely to elucidate the intraradical exchange structures in AM, in particular the arbuscules (Bonfante-Fasolo and Grippiolo 1982), it has only sporadically been used to characterize spore wall structure. Nor have other subcellular details been analyzed broadly so far; for instance, the details of nuclear division are still not known.

C. Orders and Families (For an Overview See Table 9.1)

1. Glomerales J.B. Morton and Benny (Sensu Schüßler et al. 2001b)

In this order, spore formation is exclusively glomoid, i.e., spores are formed by blastic expansion of a hyphal tip. The hyphae often remain attached to the spore, and the attachment is straight or recurved, but never with a bulbous sporogenous cell. The opening of the hyphal attachment may be closed by wall layers, a septum, or remain open; germination occurs through the attachment. The spore walls are often layered, comprising multiple lamellae. Ornamentation of the spore wall surface is usually absent; if present, it is relatively simple. This mode of spore formation, however, is also found in unrelated lineages.

The mycorrhizae usually stain strongly with trypan blue, chlorazol black, or acid fuchsin. Ovoid vesicles are often formed at later stages of colonization.

a) Glomeraceae Piroz. and Dalpé

Spore formation occurs singly, in roots or in soil, in loose clusters or in sporocarps (Fig. 9.1a). In some species, the formation of complex sporocarps occurs with peridium or hyphal

Table 9.1 Classification of the Glomeromycota

Order	Family	Approximate species number
Glomerales	Glomeraceae	108
	Claroideoglomeraceae	6
Diversisporales	Diversisporaceae	10
	Gigasporaceae	53
	Acaulosporaceae	38
	Pacisporaceae	7
Archaeosporales	Archaeosporaceae	2
	Ambisporaceae	9
	Geosiphonaceae	1
Paraglomerales	Paraglomeraceae	3
Familia incertae sedis	Entrophosporaceae	3

plexus. No flexible inner walls are found. This family seems to contain about half of all described species of the phylum, although this could not be confirmed using molecular data for many species. Field surveys have also shown that many molecular operational taxonomic units (MOTUs) belong to this lineage, which is also known as the phylogenetic group, *Glomus* Group A [see for definition Schüßler et al. (2001a) and Schwarzott et al. (2001)].

b) Claroideoglomeraceae C. Walker and A. Schüßler

In this family, spores (Fig. 9.1b), which are usually formed singly in the soil, have walls with an ephemeral outer component that sloughs off in mature spores, a characteristic that also occurs in the Glomeraceae. A semiflexible innermost component [endospore, according to Schüßler and Walker (2010)] has been reported that may, however, be difficult to distinguish from the inner lamella of a rigid spore wall. This family corresponds to *Glomus* Group B.

2. Diversisporales C. Walker and A. Schüßler

This order contains a large variety of spore morphologies and is mainly delimited based on nuclear rDNA data.

a) Gigasporaceae J.B. Morton and Benny

Species in this family form relatively large spores (diameter 120 to >1,000 μm) that develop singly in the soil and are the only infective propagules (Fig. 9.1e). A bulbous sporogenous cell is found at the hyphal attachment (gigasporoid mode of spore formation), which is usually persistent. The mycorrhizae stain uniformly dark with standard procedures; the intraradical hyphae vary considerably in width. The arbuscules often have swollen trunks. No vesicles are formed in this family. On the extraradical mycelium, characteristic thin-walled auxiliary cells of unknown function are conspicuous. No interhyphal anastomoses are formed, whereas the fungi have the ability to bridge wounded mycelium parts by end-to-end anastomoses (de la Providencia et al. 2005).

In *Gigaspora* the spore wall does not contain flexible walls, only rigid components. Spores are brightly colored (white to yellowish green unless senescent) but never hyaline. Spores germinate directly through the spore wall with the germ tube emerging from a pustulate region at the inner layers of this rigid wall. In *Scutellospora* and *Racocetra*, spores are hyaline to dark brown and possess inner flexible germinal walls, which may color deeply pink with Melzer's reagent. On this germinal wall, a permanently present germination shield is found, from which the germ tube emerges (Walker and Sanders 1986). Spores in some species have highly complex surface ornamentations.

It has been known for quite some time that *Scutellospora* is paraphyletic with respect to *Gigaspora*, the lack of germinal walls in the latter clearly being the derived condition. Oehl et al. (2008) proposed splitting *Scutellospora* into five genera and the *Gigasporaceae* into four families based on nuclear large subunit (LSU) rDNA data and the morphology of the germination shield. Because this approach relied on the interpretation of insufficiently robust phylogenetic analyses and a single, plastic morphological character only, it was later rejected by Morton and Msiska (2010). These authors proposed a classification into three genera in a single family.

b) Acaulosporaceae J.B. Morton and Benny

Spores are formed either laterally (acaulosporoid mode) or centrally within (entrophosporoid mode) the hypha terminating in a thin-walled sporiferous saccule (Fig. 9.1c) that is formed before the spore (Gerdemann and Trappe 1974). The saccule and the sporiferous hypha usually detach at spore maturity; therefore, spores of the Acaulosporaceae are mostly sessile. At the occluded points of attachment, the detached hypha leaves one (acaulosporoid type) or two (entrophosporoid type) scars on the spore wall, which may, however, be difficult to observe. The sporiferous saccules are ephemeral with thin walls. The spore color ranges from hyaline to pale golden, orange, or dark brown to black, according to the species. The spores possess ephemeral outer layers, a rigid, often laminated, structural wall, and one or two inner germinal walls with flexible components. Depending on the species, the surface of the structural wall is often ornamented, with ridges, warts, pits, or spines. The spores germinate directly through the wall with the germ tube originating from a germination orb, a round, often spiral-shaped structure formed between the germinal walls or between the germinal and the structural wall (Stürmer and Morton 1999). The mycorrhizae in the Acaulosporaceae stain with varying intensity. The vesicles formed inside the roots may be lobed, but this is not confined to this group.

The acaulosporoid and entrophosporoid mode of spore formation were once thought to be substantial enough to warrant the separation of two genera, but in fact they are derivatives of a similar process, as the two types are also found in closely related species of the Archaeosporaceae (Kaonongbua et al. 2010), which is phylogenetically quite distant from the Acaulosporaceae.

c) Pacisporaceae C. Walker, Blask., A. Schüßler and Schwarzott

This family was established for members of the Diversisporales, with spores formed in the glomoid mode but containing germinal walls (Fig. 9.1f) and a so-called germination shield (Oehl and Sieverding 2004; Walker et al. 2004). Spores are hyaline to light brown to reddish brown; structural walls are often ornamented. The detailed mycorrhizal morphology in this group is unknown as no stable and pure cultures exist.

(d) Diversisporaceae C. Walker and A. Schüßler

Spores mostly form in the glomoid mode, singly, in aggregations or in dense spore clusters (Fig. 9.1d), or sporocarps (*Diversispora*, *Redeckera*). However, in the genus *Otospora* J. Palenzuela, N. Ferrol and Oehl spore formation on a persisting ear-shaped stalk has been reported (Palenzuela et al. 2008) and *Entrophospora nevadensis* has been placed in this family (Palenzuela et al. 2010); however, both reports require additional study to validate the placement. The Diversisporaceae are well separated by rDNA phylogenies from other glomeromycotan lineages that also form glomoid spores. Previously this phylogenetic lineage was known as *Glomus* Group C.

3. Paraglomerales C. Walker and A. Schüßler

a) Paraglomeraceae J.B. Morton and D. Redecker

The four species currently known in this family form small, hyaline glomoid spores (Fig. 9.1k). They can be separated from other lineages forming glomoid spores mainly based on molecular data, i.e., nuclear rDNA and sequences of the LSU of RNA polymerase II (*rpb1*), fatty acid profiles, and antibodies (Morton and Redecker 2001). In rDNA phylogenies the Paraglomeraceae were suggested to constitute the most deeply diverging lineage of the Glomeromycota (Redecker et al. 2000b), and this conclusion has received additional support (Krüger et al. 2012). Mycorrhizae, at least in some species, stain very faintly, so that it is difficult to determine and quantify root colonization.

4. Archaeosporales C. Walker and A. Schüßler

This order constitutes a deeply divergent lineage of the phylum, comprising three families with different modes of spore formation.

a) Archaeosporaceae J.B. Morton and D. Redecker

The spores formed by the two known species in this family are acaulosporoid (Fig. 9.1j) and entrophosporoid, with thin and, thus, semiflexible layers not reacting with Melzer's reagent (Kaonongbua et al. 2010; Morton and Redecker 2001). The mycorrhizae only stain faintly. A complex germination apparatus was reported (Spain 2003) but so far has not been independently confirmed. Also, a glomoid form has been reported.

b) Ambisporaceae C. Walker, Vestberg and A. Schüßler

This family is unique in the sense that at least some species are dimorphic, that is, spores of the acaulosporoid and the glomoid type are formed on the same fungal thallus (Fig. 9.1h). Also, some fungal isolates may form only the glomoid spore type (Morton et al. 1997). Walls of glomoid spores are usually soft and pliable; therefore, the spores do not crack under pressure from the cover slip but form folds. Sometimes they are covered with a mucilaginous coat to which soil particles tend to adhere. Acaulosporoid spores may be formed on a short pedicel that may persist on the spore, giving the false impression of a glomoid spore. Their spore wall structure is complex, with two to four layers. The thick inner walls have flexible components that do not react with Melzers's reagent and do not form germination shields or orbs. Germination occurs through the opening of the pedicel. The mycorrhizae stain very weakly; occasionally vesicles have been reported.

c) Geosiphonaceae Engler and Gilg, Emend. A. Schüßler

The only species of this family, *Geosiphon pyriformis*, is unique in the phylum because it is currently not known to form AM but an endocytobiosis with cyanobacteria of the genus *Nostoc* [for a recent review, see Schüßler (2012)]. The cyanobionts are harbored within multinucleate vacuolated fungal bladders on the soil surface, which are up to 2 mm long (Fig. 9.1l). The cyanobacteria provide photosynthates to the fungal partner, which provides all necessary mineral nutrients and water to the cyanobacteria except nitrogen, which can be fixed by the cyanobacterial heterocysts. The fungus forms whitish glomoid resting spores with layered walls, singly or in loose clusters (Fig. 9.1i).

It is unknown whether the fungus also forms AM, but its endocyanobiosis clearly represents an interesting and useful model system to better understand the symbiotic interface and nutrient exchange between the Glomeromycota and their photoautotrophic partners.

Nuclear SSU rDNA data have placed this species and family firmly in the Glomeromycota as one of the basal lineages. It has been proposed that this type of symbiosis could reflect an evolutionary precursor of AM [for a recent review, see Schüßler and Walker (2011)].

5. Familia *Incertae Sedis*

a) Entrophosporaceae Oehl and Sieverd

Into this family and its only genus, *Entrophospora*, were placed species forming entrophosporoid spores that could not be assigned to either Acaulosporaceae or Archaeosporaceae. *Entrophospora infrequens* is the generic type species, and its spores (Fig. 9.1g) have a complex and characteristic wall structure comprising rigid and semiflexible components, one wall layer having pits interlocking with projections of the layer above (Hall 1977). No pure culture of this species is available, but *E. infrequens* is rather often found in mixed cultures set up from field material (so-called trap cultures), but spore production ceases after some time. The species has presented a puzzle in molecular phylogenetic studies because very diverse, putatively contaminant-derived sequences normally representing lineages with different spore morphologies were detected (Rodriguez et al. 2001). However, the origin of the sequences is unclear, and the phylogenetic position of this family and its biological background therefore remain obscure.

D. Species Concepts

Species have been described in the Glomeromycota usually as morphospecies. The size, shape, and color of spores are determined using a dissecting microscope, and hyphal attachments and the wall layer structure of the slightly cracked spores are examined in PVLG mounts at higher magnification. The reaction of spore wall components to Melzer's reagent also seems to be an important criterion (Morton 1988).

A major obstacle in studying the Glomeromycota has always been the inability to cultivate them separately from their plant host. Most often they have been propagated in open-pot cultures, which require several months to grow. Sometimes cultures are inoculated using single spores; thereby assuring that only a single species is present in the culture, but in this case special measures must be taken to achieve acceptable inoculation success. The purity of such pot cultures is difficult to maintain, and the harvested biological material always contains nonglomeromycotan microorganisms, complicating molecular analysis (Hijri et al. 2002; Walley and Germida 1996). Under these conditions, the degree of morphological and genetic variation within a species may be very difficult to assess. Monoxenic cultures on transformed roots (Bécard and Fortin 1988) offer a much higher security standard but are available only for a small fraction of the existing species.

Morphological characters to separate species are few and often difficult to observe; some species are apparently plastic in their morphology, depending on the culturing conditions and other factors. It must be emphasized that the majority of glomeromycotan species have been described not on the basis of pure cultures but using material collected from the field or mixed cultures set up from field material (trap cultures). In fact, many species have been described from obviously nonviable or degraded spores, resulting in misleading descriptions. In the strict sense, the ability to form mycorrhizae has therefore not been demonstrated but is assumed by analogy for many glomeromycotan species.

DNA sequences have been increasingly used to support (or reject) morphospecies concepts, but a stringent molecular species concept is difficult to establish. It has long been known that numerous variants of nuclear-encoded rDNA coexist within a single glomeromycotan spore (Sanders et al. 1995). Such variation was not found for the mitochondrial DNA (Raab et al. 2005), but for some other nuclear genes normally present as a single copy (Helgason et al. 2003; Koch et al. 2004), making it impossible to assign a single, unique sequence to a species. It has now been recognized that such intraorganism polymorphism is also found in other eukaryotes and has been underestimated in fungi, but in some species of the Glomeromycota it reaches exceptionally high levels (Stockinger et al. 2009, 2010). The possible contribution of pseudogenes to this polymorphism has not been determined systematically, but for the LSU rRNA gene most variants were also found in the transcriptome and indicated to be functional (Boon et al. 2010). For the rDNA Internal Transcribed Spacer (ITS) region (ITS1, 5.8S, ITS2) alone, which has been suggested as the primary DNA barcode for fungi (Schoch et al. 2012), it was shown that the intraspecific and intrasporal variation can be so high that closely related species are difficult or impossible to separate (Stockinger et al. 2009). In other fungi, molecular phylogenetic species concepts have been applied using coalescent analyses based on the criterion that species are reproductively isolated (Taylor et al. 1999). In the Glomeromycota, the genetic bases are still unclear for the great majority of lineages. Coalescent analyses cannot be applied to clonal lineages and require multilocus phylogenies, which are not yet available for the majority of glomeromycotan taxa.

Anastomosis formation could be another criterion for a biological concept of species delimitation. In *R. irregularis* (*Glomus intraradices*) hyphal cross bridges were observed between genetically distinguishable isolates at a frequency decreasing with genetic distance of the strains (Croll et al. 2009). However, in other species anastomoses only seem to occur within the same or very closely related isolates

(Giovannetti et al. 2003) or could not be observed at all (Purin and Morton 2011).

As the diversity of members of the Glomeromycota detected in environmental studies using molecular methods seems to greatly outnumber morphospecies, operative concepts were used to enumerate this diversity (e.g., Öpik et al. 2008). These concepts were based on cutoff values of sequence similarity, the definition of monophyletic groups by phylogenetic analyses, or a combination of both. However, many of these studies used exclusively the nuclear small ribosomal subunit as a marker gene, which was shown to be unsuitable for separating closely related species (Walker et al. 2007). It has become clear that cutoff values of sequence similarity cannot be generalized across families and orders. Nevertheless, molecular operational taxonomic unit (MOTU) estimates are, and will be (Hawksworth et al. 2011), highly useful as comparative proxies of biodiversity in field settings, but most authors recommend avoiding the usage of the term species in this context if MOTUs are not defined at this taxonomic level.

VII. Evolution of the Phylum

The evolutionary aspects of AMF, evolution of AM, coevolution of the symbiosis partners, and the putative impact of the AM on the colonization of land by plants has recently been reviewed in this series (Schüßler and Walker 2011). Here, some of the major points are briefly discussed.

A. Ecological Aspects

Unfortunately, not much is known about the differences in symbiotic function among the families of the Glomeromycota. Certain trends on this level were identified, for example, the differences in hyphal network architecture by the formation of anastomoses in the Glomeraceae and the absence of such networks in the Gigasporaceae (de la Providencia et al. 2005). It was also suggested that symbiotic benefits for the plant were mainly based on nutrient transport in Gigasporaceae and mainly on increased resistance against pathogens in the Glomeraceae (Klironomos et al. 2000). Different nutrient foraging behaviors have been compared among some species in the Glomerales (Jansa et al. 2005). Agricultural practice seems to have varied influence on taxon occurrence on different levels from family to species (Helgason et al. 1998; Hijri et al. 2006), which may in part be correlated with the life history strategies of species or families (Sýkorová et al. 2007).

B. Spore Structure and Ontogeny

Concerning the evolution of spore structure, more data are available. Still, as the function of many specific components of spore formation (e.g., sporiferous saccule) is unknown, it is difficult to interpret morphological evolution of spore formation, i.e., to define derived versus ancestral morphological characters. Current knowledge of glomeromycotan phylogeny allows pinpointing the following trends:

(A) The glomoid, acaulosporoid, and entrophosporoid modes of spore formation are polyphyletic. The glomoid type is particularly widespread among unrelated lineages. Glomoid and acaulosporoid types may occur in the same species, indicating that these two types of structures are nonhomologous. The switch between entrophosporoid and acaulosporoid formation seems to require only small changes in the development pattern. This may explain why in each of two very distantly related families (Acaulosporaceae and Archaeosporaceae), closely related species sharing numerous other characteristics differ only in this respect.

(B) The presence of so-called germinal walls with germination shields/orbs is restricted to the Diversisporales, where they can be found in all four spore types, but it has not yet been conclusively demonstrated whether these structures are homologous. The loss of these structures is evident in *Gigaspora*, which is clearly a derived and not a basal genus within in the family.

C. Evidence from Fossil Record and Patterns of Association with Plants

The earliest known, most widely recognized evidence for glomeromycotan fungi are 460 million-year-old fossilized glomoid spores and hyphae from Ordovician limestone (Redecker et al. 2000a). At this time, land plants had probably reached the morphological complexity of today's liverworts; therefore, it is not surprising that such plants are not well documented in the fossil record; thus a direct interaction of these fungi with the early land plants could not be shown up to now. Unequivocal evidence for embryophytes dates back to about 470 million years ago (mya), in the form of cryptospore assemblages (Rubinstein et al. 2010), and early vascular plants can be traced back about 420 million years (Stewart and Rothwell 1993).

Among the wealth of different early Devonian life forms that are exceptionally well conserved in the Rhynie Chert, dating back 400–412 mya, are the oldest known and most beautifully conserved arbuscules, the first evidence for the AM symbiosis itself (Remy et al. 1994). The fossils were detected in the rhizomes of Devonian plants such as *Aglaophyton*, with a much more advanced morphology than the putative Ordovician plants. These plants had not yet evolved roots but were colonized in their shoot cortex, illustrating the fact that roots came later than mycorrhiza, if interpreted by function and homology. In this sense, the term mycorrhiza obviously should not be used exclusively for associations involving root organs.

Besides early evidence for a number of fungal lineages, the Rhynie Chert also contained well-conserved spores of *Scutellospora*- and *Acaulospora*-like morphology and structures closely resembling germination shields (Dotzler et al. 2006, 2009). These fossils indicate that even 400 mya much of the glomeromycotan diversity on the order and family level may have been present already and that the deep lineages, such as Archaeosporaceae and Paraglomeraceae, may be considerably older. It may, however, also just indicate that character evolution of the glomeromycotan spore is more complex than previously thought, involving losses of characters previously thought to be indicative of an advanced state (Schüßler and Walker 2011).

The occurrence pattern of AM in extant plant groups indicates strongly that the ability to form this symbiosis is an ancestral character of land plants. Other types of mycorrhizae are clearly secondary associations of land plants, found exclusively in derived plant lineages and also much later in the fossil record (LePage et al. 1997). Many species of the deepest lineages (hornworts and liverworts) of land plants form associations with glomeromycotan fungi. Interestingly, recent findings indicate that extant bryophytes also form associations with zygomycetes from the *Endogone* containing clade (Bidartondo et al. 2011). This stimulates the discussion about mycorrhizal associations of early land plants, as *Endogone* most likely branches earlier than the AMF in the fungal tree of life. However, recent bryophytes also form close associations with ascomycetes and basidiomycetes, and an ancestral origin of such a symbiosis with *Endogone*-like fungi remains speculative.

The complete absence of mycorrhiza or mycorrhizalike symbioses or the presence of other types of associations than AM can be most parsimoniously explained by a loss or a switch from an ancestral state. In any case, the AM-specific plant genes and their functions are extremely conserved from bryophytes to vascular plants (Wang et al. 2010).

Taken together (Schüßler and Walker 2011), these data support the hypothesis that plants and AMF colonized the land masses together (Pirozynski and Malloch 1975), the fungi being potentially instrumental in the success of the colonization. In early terrestrial ecosystems before the formation of fertile soils and humic layers, the absorbing capacity of a fungal mycelium for nutrient uptake and transfer may have been even more crucial than today.

IX. Conclusion

There have been numerous revisions of glomeromycotan taxonomy in recent years, reflecting the steadily growing knowledge about the evolutionary relationships of these fungi. Molecular data have allowed the refinement of morphology-based approaches as well as avoidance of pitfalls by almost inevitable overinterpretation of the few morphological characters that have been used in Glomeromycotan classification. This has led to a better appreciation of the genetic diversity of arbuscular mycorrhizal fungi on all levels, from the phylum to species and populations, although a conclusive molecular species concept remains one of the major challenges for future research. There is an increasing interest in Glomeromycotan fungi as an ecologically and economically important group of organisms, for example, in the context of sustainable management of environmental resources. Future research on AM needs as a framework a robust, natural taxonomy based on the phylogenetic relationships of these fungi, which leaves space for future changes without artificial overinflation of taxa.

References

Akiyama K, Matsuzaki K, Hayashi H (2005) Plant sesquiterpenes induce hyphal branching in arbuscular mycorrhizal fungi. Nature 435:824–827

Appoloni S, Lekberg Y, Tercek M, Zabinski C, Redecker D (2008) Molecular community analysis of arbuscular mycorrhizal fungi in roots from geothermal soils in Yellowstone National Park (USA). Microb Ecol 56:649–659

Azcon-Aguilar C, Barea JM (1996) Arbuscular mycorrhizas and biological control of soil-borne plant pathogens—an overview of the mechanisms involved. Mycorrhiza 6:457–464

Bécard G, Fortin JA (1988) Early events of vesicular-arbuscular mycorrhiza formation on Ri T-DNA transformed roots. New Phytol 108:211–218

Bécard G, Pfeffer PE (1993) Status of nuclear division in arbuscular mycorrhizal fungi during in vitro development. Protoplasma 174:62–68

Bécard G, Taylor LP, Douds DD, Pfeffer PE, Doner LW (1995) Flavonoids are not necessary plant signal compounds in arbuscular mycorrhizal symbioses. Mol Plant Microbe Interact 8:252–258

Bentivenga SP, Morton JB (1996) Congruence of fatty acid methyl ester profiles and morphological characters of arbuscular mycorrhizal fungi in Gigasporaceae. Proc Natl Acad Sci USA 93:5659–5662

Berkeley M, Broome C (1873) Enumeration of the fungi of Ceylon. Part II. J Linn Soc Bot 14:29–140

Besserer A, Puech-Pagès V, Kiefer P, Gomez-Roldan V, Jauneau A, Roy S, Portais J-C, Roux C, Bécard G, Séjalon-Delmas N (2006) Strigolactones stimulate arbuscular mycorrhizal fungi by activating mitochondria. PLoS Biol 4:e226. doi:10.1371/journal.pbio.0040226

Bidartondo MI, Redecker D, Hijri I, Wiemken A, Bruns TD, Domínguez L, Sérsic A, Leake JR, Read DJ (2002) Epiparasitic plants specialized on arbuscular mycorrhizal fungi. Nature 419:389–392

Bidartondo MI, Read DJ, Trappe JM, Merckx V, Ligrone R, Duckett JG (2011) The dawn of symbiosis between plants and fungi. Biol Lett 7:574–577

Bonfante-Fasolo P, Grippiolo R (1982) Ultrastructural and cytochemical changes in the wall of a vesicular-arbuscular mycorrhizal fungus during symbiosis. Can J Bot 60:2303–2312

Boon E, Zimmerman E, Lang BF, Hijri M (2010) Intra-isolate genome variation in arbuscular mycorrhizal fungi persists in the transcriptome. J Evol Biol 23:1519–1527

Brundrett MC (2009) Mycorrhizal associations and other means of nutrition of vascular plants: understanding the global diversity of host plants by resolving conflicting information and developing reliable means of diagnosis. Plant Soil 320:37–77

Bunn RA, Zabinski CA (2003) Arbuscular mycorrhizae in thermal-influenced soils in Yellowstone National Park. West N Am Nat 63:409–415

Clapp JP, Young JPW, Merryweather JW, Fitter AH (1995) Diversity of fungal symbionts in arbuscular mycorrhizas from a natural community. New Phytol 130:259–265

Croll D, Giovannetti M, Koch AM, Sbrana C, Ehinger M, Lammers PJ, Sanders IR (2009) Nonself vegetative fusion and genetic exchange in the arbuscular mycorrhizal fungus *Glomus intraradices*. New Phytol 181:924–937

Daniell TJ, Husband R, Fitter AH, Young JPW (2001) Molecular diversity of arbuscular mycorrhizal fungi colonizing arable crops. FEMS Microbiol Lett 36:203–209

de la Providencia IE, de Souza FA, Fernández F, Delmas NS, Declerck S (2005) Arbuscular mycorrhizal fungi reveal distinct patterns of anastomosis formation and hyphal healing mechanisms between different phylogenic groups. New Phytol 165:261–271

Dickson S, Smith FA, Smith SE (2007) Structural differences in arbuscular mycorrhizal symbioses: more than 100 years after Gallaud, where next? Mycorrhiza 17:375–393

Dotzler N, Krings M, Taylor TN, Agerer R (2006) Germination shields in *Scutellospora* (Glomeromycota: Diversisporales, Gigasporaceae) from the 400 million-year-old Rhynie chert. Mycol Prog 5:178–184

Dotzler N, Walker C, Krings M, Hass H, Kerp H, Taylor TN, Agerer R (2009) Acaulosporoid glomeromycotan spores with a germination shield from the 400-million-year-old Rhynie chert. Mycol Prog 8:9–18

Frank A (1885) Über die auf Wurzelsymbiose beruhende Ernährung gewisser Bäume durch unterirdische Pilze. Ber Deut Bot Ges 3:128–145
Gallaud I (1905) Études sur les mycorrhizes endotrophes. Rev Gén Bot 17:5–48, 66–83, 123–135, 223–239, 313–325, 425–433, 479–500
Gange AC (1993) Translocation of mycorrhizal fungi by earthworms during early succession. Soil Biol Biochem 25:1021–1026
Gardes M, Bruns TD (1993) ITS primers with enhanced specificity for basidiomycetes—application to the identification of mycorrhizae and rusts. Mol Ecol 2:113–118
Gehrig H, Schüßler A, Kluge M (1996) *Geosiphon pyriforme*, a fungus forming endocytobiosis with *Nostoc* (Cyanobacteria), is an ancestral member of the Glomales: evidence by SSU rRNA analysis. J Mol Evol 43:71–81
Genre A, Chabaud M, Faccio A, Barker DG, Bonfante P (2008) Prepenetration apparatus assembly precedes and predicts the colonization patterns of arbuscular mycorrhizal fungi within the root cortex of both *Medicago truncatula* and *Daucus carota*. Plant Cell 20:1407–1420
Gerdemann JW (1955a) Wound-healing of hyphae in a phycomycetous mycorrhizal fungus. Mycologia 47:916–918
Gerdemann JW (1955b) Relation of a large soil-borne spore to phycomycetous mycorrhizal infections. Mycologia 47:429–617
Gerdemann JW, Nicolson TH (1963) Spores of mycorrhizal *Endogone* species extracted from soil by wet sieving and decanting. Trans Br Mycol Soc 46:235–244
Gerdemann JW, Trappe JM (1974) Endogonaceae in the Pacific Northwest. Mycologia Memoir 5:1–76
Gianinazzi S, Gollotte A, Binet M-N, van Tuinen D, Redecker D, Wipf D (2010) Agroecology: the key role of arbuscular mycorrhizas in ecosystem services. Mycorrhiza 20:519–530
Giovannetti M, Sbrana C, Strani P, Agnolucci M, Rinaudo V, Avio L (2003) Genetic diversity of isolates of *Glomus mosseae* from different geographic areas detected by vegetative compatibility testing and biochemical and molecular analysis. Appl Environ Microbiol 69:616–624
Hall IR (1977) Species and mycorrhizal infections of New Zealand Endogonaceae. Trans Br Mycol Soc 68:341–356
Hawksworth DL, Crous PW, Redhead SA, Reynolds DR, Samson RA, Seifert KA, Taylor JW, Samson RA, Abaci O, Asan A, Bai F, De Beer WZ, Begerow D, Boekhout T, Buzina W, Cai L, Cannon PF, Damm U, Daniel H, Demirel R, Van Diepeningen AD, Eberhardt U, Fell JW, Frisvad JC, Geml J, Glienke C, Groenwald M, Gueho-Kellermanne E, Hong S, Houbraken J, Huhndorf SM, Hyde KD, Johnston PR, Koljalg U, Kurtzman CP, Lagneau P, Levesque C, Liu X, Lombard L, Meyer W, Miller A, Minter DW, Najafzadeh MJ, Ozerskaya SM, Pennycook SR, Peterson SW, Pettersson OV, Quaedvlieg W, Robert VA, Ruibal C, Schnurer J, Schoers H, Slippers B, Spierenburg H, Taskin E, Thrane U, Uztan A, Varga J, Vasco A, Videira SI, De Vries RP, Weir BS, Yilmaz N, Yurkov A (2011) The Amsterdam declaration on fungal nomenclature. IMA Fungus 2:105–112
Helgason T, Daniell TJ, Husband R, Fitter AH, Young JPW (1998) Ploughing up the wood-wide web? Nature 394:431
Helgason T, Merryweather JW, Denison J, Wilson P, Young JPW, Fitter AH (2002) Selectivity and functional diversity in arbuscular mycorrhizas of co-occuring fungi and plants from a temperate deciduous woodland. J Ecol 90:371–384
Helgason T, Watson IJ, Young JPW (2003) Phylogeny of the Glomerales and Diversisporales (Fungi: Glomeromycota) from actin and elongation factor 1-alpha sequences. FEMS Microbiol Lett 229:127–132
Hibbett DS, Ohman A, Glotzer D, Nuhn M, Kirk P, Nilsson RH (2011) Progress in molecular and morphological taxon discovery in *Fungi* and options for formal classification of environmental sequences. Fung Biol Rev 25:38–47
Hijri M, Redecker D, Macdonald-Comber Petetot JA, Voigt K, Wöstemeyer J, Sanders IR (2002) Identification and isolation of two Ascomycete fungi from spores of the arbuscular mycorrhizal fungus *Scutellospora*. Appl Environ Microbiol 68: 4567–4573
Hijri I, Sykorova Z, Oehl F, Ineichen K, Mäder P, Wiemken A, Redecker D (2006) Communities of arbuscular mycorrhizal fungi in arable soils are not necessarily low in diversity. Mol Ecol 15: 2277–2289
Jansa J, Mozafar A, Frossard E (2005) Phosphorus acquisition strategies within arbuscular mycorrhizal fungal community of a single field site. Plant Soil 276:163–176
Janse JM (1897) Les endophytes radicaux de quelques plantes Javanaises. Annales du Jardin Botanique de Buitenzorg 14:53–212
Jany JL, Pawlowska TE (2010) Multinucleate spores contribute to evolutionary longevity of asexual *Glomeromycota*. Am Nat 175:424–435
Jin H, Pfeffer PE, Douds DD, Piotrowski E, Lammers PJ, Shachar-Hill Y (2005) The uptake, metabolism, transport and transfer of nitrogen in an arbuscular mycorrhizal symbiosis. New Phytol 168:687–696
Kaonongbua W, Morton J, Bever JD (2010) Taxonomic revision transferring species in *Kuklospora* to *Acaulospora* (Glomeromycota) and a description of *Acaulospora colliculosa* sp. nov. from field collected spores. Mycologia 102:1497–1509
Klironomos JN, McCune J, Hart M, Neville J (2000) The influence of arbuscular mycorrhizae on the relationship between plant diversity and productivity. Ecol Lett 3:137–141

Koch A, Kuhn G, Fontanillas P, Fumagalli L, Goudet J, Sanders IR (2004) High genetic variability and low local diversity in a population of arbuscular mycorrhizal fungi. Proc Natl Acad Sci USA 101:2369–2374

Koide R, Mosse B (2004) A history of research on arbuscular mycorrhiza. Mycorrhiza 14:145–163

Krüger M, Walker C, Schüßler A (2011) *Acaulospora brasiliensis* comb. nov. and *Acaulospora alpina* (*Glomeromycota*) from upland Scotland: morphology, molecular phylogeny and DNA-based detection in roots. Mycorrhiza 21:577–587

Krüger M, Krüger C, Walker C, Stockinger H, Schüßler A (2012) Phylogenetic reference data for systematics and phylotaxonomy of arbuscular mycorrhizal fungi from phylum to species level. New Phytol 193:970–984

Lee J, Young JPW (2009) The mitochondrial genome sequence of the arbuscular mycorrhizal fungus *Glomus intraradices* isolate 494 and implications for the phylogenetic placement of *Glomus*. New Phytol 183:200–211

LePage BA, Currah RS, Stockey RA, Rothwell GW (1997) Fossil ectomycorrhizae from the Middle Eocene. Am J Bot 84:410

Link HF (1809) Observationes in ordines plantarum naturales. Magazin für die neuesten Entdeckungen in der gesammten Naturkunde, Gesellschaft Naturforschender Freunde zu Berlin 3:3–42

Liu Y, Steenkamp E, Brinkmann H, Forget L, Philippe H, Lang BF (2009) Phylogenomic analyses predict sistergroup relationship of nucleariids and Fungi and paraphyly of zygomycetes with significant support. BMC Evol Biol 9:272. doi:10.1186/1471-2148-9-272

Mangan SA, Adler GH (2002) Seasonal dispersal of arbuscular mycorrhizal fungi by spiny rats in a neotropical forest. Oecologia 131:587–597

Marleau J, Dalpe Y, St-Arnaud M, Hijri M (2011) Spore development and nuclear inheritance in arbuscular mycorrhizal fungi. BMC Evol Biol 11:51

Morton JB (1988) Taxonomy of VA mycorrhizal fungi: classification, nomenclature, and identification. Mycotaxon 32:267–324

Morton JB (1995) Taxonomic and phylogenetic divergence among five *Scutellospora* species based on comparative developmental sequences. Mycologia 87:127–137

Morton JB, Benny GL (1990) Revised classification of arbuscular mycorrhizal fungi (Zygomycetes): a new order, Glomales, two new suborders, Glomineae and Gigasporineae, and two new families, Acaulosporaceae and Gigasporaceae, with an emendation of Glomaceae. Mycotaxon 37:471–491

Morton JB, Msiska Z (2010) Phylogenies from genetic and morphological characters do not support a revision of *Gigasporaceae* (Glomeromycota) into four families and five genera. Mycorrhiza 7:483–496

Morton JB, Redecker D (2001) Two new families of Glomales, Archaeosporaceae and Paraglomaceae, with two new genera *Archaeospora* and *Paraglomus*, based on concordant molecular and morphological characters. Mycologia 93:181–195

Morton JB, Bever JD, Pfleger FL (1997) Taxonomy of *Acaulospora gerdemannii* and *Glomus leptotichum*, synanamorphs of an arbuscular mycorrhizal fungus in Glomales. Mycol Res 101:625–631

Mosse B (1953) Fructifications associated with mycorrhizal strawberry roots. Nature 171:974

Nägeli C (1842) Pilze im Innern von Zellen. Linnaea 16:278–285

Nicolson TH, Gerdemann JW (1968) Mycorrhizal *Endogone* species. Mycologia 60:313–325

Oehl F, Sieverding E (2004) *Pacispora*, a new vesicular arbuscular mycorrhizal fungal genus in the Glomeromycetes. J Appl Bot 78:72–82

Oehl F, Sieverding E, Ineichen K, Mäder P, Boller T, Wiemken A (2003) Impact of land use intensity on the species diversity of arbuscular mycorrhizal fungi in agroecosystems of Central Europe. Appl Environ Microbiol 69:2816–2824

Oehl F, de Souza FA, Sieverding E (2008) Revision of *Scutellospora* and description of five new genera and three new families in the arbuscular mycorrhiza-forming Glomeromycetes. Mycotaxon 106: 311–360

Öpik M, Moora M, Liira J, Zobel M (2006) Composition of root-colonizing arbuscular mycorrhizal fungal communities in different ecosystems around the globe. J Ecol 94:778–790

Öpik M, Moora M, Zobel M, Saks U, Wheatley R, Wright F, Daniell T (2008) High diversity of arbuscular mycorrhizal fungi in a boreal herb-rich coniferous forest. New Phytol 179:867–876

Palenzuela J, Ferrol N, Boller T, Azcon-Aguilar C, Oehl F (2008) *Otospora bareai*, a new fungal species in the *Glomeromycetes* from a dolomitic shrub land in Sierra de Baza National Park (Granada, Spain). Mycologia 100:296–305

Palenzuela J, Barea JM, Ferrol N, Azcon-Aguilar C, Oehl F (2010) *Entrophospora nevadensis*, a new arbuscular mycorrhizal fungus from Sierra Nevada National Park (southeastern Spain). Mycologia 102:624–632

Peyronel B (1923) Fructification de l'endophyte à arbuscules et à vesicules des mycorhizes endotrophes. Bull Soc Mycol 39:119–126

Pirozynski KA, Malloch DW (1975) The origin of land plants: a matter of mycotrophism. BioSystems 6:153–164

Purin S, Morton J (2011) *In situ* analysis of anastomosis in representative genera of arbuscular mycorrhizal fungi. Mycorrhiza 21:505–514

Raab P, Brennwald A, Redecker D (2005) Mitochondrial large ribosomal subunit sequences are homogeneous within isolates of *Glomus* (arbuscular mycorrhizal fungi, Glomeromycota). Mycol Res 109:1315–1322

Redecker D, Raab P (2006) Phylogeny of the Glomeromycota (arbuscular mycorrhizal fungi): recent developments and new gene markers. Mycologia 98:885–895

Redecker D, Kodner R, Graham LE (2000a) Glomalean fungi from the Ordovician. Science 289:1920–1921

Redecker D, Morton JB, Bruns TD (2000b) Ancestral lineages of arbuscular mycorrhizal fungi. Mol Phylogenet Evol 14:276–284

Redecker D, Raab P, Oehl F, Camacho FJ, Courtecuisse R (2007) A novel clade of sporocarp-forming species of glomeromycotan fungi in the Diversisporales lineage. Mycol Prog 6:35–44

Remy W, Taylor TN, Hass H, Kerp H (1994) Four hundred-million-year-old vesicular arbuscular mycorrhizae. Proc Natl Acad Sci USA 91: 11841–11843

Rodriguez A, Dougall T, Dodd JC, Clapp JP (2001) The large subunit ribosomal RNA genes of *Entrophospora infrequens* comprise sequences related to two different glomalean families. New Phytol 152: 159–167

Rosendahl S (2008) Communities, populations and individuals of arbuscular mycorrhizal fungi. New Phytol 178:253–265

Rubinstein CV, Gerrienne P, De La Puente GS, Astini RA, Steemans P (2010) Early Middle Ordovician evidence for land plants in Argentina (eastern Gondwana). New Phytol 188:365–369

Sanders IR, Croll D (2010) Arbuscular mycorrhiza: the challenge to understand the genetics of the fungal partner. Annu Rev Genet 44:271–292

Sanders IR, Alt M, Groppe K, Boller T, Wiemken A (1995) Identification of ribosomal DNA polymorphism among and within spores of the Glomales: application to studies on the genetic diversity of arbuscular mycorrhizal fungal communities. New Phytol 130:419–427

Schlicht A (1889) Beitrag zur Kenntnis der Verbreitung und Bedeutung der Mycorhizen. Landwirtschaftliche Jahrbücher 18:478–506

Schoch CL, Seifert KA, Huhndorf S, Robert V, Spouge JL, Bolchacova E, Voigt K, Chen W, Levesque CA, Crous PW, Fungal Barcoding Consortium (2012) The internal transcribed spacer (ITS) as a universal DNA barcode maker for fungi. Proc Natl Acad Sci USA 109(16):6241–6246

Schüßler A (2012) The *Geosiphon–Nostoc* endosymbiosis and its role as a model for arbuscular mycorrhiza research. In: Hock B (ed) The Mycota IX—Fungal associations, 2nd edn. Springer, Berlin

Schüßler A, Walker C (2010) The *Glomeromycota*: a species list with new families and new genera. Published in libraries at The Royal Botanic Garden Edinburgh, The Royal Botanic Garden Kew, Botanische Staatssammlung Munich, and Oregon State University. Electronic version freely available online at http://www.amf-phylogeny.com. ISBN-13:978-1466388048

Schüßler A, Walker C (2011) Evolution of the 'plant-symbiotic' fungal phylum, *Glomeromycota*. In: Pöggeler S, Wöstemeyer J (eds) Evolution of fungi and fungal-like organisms. Springer, Berlin, pp 163–185

Schüßler A, Mollenhauer D, Schnepf E, Kluge M (1994) *Geosiphon pyriforme*, an endosymbiotic association of fungus and cyanobacteria: the spore structure resembles that of arbuscular mycorrhizal (AM) fungi. Bot Acta 107:36–45

Schüßler A, Gehrig H, Schwarzott D, Walker C (2001a) Analysis of partial Glomales SSU rRNA gene sequences: implications for primer design and phylogeny. Mycol Res 105:5–15

Schüßler A, Schwarzott D, Walker C (2001b) A new fungal phylum, the *Glomeromycota*: phylogeny and evolution. Mycol Res 105:1413–1421

Schüßler A, Martin H, Cohen D, Fitz M, Wipf D (2006) Characterization of a carbohydrate transporter from symbiotic glomeromycotan fungi. Nature 444:933–936

Schwarzott D, Walker C, Schüßler A (2001) *Glomus*, the largest genus of the arbuscular mycorrhizal fungi (Glomales), is non-monophyletic. Mol Phylogenet Evol 21:190–197

Simon L, Bousquet J, Levesque RC, Lalonde M (1993) Origin and diversification of endomycorrhizal fungi and coincidence with vascular land plants. Nature 363:67–69

Smith SE, Read DJ (2008) Mycorrhizal symbiosis, 3rd edn. Academic, London

Smith SE, Dickson S, Smith FA (2001) Nutrient transfer in arbuscular mycorrhizas: how are fungal and plant processes integrated? Aust J Plant Physiol 28:683–694

Sondergaard M, Laegaard S (1977) Vesicular-arbuscular mycorrhiza in some aquatic vascular plants. Nature 268:232–233

Spain JL (2003) Emendation of *Archaeospora* and of its type species, *Archaeospora trappei*. Mycotaxon 87: 109–112

Stewart WN, Rothwell GW (1993) Paleobotany and the evolution of plants, 2nd edn. Cambridge University Press, Cambridge

Stockinger H, Walker C, Schüßler A (2009) '*Glomus intraradices* DAOM197198', a model fungus in arbuscular mycorrhiza research, is not *Glomus intraradices*. New Phytol 183:1176–1187

Stockinger H, Krüger M, Schüßler A (2010) DNA barcoding of arbuscular mycorrhizal fungi. New Phytol 187:461–474

Stürmer SL (2012) A history of the taxonomy and systematics of arbuscular mycorrhizal fungi belonging to the phylum Glomeromycota. Mycorrhiza 22:247–258

Stürmer SL, Morton JB (1999) Taxonomic reinterpretation of morphological characters in Acaulo-

sporaceae based on developmental patterns. Mycologia 91:849–857
Stutz JC, Morton JB (1996) Successive pot cultures reveal high species richness of arbuscular endomycorrhizal fungi in arid ecosystems. Can J Bot 74:1883–1889
Sýkorová Z, Ineichen K, Wiemken A, Redecker D (2007) The cultivation bias: different communities of arbuscular mycorrhizal fungi detected in roots from the field, from bait plants transplanted to the field, and from a greenhouse trap experiment. Mycorrhiza 18:1–14
Taylor JW, Geiser DM, Burt A, Koufopanou V (1999) The evolutionary biology and population genetics underlying fungal strain typing. Clin Microbiol Rev 12:126–146
Tulasne LR, Tulasne C (1844) Fungi nonnulli hypogaei, novi v. minus cogniti auct. G Bot Ital 2:55–63
Walker C (1983) Taxonomic concepts in the Endogonaceae: spore wall characteristics in species descriptions. Mycotaxon 18:443–455
Walker C, Sanders FE (1986) Taxonomic concepts in the Endogonaceae: III. The separation of Scutellospora gen nov from Gigaspora Gerd and Trappe. Mycotaxon 27:169–182
Walker C, Schüßler A (2004) Nomenclatural clarifications and new taxa in the *Glomeromycota*. Mycol Res 108:981–982
Walker C, Blaszkowski J, Schwarzott D, Schüßler A (2004) *Gerdemannia* gen. nov., a genus separated from *Glomus*, and *Gerdemanniaceae* fam. nov., a new family in the *Glomeromycota*. Mycol Res 108:707–718
Walker C, Vestberg M, Demircik F, Stockinger H, Saito M, Sawaki H, Nishmura I, Schüßler A (2007) Molecular phylogeny and new taxa in the *Archaeosporales* (*Glomeromycota*): *Ambispora fennica* gen. sp nov., *Ambisporaceae* fam. nov., and emendation of *Archaeospora* and *Archaeosporaceae*. Mycol Res 111:137–153
Walley FL, Germida JJ (1996) Failure to decontaminate *Glomus clarum* NT4 spores is due to spore wall-associated bacteria. Mycorrhiza 6:43–49
Wang B, Yeun LH, Xue J-Y, Liu Y, Ane J-M, Qiu Y-L (2010) Presence of three mycorrhizal genes in the common ancestor of land plants suggests a key role of mycorrhizas in the colonization of land by plants. New Phytol 186:514–525
Young G (2008) What's on the menu? Nat Rev Microbiol 6:499

10 Pucciniomycotina

M. Catherine Aime[1], Merje Toome[1], David J. McLaughlin[2]

CONTENTS

I. Introduction

More than **8,400 species of Pucciniomycotina** have been described (Table 10.1), or more than 8 % of all described Fungi (at 98,998 spp.) (Kirk et al. 2008). Pucciniomycotina is the sister to the Ustilaginomycotina and Agaricomycotina, forming the basal lineage of Basidiomycota. **All members of the subphylum** thus far studied **have simple septal pores** lacking dolipores (septal pore swellings) and septal pore caps, which, along with predominant cell wall sugars (mannose, Prillinger et al. 1993) and disclike spindle pole bodies (McLaughlin et al. 1995; Wells 1994), distinguishes them from most other Basidiomycota. Although some Ustilaginomycotina species appear to have simple septal pores (e.g., Lutzoni et al. 2004), these are reportedly associated with membranous plates that are continuous with the plasma membrane (Bauer et al. 2006). While the position of Pucciniomycotina and the monophyly of eight of the nine classes have been established, deeper level phylogenetic relationships within the subphylum have yet to be resolved (Fig. 10.1).

Fungi belonging to Pucciniomycotina are found in a diversity of habitats, including specialized niches that are historically undersampled for Fungi. Ecologically, most discovered species are plant associates, predominantly phytopathogens but also including asymptomatic members of the phylloplane and species that form mycorrhizal associations with orchids. Others are insect and fungal pathogens, and a few are presumably saprobic. Pucciniomycotina species have been recovered from soils, freshwater and marine habitats, and the Arctic and tropical environments. They are shown to **have an array of life cycles, ranging from simple teliosporic yeasts** (Fig. 10.2) **to the elaborate five-stage life cycles of the biotrophic rust fungi** (Fig. 10.3), often regarded as the most complex organisms in Kingdom Fungi (Lutzoni et al. 2004). The number of new species and new lineages of Pucciniomycotina continues to rise, and it is predicted that much diversity within this group remains to be discovered.

[1]Department of Botany and Plant Pathology, Purdue University, West Lafayette, IN 47907, USA; e-mail: maime@purdue.edu;
[2]Department of Plant Biology, University of Minnesota, St. Paul, MN 55108, USA

Systematics and Evolution, 2nd Edition
The Mycota VII Part A
D.J. McLaughlin and J.W. Spatafora (Eds.)

Table 10.1 Pucciniomycotina: classes, orders, families, and number of species

Class	Order	Family	No. spp.
Agaricostilbomycetes R. Bauer et al.[a] 75 spp.	Agaricostilbales Oberw. & R. Bauer	Agaricostilbaceae Oberw. & R. Bauer	30
		Chionosphaeraceae Oberw. & Bandoni	34
		Kondoaceae R. Bauer et al.[a]	6
	Spiculogloeales R. Bauer et al.[a]	Spiculogloeaceae Denchev	5
Atractiellomycetes R. Bauer et al.[a] 44 spp.	Atractiellales Oberw. & Bandoni	Hoehnelomycetaceae Jülich	38
		Phleogenaceae Gäum.	1
		Saccoblastiaceae Jülich	2
	Incertae sedis		3
Classiculomycetes R. Bauer et al.[a] 2 spp.	Classiculales R. Bauer, Begerow, Oberw. & Marvanová	Classiculaceae R. Bauer, Begerow, Oberw. & Marvanová	2
Cryptomycocolacomycetes R. Bauer et al.[a] 2 spp.	Cryptomycocolacales Oberw. & R. Bauer	Cryptomycocolacaceae Oberw. & R. Bauer	2
Cystobasidiomycetes R. Bauer et al.[a] 43 spp.	Cystobasidiales R. Bauer et al.[a]	Cystobasidiaceae Gum.	11
	Erythrobasidiales R. Bauer et al.[a]	Erythrobasidiaceae Denchev	11
	Naohideales R. Bauer et al.[a]	Naohideaceae Denchev	1
	Incertae sedis		20
Microbotryomycetes R. Bauer et al.[a] 227 spp.	Heterogastridiales Oberw. & R. Bauer	Heterogastridiaceae Oberw. & R. Bauer	7
	Kriegeriales Toome & Aime	Kriegeriaceae Toome & Aime	9
	Leucosporidiales J.P. Samp. M. Weiss & R. Bauer	Camptobasidiaceae R.T. Moore	5
		Leucosporidiaceae Jülich	15
	Microbotryales R. Bauer & Oberw.	Microbotryaceae R.T. Moore	105
		Ustilentylomataceae R. Bauer & Oberw.	11
	Sporidiobolales Doweld	Sporidiobolaceae R.T. Moore	37
	Incertae sedis		39
Mixiomycetes R. Bauer et al.[a] 1 sp.	Mixiales R. Bauer et al.[a]	Mixiaceae C.L. Kramer	1
Pucciniomycetes R. Bauer et al.[a] 8016 spp.	Helicobasidiales R. Bauer et al.[a]	Helicobasidiaceae P.M. Kirk	17
	Pachnocybales Bauer et al.	Pachnocybaceae Oberw. & R. Bauer	1
	Platygloeales R.T. Moore	Eocronartiaceae Jülich	9
		Platygloeaceae Racib.	6
	Pucciniales Clem. & Shear	Chaconiaceae Cummins & Y. Hirats.	75
		Coleosporiaceae Dietel	313
		Melampsoraceae Dietel	90
		Mikronegeriaceae Cummins & Y. Hirats.	13
		Phakopsoraceae Cummins & Y. Hirats.	205
		Phragmidiaceae Corda	164
		Pileolariaceae Cummins & Y. Hirats.	34
		Pucciniaceae Chevall.	6,095
		Raveneliaceae Leppik	323
		Uncolaceae Buriticá	3
		Uropyxidaceae Cummins & Y. Hirats.	143
	Septobasidiales Couch ex Donk	*Incertae sedis*	340
		Septobasidiaceae Racib.	179
	Incertae sedis		6
Tritirachiomycetes Aime & Schell 6 spp.	Tritirachiales Aime & Schell	Tritirachiaceae Aime & Schell	6

Numbers are approximate estimates from Kirk et al. (2008), Kurtzman et al. (2011), and newly published papers cited in text
[a] R. Bauer, Begerow, J. P. Samp., M. Weiss & Oberw.

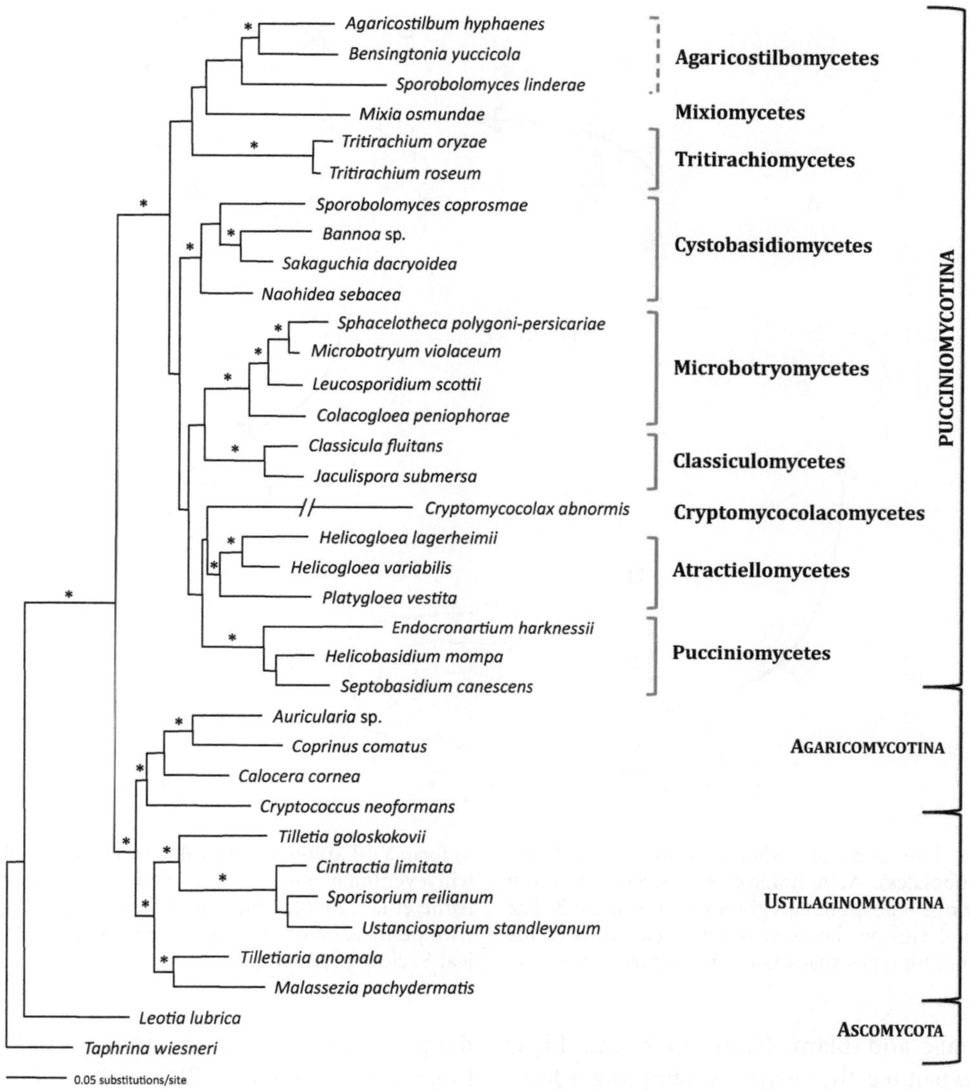

Fig. 10.1 Phylogenetic resolution of Pucciniomycotina classes. Tree based on maximum likelihood analyses of combined nuclear ribosomal small and large subunits and translation elongation factor 1-alpha DNA sequences. Ascomycota sequences included as outgroups; representative Agaricomycotina and Ustilaginomycotina sequences included to show monophyly of Pucciniomycotina. Asterisk (*) denotes nodes that have received strong (>80 %) support in the analyses of Aime et al. (2006), Padamsee et al. (2012), and Schell et al. (2011). Backbone resolution remains poor within Pucciniomycotina. Figure adapted from Schell et al. (2011)

II. Systematics of Pucciniomycotina

Most early treatments of basidiomycetes recognized one main division in the group, between those species that formed holobasidia (homobasidiomycetes) and those with phragmobasidia (heterobasidiomycetes). However, analyses of 5S ribosomal RNA sequences by Walker and Doolittle (Walker and Doolittle 1982) divided Basidiomycota into two groups, not by any traditional characters but by whether they possessed simple pores or dolipores.

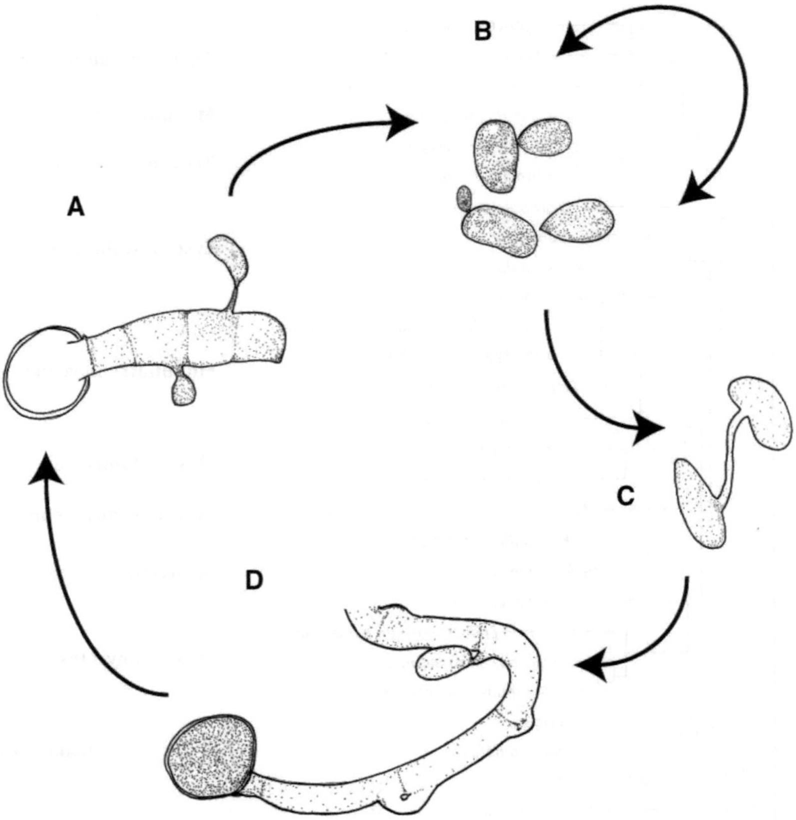

Fig. 10.2 Life cycle of *Rhodosporidium toruloides* (Sporidiobolales). A. A transversely septate basidium arises from a teliospore and gives rise to spores. B. The spores bud and persist as yeasts. C. Yeast cells of the proper mating types fuse via a thin hyphal connection to form a dikaryon. D. The dikaryon forms hyphae that will eventually give rise to teliospores. Figure from Aime et al. (2006), courtesy of D. Henk and reprinted with permission of *Mycologia*. copyright The Mycological Society of America

Gottschalk and Blanz (Gottschalk and Blanz 1985) expanded that work by sampling a large diversity of mostly basidiomycetous yeasts and taking into account the **5S RNA secondary structure** in addition to sequence, again **showing a deep division of Basidiomycota into two groups.** Those with type A secondary structure included members of the smut group *pro parte* (p.p.) (including the anther smut *Microbotryum*) and members of the heterobasidiomycetes that had simple septal pores, including members of Auriculariales p.p. and Atractiellales p.p. (Gottschalk and Blanz 1985). Species with type B secondary structure were found in most of the smut groups excepting the anther smuts, in heterobasidiomycetes with dolipore septa, and in mushroom-forming fungi (Gottschalk and Blanz 1985). The ascomycete *Taphrina deformans* was found to have a 5S secondary structure of type A, while the rust fungi, represented by four species in their analyses, were reported to have type B secondary structure (Gottschalk and Blanz 1984, 1985). Cladistic analyses by these authors of the representative 5S RNA sequences provided evidence for a basal lineage of Basidiomycota that included many yeast-forming fungi, phragmobasidiate fungi, and smutlike fungi that could be distinguished from their convergent cohorts by the absence of dolipore septa. **The group with type A secondary structure was initially referred to as the *simple***

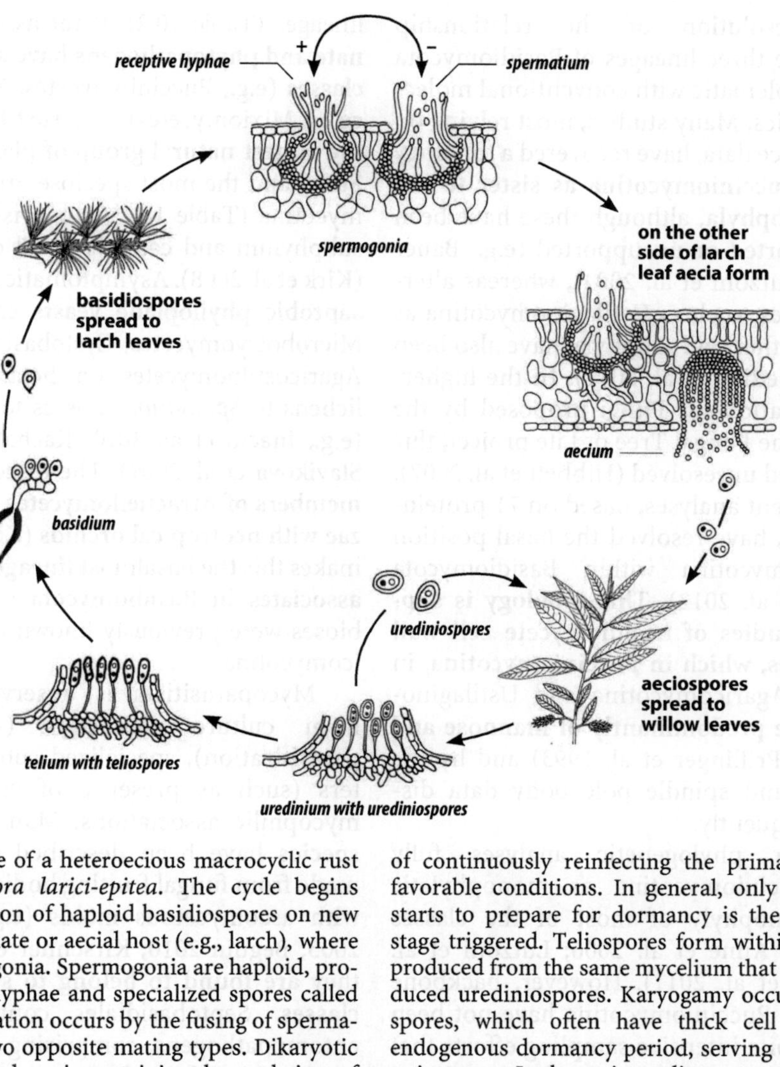

Fig. 10.3 Life cycle of a heteroecious macrocyclic rust fungus, *Melampsora larici-epitea*. The cycle begins with the germination of haploid basidiospores on new leaves of the alternate or aecial host (e.g., larch), where they form spermogonia. Spermogonia are haploid, producing receptive hyphae and specialized spores called spermatia. Fertilization occurs by the fusing of spermatia or hyphae of two opposite mating types. Dikaryotic hyphae form aecial sori containing long chains of aeciospores. Aeciospores function as wind-disseminated propagules that serve to colonize the primary or telial host (e.g., willow). Germinating aeciospores produce dikaryotic uredinia with urediniospores. The uredinial stage is the cyclic asexual stage, capable of continuously reinfecting the primary host under favorable conditions. In general, only when the host starts to prepare for dormancy is the fifth, or telial, stage triggered. Teliospores form within telia that are produced from the same mycelium that previously produced urediniospores. Karyogamy occurs in the teliospores, which often have thick cell walls and an endogenous dormancy period, serving as the overwintering stage. In the spring, teliospores act as probasidia and germinate into basidia. Meiosis occurs within basidia of the auricularioid phragmobasidium type, producing four haploid basidiospores that are forcibly discharged to a new aecial host, completing the life cycle. Figure from Toome (2010)

septate basidiomycete lineage (e.g., Nishida et al. 1995) **or Atractiellales *sensu lato*** (s.l.) (Hawksworth et al. 1995).

Subsequent analyses of small subunit ribosomal DNA (rDNA) sequence (18S rDNA) data revealed the existence of three, rather than two, major lineages of Basidiomycota (Swann and Taylor 1993, 1995). With the exception of the rust fungi, those with a type B secondary structure belonged to two lineages, Ustilaginomycetes (true smut fungi, now Ustilaginomycotina) and Hymenomycetes (mushroom-forming fungi and their relatives with dolipore septa, now Agaricomycotina). **The lineage containing the simple septate basidiomycetes with type A secondary structure and the rust fungi were united in Urediniomycetes (now Pucciniomycotina)** (Swann and Taylor 1995).

However, resolution of the relationship between these three lineages of Basidiomycota has been problematic with conventional molecular systematics. Many studies, most relying on rDNA sequence data, have recovered a topology that places **Pucciniomycotina as sister to the other two subphyla**, although these have been weakly supported or unsupported (e.g., Bauer et al. 2006; Lutzoni et al. 2004), whereas alternate topologies, such as Ustilaginomycotina as sister to the other two subphyla, have also been recovered (Medina et al. 2011). In the higher-level classification for Fungi proposed by the Assembling the Fungal Tree of Life project, this node remained unresolved (Hibbett et al. 2007). However, recent analyses, based on 71 protein-coding genes, have resolved the basal position of Pucciniomycotina within Basidiomycota (Padamsee et al. 2012). **This topology is supported by studies of basidiomycete cell wall carbohydrates, which in Pucciniomycotina**, in contrast to Agaricomycotina and Ustilaginomycotina, **are predominantly of mannose and lack xylose** (Prillinger et al. 1993) and by the septal pore and spindle pole body data discussed subsequently.

Molecular phylogenetic analyses fully support Pucciniomycotina as monophyletic and the monophyly of most of the classes therein (e.g., Aime et al. 2006; Lutzoni et al. 2004; Schell et al. 2011). However, backbone nodes within Pucciniomycotina have not been resolved despite intensive sampling efforts that included nearly the entire known generic diversity (excluding that of the rust fungi) in the subphylum (Aime et al. 2006); current research is now focused on increased locus sampling. Phylogenetic relationships within the subphylum as currently understood are presented in Fig. 10.1. The systematics of the lineages will be discussed in the classification section.

III. Diversity

A. Ecological Diversity

Pucciniomycotina species play diverse ecological roles, although these are incompletely known or can only be inferred for a number of species and lineages (Table 10.2). **Plant associations dominate and phytopathogens have arisen in several classes** (e.g., Pucciniomycetes, Microbotryomycetes, Mixiomycetes). **The rust fungi form both the largest natural group of plant pathogens** in Fungi and the most speciose order in Pucciniomycotina (Table 10.1), comprising 95 % of the subphylum and ca. 8 % of all described Fungi (Kirk et al. 2008). Asymptomatic and presumably **saprobic phylloplane yeasts** can be found in Microbotryomycetes, Cystobasidiomycetes, and Agaricostilbomycetes on hosts ranging from lichens to *Sphagnum* mosses to vascular plants (e.g., Inácio et al. 2010; Kachalkin et al. 2008; Sláviková et al. 2009). The discovery that **some members of Atractiellomycetes form mycorrhizae with neotropical orchids** (Kottke et al. 2010) makes this **the basalmost lineage of mycorrhizal associates in Basidiomycota** since these symbioses were previously known only from Agaricomycotina.

Mycoparasitism is observed or inferred from culture characters (such as self-parasitization), specialized subcellular characters (such as presence of colacosomes), or mycophilic associations. Many **mycoparasitic species** have been described from isolations made from fungal fruiting bodies or co-isolated with ascomycetous molds (e.g., Bauer et al. 2003; Beguin 2010; Kirschner et al. 2001), and they **are found to belong to several different classes**. Septobasidiales contains the only entomopathogens, comprising species that are symbiotic with scale insect colonies (Couch 1938), although the true nature of the association may be more commensal than truly parasitic (Henk and Vilgalys 2007).

Freshwater and marine yeasts can be found primarily in Cystobasidiomycetes and some Microbotryomycetes (Fell 1966; Sampaio 2004), but they also include the enigmatic fungus *Reniforma strues*, which was isolated from biofilms in a wastewater treatment plant (Pore and Sorenson 1990) and is placed *incertae sedis* within Pucciniomycotina by rDNA sequences (Aime et al. 2006). Classiculomycetes and *Cyrenella elegans* (Cystobasidiomycetes) are **aquatic hyphomycetes** that share convergent characters with other primarily ascomycetous aquatic fungi (Bauer et al. 2003; Gochenaur 1981).

Table 10.2 Pucciniomycotina: Synopsis of key ecological and morphological characters by class

Class	Ecological diversity	Asexual reproduction: Yeast state	Asexual reproduction: Conidia	Sexual reproduction: Fruiting body	Sexual reproduction: Basidia	Subcellular characters: Clamps	Septal pore associations	Spindle pole body	Other features
Agaricostilbomycetes	Mycoparasites, saprobes	+	+ (blasto-)	Various—stilboid, pustulate or none	Phragmo-, holo-	+/–	Microbodies	Multilayered disc	Tremelloid haustoria
Atractiellomycetes	Saprobes, orchid mycorrhiza	–	+	Stilboid, resupinate	Phragmo-	+	Atractosomes, microbodies	Multilayered disc	Microscala/symplechosomes
Classiculomycetes	Aquatic, mycoparasites	–	+ (triradiate)	–	Phragmo- w/subapically swollen sterigmata	+	Microbodies	n/a	Binucleate tremelloid haustoria
Cryptomycocolacomycetes	Mycoparasites	– (basidiospores may form yeast-like buds)	+	–	Holo- (w/unique development)	+	Microbodies, pore plugs	Layered disc	Colacosomes/lenticular bodies
Cystobasidiomycetes	Mycoparasites, saprobes	+	– (+in *Cyrenella*)	–	Holo-, phragmo-	+	Cystosome pore plug	n/a	Tremelloid haustoria
Microbotryomycetes	Phyto- or mycoparasites, saprobes (aquatic)	+	–/+	Various—pycnidioid, sori	Phragmo-	+	Pulley-wheel-shaped pore plug, microbodies (*Colacosiphon*)	Subgloboid with flat internalized layer	Colacosomes/lenticular bodies
Mixiomycetes	Phytoparasite	+	?	–	Unknown	–	n/a	n/a	n/a
Pucciniomycetes	Obligate phyto-, entomo- or mycoparasites (saprobe)	– (+in Septobasidiales)	+	Various—stilboid, resupinate, clavarioid, sori	Phragmo-, holo-	–	Pulley-wheel-shaped pore plug	Multilayered disc	Microscala/symplechosomes
Tritirachiomycetes	Saprobes, human pathogens, mycoparasites	–	+	–	Unknown	–	Pore plug	n/a	n/a

Because of the microscopic or cryptic nature of most of the fungi in Pucciniomycotina, their presence and ecological roles may have been overlooked in the past. For example, **sequences generated by environmental sampling** studies are providing data that **suggest the presence of unknown species of Pucciniomycotina in soil rhizospheres** (e.g., Porter et al. (2008), as uncultured basidiomycete; Stefani et al. (2010), as uncultured soil fungus), **anoxic deep-sea habitats** (e.g., Bas et al. (2007), as Urediniomycetes; Jebaraj et al. (2010), as unnamed Pucciniomycotina), **and Arctic ice** (D'Elia et al. 2009). In fact, extreme environments can harbor a diversity of psychrophilic (e.g., Libkind et al. 2005; Libkind et al. 2010; Turchetti et al. 2011), osmotolerant (e.g., Fell 1966), and toxicity-tolerant (e.g., Pohl et al. 2011) Pucciniomycotina yeasts, and such environments may prove to harbor additional untapped diversity.

B. Life Cycles

A striking feature of Pucciniomycotina is the **predominance of asexual stages within most lineages**. Some lineages, in fact, are known only from anamorphs, such as Tritirachiomycetes and, potentially, Mixiomycetes (Table 10.2). Perhaps another striking character of Pucciniomycotina is the number of unique developmental patterns and life cycles that apparently arose in what might be thought of as early experiments into basidiomycetization, culminating in the elaborate life cycles in Pucciniales wherein up to five different sporulating stages can be produced on two unrelated hosts (Fig. 10.3). Interestingly, the character of **heteroecism seems to have arisen only once in Fungi outside of Pucciniomycetes** in the unrelated chytrid genus *Coelomomyces* (Blastocladiales, Blastocladiomycetes) (Whisler et al. 1975; see James et al. 2014). The complexity of the rust life cycle is perhaps why complete life cycle data are missing for many of the species, including emerging pathogens of great agricultural significance such as *Phakopsora pachyrhizi*, *Puccinia psidii*, and *Hemileia vastatrix*. At the other extreme are simple teliosporic yeasts, such as found in Sporidiobolales (Fig. 10.2). Other life cycles will be discussed within the relevant sections to follow.

C. Morphological and Genomic Diversity

The morphological diversity in Pucciniomycotina is immense. Table 10.2 presents some salient morphological characters by class. **A diversity of sporulating forms** is exhibited in Pucciniomycotina species, **ranging from macrobasidiocarp formers to single-celled yeasts** (e.g., Fig. 10.4). To cite a few examples, when present, basidiocarps may be stipitate-capitate or stilboid, such as the fruiting bodies of *Agaricostilbum* species, resupinate, as is found in, for example, *Septobasidium* and *Helicobasidium* species, sporodochial, as in *Mycogloea* species, or, rarely, clavarioid, as in *Eocronartium muscicola*; others, such as Pucciniales and *Microbotryum* species, form spore-filled sori within their hosts.

As early basidiomycetes evolved, new mechanisms for spore formation and dispersal must have arisen, resulting in the amazing **variety of basidial morphologies** present in extant Pucciniomycotina (e.g., Figs. 10.5–11). In Cystobasidiomycetes alone basidia may be unicelled, phragmobasidia of the auricularioid type (i.e., transversely septate), elongate filamentous phragmobasidia, or two-celled with budding basidiospores, and they may germinate from probasidia, teliospores, or directly from terminal hyphal cells. Mechanisms for producing and dispersing mitospores are also diverse (e.g., Fig. 10.12). These may reproduce, for example, by budding, ballistosporic discharge from stalklike condiophores, or production of sessile conidia. Mitospores may be single-celled, multicelled and coiled (e.g., *Hobsonia* spp.), or resemble those of Ingoldian fungi with filamentous appendages adapted for water dispersal (e.g., *C. elegans*). The anamorphic yeast *Reniforma strues* has kidney-shaped cells that produce miniature reniform buds (Pore and Sorenson 1990). One unique spore developmental pattern is found

Fig. 10.4 Representatives of Pucciniomycotina. a. *Jola* cf. *javensis* (Platygloeales) fruiting on *Sematophyllum swartzii* (E. Frieders). b. *Septobasidium burtii* (Septobasidiales) fungal mat completely covering scale insects (D. Henk). c. *Eocronartium muscicola* (Platygloeales) fruiting on moss *Climacium dendroides* (E. Frieders). d. Yeast and filamentous cells of *Sporidiobolus pararoseus* (Sporidiobolales) (M.C. Aime). e. Cultures of two *Sporidiobolus* species in *S. pararoseus* clade (Sporidiobolales) (M.C. Aime). f. *Phragmidium* sp. (Pucciniales) on *Rosa rubiginosa* (M.C. Aime). Figure from Aime et al. (2006) and reprinted with permission of *Mycologia*. copyright The Mycological Society of America

within the monotypic *Mixia osmundae*, which produces hundreds of exogenous, enteroblastic spores at a time from a single saclike sporogenous cell (Nishida et al. 1995). Although the life cycle of this fungus remains to be fully described, recent genomic studies have suggested that the spores on these sporogenous cells are likely mitotic (Toome et al. 2014).

A uniting feature of Pucciniomycotina is the presence of simple septal pores that lack dolipores and septal pore caps (parenthesomes) that otherwise characterize most Basidiomycota (Celio et al. 2006). The presence of Woronin bodies in association with the septal pore is characteristic of Pezizomycotina in the Ascomycota. Although Woronin-like bodies

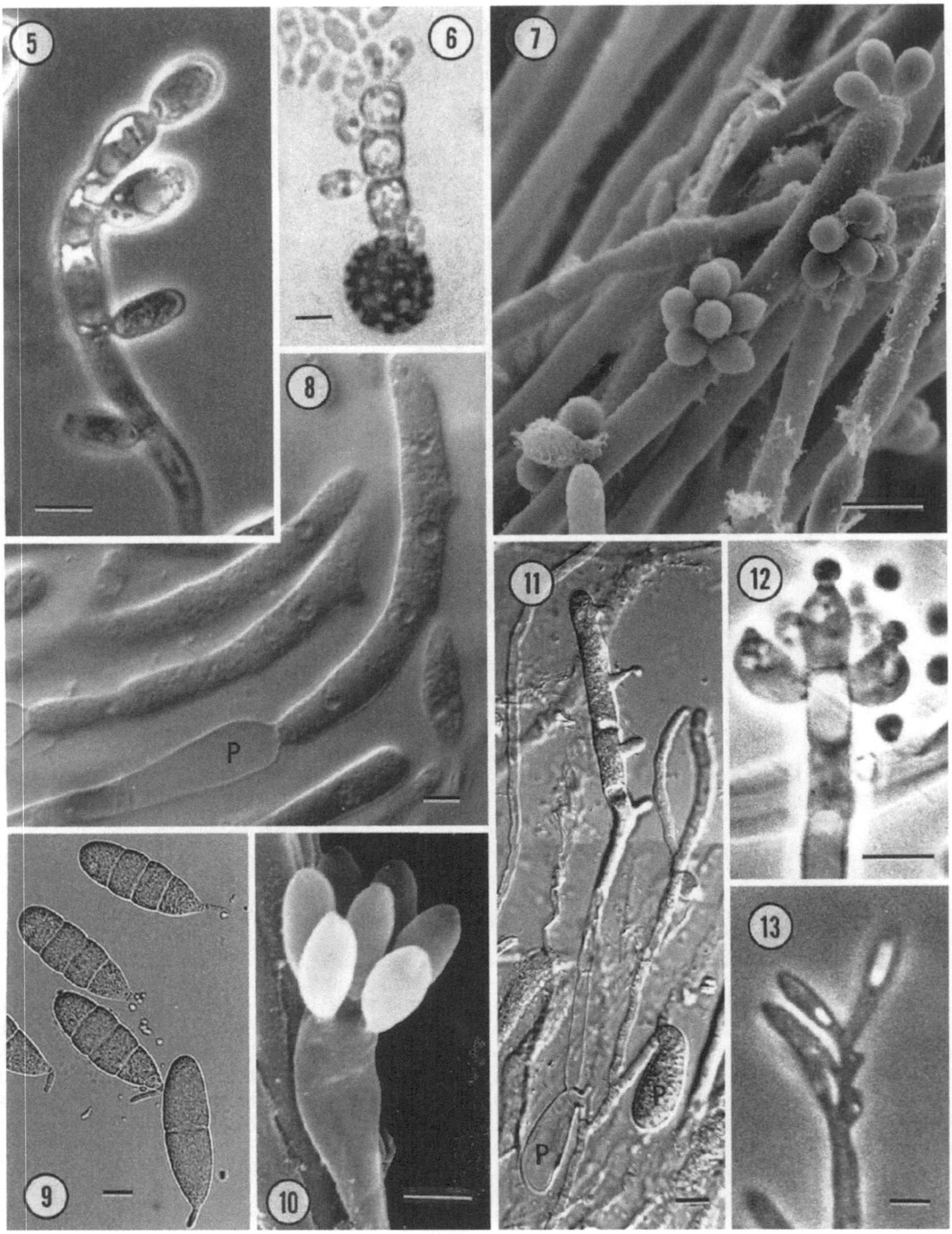

Figs. 10.5–10.13 Basidial and conidial morphology in Pucciniomycotina. **5–11.** Basidia. **5.** Gasteroid auricularioid basidium of *Atractiella* sp. with sessile basidiospores; differentiated probasidium absent (E. Swann, ECS CR27); *bar* 10 μm. **6.** Gasteroid basidium of *Microbotryum reticulatum* with teliospore, sessile basidiospores, and yeast stage (E. Swann, ECS 698); *bar* 5 μm. **7.** Gasteroid auricularioid basidium of *Agaricostilbum pulcherrimum* with multiple basidiospores on each compartment (F. Oberwinkler, F219); *bar* 5 μm. **8.** Ballistosporic auricularioid basidium of *Jola* cf. *javensis* with differentiated probasidium (*P*) (E. Frieders, EMF 004); *bar* 5 μm. **9.** Deciduous auricularioid metabasidia of *Kriegeria eriophori* prior to basidiospore production

have been reported in Agaricostilbomycetes and Cryptomycocolacomycetes (Kirschner et al. 2001; Oberwinkler and Bauer 1989, 1990), cytochemical data are needed to ascertain whether these are homologous with the similar structures in ascomycetes (Celio et al. 2006; Dhavale and Jedd 2007; Roberson et al. 2010). Additional **septal pore features may be diagnostic for some classes.** For instance, pores may be occluded by a pulley-wheel-shaped plug associated with a zone of organelle exclusion bounded by microbodies (e.g., Pucciniomycetes) (Fig. 10.14) or by a cystosome, a more or less cylindrical plug with a reticulate surface (e.g., Cystobasidiomycetes; Sampaio et al. 1999), or distinctive pore-associated microbodies may be present (e.g., Atractiellomycetes) (Fig. 10.15).

Spindle pole bodies (SPBs), organelles that organize microtubules during nuclear division, and nuclear division characters have been examined for many Pucciniomycotina (e.g., McLaughlin et al. 1995; Swann et al. 2001 and references therein). **All species** in Pucciniomycotina **have layered discoid** (although this may verge on globoid) **SPBs** (Figs. 10.16, 17). SPB morphology has not been studied in all classes, but it seems to be a diagnostic character for at least some (Celio et al. 2006) (Table 10.2). During nuclear division the SPB in many Pucciniomycotina is **more or less internalized within the nucleus,** but in the Pucciniomycetes and Atractiellomycetes it is inserted in a nuclear pore (Figs. 10.16, 17). In Pucciniomycetes, except for Pucciniales, and Atractiellomycetes the SPB is surrounded by an endoplasmic reticulum cap (Fig. 10.17), the loss of which seems to be apomorphic in Pucciniales (Fig. 10.16).

One subcellular character that seems to be synapomorphic for Atractiellomycetes is the presence of membrane complexes called **microscala or symplechosomes** (McLaughlin 1990; Oberwinkler and Bauer 1989). These **consist of stacked cisternae of endoplasmic reticulum that are regularly cross-linked by filaments** that may also connect them with mitochondria (Fig. 10.18). **Colacosomes** (sometimes referred to as lenticular bodies), subcellular organelles associated with mycoparasitism that **serve to connect the hyphal cell of the host with that of the parasite,** are found in many species (Bauer et al. 1997), especially in Cryptomycocolacomycetes and Microbotryomycetes.

Tremelloid haustoria, named for the type of haustoria formed by mycoparasitic Tremellales (Agaricomycotina), **can be found in many mycoparasitic** or presumed mycoparasitic Pucciniomycotina, although it is not known whether these structures are truly homologous with those formed in Tremellales. Nonetheless, in Classiculomycetes these haustorial cells are binucleate, rather than uninucleate as in all other studied species, and are thus diagnostic for this class (Bauer et al. 2003). The production of coenocytic hyphae in Mixiomycetes may be synapomorphic for that lineage and is otherwise rare in basidiomycetes. Another rare character in Fungi is branch origin, involving the breaking of the hyphal wall. However, this pattern occurs in both Pucciniomycetes and Atractiellomycetes (Fig. 10.19) and seems to be diagnostic for these classes (Swann et al. 2001).

Whole-genome data are lacking for most lineages of Pucciniomycotina. However, genome sizes for those known range from as small as 13 Mbp for *Mixia osmundae* to 415 Mbp for the rust fungus *Uromyces fabae*, one of the largest known in Fungi (Eilam et al. 1994; Grigoriev et al. 2012). Complete genome sequence data have been released for representatives of three classes, Pucciniomycetes, Microbotryomycetes and Mixiomycetes – *Puccinia graminis* f. sp. *tritici*, *P. triticina*, *P. striiformis*,

Figs. 10.5–10.13 (continued) (J.C. Doublés in Doublés and McLaughlin 1992); *bar* 10 μm. **10.** Gasteroid holobasidium of *Pachnocybe ferruginea* with apical basidiospores (Kleven and McLaughlin 1988); *bar* 2.5 μm. **11.** Maturing ballistosporic auricularioid basidium of *Helicogloea intermedia* with saccate lateral probasidium (*P*) and an adjacent probasidium prior to metabasidium formation (J.C. Doublés); *bar* 10 μm. **12, 13.** Conidia and conidiophores. **12.** Microconidia formation in *Atractiella* sp. (D.J. McLaughlin, DJM 969); *bar* 5 μm. **13.** Sympodial conidium formation in *Jola* cf. *javensis* (D.J. McLaughlin, DJM 739); *bar* 5 μm. **5, 6, 8, 9, 11–13.** Bright-field micrographs. **7, 10.** Scanning electron micrographs. Figures reproduced from Swann et al. (2001); collection or culture number in parentheses

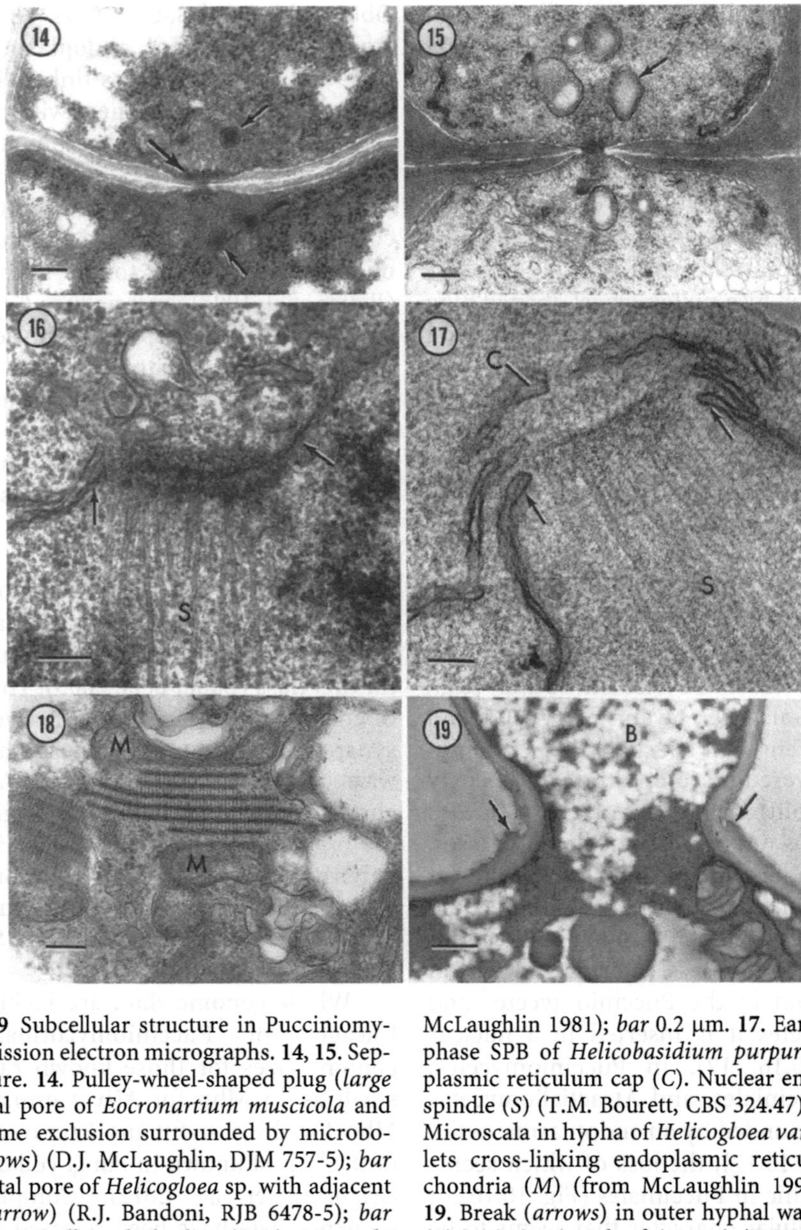

Fig. 10.14–10.19 Subcellular structure in Pucciniomycotina. Transmission electron micrographs. **14, 15.** Septal pore structure. **14.** Pulley-wheel-shaped plug (*large arrow*) in septal pore of *Eocronartium muscicola* and zone of ribosome exclusion surrounded by microbodies (*small arrows*) (D.J. McLaughlin, DJM 757-5); *bar* 0.2 µm. **15.** Septal pore of *Helicogloea* sp. with adjacent microbodies (*arrow*) (R.J. Bandoni, RJB 6478-5); *bar* 0.2 µm. **16, 17.** Spindle pole bodies (SPB). **16.** Early meiotic metaphase I SPB of *Puccinia malvacearum* inserted into a pore of the nuclear envelope (*arrows*). Spindle (*S*) (K.L. O'Donnell, in O'Donnell and McLaughlin 1981); *bar* 0.2 µm. **17.** Early mitotic metaphase SPB of *Helicobasidium purpureum* with endoplasmic reticulum cap (*C*). Nuclear envelope (*arrows*); spindle (*S*) (T.M. Bourett, CBS 324.47); *bar* 0.1 µm. **18.** Microscala in hypha of *Helicogloea variabilis* with rodlets cross-linking endoplasmic reticulum and mitochondria (*M*) (from McLaughlin 1990); *bar* 0.2 µm. **19.** Break (*arrows*) in outer hyphal wall during branch (*B*) initiation in *Jola* cf. *javensis* (D.J. McLaughlin, DJM 739 ps); *bar* 1 µm. Figures reproduced from Swann et al. (2001); collection or culture number provided in parentheses

Cronartium quercuum, and *Melampsora larici-populina* (Pucciniales), *R. graminis*, *Sporobolomyces* sp. (as *S. roseus*) (Sporidiobolales), and *Microbotryum violaceum* (Microbotryales) and *Mixia osmundae* (Mixiales) [Grigoriev et al. (2012); *Microbotryum violaceum* Sequencing Project, Broad Institute of Harvard and MIT (http://www.broadinstitute.org/); *Puccinia* Group Sequencing Project, Broad Institute of Harvard and MIT (http://www.broadinstitute.org/]. Comparative genomics of these fungi has already impacted our understanding of the molecular bases of obligate biotrophy (Duplessis et al. 2011).

D. Species Discovery and Diversity

Approximately 8,416 species of Pucciniomycotina have been described to date (Table 10.1), the majority belonging to Pucciniales. **A number of higher-level Pucciniomycotina lineages are monotypic** (e.g., Mixiomycetes, Naohideales, Pachnocybaceae) or contain less than ten described species (Tritirachiomycetes, Classiculomycetes, Cryptomycocolacomycetes).

New species discovery continues at a high rate. A recent study of the moldlike genus *Tritirachium* reassigned these fungi to Pucciniomycotina, revealing several cryptic species in the genus (Schell et al. 2011). New species and genera have been recently described from habitats not traditionally associated with Pucciniomycotina, such as soil (e.g., Bauer et al. 2009; Yurkov et al. 2011), beetle galleries (e.g., Oberwinkler et al. 2006), and extreme or toxic environments (e.g., Libkind et al. 2010; Pohl et al. 2011). The phylloplane has continued to be a rich source of species discovery, especially of yeasts in Microbotryomycetes (e.g., Golubev and Scorzetti 2010; Toome et al. 2013; Valério et al. 2008).

Even within Pucciniales, whose members, due to their importance in agriculture, have been one of the better studied groups of Fungi, new species discovery continues as molecular systematic studies show that some morphologically circumscribed species are, in fact, composed of numerous, sometimes unrelated, cryptic species. For example, in one study, rust fungi morphologically assigned to *Melampsora epitea* in the Pacific Northwest of North America were found to belong to 14 different phylospecies, of which none seems to have been previously described (Bennett et al. 2011), and investigations of the previously monotypic genus *Dasyspora* revealed it to contain at least 11 species within Central and South America (Beenken et al. 2012).

IV. Classification

Pucciniomycotina contains 9 classes **divided into 20 orders and 37 families** (Table 10.1). The systematics and composition of the three major lineages of Basidiomycota have been rapidly evolving over the last two decades, none more so than within Pucciniomycotina. **Before 2006 these fungi were known as class Urediniomycetes,** which comprised four lineages (Swann and Taylor 1995). The application of molecular systematics to fungal studies has driven the expansion of Pucciniomycotina to now include many lineages of fungi that had previously been placed within other groups. Plesiomorphic characters, such as that of a simple septal pore apparatus, led to the original assignment of fungi in classes Mixiomycetes and Tritirachiomycetes within Ascomycota (Nishida et al. 1995; Schell et al. 2011). Another potentially plesiomorphic character, that of phragmobasidia of the auricularioid type (Fig. 10.11), led to the original classification of most members of Platygloeales in Auriculariales (Agaricomycotina). Similarity in life cycles and morphology, now known to be the result of convergent evolution, led to the original classification of Microbotryomycetes within the smut fungi (Ustilaginomycotina).

Basidiomycetes with yeast states occur in all three subphyla. **Anamorphic yeasts** under previous versions of the International Code of Botanical Nomenclature **were treated separately from teleomorphic species and assigned to form genera principally based on carbon assimilation tests** (Kurtzman et al. 2011 and references therein). Numerous molecular phylogenetic studies have highlighted the artificiality of this system. For example, **species of *Sporobolomyces* occur across most of the yeast-forming Pucciniomycotina classes; species of *Rhodotorula* can be found in Ustilaginomycotina and Pucciniomycotina** (Sampaio 2004; Scorzetti et al. 2002). The type species for both *Rhodotorula* (*R. glutinus*) and *Sporobolomyces* (*S. salmonicolor*) are placed in Sporidiobolales with molecular data (Scorzetti et al. 2002). At the 2011 meeting of the International Botanical Congress changes were adopted that will discontinue the use of a dual nomenclature in Fungi (Hawksworth 2011). One challenge for the future will be to implement the changes now allowed under the new *Code* and integrate the various clades of

Sporobolomyces and *Rhodotorula* and other polyphyletic anamorphic yeast genera into a phylogenetic classification.

A. Agaricostilbomycetes

The type genus ***Agaricostilbum*** **was originally described as an anamorphic member of the Ascomycota** and later transferred to the Auriculariales (Agaricomycotina) (Wright 1970; Wright et al. 1981) before being allied in Pucciniomycotina. Agaricostilbomycetes as currently defined contains two orders, Agaricostilbales and Spiculogloeales (Bauer et al. 2006). The monophyly of both orders has been demonstrated using molecular data (e.g., Aime et al. 2006; Bauer et al. 2009), but strong support for a monophyletic Agaricostilbomycetes as currently circumscribed is lacking. Genera included in Agaricostilbales are ***Agaricostilbum*****,** ***Bensingtonia*** **(anamorphic),** ***Chionosphaera*****,** ***Cystobasidiopsis*****,** ***Kondoa*****,** ***Kurtzmanomyces*** **(anamorphic),** ***Mycogloea*** **p. p.,** ***Sterigmatomyces*** **(anamorphic), and** ***Stilbum***; *Mycogloea* p.p. and *Spiculogloea* are assigned to Spiculogloeales; anamorphic *Sporobolomyces* yeasts are found in both orders (Aime et al. 2006; Bauer et al. 2006 and references therein; Bauer et al. 2009; Kurtzman et al. 2011). *Mycogloea* s.l. is not monophyletic (Aime et al. 2006; Bauer et al. 2009), and sampling of the type species, *M. carnosa*, is needed to resolve the placement of this genus.

Together, the species of Agaricostilbomycetes possess a wide array of ecological and morphological variation. **Most species are believed to be either saprobic or mycoparasitic.** *Agaricostilbum* and *Stilbum* species are typically isolated from dead plant material (of palms in the case of *A. pulcherrimum*), and *Cystobasidiopsis* is known only from soil (Bauer et al. 2009). However, there is evidence for a mycoparasitic habit for many of the species, and many could be mycoparasitic rather than saprobic. For instance, the original description of *A. palmicola*, the type species of *Agaricostilbum*, notes that the fungus was almost always found in association with an ascomycetous *Anthostoma*-like fungus (Wright 1970). *Chionosphaera*, *Mycogloea*, and *Kondoa* species are found similarly in association with other fungi, and species of *Spiculogloea* are known mycoparasites. *Mycogloea* and *Spiculogloea* species also produce tremelloid haustorial cells such as are commonly found in other known mycoparasitic fungi, especially those in Tremellales (Agaricomycotina) (Bauer et al. 2006). Species of *Kurtzmanomyces* seem to be very rare and are known only from type cultures (Kurtzman et al. 2011), in contrast to *A. pulcherrimum*, which is pantropical in distribution. *Sterigmatomyces halophilus* is usually found in association with marine environments, and both species of *Sterigmatomyces* are osmotolerant (Fell 1966). The ecological niche of many species, however, remains unknown.

All species form a yeastlike stage, with the exception of *Cystobasidiopsis nirenbergiae* (Bauer et al. 2009), and, excepting *C. nirenbergiae*, those with known teleomorphs are dimorphic. With one exception (members of the genus *Chionosphaera*), teleomorphic species in Agaricostilbomycetes produce phragmobasidia; species of *Spiculogloea* and *Kondoa* are ballistosporic (Bauer et al. 2006), and *Kurtzmanomyces* and *Sterigmatomyces* species form ballistoconidia on a stalked conidiophore, a character otherwise not found in Pucciniomycotina (Kurtzman et al. 2011). Stilboid basidiocarps are formed in three genera (*Agaricostilbum*, *Stilbum*, and *Chionosphaera*), and minute pustulate fruiting bodies are formed by members of *Mycogloea* (Bandoni 1998). Basidia are formed directly from probasidia on hyphae in *Cystobasidiopsis* (Bauer et al. 2009).

The septal pore in *Agaricostilbum* species is associated with microbodies containing electron-dense material that has been suggested to resemble the Woronin bodies of Ascomycota (Oberwinkler and Bauer 1989). Additionally, **in studied members of the Agaricostilbaceae and Chionosphaeraceae an unusual pattern of mitosis has been documented wherein, in the yeast phage, the nucleus divides in the parent cell, rather than migrating into the bud prior**

to division (Frieders and McLaughlin 1996; McLaughlin et al. 2004; Swann et al. 2001).

B. Atractiellomycetes

This class contains a single order, Atractiellales, and **fewer than 50 species** in the genera ***Atractiella, Basidiopycnis, Helicogloea, Hobsonia*** **(anamorphic),** ***Infundibura*** **(anamorphic),** ***Leucogloea*** **(anamorphic),** ***Phleogena, Proceropycnis*** **(anamorphic), and** ***Saccoblastia.*** Atractiellales was originally erected to accommodate a number of genera and species formerly placed in Auriculariales (Agaricomycotina) and subsequently separated by the presence of simple septal pores and 5S RNA secondary structure (Gottschalk and Blanz 1985; Oberwinkler and Bandoni 1982). As in Agaricostilbomycetes, stilboid fruiting bodies are formed in *Atractiella* and *Phleogena* and basidia are phragmobasidia of the auricularioid type. However, yeast states are not known for these fungi, and anamorphic states are typically conidial. In *Hobsonia* species conidia are tightly coiled on short conidiophores, forming a minute sporodochium-like fruiting body on dead vegetation (Martin 1959).

Ultrastructurally, members possess **organelles termed microscala or symplechosomes that have no known function but seem to be synapomorphic for the class** (Bauer et al. 2006; McLaughlin 1990; Oberwinkler and Bauer 1989). Perhaps the most intriguing Pucciniomycotina discovery of recent years was that of the association of three unidentified species of Atractiellomycetes with tropical orchids, confirmed by transmission electron microscopy and molecular phylogenetics (Kottke et al. 2010). Prior to this discovery, all Atractiellomycetes were presumed saprobic, and the basalmost mycorrhizal formers known in Basidiomycota were to be found within Auriculariales. The sampling area of Kottke et al. (Kottke et al. 2010) was limited to a tropical montane rainforest in southern Ecuador; thus, it is unknown how widespread this association is. Nonetheless, the find remains significant for documenting the first known instance of a plant mutualistic association within Pucciniomycotina.

C. Classiculomycetes

This class contains a single order, Classiculales, for which **only two species are known,** ***Classicula fluitans*** **(anamorph** ***Naiadella fluitans*****) and the hyphomycete** ***Jaculispora submersa*** (teleomorph unknown) (Hudson and Ingold 1960; Marvanová and Bandoni 1987). Morphological and sequence data clearly show that *C. fluitans* and *J. submersa* form a separate lineage in Pucciniomycotina (Aime et al. 2006; Bauer et al. 2006). Both species are **aquatic and are associated with leaf litter in freshwater habitats.** Plant host preferences have been tested for *J. submersa* and suggest an affinity for oak leaves (Prokhorov and Bodyagin 2007). Additionally, there is evidence that they may be mycoparasitic; *C. fluitans* has been observed to parasitize its own hyphae in culture, and both species form tremelloid haustorial cells such as are commonly found in other known mycoparasitic fungi (Bauer et al. 2003).

In both *C. fluitans* and *J. submersa* the septal pores are surrounded by microbodies that are arranged in a circular pattern, such as are also found in Pucciniales and a few other members of Pucciniomycotina (Bauer et al. 2003). The **combination of binucleate, tremelloid haustorial cells and pore-associated microbodies, however, is unique to Classiculomycetes.** Both species are hyphal with hyaline cells. Primary septa are formed in association with nuclear division and have clamp connections; adventitious septa may also be formed independently of nuclear division and are clampless. Asexual reproduction takes place via conidia that have two to three long fusiform subapical appendages resembling the conidia of other unrelated aquatic fungi (Marvanová and Bandoni 1987), including the cystobasidiomycete *C. elegans.* The sexual stage of *C. fluitans* has been observed to occur only on the surface of water. The basidia occur in clusters and have auricularioid septation and **subapically swollen sterigmata**, the last of which is unique in Pucciniomycotina. Basidiospores are small and fusiform, which is another convergent character found in other unrelated aquatic fungi (Bauer et al. 2003).

D. Cryptomycocolacomycetes

This is a small enigmatic class with **two known species, *Cryptomycocolax abnormis* (published as *C. abnorme*) and *Colacosiphon filiformis* (anamorphic)** (Kirschner et al. 2001; Oberwinkler and Bauer 1990). Both species are apparently **rare**, having been isolated only once each from a parasitized ascomycete (*C. abnormis*) and bark beetle galleries, where it was found parasitizing a co-isolated ascomycete (*C. filiformis*). Available DNA sequence data of the nuclear ribosomal large subunit indicate that these fungi form a separate class-level lineage within Pucciniomycotina (Bauer et al. 2006). Unfortunately, type or other material of either species could not be located for additional analyses, and thus phylogenetic resolution of this lineage will not be possible until additional isolates are discovered.

Within Pucciniomycotina, the **extremely elongate holobasidia** produced by *C. abnormis* **are unique.** *C. filiformis* is described as mitosporic with elongate conidiophores (Kirschner et al. 2001), although Bauer et al. (2006) indicate that this species also produces elongate basidia of the *Cryptomycocolax* type. Species form hyaline hyphae that are clamped in *Cryptomycocolax*; basidia are produced on the host surface and undergo a unique developmental pattern (Oberwinkler and Bauer 1990). Members of Cryptomycocolacomycetes possess mycoparasitic organelles termed colacosomes, a character that is shared only with some Microbotryomycetes. Microbodies interpreted as Woronin-like bodies have been reported in association with the septal pores of both species (Kirschner et al. 2001; Oberwinkler and Bauer 1990); septal-pore-associated microbodies and pore plugs are present on some, but not all, septa of *C. abnormis* (Oberwinkler and Bauer 1990).

E. Cystobasidiomycetes

This is a small class of **predominantly yeast-like fungi**. Genera include ***Bannoa*, *Cyrenella* (anamorphic), *Cystobasidium*, *Erythrobasidium*, *Naohidea*, *Occultifur*, and *Sakaguchia*, as well as** several anamorphic yeasts currently placed in *Rhodotorula* and *Sporobolomyces* (Aime et al. 2006; Kurtzman et al. 2011), most of which are, or presumably are, mycoparasitic. For instance, *C. elegans* was isolated from a mushroom that had been submerged in freshwater (Gochenaur 1981). Species of *Cystobasidium*, *Erythrobasidium*, *Naohidea*, and *Occultifur* have been isolated from ascomycete or basidiomycete fruiting bodies. The number of known species in this class has probably tripled in the last decade, primarily because of the discovery of new yeast species from extreme habitats (e.g., Libkind et al. 2010; Pohl et al. 2011). As is also true of some Microbotryomycetes, a number of species seem to be psychrophiles (e.g., Libkind et al. 2008). The majority of Cystobasidiomycetes genera (*Bannoa*, *Cyrenella*, *Erythrobasidium*, *Naohidea*, and *Sakaguchia*) are monotypic, and most of these are known from single cultures or a single geographic locale, making it likely that a tremendous amount of undiscovered diversity exists within the class.

The variety of sexual state morphologies in this group is unusual, although not all researchers have reached similar interpretations of the structures involved, especially for *Bannoa* and *Erythrobasidium*, for which the same reproductive cells have been described as basidial (e.g., Sugiyama and Suh 1993) or conidial (Bauer et al. 2006). However, Kurtzman et al. (2011) provide convincing evidence, including the illustration of conjugation tubes and basidiospore germination, that these are indeed teleomorphic species. *Bannoa* and *Erythrobasidium* species form holobasidia; *Naohidea*, *Cystobasidium*, and *Occultifur* members form phragmobasidia of the auricularioid type, which germinate from probasidia in the latter two but not in *Naohidea*; and in *Sakaguchia dacryoidea* two- to four-celled basidia germinate from teliospores produced on short hyphal stalks (Kurtzman et al. 2011; Oberwinkler 1990; Sugiyama and Suh 1993; Yamada et al. 1994).

The bipolar multiallelic mating system of *Bannoa hahajimensis* is unique within Pucciniomycotina (Kurtzman et al. 2011). *C. elegans* is also unique among yeast species in producing subclavate tetraradiate conidia, which are found in unrelated aquatic hyphomycetes, making an aquatic habit likely for this species (Kurtzman et al. 2011). These differ from the aquatic conidia produced in Classiculomycetes in shape and number of appendages. Mycoparasitic tremelloid haustorial cells are produced by members of Cystobasidiales (Bauer et al. 2006). Ultrastructurally, septal pores of Cystobasidiales are occluded by a cystosome.

F. Microbotryomycetes

Microbotryomycetes are known for containing the model genetic organism *Microbotryum violaceum* and several ubiquitous red yeasts including *Sporidiobolus pararoseus*. Three members of Microbotryomycetes, the yeasts *Sporobolomyces* sp. (as *S. roseus*) and *Rhodotorula graminis* and the anther smut *M. violaceum*, are the only Pucciniomycotina species outside of Pucciniales to be whole-genome sequenced to date [Grigoriev et al. (2012); *Microbotryum violaceum* Sequencing Project, Broad Institute of Harvard and MIT (http://www.broadinstitute.org/)]. With more than 200 described species, it is the **second largest class in Pucciniomycotina** (Kirk et al. 2008) (Table 10.1). Five orders and seven families have been described. Genera include ***Atractocolax, Aurantiosporium, Bauerago, Camptobasidium, Colacogloea, Curvibasidium, Fulvisporium, Heterogastridium, Kriegeria, Krieglsteinera, Leucosporidium* p.p., *Liroa, Mastigobasidium, Meredithblackwellia, Microbotryum, Rhodosporidium, Sphacelotheca, Sporidiobolus, Ustilentyloma, Zundeliomyces,* and *Zymoxenogloea* (anamorphic), and numerous anamorphic yeasts placed in *Glaciozyma, Leucosporidiella, Rhodotorula,* and *Sporobolomyces,* including the type species of *Rhodotorula* and *Sporobolomyces*** (Aime et al. 2006; Bauer et al. 2006; Toome et al. 2013; Turchetti et al. 2011). A large percentage of the described genera are monotypic (e.g., *Atractocolax, Camptobasidium, Fulvisporium, Heterogastridium, Krieglsteinera, Liroa, Mastigobasidium, Meredithblackwellia,* and *Zundeliomyces*), which may be an indication of an as yet undiscovered diversity.

Microbotryum species, often referred to as the **anther smuts**, were originally classified within Ustilaginomycotina, although numerous lines of evidence now show that the smut syndrome, including an anamorphic yeast phase, gasteroid basidia, pigmented teliospores, and parasitism of plant reproductive parts, has independently evolved at least twice within Basidiomycota. Most phylogenetic analyses recover Microbotryomycetes as a monophyletic class (e.g., Aime 2006; Bauer et al. 2006), yet the backbone within the class has not been adequately resolved, and nearly 20 % of the species now classified in Microbotryomycetes have not been confidently placed to order or family (Table 10.1).

Yeast stages of this group are increasingly recovered in environmental samplings of phylloplanes, soils, and extremely cold habitats with concomitant new species discovery (e.g., Golubev and Scorzetti 2010; Kachalkin et al. 2008; Libkind et al. 2005; Toome et al. 2013; Turchetti et al. 2011; Valério et al. 2008; Yurkov et al. 2011). The tractability of many of these organisms in the laboratory has led to the development of molecular biological and genomics tools for studying genetics and gene function in Microbotryomycetes that are lacking in other Pucciniomycotina (e.g., Coelho et al. 2011; Ianiri et al. 2011). The first studies to identify mating type loci in Pucciniomycotina were conducted with a member of Microbotryomycetes (Coelho et al. 2008; Giraud et al. 2008).

Most teleomorphic species are dimorphic with haploid yeast stages and phragmobasidiate teleomorphs, with the exception of *Curvibasidium* (Bauer et al. 2006). **Colacosomes, subcellular organelles associated with mycoparasitism**, of similar appearance to those in Cryptomycocolacomycetes, **are found in many species** (Bauer et al. 1997), but otherwise there is a tremendous diversity in morphology and ecology within this class, which is discussed in detail in Bauer et al. (2006) and Swann et al. (2001). There is a range of fruiting morphologies from the simple teliosporic yeasts, e.g., *Rhodosporidium* (Fig. 10.2), to the pycnidioid fruiting bodies of *Heterogastridium* species. Ecologically, many are plant associates, either as presumably saprobic yeasts on plant surfaces or as pathogens of leaves (e.g., *Kriegeria*) and plant anthers (e.g., *Microbotryum*). *Heterogastridium* species are mycoparasites, and the presence of colacosomes in most other genera in Heterogastridiales would suggest a similar habit for these. The yeast *Camptobasidium hydrophilum* is aquatic (Marvanová and Suberkropp 1990), and several Sporidiobolales members are cosmopolitan, having been recovered from many terrestrial and marine habitats (e.g., Sampaio 2004).

G. Mixiomycetes

***Mixia osmundae* is the only species** currently known in Mixiomycetes. It was first described as an ascomycete (*Taphrina osmundae*) and remained classified within Ascomycota for more than 80 years, primarily due to superficial similarities between the sporogenous cells of *Mixia* and the asci produced by some Ascomycota. However, molecular and closer morphological studies of the sporogenous cells in the 1990s provided multiple lines of evidence that *Mixia* belongs to Basidiomycota (Nishida et al. 1995). Later phylogenetic analyses of rDNA sequences support its placement in Basidiomycota and show clearly that it is a member of Pucciniomycotina (Aime et al. 2006; Bauer et al. 2006) (Fig. 10.1).

The fungus is an **intracellular parasite of ferns in the genus *Osmunda*,** in which it causes small yellow to brown leaf spots. *Mixia* is known from *Osmunda regalis* in Japan and Taiwan and *Osmunda cinnamomea* in the USA (Kramer 1958; Mix 1947; Nishida 1911; Sugiyama and Katumoto 2008), but it is rarely found, and many aspects of its biology are unknown.

When growing within a host, *Mixia* forms intercellular coenocytic hyphal cells, forming large saclike, nonseptate, sporogenous cells on the surface of the host epidermis. **The production of coenocytic hyphae is a rare condition in Basidiomycota, and the sporogenous cells produced by *Mixia* are unique in the phylum.** Spore production is very unusual in that the spores are formed on the surface of the sporogenous cell simultaneously, creating a powdery layer on fern leaves (Nishida et al. 1995). Genome sequencing revealed that these spores are haploid and likely produced via asexual reproduction (Toome et al. 2014). In culture, *M. osmundae* forms yeastlike cells that reproduce by budding. Septal pore ultrastructure has not yet been determined for this fungus, likely due to the limited formation of septa (Bauer et al. 2006).

H. Pucciniomycetes

Pucciniomycetes is a diverse class containing **the vast majority** (ca. 8000; Kirk et al. 2008) **of Pucciniomycotina species.** Before the availability of DNA sequence data, Pucciniomycetes were placed in various positions on the fungal tree of life. For instance, based on some of their structural characters (e.g., lack of clamp connections) and parasitic life style, Pucciniales and their relatives were often thought to represent an early diverging lineage of Basidiomycota. Phylogenetic studies based on rDNA have shown that rust fungi and their closest relatives in Pucciniomycetes are a derived group within the Pucciniomycotina (Aime et al. 2006). One earlier name for this lineage is Urediniomycetidae sensu Swann et al. (2001).

Almost all of the organisms in Pucciniomycetes are **parasites of plants, insects, or other fungi.** The class contains five orders (Table 10.1), **the most speciose** of which, at ca. 7,800 species in ca. 150 genera (Kirk et al. 2008), **is Pucciniales,** or rust fungi, named for the typically rusty coloration of their urediniospores. Rust fungi are parasites of vascular plants with highly complex life cycles requiring the production of up to five different spore stages on two different host plants (Cummins and Hiratsuka 2003) (Fig. 10.3). In Fungi true obligate biotrophs, i.e., organisms that completely depend on a living host to complete their life cycle, are rare, mainly comprising the powdery mildews (Erysiphales, Ascomycota) and the rust fungi. **Species of Pucciniales cause some of the most devastating plant diseases and therefore have been studied in greater detail than other Pucciniomycotina.** However, their obligately biotrophic nature renders them recalcitrant organisms for molecular studies. Thus, family and generic concepts are predominantly morphology-based, which has led to the recognition of several artificial taxa. Comprehensive phylogenetic treatments of the order include Aime (2006), Maier et al. (2003), and Wingfield et al. (2004). Descriptions of families and genera can be found in Cummins and Hiratsuka (2003).

The remaining approximately 200 species in the Pucciniomycetes are of little economic importance. The largest among these is **Septobasidiales [*Auriculoscypha, Coccidiodictyon, Johncouchia* (anamorphic), *Ordonia, Septobasidium*, and *Uredinella*]**, of which more than 150 species are known and which contains the only entomopathogenic species in Pucciniomycotina. Members of Septobasidiales **parasitize scale insects that feed on trees**, forming dense fungal mats that cover the insects on their hosts (Couch 1938). **The next two largest orders contain species parasitic on mosses and ferns (Platygloeales, ca. 20 species in *Eocronartium, Herpobasidium, Jola, Insolibasidium, Platygloea* s.s., *Platycarpa*, and *Ptechetelium*) or parasites that alternate between plant roots and rust fungi (Helicobasidiales, ca. 17 species of *Helicobasidium*, and its *Tuberculina* anamorph)**. Some species in Helicobasidiales have been the focus of ecological studies to determine their potential as biocontrol agents against rust fungi, but very little is known about the other species in these orders. The fifth order, **Pachnocybales, contains a single species, *Pachnocybe ferruginea***, which seems to be a saprobe, having been repeatedly isolated from creosoted telephone poles, and therefore differs significantly from all other Pucciniomycetes by having a nonparasitic habit.

The **dikaryon is the dominant phase in Pucciniomycetes,** and only one of the orders, Septobasidiales, is known to have a yeast stage. However, production of asexual spores is often well developed, especially among rust fungi. Members of Pucciniomycetes lack clamp connections (Bauer et al. 2006). Basidia are of the auricularioid type, germinating from a probasidium that, in the rust fungi, is a thick-walled resting spore (i.e., teliospore). *P. ferruginea* is again the exception for Pucciniomycetes in that it produces holobasidia rather than phragmobasidia (Kropp and Corden 1986). Pucciniales are heteroecious, i.e., they alternate between two unrelated hosts during different stages of their life cycle. Members of Helicobasidiales also need to alternate between two hosts; the dikaryon parasitizes plant stems and roots, whereas the monokaryotic stage parasitizes rust fungi in the Pucciniales (Lutz et al. 2004).

The most important ultrastructural character of Pucciniomycetes is that of a **simple septal wall with a central pore that in many species has a pulley-wheel-shaped plug** (Swann et al. 2001), and the SPB is inserted into a pore of the nuclear envelope (Bauer et al. 2006). In Septobasidiales and Pachnocybales, the presence of microscala has also been reported (Swann et al. 2001).

I. Tritirachiomycetes

This class contains a single order, Tritirachiales, with **six currently known *Tritirachium* species**. Until recently the genus *Tritirachium* was placed in phylum Ascomycota, primarily due to similarities in conidiophore morphology with other mold species in subphylum Pezizomycotina. However, multigene analyses based on nuclear small and large subunits and translation elongation factor 1-alpha revealed that most species currently placed in *Tritirachium*, including the type species, belong to Pucciniomycotina (Schell et al. 2011).

All the members of Tritirachiomycetes are **anamorphic molds with no known teleomorphic stage**. Species have been isolated from dead plant roots, indoor environments, and insects (Beguin 2010; Jebaraj et al. 2010; Limber 1940; Schell et al. 2011). While the precise role of *Tritirachium* species in the environment is not known, there is strong evidence that *T. dependens* is a potentially obligate parasite of *Penicillium* and other ascomycetous species, on which it depends for certain micronutrients (Beguin 2010, as *T. egenum*). Two species, *T. oryzae* and *T. roseum*, can be causal agents of infections on human cornea and scalp (Moraes et al. 2010; Rodrigues and Laibson 1975). There is little information about the biology of *Tritirachium* species, and only those of potential medical importance have been studied in any detail. Although not originally identified as such, environmental sequences of what seem to be species of *Tritirachium* have been generated from soil clones from a rhizosphere in Canada in a study by Stefani et al. (2010) and from minimally oxygenated deep waters of the Arabian Sea reported by Jebaraj et al. (2010).

There is some overlapping of cultural and morphological characteristics between the species currently placed in *Tritirachium*; therefore, morphological observations alone may not be sufficient for diagnosing all members of this genus at the species level. At the genus level, *Tritirachium* species are hyphal in culture, producing conidiophores that branch in a characteristic zigzaglike pattern. Conidia are hyaline and single-celled. Septal pores are uniperforate with a small pore plug (Schell et al. 2011), but little else is currently known of the subcellular characters of these fungi.

V. Culturing

For the majority of Pucciniomycotina species **in Pucciniales, culturing on standard media is not possible because these are obligate plant pathogens**. Nevertheless, various methods have been developed to facilitate the multiplication of rust fungi, and uredinial spore states can be maintained on susceptible host plant tissue for a number of species. A few species of rust fungi have been successfully cultured from germinating basidiospores or hyphae from leaves; however, complex media are needed, and the growth rate of such cultures is extremely slow and limited (Kinloch and Dupper 1996; Moricca and Ragazzi 2001).

Most other known members of Pucciniomycotina are culturable and grow well on standard nutrient sources, both in liquid and on solid media. Those with forcible spore discharge can be isolated via the spore fall method by suspending the substrate (such as plant leaf) above nutrient media with antibiotics (e.g., Toome et al. 2013). This method works well for separating many mycoparasites from their fungal hosts (e.g., Langer and Oberwinkler 1998). Gasteroid species, yeasts, and anther smuts can be isolated via streak plating on antibiotic media (e.g., Kurtzman and Fell 2004). Some halotolerant species, such as *Sterigmatomyces* spp., can be isolated by exposing air to media with high sodium (up to 20 %) or glucose (up to 50 %) content (e.g., Fell 1966).

VI. Conclusion

Pucciniomycotina contains a diversity of fungi that are united in possessing simple septal pores that lack dolipores and septal pore caps. Most, but not all, produce phragmobasidia, and many have yeast states. Members now united in Pucciniomycotina were previously placed within Ascomycota and the other two subphyla (Ustilaginomycotina and Agaricomycotina) of Basidiomycota. More than 8 % of all described Fungi belong to Pucciniomycotina, whose members can be found in habitats ranging from deep oceans and Arctic ice to most terrestrial systems. Plant associations dominate, and the majority of described species are phytopathogens of vascular plants, ferns, and mosses, but other members are known as asymptomatic members of the phylloplane, entomopathogens and mycoparasites, or mycorrhizal symbionts of orchids. Life cycles range from simple teliosporic yeasts to the elaborate life cycles found in the biotrophic rust fungi. The description of new species of Pucciniomycotina has been steadily rising in the last 10 years, and it is predicted that much diversity within this group remains to be discovered.

Acknowledgements Some of the figures were previously published in Swann et al. (2001), Aime et al. (2006), and Toome (2010), and we would like to thank the individuals who produced the original images, including Drs. D. Henk, E.M. Frieders, J.C. Doublés, E. C. Swann, R.J. Bandoni, K. O'Donnell, and T.M. Bourett. This work was supported in part by National Science Foundation Grants DEB 0732968 and DEB-0732550, Assembling the Fungal Tree of Life: Resolving the Evolutionary History of the Fungi.

References

Aime MC (2006) Toward resolving family-level relationships in rust fungi (Uredinales). Mycoscience 47:112–122

Aime MC, Matheny PB, Henk DA, Frieders EM, Nilsson RH, Piepenbring M, McLaughlin DJ, Szabo LJ, Begerow D, Sampaio JP, Bauer R, Weiss M,

Oberwinkler F, Hibbett DS (2006) An overview of the higher-level classification of Pucciniomycotina based on combined analyses of nuclear large and small subunit rDNA sequences. Mycologia 98:896–905
Bandoni RJ (1998) On some species of *Mycogloea*. Mycoscience 39:31–36
Bas D, Howe A, Brown N, Barton H, Demidova M, Michelle H, Li L, Sanders H, Watkinson SC, Willcock S, Richards TA (2007) Yeast forms dominate fungal diversity in the deep oceans. Proc R Soc B 274:3069–3077
Bauer R, Oberwinkler F, Vánky K (1997) Ultrastructural markers and systematics in smut fungi and allied taxa. Can J Bot 75:1273–1314
Bauer R, Begerow D, Oberwinkler F, Marvanová L (2003) *Classicula*: the teleomorph of *Naiadella fluitans*. Mycologia 95:756–764
Bauer R, Begerow D, Sampaio JP, Weiβ M, Oberwinkler F (2006) The simple-septate basidiomycetes: a synopsis. Mycol Progress 5:41–66
Bauer R, Metzler B, Begerow D, Oberwinkler F (2009) *Cystobasidiopsis nirenbergiae*, a new agaricostilbomycete (Pucciniomycotina). Mycol Res 113:960–966
Beenken L, Zoller S, Berndt R (2012) Rust fungi on Annonaceae II: the genus *Dasyspora* Berk. & M.A. Curtis. Mycologia 104(3):659–681. doi:10.3852/11-068
Beguin H (2010) *Tritirachium egenum*, a thiamine- and siderophore-auxotrophic fungal species isolated from *Penicillium rugulosum*. FEMS Microbiol Ecol 74:165–173
Bennett C, Aime MC, Newcombe G (2011) Molecular and pathogenic variation within *Melampsora* on *Salix* in western North America reveals numerous cryptic species. Mycologia 103:1004–1018
Celio GJ, Padamsee M, Dentinger BTM, Bauer R, McLaughlin DJ (2006) Assembling the fungal tree of life: constructing the structural and biochemical database. Mycologia 98:850–859
Coelho MA, Rosa A, Rodrigues N, Fonseca A, Gonçalves P (2008) Identification of mating type genes in the bipolar basidiomycetous yeast *Rhodosporidium toruloides*: first insight into the *MAT* locus structure of the *Sporidiobolales*. Eukaryot Cell 7:1053–1061
Coelho MA, Gonçalves P, Sampaio JP (2011) Evidence for maintenance of sex determinants but not of sexual stages in red yeasts, a group of early diverged basidiomycetes. BMC Evol Biol 11:249
Couch JN (1938) The genus *Septobasidium*. University of North Carolina Press, Chapel Hill, NC
Cummins GB, Hiratsuka Y (2003) Illustrated genera of rust fungi. APS Press, St. Paul, MN
D'Elia TD, Veerapaneni R, Theraisnathan V, Rogers SO (2009) Isolation of fungi from Lake Vostok accretion ice. Mycologia 101:751–763
Dhavale T, Jedd G (2007) The fungal Woronin body. In: Howard RJ, Gow NAR (eds) The Mycota, vol VIII, 2nd edn, Biology of the fungal cell. Springer, Berlin, pp 87–96
Doublés JC, McLaughlin DJ (1992) Basidial development, life history, and the anamorph of *Kriegeria eriophori*. Mycologia 84:668–678
Duplessis S, Cuomo CA, Lin YC, Aerts A, Tisserant E, Veneault-Fourrey C, Joly DL, Hacquard S, Amselem J, Cantarel BL, Chiu R, Coutinho PM, Feau N, Field M, Frey P, Gelhaye E, Goldberg J, Grabherr MG, Kodira CD, Kohler A, Kües U, Lindquist EA, Lucas SM, Mago R, Mauceli E, Morin E, Murat C, Pangilinan JL, Park R, Pearson M, Quesneville H, Rouhier N, Sakthikumar S, Salamov AA, Schmutz J, Selles B, Shapiro H, Tanguay P, Tuskan GA, Henrissat B, Van de Peer Y, Rouzé P, Ellis JG, Dodds PN, Schein JE, Zhong S, Hamelin RC, Grigoriev IV, Szabo LJ, Martin F (2011) Obligate biotrophy features unraveled by the genomic analysis of rust fungi. Proc Natl Acad Sci USA 108:9166–9171
Eilam T, Bushnell WR, Anikster Y (1994) Relative nuclear DNA content of rust fungi estimated by flow cytometry of propidium iodide-stained pycniospores. Phytopathology 84:728–735
Fell JW (1966) *Sterigmatomyces*, a new fungal genus from marine areas. Antonie Van Leeuwenhoek 32:99–104
Frieders EM, McLaughlin DJ (1996) Mitosis in the yeast phase of *Agaricostilbum pulcherrimum* and its evolutionary significance. Can J Bot 74:1392–1406
Giraud T, Yockteng R, López-Villavicencio M, Refrégier G, Hood ME (2008) Mating system of the anther smut fungus *Microbotryum violaceum*: selfing under heterothallism. Eukaryotic Cell 7:765–775
Gochenaur SE (1981) *Cyrenella elegans* gen. et sp. nov., a dikaryotic anamorph. Mycotaxon 13:267–277
Golubev WI, Scorzetti G (2010) *Rhodotorula rosulata* sp. nov., *Rhodotorula silvestris* sp. nov. and *Rhodotorula straminea* sp. nov., novel *myo*-inositol-assimilating yeast species in the Microbotryomycetes. Int J Syst Evol Microbiol 60:2501–2506
Gottschalk M, Blanz PA (1984) Highly conserved 5S ribosomal RNA sequences in four rust fungi and atypical 5S rRNA secondary structure in *Microstroma juglandis*. Nucleic Acids Res 12:3951–3958
Gottschalk M, Blanz PA (1985) Untersuchungen an 5S ribosomalen ribonukleinsäuren als Beitrag zur Klärung von Systematik und Phylogenie der Basidiomyceten. Z Mykol 51:205–243
Grigoriev IV, Nordberg H, Shabalov I, Aerts A, Cantor M, Goodstein D, Kuo A, Minovitsky S, Nikitin R, Ohm RA, Otillar R, Poliakov A, Ratnere I, Riley R, Smirnova T, Rokhsar D, Dubchak I (2012) The genome portal of the Department of Energy Joint Genome Institute. Nucleic Acids Res 40:D26–D32
Hawksworth DL (2011) A new dawn for the naming of fungi: impacts of decisions made in Melbourne in July 2011 on the future publication and regulation of fungal names. IMA Fungus 2:155–162
Hawksworth DL, Kirk PM, Sutton BC, Pegler DN (eds) (1995) Dictionary of the fungi, 8th edn. CABI, Wallingford

Henk D, Vilgalys R (2007) Molecular phylogeny suggests a single origin of insect symbiosis in the Pucciniomycetes with support for some relationships within the genus *Septobasidium*. Am J Bot 94:1515–1526

Hibbett DS, Binder M, Bischoff JF, Blackwell M, Canon PF, Eriksson OE, Huhndorf S, James T, Kirk PM, Lücking R, Lumbsch HT, Lutzoni F, Matheny PB, McLaughlin DJ, Powell MJ, Redhead S, Schoch CL, Spatafora JW, Stalpers JA, Vilgalys R, Aime MC, Aptroot A, Bauer R, Begerow D, Benny GL, Castlebury LA, Crous PW, Dai YC, Gams W, Geiser DM, Griffith GW, Gueidan C, Hawksworth DL, Hestmark G, Hosaka K, Humber RA, Hyde KD, Ironside JE, Kõljalg U, Kurtzman CP, Larsson KH, Lichtwardt R, Longcore J, Miadlikowska J, Miller A, Moncalvo JM, Mozley-Standridge S, Oberwinkler F, Parmasto E, Reeb V, Rogers JD, Roux C, Ryvarden L, Sampaio JP, Schuessler A, Sugiyama J, Thorn RG, Tibell L, Untereiner WA, Walker C, Wang Z, Weir A, Weiss M, White MM, Winka K, Yao YJ, Zhang N (2007) A higher-level phylogenetic classification of the Fungi. Mycol Res 111:509–547

Hudson HJ, Ingold CT (1960) Aquatic hyphomycetes from Jamaica. Trans Br Mycol Soc 43:469–478

Ianiri G, Wright SAI, Castoria R, Idnurm A (2011) Development of resources for the analysis of gene function in Pucciniomycotina red yeasts. Fungal Genet Biol 48:685–695

Inácio J, Ludwig W, Spencer-Martins I, Fonseca A (2010) Assessment of phylloplane yeasts on selected Mediterranean plants by FISH with group- and species-specific oligonucleotide probes. FEMS Microbiol Ecol 71:61–72

James TY, Porter TM, Wallace Martin W (2014) Blastocladiomycota. In: McLaughlin DJ, Spatafora JW (eds) Systematics and evolution: part A, vol VII. Springer, Berlin

Jebaraj CS, Raghukumar C, Behnke A, Stoeck T (2010) Fungal diversity in oxygen-depleted regions of the Arabian Sea revealed by targeted environmental sequencing combined with cultivation. FEMS Microbiol Ecol 71:399–412

Kachalkin AV, Glushakova AM, Yurkov AM, Chernov IY (2008) Characterization of yeast groupings in the phyllosphere of *Sphagnum* mosses. Microbiology 77:474–481

Kinloch BB, Dupper GE (1996) Genetics of *Cronartium ribicola*. I. Axenic culture of haploid clones. Can J Bot 74:456–460

Kirk PM, Cannon PF, Minter DW, Stalpers JA (2008) Dictionary of the fungi, 10th edn. CABI, Wallingford

Kirschner R, Bauer R, Oberwinkler F (2001) *Colacosiphon:* a new genus described for a mycoparasitic fungus. Mycologia 93:634–644

Kleven NL, McLaughlin DJ (1988) Sporulation of the basidiomycete *Pachnocybe ferruginea* in terrestrial and aquatic environments. Mycologia 80:804–810

Kottke I, Suárez JP, Herrera P, Cruz D, Bauer R, Haug I, Garnica S (2010) Atractiellomycetes belonging to the 'rust' lineage (Pucciniomycotina) form mycorrhizae with terrestrial and epiphytic neotropical orchids. Proc Biol Sci 22:1289–1298

Kramer CL (1958) A new genus in the Protomycetaceae. Mycologia 50:916–926

Kropp BR, Corden ME (1986) Morphology and taxonomy of *Pachnocybe ferruginea*. Mycologia 78:334–342

Kurtzman CP, Fell JW (2004) Yeasts. In: Mueller GM, Bills GF, Foster MS (eds) Biodiversity of fungi: inventory and monitoring methods. Elsevier, Burlington, MA, pp 337–342

Kurtzman CP, Fell JW, Boekhout T (eds) (2011) The yeasts, a taxonomic study, vol 3, 5th edn. Elsevier, London

Langer E, Oberwinkler F (1998) *Spiculogloea occulta* (Heterobasidiomycetes) morphology and culture characters. Mycotaxon 69:249–254

Libkind D, Gadanho M, van Broock M, Sampaio JP (2005) *Sporidiobolus longiusculus* sp. nov. and *Sporobolomyces patagonicus* sp. nov., novel yeasts of the Sporidiobolales isolated from aquatic environments in Patagonia, Argentina. Int J Syst Evol Microbiol 55:503–509

Libkind D, Arts MT, van Broock M (2008) Fatty acid composition of cold-adapted carotenogenic basidiomycetous yeasts. Rev Argent Microbiol 40:193–197

Libkind D, Sampaio JP, van Broock M (2010) Cystobasidiomycetes yeasts from Patagonia (Argentina): description of *Rhodotorula meli* sp. nov. from glacial meltwater. Int J Syst Evol Microbiol 60:2251–2256

Limber DP (1940) A new form genus of the Moniliaceae. Mycologia 32:23–30

Lutz M, Bauer R, Begerow D, Oberwinkler F, Triebel D (2004) *Tuberculina*: rust relatives attack rusts. Mycologia 96:614–626

Lutzoni F, Kauff F, Cox CJ, McLaughlin D, Celio G, Dentinger B, Padamsee M, Hibbett D, James TY, Baloch E, Grube M, Reeb V, Hofstetter V, Schoch C, Arnold AE, Miadlikowska J, Spatafora J, Johnson D, Hambleton S, Crockett M, Shoemaker R, Sung GH, Lücking R, Lumbsch T, O'Donnell K, Binder M, Diederich P, Ertz D, Gueidan C, Hansen K, Harris RC, Hosaka K, Lim YW, Matheny B, Nishida H, Pfister D, Rogers J, Rossman A, Schmitt I, Sipman H, Stone J, Sugiyama J, Yahr R, Vilgalys R (2004) Assembling the fungal tree of life: progress, classification, and evolution of subcellular traits. Am J Bot 91:1446–1480

Maier W, Begerow D, Weiß M, Oberwinkler F (2003) Phylogeny of the rust fungi: an approach using nuclear large subunit ribosomal DNA sequences. Can J Bot 81:12–23

Martin GW (1959) On the genus *Hobsonia*. Brittonia 11:98–101

Marvanová L, Bandoni RJ (1987) *Naiadella fluitans* gen. et sp. nov.: a conidial basidiomycete. Mycologia 79:578–586

Marvanová L, Suberkropp K (1990) *Camptobasidium hydrophilum* and its anamorph, *Crucella subtilis*: a new heterobasidiomycete from streams. Mycologia 82:208–217

McLaughlin DJ (1990) A new cytoplasmic structure in the basidiomycete *Helicogloea*: the microscala. Exp Mycol 14:331–338

McLaughlin DJ, Frieders EM, Lü H (1995) A microscopist's view of heterobasidiomycete phylogeny. Stud Mycol 38:91–109

McLaughlin DJ, Hanson RW, Frieders EM, Swann EC, Szabo LJ (2004) Mitosis in the yeast phase of the basidiomycetes *Bensingtonia yuccicola* and *Stilbum vulgare* and its phylogenetic implications. Am J Bot 91:808–818

Medina EM, Jones GW, Fitzpatrick DA (2011) Reconstructing the fungal tree of life using phylogenomics and a preliminary investigation of the distribution of yeast prion-like protein in the fungal kingdom. J Mol Evol 73:116–133

Mix AJ (1947) *Taphrina osmundae* Nishida and *Taphrina higginsii* sp. nov. Mycologia 39:71–76

Moraes RN, Ribeiro MC, Nogueira MC, Cunha KC, Soares MM, Almeida MT (2010) First report of *Tritirachium oryzae* infection of human scalp. Mycopathologia 169:257–259

Moricca S, Ragazzi A (2001) Establishment of single-genotype axenic cultures from the haploid stage of the pine blister rust *Cronartium flaccidum*. Mycol Res 105:1527–1532

Nishida H, Ando K, Ando Y, Hirata A, Sugiyama J (1995) *Mixia osmundae*: transfer from the Ascomycota to the Basidiomycota based on evidence from molecules and morphology. Can J Bot 73 (Suppl 1):S660–S666

Nishida T (1911) A contribution to the monograph of the parasitic Exoascaceae of Japan. In: Takahashi Y (ed) Miyabe-Festschrift. Rokumeikwan, Tokyo, pp 157–212

O'Donnell KL, McLaughlin DJ (1981) Ultrastructure of meiosis in the hollyhock rust fungus, *Puccinia malvacearum*. I. Prophase I – Prometaphase I. Protoplasma 108:225–244

Oberwinkler F (1990) New genera of auricularioid heterobasidiomycetes. Rept Tottori Mycol Inst 28:113–127

Oberwinkler F, Bandoni RJ (1982) A taxonomic survey of the gasteroid, auricularioid heterobasidiomycetes. Can J Bot 60:1726–1750

Oberwinkler F, Bauer R (1989) The systematics of gasteroid, auricularioid heterobasidiomycetes. Sydowia 41:224–249

Oberwinkler F, Bauer R (1990) *Cryptomycocolax*: a new mycoparasitic heterobasidiomycete. Mycologia 82:671–692

Oberwinkler F, Kirschner R, Arenal F, Villarreal M, Rubio V, Begerow D, Bauer R (2006) Two new pycnidial members of the Atractiellales: *Basidiopycnis hyalina* and *Proceropycnis pinicola*. Mycologia 98:637–649

Padamsee M, Kumar TKA, Riley R, Binder M, Boyd A, Calvo AM, Furukawa K, Hesse C, Hohmann S, James TY, LaButti K, Lapidus A, Lindquist E, Lucas S, Miller K, Shantappa S, Grigoriev IV, Hibbett DS, McLaughlin DJ, Spatafora JW, Aime MC (2012) The genome of the xerotolerant mold *Wallemia sebi* reveals adaptations to osmotic stress and suggests cryptic sexual reproduction. Fungal Genet Biol 49:217–226

Pohl CH, Smit MS, Albertyn J (2011) *Rhodotorula bloemfonteinensis* sp. nov., *Rhodotorula eucalyptica* sp. nov., *Rhodotorula orientis* sp. nov. and *Rhodotorula pini* sp. nov., yeasts isolated from monoterpene-rich environments. Int J Syst Evol Microbiol 61:2320–2327

Pore RS, Sorenson WG (1990) *Reniforma strues*, a new yeast from wastewater. Mycologia 82:549–553

Porter TM, Skillman JE, Moncalvo JM (2008) Fruiting body and soil rDNA sampling detects complementary assemblage of Agaricomycotina (Basidiomycota Fungi) in a hemlock-dominated forest plot in southern Ontario. Mol Ecol 17:3037–3050

Prillinger H, Oberwinkler F, Umile C, Tlachac K, Bauer R, Dörfler C, Taufratzhofer E (1993) Analysis of cell wall carbohydrates (neutral sugars) from ascomycetous and basidiomycetous yeasts with and without derivatization. J Genet Appl Microbiol 39:1–34

Prokhorov V, Bodyagin V (2007) The ecology of aero-aquatic hyphomycetes. Moscow Univ Biol Sci Bull 62:15–20

Roberson RW, Abril M, Blackwell M, Letcher P, McLaughlin DJ, Mouriño-Pérez RR, Riquelme M, Uchida M (2010) Hyphal structure. In: Borkovich K, Ebbole D (eds) Cellular and molecular biology of filamentous fungi. ASM Press, Washington, DC, pp 8–24

Rodrigues MM, Laibson P (1975) Exogenous corneal ulcer caused by *Tritirachium roseum*. Am J Ophthalmol 80:804–806

Sampaio JP (2004) Diversity, phylogeny and classification of basidiomycetous yeasts. In: Agerer R, Piepenbring M, Blanz P (eds) Frontiers in basidiomycote mycology. IHW, Germany, pp 49–80

Sampaio JP, Bauer R, Begerow D, Oberwinkler F (1999) *Occultifur externus* sp. nov., a new species of simple-pored auricularioid heterobasidiomycete from plant litter in Portugal. Mycologia 91:1094–1101

Schell WA, Lee AG, Aime MC (2011) A new lineage in Pucciniomycotina: class Tritirachiomycetes, order Tritirachiales, family Tritirachiaceae. Mycologia 103:1331–1340

Scorzetti G, Fell JW, Fonseca A, Statzell-Tallman A (2002) Systematics of basidiomycetous yeasts: a comparison of large subunit D1/D2 and internal transcribed spacer rDNA regions. FEMS Yeast Res 2:495–517

Sláviková E, Vadkertiová R, Vránová D (2009) Yeasts colonizing the leaves of fruit trees. Ann Microbiol 59:419–424

Stefani FOP, Tanguay P, Pelletier G, Piché Y, Hamelin RC (2010) Impact of endochitinase-transformed white spruce on soil fungal biomass and ectendomycorrhizal symbiosis. Appl Environ Microbiol 76:2607–2614

Sugiyama J, Katumoto K (2008) Identity of the plasmodial slime mold *Phytoceratiomyxa osmundae* and the lectotypification of *Taphrina osmundae*, the basionym of *Mixia osmundae*. Mycoscience 49:192–198

Sugiyama J, Suh SO (1993) Phylogenetic analysis of basidiomycetous yeasts by means of 18S ribosomal RNA sequences: relationship of *Erythrobasidium hasegawianum* and other basidiomycetous yeast taxa. Antonie Van Leeuwenhoek 63:201–209

Swann EC, Taylor JW (1993) Higher taxa of basidiomycetes: an 18S rRNA gene perspective. Mycologia 85:923–936

Swann EC, Taylor JW (1995) Phylogenetic perspectives on basidiomycete systematics: evidence from the 18S rRNA gene. Can J Bot 73(suppl 1): S862–S868

Swann EC, Frieders EM, McLaughlin DJ (2001) Urediniomycetes. In: McLaughlin DJ, McLaughlin EG, Lemke PA (eds) The mycota, vol VII, B, Systematics and evolution. Springer, Berlin, pp 37–56

Toome M (2010) Leaf rust (*Melampsora*) on willows: ecological and plant response studies. Estonian University of Life Sciences, Tartu, Estonia, pp 1–124

Toome M, Roberson RW, Aime MC (2013) *Meredithblackwellia eburnea* gen et sp. nov., Kriegeriaceae fam. nov. and Kriegeriales ord. nov. - toward resolving higher-level classification in Microbotryomycetes. Mycologia 105:486–495

Toome M, Ohm RA, Riley RW, James TY, Lazarus KL, Henrissat B, Albu S, Boyd A, Chow J, Clum A, Heller G, Lipzen A, Nolan M, Sandor L, Zvenigorodsky N, Grigoriev IV, Spatafora JW, Aime MC (2014) Genome sequencing provides insight into the reproductive biology, nutritional mode and ploidy of the fern pathogen Mixia osmundae. New Phytol 202:554–564

Turchetti B, Hall SRT, Connell LB, Branda E, Buzzini P, Theelen B, Müller WH, Boekhout T (2011) Psychrophilic yeasts from Antarctica and European glaciers: description of *Glaciozyma* gen. nov., *Glaciozyma martinii* sp. nov. and *Glaciozyma watsonii* sp. nov. Extremophiles 15:573–586

Valério E, Gadanho M, Sampaio JP (2008) Reappraisal of the *Sporobolomyces roseus* species complex and description of *Sporidiobolus metaroseus* sp. nov. Int J Syst Evol Microbiol 58:736–774

Walker WF, Doolittle WF (1982) Redividing the basidiomycetes on the basis of 5S rRNA sequences. Nature 299:723–724

Wells K (1994) Jelly fungi, then and now! Mycologia 86:18–48

Whisler HC, Zebold SL, Shemanchuk JA (1975) Life history of *Coelomomyces psorophorae*. Proc Natl Acad Sci USA 72:693–696

Wingfield BD, Ericson L, Szaro T, Burdon JJ (2004) Phylogenetic patterns in the Uredinales. Australas Plant Pathol 33:327–335

Wright JE (1970) *Agaricostilbum*, a new genus of Deuteromycetes on palm spathes from Argentina. Mycologia 62:679–682

Wright JE, Bandoni R, Oberwinkler F (1981) *Agaricostilbum*: an auricularioid basidiomycete. Mycologia 73:880–886

Yamada Y, Maeda K, Mikata K (1994) The phylogenetic relationships of *Rhodosporidium dacryoidum* Fell, Hunter et Tallman based on the partial sequences of 18S and 26S ribosomal RNAs: the proposal of *Sakaguchia* gen. nov., a heterbasidiomycetous yeast genus. Biosci Biotechnol Biochem 58:99–103

Yurkov AM, Schäfer AM, Begerow D (2011) *Leucosporidium drummii* sp. nov. a new member of Microbotryomycetes isolated from soil. Int J Syst Evol Microbiol 62(Pt 3):728–734. doi:10.1099/ijs.0.027102-0

11 Ustilaginomycotina

D. Begerow[1], A.M. Schäfer[1], R. Kellner[2], A. Yurkov[3], M. Kemler[4], F. Oberwinkler[5], R. Bauer[5]

CONTENTS

[1]Ruhr-Universität Bochum, Geobotanik, ND 03/174, 44801 Bochum, Germany; e-mail: dominik.begerow@rub.de; angela.schaefer@rub.de
[2]Max Planck Institute for Terrestrial Microbiology, Karl von Frisch Strasse 10, 35043 Marburg, Germany; e-mail: ronny.kellner@mpi-marburg.mpg.de
[3]Leibniz Institute DSMZ-German Collection of Microorganisms and Cell Cultures, Inhoffenstraße 7B, 38124 Braunschweig, Germany; e-mail: Andrey.Yurkov@dsmz.de
[4]Centre of Excellence in Tree Health Biotechnology, Forestry and Agricultural Biotechnology Institute (FABI), University of Pretoria, Pretoria 0002, South Africa
[5]Organismische Botanik, Universität Tübingen, Auf der Morgenstelle 1, 72076 Tübingen, Germany; e-mail: franz.oberwinkler@uni-tuebingen.de; robert.bauer@uni-tuebingen.de

I. Introduction

Ustilaginomycotina comprises **115 genera with more than 1,700 species** and represents one of the three subphyla of the Basidiomycota (Bauer et al. 2006; Begerow et al. 1997; Hibbett et al. 2007; Swann and Taylor 1993). They harbour **mostly plant parasites** (Fig. 11.1a–p) **that are restricted to the geographic distribution of their hosts, encompassing tropical, temperate, and Arctic regions** (Vánky 2012).

Well-known genera in Ustilaginomycotina are *Ustilago* and *Tilletia*, which contain economically important species such as karnal bunt of wheat, loose smut of barley, and corn smut (Thomas 1989; Trione 1982; Valverde et al. 1995). In some cases where yield loss is minimal, contamination of *Tilletia* smut spores in grain can be subjected to quarantine regulations with economic implications and restrictions to international trade (Carris et al. 2006; Pascoe et al. 2005). Corn smut *Ustilago maydis* (DC.) Corda generally infects 2–5 % of plants in a corn field, although under certain conditions it can infect up to 80 % (Christensen 1963). While considered a plant pathogen in some parts of the world, the galls of *U. maydis* are appreciated as a delicacy in Mesoamerican cooking (Juarez-Montiel et al. 2011; Zepeda 2006). Besides the well-known species on crops, a huge diversity of plant parasites exist that either induce a typical smut syndrome (Fig. 11.1i–p) or present inconspicuous infections like members of Entylomatales (Fig. 11.1b), Exobasidiales (Fig. 11.1c–e), or Microstromatales (Fig. 11.1h). In addition, Ustilaginomycotina harbours some ecologically

Systematics and Evolution, 2nd Edition
The Mycota VII Part A
D.J. McLaughlin and J.W. Spatafora (Eds.)

Fig. 11.1 Diversity of sori and infections of Ustilaginomycotina. (**a**) *Doassinga callitrichis* (Liro) Vánky, R. Bauer & Begerow. (**b**) *Entyloma ficariae* A.A. Fisch. Waldh. (**c**) *Coniodictyum chevalieri* Har. & Pat. (**d**) *Laurobasidium lauri* (Geyler) Jülich, (**e**) *Exobasidium rhododendri* (Fuckel) C.E. Cramer. (**f**) *Jamesdicksonia irregularis* (Johanson) R. Bauer, Begerow, A. Nagler & Oberw. (**g**) *Gjaerumia ossifragi* (Rostrup) R. Bauer, M. Lutz & Oberw. (**h**) *Microstroma juglandis* (Berenger) Sacc. (**i**) *Tilletia controversa* J.G. Kühn. (**j**) *Thecaphora thlaspeos* (Beck) Vánky. (**k**) *Urocystis primulae* (Rostr.) Vánky. (**l**) *Ustacystis waldsteiniae* (Peck) Zundel. (**m**) *Anthracoidea sempervirentis* Vánky. (**n**) *Ustanciosporium gigantosporum* (Liro) M. Piepenbr. (**o**) *Melanotaenium endogenum* (Unger) de Bary (**p**) *Ustilago hordei* (Pers.) Lagerheim.

variable anamorphic lineages such as *Malassezia*, which colonizes human and animal skin (Begerow et al. 2000, 2006).

Among the morphologically and ecologically diverse species of Ustilaginomycotina, *U. maydis* became a model organism for studying the interaction of specific plant parasites with their hosts. Using a variety of genetic tools, it has been shown that mating is an essential prerequisite to plant infection (Kahmann and Kämper 2004). ***U. maydis* was one of the first fungal genomes** sequenced, which advanced the knowledge of fungal physiology, such as the importance of secreted proteins in signaling (Brefort et al. 2009; Kämper et al. 2006). Thus, Ustilaginomycotina is a highly valuable group for comparative genomic studies in fungal pathogens and for illuminating the evolution and functionality of host–parasite interactions (Kellner et al. 2011; Schirawski et al. 2010; Xu et al. 2007).

A. Diagnosis and Evidence for Monophyletic Origin

The Ustilaginomycotina have a distinctive cell wall composition with a dominance of glucose and an absence of xylose, which separates them from the Pucciniomycotina and Agaricomycotina (Prillinger et al. 1990, 1993). They share the type B secondary structure of 5S ribosomal RNA (rRNA) with the Agaricomycotina (Gottschalk and Blanz 1985) and the lack of parenthesomes (i.e. multilayered endoplasmic reticulum elements at the septal pores) with the Pucciniomycotina (Bauer et al. 1997, 2006). Important **synapomorphies for the Ustilaginomycotina are membranous pore caps and the presence of a characteristic host–parasite interaction zone that results from fungal exocytosis of primary interactive vesicles** (Bauer et al. 1997).

Sequence analyses support the monophyly of the Ustilaginomycotina as defined earlier but with varying statistical support in different studies. Whereas the monophyly of *Tilletia caries* (DC.) L. & C. Tul., *Ustilago hordei* (Pers.) Lagerh., and *U. maydis* had high bootstrap support with small subunit (SSU) rDNA sequence analyses (Bauer et al. 2006; Swann and Taylor 1993, 1995), the bootstrap values for the Ustilaginomycotina were lower when analysed with large subunit (LSU) rDNA sequences and increased taxon sampling (Begerow et al. 1997, 2000). In particular, bootstrap support for the Ustilaginomycotina was sensitive to the inclusion or exclusion of *Entorrhiza* sequences in the LSU data set; after several analyses and varying interpretations *Entorrhiza* was excluded from the Ustilaginomycotina (Hibbett et al. 2007; Matheny et al. 2006). To date, the phylogenetic position of *Entorrhiza* remains unresolved.

B. Smut Fungi Syndrome in Other Fungal Groups

Like the terms *agaric*, *polypore*, and *lichen*, for example, **the term *smut fungus* circumscribes the organization and life strategy of a fungus** (cf. Fig. 11.1a–p) but does not represent common ancestry. Hence, **smut fungi are nonmonophyletic** when based on the presence of a powdery spore mass. Most smut fungi are in the Ustilaginomycotina. Other smut fungi, in the Microbotryales, are members of the Pucciniomycotina (Bauer et al. 2006; Begerow et al. 1997; see Aime et al. 2014). In contrast to the Ustilaginomycotina, available data indicate that the microbotryaceous taxa *Aurantiosporium*, *Bauerago*, *Fulvisporium*, *Liroa*, *Microbotryum*, *Sphacelotheca*, *Ustilentyloma*, and *Zundeliomyces* have a type A 5S rRNA secondary structure (Gottschalk and Blanz 1985; Müller 1989), mannose as the major cell wall carbohydrate (Prillinger et al. 1991, 1993), and cellular interactions without primary interactive vesicles (Bauer et al. 1997), all of which are synapomorphies of the Pucciniomycotina. Morphologically, they are distinguishable from the phragmobasidiate members of Ustilaginomycotina by the lack of intracellular hyphae or haustoria (Bauer et al. 1997). Grouping the Microbotryales within the Pucciniomycotina rather than the Ustilaginomycotina is also supported by sequence analyses (Bauer et al. 2006; Begerow et al. 1997; Swann and Taylor 1995). However, **there are significant convergences between the microbotryaceous and**

ustilaginomycetous phragmobasidiate smut fungi. Certain taxa of both groups are similar with respect to soral morphology, teliosporogenesis, life cycle, basidial morphology, and host range (Bauer et al. 1997, 2006).

As stated previously, *Entorrhiza* has been excluded from the Ustilaginomycotina mainly based on molecular phylogenetic analyses (Hibbett et al. 2007). Early studies using a smaller number of taxa placed *Entorrhiza* species basal to other Ustilaginomycotina with low bootstrap support (Begerow et al. 1997). Later studies questioned this position, and, depending on species sampling and outgroup selection, the position of *Entorrhiza* remains more or less unresolved (Begerow et al. 2006; Matheny et al. 2006). As long as a thorough multigene analysis is lacking, we follow the concept of Hibbett et al. (2007) and exclude *Entorrhiza* from the Ustilaginomycotina.

Interestingly, even non-basidiomycetous fungi can cause diseases with the formation of thick-walled propagules convergent to those of smut fungi. Species of *Schroeteria* Winter, for example, look superficially similar to Ustilaginomycotina (Vánky 1981) but belong to the Ascomycota (Nagler et al. 1989). Leaf spots similar to sori of *Entyloma* can be formed by representatives of the Protomycetales (Reddy and Kramer 1975), which belong to the Taphrinomycotina (Sugiyama et al. 2006; see Kurtzman and Sugiyama (2014), Chap. 1 Vol. VII, Part B) and produce ascospores in their synasci (Preece and Hicks 2001).

C. Hosts and Their Role in Species Definition

The Ustilaginomycotina, unlike the Pucciniomycotina and Agaricomycotina, generally are ecologically well characterized by parasitism. Besides some anamorphic taxa, which will be discussed in more detail later, all members of Ustilaginomycotina are plant parasites. Aside from *Exoteliospora* on ferns (Bauer et al. 1999b), two species of *Melaniella* on spike mosses (Bauer et al. 1999a), and two species of *Uleiella* on conifers (Butin and Peredo 1986; Schröter 1894), all other plant parasitic members of Ustilaginomycotina parasitize angiosperms with a high proportion of species on monocots, especially Poaceae and Cyperaceae. **The majority of the roughly 1,710 species occur on Poaceae** (45 %) **or on Cyperaceae** (13 %). The 121 ustilaginomycetous genera occurring on angiosperms include 72 genera that are exclusively found on monocots and 31 exclusively on dicots (mainly eudicots); 4 comprise species that parasitize both monocots and dicots. The genera found exlusively on monocots occur mainly on Poaceae (22) and on Cyperaceae (20, see below). Concerning the hosts, there are two remarkable points. (1) With a few exceptions, the teliospore-forming species of Ustilaginomycotina parasitize nonwoody herbs, whereas those without teliospores prefer woody trees or bushes. However, almost all species sporulate on parenchymatic tissues of the hosts. (2) Two of the angiosperm families with the highest number of species, the Orchidaceae with ca. 20,000 species and the Poaceae with ca. 10,000 species, play quite different roles for the Ustilaginomycotina. There are no known smut species on Orchidaceae, while the Poaceae are the most important host family of Ustilaginomycotina. **Grass smuts have** obviously **adapted to the ecology of their host group by wind-borne teliospores or basidiospores** and are thereby able to infect hosts that often occur in extensive, but often disconnected, host populations.

Host range used to play an important role in species definition. Many species, for instance in the genera *Entyloma*, *Melanotaenium*, and *Urocystis* (Vánky 1994), have few defining morphological characters, which, until the advent of ultrastructural techniques, were mainly limited to spore ornamentation and spore size. Therefore, host information has long been used in the delimitation of smut species as an additional defining characteristic. Different authors gave host specificity different emphases. Savile (1947), for instance, accepted only two species in *Entyloma* and lumped many already described species into these, whereas Vánky (2012) applied a narrower species definition and recognized 163 species based on spore morphology and host. Besides morphological and ecological concepts, phylogenetic species definitions have attracted much attention in recent years [for a review see Cai et al. (2011)], and phylogenetic approaches have in general confirmed the latter position (e.g. Begerow et al. 2002, 2004). These studies question the roles of host interaction and host range in maintaining species integrity of smuts. The species concept of smut fungi is especially perplexing because not only can closely related species [e.g. *Tilletia controversa* J.G. Kühn and *T. caries* (DC.) Tul. & C. Tul] hybridize under laboratory conditions (Trail and Mills 1990), but hybridization can even be observed between species that had their own evolutionary trajectory for millions of years (Kellner et al. 2011). It is unknown how gene flow is prevented in nature and how species integrity is main-

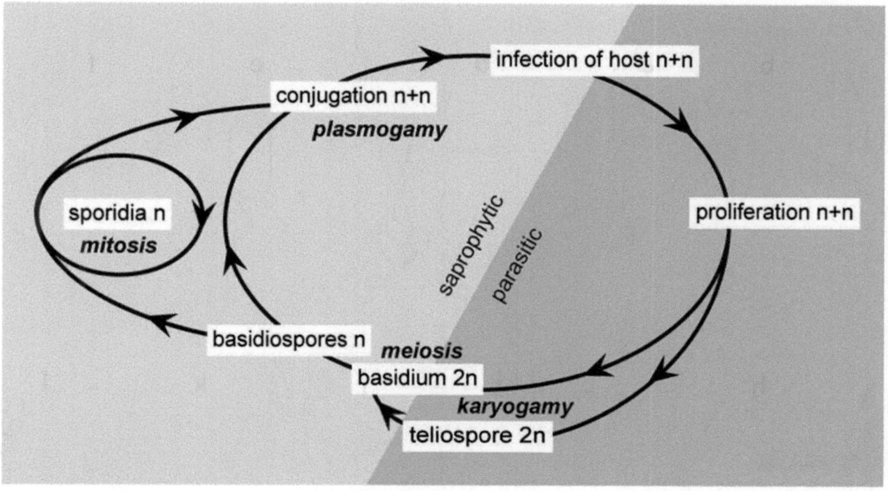

Fig. 11.2 Generalized life cycle of Ustilaginomycotina

tained for the Ustilaginomycotina, but in microbotryaceous smuts, hybrid inviability was shown to select against hybrids (De Vienne et al. 2009).

In *U. maydis*, sorus formation is initiated by a combination of parasite and host effectors. To develop teliospores, *U. maydis* specifically alters plant expression, initiating different expression profiles in different host tissues (Skibbe et al. 2010). These experiments demonstrated that the interaction between smuts and their hosts is extremely tight at the molecular level, which suggests that there are strong factors in maintaining boundaries between parasite species. Thus, species concepts incorporating host information, as applied by smut fungal taxonomists for the last century, have a biological basis, which could explain such narrow host ranges in smut fungi (Cai et al. 2011).

II. Life Cycle

Species of Ustilaginomycotina share a similar **dimorphic life cycle comprised of a saprobic haploid phase and a parasitic dikaryotic phase** (Fig. 11.2) (Brefeld 1883; de Bary 1884; Sampson 1939). The haploid phase is initiated usually by the formation of basidiospores following meiosis of the diploid nucleus in the basidium and ends with the conjugation of compatible haploid cells to produce dikaryotic, infectious hyphae. The dikaryotic phase ultimately results in the production of probasidia (i.e. often teliospores) or basidia (Fig. 11.3a–o).

Almost all Ustilaginomycotina sporulate on or in parenchymatic tissues of their hosts. In the majority of the Ustilaginomycotina, **the young basidium becomes thick-walled and at maturity separates from the sorus to function as a dispersal agent, the teliospore.** Teliospores are usually the most conspicuous stage in a smut's life cycle, representing the smut syndrome (cf. Fig. 11.1a–p). Most of the Ustilaginomycotina are dimorphic and produce a yeast or yeast-like stage in the haploid phase and form hyphae during the parasitic phase. However, there are several variations from this generalized life cycle, e.g. the occurrence of homothallism in *Anthracoidea* (Kukkonen and Raudaskoski 1964) and *Exobasidium* (Sundström 1964), the **lack of teliospores** in the Microstromatales, Exobasidiales and Ceraceosorales (Begerow et al. 2001, 2002, 2006), or even the switch to a complete anamorphic life style as assumed for *Malassezia* (Boekhout et al. 2010) and other anamorphic genera.

A. Saprobic Phase

Members of Ustilaginomycotina can survive outside their hosts during a free-living asexual state, the saprobic phase. **Many representatives are readily obtained from nature as predominantly unicellular budding states, called yeasts or sporidia**, e.g. species in *Ustilago*, *Microstroma*, and *Malassezia* (Begerow et al.

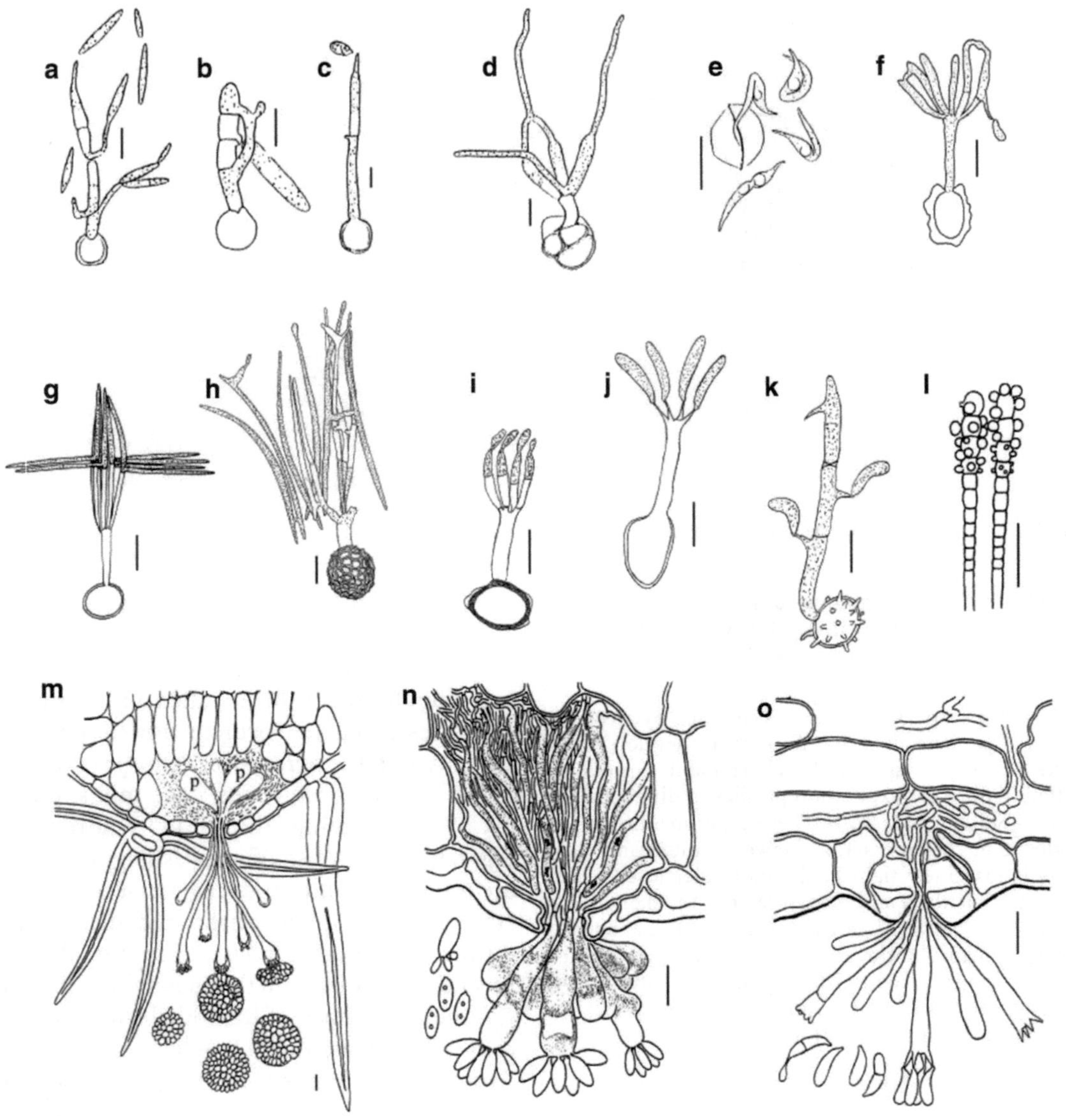

Fig. 11.3 Basidia of Ustilaginomycotina. *Bar*=10 µm. (**a**) *Ustilago maydis* (DC.) Corda. (**b**) *Cintractia axicola* (Berk.) Cornu. (**c**) *Anthracoidea altiphila* Vánky & M. Piepenb. (**d**) *Urocystis ranunculi* (Lib.) Moesz. (**e**) *Mycosyrinx cissi* (DC.) G. Beck. (**f**) *Entyloma microsporum* (Unger) Schröter. (**g**) *Rhamphospora nymphaeae* D.D. Cunn. (**h**) *Tilletia caries* (DC.) Tul. & C. Tul. (**i**) *Eballistra brachiariae* (Viégas) R. Bauer, Begerow, A. Nagler & Oberw. (**j**) *Jamesdicksonia dactylidis* (Pass.) R. Bauer, Begerow, A. Nagler & Oberw. (**k**) *Tilletiaria anomala* Bandoni & Johri. (**l**) *Graphiola phoenicis* (Moug.) Poiteau. (**m**) *Volvocisporium triumfetticola* (M.S. Patil) Begerow, R. Bauer & Oberw. (**n**) *Microstroma juglandis* (Berenger) Sacc. (**o**) *Exobasidium oxycocci* Rostr.

2000). Additionally, some smut fungi (e.g. *Exobasidium* and *Georgefischeria*) produce ballistospores to actively discharge basidiospores or secondary spores (Begerow et al. 2000). Hyphal growth is present in the saprobic phase of some members of Ustilaginomycotina; in many cases, a clear distinction between unicellular, yeast-like, pseudohyphal, and hyphal proliferation is impossible because budding cells (blastoconidia) often originate from hyphae and vice versa. This yeast–hyphal dimorphism occurs in many lineages of the Basidiomycota and might be a distinctive feature of parasitic lineages (Sampaio 2004).

In members of Ustilaginomycotina, yeast and yeast-like states are known from five orders: Ustilaginales, Entylomatales, Exobasidiales, Georgefischeriales, and Microstromatales (Begerow et al. 2000; Boekhout et al. 2011; Sampaio 2004). In the order Urocystidales, saprobic yeast-like growth of secondary sporidia was observed in some *Thecaphora* species (Vánky et al. 2008a) and in *Urocystis cepulae* Frost (Whitehead 1921). No such yeast states are known from the Doassansiales and Tilletiales.

Multiplication and propagation as yeast and yeast-like states are likely to be advantageous for survival and dispersal, and actually **some taxa are known from their asexual states only**, namely *Pseudozyma*, *Tilletiopsis*, *Sympodiomycopsis*, *Meira*, *Acaromyces*, *Jaminaea*, *Malassezia*, and probably *Quambalaria*. Members of these genera have mostly been isolated from various substrates during analyses of yeast communities in specific habitats (de Beer et al. 2006; Kurtzman et al. 2011).

Despite the economic relevance of smut infections caused by *Ustilago*, *Quambalaria*, and many others, and the rather high frequency of occurrence, little is known about the distribution of free-living yeast states. Assimilation tests, which are routinely performed for fungi historically treated as yeasts (*Pseudozyma*, *Sympodiomycopsis*, *Rhodotorula*), reveal the abilities of free-living states of Ustilaginomycotina to utilize a broad spectrum of plant-related carbohydrates, like sucrose, cellobiose, trehalose, L-arabinose, D-xylose, and some polyols (Kurtzman et al. 2011). Additionally, the capability of species of Ustilaginales (*Sporisorium*, *Ustilago*, *Farysia*, *Farysyzyma*, *Pseudozyma*), Entylomatales (*Entyloma*, *Tilletiopsis*), and Microstromatales (*Sympodiomycopsis*, *Rhodotorula*) to break down and assimilate low-weight aromatic molecules has been demonstrated (Sampaio 1999). Most of the tested cultures were able to use intermediates of lignin degradation, such as protocatechuic, *p*-coumaric acid, vanillic, and *p*-hydroxybenzoic acids (Sampaio 1999; Subba Rao et al. 1971). This adaptation seems especially interesting for dimorphic plant parasites because it might enable active degradation of cell walls, thereby allowing survival on decaying plant material. Besides the use of ligno-cellulosic derivates, the utilization of several non-conventional carbon sources of plant origin by species of Ustilaginomycotina has been reported, e.g. *Tilletiopsis washingtonensis* Nyland assimilates diverse volatile organic carbon (VOC) sources present in ripe apples (Vishniac et al. 1997). Interestingly, one component of VOC (butyl acetate), successfully utilized by *T. washingtonensis*, stimulates germination of grey mould (*Botrytis cinerea* Pers.) conidia, and the consumption of gaseous carbon products by *T. washingtonensis* decreases the development of moulds on apples (Filonow 2001). Members of Entylomatales display growth on gentisic acid (Sampaio 1999), a compound involved in regulating the defense responses of plants (Bellés et al. 2006). Members of Entylomatales and Microstromatales are able to grow on gallic acid (Sampaio 1999), a widely distributed tannin often accumulated in substantial quantities in plant material (Haslam and Cai 1994). Furthermore, the capability of some species of *Tilletiopsis*, *Pseudozyma*, and *Ustilago* to secrete enzymes, such as lipase, amylase, glucoamylase, cutinase, protease, pectinase, and xylanase, has been reported (Boekhout et al. 2006, 2011; Geiser et al. 2013; Trindade et al. 2002; Urquhart and Punja 2002).

Several interesting physiological adaptations seem to facilitate saprobic growth and survival in natural habitats. Cold tolerance is a common trait among basidiomycetous yeasts, which successfully colonize extremely cold habitats, including glaciers (Branda et al. 2010) and high-altitude regions (Connell et al. 2008; Vishniac 2006). Low temperatures also favour the development of various species of *Tilletiopsis* and anamorphs of *Entyloma* (Boekhout et al. 2006 and references therein). Extensive growth of *Tilletiopsis* spp. on apple surfaces under low oxygen concentration was reported recently (Boekhout et al. 2006). Although it is not yet clear whether this ability provides any advantage in colonizing plant substrates, several yeasts (e.g. *Meira* spp., *Pseudozyma* spp.) were reported from inside plant tissues (Abdel-Motaal et al. 2009; Gerson et al. 2008; Paz et al. 2007; Posada and Vega 2005; Takahashi et al. 2011; Tanaka et al. 2008; Yasuda et al. 2006).

The secretion of antibiotic compounds, killer toxins (proteins), and glycolipids could give yeasts a competitive advantage against other microorganisms. Glycolipids are modified long-chain fatty acids that are active against diverse groups of fungi (Golubev 2007; Mimee et al. 2005; Teichmann et al. 2007), bacteria (Kitamoto et al. 1993), and insects (Gerson et al. 2008). Antagonistic reactions towards other fungi were reported for *Acaromyces ingoldii* Boekhout, Scorzetti, Gerson & Sztejnb. and several species of the genera *Meira*, *Pseudozyma*, *Tilletiopsis*, and *Sympodiomycopsis* (Boekhout 2011; Gerson et al. 2008; Golubev 2006, 2007; Golubev et al. 2008). Consequently, some Ustilaginomycotina yeast species might even have evolved a mycoparasitic life style, as has been suggested for *T. pallescens* Gokhale, which was repeatedly isolated from basidiocarps of other fungi (Boekhout 2011). Recently, two asexual genera, *Meira* and *Acaromyces*, were found to cause the mortality of citrus mite pests (Paz et al. 2007). Although these fungi grew on mite cadavers, the capability of cell-free extracts from cultures to kill mites suggests the toxic nature of fungal secretions rather than a parasitic life style. Interestingly, cell-free extracts effectively suppressed the growth of several plant pathogens, including moulds, mildew, and soil-borne fungi (Kushnir et al. 2011).

It is not surprising that **saprobic states of Ustilaginomycotina were found on different plant-related substrates** (Begerow et al. 2000; Fonseca and Inácio 2006; Sampaio 2004). In some cases saprobic and parasitic states co-exist in the same natural habitat; however, a considerable number of species were isolated from distinct substrates (water, nectar, and fruits) or from plants totally unrelated to the known hosts. Yeasts of the genus *Farysizyma*, probably the anamorphic stage of *Farysia*, which parasitizes Cyperaceae, have been recovered from leaves of unrelated plant species of Bromeliaceae and Cistaceae, strawberry fruits, and nectar (Inácio et al. 2008). Other substrates, i.e. water, fruit pulps and flowers, also yielded saprobic states of Ustilaginomycotina (Fell et al. 2011; Inácio et al. 2008; Liou et al. 2009; Seo et al. 2007; Trindade et al. 2002; Wang et al. 2006). Although some authors reported the isolation of *Pseudozyma* yeasts from clinical samples, **invasive disease caused by these fungi are very unusual in humans** (Lin et al. 2008; Sugita et al. 2003), and only yeasts of the genus *Malassezia* are considered to be part of the normal skin mycobiota of warm-blooded vertebrates (Findley et al. 2013). However, in many circumstances they have been reported to cause various types of skin diseases like pityriasis versicolor, seborrheic dermatitis, and folliculitis (Boekhout et al. 2010).

Finally, the dual nomenclature introduced for anamorphic strains and species remains problematic because some of them represent the anamorphic stage of a well-known teleomorph (Begerow et al. 2000; de Beer et al. 2006). The application of the new rules provided by the Melbourne Code will allow phylogenetic species recognition, and it is hoped that some of the systematic problems will be resolved in the near future (Hawksworth 2011; Hawksworth et al. 2011), but the integration of anamorphic and teleomorphic systematics and nomenclature remains a challenge.

B. Parasitic Phase

The parasitic phase in Ustilaginomycotina is initiated by the **mating process,** which induces a morphological and physiological transition from saprophytic yeast cells to biotrophic filaments (Fig. 11.2) (Kahmann and Kämper 2004; Kellner et al. 2011; Snetselaar and Mims 1992). The genetic and developmental basis of the infection process and the host–parasite interaction have been studied best in the model organism *U. maydis* and will not be reviewed in detail [for a more detailed view see Brefort et al. (2009), Kahmann and Kämper (2004) and Vollmeister et al. (2012)]. **To form an infectious dikaryotic hypha, two compatible haploid sporidia must recognize each other and fuse.** In *U. maydis* the cell cycle arrests during mating until after host penetration (Garcia-Muse et al. 2003). Penetration is achieved via non-melanized appressoria at the tip of elongated dikaryotic cells and might additionally be aided by the secretion of lytic enzymes (Schirawski et al. 2005).

The subsequent steps of infection depend on the ability of the fungus to establish an **intimate interaction with its specific host** (Fig. 11.4a–e). This is **mediated by the vesicle-based exocytosis** (Bauer et al. 1997) **of secreted effector proteins that interfere with plant defenses** (Brefort et al. 2009) and host-specific metabolic processes (Djamei et al. 2011). Depending on the respective ustilaginomycetous group, hyphae grow and proliferate either only intercellularly or both intercellularly and intracellularly (Fig. 11.4a–e) (Bauer et al. 1997). Intracellular hyphae are tightly surrounded by the plant plasma membrane and develop a characteristic vesicular matrix through the accumulation of secreted deposits (Bauer et al. 1997). **Members of Doassansiales, Entylomatales, and Exobasidiales develop a characteristic interaction apparatus** (Fig. 11.4b–d), **while other groups of Ustilaginomycotina interact with their host either via small evagination zones** (Fig. 11.4a) **or along the whole hyphae** (Fig. 11.4e). These hyphae are usually not restricted to specific entrance or exit sites of host cells and, therefore, can passage from cell to cell (Bauer et al. 1997). In the Ustilaginaceae, hyphae grow directly to plant vascular bundle cells and proliferate throughout the host in

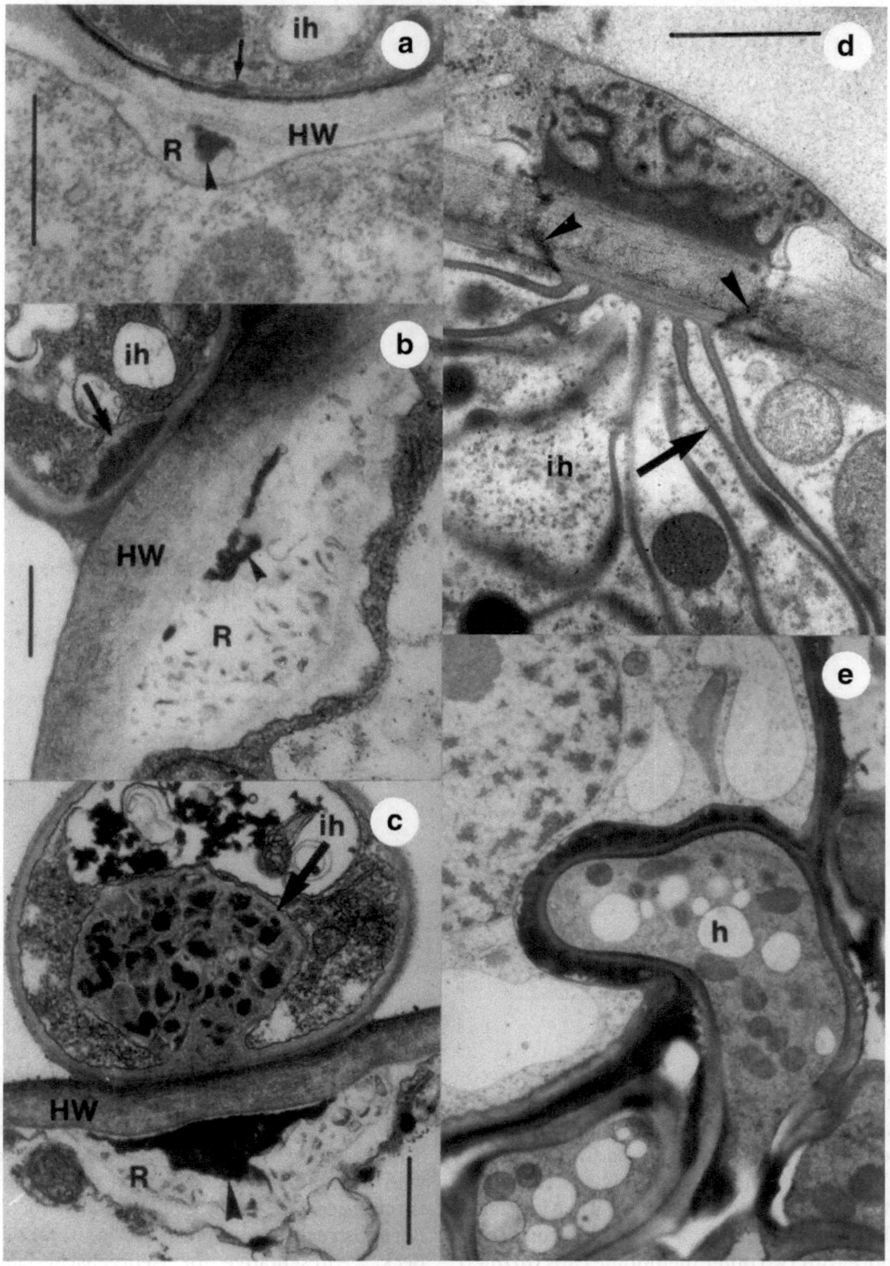

Fig. 11.4 Cellular interactions in Ustilaginomycotina. Material illustrated in (**d**) and (**e**) was prepared using freeze substitution. *Bar*=0.5 μm. (**a**)–(**d**) Local interaction zones, representative of Exobasidiomycetes. (**a**) Local interaction zone without interaction apparatus, representative of Georgefischeriales, Tilletiales, and Microstromatales. Intercellular hypha (ih) of *Conidiosporomyces ayresii* (Berk.) Vánky with secretion profile of one primary interactive vesicle (*arrow*) in contact with host cell wall (HW). Note electron-opaque deposits at host side (*arrowhead*). Host response to infection is visible at R. (**b**) Local interaction zone with simple interaction apparatus, representative of Entylomatales and Ceroceosorales. Intercellular hypha (ih) of *Entyloma hieracii* H. & P. Sydow in contact with host cell wall (HW) showing exocytosis profile of simple interaction apparatus (*arrow*). Note electron-opaque deposit at host side (*arrowhead*). Host response to infection is visible at R. (**c**) Local interaction zone with complex interaction apparatus containing cytoplasmic compartments, representative for Doassansiales. Intercellular hypha (ih) of *Doassinga callitrichis* (Liro) Vánky, R. Bauer & Begerow in contact with host cell wall (HW) showing exocytosis profile of one complex interaction apparatus (*arrow*).

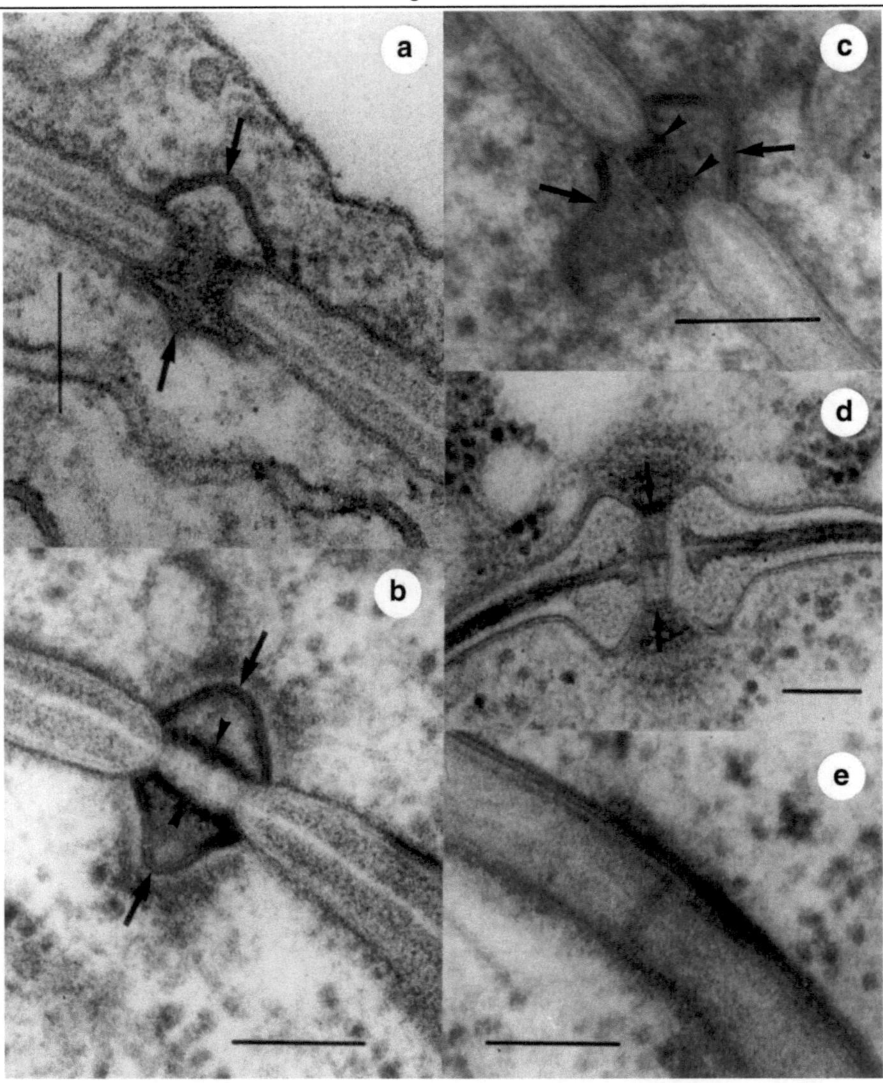

Fig. 11.5 Septation of soral hyphae in Ustilaginomycotina. Material illustrated in (**b**)–(**e**) was prepared using freeze substitution. *Scale bars*=0.1 µm. (**a**) Simple pore with two membrane caps (*arrows*) of *Doassinga callitrichis* (Liro) Vánky, R. Bauer & Begerow, representative of Melanotaeniaceae, Ceraceosorales, Doassansiales, Entylomatales, Exobasidiales, and Microstromatales. (**b**) Simple pore with two outer membrane caps (*arrows*) and two inner nonmembranous plates (*arrowheads*) of *Ustacystis waldsteiniae* (Peck) Zundel, representative of Urocystidiaceae, Floromycetaceae, and Doassansiopsidaceae. (**c**) Simple pore with two membrane caps (*arrows*) and sectioned tube in pore channel (*arrowheads*), representative of Exobasidiales. (**d**) Dolipore with membrane bands (*arrows*) of *Tilletia barclayana* (Bref.) Sacc. & P. Sydow, representative for Tilletiales. (**e**) Poroid structure in septum of *Mycosyrinx cissi* (DC.) G. Beck, representative for Georgefischeriales and core group Ustilaginales

Fig. 11.4 (continued) The interaction apparatus and its intercisternal space are excluded from the cytoplasm. Note electron-opaque deposit at host side (*arrowhead*). Host response to infection is visible at R. (**d**) Local interaction zone with complex interaction apparatus producing interaction tube, representative of Exobasidiales. Intercellular hypha (ih) of *Exobasidium pachysporum* Nannf. with interaction apparatus (*arrow*). Note sectioned interaction tube (arrowheads) at adjacent cell walls of parasitic and host cell. (**e**) Enlarged interaction zone between upper left plant cell and haustorium (h), representative of Ustilaginomycetes. The haustorium (h) of *Ustacystis waldsteiniae* (Peck) Zundel is surrounded by an electron-opaque matrix within host cell

close proximity to the vascular system until they reach their sporulation sites (Doehlemann et al. 2009). During proliferation, fungal hyphae branch and undergo mitosis and septation (cf. Fig. 11.5a–e). Members of Doassansiales, Ustilaginales, and some species of *Exobasidium* develop clamps to coordinate the synchronized division of the two nuclei. In *U. maydis* clamp primordia are formed at the tip of the growing hyphae (Scherer et al. 2006). However, clamp-like structures are observed in many species of Ustilaginomycotina. Whilst some clamps give rise to hyphal branches, others seem to correspond to fusion bridges (Fischer and Holton 1957), which ensure the migration of nuclei rather than coordinating dikaryotic mitoses.

Proliferation in the host is followed either by the direct formation of basidia, as observed in Microstomatales, Exobasidiales, and Ceraceosorales (Fig. 11.3l–o), or by the production of teliospores, which are clustered in sori (e.g. Fig. 11.1b, c, f, j–p). **Sporogenesis of teliospores often occurs in distinct organs of a plant, including roots, stems, leaves, inflorescences, anthers, ovaries, and seeds** (Fig. 11.1) (Vánky 2012). In this process biotrophic hyphae aggregate, septate, and finally differentiate into teliospores. However, teliospore formation is variable among members of Ustilaginomycotina, and propagation units range from single spores to large spore balls, which may or may not include sterile cells (Piepenbring et al. 1998). This variability can even be observed between closely related species, e.g. in *Urocystis* and *Thecaphora* (Vánky et al. 2008a). In *Ustilago* nearly all hyphae disarticulate and form teliospores in a matrix resulting from gelatinization of hyphal cell walls (Snetselaar and Mims 1994), whereas teliospores in *Rhamphospora* are formed terminally and without recognizable gelatinization (Piepenbring et al. 1998). Usually, sporogenesis occurs intercellularly either in preformed intercellular spaces or in cavities of disintegrated host cells (Luttrell 1987). The release of teliospores does not depend on living host tissue since *Schizonella* and some species of *Ustilago* sporulate within disintegrated host tissues, and species of *Clintamra*, *Exoteliospora*, and *Orphanomyces* even develop their teliospores externally to the host tissue (Piepenbring et al. 1998; Vánky 1987). Some of the varying morphological traits of soral formation or spore characteristics for ustilaginomycetous families are summarized in Fig. 11.8.

Besides the majority of Ustilaginomycota, which parasitize their host in the dikaryotic phase, there are a few examples of specific haploid yeast parasites. The most prominent ones certainly belong to the genus *Malassezia*, in which the anamorphic lipophylic yeast species specifically feed on the skin of warm-blooded animals, where they are involved in common skin diseases (Xu et al. 2007). To date, there are more species described with different specific host substrates, i.e. the mite parasitic species of *Meira* and *Acaromyces* (Gerson et al. 2008).

III. Classification System

Beginning with Tulasne and Tulasne (1847), the smut fungi have traditionally been divided into phragmobasidiate Ustilaginaceae or Ustilaginales and holobasidiate Tilletiaceae or Tilletiales (e.g. Kreisel 1969; Oberwinkler 1987). Durán (1973) and Vánky (1987) discussed the difficulties associated with smut classification in detail but did not list higher taxa in the group. Consequently, Vánky (1987) treated all smut fungi in a single order, Ustilaginales, with one family, Ustilaginaceae. Other plant parasites like *Exobasidium*, *Graphiola*, and *Microstroma* are treated in other families and orders (Hennings 1900) and are included in Ustilaginomycotina on the basis of ultrastructural characters (Bauer et al. 1997).

The **classification** proposed below **is based predominantly on characteristics of host–parasite interactions, the septal pore apparatus** (Fig. 11.6) (Bauer et al. 1997), **and LSU sequence analyses** (Fig. 11.7) (Begerow et al. 1997, 2006). However, the system is still under discussion because many groups are still poorly studied. As mentioned previously, the position of *Entorrhiza* within Basidiomycota is questionable based on molecular data, and the genus lacks some typical morphological features of Ustilaginomycotina (e.g. it does not possess membranous pore caps) (Bauer et al. 1997). The phylogenetic relationships among the different families of Ustilaginales and Urocystidales could only be clarified by molecular data and revealed the convergent evolution of several characters, e.g. the loss of septal pores or the development of intracellular hyphae (Begerow et al. 2006). Although the types of

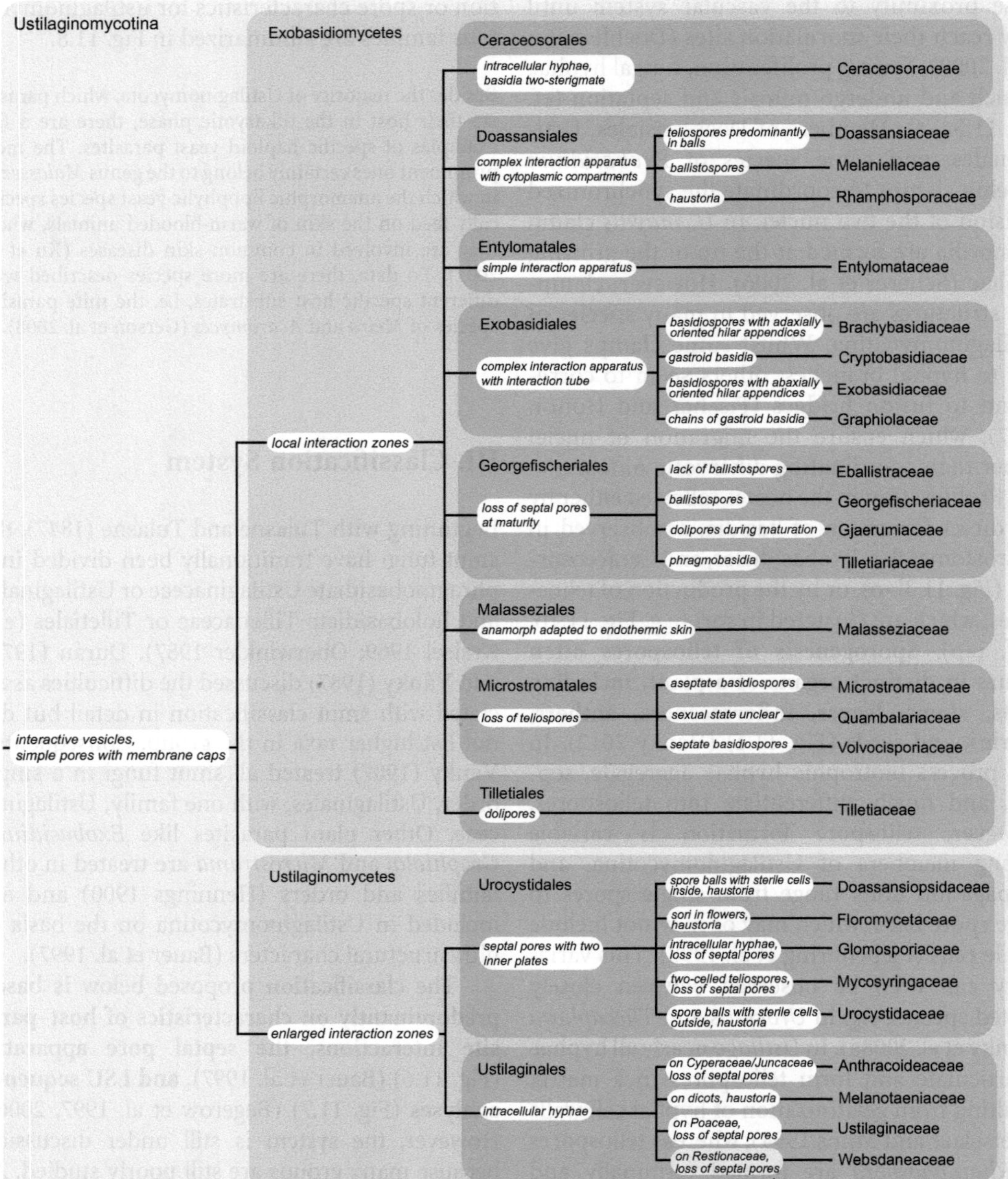

Fig. 11.6 Systematic overview of Ustilaginomycotina integrating morphological, anatomical, ecological, and molecular analyses. Characters on branches represent relevant markers reflecting apomorphies in some cases

basidial development are quite different among the various families of Exobasidiales or the families of Microstromatales, the relationships between the respective families within these orders are not always clear, and some of them are difficult to separate from each other without molecular data. Unresolved phylogenetic relationships are discussed with the respective groups in the next section. The fundamental characters used in classifying the Ustilaginomycotina were discussed in detail by Bauer et al. (1995a, b, 1997) and are therefore only briefly summarized here.

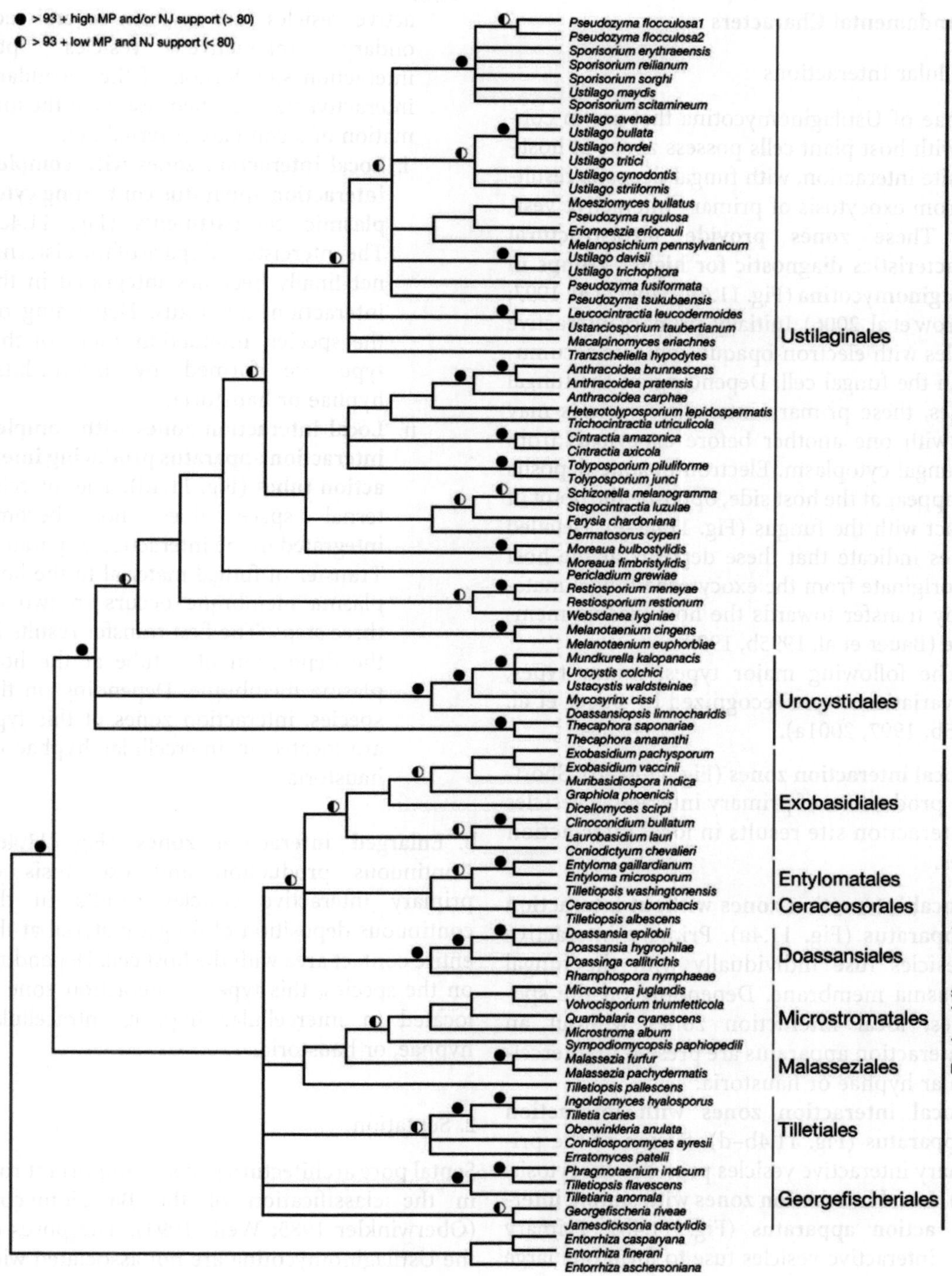

Fig. 11.7 Supertree topology from parsimony ratchet analysis (10,000 iterations) of matrix generated out of four neighbour-joining topologies (LSU, ITS, atp6, and β-tubulin genes). *Circles* next to branches summarize posterior probabilities of Bayesian analysis and bootstrap values of maximum parsimony and neighbour-joining analyses, which were based on a concatenated alignment [modified from Begerow et al. (2006)]

A. Fundamental Characters

1. Cellular Interactions

Hyphae of Ustilaginomycotina that are in contact with host plant cells possess zones of host–parasite interaction, with fungal deposits resulting from exocytosis of primary interactive vesicles. **These zones provide ultrastructural characteristics diagnostic for higher groups in Ustilaginomycotina** (Fig. 11.6) (Bauer et al. 1997; Begerow et al. 2006). Initially, primary interactive vesicles with electron-opaque contents accumulate in the fungal cell. Depending on the fungal species, these primary interactive vesicles may fuse with one another before exocytosis from the fungal cytoplasm. Electron-opaque deposits also appear at the host side, opposite the point of contact with the fungus (Fig. 11.4a–e). Detailed studies indicate that these deposits at the host side originate from the exocytosed fungal material by transfer towards the host plasma membrane (Bauer et al. 1995b, 1997).

The following major types, minor types, and variations were recognized by Bauer et al. (1995b, 1997, 2001a).

a. **Local interaction zones** (Fig. 11.4a–d). Short-term production of primary interactive vesicles at interaction site results in local inter-action zones.

1. **Local interaction zones without interaction apparatus** (Fig. 11.4a). Primary interactive vesicles fuse individually with the fungal plasma membrane. Depending on the species, local interaction zones without an interaction apparatus are present in intercellular hyphae or haustoria.
2. **Local interaction zones with interaction apparatus** (Fig. 11.4b–d). Fusion of the primary interactive vesicles precedes exocytosis.
   a) **Local interaction zones with simple interaction apparatus** (Fig. 11.4b). Primary interactive vesicles fuse to form one large secondary interactive vesicle per interaction site. Depending on the species, interaction zones of this type are located in intercellular or intracellular hyphae.
   b) **Local interaction zones with complex interaction apparatus** (Fig. 11.4c, d). Numerous primary interactive vesicles fuse to form several secondary interactive vesicles per interaction site. Fusion of the secondary interactive vesicles then results in the formation of a complex cisternal net.
      i. **Local interaction zones with complex interaction apparatus containing cytoplasmic compartments** (Fig. 11.4c). The intercisternal space of the cisternal net finally becomes integrated in the interaction apparatus. Depending on the species, interaction zones of this type are formed by intercellular hyphae or haustoria.
      ii. **Local interaction zones with complex interaction apparatus producing interaction tubes** (Fig. 11.4d). The intercisternal space does not become integrated in the interaction apparatus. Transfer of fungal material to the host plasma membrane occurs in two or three steps. The first transfer results in the deposition of a tube at the host plasma membrane. Depending on the species, interaction zones of this type are located in intercellular hyphae or haustoria.

b. **Enlarged interaction zones** (Fig. 11.4e). Continuous production and exocytosis of primary interactive vesicles results in the continuous deposition of fungal material at the entire contact area with the host cell. Depending on the species, this type of interaction zone is located in intercellular hyphae, intracellular hyphae, or haustoria.

2. Septation

Septal pore architecture plays an important role in the classification of the Basidiomycota (Oberwinkler 1985; Wells 1994). The pores of the Ustilaginomycotina are not associated with differentiated, multilayered caps or sacs derived from the endoplasmic reticulum. The septa produced in the saprobic phase of the dimorphic species of the Ustilaginomycotina are usually devoid of distinct septal pores. **Septa in soral hyphae of the Ustilaginomycotina either have pores with membrane caps or**

are poreless. Five types of septation of soral hyphae were recognized by Bauer et al. (1997): (1) presence of simple pores with two tripartite membrane caps (Fig. 11.5a), (2) presence of simple pores with two outer tripartite membrane caps and two inner nonmembranous plates (Fig. 11.5b) (see also Bauer et al. 1995a), (3) presence of simple pores with two outer tripartite membrane caps and a tube in the pore channel (Fig. 11.5c), (4) presence of dolipores with membrane bands (Fig. 11.5d) (see also Roberson and Luttrell 1989), and (5) septa without distinct pores (Fig. 11.5e) designated as poroid or poreless septa.

B. Overview

In what follows, an overview of the taxa included in the Ustilaginomycotina is given. The system is based on a review of available studies. Discrepancies with other taxa proposed in the past are discussed subsequently. Compared to a previous overview (Bauer et al. 2001b), we have included several new genera, Malasseziales and Ceraceosorales, and excluded *Entorrhiza*, mainly based on the results of molecular analyses (Hibbett et al. 2007). Host families are indicated if the host range of the respective genera comprises one or two families. Following a unification of anamorphic and teleomorphic taxonomies (Hawksworth et al. 2011), the anamorphic species are ascribed to higher teleomorphic taxa based on molecular data (Fig 11.7) (Boekhout et al. 2011). Numbers in parentheses indicate the known species of each genus (Boekhout et al. 2011; Kirk et al. 2008; Vánky 2012; Chamnanpa et al. 2013; Denchev and Denchev 2011; Lutz et al. 2012; Savchenko et al. 2013).

Ustilaginomycotina R. Bauer, Begerow, J.P. Samp., M. Weiss & Oberw.

I. Exobasidiomycetes: Begerow, M. Stoll, R. Bauer
- a. Ceraceosorales:: Begerow, M. Stoll & R. Bauer
  - i. Ceraceosoraceae Denchev & R.T. Moore
    - *Ceraceosorus* B.K. Bakshi on Malvaceae (1)
- b. Doassansiales R. Bauer & Oberw.
  - i. Melaniellaceae R. Bauer, Vánky, Begerow & Oberw.
    - *Melaniella* R. Bauer, Vánky, Begerow & Oberw. on Selaginellaceae (2)
  - ii. Doassansiaceae (Azb. & Karat.) R.T. Moore emend. R. Bauer & Oberw.
    - *Burrillia* Setchell on monocots (4)
    - *Doassansia* Cornu on mono- and eudicots (12)
    - *Doassinga* Vánky, R. Bauer & Begerow on Plantaginaceae (1)
    - *Entylomaster* Vánky & R.G. Shivas on Araceae (2)
    - *Heterodoassansia* Vánky on mono- and eudicots (8)
    - *Nannfeldtiomyces* Vánky on Typhaceae (2)
    - *Narasimhania* Thirum. & Pavgi emend Vánky on Alismataceae (1)
    - *Pseudodermatosorus* Vánky on Alismataceae (2)
    - *Pseudodoassania* (Setchell) Vánky on Alismataceae (2)
    - *Pseudotracya* Vánky on Hydrocharitaceae (1)
    - *Tracya* H. & P. Sydow on Hydrocharitaceae and Araceae (2)
  - iii. Rhamphosporaceae R. Bauer & Oberw.
    - *Rhamphospora* D.D. Cunn. on Nymphaeaceae (1)
- c. Entylomatales R. Bauer & Oberw.
  - i. Entylomataceae R. Bauer & Oberw.
    - *Entyloma* de Bary on eudicots (163)
    - *Tilletiopsis* Derx pro parte (anamorphic) (3)
- d. Exobasidiales P. Henn. emend. R. Bauer & Oberw.
  - i. Brachybasidiaceae Gäum.
    - *Brachybasidium* Gäumann on Arecaceae (1)
    - *Dicellomyces* L. S. Olive on monocots (4)
    - *Exobasidiellum* Donk on Poaceae (2)
    - *Kordyana* Racib. on Commelinaceae (5)
    - *Meira* Boekhout, Scorzetti, Gerson & Sztejnb. (anamorphic) (4)
    - *Proliferobasidium* L.J. Cunn. on Heliconiaceae (1)
  - ii. Cryptobasidiaceae Malençon ex Donk
    - *Acaromyces* Boekhout, Scorzetti, Gerson & Sztejnb. (anamorphic) (2)
    - *Botryoconis* H. & P. Sydow on Lauraceae (5)
    - *Clinoconidium* Pat. on Lauraceae (2)
    - *Coniodictyum* Har. & Pat. on Rhamnaceae (1)
    - *Drepanoconis* Schröter & P. Henn. on Lauraceae (3)
    - *Laurobasidium* Jülich on Lauraceae (1)
  - iii. Exobasidiaceae P. Henn.
    - *Arcticomyces* Savile on Saxifragaceae (1)
    - *Austrobasidium* Palfner on Hydrangeaceae (1)
    - *Exobasidium* Woronin on Ericales (50)
    - *Muribasidiospora* Kamat & Rajendren on Anacardiaceae and Ulmaceae (3)

iv. Graphiolaceae E. Fischer
Graphiola Poiteau on Arecaceae (5)
Stylina H. Sydow on Arecaceae (1)
e. Georgefischeriales R. Bauer, Begerow & Oberw.
i. Georgefischeriaceae R. Bauer, Begerow & Oberw.
Georgefischeria Thirum. & Narash. emend. Gandhe on Convolvulaceae (4)
Jamesdicksonia Thirum., Pavgi & Payak on Cyperaceae and Poaceae (16)
ii. Gjaerumiaceae R. Bauer, M. Lutz & Oberw.
Gjaerumia R. Bauer, M. Lutz & Oberw. on Asparagaceae, Melanthiaceae and Xanthorrhoeaceae (3)
Tilletiopsis Derx pro parte (anamorphic) (2)
iii. Tilletiariaceae Moore
Phragmotaenium R. Bauer, Begerow, A. Nagler & Oberw. on Poaceae (1)
Tilletiaria Bandoni & Johri (1)
Tilletiopsis Derx pro parte (anamorphic) (4)
Tolyposporella Atkinson on Poaceae (6)
iv. Eballistraceae R. Bauer, Begerow, A. Nagler & Oberw.
Eballistra R. Bauer, Begerow, A. Nagler & Oberw. on Poaceae (3)
f. Malasseziales Moore emend. Begerow, R. Bauer & Boekhout
Malassezia Baill. (anamorphic) (14)
g. Microstromatales R. Bauer & Oberw.
i. Microstromataceae Jülich
Microstroma Niessl on Juglandaceae, Fabaceae and Fagaceae (4)
Rhodotorula F.C. Harrison pro parte (anamorphic) (3)
ii. Volvocisporiaceae Begerow, R. Bauer & Oberw.
Volvocisporium Begerow, R. Bauer & Oberw. on Malvaceae (2)
iii. Quambalariaceae Z.W. de Beer, Begerow & R. Bauer
Quambalaria J.A. Simpson on Myrtaceae (6)
Jaminaea Sipiczki & Kajdacsi (anamorphic) (2)
iv. Microstromatales *incertae sedis*
Sympodiomycopsis Sugiy., Tokuoka & Komag. (anamorphic) (2)
h. Tilletiales Kreisel ex R. Bauer & Oberw.
i. Tilletiaceae Tul. & C. Tul. emend. R. Bauer & Oberw.
Conidiosporomyces Vánky on Poaceae (3)
Erratomyces M. Piepenbr. & R. Bauer on Fabaceae (5)
Ingoldiomyces Vánky on Poaceae (1)
Neovossia Körn. on Poaceae (1)
Oberwinkleria Vánky & R. Bauer on Poaceae (1)
Salmacisia D.R. Huff & A. Chandra on Poaceae (1)
Tilletia L. & C. Tul. on Poaceae (179)
i. Exobasidiomycetes *incertae sedis*
Tilletiopsis albescens Gokhale (anamorphic)
Tilletiopsis pallescens Gokhale (anamorphic)
II. Ustilaginomycetes R. Bauer, Oberw. & Vánky
a. Urocystidales R. Bauer & Oberw.
i. Doassansiopsidaceae Begerow, R. Bauer & Oberw.
Doassansiopsis (Setchell) Dietel on mono- and dicots
ii. Floromycetaceae M. Lutz, R. Bauer & Vánky
Antherospora R. Bauer, M. Lutz, Begerow, Piątek & Vánky on Asparagaceae (8)
Floromyces Vánky, M. Lutz & R. Bauer on Asparagaceae (1)
iii. Glomosporiaceae Cifferi emend. Begerow, R. Bauer & Oberw.
Thecaphora Fingerh. (including *Glomosporium*, *Kochmania*, *Tothiella*, *Sorosporium*) on eudicots (61)
iv. Mycosyringaceae R. Bauer & Oberw.
Mycosyrinx Beck on Vitaceae (4)
v. Urocystidaceae Begerow, R. Bauer & Oberw.
Flamingomyces R. Bauer, M. Lutz, Piątek, Vánky & Oberw. on Ruppiacae (1)
Melanoxa M. Lutz, Vánky & R. Bauer on Oxalidaceae (2)
Melanustilospora Denchev on Araceae (2)
Mundkurella Thirum. on Araliaceae (5)
Urocystis Rabenh. ex Fuckel on mono- and eudicots (165)
Ustacystis Zundel on Rosaceae (1)
Vankya Ershad on Liliaceae (3)
b. Ustilaginales Clinton emend. R. Bauer & Oberw.
i. Anthracoideaceae Denchev
Anthracoidea Brefeld on Cyperaceae (101)
Cintractia Cornu on Cyperaceae (13)
Dermatosorus Sawada ex Ling on Cyperaceae (6)
Farysia Racib. on Cyperaceae (21)
Farysizyma A. Fonseca (anamorphic) (4)
Heterotolyposporium Vánky on Cyperaceae (1)
Leucocintractia M. Piepenbr., Begerow & Oberw. on Cyperaceae (4)
Moreaua T. N. Liou & H. C. Cheng on Cyperaceae (36)
Parvulago R. Bauer, M. Lutz, M. Piątek, Vánky & Oberw. on Cyperaceae (1)
Pilocintractia Vánky on Cyperaceae (2)
Planetella Savile on Cyperaceae (1)
Portalia V. Gonzáles, Vánky & G. Platas on Cyperaceae (1)
Schizonella Schröter on Cyperaceae (6)
Shivasia Vánky, M. Lutz & M. Piątek on Cyperaceae (1)
Stegocintractia M. Piepenbr., Begerow & Oberw. on Juncaceae (6)
Tolyposporium Woronin ex Schröter on Juncaceae (5)

Trichocintractia M. Piepenbr. on Cyperaceae (1)
Ustanciosporium Vánky emend. M. Piepenbr. on Cyperaceae (21)

ii. Melanotaeniaceae Begerow, R. Bauer & Oberw.
Exoteliospora R. Bauer, Oberw. & Vánky on Osmundaceae (1)
Melanotaenium de Bary on eudicots (9)
Yelsemia Walker on eudicots (4)

iii. Ustilaginaceae Tul. & C. Tul. emend. R. Bauer & Oberw.
Anomalomyces Vánky, M. Lutz & R.G. Shivas on Poaceae (1)
Anthracocystis Bref. on Poaceae (124)
Eriomoeszia Vánky on Eriocaulaceae (1)
Franzpetrakia Thirum. & Pavgi emend. Guo, Vánky & Mordue on Poaceae (3)
Langdonia McTaggart & R.G. Shivas on Poaceae (8)
Macalpinomyces Langdon & Full. emend. Vánky on Poaceae (41)
Melanopsichium G. Beck on Polygonaceae (2)
Moesziomyces Vánky on Poaceae (1)
Pericladium Pass. on Malvaceae (3)
Pseudozyma Bandoni emend. Boekhout (anamorphic) (16)
Sporisorium Ehrenb. on Poaceae (195)
Stollia McTaggart & R.G. Shivas on Poaceae (5)
Tranzscheliella Lavrov on Poaceae (17)
Triodiomyces McTaggart & R.G. Shivas on Poaceae (5)
Tubisorus Vánky & M. Lutz on Poaceae (1)
Ustilago (Pers.) Roussel on Poaceae (167)

iv. Websdaneaceae Vánky
Restiosporium Vánky on Anarthriaceae and Restionaceae (23)
Websdanea Vánky on Anarthriaceae (1)

v. Ustilaginales *incertae sedis*:
Ahmadiago Vánky on Euphorbiaceae (1)
Centrolepidosporium R.G. Shivas & Vánky on Centrolepidaceae (1)
Cintractiella K.B. Boedijn emend. M. Piepenbr. on Cyperaceae (2)
Clintamra Cordas & Durán on Asparagaceae (1)
Eriocaulago Vánky on Eriocaulaceae (2)
Eriosporium Vánky on Eriocaulaceae (2)
Farysporium Vánky on Cyperaceae (1)
Geminago Vánky & R. Bauer on Malvaceae (1)
Kuntzeomyces P. Henn. ex Sacc. & P. Sydow on Cyperaceae (2)
Orphanomyces Savile on Cyperaceae (3)
Testicularia Klotzsch on Cyperaceae (3)
Uleiella Schröter on Araucariaceae (2)

C. Description

Within Ustilaginomycotina **two major groups are evident in the dendrograms resulting from ultrastructural and LSU rDNA sequence analyses** (Figs. 11.6 and 11.7) (Bauer et al. 1997; Begerow et al. 2006). Though the monophyly of the Ustilaginomycetes is well supported, this is not always the case with the Exobasidiomycetes (Hibbett et al. 2007). However, in the absence of additional studies, we follow the earlier interpretations and retain Ceraceosorales and Malasseziales as part of the Exobasidiomycetes (Begerow et al. 2000, 2006). Although many morphological characters of sori and teliospores are not consistent at higher taxonomic levels, an overview at the family level is included in Fig. 11.8.

1. Exobasidiomycetes

Exobasidiomycetes represents the sister group of Ustilaginomycetes (Bauer et al. 1997; Begerow et al. 1997, 2006; Hibbett et al. 2007). The members of Exobasidiomycetes and Ustilaginomycetes share the **presence of membrane caps or bands at the septal pores** (Fig. 11.6). However, taxa with poreless septa evolved in both groups. **Exobasidiomycetes differs from Ustilaginomycetes in the formation of local interaction zones** (Fig. 11.5). Except for Tilletiariaceae (Fig. 11.3k), all **members of Exobasidiomycetes are holobasidiate** (Fig. 11.3f–j, l–o). Among the basidiomycetes, the formation of ballistosporic holobasidia, in which the **hilar appendices of the basidiospores are oriented abaxially** (sterigmata turned outwards; basidiospores inwards) (Fig. 11.3j, o), is restricted to Exobasidiomycetes. This type of basidium is common in Exobasidiales (Fig. 11.3o) but also occurs in species of Doassansiales, Georgefischeriales (Fig. 11.3j), and Tilletiales (Goates and Hoffmann 1986). Therefore, the *Exobasidium* basidium with the specific orientation of the ballistosporic basidiospores may represent an apomorphy for Exobasidiomycetes.

Teliospores are absent or present within the Exobasidiomycetes. **Formation of teliospore**

Taxon	Sorus location	Teliospores[a]	Spore balls	Sterile cells[b]	Peridia	Holo-/phragmobasidia	Haustoria/intracellular hyphae	Septal pores	Host preferences[c,d]
Exobasidiomycetes									
Ceraceosorales									
Ceraceosoraceae	Veg. organs	○	○	○	○	●	●	●	Malvaceae*
Doassansiales									
Doassansiaceae	Veg. organs	●	◐	◐	○	●	○	●	Mono- and eudicots
Melaniellaceae	Veg. organs	●	○	○	○	●	○	●	Selaginella*
Rhamphosporaceae	Veg. organs	●	○	○	○	●	●	●	Nymphaeaceae*
Entylomatales									
Entylomataceae	Veg. organs	●	○	○	○	●	○	●	Eudicots
Exobasidiales									
Brachybasidiaceae	Veg. organs	○	○	○	○	●	●	●	Monocots
Cryptobasidiaceae	Veg. organs	○	○	○	○	●	●	●	Lauraceae
Exobasidiaceae	Veg. organs	○	○	○	○	●	◐	●	Ericales
Graphiolaceae	Veg. organs	○	○	○	○	●	●	●	Arecaceae
Georgefischeriales									
Eballistraceae	Veg. organs	●	○	○	○	●	○	○	Poaceae
Georgefischeriaceae	Veg. organs	●	○	○	○	●	○	○	Poales, Convolvulaceae
Gjaerumiaceae	Veg. organs	●	○	○	○	●	○	○	Asparagales, Liliales*
Tilletiariaceae	Veg. organs	●	○	○	○	○	○	○	Poaceae
Malasseziales									
Malasseziaceae									Human skin
Microstromatales									
Microstromataceae	Veg. organs	○	○	○	○	●	○	●	Eudicots
Quambalariaceae	Veg. organs	○	○	○	○		○	●	Myrtaceae*
Volvocisporiaceae	Veg. organs	○	○	○	○	●	○	●	Malvaceae*
Tilletiales									
Tilletiaceae	Flower/veg. organs	●	○	○	○	●	○	●	Poaceae, Fabaceae
Ustilaginomycetes									
Urocystidales									
Doassansiopsidaceae	Veg. organs	●	●	●	○	●	●	●	Mono- and dicots
Floromycetaceae	Flower	●	◐	○	○	○	●	●	Asparagales
Glomosporiaceae	Veg. organs	●	◐	◐	○	●	●	○	Eudicots
Mycosyringaceae	Witch's broom	●	●	○	○		○	○	Vitaceae*
Urocystidaceae	Flower/veg. organs	●	◐	●	○	◐	●	●	Mono- and eudicots
Ustilaginales									
Anthracoideaceae	Flower/veg. organs	●	◐	○	◐	○	●	○	Cyperaceae/Juncaceae
Melanotaeniaceae	Veg. organs	●	○	○	○	○	●	●	Eudicots, Osmundaceae
Ustilaginaceae	Flower/veg. organs	●	◐	○	◐	○	●	○	Poaceae
Websdaneaceae	Flower	●	◐	○	◐	○	●	○	Anarthriaceae, Restionaceae

Fig. 11.8 Summary of character states of Ustilaginomycotina families. [a]*Filled circle*: presence of a character or holobasidia; *empty circle*: absence of a character or presence of phragmobasidia; *half-filled circle*: mixture of characters in respective groups. [b]Sterile cells in spore balls. [c]Host preferences are identified for families with more than 90 % of their members parasitizing the respective plant taxon. [d]*Asterisk*: families of unclear preference due to small species sampling

balls only occurs in Doassansiaceae and in *Tolyposporella*. Smut fungi among the Exobasidiomycetes show terminal or intercalary teliospore formation (Roberson and Luttrell 1987; Trione et al. 1989). A gelatinization of hyphal walls preceding teliospore formation is either lacking or not clearly recognizable.

Currently we include eight orders on the basis of ultrastructural characters and molecular phylogenetic data within Exobasidiomycetes (Fig. 11.6). A superorder, Exobasidianae, including Entylomatales, Doassansiales, and Exobasidiales, was proposed based on the apomorphy of a complex interaction apparatus (Bauer et al. 1997). This grouping is highly sensitive to sampling in molecular analyses (Fig. 11.7), and therefore we follow the system of Hibbett et al. (2007). Anamorphic species without affiliation to a teleomorph have been assigned to *Tilletiopsis*, although the genus is non-monophyletic (Begerow et al. 2000). However, some anamorphic species or lineages have been named according to a unique ecology as in *Meira* and *Jaminaea*. The **phylogenetic positions of Malasseziales and Ceraceosorales are controversial**, and some authors have proposed a treatment as *incertae sedis* (Hibbett et al. 2007). Although apomorphic exobasidiomycetous morphological features like septal pore caps and local interaction zones are lacking, at least in Malasseziales, we follow the proposal of Begerow et al. (2006) based on molecular data (Fig. 11.7). All orders are presented alphabetically without additional hierarchy.

a) Ceraceosorales

Within Exobasidiomycetes, the **Ceraceosorales are characterized by intracellular hyphae with a simple interaction apparatus** (Fig. 11.6) (Begerow et al. 2006). The **septal pores in *Ceraceosorus bombacis*** (B.K. Bakshi) B.K. Bakshi **are simple and enclosed by membrane caps at both sides** (Fig. 11.5a), as seen in Melanotaeniaceae of the Ustilaginomycetes and in Microstromatales, Entylomatales, Doassansiales, and Exobasidiales of the Exobasidiomycetes (Bauer et al. 1997; Begerow et al. 2006). In *Ceraceosorus* and in Brachybasidiaceae, basidia protrude through stomata or emerge from the disintegrated epidermis. The basidia are elongated, basally thick-walled, and two-sterigmate and form ballistosporic basidiospores with an adaxial orientation of the hilar appendices in both groups (Begerow et al. 2002; Cunningham et al. 1976). Like Brachybasidiaceae and Exobasidiomycetes in general, *Ceraceosorus* produces local interaction zones (Begerow et al. 2006). However, molecular data do not support a closer relationship between Exobasidiales and Ceraceosorales (Fig. 11.7). The other Exobasidiomycetes lacking an interaction apparatus or establishing a simple interaction apparatus, such as Entylomatales, Georgefischeriales, Microstromatales, and Tilletiales, do not form intracellular hyphae or haustoria (Bauer et al. 1997; Begerow et al. 2006). Thus, *C. bombacis* (B.K. Bakshi) B.K. Bakshi seems to be isolated, and the monotypic order seems to be justified.

b) Doassansiales

A **complex interaction apparatus, including cytoplasmic compartments**, characterizes this order (Figs. 11.4c and 11.6) (Bauer et al. 1997). The studied **species of this group have parasitic hyphae with clamps.** They are teliosporic and dimorphic and do not form ballistoconidia in the haploid phase. The teliospore germinates with holobasidia (Fig. 11.3g) (Bauer et al. 1999a; Vánky et al. 1998). The members of *Burrillia*, *Doassansia*, *Entylomaster*, *Heterodoassansia*, *Nannfeldtiomyces*, *Narasimhania*, *Pseudodoassansia*, *Pseudodermatosporus*, *Pseudotracya*, and *Tracya* have complex spore balls (Vánky 1987, 2012), whereas *Doassinga* (Fig. 11.1a), *Melaniella*, and *Rhamphospora* produce single spores (Vánky 1994; Vánky et al. 1998). **The spore balls differ in the occurrence of sterile cells within the spore ball.** In addition, teliospores are darkly coloured in *Melaniella* and lightly coloured in *Doassinga*, *Rhamphospora*, and genera with complex teliospore balls. The hosts of the Doassansiales are systematically diverse, comprising spike mosses (Selaginellaceae) and various monocots as well as eudicots.

However, members of **Doassansiales are ecologically well characterized by their occurrence on paludal or aquatic plants, or at least on plants of moist habitats.** They apparently evolved in the ecological niche of aquatic plants and developed complex spore balls and more or less sigmoid basidiospores in adaptation to water dispersal (Fig. 11.3g) (Bauer et al. 1997). Interestingly, the species of *Doassansiopsis* in Urocystidales likewise parasitize aquatic plants and possess similar complex spore balls. Thus, *Doassansiopsis* and Doassansiales are excellent examples of the independent, convergent evolution of similar structures under the same environmental condition.

The order comprises three families. Ultrastructural and LSU sequence analyses revealed a basal dichotomy between Melaniellaceae presenting pigmented spores and Rhamphosporeaceae and Doassansiaceae showing hyaline teliospores (Bauer et al. 1999a; Begerow et al. 1997). In contrast to members of Doassansiaceae, *Rhamphospora nymphaeae* D. Cunn., the only species placed in the Rhamphosporaceae, forms highly branched haustoria (Fig. 11.8) (Bauer et al. 1997).

c) Entylomatales

Entylomatales is **characterized by the presence of a simple interaction apparatus** at the interaction sites (Fig. 11.4b) **and simple hyaline spores as well as simple septal pores** (Fig. 11.6) (Bauer et al. 1997). This group comprises only species of *Entyloma* occurring on eudicots (Fig. 11.1b), with the type species of *Entyloma*, *Entyloma microsporum* (Unger) Schröter (Fig. 11.3f), as well as anamorphic *Tilletiopsis* species. Ultrastructural and LSU sequence analyses revealed that the genus *Entyloma* was polyphyletic and that the previous "*Entyloma*" species occurring on monocots belonged to Georgefischeriales (designated as *Eballistra* or *Jamesdicksonia*) (Figs. 11.3i, j and 11.7) (Bauer et al. 1997; Begerow et al. 1997, 2002).

Although the species in the genus *Entyloma* are morphologically very similar, systematic analyses supported numerous host-specific species (Boekhout et al. 2006). The **majority of *Entyloma* species parasitize plant families in Ranunculales or Asteridae**, whereby the members of Ranunculales seem to be the older host group as its parasites are paraphyletic and show longer branch lengths (Begerow et al. 2002). Within Asteridae (including one *Entyloma* species on Saxifragaceae) an "explosive" radiation seems to have occurred, most likely caused by a rapid succession of **host jumps rather than cocladogenesis** (Begerow et al. 2002). This is supported by the fact that *Entyloma* species on closely related host groups are not necessarily closely related to each other. Additionally, the much longer branch lengths in Asteridae hosts indicate that their interaction with *Entyloma* is younger than the radiation of the host group (Begerow et al. 2002). Finally, the inclusion of an *Entyloma* species on *Chrysosplenium* (Saxifragaceae) supports this view of host shifts as a likely explanation for the observed host range patterns (Begerow et al. 2002).

The **anamorphic *Tilletiopsis*** species, which have been assumed to be the sister group to *Entyloma* (Fig. 11.7) (Begerow et al. 2002), are now known to have evolved independently several times within the genus *Entyloma* (Boekhout et al. 2006). Research in this group has recently gained importance because so-called white haze, a post-harvest disorder of apples, has been associated with the proliferation of pseudomycelia of various *Tilletiopsis* species. This cosmetic disorder was first described as problematic under low-oxygen storage conditions but was demonstrated to additionally occur on fruits in the field. The increase in observations in the last decade is correlated mainly with an increase in humidity and new cultivation procedures in this time frame (Baric et al. 2010).

d) Exobasidiales

Members of Exobasidiales are characterized by the presence of **interaction tubes produced by a complex interaction apparatus** (Fig. 11.4d) (Bauer et al. 1997) and **septal pores with membranous caps and an additional tube inside** (Figs. 11.5c and 11.6). The monophyly of this group is also well supported by molecular data (Fig. 11.7) (Begerow et al. 2002, 2006). Members of Exobasidiales are holobasidiate and dimorphic (Fig. 11.3l–o). They do not form teliospores in the parasitic phase or ballistoconidia in the saprobic phase. In most species, the basidiospores become septate during germination. **Hosts are mono- and eudicots.** The sori appear on leaves, fruits, and stems (Figs. 11.1a–c and 8). We currently recognize four families in this order (Fig. 11.6) (Begerow et al. 2002; Hibbett et al. 2007).

The Brachybasidiaceae sporulate on the surface of host organs. The basidia protrude through stomata or emerge from the disintegrated epidermis. The basidia are elongated and ballistosporic and have two sterigmata. The basidiospores are thin-walled. Available data indicate that the hilar appendices of the basidiospores are oriented adaxially at the apex of the basidia (see Figs. 2, 6, 13, 17 in Cunningham et al. 1976; Fig. 1-G in Ingold 1985; Figs. 1.10-2, 1.10-3 in Oberwinkler 1982; Fig. 4 in Oberwinkler 1993). *Brachybasidium pinangae* Gäumann, *Dicellomyces gloeosporus* Olive, and *Proliferobasidium heliconiae* Cunningham form persistent probasidia that are arranged in delimited fructifications. The species of Brachybasidiaceae live predominantly on monocots (Cunningham et al. 1976; Gäumann 1922; Oberwinkler 1978, 1982, 1993; Olive 1945). Molecular analyses placed the anamorphic genus *Meira*, isolated from pear fruits, into this family, although teleomorphic stages are unknown (Rush and Aime 2013; Yasuda et al. 2006).

The non-smut family Cryptobasidiaceae (Fig. 11.1c, d) contrasts with Brachybasidiaceae and Exobasidiaceae because it sporulates internally by producing holobasidia in peripheral lacunae of the host galls (Fig. 11.6). During maturation, the galls rupture and liberate the basidiospore mass. The basidia are gastroid and lack sterigmata. The basidiospores are usually thick-walled, resembling the urediniospores of rust fungi or the teliospores of smut fungi. In addition, old fructifications often resemble smut sori. These characters may explain why some members of this group were described as smut fungi (see above), whilst others were originally described as rusts [e.g. *Clinoconidium farinosum* (P. Henn.) Pat. as *Uredo farinosa* P. Henn.]. In contrast to other members of Cryptobasidiaceae, *Laurobasidium lauri* (Geyler) Jülich (Fig. 11.1d) sporulates on the surface of host organs. Additionally, the basidia of this species resemble those of *Exobasidium* but are gastroid, as in other members of Cryptobasidiaceae [for a detailed discussion see Begerow et al. (2002)]. Thus, *Laurobasidium* may occupy a systematic position at the base of the Cryptobasidiaceae and intermediate between Cryptobasidiaceae and other Exobasidiales, although this is not supported by molecular analyses so far (Begerow et al. 2002). Except for *Coniodictyum* (Fig. 11.1c), the host range of Cryptobasidiaceae is restricted to laurels. Cryptobasidiaceae species are known only from Japan, Africa, and South America (Donk 1956; Hendrichs et al. 2003; Lendner 1920; Malençon 1953; Maublanc 1914; Oberwinkler 1978, 1982, 1993; Piepenbring et al. 2010; Sydow 1926). In molecular studies the anamorphic genus *Acaromyces* isolated from mites also clusters within Cryptobasidiacae (Boekhout et al. 2003).

Exobasidiaceae species are morphologically similar to those of Brachybasidiaceae. Like members of Brachybasidiaceae, Exobasidiaceae species sporulate through stomata or from the disintegrated epidermis (Mims and Richardson 2007), the basidia are elongated and ballistosporic, and the basidiospores are thin-walled. In contrast to the Brachybasidiaceae, the hilar appendices of the basidiospores are oriented abaxially at the apex of the basidia (Fig. 11.3o) (Oberwinkler 1977, 1978, 1982). In most Exobasidiaceae species, the number of sterigmata per basidium is not fixed, varying from two to eight, with four as the most frequent number. Only a few species form generally two-sterigmate basidia. Exobasidiaceae comprises *Arcticomyces*, *Austrobasidium*, *Exobasidium*, and *Muribasidiospora* (Begerow et al. 2002). The members of this family occur on eudicots predominantly on Ericaceae (Fig. 11.1e) (Hennings 1900; Mims et al. 1987; Nannfeldt 1981; Oberwinkler 1977, 1978, 1982, 1993; Piepenbring et al. 2010; Rajendren 1968).

The Graphiolaceae are parasites of palms. Fructification of the Graphiolaceae starts between chlorenchyma and hypodermal tissue (Cole 1983). During differentiation of the cupulate to cylindrical basidiocarp, the epidermis ruptures and globose basidia are produced in chains by disarticulation of sporogenous hyphae within the basidiocarps (Fig. 11.3l). The passively released basidiospores arise laterally on the basidia (Fischer 1921, 1922; Oberwinkler et al. 1982). Haustoria are constricted at the point of penetration and consist of a clamped basal body (see Fig. 11.3 in Oberwinkler et al. 1982; Bauer et al. 1997; Begerow et al. 2002).

e) Georgefischeriales

Among the Exobasidiomycetes, this group is characterized by the presence of **poreless septa in soral hyphae** (Fig. 11.6). The Georgefischeriales species have a dimorphic life cycle and form teliospores. They interact with their respective hosts via **local interaction zones without an interaction apparatus** (Bauer et al. 1997, 2001a, 2005). Haustoria or intracellular hyphae are lacking. The Georgefischeriales sporulate in vegetative parts of the hosts, predominantly in leaves (Fig. 11.1f, g). Teliospores are yellow to brown in species of *Georgefischeria* and darkly coloured in other taxa. The teliospore masses are usually not powdery, and host tissues are not fractured to expose the sori (Bauer et al. 1997, 2001a, 2005). The order is divided into four families, Georgefischeriaceae, Gjaerumiaceae, Tilletiariaceae, and Eballistraceae (Fig. 11.8).

Except for *Georgefischeria*, with its four species on Convolvulaceae and the species of *Gjaerumia* on several monocot families, the **Georgefischeriales occur on Poales.** Because *Tilletiaria anomala* Bandoni & B.N. Johri appeared in a plate over which a polypore growing on decaying wood had been suspended (Bandoni and Johri 1972), nothing is known of its ecology. Most recently, *T. anomala* was found in the intercellular spaces of rice plants, **indicating an endophytic life style** (Takahashi et al. 2011). In this study, other grass parasites, including *Ustilago* and *Tilletia*, were also found in the intercellular spaces, and, like *T. anomala*, smut fungi occasionally form teliospores and basidia in culture (Fig. 11.3k) (Bauer et al.

1997, 2005). It is conceivable that *T. anomala* is a phytoparasite, probably on grasses.

The molecular phylogenies of this group (Bauer et al. 2005) correlate well with the family concept proposed by Bauer et al. (2001a) (Figs. 11.6 and 11.8). Species of Georgefischeriaceae, Gjaerumiaceae, and Eballistraceae are characterized by holobasidia (Fig. 11.3i, j), whereas species of Tilletiariaceae are phragmobasidiate (Fig. 11.3k) (Bandoni and Johri 1972; Bauer et al. 2001a, 2005). The basidiospores of Georgefischeriaceae, Gjaerumiaceae, and Tilletiariaceae form *Tilletiopsis*-like pseudohyphal anamorphs that produce ballistoconidia (Bandoni and Johri 1972; Bauer et al. 2005). Members of Eballistraceae do not produce ballistoconidia but form budding yeasts, which are spherical to ellipsoidal in form (Singh and Pavgi 1973).

Noteworthy is the occurrence of dolipores in young septal pores of Gjaerumiaceae. So far, within the Exobasidiomycetes dolipores are only known from members of the Tilletiales. However, in contrast to members of this group, the pores of members of Gjaerumiaceae are closed during teliosporogenesis (Bauer et al. 2005).

Molecular analyses also revealed that several anamorphic species cluster within the Georgefischeriales. The current taxonomy of these species assigned to *Tilletiopsis* awaits revision (Boekhout et al. 2006).

f) Malasseziales

The **anamorphic genus *Malassezia* comprises medically important, lipophylic yeasts** that constitute part of the fungal microflora on the skin of warm-blooded animals (Guého et al. 1998; Findley et al. 2013). It has been placed within the Exobasidiomycetes based on molecular studies (Begerow et al. 2000, 2006). *Malassezia* has been found to be associated with a variety of pathological conditions in humans, including **pityriasis versicolor, seborrheic dermatitis, folliculitis,** and **systemic infections** (Gueho et al. 1998). The cell wall of *Malassezia* yeasts is thick and multilamellate and reveals a unique substructure with an electron-opaque, helicoidal band that corresponds to a helicoidal evagination of the plasma membrane (Guého-Kellermann et al. 2010). The sexual phase of *Malassezia* is unknown, although genetic analyses revealed intact mating genes (Xu et al. 2007). The position of *Malassezia* in the Exobasidiomycetes is surprising and suggests that *Malassezia* species either are phytoparasitic in the dikaryophase or originated at least from plant parasites.

g) Microstromatales

Among the Exobasidiomycetes, the **Microstromatales are characterized by the presence of simple pores and local interaction zones without an interaction apparatus** (Fig. 11.6) (Bauer et al. 1997). Teliospores are lacking. Hosts are often **woody plants**, which is similar to the ecology of Exobasidiales. Though only a few species were initially placed in this order, three families are currently recognized: Microstromataceae, Volvocisporiaceae, and Quambalariaceae (Fig. 11.8) (Begerow et al. 2001; de Beer et al. 2006).

In Microstromataceae the young basidia protrude through the stomata and sporulate on the leaf surface (Figs. 11.1h and 11.3n) (Oberwinkler 1978; Patil 1977). They are not teliosporic, and sori are mostly less than a few millimetres in diameter (Fig. 11.1h). They are characterized by single-celled, hyaline basidiospores and infect mainly trees and bushes of various eudicot families, mainly Juglandaceae, Fabaceae, and Fagaceae (Begerow et al. 2001). In culture they form budding yeasts without ballistoconidia and pseudohyphae. In contrast to most Ustilaginomycetes, the yeast cells are more or less spherical in form.

Volvocisporiaceae are characterized by large and highly septate basidiospores (Fig. 11.3m) and are known from only two species (Begerow et al. 2001; Ritschel et al. 2008). They share the ultrastructural morphology of simple septal pores and local interaction zones with all members of Microstromatales, but they are clearly separated from other families by molecular means (Ritschel et al. 2008).

In contrast to Microstromataceae and Volvocisporiaceae, members of Quambalariaceae possess septal pores with swellings resembling dolipores of other groups (Fig. 11.6) (de Beer et al. 2006). They comprise pathogens of *Eucalyptus* and *Corymbia*, and so far, almost all host taxa are native to Australia, which suggests Australia as the centre of diversity (de Beer et al. 2006; Pegg et al. 2009). Although the development of conidiophores through stomata looks very similar to basidia of *Microstroma* sporulations, meiosis has not been observed and the sexual state remains unclear (Pegg et al. 2009).

The known species of Microstromatales may only represent the so-called tip of the iceberg for this group. Most of them are difficult to detect in nature and could easily be overlooked. Additionally, several yeasts belonging to this group have been isolated, and their affiliation is not always clear. Beecause yeast anamorphs are common in *Microstroma*, addititional surveys are needed to recognize more taxa and teleomorphs. Though the included "*Rhodotorula*" species seem to be anamorphic stages of *Microstroma*, *Sympodiomycopsis* spp., and *Jaminaea angkorensis* Sipiczki and Kajdacsi, they seem to lack close relation to any studied species (Begerow et al. 2001; Mahdi et al. 2008; Sipiczki and Kajdacsi 2009).

h) Tilletiales

The presence of **dolipores in the mature septa** (Fig. 11.5d) characterizes the Tilletiales among the Exobasidiomycetes (Fig. 11.6) (Bauer et al. 1997). In contrast to all other groups of the Exobasidiomycetes, the Tilletiales are not known to be dimorphic. **They form local interaction zones without an interaction apparatus** (Fig. 11.4a), **and their hyphal anamorphs regularly produce ballistoconidia** (e. g., Carris et al. 2006; Ingold 1987b, 1997). Among all the smut fungi studied in culture, only the members of Tilletiales present distinct pores in the septa of saprobic hyphae.

Members of Tilletiales lack haustoria and intracellular hyphae (Fig. 11.8). The **teliospores** are darkly pigmented and often ornamented. Moreover, these teliospores are usually much larger than those of other groups of the Ustilaginomycotina, and they are never arranged in balls (Fig. 11.8). The teliospores of some species produce trimethylamine, which causes a foul smell in the spores. Seven genera are described in this family, six of which exclusively parasitize **Poaceae**. The genus *Erratomyces* is solely parasitic on Fabaceae. Sori are formed in ovaries of the hosts in the majority of species (Fig. 11.1i); only a few species of *Tilletia* and *Erratomyces* form teliospores in vegetative host organs (Castlebury et al. 2005; Piepenbring and Bauer 1997; Vánky 1994, 2012; Vánky and Bauer 1992, 1995, 1996). The teliospores germinate with holobasidia, producing terminal basidiospores, which often conjugate and give rise to infectious hyphae (Ingold 1989b; Vánky 2012).

Some species of *Tilletia* are economically important. *Tilletia caries* (DC) Tul. & C. Tul. and *T. controversa* J. G. Kühn on wheat and *Tilletia horrida* Takah. on rice can cause heavy losses in grain production (Carris et al. 2006; Mathre 1996; Trione 1982). In India and the American tropics the angular black spot disease on leaves of beans is caused by *Erratomyces patelii* (Pavgi & Thirum.) M. Piepenbr. & Bauer (Piepenbring and Bauer 1997).

Within Tilletiales the taxonomy is far from resolved. Molecular data especially provided controversial results for morphology-based classification (Castlebury et al. 2005). Species concepts and species delimitations are still in discussion (Cai et al. 2011). Additionally, the discovery and addition of new species might change the taxonomic concept (Bao et al. 2010; Shivas 2009).

Remarkably, *Salmacisia buchloëana* (Kellerm. & Swingle) D.R. Huff & Amb. Chandra parasitizing the buffalograss *Buchloë dactyloides* (Nutt.) Englem induces the development of ovaries in male flowers, which leads to hermaphroditism and castration of its host plant (Chandra and Huff 2008). Alteration of host reproductive structures evolved at least three times independently within smut fungi, as seen in *Salmacisia*, *Microbotryum* spp. and *Thecaphora oxalidis* (Ellis & Tracy) M. Lutz, R. Bauer & Piątek (Roets et al. 2008; Schäfer et al. 2010).

2. Ustilaginomycetes

The presence of **enlarged interaction zones** (Fig. 11.4e) **characterizes this group** (Fig. 11.6) (Bauer et al. 1997). The members of the **Ustilaginomycetes are teliosporic, gastroid, and dimorphic**. The species isolated in the anamorphic phase are usually placed in the genus *Pseudozyma*. However, for some members closely related to *Farysia* a new genus, *Farysizyma*, has been proposed (Inácio et al. 2008). Based on the new regulations of dual nomenclature, they should be included in *Farysia* (Hawksworth 2011; Hawksworth et al. 2011)

Morphologically and ecologically, members of the Ustilaginomycetes are diverse (Fig. 11.1j–p)

(Vánky 1987, 1994, 2012), but both ultrastructural and LSU sequence analyses unite them (Figs. 11.6 and 11.7) (Bauer et al. 1997; Begerow et al. 1997, 2006). Two orders are recognized.

a) Urocystidales

As part of Ustilaginomycetes the Urocystidales were originally **characterized by the presence of haustoria and pores** in the septa of soral hyphae (Bauer et al. 1997). The morphological characterization has discrepancies with molecular data, the latter supporting the inclusion of five families: Doassansiopsidaceae, Floromycetaceae, Glomosporiaceae, Mycosyringaceae, and Urocystidaceae (Fig. 11.6) (Begerow et al. 2006; Vánky et al. 2008b). Doassansiopsidaceae, Floromycetaceae, and Urocystidaceae are characterized by the presence of haustoria and pores in the septa of soral hyphae (Bauer et al. 1997), but these characters are missing in the mature infection stuctures of Mycosyringaceae and Glomosporiaceae (Begerow et al. 2006; Vánky 1996). Additionally, molecular studies do not support the monophyly of Urocystidaceae, Doassansiopsidaceae, or Melanotaeniaeae, which are characterized by the same combination of haustoria and septal pores (Fig. 11.6). Therefore, Melanotaeniaceae is no longer part of the Urocystidales but is in the Ustilaginales (Begerow et al. 2006).

Doassansiopsidaceae shares with Urocystidaceae and Floromycetaceae an essentially identical septal pore apparatus (Fig. 11.5b) (Bauer et al. 1997). It is composed of a simple pore with two outer tripartite membrane caps and two inner non-membranous plates (Fig. 11.5b) (Bauer et al. 1995a, 1997, 2008; Vánky et al. 2008b). The species of *Doassansiopsis*, the only genus of Doassansiopsidaceae, possess complex teliospore balls. A central mass of pseudoparenchymatous cells is surrounded by a layer of firmly adhering, lightly coloured teliospores and an external cortex of sterile cells (Piątek et al. 2008; Vánky 1987). *Doassansiopsis* species form gastroid holobasidia and yeast anamorphs without ballistoconidia. The position of *Doassansiopsis* in Urocystidales is surprising. Based on teliospore ball morphology and the parasitism of aquatic plants, *Doassansiopsis* was grouped with *Burillia*, *Doassansia*, *Heterodoassansia*, *Nannfeldtiomyces*, *Narasimhania*, *Pseudodoassansia*, and *Tracya* (Vánky 1987, 1994). However, both ultrastructural and molecular analyses show that *Doassansiopsis* is not closely related to the other complex teliospore-ball-forming taxa (Fig. 11.8) (Bauer et al. 1997; Begerow et al. 1997, 2006).

Floromycetaceae includes species that parasitize various members of Asparagaceae. Within this family, haustoria and septal pores in soral hyphae are present. The genus *Antherospora* forms single spores in the anthers of the host plant (Bauer et al. 2008), whereas *Floromyces* forms spore balls in flowers. The germination of the teliospores of both genera results in phragmobasidia with sterigmata (Vánky et al. 2008b).

The family Glomosporiaceae experienced a reclassification on the basis of molecular data (Begerow et al. 2006). Originally it was included in the Ustilaginales because intracellular hyphae are formed in the host interaction (Bauer et al. 1997). *Glomosporium* and *Tothiella* were identified as synonyms of *Thecaphora* (Vánky et al. 2007, 2008a). *Thecaphora* species parasitize eudicots and display light brown teliospore balls that differ in the amount of spores (Fig. 11.1j). In the majority of species, these spore balls only consist of fertile cells, in contrast to other families within Urocystidales (Fig. 11.8). The balls vary in their integrity; in some species the balls are strongly agglutinated, whilst in other species they separate easily. Moreover, there are species that have single spores, for example *T. thlaspeos* (Beck) Vánky (Vánky et al. 2007, 2008a). Teliospore germination among species of *Thecaphora* is variable, ranging from true holobasidia to aseptate or septate hyphae that sometimes bear basidiospores (Ingold 1987c; Kochman 1939; Nagler 1986; Piepenbring and Bauer 1995). We interpret these hyphal germinations as atypical germinations resulting possibly from non-optimal environmental conditions. For example, both germination types (i.e. phragmo- and holobasidia) have been reported for *Thecaphora haumanii* Speg. (Piepenbring and Bauer 1995).

Mycosyringaceae is represented by a single genus, *Mycosyrinx*. Its host range is restricted to members of Vitaceae (Vánky 1996, 2012). The teliospores come in pairs. Germination, only known from *M. cissi* (DC.) G. Beck, results directly in basidiospores with a sigmoid shape (Fig. 11.3e) (Piepenbring and Bauer 1995; Vánky 1996). The basidia seem to be small or reduced, and the meiosporangium is represented by the teliospore. The fungus does not form haustoria or intracellular hyphae in host cells (Bauer et al. 1997). At maturity, soral hyphae lack septal pores (Fig. 11.5e) (Begerow et al. 2006).

Urocystidaceae comprises morphologically diverse species with coloured teliospores in flowers or leaves and stems (Fig. 11.1k, l). The genera *Flamingomyces*, *Melanustilospora*, and *Vankya* have single teliospores. The separation of the genera is based on the results of morphological or molecular data (Bauer et al. 2007;

Denchev 2003; Ershad 2000). The genus *Melanoxa* also has single teliospores, but the wall of the teliospores is multilamellate (Lutz et al. 2011). *Mundkurella* is characterized by one- to four-celled teliospores, and *Urocystis* and *Ustacystis* by teliospores that are united in balls with fertile and sterile cells (Vánky 1987, 1994, 2012). The teliospore germination within Urocystidaceae is also diverse. *Flamingomyces* germinates with a single hypha; *Mundkurella*, *Ustacystis*, and *Vankya* (Vánky 2012; Zundel 1945) germinate with phragmobasidia, whereas *Urocystis* germinates with holobasidia (Fig. 11.3d) (Ingold 1999). The members of Urocystidaceae form a yeast-like anamorph without ballistoconidia.

With the advent of molecular systematics, the evolutionary trends in Urocystidales became less obvious. Urocystidales includes **sporeball-forming as well as single-spore-bearing taxa,** and neither sporeball formation nor basidial morphology provides a clear distinction between the different lineages as in the Georgefischeriales or Exobasidiales (Bauer et al. 2001a; Begerow et al. 2002). Given the size and diversity of the group, further studies are needed to understand the ecology and evolution that resulted in morphological variation during the radiations within Urocystidales.

b) Ustilaginales

Poreless septa characterize the Ustilaginales in general (Figs. 11.5e and 11.6). Most of the species sporulate in the reproductive parts of their hosts (Fig. 11.1m–p), and teliosporogenesis occurs by disarticulation. A prominent **gelatinization of hyphal walls** usually **precedes teliospore formation** (Luttrell 1987; Mims and Snetselaar 1991; Mims et al. 1992; Snetselaar and Mims 1994; Snetselaar and Tiffany 1990). They have darkly coloured teliospores and usually germinate with four-celled phragmobasidia (Fig. 11.3a–c). Depending on the species and sometimes on the environmental conditions, **phragmobasidia vary in morphology** (Ingold 1983, 1987a, 1989a, 1989c). Previously, a basal dichotomy in Ustilaginales was accepted at the family level, i.e. Glomosporiaceae and Ustilaginaceae. The system according to Bauer et al. (1997) was based on morphological and anatomical apomorphies and suggested a subdivision into Glomosporiaceae, Mycosyringaceae, and Ustilaginaceae (including Anthracoideaceae and Websdaneaceae) (Bauer et al. 1997). However, this grouping was incongruent with molecular phylogenies favouring a dichotomy between Melanotaeniaceae and Ustilaginaceae in the Ustilaginales and Glomosporiaceae and Mycosyringaceae as part of the Urocystidales (Begerow et al. 1997). The split of Ustilaginaceae s.l. on Poales in favour of three families on different plant families suggests a host specificity of monophyletic lineages, which is not supported by most phylogenetic analyses (Begerow et al. 1997, 2000; Stoll et al. 2005). Thus, the **systematics of Ustilaginales is far from settled,** and our grouping reflects ongoing discussion. **Based on a combination of morphology, host specificity,** and **LSU sequence analyses the Ustilaginales are grouped into four families** (Figs. 11.6 and 11.7) (Begerow et al. 2006). Several additional, mostly monotypic, families have been proposed based on either morphological specialities or host range (Denchev 1997; Vánky 2000, 2001, 2003). For some species like *Melanopsichium* or *Dermatosorus* it can be shown that the proposed apomorphies do not provide additional systematic information (cf. Fig. 11.8), and therefore we follow the concept proposed by Begerow et al. (2006). The families of the Ustilaginales are characterized by host specificity on the family level or higher, i.e. eudicots for the Melanotaeniaceae, Anathriaceae and Restionaceae for the Websdaneaceae, Cyperaceae and Juncaceae for the Anthracoideaceae, and Poaceae for the Ustilaginaceae, thereby ignoring the fact that host jumps to distantly related hosts occurred several times, e.g. *Melanopsichium* or *Pericladium.* Vánky (2011) argued on the basis of a germination that resembles holobasidia and the isolated molecular position of *Pericladium* to establish a new family, Pericladiaceae. However, as long as a comprehensive molecular analysis presenting clear family concepts for the whole order is still lacking, we treat several genera in a preliminary state as *incertae sedis.* At the present state of knowledge, we propose the following families (Fig. 11.6).

The first family, which was excluded from Ustilaginaceae sensu Bauer et al. 1997, was Anthracoideaceae (Denchev 1997). Species of *Anthracoidea* present a unique type of two-celled basidia (Fig. 11.3c) and almost exclusively parasitize species of *Carex*. They exhibit an expanding element in their LSU sequence, which complicates their alignment with other smut species (Hendrichs et al. 2005). In molecular analyses there is no clear separation between *Anthracoidea* species and *Cintractia*-like smuts (Figs. 11.1n and 11.3b). Therefore, one family of smuts on Juncaceae and Cyperaceae was proposed (Begerow et al. 2006). In addition to the common host group, they are morphologically and ecologically similar, often presenting a whitish peridium in immature sori, which are produced in flowers or inflorescences (Fig. 11.8). Based on molecular data, members of *Anthracoidea*, *Cintractia*, *Dermatosorus*, *Farysia*, *Farysizyma*, *Heterotolyposporium*, *Leucocintractia*, *Moreaua*, *Parvulago*, *Pilocintractia*, *Planetlla*, *Portalia*, *Schizonella*, *Stegocintractia*, *Tolyposporium*, *Trichocintractia*, and *Ustanciosporium* are included (Begerow et al. 2006). Consequently, Cintractiaceae, Dermatosoraceae, and Farysiaceae (Vánky 2001) are rejected because they are interspersed in Anthracoideaceae.

Melanotaeniaceae is represented by *Melanotaenium* on eudicots (Fig. 11.1o) and *Exoteliospora* on *Osmunda*. Previous members on Poacecae have been excluded based on morphological and molecular data and are now part of the Georgefischeriales (Begerow et al. 2001). In contrast to the other three families, members of Melanotaeniaceae are characterized by simple septal pores with membrane caps and by the development of haustoria (Fig. 11.6) (Bauer et al. 1997; Begerow et al. 2006).

Ustilaginaceae comprises the large genera *Ustilago* and *Sporisorium* and several smaller genera of species previously treated as *Ustilago* or *Sporisorium*, representing a large *Ustilago*–*Sporisorium*–*Macalpinomyces* complex with more than 550 species. Except for *Eriomoeszia*, *Melanopsichium*, and *Pericladium*, all species parasitize Poaceae. *Melanopsichium pennsylvanicum* Hirschh., which occurs on Polygonaceae, is well embedded in the supported group of the Ustilaginaceae (Fig. 11.7). This indicates that jumps to distantly related hosts occasionally occur. However, no further radiations on Polygonaceae took place, which supports the important adaptation of the Ustilaginaceae to hosts of Poaceae. Several molecular studies have shown that the separation of *Ustilago* and *Sporisorium* is very difficult on the basis of hitherto used features (Stoll et al. 2003, 2005). Some genera have been proposed to accommodate species with clear apomorphies like *Eriomoeszia*, *Anomalomyces*, *Portalia*, or *Tubisorus* (Gonzales et al. 2007; Vánky 2005; Vánky and Lutz 2011; Vánky et al. 2006), but a clear structure of the group was lacking. Most recently, a four-gene phylogeny, combined with detailed studies on sorus morphology, revealed some monophyletic groups that could be excluded from the large *Ustilago*–*Sporisorium*–*Macalpinomyces* complex (McTaggart et al. 2012a, b). Based on these data, *Anthracocystis* was reestablished and *Langdonia*, *Stollia*, and *Triodiomyces* were newly described to accommodate the well-characterized monophyletic groups, together with *Ustilago* and *Sporisorium* (McTaggart et al. 2012c). Besides the host specificity of some groups, the genera are based mainly on characteristics of teliospores and sori, e.g. teliospores free or united in balls and the presence or absence of peridia, columellae, sterile cells, or sterile hyphae (see Vánky 1987, 1994). The sori of *Sporisorium* species are also covered by peridia, but these can be composed of host tissue or fungal hyphae. The teliospores are free or arranged in balls. Teliospore balls and special soral structures are lacking in *Ustilago* species, whose simple teliospores develop by replacing host organs, at least partially.

Websdaneaceae includes *Websdanea* and *Restiosporium*, both of which occur on Anarthriaceae and Restionaceae. This group is well supported in several molecular phylogenetic analyses (Begerow et al. 2006). Morphologically, they are very similar to members of Anthracoideaceae, but LSU sequence data support a sister group relationship with the other members of Ustilaginales on grasses and grass-like hosts (Fig. 11.7).

Based mainly on host relationships, Vánky (2001) created the Clintamraceae for *Clintamra*, Geminaginaceae for *Geminago* and Uleiellaceae for *Uleiella*. Unfortunately, molecular data are not available for these genera, and it is unclear whether these genera represent recent or ancient host jumps. Because there is no other support for these families at the moment, we treat them as *incertae sedis*, together with other genera lacking molecular data and clear morphological characteristics to place them in one of the described groups.

IV. Conclusions

The history of smut systematics dates back to the brothers Tulasne, who separated holobasidiate and phragmobasidiate groups for the first time (Tulasne and Tulasne 1847). This grouping was consistent for more than 100 years and to our knowledge was never questioned. Differences in the sugar composition of the cell wall of *Ustilago* and *Microbotryum* yeasts implicated a separation of smuts on monocots and dicots (Prillinger et al. 1993), but subsequent data did not support this hypothesis. The discovery of **ultrastructural markers in the host–parasite interaction and septal formation provided apomorphic characters to delimit monophyletic groups** which were supported by molecular analyses (Figs. 11.6 and 11.7) (Bauer et al. 1997; Begerow et al. 1997, 2006). Thus, the analysis of two characters remains to be discussed: basidia and host specificity. Their analyses reveal novel conclusions about the evolution of Ustilaginomycotina.

A. Basidia

The evolutionary transitions to the basidium of the Ustilaginomycotina are unknown. Nevertheless, a tentative sequence can be outlined from the distribution of the basidial types among the different groups. While the Agaricomycotina are dominated by the presence of holobasidia, and the Pucciniomycotina have almost exclusively phragmobasidia, the Ustilaginomycotina are somewhat intermediate, having both types in several groups (Fig. 11.3). The monophyly of the Agaricomycotina with the Ustilaginomycotina and the group's common ancestor with the Pucciniomycotina suggest that the **plesiomorphic state of the basidium was phragmobasidiate.** In the Ustilaginomycetes and Exobasidiomycetes, however, phragmobasidia occur only in the Anthracoideaceae, Urocystidaceae, Ustilaginaceae, Tilletiariaceae, and Websdaneaceae. Except for a few species, the phragmobasidial taxa of the Ustilaginomycetes and Exobasidiomycetes are concentrated in a single monophyletic group of the Ustilaginales, whereas the holobasidial taxa are distributed throughout all orders of the Ustilaginomycetes and Exobasidiomycetes. In addition, the early-diverged lineages of the Ustilaginomycetes and Exobasidiomycetes, i.e. Melanotaeniaceae, Glomosporiaceae, Tilletiales, and Exobasidiales, are holobasidiate. This distribution of basidial types supports a holobasidiate ancestor of the Ustilaginomycetes and Exobasidiomycetes. Consequently, the **septation of the basidia in several families must be interpreted as the result of convergent evolution.** Apart from septation, the hilar appendices responsible for the active discharge of basidiospores are unevenly distributed. Though they are present in several families of the Exobasidiomycetes showing various orientations, they seem to be absent in the Ustilaginomycetes. Hence, the **gastroid basidia** of Ustilaginomycetes might represent an apomorphy of this monophyletic group.

B. Host Specificity

Following the reorganization of the Ustilaginomycotina systematics on the basis of phylogenetic data, it became evident that **most species were highly host-specific.** Moreover, **monophyletic lineages are often restricted to monophyletic host groups** (Begerow et al. 2004). As partly discussed earlier, within the Ustilaginomycotina there are evident examples of **coevolution** with angiosperm lineages (e.g. Tilletiales, Georgefischeriales, and the *Ustilago-Sporisorium* complex with Poaceae, *Graphiola* with palms, *Anthracoidea* with Cyperaceae, *Mycosyrinx* with Vitaceae, *Exobasidium* with Ericales). On the other hand, Doassansiales and *Doassansiopsis* are two excellent examples of evolution within a given ecosystem.

As a whole, the host range of Ustilaginomycotina is restricted to angiosperms, with a few exceptions on gymnosperms and ferns, which are regarded as the result of host jumps (Bauer et al. 1997; Begerow et al. 2004). Most Ustilaginomycotina members are parasites of monocots, especially members of Poaceae and Cyperaceae. This host distribution suggests that the **Ustilaginomycotina** species **may have evolved as pathogens, either on early angiosperms or on early monocots,** with subsequent jumps to eudicots. Given the relative age of the stem group of Ustilaginomycotina of at least 300 million years (Taylor and Berbee 2006), it seems most likely that ancestral lineages date back before the radiation of angiosperms. Thus, the present specificity of some lineages could be the result of massive extinctions during evolution. In contrast, broad host ranges of some groups, e.g. the Georgefischeriales on Poaceae with a few species on Convolvulaceae and Cyperaceae, the Tilletiales on Poaceae with five species on Leguminosae, and the Ustilaginaceae on monocots with a few genera on eudicots, indicate not only that the ancestors of Ustilaginomycotina have undergone periods of parallel evolution with their hosts, but host jumps may have stimulated the evolution of a large number of taxa.

Thus, **host specificity** seen in genera like *Ustilago* and *Tilletia* on Poaceae (Stoll et al. 2005) or *Entyloma* on asterids (Begerow et al. 2002) might be a **result of adaptive radiation**. Branch lengths of molecular analyses suggest radiations in these genera, which are younger than some 100–200 million years old.

C. Evolutionary Trends

Finally, our analyses suggest the following evolutionary trends within Ustilaginomycotina:

- Cellular interactions from simple to complex forms;
- Multiple convergent evolution of intracellular fungal elements;
- Multiple convergent evolution of spore balls;
- Repeated loss of septal pores in senescent hyphae;
- Repeated loss of teliospores as propagules;
- Multiple convergent evolution of gastroid taxa;
- Repeated loss of ballistosporic mechanism;
- Repeated change of sorus location from vegetative organs to flowers;
- Multiple convergent evolution of sporulation in anthers;
- Coevolution with host groups, but also with ecosystems;
- Repeated jumps to unrelated hosts.

Acknowledgements We thank Alistair McTaggart (Agri-Science Queensland, Australia) for his critical reading of the manuscript, Lori Carris (Washington State University, USA) and Hans Henrik Bruun (Center for Macroecology, Evolution and Climate, Denmark) for providing photographs of *T. controversa* and *M. endogenum* for Fig. 11.1, and Meike Piepenbring (Goethe Universität, Germany) for several drawings for Fig. 11.3. Kálmán Vánky (HUV, Germany) is acknowledged for many specimens and the Deutsche Forschungsgemeinschaft for financial support.

References

Abdel-Motaal FF, El-Zayat SA, Kosaka Y, El-Sayed MA, Nassar MSM, Ito S (2009) Four novel ustilaginomycetous anamorphic yeast species isolated as endophytes from the medicinal plant *Hyoscyamus muticus*. Asian J Plant Sci 8:526–535

Aime MC, Toome M, McLaughlin DJ (2014) Pucciniomycotina. In: McLaughlin DJ, Spatafora JW (eds) Systematics and evolution. Springer, Heidelberg

Bandoni RJ, Johri BN (1972) *Tilletiaria*: a new genus in the Ustilaginales. Can J Bot 50:39–43

Bao X, Carris LM, Huang G (2010) *Tilletia puccinelliae*, a new species of reticulate-spored bunt fungus infecting *Puccinellia distans*. Mycologia 120:613–623. doi:10.3852/09-135

Baric S, Lindner L, Marschall K, Dalla Via J (2010) Haplotype diversity of *Tilletiopsis* spp. causing white haze in apple orchards in Northern Italy. Plant Pathol 59:535–541. doi:10.1111/j.1365-3059.2009.02217.x

Bauer R, Mendgen K, Oberwinkler F (1995a) Septal pore apparatus of the smut *Ustacystis waldsteiniae*. Mycologia 87:18–24

Bauer R, Mendgen K, Oberwinkler F (1995b) Cellular interaction of the smut fungus *Ustacystis waldsteiniae*. Can J Bot 73:867–883

Bauer R, Oberwinkler F, Vánky K (1997) Ultrastructural markers and systematics in smut fungi and allied taxa. Can J Bot 75:1273–1314

Bauer R, Oberwinkler F, Vánky K (1999a) Ustilaginomycetes on *Osmunda*. Mycologia 91:669–675

Bauer R, Vánky K, Begerow D, Oberwinkler F (1999b) Ustilaginomycetes on *Selaginella*. Mycologia 91:475–484

Bauer R, Begerow D, Nagler A, Oberwinkler F (2001a) The Georgefischeriales: a phylogenetic hypothesis. Mycol Res 105:416–424

Bauer R, Begerow D, Oberwinkler F, Piepenbring M, Berbee ML (2001b) Ustilaginomycetes. In: McLaughlin DJ, McLaughlin EG, Lemke PA (eds)

The Mycota, vol 7B, Systematics and evolution. Springer, Berlin, pp 57–83
Bauer R, Lutz M, Oberwinkler F (2005) *Gjaerumia*, a new genus in the Georgefischeriales (Ustilaginomycetes). Mycol Res 109:1250–1258
Bauer R, Begerow D, Sampaio JP, Weiss M, Oberwinkler F (2006) The simple-septate basidiomycetes: a synopsis. Mycol Prog 5:41–66
Bauer R, Lutz M, Piątek M, Vánky K, Oberwinkler F (2007) *Flamingomyces* and *Parvulago*, new genera of marine smut fungi (Ustilaginomycotina). Mycol Res 111:1199–1206
Bauer R, Lutz M, Begerow D, Piątek M, Vánky K, Bacigálová K, Oberwinkler F (2008) Anther smut fungi on monocots. Mycol Res 112:1297–1306
Begerow D, Bauer R, Oberwinkler F (1997) Phylogenetic studies on nuclear large subunit ribosomal DNA sequences of smut fungi and related taxa. Can J Bot 75:2045–2056
Begerow D, Bauer R, Boekhout T (2000) Phylogenetic placements of ustilaginomycetous anamorphs as deduced from nuclear LSU rDNA sequences. Mycol Res 104:53–60
Begerow D, Bauer R, Oberwinkler F (2001) *Muribasidiospora*: Microstromatales or Exobasidiales? Mycol Res 105:798–810
Begerow D, Bauer R, Oberwinkler F (2002) The Exobasidiales: an evolutionary hypothesis. Mycol Prog 1:187–199
Begerow D, Göker M, Lutz M, Stoll M (2004) On the evolution of smut fungi on their hosts. In: Agerer R, Piepenbring M, Blanz P (eds) Frontiers in basidiomycote mycology. IHW, Echingen, German, pp 81–98
Begerow D, Stoll M, Bauer R (2006) A phylogenetic hypothesis of Ustilaginomycotina based on multiple gene analyses and morphological data. Mycologia 98:906–916
Bellés JM, Garro R, Pallás V, Fayos J, Rodrigo I, Conejero V (2006) Accumulation of gentisic acid as associated with systemic infections but not with the hypersensitive response in plant-pathogen interactions. Planta 223:500–511
Boekhout T (2011) *Tilletiopsis* Derx ex Derx (1930). In: Kurtzman CP, Fell JW, Boekhout T (eds) The yeasts: a taxonomic study, 5th edn. Elsevier, London, pp 2003–2014
Boekhout T, Theelen B, Houbraken J, Robert V, Scorzetti G, Gafni A, Gerson U, Sztejnberg A (2003) Novel anamorphic mite-associated fungi belonging to the Ustilaginomycetes: *Meira geulakonigii* gen. nov., sp. nov., *Meira argovae* sp. nov. and *Acaromyces ingoldii* gen. nov., sp. nov. Int J Syst Evol Microbiol 53:1655–1664
Boekhout T, Gildemacher P, Theelen B, Müller WH, Heijne B, Lutz M (2006) Extensive colonization of apples by smut anamorphs causes a new postharvest disorder. FEMS Yeast Res 6:63–76
Boekhout T, Guého E, Mayser P, Velegraki A (2010) *Malassezia* and the skin. Science and clinical practice. Springer, Berlin
Boekhout T, Fonseca A, Sampaio JP, Bandoni RJ, Kwon-Chung KJ (2011) Discussion of teleomorphic and anamorphic basidiomycetous yeasts. In: Kurtzman CP, Fell JW, Boekhout T (eds) The yeasts: a taxonomic study, 5th edn. Elsevier, London, pp 1339–1372
Branda E, Turchetti B, Diolaiuti G, Pecci M, Smiraglia C, Buzzini P (2010) Yeast and yeast-like diversity in the southernmost glacier of Europe (Calderone Glacier, Apennines, Italy). FEMS Microbiol Ecol 72:354–369
Brefeld O (1883) Botanische Untersuchungen über Hefepilze. 5. Die Brandpilze I (Ustilagineen). A. Felix, Leipzig, Germany
Brefort T, Doehlemann G, Mendoza-Mendoza A, Reissmann S, Djamei A, Kahmann R (2009) *Ustilago maydis* as a pathogen. Annu Rev Phytopathol 47:423–445
Butin H, Peredo HL (1986) Hongos parásitos en coníferas de América del Sur, con especial referencia a Chile. Biblioth Mycol 101:1–100
Cai L, Giraud T, Zhang N, Begerow D, Cai G, Shivas RG (2011) The evolution of species concepts and species recognition criteria in plant pathogenic fungi. Fungal Divers 50:121–133
Carris LM, Castlebury LA, Goates BJ (2006) Nonsystemic bunt fungi-*Tilletia indica* and *T. horrida*: a review of history, systematics, and biology. Annu Rev Phytopathol 44:113–133
Castlebury LA, Carris LM, Vánky K (2005) Phylogenetic analysis of *Tilletia* and allied genera in order Tilletiales (Ustilaginomycetes; Exobasidiomycetidae) based on large subunit nuclear rDNA sequences. Mycologia 97:888–900
Chamnanpa T, Limtong P, Srisuk N, Limtong S (2013) *Pseudozyma vetiver* sp. nov., a novel anamorphic ustilaginomycetous yeast species isolated from the phylloplane in Thailand. Antonie Von Leeuwenhoek 104(5):637–44
Chandra A, Huff DR (2008) *Salmacisia*, a new genus of Tilletiales: reclassification of *Tilletia buchloeana* causing induced hermaphroditism in buffalograss. Mycologia 100:81–93
Christensen JJ (1963) Corn smut caused by *Ustilago maydis*. Am Phytopathol Soc Monogr 2, 41 pp
Cole GT (1983) *Graphiola phoenicis*: a taxonomic enigma. Mycologia 75:93–116
Connell L, Redman R, Craig S, Scorzetti G, Iszard M, Rodriguez R (2008) Diversity of soil yeasts isolated from South Victoria Land, Antarctica. Microb Ecol 56:448–459
Cunningham JL, Bakshi BK, Lentz PL (1976) Two new genera of leaf-parasitic fungi (Basidiomycetidae: Brachybasidiaceae). Mycologia 68:640–654

De Bary A (1884) Vergleichende Morphologie und Biologie der Pilze, Mycetozoen und Bacterien. Verlag W. Engelmann, Leipzig, Germany

De Beer ZW, Begerow D, Bauer R, Pegg GS, Crous PW, Wingfield MJ (2006) Phylogeny of the Quambalariaceae fam nov., including important eucalyptus pathogens in South Africa and Australia. Stud Mycol 55:289–298

De Vienne DM, Refrégier G, Hood ME, Guigue A, Devier B, Vercken E, Smadja C, Deseille A, Giraud T (2009) Hybrid sterility and inviability in the parasitic fungal species complex *Microbotryum*. J Evol Biol 22:683–698

Denchev CM (1997) Anthracoideaceae, a new family in the Ustilaginales. Mycotaxon 64:411–417

Denchev CM (2003) *Melanustilospora*, a new genus in the Urocystales (smut fungi). Mycotaxon 87:475–477

Denchev CM, Denchev TT (2011) *Anthracoidea melanostachyae* sp. nov. (*Anthracoideaceae*). Mycol Balc 8:153–155

Djamei A, Schipper K, Rabe F, Ghosh A, Vincon V, Kahnt J, Osorio S, Tohge T, Fernie AR, Feussner I, Feussner K, Meinicke P, Stierhof YD, Schwarz H, Macek B, Mann M, Kahmann R (2011) Metabolic priming by a secreted fungal effector. Nature 478:395. doi:10.1038/nature10454

Doehlemann G, van der Linde K, Aßmann D, Schwammbach D, Hof A, Mohanty A, Jackson D, Kahmann R (2009) Pep1, a secreted effector protein of *Ustilago maydis*, is required for successful invasion of plant cells. PLoS Pathog 5(2):e1000290. doi:10.1371/journal.ppat.1000290

Donk MA (1956) The generic names proposed for hymenomycetes-VI. Brachybasidiaceae, Cryptobasidiaceae, Exobasidiaceae. Reinwardtia 4:113–118

Durán R (1973) Ustilaginales. In: Ainsworth GC, Sparrow FK, Sussman AS (eds) The fungi, vol 4B. Academic, New York, pp 281–300

Ershad D (2000) *Vankya*, a new genus of smut fungi. Rostaniha 1:65–72

Fell JW, Statzell-Tallman A, Scorzetti G, Gutiérrez MH (2011) Five new species of yeasts from fresh water and marine habitats in the Florida Everglades. Antonie Van Leeuwenhoek 99:533–549

Filonow AB (2001) Butyl acetate and yeasts interact in adhesion and germination of *Botrytis cinerea* conidia in vitro and in fungal decay of golden delicious apple. J Chem Ecol 27:831–844

Findley K, Oh J, Yang J, Conlan S, Deming C, Meyer JA, Schoenfeld D, Nomicos E, Park M, NIH Intramural Sequencing Center Comparative Sequencing Program, Kong HH, Segre JA (2013) Topographic diversity of fungal and bacterial communities in human skin. Nature 498:367–370

Fischer E (1921) Zur Kenntnis von *Graphiola* and *Farysia*. Ann Mycol 18:118–197

Fischer E (1922) Weitere Beiträge zur Kenntnis der Gattung *Graphiola*. Ann Mycol 20:228–237

Fischer GW, Holton CS (1957) Biology and control of the smut fungi. Ronald Press, New York

Fonseca Á, Inácio J (2006) Phylloplane yeasts. In: Peter G, Rosa C (eds) Biodiversity and ecophysiology of yeasts. Springer, Berlin, pp 263–301

Garcia-Muse T, Steinberg G, Perez-Martin J (2003) Pheromone-induced G2 arrest in the phytopathogenic fungus *Ustilago maydis*. Eurkaryot Cell 2:494–500. doi:10.1128/EC.2.3.494-500.2003

Gäumann E (1922) Über die Gattung *Kordyana* Rac. Ann Mycol 20:257–271

Geiser E, Wierckx N, Zimmermann M, Blank LM (2013) Identification of an endo-1, 4-beta-xylanase of *Ustilago maydis*. BMC Biotechnol 13:59

Gerson U, Gafni A, Paz Z, Sztejnberg A (2008) A tale of three acaropathogenic fungi in Israel: *Hirsutella*, *Meira* and *Acaromyces*. Exp Appl Acarol 46:183–194

Goates BJ, Hoffmann JA (1986) Formation and discharge of secondary sporidia of the bunt fungus, *Tilletia foetida*. Mycologia 78:371–379

Golubev WI (2006) Antagonistic interactions among yeasts. In: Peter G, Rosa C (eds) Biodiversity and ecophysiology of yeasts. Springer, Berlin, pp 197–219

Golubev WI (2007) Mycocinogeny in smut yeast-like fungi of the genus *Pseudozyma*. Microbiology 76:719–722

Golubev WI, Kulakovskaya TV, Shashkov AS, Kulakovskaya EV, Golubev NV (2008) Antifungal cellobiose lipid secreted by the epiphytic yeast *Pseudozyma graminicola*. Microbiology 77:171–175

Gonzales V, Vánky K, Platas G, Lutz M (2007) *Portalia* gen. nov (Ustilaginomycotina). Fungal Divers 27:45–59

Gottschalk M, Blanz PA (1985) Untersuchungen an 5S ribosomalen Ribonucleinsäuren als Beitrag zur Klärung von Systematik und Phylogenie der Basidiomyceten. Z Mycol 51:205–243

Guého E, Boekhout T, Ashbee HR, Guillot J, Van Belkum A, Faergemann J (1998) The role of *Malassezia* species in the ecology of human skin and as pathogens. Med Mycol 36:220–229

Guého-Kellermann E, Boekhout T, Begerow D (2010) Biodiversity, phylogeny and ultrastructure. In: Boekhout T, Guého E, Mayser P, Velegraki A (eds) *Malassezia* and the skin. Science and clinical practice. Springer, Berlin, pp 17–63

Haslam E, Cai Y (1994) Plant polyphenols (vegetable tannins): gallic acid metabolism. Nat Prod Rep 11:41–66

Hawksworth DL (2011) A new dawn for the naming of fungi: impacts of decisions made in Melbourne in July 2011 on the future publication and regulation of fungal names. Mycokeys 1:7–20

Hawksworth DL, Crous PW, Redhead SA, Reynolds DR, Samson RA, Seifert KA, Taylor JW, Wingfield MJ,

Abaci Ö, Aime C, Asan A, Bai F-Y, de Beer ZW, Begerow D, Berikten D, Boekhout T, Buchanan PK, Burgess TI, Buzina W, Cai L, Cannon PF, Crane JL, Damm U, Daniel H-M, van Diepeningen AD, Druzhinina IS, Dyer PS, Eberhardt U, Fell JW, Frisvad JC, Geiser DM, Geml J, Glienke C, Gräfenhan T, Groenewald JZ, Groenewald M, de Gruyter JZ, Guého-Kellermann E, Hibbett DS, Hong S-B, de Hoog GS, Houbraken JAMP, Huhndorf SM, Hyde KD, Ismail A, Johnston PR, Kadaifciler DG, Kirk PM, Kõljalg U, Kurtzman CP, Lagneau P-E, Lévesque CA et al (2011) The Amsterdam Declaration on Fungal Nomenclature. IMA Fungus 2:105–112. doi:10.5598/imafungus.2011.02.01.14

Hendrichs M, Bauer R, Oberwinkler F (2003) The Cryptobasidiaceae of tropical Central and South America. Sydowia 55:33–64

Hendrichs M, Begerow D, Bauer R, Oberwinkler F (2005) The genus *Anthracoidea* (Basidiomycota, Ustilaginales): a molecular phylogenetic approach using LSU rDNA sequences. Mycol Res 109:31–40

Hennings P (1900) Exobasidiineae. In: Engler A, Prantl K (eds) Die natürlichen Pflanzenfamilien 1,1. Verlag von W. Engelmann, Leipzig, Germany, pp 103–105

Hibbett DS, Binder M, Bischoff JF, Blackwell M, Cannon PF, Eriksson OE, Huhndorf S, James T, Kirk PM, Lucking R, Lumbsch TH, Lutzoni F, Matheny PB, McLaughlin DJ, Powell MJ, Redhead S, Schoch CL, Spatafora JW, Stalpers JA, Vilgalys R, Aime MC, Aptroot A, Bauer R, Begerow D, Benny GL, Castlebury LA, Crous PW, Dai YC, Gams W, Geiser DM, Griffith GW, Gueidan C, Hawksworth DL, Hestmark G, Hosaka K, Humber RA, Hyde KD, Ironside JE, Kõljalg U, Kurtzman CP, Larsson KH, Lichtwardt R, Longcore J, Miadlikowska J, Miller A, Moncalvo JM, Mozley-Standridge S, Oberwinkler F, Parmasto E, Reeb V, Rogers JD, Roux C, Ryvarden L, Sampaio P, Schussler A, Sugiyama J, Thorn RG, Tibell L, Untereiner WA, Walker C, Wang Z, Weir A, Weiss M, White MM, Winka K, Yao YJ, Zhang N (2007) A higher-level phylogenetic classification of the Fungi. Mycol Res 111:509–547

Inácio J, Landell MF, Valente P, Wang PH, Wang YT, Yang SH, Manson JS, Lachance MA, Rosa CA, Fonseca Á (2008) *Farysizyma* gen. nov., an anamorphic genus in the Ustilaginales to accommodate three novel epiphytic basidiomycetous yeast species from America, Europe and Asia. FEMS Yeast Res 8:499–508

Ingold CT (1983) The basidium in *Ustilago*. Trans Br Mycol Soc 81:573–584

Ingold CT (1985) *Dicellomyces scirpi*: its conidial stage and taxonomic position. Trans Br Mycol Soc 84:542–545

Ingold CT (1987a) Germination of teliospores in certain smuts. Trans Br Mycol Soc 88:355–363

Ingold CT (1987b) Ballistospores and blastic conidia of *Tilletia ayresii*, and comparison with those of *T. tritici* and *Entyloma ficariae*. Trans Br Mycol Soc 88:75–82

Ingold CT (1987c) Aerial sporidia of *Ustilago hypodytes* and *Sorosporium saponariae*. Trans Br Mycol Soc 89:471–475

Ingold CT (1989a) Basidium development in some species of *Ustilago*. Mycol Res 93:405–412

Ingold CT (1989b) Note on the basidium in the Tilletiaceae. Mycol Res 93:387–389

Ingold CT (1989c) The basidium of *Anthracoidea inclusa* in relation to smut taxonomy. Mycol Res 92:245–246

Ingold CT (1997) Teliospore germination in *Tilletia opaca* and *T. sumatii* and the nature of the tilletiaceous basidium. Mycol Res 101:281–284

Ingold CT (1999) Two types of basidia in *Urocystis hypoxis* and the implications for smut taxonomy. Mycol Res 103:18–20

Juarez-Montiel M, Ruiloba de Leon S, Chavez-Camarillo G, Hernandez-Rodriguez C, Villa-Tanaca L (2011) Huitlacoche (corn smut), caused by the phytopathogenic fungus *Ustilago maydis*, as a functional food. Rev Iberoam Micol 28:69–73. doi:10.1016/j.riam.2011.01.001

Kahmann R, Kämper J (2004) *Ustilago maydis*: how its biology relates to pathogenic development. New Phytol 164:31–42

Kämper J, Kahmann R, Bölker M, Ma LJ, Brefort T, Saville BJ, Banuett F, Kronstad JW, Gold SE, Muller O, Perlin MH, Wosten HA, De Vries R, Ruiz-Herrera J, Reynaga-Pena CG, Snetselaar K, Mccann M, Perez-Martin J, Feldbrügge M, Basse CW, Steinberg G, Ibeas JI, Holloman W, Guzman P, Farman M, Stajich JE, Sentandreu R, Gonzalez-Prieto JM, Kennell JC, Molina L, Schirawski J, Mendoza-Mendoza A, Greilinger D, Munch K, Rossel N, Scherer M, Vranes M, Ladendorf O, Vincon V, Fuchs U, Sandrock B, Meng S, Ho EC, Cahill MJ, Boyce KJ, Klose J, Klosterman SJ, Deelstra HJ, Ortiz-Castellanos L, Li W, Sanchez-Alonso P, Schreier PH, Hauser-Hahn I, Vaupel M, Koopmann E, Friedrich G, Voss H, Schluter T, Margolis J, Platt D, Swimmer C, Gnirke A, Chen F, Vysotskaia V, Mannhaupt G, Guldener U, Munsterkotter M, Haase D, Oesterheld M, Mewes HW, Mauceli EW, Decaprio D, Wade CM, Butler J, Young S, Jaffe DB, Calvo S, Nusbaum C, Galagan J, Birren BW (2006) Insights from the genome of the biotrophic fungal plant pathogen *Ustilago maydis*. Nature 444:97–101

Kellner R, Vollmeister E, Feldbrügge M, Begerow D (2011) Interspecific sex in grass smuts and the genetic diversity of their pheromone-receptor system. PLoS Genet 7:e1002436. doi:10.1371/journal.pgen.1002436

Kirk PM, Cannon PF, Minter DW, Stalpers JA (2008) Dictionary of fungi, 10th edn. CABI, Wallingford

Kitamoto D, Yanagishita H, Shinbo T, Nakane T, Kamisawa C, Nakahara T (1993) Surface active properties and antimicrobial activities of mannosylerythritol lipids as biosurfactants produced by *Candida antarctica*. J Biotechnol 29:91–96

Kochman J (1939) Beitrag zur Kenntnis der Brandpilzflora Polens II. Acta Soc Bot Pol 16:53–67

Kreisel H (1969) Grundzüge eines natürlichen Systems der Pilze. Cramer, Vaduz

Kukkonen I, Raudaskoski M (1964) Studies on the probable homothallism and pseudohomothallism in the genus *Anthracoidea*. Ann Bot Fenn 1:257–271

Kurtzman CP, Fell JW, Boekhout T (2011) The yeasts: a taxonomic study. Elsevier, London

Kurtzman CP, Sugiyama J (2014) Taphrinomycotina and Saccharomycotina. In: McLaughlin DJ, Spatafora JW (eds) The Mycota, vol. 7B, 2nd edn, Systematics and evolution. Springer, Berlin

Kushnir L, Paz Z, Gerson U, Sztejnberg A (2011) The effect of three basidiomycetous fungal species on soil-borne, foliage and fruit-damaging phytopathogens in laboratory experiments. BioControl 56:799–810

Lendner A (1920) Un champignon parasite sur une Lauracée du genre *Ocotea*. Bull Soc Bot Genève 12:122–128

Lin SS, Pranikoff T, Smith SF, Brandt ME, Gilbert K, Palavecino EL, Shetty AK (2008) Central venous catheter infection associated with *Pseudozyma aphidis* in a child with short gut syndrome. J Med Microbiol 57:516–518

Liou GY, Wei YH, Lin SJ, Wen CY, Lee FL (2009) *Pseudozyma pruni* sp. nov., a novel ustilaginomycetous anamorphic fungus from flowers in Taiwan. Int J Syst Evol Microbiol 59:1813–1817

Luttrell ES (1987) Relations of hyphae to host cells in smut galls caused by species of *Tilletia*, *Tolyposporium*, and *Ustilago*. Can J Bot 65:2581–2591

Lutz M, Vánky K, Bauer R (2011) *Melanoxa*, a new genus in the Urocystidales (Ustilaginomycotina). Mycol Prog 11:149–158. doi:10.1007/s11557-010-0737-7

Lutz M, Vánky K, Piątek M (2012) Shivasia gen. nov. for the Australasian smut Ustilago solida that historically shifted through five different genera. IMA Fungus 3:143–154

Mahdi LE, Statzell-Tallman A, Fell JW, Brown MV, Donachie SP (2008) *Sympodiomycopsis lanaiensis* sp. nov., a basidiomycetous yeast (Ustilaginomycotina: Microstromatales) from marine driftwood in Hawai'i. FEMS Yeast Res 8:1357–1363

Malençon G (1953) Le *Coniodyctium chevalieri* Har. et Pat., sa nature et ses affinitiés. Bull Soc Myc Fr 69:77–100

Matheny PB, Gossmann JA, Zalar P, Kumar TKA, Hibbett DS (2006) Resolving the phylogenetic position of the Wallemiomycetes: an enigmatic major lineage of Basidiomycota. Can J Bot 84: 1794–1805

Mathre DE (1996) Dwarf bunt: politics, identification, and biology. Annu Rev Phytopathol 34:67–85

Maublanc MA (1914) Les genres *Drepanoconis* Schr. et Henn. et *Clinoconidium* Pat.: leur structure et leur place dans la classification. Bull Soc Myc Fr 30:444–446

McTaggart AR, Shivas RG, Geering ADW, Callaghan B, Vánky K, Scharaschkin T (2012a) Soral synapomorphies are significant for the systematics of the *Ustilago-Sporisorium-Macalpinomyces* complex (*Ustilaginaceae*). Persoonia 29:63–77

McTaggart AR, Shivas RG, Geering ADW, Vánky K, Scharaschkin T (2012b) A review of the *Ustilago-Sporisorium-Macalpinomyces* complex. Persoonia 29:55–62

McTaggart AR, Shivas RG, Geering ADW, Vánky K, Scharaschkin T (2012c) Taxonomic revision of *Ustilago*, *Sporisorium* and *Macalpinomyces*. Persoonia 29:116–132

Mimee B, Labbé C, Pelletier R, Bélanger RR (2005) Antifungal activity of flocculosin, a novel glycolipid isolated from *Pseudozyma flocculosa*. Antimicrob Agents Chemother 49:1597–1599

Mims CW, Richardson EA (2007) Light and electron microscopic observation of the infection of *Camellia sasanqua* by the fungus *Exobasidium camelliae* var. *gracilis*. Can J Bot 85:175–183

Mims CW, Snetselaar KM (1991) Teliospore maturation in the smut fungus *Sporisorium sorghi*: an ultrastructural study using freeze substitution fixation. Bot Gaz 152:1–7

Mims CW, Richardson EA, Roberson RW (1987) Ultrastructure of basidium and basidiospore development in three species of the fungus *Exobasidium*. Can J Bot 65:1236–1244

Mims CW, Snetselaar KM, Richardson EA (1992) Ultrastructure of the leaf stripe smut fungus *Ustilago striiformis*: host-pathogen relationship and teliospore development. Int J Plant Sci 153:289–290

Müller B (1989) Chemotaxonomische Untersuchungen an Basidiomycetenhefen. Thesis, University of Tübingen, Tübingen, Germany

Nagler A (1986) Untersuchungen zur Gattungsabgrenzung von *Ginanniella* Ciferri und *Urocystis* Rabenhorst sowie zur Ontogenie von *Thecaphora seminis-convolvuli* (Desm.) Ito. Thesis, University of Tübingen, Tübingen, Germany

Nagler A, Bauer R, Berbee M, Vánky K, Oberwinkler F (1989) Light and electron microscopic studies of *Schroeteria delastrina* and *S. poeltii*. Mycologia 81:884–895

Nannfeldt JA (1981) *Exobasidium*, a taxonomic reassessment applied to the European species. Symb Bot Upsal 23:1–71

Oberwinkler F (1977) Das neue System der Basidiomyceten. In: Frey W, Hurka H, Oberwinkler F (eds) Beiträge zur Biologie der niederen Pflanzen. Gustav Fischer, Stuttgart, pp 59–105

Oberwinkler F (1978) Was ist ein Basidiomycet? Z Mykol 44:13–29
Oberwinkler F (1982) The significance of the morphology of the basidium in the phylogeny of basidiomycetes. In: Wells K, Wells EK (eds) Basidium and basidiocarp. Springer, Berlin, Heidelberg, New York, pp 9–35
Oberwinkler F (1985) Zur Evolution und Systematik der Basidiomyceten. Bot Jahrb Syst 107:541–580
Oberwinkler F (1987) Heterobasidiomycetes with ontogenetic yeast stages-systematic and phylogenetic aspects. Stud Mycol 30:61–74
Oberwinkler F (1993) Diversity and phylogenetic importance of tropical heterobasidiomycetes. In: Isaac S, Frankland JC, Watling R, Whalley AJS (eds) Aspects of tropical mycology. Cambridge University Press, Cambridge, UK, pp 121–147
Oberwinkler F, Bandoni RJ, Blanz P, Deml G, Kisimova-Horovitz L (1982) Graphiolales: basidiomycetes parasitic on palms. Plant Systemat Evol 140:251–277
Olive LS (1945) A new *Dacrymyces*-like parasite of *Arundinaria*. Mycologia 37:543–552
Pascoe IG, Priest MJ, Shivas RG, Cunnington JH (2005) Ustilospores of *Tilletia ehrhartae*, a smut of *Ehrharta calycina*, are common contaminants of Australian wheat grain, and potential source of confusion with *Tilletia indica*, the cause of Karnal bunt of wheat. Plant Pathol 54:161–168
Patil MS (1977) A new species of *Muribasidiospora* from Kolhapur. Kavaka 5:31–33
Paz Z, Burdman S, Gerson U, Sztejnberg A (2007) Antagonistic effects of the endophytic fungus *Meira geulakonigii* on the citrus rust mite *Phyllocoptruta oleivora*. J Appl Microbiol 103:2570–2579
Pegg GS, Webb RI, Carnegie AJ, Wingfield MJ, Drenth A (2009) Infection and disease development of *Quambalaria* spp. on *Corymbia* and *Eucalyptus* species. Plant Pathol 58:642–654
Piątek M, Vánky K, Mossebo DC, Piątek J (2008) *Doassansiopsis caldesiae* sp. nov. and *Doassansiopsis tomasii*: two remarkable smut fungi from Cameroon. Mycologia 100:662–672
Piepenbring M, Bauer R (1995) Noteworthy germinations of some Costa Rican Ustilaginales. Mycol Res 99:853–858
Piepenbring M, Bauer R (1997) *Erratomyces*, a new genus of Tilletiales with species on Leguminosae. Mycologia 89:924–936
Piepenbring M, Bauer R, Oberwinkler F (1998) Teliospores of smut fungi: general aspects of teliospore walls and sporogenesis. Protoplasma 204: 155–169
Piepenbring M, Espinoza J, Saldana L, Caceres O (2010) New records, host plants, morphological and molecular data of Exobasidiales (Basidiomycota) from Panama. Nova Hedwigia 19:231–242. doi:10.1127/0029-5035/2010/0091-0231
Posada F, Vega FE (2005) Coffee endophytes pathogenic to the coffee berry borer. In: 38th annual meeting of the society for invertebrate pathology, program and abstracts, 7–11 Aug 2005, Anchorage, AK
Preece TF, Hick AJ (2001) An introduction to the Protomycetales: *Burenia inundata* on *Apium nodiflorum* and *Protomyces macrosporus* on *Anthriscus sylvestris*. Mycologist 15:119–125
Prillinger H, Dörfler C, Laaser G, Hauska G (1990) Ein Beitrag zur Systematik und Entwicklungsbiologie höherer Pilze. Hefe-Typen der Basidiomyceten. Teil III. *Ustilago*-Typ. Z Mykol 56:251–278
Prillinger H, Deml G, Dörfler C, Laaser G, Lockau W (1991) Ein Beitrag zur Systematik und Entwicklungsbiologie höherer Pilze: Hefe-Typen der Basidiomyceten. Teil II. *Microbotryum*-Typ. Bot Acta 104:5–17
Prillinger H, Oberwinkler F, Umile C, Tlachac K, Bauer R, Dörfler C, Taufratzhofer E (1993) Analysis of cell wall carbohydrates (neutral sugars) from ascomycetous and basidiomycetous yeasts with and without derivatization. J General Appl Microbiol 39:1–34
Rajendren RB (1968) *Muribasidiospora*-a new genus of the Exobasidiaceae. Mycopathologia 36:218–222
Reddy MS, Kramer CL (1975) A taxonomic revision of the Protomycetales. Mycotaxon 3:1–50
Ritschel A, Begerow D, Oberwinkler F (2008) A new species of *Volvocisporium* from Namibia, *V. grewiae* sp. nov. (Microstromatales, Ustilaginomycota). Mycol Prog 7:1–5
Roberson RW, Luttrell ES (1987) Ultrastructure of teliospore ontogeny in *Tilletia indica*. Mycologia 79:753–763
Roberson RW, Luttrell ES (1989) Dolipore septa in *Tilletia*. Mycologia 81:650–652
Roets F, Dreyer LL, Wingfield MJ, Begerow D (2008) *Thecaphora capensis* sp. nov., an unusual new anther smut on *Oxalis* in South Africa. Persoonia 21:47–152
Rush TA, Aime MC (2013) The genus *Meira*: phylogenetic placement and description of a new species. Antonie Van Leeuwenhoek 103:1–10
Sampaio JP (1999) Utilization of low molecular weight aromatic compounds by heterobasidiomycetous yeasts: taxonomic implications. Can J Microbiol 45:491–512
Sampaio JP (2004) Diversity, phylogeny and classification of basidiomycetous yeasts. In: Agerer R, Piepenbring M, Blanz P (eds) Frontiers in basidiomycote mycology. IHW, Eching, Germany, pp 49–80
Sampson K (1939) Life cycles of smut fungi. Trans Br Mycol Soc 23:1–23
Savchenko KG, Lutz M, Piątek M, Heluta VP, Nevo E (2013) *Anthracoidea caricis-meadii* is a new North American smut fungis *Carex* sect. Paniceae. Mycologia 105:181–193
Savile DBO (1947) A study of the species of *Entyloma* on North American composites. Can J Res 25:105–120

Schäfer AM, Kemler M, Bauer R, Begerow D (2010) The illustrated life cycle of *Microbotryum* on the host plant *Silene latifolia*. Can J Bot 88:875–885

Scherer M, Heimel K, Starke V, Kämper J (2006) The Clp1 protein is required for clamp formation and pathogenic development of *Ustilago maydis*. Plant Cell 18:2388–2401

Schirawski J, Böhnert HU, Steinberg G, Snetselaar K, Adamikowa L, Kahmann R (2005) Endoplasmic reticulum glucosidase II is required for pathogenicity of *Ustilago maydis*. Plant Cell 17:3532–3543

Schirawski J, Mannhaupt G, Münch K, Brefort T, Schipper K, Doehlemann G, Di Stasio M, Rössel N, Mendoza-Mendoza A, Pester D, Müller O, Winterberg B, Meyer E, Ghareeb H, Wollenberg T, Münsterkötter M, Wong P, Walter M, Stukenbrock E, Güldener U, Kahmann R (2010) Pathogenicity determinants in smut fungi revealed by genome comparison. Science 330:1546–1548. doi:10.1126/science.1195330

Schröter J (1894) *Uleiella* gen. nov. In: Pazschke, Sammlungen. Hedwigia Beibl 33:64–66

Seo HS, Um HJ, Min J, Rhee SK, Cho TJ, Kim YH, Lee J (2007) *Pseudozyma jejuensis* sp. nov., a novel cutinolytic ustilaginomycetous yeast species that is able to degrade plastic waste. FEMS Yeast Res 7:1035–1045

Shivas RG (2009) Three new species of *Tilletia* on native grasses from Australia. Australas Plant Pathol 38:128–131

Singh RA, Pavgi MS (1973) Morphology, cytology and development of *Melanotaenium brachiariae*. Cytologia 38:455–466

Sipiczki M, Kajdacsi E (2009) *Jaminaea angkorensis* gen. nov., sp. nov., a novel anamorphic fungus containing an S943 nuclear small-subunit rRNA group IB intron represents a basal branch of Microstromatales. Int J Syst Evol Microbiol 59:914–920

Skibbe DS, Doehlemann G, Fernandes J, Walbot V (2010) Maize tumors caused by *Ustilago maydis* require organ-specific genes in host and pathogen. Science 328:89. doi:10.1126/science.1185775

Snetselaar KM, Mims CW (1992) Sporidial fusion and infection of maize seedlings by the smut fungus *Ustilago maydis*. Mycologia 84:193–203

Snetselaar KM, Mims CW (1994) Light and electron microscopy of *Ustilago maydis* hyphae in maize. Mycol Res 98:347–355

Snetselaar KM, Tiffany LH (1990) Light and electron microscopy of sorus development in *Sorosporium provinciale*, a smut of big bluestem. Mycologia 82:480–492

Stoll M, Piepenbring M, Begerow D, Oberwinkler F (2003) Molecular phylogeny of *Ustilago* and *Sporisorium* species (Basidiomycota, Ustilaginales) based on internal transcribed spacer (ITS) sequences. Can J Bot 81:976–984

Stoll M, Begerow D, Oberwinkler F (2005) Molecular phylogeny of *Ustilago*, *Sporisorium*, and related taxa based on combined analyses of rDNA sequences. Mycol Res 109:342–356

Subba Rao PV, Fritig B, Vose JR, Towers GHN (1971) An aromatic 3,4-oxygenase from *Tilletiopsis washingtonensis*—oxidation of 3,4-dihydroxyphenylacetic acid to β-carboxymethylmuconolactone. Phytochemistry 10:51–56

Sugita T, Takashima M, Poonwan N, Mekha N, Malaithao K, Thungmuthasawat B, Prasarn S, Luangsook P, Kudo T (2003) The first isolation of ustilaginomycetous anamorphic yeasts, *Pseudozyma* species, from patients blood and a description of two new species: *P. parantarctica* and *P. thailandica*. Microbiol Immunol 47:183–190

Sugiyama J, Hosaka K, Suh SO (2006) Early diverging Ascomycota: phylogenetic divergence and related evolutionary enigmas. Mycologia 98:996–1005. doi:10.3852/mycologia.98.6.996

Sundström KR (1964) Studies of the physiology, morphology and serology of *Exobasidium*. Symb Bot Ups 18:3, 89 p

Swann EC, Taylor JW (1993) Higher taxa of basidiomycetes: an 18S rRNA gene perspective. Mycologia 85:923–936

Swann EC, Taylor JW (1995) Phylogenetic diversity of yeast-producing basidiomycetes. Mycol Res 99:1205–1210

Sydow H (1926) Fungi in itinere costaricensi collecti. Pars secunda. Ann Mycol 24:283–288

Takahashi H, Sekiguchi H, Ito T, Sasahara M, Hatanaka N, Ohba A, Hase S, Ando S, Hasegawa H, Takenaka S (2011) Microbial community profiles in intercellular fluid of rice. J Gen Plant Pathol 77:121–131

Tanaka E, Shimizu K, Imanishi Y, Yasuda F, Tanaka C (2008) Isolation of basidiomycetous anamorphic yeast-like fungus *Meira argovae* found on Japanese bamboo. Mycoscience 49:329–333

Taylor JW, Berbee ML (2006) Dating divergences in the Fungal Tree of Life: review and new analyses. Mycologia 98:838–849

Teichmann B, Linne U, Hewald S, Marahiel MA, Bölker M (2007) A biosynthetic gene cluster for a secreted cellobiose lipid with antifungal activity from *Ustilago maydis*. Mol Microbiol 66:525–533

Thomas PL (1989) Barley smuts in the prairie provinces of Canada, 1983-1988. Can J Phytopathol 11:133–136

Trail F, Mills D (1990) Growth of haploid *Tilletia* strains in planta and genetic analysis of a cross of *Tilletia caries* X *Tilletia controversa*. Phytopathology 80:367–370

Trindade RC, Resende MA, Silva CM, Rosa CA (2002) Yeasts associated with fresh and frozen pulps of Brazilian tropical fruits. Syst Appl Microbiol 25:294–300

Trione EJ (1982) Dwarf bunt of wheat and its importance in international wheat trade. Plant Dis 66:1083–1088

Trione EJ, Hess WM, Stockwell VO (1989) Growth and sporulation of the dikaryons of the dwarf bunt fungus in wheat plants and in culture. Can J Bot 67:1671–1680

Tulasne L, Tulasne C (1847) Mémoire sur les Ustilaginées comparées Uredinées. Ann Sci Nat Bot 3:12–127

Urquhart EJ, Punja ZK (2002) Hydrolytic enzymes and antifungal compounds produced by *Tilletiopsis* species, phyllosphere yeasts that are antagonists of powdery mildew fungi. Can J Microbiol 48:219–229

Valverde ME, Paredes-Lópes O, Pataky JK, Guevara-Lara F (1995) Huitlacoche (*Ustilago maydis*) as a food source—biology, composition, and production. CRC Crit Rev Food Sci Nutr 35:191–229

Vánky K (1981) The genus *Schroeteria* Winter (Ustilaginales). Sydowia 34:157–166

Vánky K (1987) Illustrated genera of smut fungi. Cryptogam Stud 1:1–159

Vánky K (1994) European smut fungi. Gustav Fischer, Stuttgart

Vánky K (1996) *Mycosyrinx* and other pair-spored Ustilaginales. Mycoscience 37:173–185

Vánky K (2000) New taxa of Ustilaginomycetes. Mycotaxon 74:343–356

Vánky K (2001) The emended Ustilaginaceae of the modern classificatory system for smut fungi. Fungal Divers 6:131–147

Vánky K (2003) Cintractiellaceae fam. nov. (Ustilaginomycetes). Fungal Divers 13:167–173

Vánky K (2005) The smut fungi (Ustilaginomycetes) of Eriocaulaceae. I. *Eriomoeszia* gen. nov. Mycol Balc 2:105–111

Vánky K (2011) The genus *Pericladium* (Ustilaginales). *Pericladiaceae* fam. nov. Mycol Balc 8:147–152

Vánky K (2012) Smut fungi of the world. APS, St. Paul, MN

Vánky K, Bauer R (1992) *Conidiosporomyces*, a new genus of Ustilaginales. Mycotaxon 43:426–435

Vánky K, Bauer R (1995) *Oberwinkleria*, a new genus of Ustilaginales. Mycotaxon 53:361–368

Vánky K, Bauer R (1996) *Ingoldiomyces*, a new genus of Ustilaginales. Mycotaxon 59:277–287

Vánky K, Lutz M (2011) *Tubisorus*, a new genus of smut fungi (Ustilaginomycetes) for *Sorosporium pachycarpum*. Mycol Blac 8:129–135

Vánky K, Bauer R, Begerow D (1998) *Doassinga*, a new genus of Doassansiales. Mycologia 90: 964–970

Vánky K, Lutz M, Shivas RG (2006) *Anomalomyces panici*, new genus and species of Ustilaginomycetes from Australia. Mycol Balc 3:119–126

Vánky K, Lutz M, Bauer R (2007) Revision of some *Thecaphora* species (Ustilaginomycotina) on Caryophyllaceae. Mycol Res 111:1207–1219

Vánky K, Lutz M, Bauer R (2008a) About the genus *Thecaphora* (Glomosporiaceae) and its new synonyms. Mycol Prog 7:31–39

Vánky K, Lutz M, Bauer R (2008b) *Floromyces*, a new genus of Ustilaginomycotina. Mycotaxon 104:171–184

Vishniac HS (2006) A multivariate analysis of soil yeasts isolated from a latitudinal gradient. Microb Ecol 52:90–103

Vishniac HS, Anderson JA, Filonow AB (1997) Assimilation of volatiles from ripe apples by *Sporidiobolus salmonicolor* and *Tilletiopsis washingtonensis*. Antonie Van Leeuwenhoek 72:201–207

Vollmeister E, Schipper K, Baumann S, Haag C, Pohlmann T, Stock J, Feldbrügge M (2012) Fungal development of the plant pathogen *Ustilago maydis*. FEMS Microbiol Rev 36:59–77. doi:10.1111/j.15746976.2011.00296.x

Wang QM, Jia JH, Bai FY (2006) *Pseudozyma hubeiensis* sp. nov. and *Pseudozyma shanxiensis* sp. nov., novel ustilaginomycetous anamorphic yeast species from plant leaves. Int J Syst Evol Microbiol 56:289–293

Wells K (1994) Jelly fungi, then and now! Mycologia 86:18–48

Whitehead T (1921) On the life history and morphology of *Urocystis cepulae*. Trans Br Mycol Soc 7:65–71, plate II

Xu J, Saunders CW, Hu P, Grant RA, Boekhout T, Kuramae EE, Kronstad JW, DeAngelis YM, Reeder NL, Johnstone KR, Leland M, Fieno AM, Begley WM, Sun Y, Lacey MP, Chaudhary T, Keough T, Chu L, Sears R, Yuan B, Dawson TL (2007) Dandruff-associated *Malassezia* genomes reveal convergent and divergent virulence traits shared with plant and human fungal pathogens. Proc Natl Acad Sci USA 104:18730–18735

Yasuda F, Yamagishi D, Hajime Akamatsu H, Izawa H, Kodama M, Otani H (2006) *Meira nashicola* sp. nov., a novel basidiomycetous, anamorphic yeast-like fungus isolated from Japanese pear fruit with reddish stain. Mycoscience 47:36–40

Zepeda L (2006) The Huitlacoche project: a tale of smut and gold. Renew Agr Food Syst 21:224–226. doi:10.1079/RAF2006153

Zundel GL (1945) A change in generic name. Mycologia 37:795–796

12 Tremellomycetes and Related Groups

MICHAEL WEISS[1], ROBERT BAUER[1], JOSÉ PAULO SAMPAIO[2], FRANZ OBERWINKLER[1]

CONTENTS

I. Introduction

Tremellomycetes is a fungal group full of contrasts. It includes jelly fungi with conspicuous macroscopic basidiomes, such as some species of *Tremella*, as well as macroscopically invisible inhabitants of other fungal fruiting bodies and a plethora of species known so far only as asexual yeasts. Tremellomycetes may be beneficial to humans, as exemplified by the production of edible *Tremella* fruiting bodies, which increased in China alone from 100 MT in 1998 to more than 250,000 MT in 2007 (Chang and Wasser 2012), or extremely harmful, such as the systemic human pathogen *Cryptococcus neoformans*. The systematics and taxonomy of many species now contained in Tremellomycetes have significantly changed during the past three decades and are about to change again as a result of changes in the taxonomic treatment of anamorph forms in the International Code of Nomenclature for algae, fungi, and plants (McNeill et al. 2012). An integrated systematic view of the Tremellomycetes has been hampered by the fact that the anamorphic taxa, i.e., the yeasts, and the basidiome-forming dimorphic taxa have traditionally been studied by different scientific communities. Recently, the group has been discussed in more integrative treatments (e.g., Boekhout et al. 2011; Millanes et al. 2011; Sampaio 2004).

Since the last edition of *The Mycota*, key systematic concepts in the Basidiomycota have changed conspicuously. While the tremellomycetous groups were then treated in a separate *Heterobasidiomycetes* chapter (Wells and Bandoni 2001), "Heterobasidiomycetes" is no

Dedicated to the memory of Robert Joseph Bandoni (1926–2009)

[1]Fachbereich Biologie, Eberhard-Karls-Universität Tübingen, Auf der Morgenstelle 1, 72076 Tübingen, Germany; e-mail: michael.weiss@uni-tuebingen.de; ;
[2]Centro de Recursos Microbiológicos, Departamento de Ciências da Vida, Universidade Nova de Lisboa, 2829-516 Caparica, Portugal

Systematics and Evolution, 2nd Edition
The Mycota VII Part A
D.J. McLaughlin and J.W. Spatafora (Eds.)

longer considered a monophyletic group (Weiß et al. 2004); for the present edition its members are discussed in this chapter, in Agaricomycetes (*Tulasnella*, *Ceratobasidium* and relatives, Auriculariales, Sebacinales; see Hibbett et al. 2014), and in Dacrymycetes (see Oberwinkler 2014).

In this overview we provide an introduction to the taxonomy, morphology, ecology, and phylogenetic relationships of the Tremellomycetes, including a phylogenetic tree that covers the vast majority of species of this group for which molecular data [nuclear rDNA coding for the D1/D2 regions of the large ribosomal subunit (LSU)] are available, using type or ex-type sequences wherever possible. It illustrates both the phylogenetic resolution presently available in the Tremellomycetes and the degree to which current taxonomy matches the phylogenetic relationships in this group. Considering the impressive progress in genome sequencing and phylogenomics we anticipate that at least the higher-level relationships will be much better resolved in the near future.

A. Historical Concepts

The genus *Tremella* was validly described by Persoon (1794). Some years later (Fries 1821) the genus was the basis for the family Tremellaceae (as "Tremellini", including also *Dacrymyces*), and for the order Tremellales (as "Tremellinae")—one of the six orders that Fries described in his "Hymenomycetes"—which roughly corresponds to what today are called jelly fungi. Since the acceptance of basidial morphology as a key character in the systematics of the basidiomycetes (Brefeld 1888; Patouillard 1887; Tulasne 1853), Tremellaceae/Tremellales have often been used as the taxon containing all hymenomycetes with longitudinally septate basidia—as opposed to the Auriculariaceae/Auriculariales, which according to these concepts included the taxa with transversely septate basidia [see Bandoni (1984) for a systematic treatment of the taxonomic history].

B. Modern View

This concept was challenged by Bandoni (1984), who redefined Tremellales and Auriculariales based on ultrastructural characters, the nature of the haploid states, and trophic strategies, rather than on basidial morphology. This alternative concept has been largely confirmed by molecular data (e.g., Swann and Taylor 1995; Weiß and Oberwinkler 2001) and is currently widely accepted (Hibbett et al. 2007).

Particular taxonomic problems in the Tremellomycetes to be solved in the future include the obvious nonmonophyly of established morphogenera, such as *Tremella*, and the question of how to best treat originally anamorphic genera, such as *Bullera* and *Cryptococcus*, in a modern nomenclature that no longer gives priority to generic names based on teleomorphs (Hawksworth 2011; McNeill et al. 2012). Since it is too early to solve these questions in this text, here we still adopt some widely used names that are likely to change in the near future.

II. Morphology and Anatomy

A. Basidiocarps

Basidiocarps (Fig. 12.1) are known from species of Tremellales, Holtermanniales, and Filobasidiales (*Syzygospora*). In species of Tremellales, basidiocarps are mostly of a **gelatinous consistency**. Many species can undergo prolonged phases of exsiccation, reviving when rehydrated, with renewed growth and production of conidia and/or basidiospores (Wells and Bandoni 2001). They are thus well adapted to habitats on dead wood, on which the more exposed species, for example, *Tremella*, are often found. Basidiocarp forms vary from **pustulate**, for example, *Tremella* spp., *Tetragoniomyces*, or *Sirobasidium*, to **cushion-shaped, lobose-cerebriform**, for example, *Tremella mesenterica* (Fig. 12.1a, b), to foliose, for example, *Tremella foliacea* (Fig. 12.1g) and *Tremella fuciformis* (Fig. 12.1h). Often they originate from a host fungus that they obviously parasitize (see below). Mature basidiocarps may even

Fig. 12.1 (**a–e**) *Tremella mesenterica*. (**a**) Young basidiocarps on *Peniophora laeta* growing on *Carpinus betulus*, *bar*=2 mm. (**b**) Mature basidiocarp, *bar*=5 mm. (**c**) Part of hymenium with basidia, *bar*=10 µm. (**d**) Yeast budding, *bar*=10 µm. (**e**) Basidiospore with secondary spore, *bar*=5 µm. (**f**) *Tremella encephala* showing whitish core with hyphal mixture of host *Stereum sanguinolentum* and mycoparasite, *bar*=1 cm. (**g**) *Tremella foliacea*, *bar*=2 cm. (**h**) *Tremella fuciformis*, *bar*=3 cm. (**i**) *Tremella aurantia*, *bar*=3 cm. (**j**) *Sirobasidium magnum*, bar 3 cm. (**k–m**) *Syzygospora pallida*. (**k**) Pustular basidiocarps emerging from host *Phanerochaete cremea*, *bar*=2 mm. (**l**) Basidium and conidiophores, *bar*=10 µm. (**m**) Budding yeasts, *bar*=10 µm. (**n, o**) *Tetragoniomyces uliginosus*. (**n**) Basidiocarps in culture, *bar*=1 mm. (**o**) Germinating basidium, *bar*=10 µm

show a central core composed of hyphae of host and mycoparasite, as in *Tremella encephala* (Fig. 12.1f). Basidiocarps in the Holtermanniales are tough-gelatinous, with a clavarioid appearance.

Numerous teleomorphic species in the Tremellomycetes apparently lack basidiocarps. Such species grow **intrahymenially** in their fungal hosts, either without causing any macroscopic symptoms, such as *Tremella giraffa*, *Tremella obscura*, and *Tremella penetrans*, or inducing galls on their hosts, for example, lichenicolous species of *Tremella* or *Biatoropsis usnearum*. Sexual stages in some species of Tremellales, for example, *Bulleribasidium* (Fig. 12.2e), *Filobasidiella* (Fig. 12.3a), *Kwoniella*, and *Rhynchogastrema* (Fig. 12.2f), as well as all known sexual stages in Cystofilobasidiales, are known only from pure cultures.

B. Micromorphology

Most species in the Tremellomycetes grow as **yeasts** in their haploid stages (Figs. 12.2 and 12.3). Such yeast stages may proliferate by budding, but they may also produce **ballistoconidia** that are morphologically and functionally similar to basidiospores. Yeast cells are generally globoid to ellipsoid but may also be elongate, as in *Carcinomyces*. Diploid stages are generally filamentous, with clamped hyphae.

There is conspicuous variation in basidial morphology, which has been one of the most important characters used in traditional morphogeneric concepts. The **basidia** of the species of *Tremella* are usually longitudinally septate (so-called *tremelloid* basidia), with the basidial compartments protruding into elongated tubes, designated as *epibasidia* by some authors (Wells and Bandoni 2001), that pervade the often gelatinous matrix of their own or the host basidiome and, finally, apically bear a sterigma, from which the mostly globular basidiospores are actively discharged into the air (Fig. 12.4). There are, however, numerous variations.

First, **tremelloid basidia** are only known in the Tremellales and in *Holtermannia*. In some other species of Tremellales the basidial compartments may be arranged in a more linear order, with **transverse or oblique basidial septa**, as in *Auriculibuller*, *Bulleromyces*, and *Papiliotrema* (Fig. 12.2b). Development of the basidial compartments is often strongly desynchronized (Wells and Bandoni 2001), and basidial compartments may detach in some species, for example in *Sirobasidium* (Fig. 12.2h), before giving rise to a ballistospore (Bandoni 1984). Basidial septation may also be lacking, resulting in **holobasidia**, as in *Carcinomyces* (Fig. 12.3d) and *Filobasidiella* (Fig. 12.3a) (Tremellales); *Filobasidium* (Fig. 12.3b) and *Syzygospora* (Fig. 12.3e, f) (Filobasidiales); *Cystofilobasidium* (Fig. 12.3c), and *Xanthophyllomyces* (Cystofilobasidiales). In some species a **partial apical septation** in holobasidia has been reported, for example in *Rhynchogastrema* (Fig. 12.2f) and *Syzygospora* (Metzler et al. 1989; Oberwinkler and Lowy 1981). Obviously, a transition from phragmobasidia to holobasidia has occurred independently several times in the Tremellomycetes (Millanes et al. 2011).

Second, there are also exceptions concerning the development and arrangement of basidia. While basidia usually appear singly or in clusters proliferating from subbasidial clamps, for example in basidiomes of *Tremella*, **basidial chains** can be observed in species of *Sirobasidium* (Fig. 12.2h) and, to a lesser degree, in *Sirotrema*. In these species, basidia proliferate basipetally, starting from an apical basidium. In *Cystofilobasidiales* basidia arise from **teliospores** (Fig. 12.3c).

Third, tremellomycete species differ concerning the release and functioning of **basidiospores**. In most teleomorphic species basidiospores are actively discharged from sterigmata. In *Sirobasidium*, on the other hand, basidia give rise to passively released fusoid basidiospores (Fig. 12.2h) [alternatively designated as epibasidia (Wells and Bandoni 2001)] that may proliferate by budding or by the formation of secondary spores (Bandoni 1984). Some other species in the Tremellomycetes, such as the phragmobasidiate species of *Kwoniella* and the holobasidiate species of *Carcinomyces*, produce sessile basidiospores.

A feasible concept uniting the heterogeneity in basidial morphology observed in the Tremellomycetes has been proposed by Bandoni

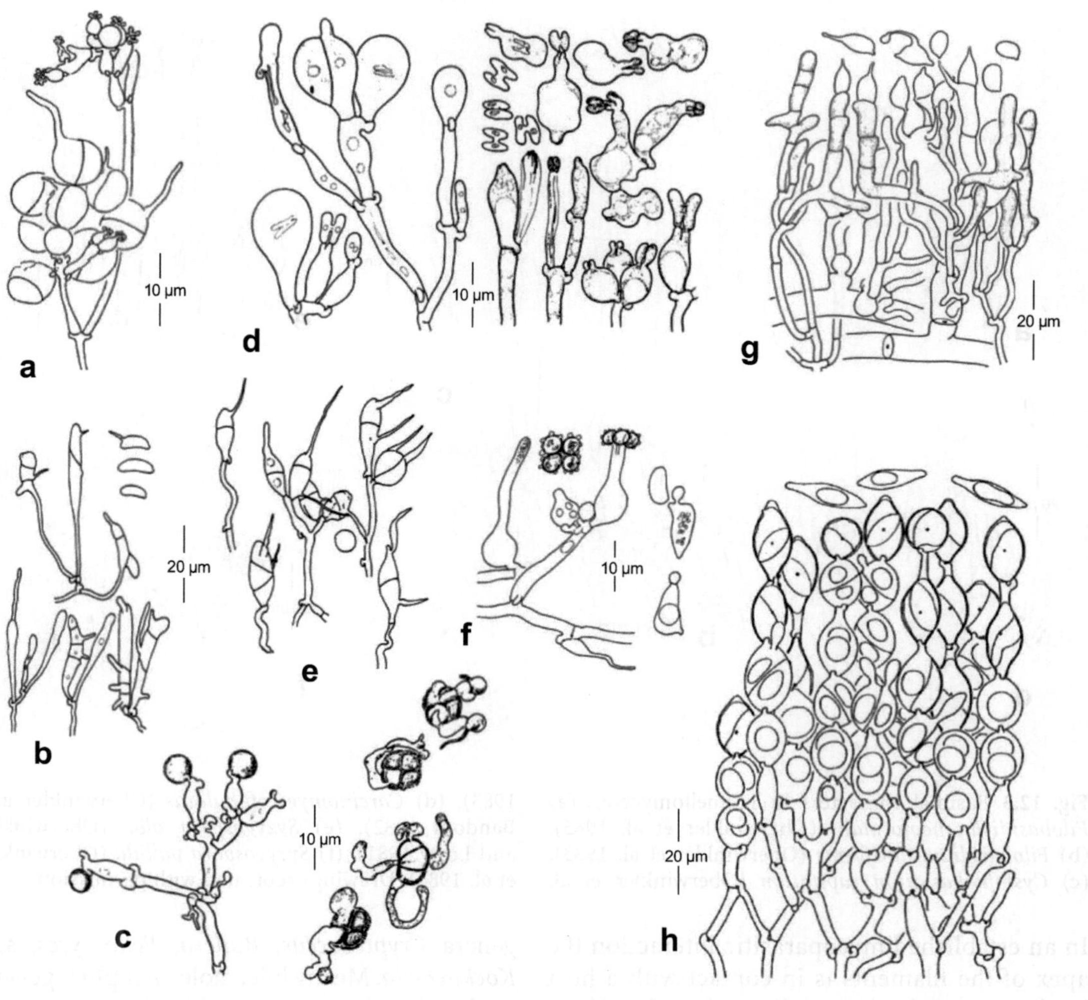

Fig. 12.2 Basidial characters in Tremellomycetes. (**a**) *Cuniculitrema polymorpha* (Kirschner et al. 2001). (**b**) *Papiliotrema bandonii* (Sampaio et al. 2002). (**c**) *Tetragoniomyces uliginosus* (Oberwinkler and Bandoni 1981). (**d**) *Trimorphomyces papilionaceus* (Oberwinkler and Bandoni 1983). (**e**) *Bulleribasidium oberjochense* (Sampaio et al. 2002). (**f**) *Rhynchogastrema coronatum* (Metzler et al. 1989). (**g**) *Phragmoxenidium mycophilum* (Oberwinkler et al. 1990). (**h**) *Sirobasidium magnum* (Chen 1998). Drawings reprinted with permission

(1984), who suggested that the basidial "compartments" themselves may actually be meiotic products (**endospores**) that in most species form a germtube (the so-called epibasidium) to produce a secondary spore (basidiospore in common terminology). Longitudinal, transverse, or oblique septation of the basidium then may simply result from a varying arrangement of the primary spores (endospores) within the basidium.

Teliospores, i.e., one-celled conidia that give rise to basidia after a resting period, are only known from species of Cystofilobasidiales. These structures provide an eloquent example of convergent evolution as they are present in various distantly related groups of basidiomycetes, such as the rust and the smut fungi.

Many presumably mycoparasitic species of Tremellomycetes feature a characteristic tremelloid haustorial type in their filamentous stages (e.g., Chen 1998; Oberwinkler and Bandoni 1981; Zugmaier et al. 1994) (Fig. 12.2c, f, g). **Tremelloid haustoria** arise from clamp connections and consist of single cells that are globular or short clavate at the base and extend into one or more narrow filaments (Fig. 12.4).

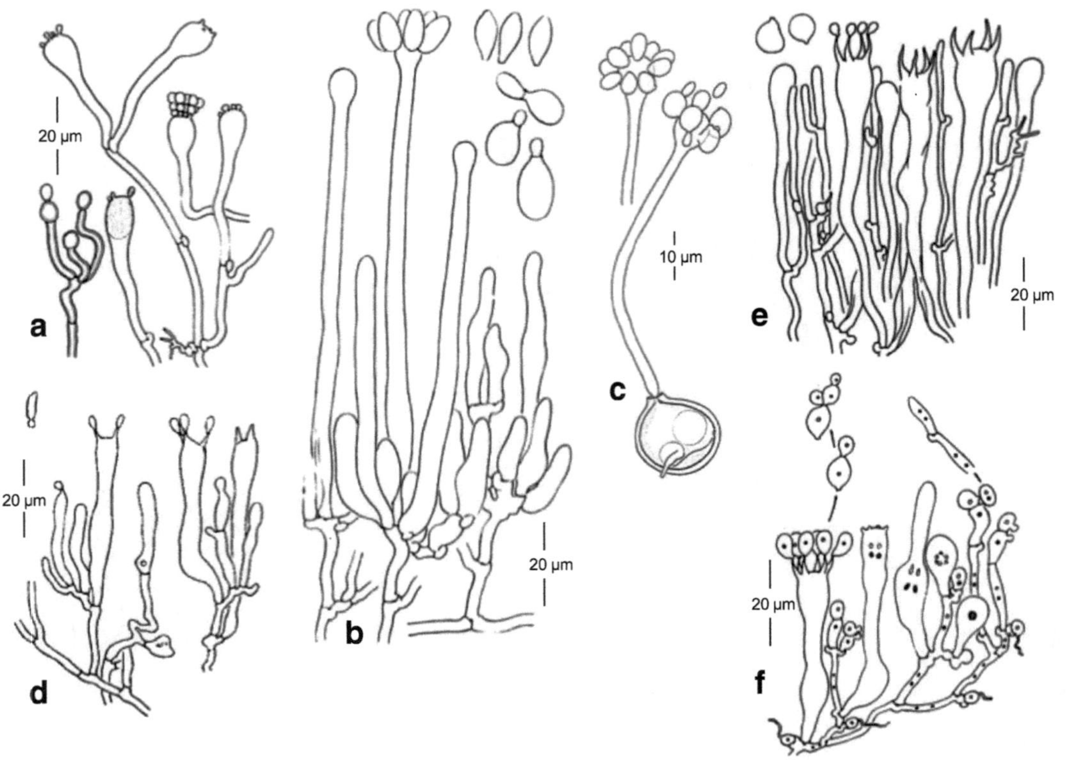

Fig. 12.3 Basidial characters in Tremellomycetes. (**a**) *Filobasidiella neoformans* (Oberwinkler et al. 1983). (**b**) *Filobasidium floriforme* (Oberwinkler et al. 1983). (**c**) *Cystofilobasidium capitatum* (Oberwinkler et al. 1983). (**d**) *Carcinomyes effibulatus* (Oberwinkler and Bandoni 1982). (**e**) *Syzygospora alba* (Oberwinkler and Lowy 1981). (**f**) *Syzygospora pallida* (Oberwinkler et al. 1984). Drawings reprinted with permission

In an established mycoparasitic interaction the apex of the filaments is in contact with a host hypha (see subsequent discussion for ultrastructural details).

Several types of conidia have been observed in the Tremellomycetes. Globular **blastoconidia** are occasionally found in Tremellales fruiting bodies [e.g., in *T. mesenterica* (Fig. 12.4), where ample production of blastoconidia creates the characteristic orange color of the fruiting bodies], before or synchronously with the production of basidia and basidiospores. Production of blastoconidia on elongated stalks is known from species of *Fellomyces* and *Cuniculitrema*. **Arthroconidia** are typical of most Trichosporonales species but can also be found in other species, for example, in *Guehomyces* and *Tausonia* (Cystofilobasidiales). Many species in the Tremellomycetes also form **ballistoconidia.**

The formation or absence of ballistoconidia was used in earlier classifications to separate the genera *Cryptococcus*, *Bullera*, *Fellomyces*, and *Kockovaella*. Meanwhile, molecular phylogenetic studies have shown that this character is not useful for circumscribing monophyletic genera (Boekhout et al. 2011). Consequently, genera such as *Derxomyces*, *Dioszegia*, and *Hannaella* have been proposed for monophyletic groups that contain both species with or without the formation of ballistoconidia. However, the footprints of the old classification marker "presence/absence of ballistoconida" are still visible in the current tree of the Tremellomycetes (Fig. 12.7).

Zygoconidia, i.e., dikaryotic H-shaped conidia, are known from several distantly related taxa, such as *Carcinomyces*, *Papiliotrema*, and *Trimorphomyces* (Fig. 12.2d) (Tremellales), as well as from *Syzygospora* (Filobasidiales) (e.g., Oberwinkler and Bandoni 1983; Oberwinkler and Lowy 1981; Sampaio et al. 2002).

Finally, four-spined **asteroconidia** have been observed in some lichenicolous species

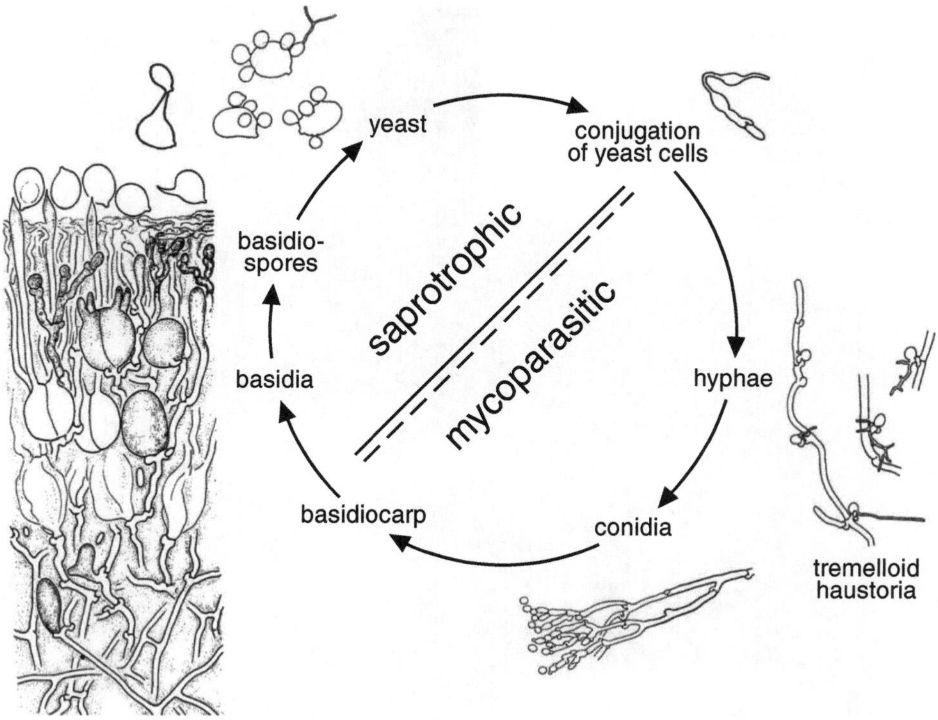

Fig. 12.4 Life cycle of *Tremella mesenterica*

of *Tremella* (Diederich 1996; Millanes et al. 2011).

C. Ultrastructure

Septal pores in the Tremellomycetes are **dolipores** that, except for members of Cystofilobasidiales (Oberwinkler et al. 1983; Wells 1994; R. Bauer, unpublished), are surrounded at both sides by **sacculate caps** arranged in hemispherical outlines (Fig. 12.5) (Berbee and Wells 1988). In three-dimensional configurations these saccules represent fingerlike extensions of the endoplasmic reticulum surrounding the pore on either side (as is visible in one saccule illustrated in Fig. 12.5b), in which the intracisternal surface of the membrane is accompanied by an additional electron-opaque nonmembranous layer (Fig. 12.5b). In cross or oblique sections, these fingerlike extensions are mapped as saccules with abseptal openings (Fig. 12.5b, c). Saccular parenthesomal elements may, however, be missing in some pores of a studied specimen (R. Bauer, unpublished; Padamsee et al. 2012), which may help to explain some inconsistencies documented in the literature.

Spindle pole bodies of the studied species in Tremellales are biglobular during prophase (Berbee and Wells 1988), a character state that supports the inclusion of the Tremellomycetes in Agaricomycotina.

The cellular interaction between species of the Tremellomycetes and their presumed host fungi occurs via the formation of **tremelloid haustoria** (Figs. 12.2c, f, g and 12.4; see previous discussion). The haustorial filaments are capable of fusing with host cells via pores of roughly 15 nm in diameter, where plasma membranes of both fungi are continuous with each other (Fig. 12.6). This yields a direct cytoplasmic contact; however, the size of these **fusion channels** prevents an exchange of organelles, including ribosomes, between the interacting organisms (Bauer and Oberwinkler 1990a; Oberwinkler et al. 1984; Zugmaier et al. 1994). While in

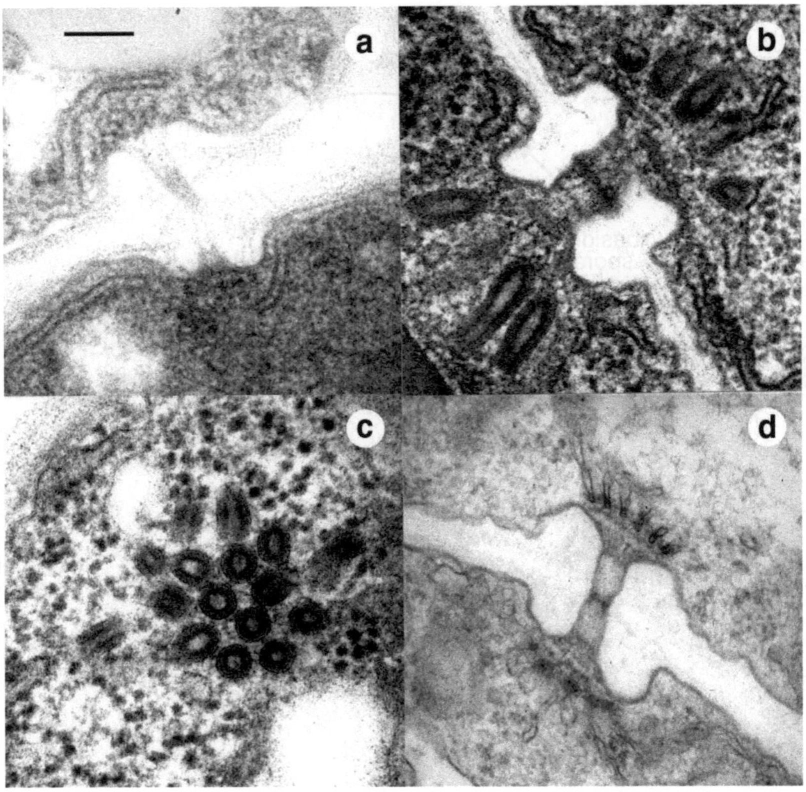

Fig. 12.5 Septal pore architecture in Tremellomycetes. *Bar*=0.1 μm in (**a–c**), 0.2 μm in (**d**). (**a**) Dolipore of *Cystofilobasidium ferigula* without specialized multilamellate caps, representative of Cystofilobasidiales. Note that pore is surrounded at each side by a more or less dome-shaped ER cisterna. (**b–d**) Dolipores surrounded at each side by many multilamellate cupulate cap elements, representative of Tremellomycetes [except for Cystofilobasidiales; see (**a**)] and *Wallemia*. (**b**, **c**) *Tremella* sp. Cupulate cap elements are sectioned longitudinally in (**b**), transversally in (**c**). Continuity between ER and saccules is visible for one of upper saccules in (**b**). (**d**) *Wallemia sebi* (**a**) reprinted from Sampaio et al. (2001) with permission

most studied species of the Tremellomycetes a haustorial filament forms only one fusion channel (Fig. 12.6a, b), in *Syzygospora pallida* a single haustorial filament may form several protrusions into the host cytoplasm, resulting in numerous fusion channels per filament (Fig. 12.6c, d) (Bauer 2004; Bauer and Oberwinkler 1990b; Oberwinkler et al. 1984).

III. Life Cycles

A. Dimorphism

Dimorphism, i.e., differing morphological organization of different life stages, is a characteristic trait in most species of the Tremellomycetes for which a teleomorph is known. A typical life cycle of a *Tremella* species is illustrated in Fig. 12.4. In these species, basidiospores germinate by budding to establish a **haploid yeast stage.** Since this stage can easily be maintained in pure culture on standard media, it is assumed that the yeast stage is saprotrophic.

Conjugation of compatible yeast cells initiates a dikaryotic hyphal stage, which is considered mycoparasitic in many species based on two lines of evidence. First, axenic cultivation of this stage has seldom been reported (Zugmaier and Oberwinkler 1995; Zugmaier et al. 1994). Second, tremelloid haustoria attached to hyphae of other fungal species are often observed microscopically

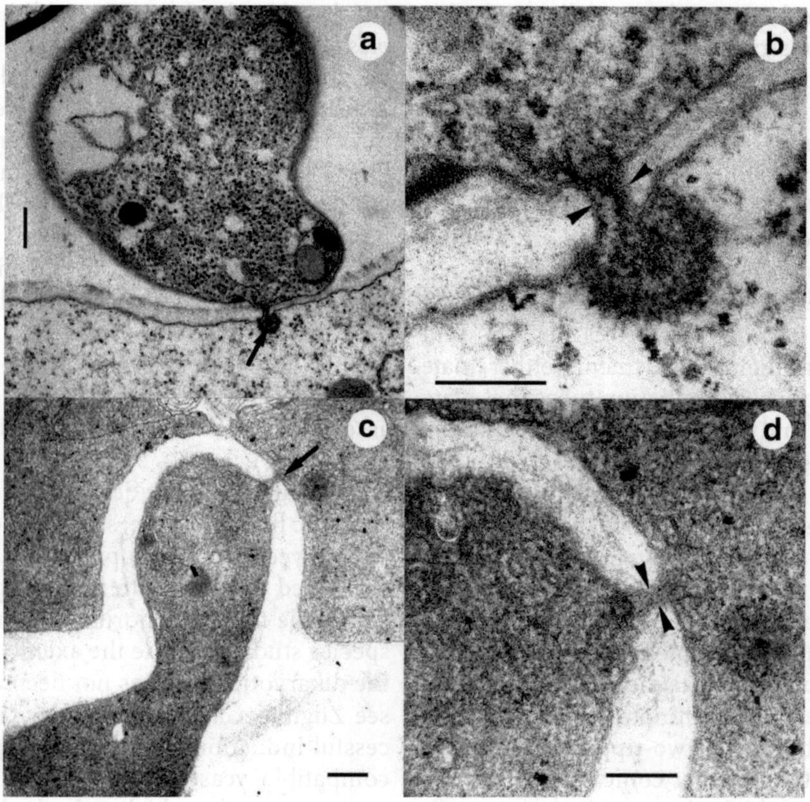

Fig. 12.6 Mycoparasitic interaction stages of some Tremellomycetes. *Bars*=0.2 μm in (**a, c**) and 0.1 μm in (**b, d**). (**a, b**) Haustorial filament of *Tetragoniomyces uliginosus* (*upper cell*) attached to cell of *Rhizoctonia* sp. (**a**) Note medianly sectioned micropore (*arrow*) connecting haustorial apex with host cell. (**b**) Detail from (**a**) Note that pore membrane (*arrowheads*) is continuous with plasma membranes of both cells. (**c, d**) Haustorial filament of *Syzygospora pallida* penetrating a cell of *Phanerochaete cremea*. (**c**) One of several micropores connecting haustorial filament with host cell is medianly sectioned (*arrow*). (**d**) Detail from (c). Note that pore membrane (*arrowheads*) is continuous with plasma membranes of both cells. (**a, b**) from Bauer and Oberwinkler (1990a), (**c, d**) from Bauer and Oberwinkler (1990b)

(see previous discussion). **Dikaryotic hyphae** may constitute a fruiting body that ultimately produces basidia and basidiospores or conidiogenous hyphae, giving rise to conidia. In species lacking a fruiting body, dikaryotic hyphae grow inside a host fruiting body and finally sporulate at its surface.

Holtermanniella mycelialis has been reported to be dimorphic and haploid (Golubev and Golubev 2003). In this species, after some days of cultivation, yeast colonies build clamped hyphae with tremelloid haustoria and release blastoconidia. Basidia have not been observed in this species, which may represent an anamorph of a *Holtermannia* species, where a teleomorph is possibly induced in the presence of a particular fungal host.

B. Deviance from Dimorphism

The designation of a species of the Tremellomycetes (typically a yeast) as monomorphic should always be considered as being preliminary. There are instances where a filamentous stage was obtained by mating compatible strains long after the first description of the yeast stage, for example, in *Bullera*/*Bulleromyces*, *Cryptococcus*/*Filobasidiella*, and *Cryptococcus*/*Kwoniella*. More recently, genomic methods have been used to predict and ultimately demonstrate sexuality in fungi that were previously considered asexual (Metin et al. 2010; O'Gorman et al. 2009). Consequently, many other inconspicuous teleomorphs may still await detection and description (Metin et al. 2010).

Some species lack a yeast stage. In *Tetragoniomyces uliginosus* (Fig. 12.2c) basidia do not produce external basidiospores. Instead, the thick-walled basidia themselves detach, and compatible basidial compartments either mate directly or produce germination tubes that mate (Oberwinkler and Bandoni 1981), inducing the next hyphal generation. Yeast stages are also unknown for many species of Trichosporonales and in *Filobasidiella depauperata.*

Trimorphomyces papilionaceus (Tremellales) is the only known species of the Tremellomycetes that has a dikaryotic yeast stage in addition to the usual haploid yeast stage, which arises from budding basidiospores. Here, the dikaryotic yeast cells initiate from dikaryotic zygoconidia borne on two-tipped conidiogenous cells located in either conidiomata or fruiting bodies in which the conidiogenous cells occur together with basidia (Fig. 12.2d). In the presence of a suitable fungal host, zygoconidia alternatively germinate with hyphae that form clamps and tremelloid haustoria.

The life cycle of *Itersonilia perplexans* comprises clamped dikaryotic hyphae, short unclamped monokaryotic hyphae, monokaryotic yeast cells, chlamydosporelike resting cells, and ballistoconidia (Boekhout 2011; F. Oberwinkler, unpublished).

IV. Ecology

A. Mycoparasitism

That a mycoparasitic lifestyle is a distinctive feature of the teleomorphic stages for many, if not all, members of the Tremellomycetes has been deduced from obvious host specificity, from morphological evidence, such as the presence of hyphae of putative host fungi growing inside fruiting bodies of Tremellomycetes, or from the presence of tremelloid haustoria, which may attach to host hyphae and establish minute cytoplasm-to-cytoplasm contacts (see previous discussion; Bandoni 1984; Bauer and Oberwinkler 1990a; Zugmaier et al. 1994). However, a flux of carbon compounds or other nutrients from a fungal host species to a tremellomycete has not yet been demonstrated. Apparently, the mycoparasitic potential is initiated with the transition from the monokaryotic to the dikaryotic life stage. Molecular mechanisms, such as host recognition, are still unknown.

That hyphal stages of some phylogenetically close species of the Tremellomycetes are associated with fungi that are closely related inter se (Fig. 12.7: 22) may be taken as an additional piece of evidence in favor of a mycoparasitic lifestyle. Here, strongly dependent tremellomycetous mycoparasites may have coevolved together with their fungal hosts.

While for the majority of Tremellomycetes species studied to date the axenic cultivation of the dikaryotic stage has not been achieved [but see Zugmaier and Oberwinkler (1995)], a successful induction of the teleomorph by mating compatible yeast cells has been reported for some species for which only the haploid stage, i.e., the yeast in most cases, was known previously. These include *Filobasidiella neoformans*, *Bulleromyces* and *Kwoniella* (Tremellales), and *Cystofilobasidium* (Cystofilobasidiales).

B. Tremellomycetous Yeasts

Apparently, all of the known yeast stages in the Tremellomycetes can be cultured axenically in standard media. Tremellomycetous yeasts are ubiquitous elements of terrestrial and aquatic ecosystems and have been reported from Antarctic soils as well as from hydrothermal oceanic vents. They have been isolated from sources as diverse as the surface of land plants, including flowers and tree bark, from freshwater and seawater samples, from clinical specimens, and from animals or their excrements [see Kurtzman et al. (2011)]. Some species seem to occupy rather diverse niches, for example, *Cryptococcus curvatus* has been reported mainly from medical sources and from food products but was also shown to be

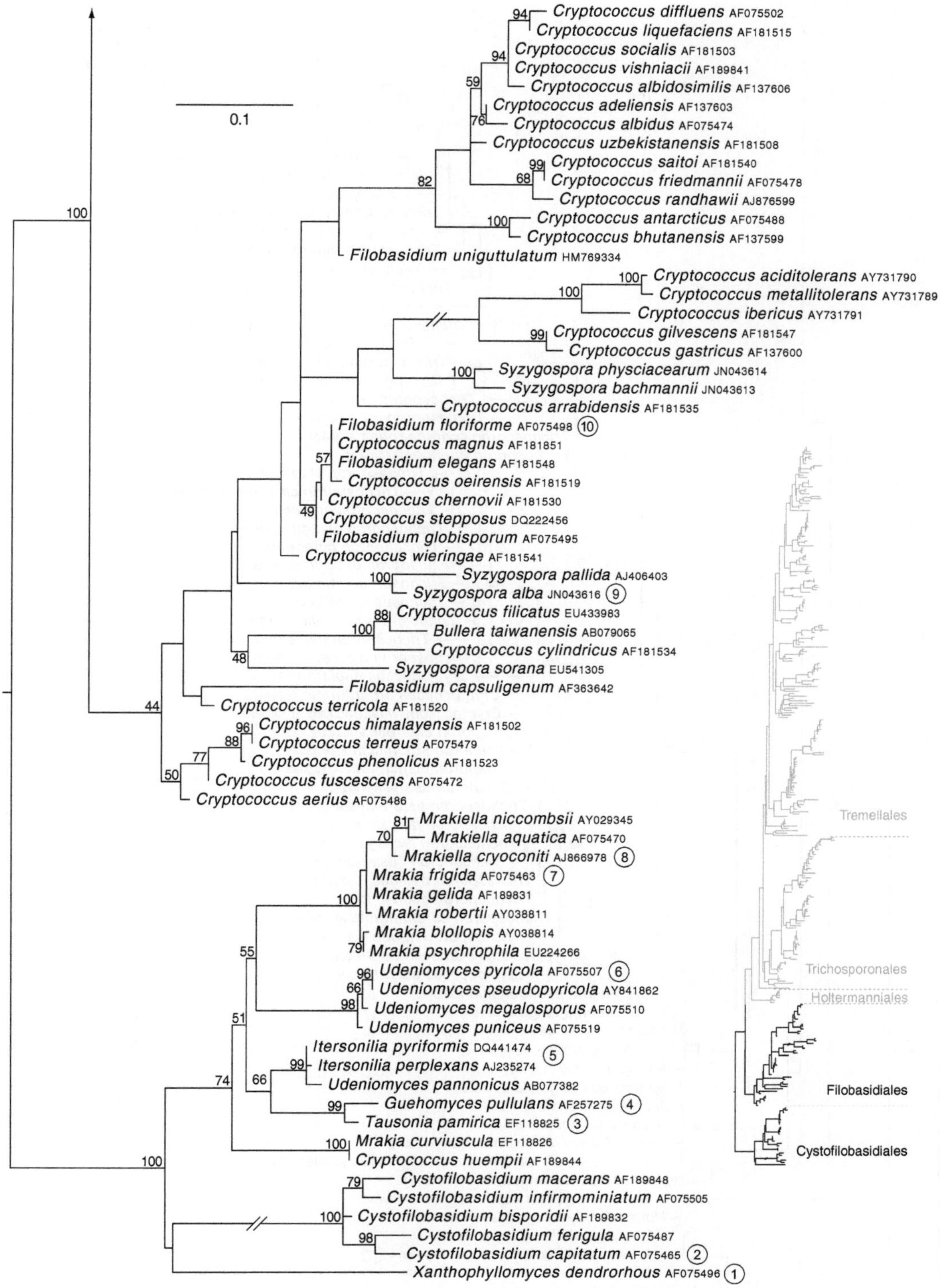

Fig. 12.7 (continued)

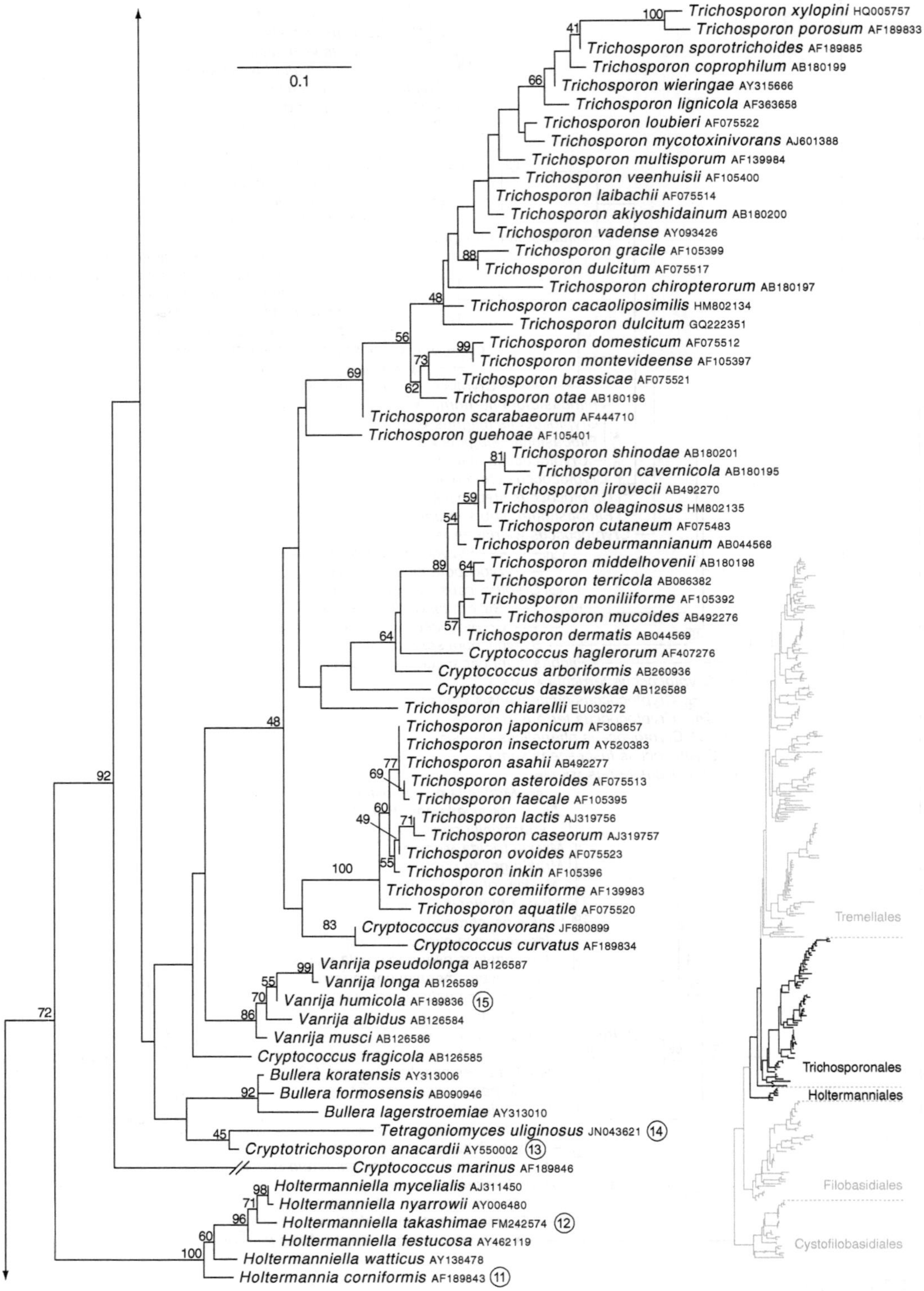

Fig. 12.7 (continued)

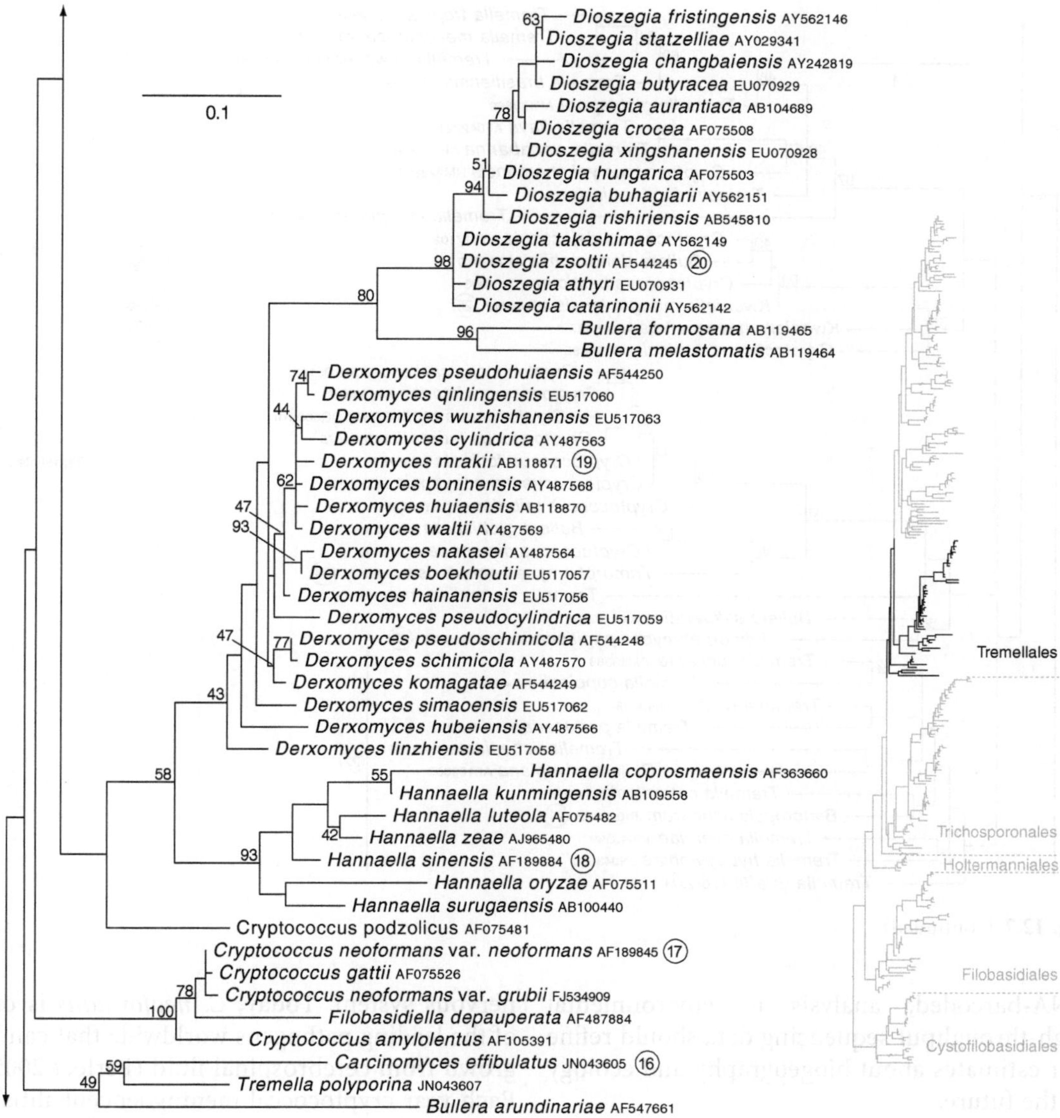

Fig. 12.7 (continued)

the dominant microbial eukaryote in sediments of a methane seep at a water depth of 640 m in the Pacific Ocean (Takishita et al. 2006). For some species ecological trends are visible. Psychrophilic species, such as members of *Mrakia*, have been isolated in Antarctica or Greenland or from glaciers but were also reported from refrigerated food (Fell 2011). *Bullera alba* is frequently isolated from the phylloplane (Sampaio 2004). *Xanthophyllomyces dendrorhous* is known from the sap of various tree species (David-Palma et al. 2014; Fell et al. 2011). Some species of *Fellomyces* have only been found on lichen thalli (Lopandic et al. 2011).

For many, if not most, of the known species, however, data are still too sparse to estimate distribution and ecology with confidence. Additionally, it may be problematic to integrate data based on morphological and physiological species determination with data based on sequence-based identification. Since all known tremellomycetous yeast species have been

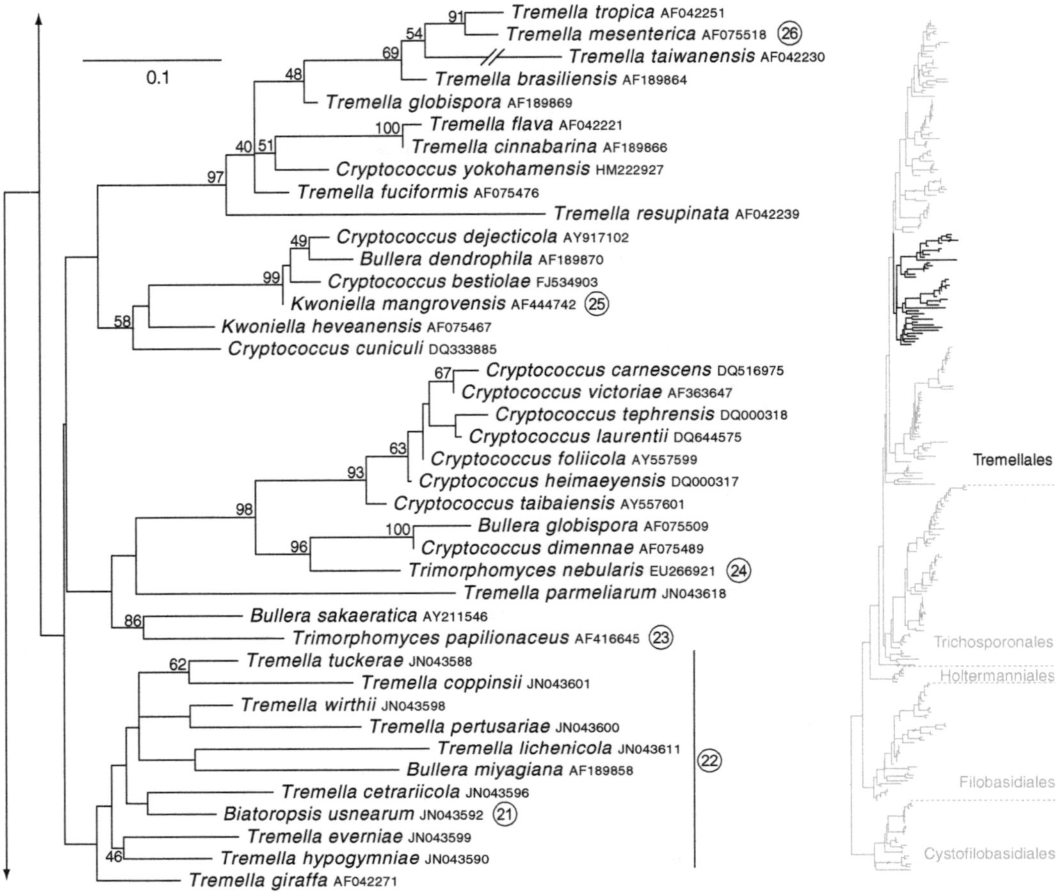

Fig. 12.7 (continued)

DNA-barcoded, analysis of environmental high-throughput sequencing data should refine our estimates about biogeography and ecology in the future.

C. Animal and Human Pathogens

C. neoformans, the yeast stage of *F. neoformans*, is an opportunistic pathogen in immunocompromised humans and animals around the world. The fungus is able to infest immunocompetent individuals without causing noticeable disease symptoms. However, in immunocompromised individuals, for example, those with an HIV infection, it may disseminate from a local infection to any organ of a patient and in particular invade the central nervous system. Today, *C. neoformans* is one of the leading pathogens worldwide that can be grown from cerebrospinal fluid (Perfect 2005). Each year cryptococcal meningoencephalitis is diagnosed in nearly a million individuals and accounts for more than 600,000 deaths (Park et al. 2009). Even if treated with state-of-the-art therapy, cryptococcosis is fatal in ca. 20 % of cases (Desnos-Ollivier et al. 2010). As the closely related *Cryptococcus gattii* (anamorph of *Filobasidiella bacillispora*), *C. neoformans* apparently has a wide spectrum of potential host taxa, including both vertebrate and invertebrate species and even protozoans [see Kwon-Chung (2011)].

The genus *Trichosporon* contains many known pathogens of animals and humans, and more than 30 % of *Trichosporon* species were

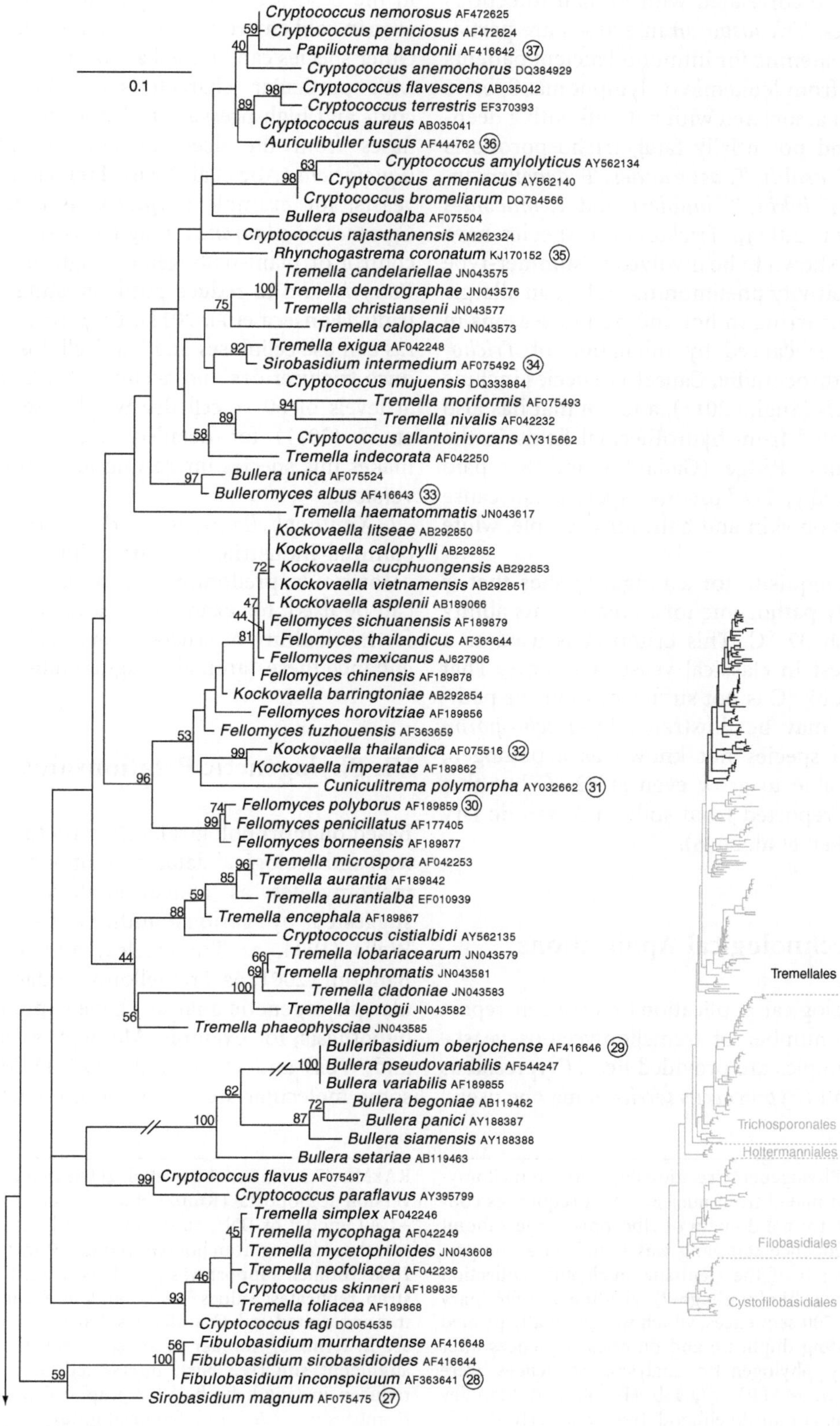

Fig. 12.7 (continued)

found to be correlated with human infections or allergies. *Trichosporon* infections are particularly threatening for immunodeficient patients suffering from leukemia or lymphoma. Species often seen associated with patients with a deep-seated, and potentially fatal, trichosporonosis include *T. asahii*, *T. asterioides*, *T. debeurmannianum*, *T. inkin*, *T. loubieri*, and *T. mucoides* [see Sugita (2011)]. *Trichosporon* species have also been shown to be involved in summer-type hypersensitivity pneumonitis (SHP), an allergic disease occurring in hot and humid seasons in Asia that is caused by inhalation of *Trichosporon* arthroconidia. Causative species include *T. dermatis* (Sugita 2011), a taxon that has also been isolated from hydrothermal fields in the Mid-Atlantic Ridge (Gadanho and Sampaio 2005). Finally, *Trichosporon* species can cause infections on skin and hair, for example, white piedra.

A prerequisite for a fungal species that is potentially pathogenic for humans is its ability to grow at 37 °C. This criterion is used as a routine test in classical yeast taxonomy. That growth at 37 °C is not sufficient to prove pathogenicity may be illustrated by *Trichosporon louberi*, a species not known as a pathogen, which is able to grow even at 42 °C but that has been reported from soils in Antarctic Dry Valleys (Fell et al. 2006).

V. Biotechnological Applications

Biotechnological applications have been reported for a number of tremellomycetous yeasts. Some examples are provided here. *Cryptococcus terreus* and *Cryptococcus terricola* may be useful in the biodegradation of phenolic compounds, even in cold environments (Bergauer et al. 2005). Other species can be used as sources of enzymes with particular characteristics, for example, cold- and high-pressure-tolerant polygalacturonases from the deep-sea yeast *Cryptococcus liquefaciens* (Abe et al. 2006). Tremellomycetous yeasts, for example *Cryptococcus albidus* and *Cryptococcus laurentii*, may be used in the biocontrol of plant-pathogenic fungi, for example *Botrytis*, and to reduce postharvesting decay of fruits (Fonseca et al. 2011). *Cryptococcus curvatus* can use celluloses and hemicelluloses to produce triglycerides and accumulates these lipids at levels of 60 % cell dry weight [see Fonseca et al. (2011) for detailed references], which makes this species interesting for biomass conversion.

Xanthophyllomyces dendrorhous is cultured industrially for its ability to produce carotenoids, predominantly astaxanthin, which can be used, for example, as a dietary additive for mariculture of crustaceans or salmonids to enhance these "animals" pigmentation.

VI. Phylogenetic Relationships

Based on morphological, ultrastructural, chemical, and ecological data, the monophyly of Tremellomycetes as conceived here has been suggested by various authors, for example, Wells (1994, as Tremellales) and Wells and Bandoni (2001, as Tremellomycetidae). Molecular phylogenetic analyses have supported this hypothesis, for example, Matheny et al. (2006) and Weiß and Oberwinkler (2001). However, some molecular studies based on nrDNA have

Fig. 12.7 Phylogenetic relationships in Tremellomycetes, as estimated from nuclear rDNA sequences coding for 5′ terminal domain of ribosomal large subunit (nLSU). Sequence sampling was based on a comprehensive search of the GenBank nucleotide collection (http://www.ncbi.nlm.nih.gov/), yielding a preliminary set of ca. 1,700 sequences, which was gradually pruned by eliminating duplicate and dubious sequences after preliminary phylogenetic analyses. Sequences were aligned with MAFFT v7.045b (Katoh and Standley 2013), a maximum-likelihood tree was derived with RAxML v7.3.2 (Stamatakis et al. 2008) in a parallelized version at Bioportal (Kumar et al. 2009) using the GTR+CAT model of DNA substitution and with heuristic searches starting from bootstrap trees (Stamatakis et al. 2008). Branch support was calculated from 1,000 bootstrap replicates; values below 40 % are omitted. The tree was rooted with Cystofilobasidiales. Branch lengths are in terms of number of expected substitutions per alignment site (see bar); intersected branches were reduced in length by half for graphical presentation. Numbers in *circles* are referenced in text

yielded alternative topologies, in which tremellomycetous taxa form a grade, with a more basal Cystofilobasidiales separated from the remainig taxa (Bauer et al. 2006; Matheny et al. 2006; Millanes et al. 2011). Within Agaricomycotina, Tremellomycetes obtains a basal position (Floudas et al. 2012; James et al. 2006; Weiß et al. 2004).

The phylogenetic tree derived for this review from a comprehensive sampling of nrLSU data is shown in Fig. 12.7. We did not test whether or not Cystofilobasidiales is part of a monophyletic Tremellomycetes and so did not include any outgroup sequences, which increased alignment quality. Consistent with the current literature, our tree was rooted with Cystofilobasidiales. Filobasidiales branches next, followed by Holtermanniales, which is consistent with Millanes et al. (2011) and Wuczkowski et al. (2011), but in contrast to the analysis by Boekhout et al. (2011), where a sister-group relationship of Holtermanniales and Filobasidiales received high bootstrap support. The most basal branch in the remaining subtree is occupied by *Cryptococcus marinus*, a species that was found in an isolated position in several studies, for example, Boekhout et al. (2011) and Scorzetti et al. (2002). Trichosporonales and Tremellales appear as sister groups, consistent with Boekhout et al. (2011) but in contrast to other analyses where Trichosporonales cluster nested within Tremellales (Millanes et al. 2011; Sampaio 2004).

VII. Taxonomy

A. Taxonomy in Flow

Among all groups of Agaricomycotina, Tremellomycetes is particularly prone to future taxonomic changes. First, molecular phylogenetic studies strongly suggest that *Tremella*, the largest teleomorphic genus in this group, is nonmonophyletic (Fig. 12.7) (Boekhout et al. 2011; Millanes et al. 2011). The same is true for the main anamorphic genera, *Bullera* and *Cryptococcus* (Fig. 12.7). Accordingly, segregation of subgroups of these catch-all genera is being or already has been implemented (e.g., *Derxomyces*, *Dioszegia*, *Hannaella*, *Vanrija*). Second, the International Code of Nomenclature for algae, fungi, and plants (ICN) (McNeill et al. 2012) has abandoned taxonomic priority for teleomorphic stages, principally rendering obsolete taxa that had been established for teleomorphic stages detected in groups that formerly only contained anamorphic species, for example, *Bulleromyces*, *Bulleribasidium*, and *Cuniculitrema*. Likewise, it is no longer necessary to keep genera for anamorphs in originally solely teleomorphic groups, for example, *Holtermanniella* and *Mrakiella*. Some of these more recently created names for teleomorphic or anamorphic genera may be used to define appropriate monophyletic subgroups in the future. Ongoing discussions in the mycological community will yield proposals about which of the competing names to conserve or abandon.

B. Taxonomic Synopsis

What follows is a synopsis of generic names in the Tremellomycetes that are currently in use. As of this writing, questions regarding the taxonomic priority of names versus names to be conserved that have emerged as a result of the ICN (McNeill et al. 2012) (see the discussion in the previous section) have not been resolved. Therefore, we include both anamorph- and teleomorph-derived names for groups in which one or the other will probably be eliminated in the future. Taxa typified with a teleomorph are designated by an asterisk (*).

Cystofilobasidiales Fell, Roeijmans & Boekhout 1999

Cryptococcus Vuill. 1901 p.pte (type *C. neoformans*) (Fig. 12.7: 17)
*Cystofilobasidium** Oberw. & Bandoni 1983 (type *C. capitatum*) (Fig. 12.7: 2)
Guehomyces Fell & Scorzetti 2004 (type *G. pullulans*) (Fig. 12.7: 4)
Itersonilia Derx 1948 (type *I. perplexans*) (Fig. 12.7: 5)
*Mrakia** Y. Yamada & Komag. 1987 (type *M. frigida*) (Fig. 12.7: 7)
Mrakiella Margesin & Fell 2008 (type *M. cryoconiti*) (Fig. 12.7: 8)
Phaffia M.W. Mill., Yoney. & Soneda 1976 (type *P. rhodozyma*) (Fig. 12.7: 1)
Tausonia Babeva 1998 (type *M. pamirica*) (Fig. 12.7: 3)
Udeniomyces Nakase & Takem. 1992 (type *U. pyricola*) (Fig. 12.7: 6)

*Xanthophyllomyces** Golubev 1995 (type *X. dendrorhous*; teleomorph of *Phaffia rhodozyma*) (Fig. 12.7: 1)

Filobasidiales Jülich 1981

Bullera Derx 1930 p.pte (type *B. alba*) (Fig. 12.7: 33)
Cryptococcus Vuill. 1901 p.pte (type *C. neoformans*) (Fig. 12.7: 17)
*Filobasidium** L.S. Olive 1968 (type *F. floriforme*) (Fig. 12.7: 10)
*Syzygospora** G.W. Martin 1937 (type *S. alba*) (Fig. 12.7: 9)

Holtermanniales

*Holtermannia** Sacc. & Traverso 1910 (type *H. pinguis*) (Fig. 12.7: 11)
Holtermanniella Libkind, Wuczkowski, Turchetti & Boekhout 2010 (type *H. takashimae*) (Fig. 12.7: 12)

Trichosporonales Boekhout & Fell 2001

Bullera Derx 1930 p.pte (type *B. alba*) (Fig. 12.7: 33)
Cryptococcus Vuill. 1901 p.pte (type *C. neoformans*) (Fig. 12.7: 17)
Cryptotrichosporon Okoli & Boekhout 2007 (type *C. anacardii*) (Fig. 12.7: 13)
*Tetragoniomyces** Oberw. & Bandoni 1981 (type *T. uliginosus*) (Fig. 12.7: 14)
Trichosporon Behrend 1890 (type *T. beigelii*)
Vanrija R.T. Moore 1980 (type *V. humicola*) (Fig. 12.7: 15)

Tremellales Fr. 1821

*Auriculibuller** J.P. Samp. & Fonseca 2004 (type *A. fuscus*) (Fig. 12.7: 36)
Bandoniozyma P. Valente, Pagnocca, C.A. Rosa, C.F. Lee, S.O. Suh, M. Blackw., G. Péter & Fell 2012 (type *B. noutii*)
*Biatoropsis** Räsänen 1934 (type *B. usnearum*) (Fig. 12.7: 21)
Bullera Derx 1930 p.pte (type *B. alba*) (Fig. 12.7: 33)
*Bulleribasidium** J.P. Samp., M. Weiß & R. Bauer 2002 (type *B. oberjochense*) (Fig. 12.7: 29)
*Bulleromyces** Boekhout & A. Fonseca 1991 (type *B. albus*) (Fig. 12.7: 33)
*Carcinomyces** Oberw. & Bandoni 1982 (type *C. mycetophilus*) (Fig. 12.7: 16)
Cryptococcus Vuill. 1901 p.pte (type *C. neoformans*) (Fig. 12.7: 17)
*Cuniculitrema** J.P. Samp. & R. Kirschner 2001 (type *C. polymorpha*; teleomorph of *Sterigmatosporidium polymorphum*) (Fig. 12.7: 31)
Derxomyces F.Y. Bai & Q.M. Wang 2008 (type *D. mrakii*) (Fig. 12.7: 19)
Dioszegia Zsolt 1957 (type *D. hungarica*) (Fig. 12.7: 20)
Fellomyces Y. Yamada & I. Banno 1984 (type *F. polyborus*) (Fig. 12.7: 30)
*Fibulobasidium** Bandoni 1979 (type *F. inconspicuum*) (Fig. 12.7: 28)
*Filobasidiella** Kwon-Chung 1976 (type *F. neoformans*, teleomorph of *Cryptococcus neoformans*) (Fig. 12.7: 17)
Hannaella F.Y. Bai & Q.M. Wang 2008 (type *H. sinensis*) (Fig. 12.7: 18)
Kockovaella Nakase, I. Banno & Y. Yamada 1991 (type *K. thailandica*) (Fig. 12.7: 32)
*Kwoniella** Statzell-Tallman, Belloch & J.W. Fell 2008 (type *K. mangrovensis*) (Fig. 12.7: 25)
*Papiliotrema** J.P. Samp., M. Weiß & R. Bauer 2002 (type *P. bandonii*) (Fig. 12.7: 37)
*Phragmoxenidium** Oberw. 1990 (type *Phragmoxenidium mycophilum*; no DNA data available)
*Phyllogloea** Lowy 1961 (type *P. singeri*; no DNA data available)
*Rhynchogastrema** B. Metzler & Oberw. 1989 (type *R. coronatum*) (Fig. 12.7: 35)
*Sigmogloea** Bandoni & J.C. Krug 2000 (type *S. tremelloidea*; no DNA data available)
*Sirobasidium** Lagerh. & Pat. 1892 (type *S. sanguineum*) (Fig. 12.7: 27, 34)
*Sirotrema** Bandoni 1986 (type *S. pusilla*; no DNA data available)
Sterigmatosporidium G. Kraep. & U. Schulze 1983 (type *S. polymorphum*, the anamorph of *Cuniculitrema p.*) (Fig. 12.7: 31)
*Tremella** Pers. 1794 (type *T. mesenterica*) (Fig. 12.7: 26)
Tremellina Bandoni 1986 (type *T. pyrenophila*; no DNA data available)
*Trimorphomyces** Bandoni & Oberw. 1983 (type *T. papilionaceus*) (Fig. 12.7: 23, 24)
*Xenolachne** D.P. Rogers 1947 (type *X. flagellifera*; no DNA data available)

Incertae sedis

Dictyotremella Kobayashi 1971 (type *Dictyotremella novoguineensis*; no DNA data available)
Heteromycophaga P. Roberts 1997 (type *H. glandulosae*; no DNA data available)
Hyalococcus Schroeter 1889 (no DNA data available)
Neotremella Lowy 1979 (type *N. guzmanii*; no DNA data available)
Bartheletia G. Arnaud ex Scheuer, R. Bauer, M. Lutz, Stabentheiner, Melnik & Grube 2008 (type *B. paradoxa*)
Wallemia Johan-Olsen 1887 (type *W. ichthyophaga*)

C. Key Groups

1. Cystofilobasidiales

Whether this is the most basal group in the Tremellomycetes or whether Cystofilobasidiales should be excluded from Tremellomycetes in order to assure its monophyly has not yet been answered with certainty (see preceding Sect. VI). According to molecular phylogenetic analyses [e.g., Boekhout et al. (2011); this study, Fig. 12.7] the order splits into *Cystofilobasidium*, *Xanthophyllomyces* (species of both have slender holobasidia), and a clade containing *Mrakia*/*Mrakiella* and several anamorphic species of *Tausonia*/*Guehomyces*, *Itersonilia*, and *Udeniomyces*. A characteristic trait of the order, which is absent in all other groups of the Tremellomycetes, is the formation of teliospores, which can be observed in species of *Cystofilobasidium* and *Mrakia*. Teleomorphs have holobasidia producing sessile basidiospores. Dolipores lack parenthesomes (Fig. 12.5a); basidiomes are not known in this group. Species of *Cystofilobasidium* and *Xanthophyllomyces* produce carotenoids, a trait that is commercially used in *X. dendrorhous*, where the carotenoid astaxanthin produced by an optimized strain is used in industrial mariculture (Johnson and Schroeder 1995). Biogeographically, some species of the Cystofilobasidiales, for example, *Cystofilobasidium bisporidii*, and the species of *Mrakia*/*Mrakiella* are clearly cold-adapted and have been found in Arctic environments.

2. Filobasidiales

Filobasidiales contains a taxonomically heterogeneous assemblage of species. Teleomorphic species have been assigned to the morphogenera *Filobasidium* and *Syzygospora*, neither of which seems to represent a monophyletic taxon in its current circumscription (Fig. 12.7) (Boekhout et al. 2011; Millanes et al. 2011). Teleomorphic species have holobasidia; spores are sessile in most species. Species of *Filobasidium* have characteristically elongate slender basidia bearing apically a whorl of sessile basidiospores (Fig. 12.2b). Macroscopically visible fruiting bodies may be present (*Syzygospora alba*, *Syzygospora pallida*) (Fig. 12.1k) or absent. Parenthesomes are lacking in *S. pallida* (Oberwinkler et al. 1984). Species of *Syzygospora* parasitize fruiting bodies of asco- or basidiomycetes or lichen thalli (Diederich 1996; Oberwinkler et al. 1984). The ecology of most other Filobasidiales species is not known; strains have been isolated from different sources such as plants, animals, or soils.

3. Holtermanniales

Holtermanniales is the most understudied order in the Tremellomycetes. It currently contains the teleomorphic species of *Holtermannia* and some yeast species, for which the genus *Holtermanniella* was recently established (Wuczkowski et al. 2011). The only *Holtermannia* species that has been cultured and sequenced is *Holtermannia corniformis*, with small clavarioid and anatomically complex basidiomes reminiscent of *Calocera* (Bandoni et al. 2011) and tremelloid basidia. Since *H. corniformis* grows on ascomycetous stromata on dead wood and possesses tremelloid haustoria, it is probably a mycoparasitic species. Based on the available morphological data the other six described species of *Holtermannia* do not seem to be closely related to *H. corniformis* (Bandoni et al. 2011); thus, detailed morphological studies and analyses of sequence data are needed to clarify their phylogenetic position. While species of *Holtermannia* are only known from Southeast Asia and Brazil (Kirk et al. 2008), *Holtermanniella* species have been reported from Europe and North America.

4. Trichosporonales

This order nearly exclusively comprises anamorphic species, most of which are characterized by the formation of hyphae and arthroconidia (*Trichosporon*) and the lack of a yeast stage. If merged with some yeast species that probably secondarily lost the ability to form arthroconidia and are still classified in *Cryptococcus*, *Trichosporon* may represent a monophyletic group (Fig. 12.7). Roughly one-third of all described *Trichosporon* species are associated with human infections or allergic diseases (Sugita 2011).

Following Fonseca et al. (2011) and Sugita (2011) we give *Vanrija* (Moore 1980) nomenclatural priority over *Asterotremella* (Prillinger et al. 2007) for a monophyletic group of yeasts closely related to *Trichosporon* that lack arthroconidia and formerly were classified in *Cryptococcus* (*humicola* group) (Fig. 12.7: 15). Thus, we restrict the original concept of *Vanrija* (Moore 1980, 1987) to the *humicola* clade of the Trichosporonales and add some species that were described later. Our emended concept includes the new combinations *Vanrija albida* (C. Ramírez) M. Weiß, *Vanrija longa* (M. Takash., Sugita, Shinoda & Nakase) M. Weiß, *Vanrija musci* (M. Takash., Sugita, Shinoda & Nakase) M. Weiß, and *Vanrija pseudolonga* (M. Takash., Sugita, Shinoda & Nakase) M. Weiß based on *Sporobolomyces albidus* C. Ramírez (Ramírez Gómez 1957, p. 238), *Cryptococcus longus* M. Takash., Sugita, Shinoda & Nakase (Takashima et al. 2001, p. 2207), *Cryptococcus musci* M. Takash., Sugita, Shinoda & Nakase (Takashima et al. 2001, p. 2207), and *Cryptococcus pseudolongus* M. Takash., Sugita, Shinoda & Nakase (Takashima et al. 2001, p. 2208), respectively.

Interestingly, in our molecular phylogenetic analysis, as well as in Millanes et al. (2011), *Tetragoniomyces uliginosus* (Fig. 12.7: 14) seems to be a basal member of the Trichosporonales. If this position is verified in future analyses, this species would be the only member of this order for which a sexual stage is known. Like most species of *Trichosporon*, but in contrast to the majority of species in Tremellomycetes, *Tetragoniomyces* lacks a yeast stage. *Tetragoniomyces* basidia detach, and compatible basidial compartments fuse either directly or via germination tubes to establish a new dikaryotic hyphal cell (Fig. 12.2c).

5. Tremellales

Tremellales is the largest group in the Tremellomycetes and shows a high diversity of features regarding life cycles and morphology. Within the Tremellomycetes, Tremellales harbors most of the teleomorphs with conspicuous basidiocarps, most of which are still classified in the genus *Tremella* [ca. 90 species; Kirk et al. (2008)]. However, according to molecular phylogenetic analyses (e.g., Boekhout et al. 2011; Millanes et al. 2011; this study) (Fig. 12.7), *Tremella* seems to be polyphyletic and, consequently, will have to be split into monophyletic subgroups in future classifications. Obviously it will also be necessary to include anamorphic yeast species (in current taxonomy mostly still assigned to *Cryptococcus* or *Bullera*) in most of these subgroups to render them monophyletic. Judging from published sequence data, yeast stages of sequenced *Tremella* species have not yet been isolated from environmental samples; however, since most described species of *Tremella* are still without sequence data, this may be a preliminary observation.

Other teleomorphic genera of the Tremellales with sequenced members include *Auriculibuller*, *Biatoropsis*, *Bulleribasidium*, *Bulleromyces*, *Carcinomyces*, *Cuniculitrema*, *Fibulobasidium*, *Filobasidiella*, *Kwoniella*, *Papiliotrema*, *Rhynchogastrema*, *Sirobasidium*, and *Trimorphomyces*. In current molecular phylogenetic analyses (e.g., Boekhout et al. 2011; Millanes et al. 2011; this study) (Fig. 12.7), these taxa appear scattered over the Tremellales tree. Since backbone resolution is still poor, we will not speculate about phylogenetic relationships here. Basidial morphology in these taxa varies from longitudinally to obliquely to transversely or irregularly septate to nonseptate (see Micromorphology in Sect. II). Species of *Carcinomyces* and *Filobasidiella* have holobasidia; basidia in *Rhynchogastrema* are apically partially septate. In many instances teleomorphic species appear closely related to yeast species for which teleomorphic stages have not yet been observed.

Some parts of the tree contain monophyletic clades that are currently exclusively composed of yeast species. Some of these have recently been transferred from *Cryptococcus* or *Bullera* into genera of their own, for example, *Dioszegia* and *Hannaella*.

Of the various families that have been proposed in Tremellales in the past, only two, Cuniculitremaceae and Sirobasidiaceae, seem to represent monophyletic groups. A typical feature present in Cuniculitremaceae (*Fellomyces*, *Kockovaella*, *Cuniculitrema*) is the production of ballistoconidia on elongate conidiophores. Members of Sirobasidiaceae (*Fibulobasidium*, *Sirobasidium*) form cylindrical to fusiform and passively released basidiospores that are possibly homologous to the epibasidial tubes in *Tremella* (Wells and Bandoni 2001).

The production of basidiospores in chains is a unique feature of the species of *Filobasidiella*. Teleomorphs of this genus have never been reported from natural environments and are only known from in vitro fusion of compatible yeast strains. Since *C. neoformans*, the most virulent pathogen in the Tremellomycetes (see Animal and Human Pathogens in Sect. IV), is the type species of its genus, *Cryptococcus* may in the future be restricted to species now known as *Filobasidiella*.

In recent molecular phylogenetic studies (Millanes et al. 2011; this study) (Fig. 12.7: 16), *Carcinomyces effibulatus*, a holobasidiate species parasitizing the agaric *Gymnopus dryophilus* and inducing the formation of characteristic tumors in the host basidiomes, was found to be included in Tremellales. Preliminary sequence data (M. Weiß, unpublished) suggest that this taxon is conspecific with *Carcinomyces mycetophilus* and *Carcinomyces tumefaciens*. We thus reinstall *Carcinomyces* (Oberwinkler and Bandoni 1982), a genus apparently well separated from species of *Syzygospora* (Filobasidiales) (Fig. 12.7, part 1), with which *Carcinomyces* had been merged earlier (Ginns 1986).

D. Possibly Related Taxa *Incertae Sedis*

1. *Bartheletia*

In molecular phylogenetic analyses Scheuer et al. (2008) found that *Bartheletia paradoxa*, the only species of *Bartheletia*, is a member of the Agaricomycotina, but they were unable to assign it to any particular subgroup of this subphylum. *Bartheletia paradoxa* is a dimorphic fungus that rapidly develops on fallen *Ginkgo biloba* leaves in its filamentous conidiogenous anamorphic state in autumn. Later, resting teliospores are formed that germinate a year later into longitudinally septate basidia. Since all other teliospore-generating taxa in the Agaricomycotina belong to the *Cystofilobasidiales*, Scheuer et al. (2008) speculated about a possible relationship of *Bartheletia* to the Tremellomycetes. However, unlike most members of the Tremellomycetes, *B. paradoxa* lacks a yeast stage. In addition, that fungus differs from all other known members of the Agaricomycotina by the absence of dolipores. Instead, *Bartheletia* has multiple plasmodesmalike perforations in its hyphal septa (Scheuer et al. 2008).

2. *Wallemia*

The hyphomycete genus *Wallemia* includes three described species that can tolerate osmotic stress and are hence regularly detected as contaminants of low-moisture foods (Zalar et al. 2005). Ultrastructural data have shown the basidiomycetous nature of *Wallemia sebi* (Moore 1986). However, molecular phylogenetic analyses could not unambiguously assign it to any of the major basidiomycetous clades (Matheny et al. 2006). Recent phylogenomic studies with limited taxon sampling placed *Wallemia* in a basal position within the Agaricomycotina (Padamsee et al. 2012; Zajc et al. 2013). The septal pore apparatus of *W. sebi* resembles that of Tremellales (Fig. 12.5d) (Padamsee et al. 2012).

VIII. Conclusions

We have provided an overview of Tremellomycetes, a basal group in the Agaricomycotina. For most of its species, knowledge of ecology and phylogeography is still sparse. A particular problem for the taxonomy of this group is the fact that teleomorphs and anamorphs have mostly been studied by different scientific communities using different taxonomic methods. Polymerase chain reaction–based advances in molecular biology have triggered an integration of these two taxonomic approaches into a consistent classification system, yet a sound phylogenetic classification is only just emerging. We expect that most of the open systematic questions will be solved in the near future by phylogenomic analyses, when more genomes in the Tremellomycetes will be available (as of writing this chapter, genome data are only available for *C. neoformans* and *T. mesenterica*). We hope that, along with the progress in molecular techniques and data, a rising number of mycologists will be interested in and capable of studying these fascinating fungi in the field, to shed light on the biodiversity still unknown.

References

Abe F, Minegishi H, Miura T, Nagahama T, Usami R, Horikoshi K (2006) Characterization of cold- and high-pressure-active polygalacturonases from a deep-sea yeast, *Cryptococcus liquefaciens* strain N6. Biosci Biotechnol Biochem 70:296–299

Bandoni RJ (1984) The Tremellales and Auriculariales: an alternative classification. Trans Mycol Soc Jpn 25:489–530

Bandoni RJ, Boekhout T, Sampaio JP (2011) *Holtermannia* (Saccardo & Traverso 1910). In: Kurtzman CP, Fell JW, Boekhout T (eds) The yeasts: a taxonomic study. Elsevier, Amsterdam, pp 1467–1470

Bauer R (2004) Basidiomycetous interfungal cellular interactions—a synopsis. In: Agerer R, Piepenbring M, Blanz P (eds) Frontiers in basidiomycote mycology. IHW, Eching, pp 325–337

Bauer R, Oberwinkler F (1990a) Direct cytoplasm-cytoplasm connection: an unusual host-parasite interaction of the tremelloid mycoparasite *Tetragoniomyces uliginosus*. Protoplasma 154:157–160

Bauer R, Oberwinkler F (1990b) Haustoria of the mycoparasitic heterobasidiomycete *Christiansenia pallida*. Cytologia 55:419–424

Bauer R, Begerow D, Sampaio JP, Weiß M, Oberwinkler F (2006) The simple-septate basidiomycetes: a synopsis. Mycol Prog 5:41–66

Berbee ML, Wells K (1988) Ultrastructural studies of mitosis and the septal pore apparatus in *Tremella globospora*. Mycologia 80:479–492

Bergauer P, Fonteyne PA, Nolard N, Schinner F, Margesin R (2005) Biodegradation of phenol and phenol-related compounds by psychrophilic and cold-tolerant alpine yeasts. Chemosphere 59:909–918

Boekhout T (2011) *Itersonilia* Derx (1948). In: Kurtzman CP, Fell JW, Boekhout T (eds) The yeasts: a taxonomic study. Elsevier, Amsterdam, pp 1777–1780

Boekhout T, Fonseca Á, Sampaio JP, Bandoni RJ, Fell JW, Kwon-Chung KJ (2011) Discussion of teleomorphic and anamorphic basidiomycetous yeasts. In: Kurtzman CP, Fell JW, Boekhout T (eds) The yeasts: a taxonomic study. Elsevier, Amsterdam, pp 1339–1372

Brefeld O (1888) Basidiomyceten II. Protobasidiomyceten, vol VII. Untersuchungen aus dem Gesammtgebiete der Mykologie. Arthur Felix, Leipzig

Chang ST, Wasser SP (2012) The role of culinary-medicinal mushrooms on human welfare with a pyramid model for human health. Int J Med Mushrooms 14:95–134

Chen C-J (1998) Morphological and molecular studies in the genus *Tremella*. Bibl Mycol 174:1–225

David-Palma M, Libkind D, Sampaio JP (2014) Global distribution, diversity hot spots and niche transitions of an astaxanthin-producing eukaryotic microbe. Mol Ecol 23:921–932

Desnos-Ollivier M, Patel S, Spaulding AR, Charlier C, Garcia-Hermoso D, Nielsen K, Dromer F (2010) Mixed infections and in vivo evolution in the human fungal pathogen *Cryptococcus neoformans*. mBio 1:e00091-10

Diederich P (1996) The lichenicolous heterobasidiomycetes. Bibl Lichenol 61:1–198

Fell JW (2011) *Mrakia* Y. Yamada & Komagata (1987). In: Kurtzman CP, Fell JW, Boekhout T (eds) The yeasts: a taxonomic study. Elsevier, Amsterdam, pp 1503–1510

Fell JW, Scorzetti G, Connell L, Craig S (2006) Biodiversity of micro-eukaryotes in Antarctic dry valley soils with < 5 % soil moisture. Soil Biol Biochem 38:3107–3119

Fell JW, Johnson EA, Scorzetti G (2011) *Xanthophyllomyces* Golubev (1995). In: Kurtzman CP, Fell JW, Boekhout T (eds) The yeasts: a taxonomic study. Elsevier, Amsterdam, pp 1595–1599

Floudas D, Binder M, Riley R, Barry K, Blanchette RA, Henrissat B, Martinez AT, Otillar R, Spatafora JW, Yadav JS, Aerts A, Benoit I, Boyd A, Carlson A, Copeland A, Coutinho PM, de Vries RP, Ferreira P, Findley K, Foster B, Gaskell J, Glotzer D, Gorecki P, Heitman J, Hesse C, Hori C, Igarashi K, Jurgens JA, Kallen N, Kersten P, Kohler A, Kues U, Kumar TKA, Kuo A, LaButti K, Larrondo LF, Lindquist E, Ling A, Lombard V, Lucas S, Lundell T, Martin R, McLaughlin DJ, Morgenstern I, Morin E, Murat C, Nagy LG, Nolan M, Ohm RA, Patyshakuliyeva A, Rokas A, Ruiz-Duenas FJ, Sabat G, Salamov A, Samejima M, Schmutz J, Slot JC, John FS, Stenlid J, Sun H, Sun S, Syed K, Tsang A, Wiebenga A, Young D, Pisabarro A, Eastwood DC, Martin F, Cullen D, Grigoriev IV, Hibbett DS (2012) The paleozoic origin of enzymatic lignin decomposition reconstructed from 31 fungal genomes. Science 336:1715–1719

Fonseca Á, Boekhout T, Fell JW (2011) *Cryptococcus* Vuillemin (1901). In: Kurtzman CP, Fell JW, Boekhout T (eds) The yeasts: a taxonomic study. Elsevier, Amsterdam, pp 1661–1737

Fries EM (1821) Systema mycologicum I. Lundae

Gadanho M, Sampaio JP (2005) Occurrence and diversity of yeasts in the mid-atlantic ridge hydrothermal fields near the Azores Archipelago. Microb Ecol 50:408–417

Ginns J (1986) The genus *Syzygospora* (Heterobasidiomycetes: Syzygosporaceae). Mycologia 78:619–636

Golubev WI, Golubev NW (2003) A new basidiomycetous yeast species, *Cryptococcus mycelialis*, related to *Holtermannia* Saccardo et Traverso. Microbiology (Moscow) 72:728–732

Hawksworth DL (2011) A new dawn for the naming of fungi: impacts of decisions made in Melbourne in July 2011 on the future publication and regulation of fungal names. IMA Fungus 2:155–162

Hibbett DS, Binder M, Bischoff JF, Blackwell M, Cannon PF, Eriksson OE, Huhndorf S, James T, Kirk PM,

Lücking R, Lumbsch T, Lutzoni F, Matheny PB, McLaughlin DJ, Powell MJ, Redhead S, Schoch CL, Spatafora JW, Stalpers JA, Vilgalys R, Aime MC, Aptroot A, Bauer R, Begerow D, Benny GL, Castlebury LA, Crous PW, Dai Y-C, Gams W, Geiser DM, Griffith GW, Gueidan C, Hawksworth DL, Hestmark G, Hosaka K, Humber RA, Hyde K, Ironside JE, Kõljalg U, Kurtzman CP, Larsson K-H, Lichtwardt R, Longcore J, Miadlikowska J, Miller A, Moncalvo J-M, Mozley-Standridge S, Oberwinkler F, Parmasto E, Reeb V, Rogers JD, Roux C, Ryvarden L, Sampaio JP, Schüssler A, Sugiyama J, Thorn RG, Tibell L, Untereiner W, Walker C, Wang Z, Weir A, Weiß M, White MM, Winka K, Yao Y-J, Zhang N (2007) A higher-level phylogenetic classification of the Fungi. Mycol Res 111:509–547

Hibbett DS, Bauer R, Binder M, Giachini AJ, Hosaka K, Justo A, Larsson E, Larsson KH, Lawrey JD, Miettinen O, Nagy L, Nilsson RH, Weiß M, Thorn RG (2014) Agaricomycetes, chapter 14. In: McLaughlin DJ, Spatafora JW (eds) The mycota, Vol. VII, Part B: systematics and evolution. Springer, Heidelberg

James TY, Kauff F, Schoch CL, Matheny PB, Hofstetter V, Cox CJ, Celio G, Gueidan C, Fraker E, Miadlikowska J, Lumbsch HT, Rauhut A, Reeb V, Arnold AE, Amtoft A, Stajich JE, Hosaka K, Sung GH, Johnson D, O'Rourke B, Crockett M, Binder M, Curtis JM, Slot JC, Wang Z, Wilson AW, Schussler A, Longcore JE, O'Donnell K, Mozley-Standridge S, Porter D, Letcher PM, Powell MJ, Taylor JW, White MM, Griffith GW, Davies DR, Humber RA, Morton JB, Sugiyama J, Rossman AY, Rogers JD, Pfister DH, Hewitt D, Hansen K, Hambleton S, Shoemaker RA, Kohlmeyer J, Volkmann-Kohlmeyer B, Spotts RA, Serdani M, Crous PW, Hughes KW, Matsuura K, Langer E, Langer G, Untereiner WA, Lucking R, Budel B, Geiser DM, Aptroot A, Diederich P, Schmitt I, Schultz M, Yahr R, Hibbett DS, Lutzoni F, McLaughlin DJ, Spatafora JW, Vilgalys R (2006) Reconstructing the early evolution of Fungi using a six-gene phylogeny. Nature 443:818–822

Johnson EA, Schroeder W (1995) Astaxanthin from the yeast *Phaffia rhodozyma*. Stud Mycol 38:81–90

Katoh K, Standley DM (2013) MAFFT multiple sequence alignment software Version 7: improvements in performance and usability. Mol Biol Evol 30:772–780

Kirk PM, Cannon PF, Minter DW, Stalpers JA (2008) Dictionary of the Fungi, 10th edn. CAB International, Wallingford

Kirschner R, Sampaio JP, Gadanho M, Weiß M, Oberwinkler F (2001) *Cuniculitrema polymorpha* (Tremellales, gen. nov. and sp. nov.), a heterobasidiomycete vectored by bark beetles, which is the teleomorph of *Sterigmatosporidium polymorphum*. Antonie Van Leeuwenhoek 80:149–161

Kumar S, Skjaeveland A, Orr RJS, Enger P, Ruden T, Mevik BH, Burki F, Botnen A, Shalchian-Tabrizi K (2009) AIR: a batch-oriented web program package for construction of supermatrices ready for phylogenomic analyses. BMC Bioinformatics 10:357

Kurtzman CP, Fell JW, Boekhout T (eds) (2011) The yeasts: a taxonomic study. Elsevier, Amsterdam

Kwon-Chung KJ (2011) *Filobasidiella* Kwon-Chung (1975). In: Kurtzman CP, Fell JW, Boekhout T (eds) The yeasts: a taxonomic study. Elsevier, Amsterdam, pp 1443–1455

Lopandic K, Prillinger H, Wuczkowski M (2011) *Fellomyces* Y. Yamada & Banno (1984). In: Kurtzman CP, Fell JW, Boekhout T (eds) The yeasts: a taxonomic study. Elsevier, Amsterdam, pp 1759–1772

Matheny PB, Gossmann JA, Zalar P, Arun Kumar TK, Hibbett DS (2006) Resolving the phylogenetic position of the Wallemiomycetes: an enigmatic major lineage of Basidiomycota. Can J Bot 84:1794–1805

McNeill J, Barrie FR, Buck WR, Demoulin V, Greuter W, Hawksworth DL, Herendeen PS, Knapp S, Marhold K, Prado J, Prud'homme van Reine WF, Smith GF, Wiersema JH, Turland NJ (2012) International code of nomenclature for algae, fungi, and plants (Melbourne Code). Regnum Vegetabile, vol 154. Koeltz Scientific Books

Metin B, Findley K, Heitman J (2010) The mating type locus (MAT) and sexual reproduction of *Cryptococcus heveanensis*: insights into the evolution of sex and sex-determining chromosomal regions in Fungi. PLoS Genet 6:e1000961

Metzler B, Oberwinkler F, Petzold H (1989) *Rhynchogastrema* gen. nov. and Rhynchogastremaceae fam. nov. (Tremellales). Syst Appl Microbiol 12:280–287

Millanes AM, Diederich P, Ekman S, Wedin M (2011) Phylogeny and character evolution in the jelly fungi (Tremellomycetes, Basidiomycota, Fungi). Mol Phylogenet Evol 61:12–28

Moore RT (1980) Taxonomic proposals for the classification of marine yeasts and other yeast-like fungi including the smuts. Bot Mar 23:361–374

Moore RT (1986) A note on *Wallemia sebi*. Antonie Van Leeuwenhoek 52:183–187

Moore RT (1987) Additions to the genus *Vanrija*. Bibl Mycol 108:167–173

O'Gorman CM, Fuller HT, Dyer PS (2009) Discovery of a sexual cycle in the opportunistic fungal pathogen *Aspergillus fumigatus*. Nature 457:471–474

Oberwinkler F (2014) Dacrymycetes, chapter 13. In: McLaughlin DJ, Spatafora JW (eds) The mycota, Vol. VII, Part B: systematics and evolution. Springer, Heidelberg

Oberwinkler F, Bandoni RJ (1981) *Tetragoniomyces* gen. nov. and Tetragoniomycetaceae fam. nov. (Tremellales). Can J Bot 59:1034–1040

Oberwinkler F, Bandoni R (1982) Carcinomycetaceae: a new family in the Heterobasidiomycetes. Nord J Bot 2:501–516

Oberwinkler F, Bandoni RJ (1983) *Trimorphomyces*: a new genus in the Tremellaceae. Syst Appl Microbiol 4:105–113

Oberwinkler F, Lowy B (1981) *Syzygospora alba*, a mycoparasitic heterobasidiomycete. Mycologia 73:1108–1115

Oberwinkler F, Bandoni R, Blanz P, Kisimova-Horovitz L (1983) *Cystofilobasidium*: a new genus in the Filobasidiaceae. Syst Appl Microbiol 4:114–122

Oberwinkler F, Bandoni RJ, Bauer R, Deml G, Kisimova-Horovitz L (1984) The life-history of *Christiansenia pallida*, a dimorphic, mycoparasitic heterobasidiomycete. Mycologia 76:9–22

Oberwinkler F, Bauer R, Schneller J (1990) *Phragmoxenidium mycophilum* sp. nov., an unusual mycoparasitic heterobasidiomycete. Syst Appl Microbiol 13:186–191

Padamsee M, Kumar TKA, Riley R, Binder M, Boyd A, Calvo AM, Furukawa K, Hesse C, Hohmann S, James TY, LaButti K, Lapidus A, Lindquist E, Lucas S, Miller K, Shantappa S, Grigoriev IV, Hibbett DS, McLaughlin DJ, Spatafora JW, Aime MC (2012) The genome of the xerotolerant mold *Wallemia sebi* reveals adaptations to osmotic stress and suggests cryptic sexual reproduction. Fungal Genet Biol 49:217–226

Park BJ, Wannemuehler KA, Marston BJ, Govender N, Pappas PG, Chiller TA (2009) Estimation of the current global burden of cryptococcal meningitis among persons living with HIV/AIDS. AIDS 23:525–530

Patouillard N (1887) Les Hyménomycètes d'Europe. Paris

Perfect JR (2005) *Cryptococcus neoformans*: a sugar-coated killer with designer genes. FEMS Immunol Med Microbiol 45:395–404

Persoon CH (1794) Dispositio methodica fungorum. Neues Magazin für die Botanik 1:81–128

Prillinger H, Lopandic K, Sugita T, Wuczkowski M (2007) *Asterotremella* gen. nov. *albida*, an anamorphic tremelloid yeast isolated from the agarics *Asterophora lycoperdoides* and *Asterophora parasitica*. J Gen Appl Microbiol 53:167–175

Ramírez Gómez C (1957) Contribución al estudio de la ecología de las levaduras. I. Estudio de levaduras aisladas de hongos carnosos. Microbiol Esp 10:215–247

Sampaio JP (2004) Diversity, phylogeny and classification of basidiomycetous yeasts. In: Agerer R, Piepenbring M, Blanz P (eds) Frontiers in basidiomycote mycology. IHW, Eching, pp 49–80

Sampaio JP, Gadanho M, Bauer R (2001) Taxonomic studies on the genus Cystofilobasidium: description of *Cystofilobasidium ferigula* sp. nov. and clarification of the status of Cystofilobasidium lari-marini. Int J Syst Evol Microbiol 51:221–229

Sampaio JP, Weiß M, Gadanho M, Bauer R (2002) New taxa in the Tremellales: *Bulleribasidium oberjochense* gen. et sp. nov., *Papiliotrema bandonii* gen. et sp. nov. and *Fibulobasidium murrhardtense* sp. nov. Mycologia 94:873–887

Scheuer C, Bauer R, Lutz M, Stabentheiner E, Mel'nikd VA, Grube M (2008) *Bartheletia paradoxa* is a living fossil on *Ginkgo* leaf litter with a unique septal structure in the Basidiomycota. Mycol Res 112:1265–1279

Scorzetti G, Fell JW, Fonseca A, Statzell-Tallman A (2002) Systematics of basidiomycetous yeasts: a comparison of large subunit D1/D2 and internal transcribed spacer rDNA regions. FEMS Yeast Res 2:495–517

Stamatakis A, Hoover P, Rougemont J (2008) A rapid bootstrap algorithm for the RAxML web servers. Syst Biol 57:758–771

Sugita T (2011) *Trichosporon* Behrend 1890. In: Kurtzman CP, Fell JW, Boekhout T (eds) The yeasts: a taxonomic study. Elsevier, Amsterdam, pp 2015–2061

Swann EC, Taylor JW (1995) Phylogenetic perspectives on basidiomycete systematics: evidence from the 18S rRNA gene. Can J Bot 73:S862–S868

Takashima M, Sugita T, Shinoda T, Nakase T (2001) Reclassification of the *Cryptococcus humicola* complex. Int J Syst Evol Microbiol 51:2199–2210

Takishita K, Tsuchiya M, Reimer JD, Maruyama T (2006) Molecular evidence demonstrating the basidiomycetous fungus *Cryptococcus curvatus* is the dominant microbial eukaryote in sediment at the Kuroshima Knoll methane seep. Extremophiles 10:165–169

Tulasne LR (1853) Observations sur l'organisation des Trémellinées. Ann Sci Nat Bot 19:193–231

Weiß M, Oberwinkler F (2001) Phylogenetic relationships in Auriculariales and related groups—hypotheses derived from nuclear ribosomal DNA sequences. Mycol Res 105:403–415

Weiß M, Bauer R, Begerow D (2004) Spotlights on heterobasidiomycetes. In: Agerer R, Piepenbring M, Blanz P (eds) Frontiers in basidiomycote mycology. IHW, Eching, pp 7–48

Wells K (1994) Jelly fungi, then and now! Mycologia 86:18–48

Wells K, Bandoni R (2001) Heterobasidiomycetes. In: McLaughlin DJ, McLaughlin EG, Lemke PA (eds) The mycota, Vol. VII, Part B: systematics and evolution. Springer, Berlin, pp 85–120

Wuczkowski M, Passoth V, Turchetti B, Andersson AC, Olstorpe M, Laitila A, Theelen B, van Broock M, Buzzini P, Prillinger H, Sterflinger K, Schnurer J, Boekhout T, Libkind D (2011) Description of *Holtermanniella* gen. nov., including *Holtermanniella takashimae* sp. nov. and four new combinations, and proposal of the order Holtermanniales to accommodate tremellomycetous yeasts of the *Holtermannia* clade. Int J Syst Evol Microbiol 61:680–689

Zajc J, Liu Y, Dai W, Yang Z, Hu J, Constinčar C, Gunde-Cimerman N (2013) Genome and transcriptome sequencing of the halophilic fungus

Wallemia ichthyophaga: haloadaptations present and absent. BMC Genomics 14:617
Zalar P, de Hoog GS, Schroers H-J, Frank JM, Gunde-Cimerman N (2005) Taxonomy and phylogeny of the xerophilic genus *Wallemia* (Wallemiomycetes and Wallemiales, cl. et ord. nov.). Antonie Van Leeuwenhoek 87:311–328
Zugmaier W, Oberwinkler F (1995) Tremelloid haustorial cells with haustorial filaments and potential host-range of *Tremella mesenterica*. Nord J Bot 15:207–213
Zugmaier W, Bauer R, Oberwinkler F (1994) Mycoparasitism of some *Tremella* species. Mycologia 86:49–56

13 Dacrymycetes

Franz Oberwinkler[1]

CONTENTS

I. Introduction

Dacrymycetes is a well-defined class in the Agaricomycotina, Basidiomycota. **The most important characteristic of this class is its unique basidial ontogeny and morphology** (Fig. 13.1d). Young basidia grow to become slightly clavate and expand apically to produce two thick, cylindrical, and long sterigmata that taper abruptly to form spicula on which large asymmetrically attached spores develop. Between the sterigmata the original apex of the young basidium remains visible, a characteristic that is morphologically rather distinctive. Most species have curved-cylindrical basidiospores that are typically transversely septate when mature (Fig. 13.1e–g). Basidiospores commonly germinate with microconidia (Fig. 13.1e, f), but germ-tube formation is also widespread (Fig. 13.1g). The production of secondary spores is not known in Dacrymycetales. Several species of *Dacrymyces*, *Femsjonia*, *Guepiniopsis*, and *Calocera*, which have been studied extensively in pure culture, produce **limited yeast colonies** that originate from microconidia (Fig. 13.1f). Microconidium formation also occurs on hyphae (Fig. 13.1h) and functions as additional asexual reproduction. In *Dacrymyces stillatus*, hyphal fragmentation, also considered a conidial stage, is rather common and serves as an effective dispersal mechanism (Fig. 13.1i). Hyphal walls partly gelatinize, thereby forming a soft waxy to gelatinous, sometimes tough, consistency. **Basidiocarp morphology** varies from strictly corticioid to pustulate, cupulate, cyphelloid, stalked-capitate, or clavarioid. The abhymenial surfaces often produce strongly differentiated so-called marginal hairs, which are terminal hyphal cells that are characteristic of certain taxa. Cystidia are absent in the hymenia, but conspicuously branched dikaryophyses often occur. Dacrymycetes species form predominantly gelatinous, **mostly bright yellowish to orange basidiocarps** that are pigmented by carotenoids. Most fructifications have rootlike bases in the wood; rarely are they broadly attached to the substrate. An intensive brown rot is associated with growth in coniferous and angiosperm wood, but lignin decomposition may also occur.

[1]Institut für Evolution und Ökologie, Universität Tübingen, Auf der Morgenstelle 1, 72076 Tübingen, Germany; e-mail: franz.oberwinkler@uni-tuebingen.de

Systematics and Evolution, 2nd Edition
The Mycota VII Part A
D.J. McLaughlin and J.W. Spatafora (Eds.)

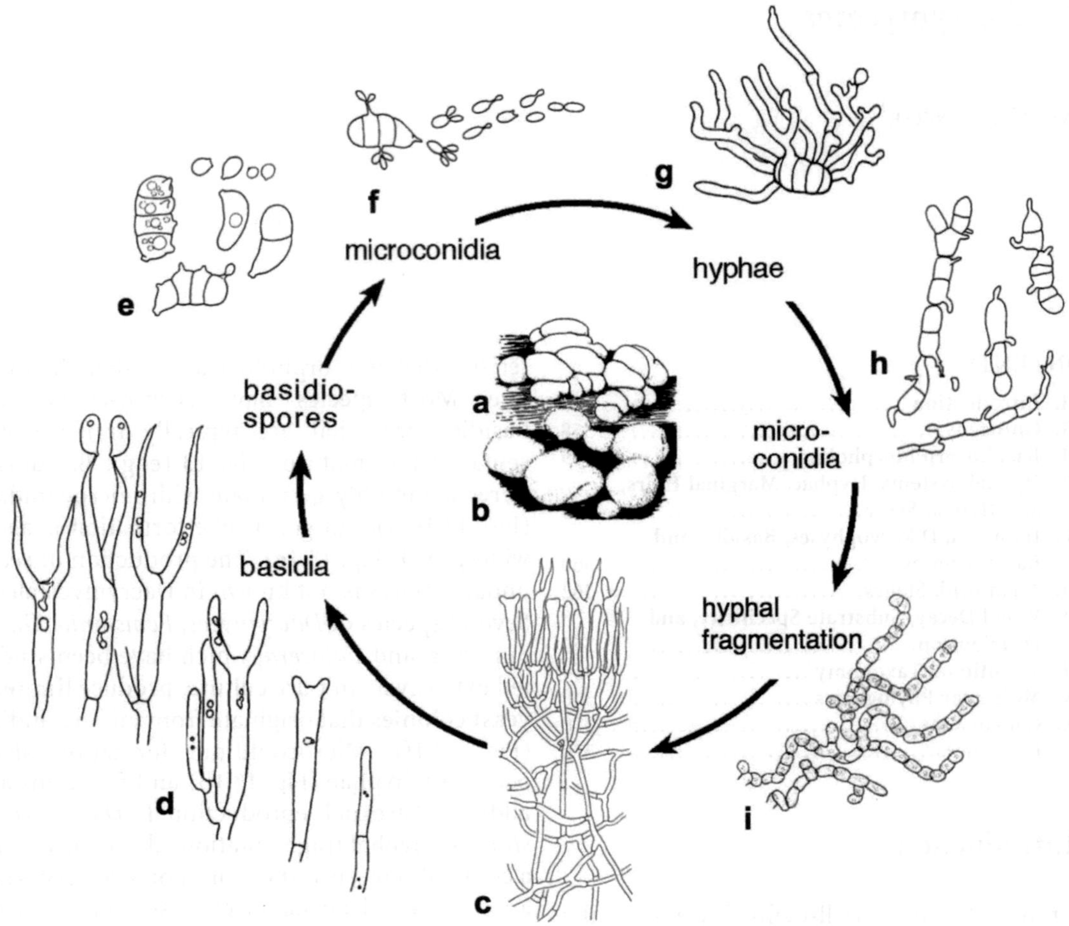

Fig. 13.1 Life cycle of *Dacrymyces stillatus*. (**a**) Basidiocarps (**b**) Fructification with fragmenting hyphae (**i**). (**c**) Detail of hymenium and subhymenium. (**d**) Basidial ontogeny showing stages of nuclear divisions in basidia. (**e**) Basidiospores and spore germination. (**f**) Yeastlike budding of microconidia. (**g**) Spore germination with hyphae, illustrated from *D. palmatus* but also occurring in *D. stillatus*. (**h**) Fragmented hyphae producing microconidia. (**i**) Short-celled fragmentation of peripheral hyphae from anamorph fructification (**b**). Figures not to scale; from Oberwinkler (2012)

II. Ontogeny

The life cycle of *D. stillatus* (Fig. 13.1) is taken as a representative example of the Dacrymycetes to illustrate the most important developmental stages. The morphology of basidiocarps (Figs. 13.2 and 13.5), however, is rather varied. Usually, the hyphal context of the trama and the subhymenium (Figs. 13.1c and 13.3a) is gelatinous due to the gelatinizing outer hyphal walls and depends on the water content of the fructifications. Hymenia are single-layered in young stages and may develop into mutlilayered thickening ones in older basidiocarps (Fig. 13.1c). **A typical basidial ontogeny** is illustrated in Fig. 13.1d. Only in rare cases are basidia three- or one-sterigmate, the latter as in *Unilacryma unispora* (Fig. 13.8e). In most cases, mature basidiospores are transversely septate (Fig. 13.1e) and germinate with microconidia that appear to reproduce by budding (Fig. 13.1e), however limited in time and space. Spore germination with hyphae, illustrated by *Dacrymces palmatus* but also occurring in *D. stillatus*, is common, and microconidia can develop on these haploid hyphae (Fig. 13.1g). Fragmented hyphae are also

Fig. 13.2 Basidiocarps and anamorph fructification (**b**) of Dacrymycetales species. (**a, b**) *Dacrymyces stillatus*. (**a**) Basidial stage with hymenium covering whole upper side. (**b**) Conidial stage with hyphal fragmentation in whole fructification. (**c**) *Dacrymyces variisporus*, well-developed basidiocarp. (**d**) *Dacrymyces palmatus*, hymenium on upper, folded surface of basidiocarp. (**e**) *Ditiola haasii*; note hyaline basidiocarps without carotenoids, hymenia marked by light grayish and slightly rough lower surfaces. (**f**) *Dacrymyces chrysospermus*, hymenium covering folded upper surface of basidiocarp. (**g**) *Dacryopinax spathularia*, hymenia on underside of bent basidiocarps. (**h**) *Heterotextus alpinus*, conelike basidiocarps with hymenia on flattened underside.(**i**) *Guepiniopsis buccina*, hymenia in cups, geotropically positively oriented. (**j**) *Dacryonaema rufum*, young basidiocarps on exposed, hard coniferous wood together with crustose lichens. (**k**) *Calocera cornea* with mostly unforked clavarioid basidiocarps on hardwood. (**l**) *Calocera viscosa*, coralloid basidiocarp with long root inserted in coniferous wood. *Bars*=5 mm. All figure originals F. Oberwinkler

capable of producing microconidia (Fig. 13.1h). **Fragmenting hyphae** (Fig. 13.1i) normally occur in asexual fructifications but occasionally in basidiocarps (Fig. 13.1a, b). Asexual fructifications are common in *D. stillatus* but rare or lacking in other species of the Dacrymycetes.

III. Basidiocarp Morphology

Basidiocarp morphology is distinct in most Dacrymycetes species (Figs. 13.2 and 13.5), but variation during ontogenetic development must be considered. The **traditional generic concept** is based predominantly on the morphology of basidiocarps (Fig. 13.5). In contrast to all other Dacrymycetes, species of *Cerinomyces* (Fig. 13.8a) have corticioid (resupinate) growth without definite margins. The flat to slightly cupulate fructifications of *Arrhytidia* have rooting bases and distinct margins. The basidiocarps of other cupulate genera (Fig. 13.5), for example, *Ditiola*, *Heterotextus*, *Femsjonia*, *Guepiniopsis*, and *Dacryopinax*, can be

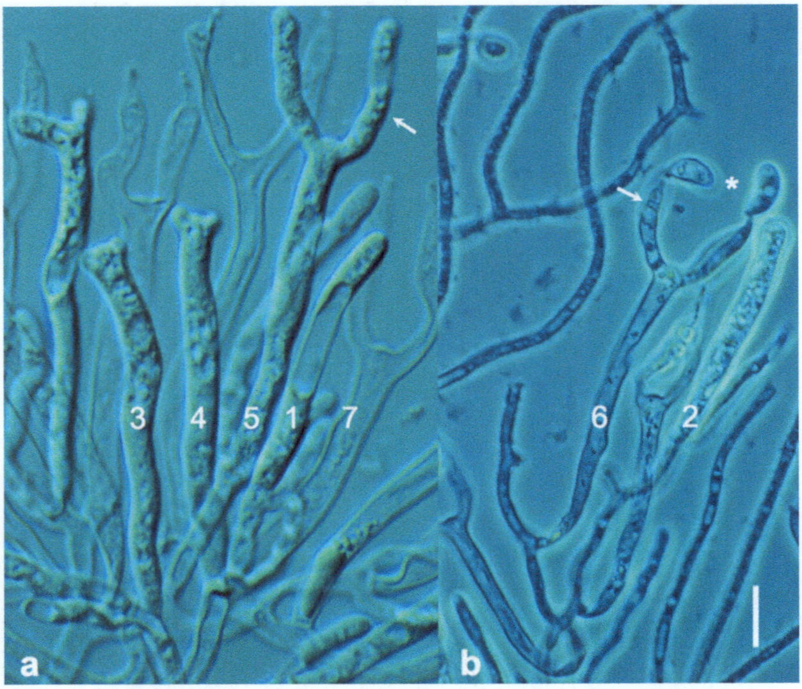

Fig. 13.3 Light microscopy of *Dacrymyces stillatus* showing sequence of ontogenetic steps, 1–7, in basidial development. (**a**) Nomarski contrast. (**b**) Transmitted bright-field microscopy. Note long basidial bodies in which nuclear divisions occur and two long sterigmata (*arrows*) with terminally narrowing spicula on which basidiospores (*asterisk*) develop. *Bar*=10 μm. Originals F. Oberwinkler

characterized by growth direction, stipe morphology, and cell differentiations of marginal hairs. Other stalked species are grouped in *Dacryonaema* with globose fertile parts, *Calocera* with simple or forked clavarioid basidiocarps, and *Dacryomitra* with a minute morchelloid habit.

IV. Hyphal Systems, Hyphae, Marginal Hairs, and Hyphal Septa

The reason for the development of jelly fructifications with a soft or tough context in Dacrymycetes is a **strong tendency of outer hyphal wall layers to gelatinize**. In addition, thin-walled and hyaline hyphae are rather common, but various kinds of wall thickenings and carotenoid pigmentations also occur. In particular, terminal cells or cell chains of marginal hyphae are often very characteristically structured and therefore used for traditional generic delimitations, as in *Dacryonaema*, *Guepiniopsis*, and *Heterotextus* (Fig. 13.7). Hyphal clamps are present or lacking and are thus consistent specific characters. All dacrymycetaceous species studied so far show dolipores with continuous parenthesomes, except for a tiny central pore (Fig. 13.4).

V. Hymenia, Dikaryophyses, Basidia, and Basidiospores

Hymenial surfaces in dacrymycetaceous species are normally smooth. However, well-developed hymenia may become cerebriform to irregularly flabellate. Luxuriantly growing hymenia will alter the original shape of basidiocarps considerably.

Usually, the **hymenial layer** is composed exclusively of basidia in different developmental stages (Fig. 13.3). Some species, like *Dacrymyces estonicus*, possess slender, unbranched

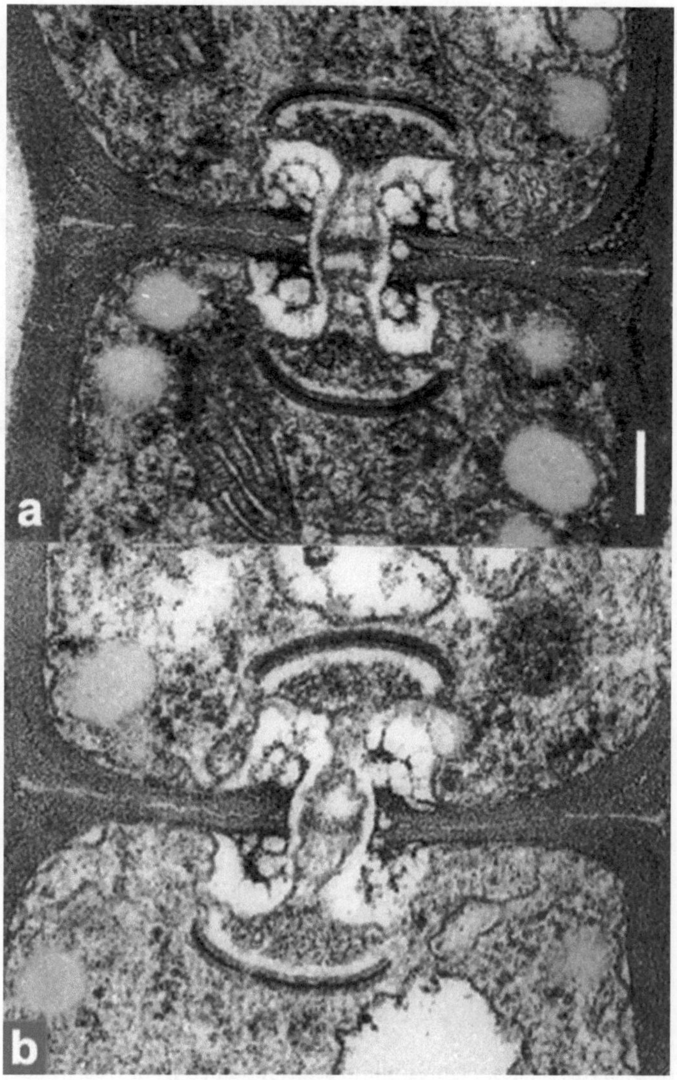

Fig. 13.4 (**a**, **b**) Transmission electron micrographs showing dolipores of *Dacrymyces stillatus* with continuous parenthesomes on both sides of pores and central bandings of different electron-dense structures inside pores. (**a**) Upper parenthesome contains central pore. *Bar*=0.5 μm. Originals F. Oberwinkler

dikaryophyses (hyphidia). Branched dikaryophyses are known, for example, in *Dacrymyces enatus* and *Dacrymyces ovisporus* (Fig. 13.8d). Conspicuously branched dikaryophyses, dendrohyphidia, are found in *Dacrymyces dendrocalami* (Fig. 13.8b), *Dacrymyces macnabbii* (Reid 1974), *Dacrymyces paraphysatus*, and *U. unispora* (Fig. 13.8e).

As mentioned previously, basidial ontogeny and morphology are the most distinctive characters of Dacrymycetes (Figs. 13.1, 13.3, and 13.7). Deviations, as in *U. unispora*, are very rare. *D. ovisporus* (Fig. 13.8d) clearly bridges the gap to typical dacrymycetaceous basidia.

In addition, the majority of **basidiospores** share common characters, i.e., transverse septation in mature stages and germination with microconidia or hyphae but not with secondary spores (Fig. 13.9b–f, h). Exceptionally, septation of basidiospores is lacking, as in *Cerinomyces crustulinus* (Fig. 13.9a), *Cerinomyces canadensis*, and *Cerinomyces pallidus*. In addition, most species share cylindrical to allantoid basidiospores, while subglobose ones are restricted to

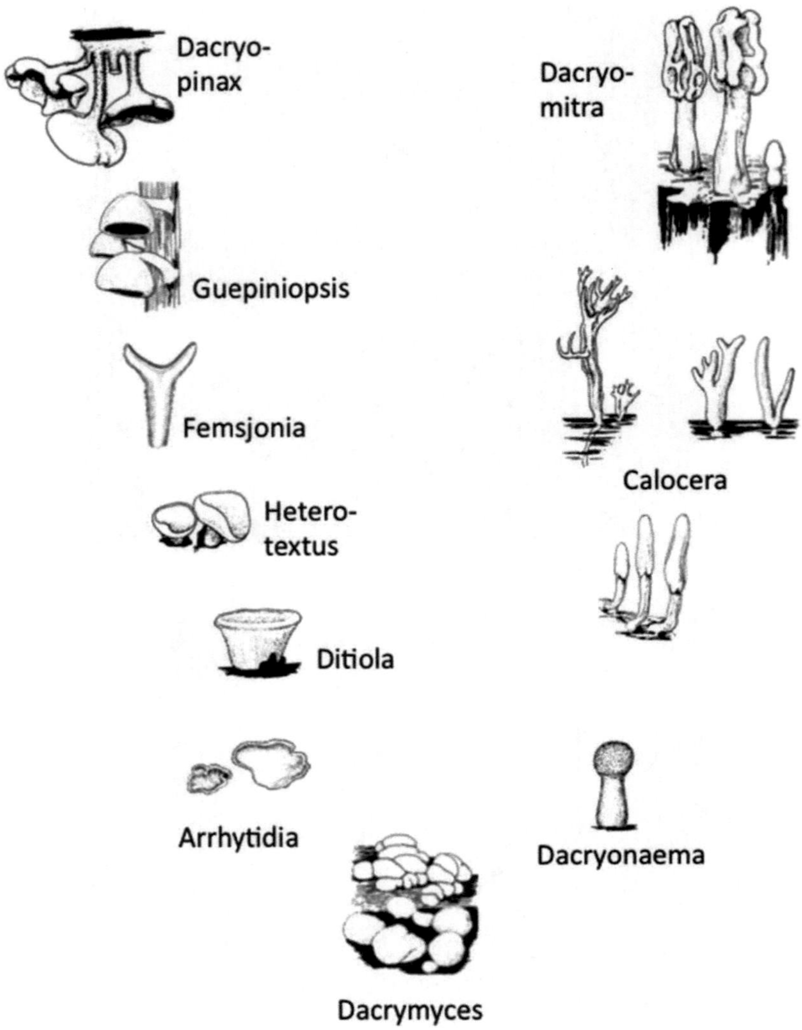

Fig. 13.5 Basidiocarps of Dacrymycetales. Figures not to scale, modified from Oberwinkler (2012)

D. ovisporus and *U. unispora* (Figs. 13.8e and 13.9d). The latter species have oblique to muriform spore septations.

VI. Anamorph Stages

Microconidial formation during germination of basidiospores and on monokaryotic hyphae seems to be an **effective dispersal strategy** in the haplophase (Fig. 13.9e–h). However, this requires the fusion of compatible cells to establish the dikaryophase. Disarticulation of dikaryotic cells along septa to form arthroconidia is characteristic of few dacrymycetaceous taxa. Though it is a very significant distribution mechanism in *D. stillatus*, even in the closely related *Dacrymyces minor*, arthrospore production is not known.

Recently, **additional anamorph fructifications** of Dacrymycetes have been described. *Dacryoscyphus chrysochilus* (Fig. 13.6) was introduced by Kirschner and Yang (2005), and two anamorphic *Dacrymyces* species, *D. pinacearum* and *D. subarcticus*, proposed by Shirouzu and Tokumasu (Shirouzu et al. 2009), were transferred to *Dacryoscyphus* (Kirschner et al. 2010).

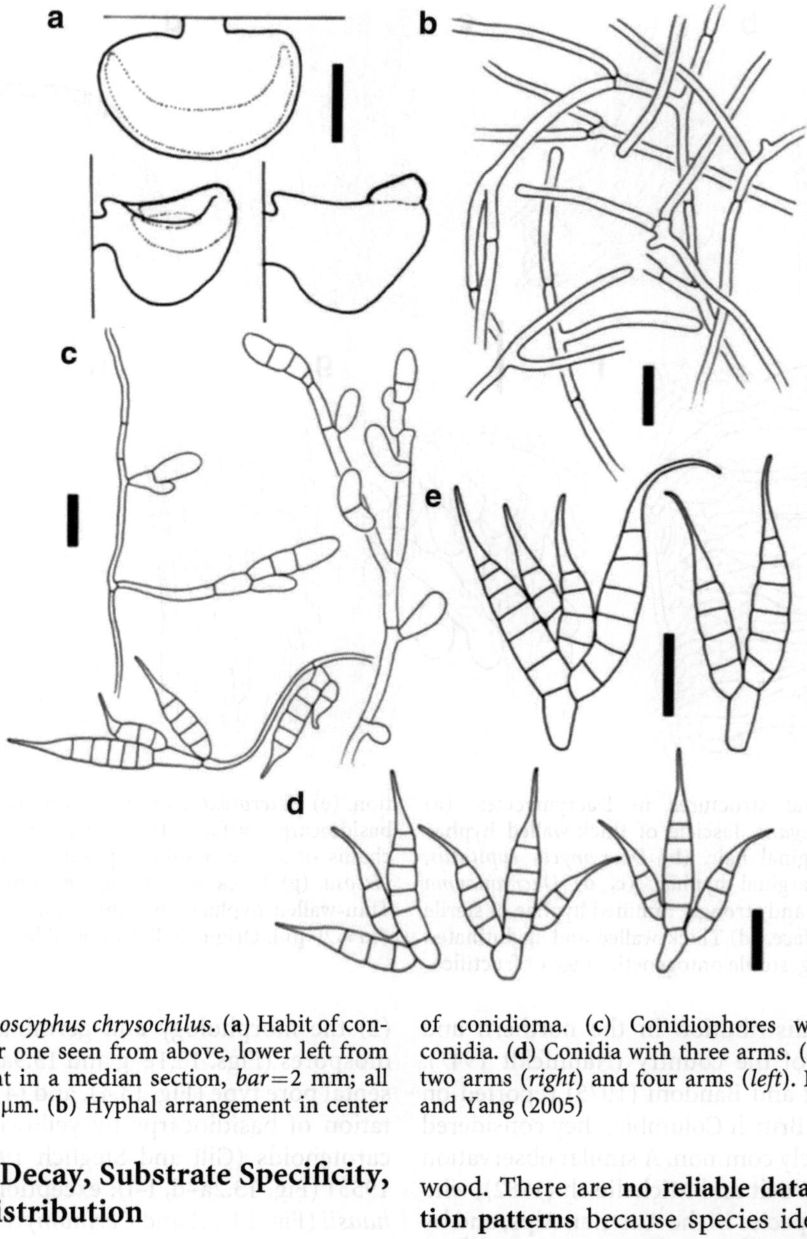

Fig. 13.6 *Dacryoscyphus chrysochilus.* (**a**) Habit of conidiomata; upper one seen from above, lower left from side, lower right in a median section, *bar*=2 mm; all other *bars*=10 μm. (**b**) Hyphal arrangement in center of conidioma. (**c**) Conidiophores with developing conidia. (**d**) Conidia with three arms. (**e**) Conidia with two arms (*right*) and four arms (*left*). From Kirschner and Yang (2005)

VII. Wood Decay, Substrate Specificity, and Distribution

All Dacrymycetes species grow on wood, and presumably all are more or less **strong brown-rot fungi** in coniferous and angiosperm wood, but lignin decomposition may also occur (Seifert 1983). Brown rot is considered an evolutionarily old form of wood decay (Floudas et al. 2012).

Calocera cornea (Fig. 13.2k) preferably grows on angiosperm wood, while *Calocera viscosa* (Fig. 13.2l) is widespread on coniferous wood. There are **no reliable data on distribution patterns** because species identification is unclear in many cases. *D. stillatus* (Fig. 13.2a, b) is considered a species of cosmopolitan distribution (McNabb 1973), while *D. dendrocalami* (Fig. 13.8b) is only known from Taiwan and Japan (Shirouzu et al. 2009), and *Ditiola haasii* (Fig. 13.2e) from the Northern Alps (Oberwinkler 1989). Data on both subtrate specificity and species distribution suffer considerably because of inadequate sampling. An example of this is *Dacryonaema rufum* (Figs. 13.2j and 13.7c, d), a species originally described in Sweden

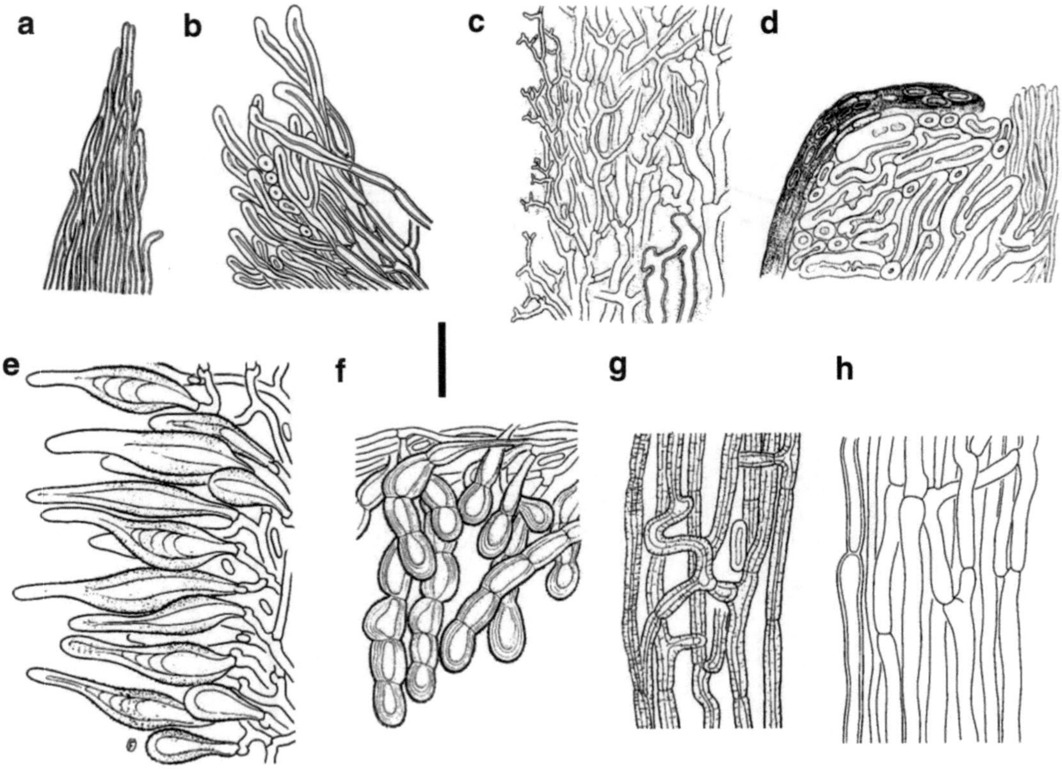

Fig. 13.7 Hyphal structures in Dacrymycetes. (**a**) *Dacryopinax elegans*, fascicle of thick-walled hyphae forming a marginal hair. (**b**) *Dacrymyces capitatus*, thick-walled marginal hyphae. (**c, d**) *Dacryonaema rufum*. (**c**) Thin and strongly ramified hyphae of sterile basidiocarp surface. (**d**) Thick-walled and agglutinated hyphae of young, sterile ontogenetic stage of fructification. (**e**) *Heterotextus miltinus*, terminal cells of sterile basidiocarp surface. (**f**) *Guepiniopsis buccina*, cell chains of sterile basidiocarp surface. (**g, h**) *Calocera viscosa*. (**g**) Thick-walled hyphae from outer part. (**h**) Thin-walled hyphae from inner part of rooting base. *Bar*=20 μm. Originals F. Oberwinkler

with a wide distribution in the northern and central parts of the country (Nannfeldt 1947). When Brough and Bandoni (1975) reported on the species in British Columbia, they considered it to be relatively common. A similar observation was made by Poelt and Michelitsch (1982), who collected the species in the Austrian Alps, and by the present author, who recorded it in the Bavarian Alps (unpublished).

VIII. Traditional Taxonomy

Dacrymycetales represents a natural taxon that is suppported by the stability of several **important characteristics:** (1) the ontogeny and morphology of the basidium (Fig. 13.1d), (2) the morphology and germination of basidiospores (Figs. 13.1e–g and 13.9a–f, h), (3) the septal pore type (Fig. 13.4), and (4) the pigmentation of basidiocarps by yellowish to orange carotenoids (Gill and Steglich 1987; Goodwin 1953) (Fig. 13.2a–d, f–l); exceptions are *Ditiola haasii* (Fig. 13.2e) and *Cerinomyces* species, and (5) the predominantly brown-rotting wood decay. In a few earlier publications (Donk 1966; Eriksson and Ryvarden 1973), species of the genus *Cerinomyces* were considered to be intermediate between the Dacrymycetales and the Aphyllophorales. However, all molecularly based phylogenetic hypotheses that include *Cerinomyces* spp. cluster them with Dacrymycetales.

What follows is an account of the **accepted genera** of Dacrymycetaceae (cf. Fig. 13.5).

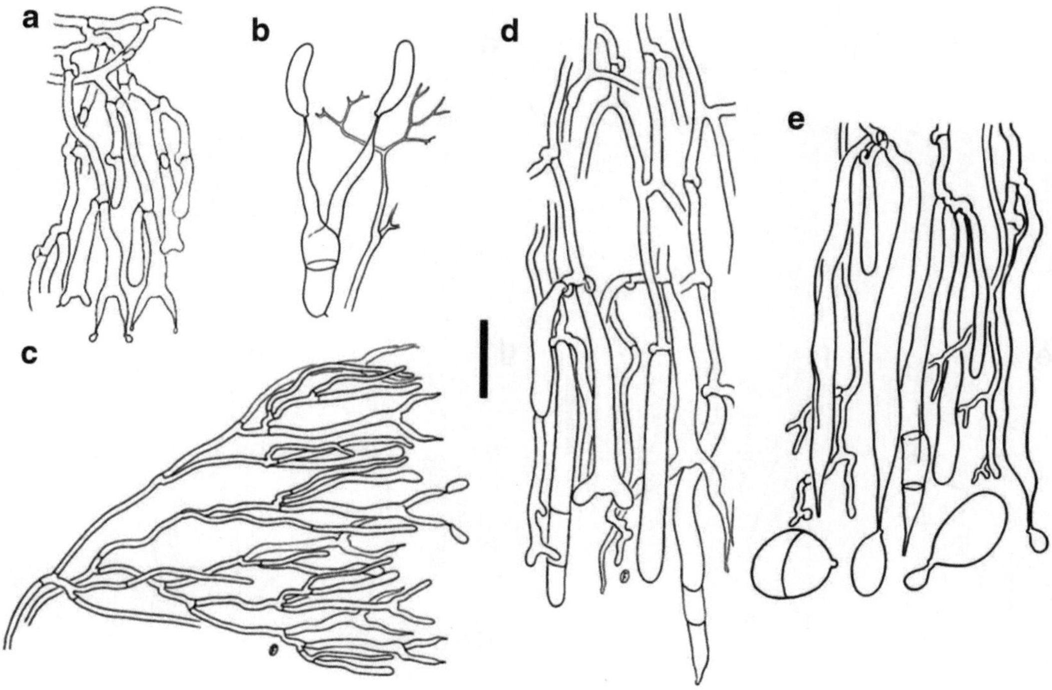

Fig. 13.8 Parts of hymenia, basidia, basidiospores, and dikaryophyses in Dacrymycetes. (**a**) *Cerinomyces pallidus*, part of hymenium with basidia in different developmental stages. (**b**) *Dacrymyces dendrocalami*, basidium with adventitious septum and two attached basidiospores, and dikaryophysis. (**c**) *Calocera viscosa*, part of hymenium with basidia in different developmental stages. (**d**) *Dacrymyces ovisporus*, part of hymenium with two bisterigmate, two unisterigmate basidia with adventitious septa, and two young basidia. (**e**) *Unilacryma unispora*, part of hymenium with three unisterigmate basidia, two basidiospores, one transversely septate, another with an initial stage of germination. *Bar*=20 μm. Originals F. Oberwinkler

Brefeld (1888) introduced the Dacrymycetaceae and recognized four genera. A generic survey of Martin and Fischer (1933) posited nine genera, but Neuhoff (1936) accepted only two. The recognition of nine genera of north-central North American Dacrymycetes by Martin (1952) was again addressed by Kennedy (1958). A comprehensive study of Dacrymycetes on a worldwide scope was carried out by McNabb (1964, 1965a, b, c, d, e, 1966, 1973) and dealt with eight genera. A key to identifying nine genera of Dacrymycetales was provided by McNabb and Talbot (1973). Twelve genera were treated by Oberwinkler (1994).

Martin (1949) erected the genus ***Cerinomyces*** with three species and *C. pallidus* as the type (Fig. 13.8a). Originally, the genus comprised corticioid and non-orange-colored species with nonseptate basidiospores. This narrow scope of the genus was broadened by McNabb (1964) and Ginns (1982) to include species with orange basidiocarps and septate basidiospores.

Based on micromorphological characters, Eriksson (1958) tentatively placed *Cerinomyces* in Corticiaceae, and Eriksson and Ryvarden (1973) considered it as occupying an intermediate position between Dacrymycetaceae and Corticiaceae. McNabb (1964) and Donk (1972) posited a close relationship between *Cerinomyces* and the two-sterigmate *Clavulicium*. However, Martin (1952), Kennedy (1958a), Parmasto (1961), McNabb (1964), and Oberwinkler (1994) kept *Cerinomyces* in Dacrymycetaceae.

The type species of ***Dacrymyces***, *D. stillatus* (Figs. 13.1, 13.2a, b, and 13.3–13.5), is widespread and very common on coniferous wood in northern temperate regions. It is commonly associated with anamorphic fructifications. Thus on this basis, together with micromorphological features, by one classification

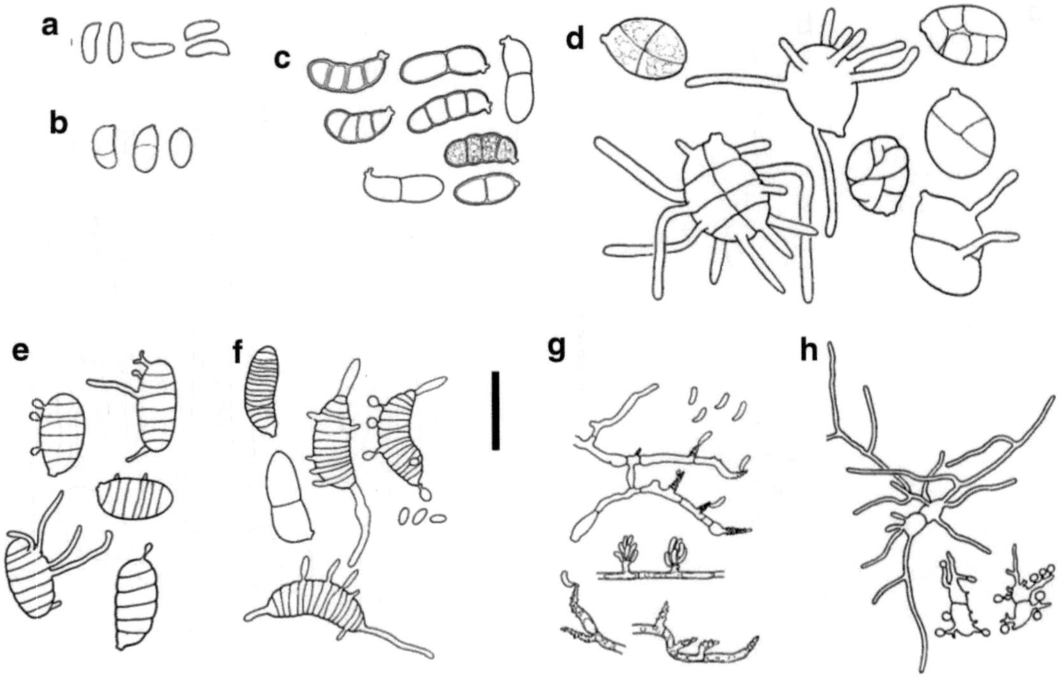

Fig. 13.9 Germination of basidiospores and microconidia in Dacrymycetes. (**a**) *Cerinomyces crustulinus*, unseptate basidiospores. (**b**) *Dacrymyces minor*, basidiospores with one septum each or unseptate. (**c**) *Dacryomitra pusilla*, slightly thick-walled basidiospores, all septate. (**d**) *Dacrymyces ovisporus*, subglobose basidiospores, septate, except for one spore; three germinating with hyphae. (**e**) *Dacrymyces estonicus*, multiseptate basidiospores, three spores germinating with hyphae, two germinating with microconidia. (**f, g**) *Femsjonia peziziformis*. (**f**) Young basidiospore with one septum, four old multiseptate spores, three germinating with hyphae or microconidia. (**g**) Haploid hyphae producing microconidia. (**h**) *Calocera viscosa*, three basidiospores, each with one transverse septum; one spore germinating with hyphae, two with microconidia. *Bar*=20 μm. Originals F. Oberwinkler

scheme, *Dacrymyces* s. str. includes *Dacrymyces aquaticus* (Bandoni and Hughes 1984) and possibly *Dacrymyces capitatus* and *Dacrymyces minor* (Oberwinkler 1994). McNabb (1973) included 30 species in his monograph on the genus and recognized two subgenera, *Dacrymyces* and *Turbinaster*, that had already been introduced by Kobayashi (1939). Species of subgenus *Dacrymyces* have pustular basidiocarps with amphigeneous hymenia; those of subgenus *Turbinaster* are turbinate to pezizoid with hymenia restricted to the apical, disclike parts, a distinction that conforms to practical classificatorial principles but possesses a questionable systematic meaning. A detailed study on selected species of *Dacrymyces*, with comprehensive micromorphological illustrations, was conducted by Göttel (1983).

D. ovisporus, described by Brefeld (1888), has subglobose to ovoid basidiospores, with cruciform to muriform septations in mature stages (Fig. 13.9d), and a mixture of two- and one-sterigmate basidia. Oberwinkler (1994) restudied the type of *Platygloea unispora* (Olive 1947) and considered it to be closely related to *D. ovisporus*, deviating from it by a high percentage of unisterigmate basidia but identical in basidiocarp and spore morphology and germination and in simple dikaryophyses with short ramifications.

Arrhytidia flava, the type species of ***Arrhytidia*** (Berkeley and Curtis 1849), has flat discoid basidiocarps attached to the wood by rootlike bases and fimbriate margins composed of hyphae with morphologically distinct terminal cells. Berkeley (1860) himself was uncertain about the taxonomic meaning of the new genus, and later workers did not arrive at a definitive solution because of insufficient documentation of the micromorphological characters. Coker (1928) synonymized *A. flava* with *Dacrymyces corticioides*, and Kennedy (1958) was of the opinion that the type species might be identical to *A. involuta* sensu Coker and restricted the genus to the latter species. Lloyd (1919), Brasfield (1938), Martin (1949), Donk (1966), and McNabb (1973) did not recognize *Arrhytidia* as being generically different from *Dacrymyces*, a view that was critically questioned by Oberwinkler (1994).

A detailed study of ***Ditiola radicata***, the type species of the genus, clearly revealed the dacrymycetaceous nature of its basidia (Lindau 1894). The prominent rooting base and the capitate to discoid basidiocarps were used to circumscribe the genus (Kennedy 1964; Kobayashi 1939; McNabb 1966; Oberwinkler 1994), though such character combinations also occur in *Dacryopinax*, *Femsjonia*, *Guepiniopsis*, and *Heterotextus*. However, in these genera, marginal hyphae of the sterile basidiocarp surfaces have rather characteristic and distinctive morphological features in comparison with *D. radicata*. Two species were recognized by McNabb (1966) in his monograph on the genus. A nonpigmented species, *Ditiola haasii* (Fig. 13.2e), was described and placed in *Ditiola* by Oberwinkler (1989).

The cortical hairs of ***Heterotextus*** species are thick-walled, basally swollen, and apically bluntly beaked (Fig. 13.7e), which makes them easily distinguishable from other dacrymycetous genera (cf. Fig. 13.7). However, in *Dacrymyces suecicus* the structure of the marginal hyphae is similar to that found in *Heterotextus* (McNabb 1973), indicating that it should be included in *Heterotextus* (Oberwinkler 1994). In his monograph on the genus, McNabb (1965d) accepted four species.

According to the generic concept of McNabb (1965e), ***Femsjonia*** comprises two turbinate to pezizoid species with thick-walled marginal hairs and internal hyphae bearing conspicuous clamp connections. Its allantoid basidiospores have one to many septa and germinate with microconidia (Fig. 13.9f). Microconidia can also develop on haploid hyphae (Fig. 13.9g). Since the publication of McNabb's work, three additional species from China have been described (Liu et al. 1988; Zang 1983).

In his monograph on *Guepiniopsis*, McNabb (1965c) accepted only the type species ***Guepiniopsis buccina*** (Figs. 13.2i, 13.5, and 13.7f), which is defined by its catenulate marginal hairs with stout and thick-walled cells, with the walls often being characteristically layered. The infrageneric taxonomy is discussed by Oberwinkler (1994).

The cyphelloid basidiocarps of the type species ***Dacryopinax elegans*** consist of thick-walled hyphae, except for the hymenium. In addition, the fascicles of the marginal hairs have the same hyphal composition (Fig. 13.7a). The walls of the basidiospores and spore septa of *Dacryopinax* species are also conspicuously thick-walled. McNabb (1965b) broadened the scope of the genus and included six additional species, thereby creating a morphologically heterogeneous assemblage. One of the strongly deviating species is *Dacryopinax spathularia* (Fig. 13.2h), a fungus with a widespread distribution in the tropics.

The ontogeny and morphology of ***Dacryonaema rufum*** (Nannfeldt 1947), the type and single species of the genus, are unique. There is a primordial conelike, sterile stage (Figs. 13.2j and 13.7c, d), obviously well adapted to extremely dry environmental conditions. The fertile part of the fructification develops into a globose capitulum (Fig. 13.5).

Commonly, all dacrymycetaceous fungi with clavarioid basidiocarps are included in ***Calocera*** (Figs. 13.2k, l, 13.5, 13.7g, h, 13.8c, and 13.9h), indicating acceptance of the generic concept of McNabb (1965a) that considered *Corynoides*, *Dacryomitra*, and *Caloposis* to be synonymous. The type species, *C. viscosa*, has a large rooting base (Fig. 13.2l) with a dimorphic hyphal arrangement in three zones (Fig. 13.7g, h) that elongates into the fructification. Well-developed basidiocarps are conspicuously ramified, a character allowing for easy recognition of the species, and the amphigeneous hymenium (Fig. 13.8c) is not markedly separated from the sterile base. Basidiospores germinate with either microconidia or hyphae (Fig. 13.9h). McNabb (1965a) recognized 11 species, and 3 additional ones from China were described by Liu et al. (1988).

Dacryomitra pusilla, the type species of the genus, was introduced by Tulasne and Tulasne (1872) to accommodate a tiny, morel-like fungus with a sterile stalk and a morchelloid hymenium (Fig. 13.5). In contrast to most American researchers, McNabb (1965a), Donk (1966), and Reid (1974) did not accept the genus.

Kirschner and Yang (2005) introduced ***Dacryoscyphus*** *chrysochilus*, an anamorphic species with cupulate conidiomata and staurosporous conidia,that grows on the dead twigs of *Rhododendron* sp. in southwest China (Fig. 13.6). A molecular phylogeny derived from partial LSU rDNA clustered the new species with the Dacrymycetes. In addition, dolipores with continuous parenthesomes supported this systematic position. Based on comparative morphology, Kirschner et al. (2010) included the anamorphic *D. pinacearum* and *Dacrymyces subarcticus* (Shirouzu et al. 2009) in *Dacryoscyphus.*

IX. Molecular Phylogenies

In a study on the phylogenetic relationship of the Auriculariales, Weiß and Oberwinkler (2001) included representative taxa of Agaricomycotina, inclusive of nine Dacrymycetes species. These clustered in a **well-supported monophyletic group**. Similar results were obtained when members of Sebacinales were studied (Weiß et al. 2004), but only a few members of the Dacrymycetes were included. In proposing *Dacryoscyphus*, Kirschner and Yang (2005) also documented the monophyletic clade of dacrymycetaceous fungi.

An **age of approximately 400 million years** was calculated for Dacrymycetes in tracing the origin of wood decay fungi (Floudas et al. 2012).

Attempts to explore **evolutionary trends** within Dacrymycetes were seriously hampered by inadequate sampling. At present, species with clamps seem to represent basal relationships, while the loss of clamps seems to indicate derived evolutionary stages (Shirouzu et al. 2013). The morphological characters of basidiocarps, basidia, basidiospore septa, and sterile marginal hyphae could not be interpreted in meaningful ancestral state reconstructions.

In an analysis that included the type species ***Cerinomyces*** *pallidus* as well as *C. crustulinus*, *C. albosporus*, and *C. canadensis*, these species clustered in one clade together with *Dacrymyces punctiformis* (Shirouzu et al. 2009) (Fig. 13.10). However, in an extended sampling together with *C. ceraceus*, *C. grandinioides*, and *C. lagerheimii*, the latter species represented a separate cluster, clearly distinct from the type species group (Shirouzu et al. 2013) (Fig. 13.10). Since a detailed comparative micromorphology of *Cerinomyces* species is lacking, the two independent clades cannot be characterized by additional features.

Shirouzu et al. (2007) were the first to document the polyphyletic assemblage of *Dacrymyces* species using the 28S rRNA gene D1/D2 region. The type species, *D. stillatus*, clustered with *D. minor*, i.e., ***Dacrymyces* s. str.**, with *G. buccina* as sister group. This finding was confirmed in extended samplings with Japanese Dacrymycetes (Shirouzu et al. 2009) and in a comprehensive study on the phylogeny of Dacrymycetes (Shirouzu et al. 2013) (Fig. 13.10). Unfortunately, *D. aquaticus* (Bandoni and Hughes 1984) could not yet be included in molecular analyses.

The minute ***Dacrymyces dendrocalami*** (Oberwinkler and Tschen 1989) has stout basidia, often with adventitious septa, and strongly branched dikaryophyses (Fig. 13.8b). Basidia and basidiospores often are thick-walled in mature stages. This species clusters with *D. adpressus* in its own clade (Shirouzu et al. 2013) (Fig. 13.10). The micromorphological data available for the latter species (McNabb 1973; Shirouzu et al. 2009) do not allow a detailed comparison with *D. dendrocalami.*

As discussed earlier, the type species of ***Femsjonia*** and ***Heterotextus*** differ considerably in structural characters of marginal hairs and the hyphal context of the basidiocarps. Nevertheless, they group as sister taxa in the molecular phylogeny (Fig. 13.10) of Shirouzu et al. (2013).

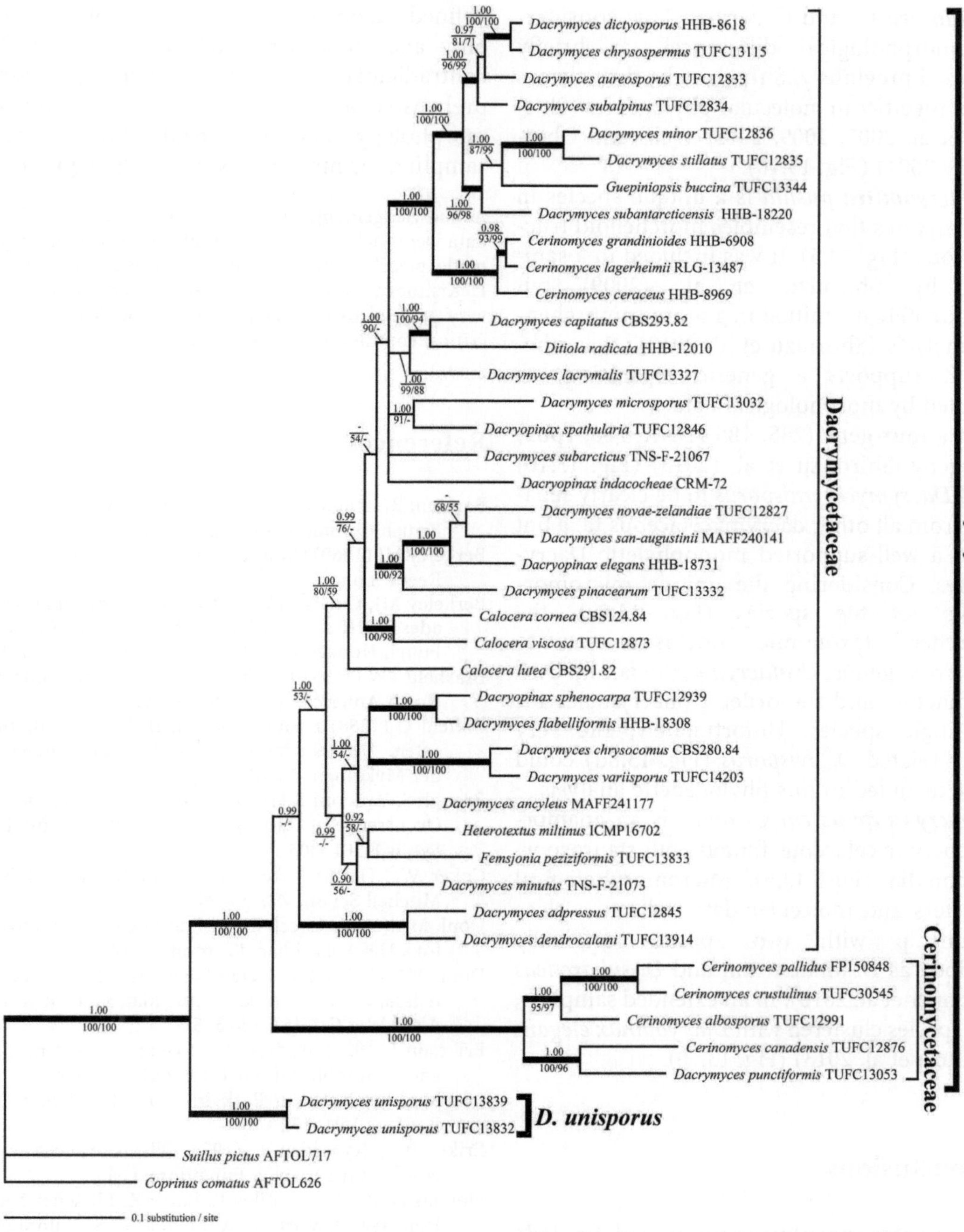

Fig. 13.10 Bayesian tree of Dacrymycetes based on multigene sequences (3,163 bp). Bayesian posterior probabilities PP ≥ 0.90 shown above branches. Maximum likelihood bootstrapping MLBP ≥ 50 %/maximum parsimony bootstrap proportions MPBP ≥ 50 % below branches. Bayesian PP ≥ 0.95, MLBP ≥ 90 %, and MPBP ≥ 90 % indicated by thickened branches. From Shirouzu et al. (2012)

The infrageneric taxonomy of ***Guepiniopsis*** has been discussed by Oberwinkler (1994), who considered *G. chrysocoma* as an additional species and *Dacrymyces minuta* (Olive 1954) to be closely related. However, in the molecular phylogenies of Shirouzu et al. (2007, 2009, 2013), *G. buccina* is a sister of *Dacrymyces* s. str., i.e., *D. stillatus* and *D. minor* (Fig. 13.10).

Calocera v. and *C. cornea* show considerable morphological differences, as briefly explained previously. Surprisingly, they always cluster together in molecular phylogenies (Shirouzu et al. 2007, 2009, 2013; Weiß and Oberwinkler 2001) (Fig. 13.10).

***Dacryomitra** pusilla* is a unique species in Dacrymycetes that resembles morchelloid fructifications (Fig. 13.5). It was included in a sampling by Shirouzu et al. (2009) but unfortunately is omitted in a more comprehensive analysis (Shirouzu et al. 2013). Nevertheless, it supports a generic separation, as indicated by morphological data.

In a four-gene (28S, 18S rDNA, ITS, rpb2) phylogeny, Shirouzu et al. (2013) (Fig. 13.10) found *Dacrymyces unisporus* to be clearly separated from all other dacrymycetaceous taxa but within a well-supported monophyletic Dacrymycetes. Considering the unique micromorphology of the species (Fig. 13.8e), the researchers' taxonomic conclusion was to erect a new genus, ***Unilacryma***, the family Unilacrymaceae, and the order Unilacrymales for this single species. Unfortunately, the very closely related *D. ovisporus* (Fig. 13.8d) could not be included in this phylogenetic analysis.

***Dacryoscyphus** chrysocomus* is an anamorphic dacrymycetaceous fungus with staurosporous conidia (Fig. 13.6). Micromorphological characters and molecular data indicate a close relationship with two species, originally described as *D. pinacearum* and *D. subarcticus* (Kirschner et al. 2010). In an extended sampling, these species clustered with *Dacryopinax elegans* (Shirouzu et al. 2013) (Fig. 13.10).

X. Conclusions

Dacrymycetes constitutes a **monophyletic clade in the Agaricomycotina**. The ontogeny and morphology of the basidia and basidiospores of this clade's members are the best indicators of their natural relationship. Hyphal context, basidiocarp pigmentation, and brown-rot wood decay are additional important characters. All molecular phylogenies available support the notion of Dacrymycetes as a well-defined taxon. In contrast, traditional taxonomy and molecular phylogenies are mostly contradictory. To solve this dilemma, a comprehensive and detailed comparative micromorphology and a considerably enlarged sampling for molecular studies are required.

Acknowledgements In this review many unpublished data were included that accumulated over a long time, made possible through cultural work carried out by L. Kisimova-Horowitz in my former laboratory. I am very grateful to Paul Blanz and Roland Kirschner for critical comments on this manuscript.

References

Bandoni RJ, Hughes GC (1984) A new *Dacrymyces* from British Columbia. Mycologia 76:63–66

Berkeley MJ (1860) Outlines of British fungology. Lovell Reeve, London

Berkeley MJ, Curtis MA (1849) Decades of fungi. Decades XXIII and XXIV. North and South Carolina Fungi. Hooker's J Bot Kew Gard Misc 1:234–239

Brasfield TW (1938) The Dacrymycetaceae of temperate North America. Am Midl Nat 20:211–235

Brefeld O (1888) Basidiomyceten. II. Protobasidiomyceten. Untersuchungen aus dem Gesamtgebiete der Mykologie 7:1–178

Brough S, Bandoni R (1975) Epistolae mycologicae. VI. Occurrence of Dacryonaema in British Columbia. Syesis 8:301–303

Coker WC (1928) Notes on Basidiomycetes. J Elisha Mitchell Sci Soc 43:233–242

Donk MA (1966) Check list of European hymenomycetous Heterobasidiae. Persoonia 4:145–335

Donk MA (1972) The Heterobasidiomycetes: a reconnaissance -I. A restricted emendation. Proc K Ned Akad Wet C 75(5):365–375

Eriksson J (1958) Studies in the Heterobasidiomycetes and Homobasidiomycetes-Aphyllophorales of Muddus National Park in North Sweden. Symb Bot Ups 16:1–172

Eriksson J, Ryvarden L (1973) The Corticiaceae of North Europe, vol 2. Fungiflora, Oslo

Floudas D, Binder M, Riley R, Barry K, Blanchette RA, Henrissat B, Martínez AT, Otillar R, Spatafora JW, Yadav JS, Aerts A, Benoit I, Boyd A, Carlson A, Copeland A, Coutinho PM, de Vries R, Ferreira P, Findley K, Foster B, Gaskell J, Glotzer D, Górecki P, Heitman J, Hesse C, Hori C, Igarashi K, Jurgens JA, Kallen N, Kersten P, Kohler A, Kües U, Kumar TKA, Kuo A, LaButti K, Karrondo LF, Lindquist E, Ling A, Lombard V, Lucas S, Lundell T, Martin R, McLaughlin DJ, Morgenstern I, Morin E, Murat C, Nagy LG, Nolan M, Ohm RA, Patyshakuliyeva A, Rokas A, Ruiz-Dueñas FJ, Sabat G, Salamov A,

Samejima M, Schmutz J, Slot JC, John FS, Stenlid J, Sun H, Sun S, Syed K, Tsang A, Wiebenga A, Young D, Pisabarro A, Eastwood DC, Martin F, Cullen D, Grigoriev IV, Hibbett DS (2012) The paleozoic origin of enzymatic lignin decomposition reconstructed from 31 fungal genomes. Science 336:1715–1719

Gill M, Steglich W (1987) Pigments of fungi (Macromycetes). Prog Chem Org Nat Prod 51:1–317

Ginns J (1982) *Cerinomyces ceraceus* sp. nov. and the similar *C. grandinioides* and *C. lagerheimii*. Can J Bot 60:519–524

Goodwin TW (1953) Studies in carotenogenesis. 8. The carotenoids present in the basidiomycete *Dacryomyces stillatus*. Biochem J 53:538–540

Göttel G (1983) Untersuchungen zur Systematik der Gattung *Dacrymyces* Nees per Fr. (Basidiomycetes). Dissertation, University of Tübingen, Tübingen, Germany

Kennedy LL (1958) The genera of the Dacrymycetaceae. Mycologia 50:874–895

Kennedy LL (1964) The genus *Ditiola*. Mycologia 56:289–308

Kirschner R, Yang ZL (2005) *Dacryoscyphus chrysochilus*, a new staurosporous anamorph with cupulate conidiomata from China and with affinities to the Dacrymycetales (Basidiomycota). Antonie Van Leeuwenhoek 87:329–337

Kirschner R, Yang ZL, Zhao Q, Feng B (2010) *Ovipoculum album*, a new anamorph with gelatinous cupulate bulbilliferous conidiomata from China and with affinities to the Auriculariales (Basidiomycota). Fungal Divers 43:55–65

Kobayashi Y (1939) On the Dacrymyces group. Sc Rep Tokyo Bunrika Daigaku Sect B 4:705–728

Lindau G (1894) Über Bau und systematische Stellung von *Ditiola radicata* (Alb. et Schwein.) Fr. Hedwigia 33:234–240

Liu B, Fan L, Tao K (1988) Five new species of Dacrymycetaceae from China. Acta Mycol Sin 7:1–6

Lloyd CG (1919) Mycological notes at the Lloyd library and museum. 61:877–903

Martin GW (1949) The genus *Ceracea* Cragin. Mycologia 41:47–86

Martin GW (1952) Revision of the North Central Tremellales. Univ Iowa Stud Nat Hist 19:1–122

Martin GW, Fisher MC (1933) The genera of the Dacrymycetaceae. Univ Iowa Stud Nat Hist 25:8–13

McNabb RFR (1964) Taxonomic studies in the Dacrymycetaceae. I. *Cerinomyces* Martin. NZ J Bot 2:424–425

McNabb RFR (1965a) Taxonomic studies in the Dacrymycetaceae. II. *Calocera* (Fries) Fries. NZ J Bot 3:31–58

McNabb RFR (1965b) Taxonomic studies in the Dacrymycetaceae. III. *Dacryopinax* Nees ex Fries. NZ J Bot 3:59–72

McNabb RFR (1965c) Taxonomic studies in the Dacrymycetaceae. IV. *Guepiniopsis* Patouillard. NZ J Bot 3:759–169

McNabb RFR (1965d) Taxonomic studies in the Dacrymycetaceae. V. *Heterotextus* Lloyd. NZ J Bot 3:275–222

McNabb RFR (1965e) Taxonomic studies in the Dacrymycetaceae. VI. *Femsjonia* Nees ex Fries. NZ J Bot 3:223–228

McNabb RFR (1966) Taxonomic studies on the Dacrymycetaceae VII. *Ditiola* Fries. N Z J Bot 4:546–558

McNabb RFR (1973) Taxonomic studies in the Dacrymycetaceae. VIII. *Dacrymyces* Nees ex Fries. NZ J Bot 11:461–524

McNabb RFR, Talbot PHB (1973) Holobasidiomycetidae: Exobasidiales, Brachybasidiales, Dacrymycetales. In: Ainsworth CC (ed) The Fungi, vol IV B. pp 317–325

Nannfeldt JA (1947) *Sphaeronema rufum* Fr., a misunderstood member of the Dacrymycetaceae. Svensk Bot Tidskr 47:177–180

Neuhoff W (1936) Die Gallertpilze Schwedens (Tremellaceae, Dacrymycetaceae, Tulasnellaceae, Auriculariaceae). Ark Bot 28 A:1–57

Oberwinkler F (1989) *Ditiola haasii* sp. nov., eine neue Art der Dacrymycetales. Z Mykol 55:797–207

Oberwinkler F (1994) Genera in monophyletic group: the Dacrymycetales. Mycol Helvet 6:35–72

Oberwinkler F (2012) Evolutionary trends in Basidiomycota. Stapfia 96:45–104

Oberwinkler F, Tschen J (1989) A new *Dacrymyces* species from Taiwan. Trans Mycol Soc Jpn 30:349–356

Olive LS (1947) Notes on the Tremellales of Georgia. Mycologia 39:90–108

Olive LS (1954) New or noteworthy species of Tremellales from the Southern Appalachians. Bull Torrey Bot Cl III 81:329–339

Parmasto E (1961) A preliminary review of the genus *Cerinomyces* Martin in the U.S.S.R. Acad Sci Estonian SSR Inst Zool Bot 10:231–235

Poelt J, Michelitsch S (1982) *Dacryonaema rufum*, ein übersehener, aber weit verbreiteter Pilz der Alpen. Mitt nat Ver Steiermark 112:141–144

Reid DA (1974) A monograph of the British Dacrymycetales. Trans Br Mycol Soc 62:433–494

Seifert KA (1983) Decay of wood by the Dacrymycetales. Mycologia 75:1011–1018

Shirouzu T, Hirose D, Tokumasu S (2007) Sequence analyses of the 28S rRNA gene D1/D2 region suggest *Dacrymyces* (Heterobasidiomycetes, Dacrymycetales) is polyphyletic. Mycoscience 48:388–394

Shirouzu T, Hirose D, Tokumasu S (2009) Taxonomic studies of the Japanese Dacrymycetes. Persoonia 23:16–34

Shirouzu T, Hirose D, Oberwinkler F, Shimomura N, Maekawa N, Tokumasu S (2013) Combined molecular and morphological data for improving phylogenetic hypotheses in Dacrymycetes. Mycologia 105:1110–1125

Tulasne M, Tulasne LR (1872) Nouvelles notes sur les Fungi Tremellini et leurs alliés. Ann Sci Nat 5. Ser, Bot 15:193–235

Weiß M, Oberwinkler F (2001) Phylogenetic relationships in Auriculariales and related groups — hypotheses derived from nuclear ribosomal DNA sequences. Mycol Res 105:403–415

Weiß M, Selosse M-A, Rexer K-H, Urban A, Oberwinkler F (2004) Sebacinales: a hitherto overlooked cosm of heterobasidiomycetes with a broad mycorrhizal potential. Mycol Res 108: 1003–1010

Zang M (1983) Notes on the genus *Femsjonia* in China. Mycologia 75:468–471

14 Agaricomycetes

D.S. Hibbett[1], R. Bauer[2], M. Binder[3], A.J. Giachini[4], K. Hosaka[5], A. Justo[1], E. Larsson[6], K.H. Larsson[7], J.D. Lawrey[8], O. Miettinen[1,9], L.G. Nagy[1], R.H. Nilsson[6], M. Weiss[10], R.G. Thorn[11]

CONTENTS

[1]Biology Department, Clark University, Worcester, MA, USA; e-mail: dhibbett@clarku.edu
[2]Institut für Evolution und Ökologie, Universität Tübingen, Auf der Morgenstelle 1, 72076 Tübingen, Germany
[3]Centraalbureau voor Schimmelcultures, Uppsalalaan 8, Utrecht 3584 CT, Netherlands
[4]Departamento de Microbiologia e Parasitologia, Centro de Ciencias Biologicas, Universidade Federal de Santa Catarina, Caixa Postal 476, SC 88040 900 Florianopolis, Brazil
[5]Department of Botany, National Museum of Nature and Science, 4-1-1 Amakubo, Tsukuba, Ibaraki 305-0005, Japan
[6]Department of Biological and Environmental Sciences, University of Gothenburg, Box 100, 405 30 Gothenburg, Sweden
[7]Natural History Museum, P.O. Box 1172 Blindern, 0318 Oslo, Norway
[8]Environmental Science and Policy Department, George Mason University, 4400 University Drive, Fairfax, VA, USA
[9]Botanical Museum, University of Helsinki, P.O. Box 7, 00014 Helsinki, Finland
[10]Fachbereich Biologie, Universität Tübingen, Auf der Morgenstelle 1, 72076 Tübingen, Germany
[11]Department of Biology, University of Western Ontario, 1151 Richmond St. North, London, ON, Canada N6A 5B7

I. Introduction

Agaricomycetes is a clade of Basidiomycota that contains ca. 21,000 described species, which is one-fifth of all known Fungi (Kirk et al. 2008). However, new taxa are continually being described, and molecular ecologists routinely detect DNA sequences of Agaricomycetes that cannot be referred to known species, suggesting that the actual diversity of the group far exceeds the current catalog (Blackwell 2011; Hibbett et al. 2011). Many members of Agaricomycetes produce conspicuous fruiting bodies that are popular subjects for artists and amateur naturalists (Petersen 2012). In addition, most edible mushrooms are Agaricomycetes, including cultivated saprotrophs, such as *Agaricus bisporus* (champignon), *Pleurotus ostreatus* (oyster mushroom), and *Lentinula edodes* (shiitake), and wild-collected ectomycorrhizal (ECM) species, such as *Boletus edulis* (porcini), *Cantharellus cibarius* (chanterelle), and *Tricholoma matsutake* (matsutake). Psychoactive taxa, particularly species of *Psilocybe*, have

Systematics and Evolution, 2nd Edition
The Mycota VII Part A
D.J. McLaughlin and J.W. Spatafora (Eds.)

been used both as recreational drugs and religious sacraments (Heim and Wasson 1958). Other members of Agaricomycetes are toxic, with effects that range from gastrointestinal distress, caused by diverse taxa such as *Chlorophyllum molybdites*, to life-threatening amatoxin poisoning, caused by *Amanita phalloides*, *Galerina autumnalis*, and others (Benjamin 1995). The toxic compound phalloidin (from *A. phalloides*) binds to actin, making it useful as a component of fluorescent stains for visualizing the cytoskeleton.

Agaricomycetes are not common as human pathogens, although *Schizophyllum commune*, which normally occurs as a wood-decay fungus, is known to cause serious infections of lungs and other organs (Sigler et al. 1995). Several Agaricomycetes have been important as model systems in studies of fungal mating genetics and development (*S. commune*, *Coprinopsis cinerea*) (Ohm et al. 2010; Raper and Miles 1958; Stajich et al. 2010) and the biochemistry of wood decay (*Phanerochaete chrysosporium*, *Postia placenta*, and others) (Martinez et al. 2004, 2009). Finally, there is interest in uses of Agaricomycetes in industrial bioconversion processes and bioremediation (Ruiz-Dueñas and Martínez 2009).

Most of the taxa now classified in the Agaricomycetes were included in a chapter on Homobasidiomycetes in the previous edition of *The Mycota* (Hibbett and Thorn 2001). Eight informally named clades (e.g., euagarics clade, russuloid clade) were proposed, based almost entirely on analyses of ribosomal RNA (rRNA) gene sequences. A separate chapter treated Heterobasidiomycetes (Wells and Bandoni 2001), which included jelly fungi and others with mostly septate basidia (Weiß et al. 2004a). Today, Agaricomycetes is recognized as one of four major clades of Agaricomycotina, the others being the Dacrymycetes (see Oberwinkler 2014), Tremellomycetes (see Weiß et al. 2014), and Wallemiomycetes (Fig. 14.1) (Hibbett 2006; Padamsee et al. 2012). The 2007 AFTOL classification of Fungi (Hibbett et al. 2007) included 17 orders of Agaricomycetes, three of which contain species formerly classified as Heterobasidiomycetes, namely Auriculariales, Sebacinales, and Cantharellales pro parte (i.e., Ceratobasidiaceae and Tulasnellaceae). Since 2007, three new orders of Agaricomycetes have been proposed: Amylocorticiales, Jaapiales, and Lepidostromatales (Binder et al. 2010; Hodkinson et al. 2013). This chapter provides a phylogenetic overview of Agaricomycetes, emphasizing recent molecular studies that address the diversity and phylogenetic relationships of each order (of course, clades of Agaricomycetes classified as orders are simply mutually exclusive groups; they are not necessarily equivalent in age, number of species, or phenotypic diversity).

A. Higher-Level Relationships

All currently recognized orders of Agaricomycetes have been resolved as monophyletic in at least one analysis of rRNA genes, but support for some groups has been weak or absent, in part because of elevated rates of evolution in nuclear rRNA (nrRNA) genes in certain Cantharellales and other lineages (Binder and Hibbett 2002; Binder et al. 2005; Hibbett et al. 1997b; Moncalvo et al. 2006). Genes encoding proteins, such as subunits 1 and 2 of RNA polymerase II (*rpb1*, *rpb2*), mitochondrial ATPase subunit 6 (*atp6*), and translation elongation factor 1-α (*tef1*), started to be used in fungal molecular systematics in the late twentieth century (Kretzer and Bruns 1999; Liu et al. 1999; O'Donnell et al. 2001), and by 2006 a 6-gene, 200-species, kingdom-wide fungal phylogeny had been produced that included 37 species of Agaricomycetes (James et al. 2006). The first in-depth study of Agaricomycetes combining rRNA and protein-coding genes was that of Matheny et al. (2007), who analyzed a 6.6 kb data set of *rbp2*, *tef1*, and nrRNA genes in 146 species (119 species of Agaricomycetes). This was the first analysis to provide strong support for the monophyly of Polyporales (which had been weakly supported in rRNA analyses), and it suggested that the Sebacinales, Cantharellales, Auriculariales, and Phallomycetidae formed a paraphyletic assemblage, within which a clade containing the remaining Agaricomycetes is nested.

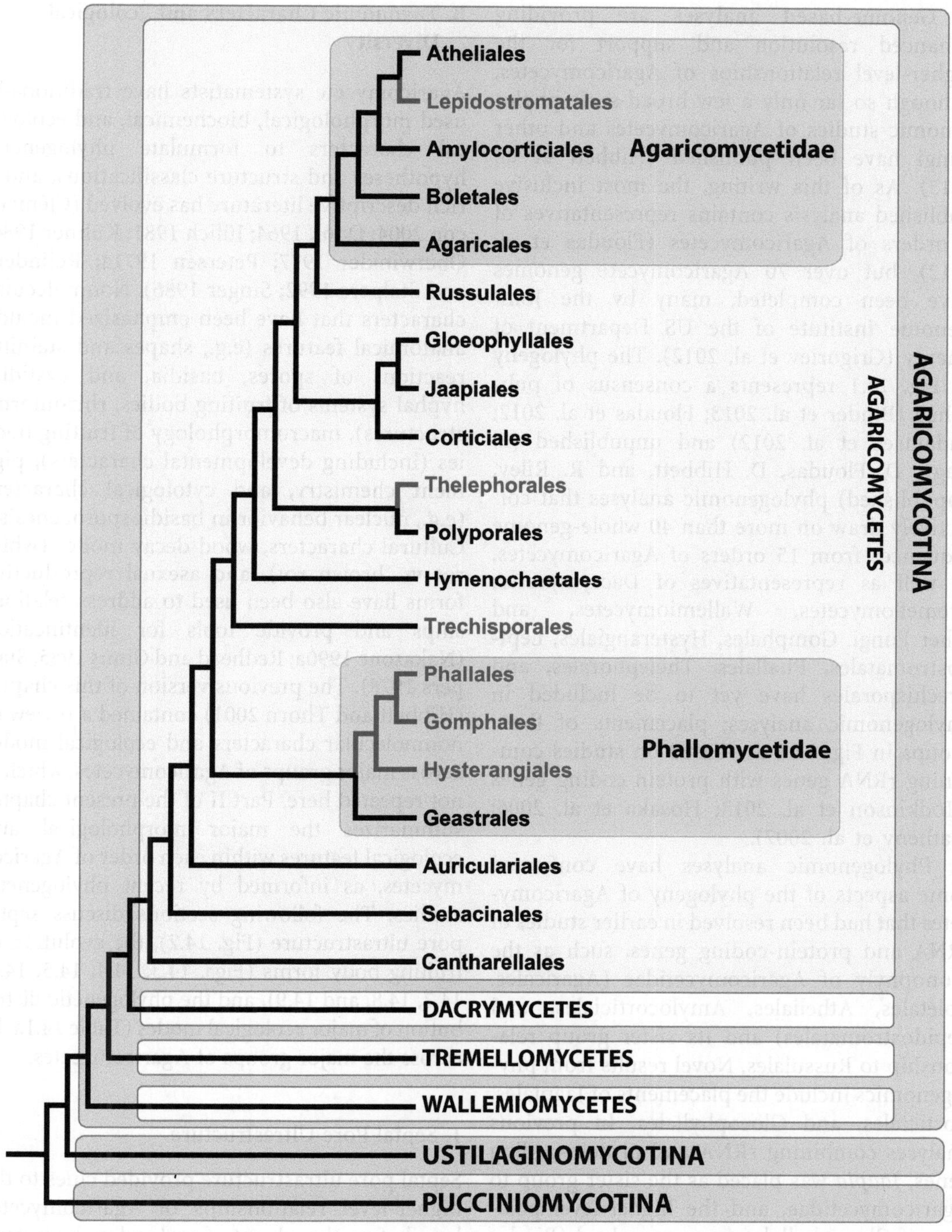

Fig. 14.1 Higher-level phylogenetic relationships of major groups of Agaricomycetes and other Basidiomycota. The major topology (*black lines* and text) is based on published Floudas et al. 2012; Padamsee et al. 2012) and unpublished (L. Nagy, D. Floudas, R. Riley, D. Hibbett et al., unpublished) phylogenomic analyses. Names in *gray* represent groups that have not been included in phylogenomic analyses; placements of these taxa are based on studies combining rRNA genes with 2–3 protein-coding genes (Hodkinson et al. 2013; Hosaka et al. 2006; Matheny et al. 2007)

Genome-based analyses are providing enhanced resolution and support for the higher-level relationships of Agaricomycetes, although so far only a few broad-scale phylogenomic studies of Agaricomycetes and other Fungi have been published (Hibbett et al. 2013). As of this writing, the most inclusive published analysis contains representatives of 8 orders of Agaricomycetes (Floudas et al. 2012), but over 70 Agaricomycete genomes have been completed, many by the Joint Genome Institute of the US Department of Energy (Grigoriev et al. 2012). The phylogeny in Fig. 14.1 represents a consensus of published (Binder et al. 2013; Floudas et al. 2012; Padamsee et al. 2012) and unpublished (L. Nagy, D. Floudas, D. Hibbett, and R. Riley, unpublished) phylogenomic analyses that collectively draw on more than 40 whole-genome sequences from 15 orders of Agaricomycetes, as well as representatives of Dacrymycetes, Tremellomycetes, Wallemiomycetes, and other Fungi. Gomphales, Hysterangiales, Lepidostromatales, Phallales, Thelephorales, and Trechisporales have yet to be included in phylogenomic analyses; placements of these groups in Fig. 14.1 are based on studies combining rRNA genes with protein-coding genes (Hodkinson et al. 2013; Hosaka et al. 2006; Matheny et al. 2007).

Phylogenomic analyses have confirmed some aspects of the phylogeny of Agaricomycetes that had been resolved in earlier studies of rRNA and protein-coding genes, such as the monophyly of Agaricomycetidae (Agaricales, Boletales, Atheliales, Amylocorticiales, and Lepidostromatales) and its sister group relationship to Russulales. Novel results from phylogenomics include the placements of Jaapiales, Corticiales, and Gloeophyllales. In previous analyses combining rRNA and protein-coding genes, *Jaapia* was placed as the sister group to Agaricomycetidae, and the higher-level position of Gloeophyllales was unresolved (Binder et al. 2010; Garcia-Sandoval et al. 2011). Recent phylogenomic analyses indicate that Jaapiales, Gloeophyllales, and Corticiales form a strongly supported clade, but its higher-level position is ambiguous.

B. Taxonomic Characters and Ecological Diversity

Agaricomycete systematists have traditionally used morphological, biochemical, and ecological characters to formulate phylogenetic hypotheses and structure classifications, and a rich descriptive literature has evolved (Clémençon 2004; Donk 1964; Jülich 1981; Kühner 1984; Oberwinkler 1977; Petersen 1971a; Reijnders and Stalpers 1992; Singer 1986). Nonmolecular characters that have been emphasized include anatomical features (e.g., shapes and staining reactions of spores, basidia, and cystidia, hyphal systems of fruiting bodies, rhizomorph structures), macromorphology of fruiting bodies (including developmental characters), pigment chemistry, and cytological characters (e.g., nuclear behavior in basidiosporogenesis). Cultural characters, wood-decay modes (white rot vs. brown rot), and asexual reproductive forms have also been used to address relationships and provide tools for identification (Nakasone 1990a; Redhead and Ginns 1985; Stalpers 1978). The previous version of this chapter (Hibbett and Thorn 2001) contained a review of nonmolecular characters and ecological modes across major groups of Agaricomycetes, which is not repeated here. Part II of the present chapter summarizes the major morphological and ecological features within each order of Agaricomycetes, as informed by recent phylogenetic studies. The following sections discuss septal pore ultrastructure (Fig. 14.2), the evolution of fruiting body forms (Figs. 14.3, 14.4, 14.5, 14.6, 14.7, 14.8, and 14.9), and the phylogenetic distribution of major ecological modes (Table 14.1a, b) across the major groups of Agaricomycetes.

1. Septal Pore Ultrastructure

Septal pore ultrastructure provided clues to the higher-level relationships of Agaricomycetes long before the advent of molecular characters. The union of Dacrymycetes and Agaricomycetes is supported by their shared possession of dolipores that are surrounded at each side by a more or less dome-shaped modified ER (endoplasmic reticulum) cisterna, the so-called

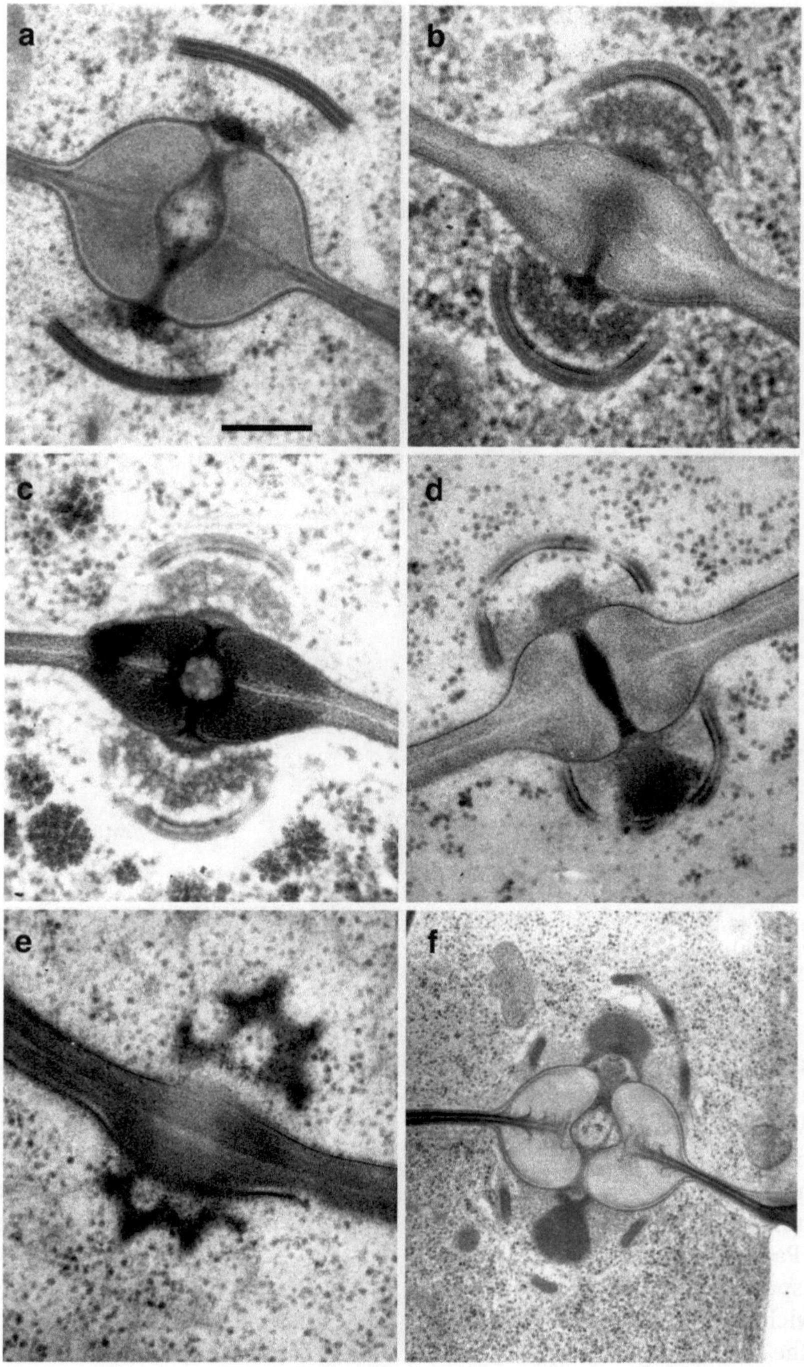

Fig. 14.2 Dolipores of Agaricomycetes in transverse (**a–d**, **f**) or tangential sections (**e**) through the septal pore caps (SPCs). Material was prepared by freeze substitution. *Bar* =0.2 μm in (**a**)–(**e**), and 0.5 μm in (**f**). (**a**) *Tulasnella* sp. (**b, c**) *Sphaerobolus* sp. Note the imperforate SPCs in (**b**) and the perforate SPCs in (**c**). (**d, e**) *Schizophyllum commune*. The regularly arranged perforations are especially visible in (**e**). (**f**) *Ceratobasidium* sp.

Fig. 14.3 Sebacinales (**a**, **b**), Auriculariales (**c**, **d**), and Cantharellales (**e**, **f**). (**a**) *Sebacina epigaea*. (**b**) *Craterocolla cerasi*. (**c**) *Pseudohydnum gelatinosum*. (**d**) *Exidia truncata*. (**e**) *Craterellus tubaeformis*. (**f**) *Clavulina cristata*. Photos by Michael Wood (**a**, **e**, **f**; http://www.mykoweb.com) and Jaroslav Maly (**b–d**)

septal pore cap (SPC), also termed the parenthesome. In contrast, the SPCs of Tremellomycetes and Wallemiomycetes are composed of saccules or fingerlike projections arising from the endoplasmic reticulum (SPCs are also occasionally absent in both groups) (Wells and Bandoni 2001; Padamsee et al. 2012). The intracisternal space of the SPCs of Agaricomycetes is sandwiched by a fine electron-opaque layer so that the SPCs altogether appear nine-lamellate in optimal sections (the double-sectioned cisternal membrane with three layers each and the intracisternal lumen with three layers). Variation in the perforation of SPCs within Agaricomycetes was summarized by van Driel et al. (2009, and references therein). The following account completes and corrects this synopsis with new observations (by R. Bauer) using serial sections. Accordingly, within the Dacrymycetes/Agaricomycetes union, four types of SPC are evident.

Imperforate to uni-perforate SPCs (Fig. 14.2a). Probably depending on the section, no or only one median perforation of ca. 100 nm appears in one section of the series. This type is typical for Dacrymycetes and within Agaricomycetes for Sebacinales, Auriculariales, Trechisporales, the *Botryobasidium* and *Tulasnella* clades within Cantharellales, and the Hymenochaetaceae, *Hyphodontia*, *Coltricia*, *Kneiffiella*, and *Trichaptum* clades within Hymenochaetales [for the taxa and clades see Hibbett (2006) and van Driel et al. (2009)].

Fig. 14.4 Phallomycetidae, including Phallales (**a**), Geastrales (**b**), Hysterangiales (**c**), and Gomphales (**d**, **e**). (**a**) *Clathrus ruber*. (**b**) *Geastrum saccatum*. (**c**) *Hysterangium coriaceum*. (**d**) *Turbinellus floccosus*. (**e**) *Kavinia himantioides*. Photos by Michael Wood (**a–d**; http://www.mykoweb.com) and Otto Miettinen (**e**)

Unstable perforate SPCs (Fig. 14.2b, c). In all members of the Phallomycetidae studied, such as *Aseroe* sp., *Clathrus archeri*, *Geastrum* spp., *Gomphus clavatus*, *Phallus impudicus*, *Ramaria* spp., and *Sphaerobolus* sp., there are predominantly imperforate SPCs, but in each species, often in the same sections, there are also dolipores with perforate SPCs. Usually, the intracisternal layering of the imperforate SPCs appears somehow incomplete. These observations suggest that the perforation formation in the Phallomycetidae begins later in comparison to that in the other accordant groups. Accordingly, the SPCs in this group may represent a unique type.

Regularly perforate SPCs with openings of roughly 100 nm (Fig. 14.2d, e). Usually, the openings are more or less hexagonally arranged. This type is realized in the Agaricomycetidae and related groups (i.e., Russulales, Corticiales, Gloeophyllales, Polyporales, and Thelephorales), the *Rickenella* and *Peniophorella praetermissa* clades within the Hymenochaetales, and the core cantharelloid group within Cantharellales [for the taxa and clades see Hibbett (2006) and van Driel et al. (2009)].

Irregularly perforate SPCs with a few large openings of several hundred nanometers (Fig. 14.2f). This type characterizes the Ceratobasidiales clade within the Cantharellales.

Fig. 14.5 Hymenochaetales (**a**–**e**) and Trechisporales (**f**–**h**). (**a**) *Hymenochaete* sp. (**b**) *Coltricia perennis*; (**c**) *Alloclavaria purpurea*. (**d**) *Rickenella fibula*. (**e**) *Xylodon* (=*Hyphodontia*) *crustosus*. (**f**) *Trechispora stevensonii*, with pustulate anamorphic regions. (**g**) *Trechispora hymenocystis*. (**h**) *Scytinopogon angulisporus*. Photos by Otto Miettinen (**a**, **b**, **e**–**g**), Ellen Larsson (**c**), Lasse Kosonen (**d**), and Nourou Yourou (**h**)

2. Fruiting Bodies

The **morphological diversity** of fruiting bodies in Agaricomycetes is unparalleled in any other clade of Fungi. Agaricomycete fruiting bodies include complex, developmentally integrated forms, such as stinkhorns (e.g., *Clathrus ruber*, Phallales) and veiled agarics (e.g., *Lepiota lilacina*, Agaricales), as well as relatively simple corticioid forms (e.g., *Trechispora stevensonii*, Trechisporales). Agaricomycete fruiting bodies range over several orders of magnitude in size, from tiny cyphelloid forms, such as *Henningsomyces candidus* or

Fig. 14.6 Polyporales. (**a**) *Panus* sp. (**b**) *Perenniporia ochroleuca.* (**c**) *Phlebia femsjoeensis.* (**d**) *Mycoacia aurea.* (**e**) *Polyporus* sp. (**f**) *Podoscypha* sp. Photos by Otto Miettinen (**a**, **b**, **d**–**f**) and O. Manninen (**c**)

the minute red algal parasite *Mycaureola dilseae* (Agaricales), which are often less than 1 mm in diameter (Binder et al. 2006; Bodensteiner et al. 2004), to giant polypores, like *Bridgeoporus nobilissimus* and *Phellinus ellipsoideus* (Hymenochaetales), which can be more than 1 m in diameter (Burdsall et al. 1996; Dai and Cui 2011; Redberg et al. 2003). They may be ephemeral, with deliquescent forms such as *Coprinopsis cinerea* (Agaricales) appearing and disappearing over a few hours, or perennial, with woody "conks" like *Fomes fomentarius* (Polyporales) persisting for years.

The phylogenetic distribution of fruiting body forms and hymenophore configurations across orders of Agaricomycetes was reviewed by Hibbett (2007). All of the major morphotypes (e.g., pileate-stipitate, coralloid, polyporoid,

Fig. 14.7 Gloeophyllales (**a**, **b**) and Thelephorales (**c**, **d**). (**a**) *Neolentinus ponderosus*. (**b**) *Veluticeps fimbriata*. (**c**) *Hydnellum aurantiacum*. (**d**) *Polyozellus m*ultiplex. Photos by Michael Wood (http://www.mykoweb.com)

Fig. 14.8 Russulales. (**a**) *Lactarius subcircellatus*. (**b**) *Laurilia sulcata*. (**c**) *Artomyces pyxidatus*. (**d**, **e**). *Macowanites vinaceodorus*. Photos by Ellen Larsson (**a–c**) and M. Jeppson (**d**, **e**)

corticioid, and gasteroid forms) have evolved repeatedly. Several orders are composed exclusively (or almost exclusively) of resupinate forms (Amylocorticiales, Atheliales, Corticiales, Jaapiales, Trechisporales, and Lepidostromatales), and this is the only morphotype that is known in all orders of Agaricomycetes (except Geastrales, Hysterangiales, and Phallales). Analyses of character evolution using likelihood and parsimony methods suggest that corticioid forms may represent the plesiomorphic condition in Agaricomycetes, with many independent origins of pileate-erect fruiting bodies, as well as reversals to resupinate forms (Hibbett 2004; Hibbett and Binder 2002). These analyses should be interpreted with caution because they are based on limited taxon samples and simplistic models of

Fig. 14.9 Agaricomycetidae, including Agaricales (**a, c**), Amylocorticales (**b**), Atheliales (**d**), and Boletales (**e, f**). (**a**) *Lepiota lilacea*. (**b**) *Anomoporia bombycina*. (**c**) *Lycoperdon perlatum*. (**d**) *Athelia salicum*. (**e**) *Boletus lupinus*. (**f**) *Serpula lacrymans*. Photos by Guillermo Muñoz (**a, c, e**), Yu-Cheng Dai [**b**, reprinted from Binder et al. (2010), with permission of the Mycological Society of America], Paul Diederich (**d**), and Doris Haas (**f**)

character evolution (Hibbett 2007). Nonetheless, it is plausible that the ancestor of the Agaricomycetes may have been a corticioid fungus. Alternatively, the paraphyletic arrangement of Tremellomycetes and Dacrymycetes within Agaricomycotina could indicate that the ancestor was a jelly fungus.

3. Ecological Roles

Agaricomycetes function as **saprotrophs, pathogens,** and **mutualists**. The group contains the major concentration of ECM taxa, as well as white-rot and brown-rot decayers of massive woody substrates. The distribution of major ecological roles across the orders of

Table 14.1a Nutritional modes in Agaricomycete orders, with examplar genera (Saprotrophs[a])

	White rot	Brown rot	Other/uncertain[b]
Agaricales	***Pleurotus***	***Fistulina***	***Coprinopsis***
Boletales		***Serpula***	
Amylocorticiales	***Plicaturopsis***	*Anomoporia*	
Atheliales	*Athelopsis*		*Athelopsis*
Lepidostromatales			
Polyporales	***Phanerochaete***	***Postia***	
Russulales	***Stereum***		
Thelephorales	*Lenzitopsis*		
Gloeophyllales		***Gloeophyllum***	*Boreostereum*
Jaapiales			***Jaapia***
Corticiales	***Punctularia***		
Hymenochaetales	***Trichaptum***	*Bridgeoporus?*	*Tubulicrinis*
Trechisporales	***Sistotremastrum***		*Porpomyces*
Hysterangiales			*Phallogaster*
Gomphales	*Lentaria*		*Kavinia*
Phallales	*Phallus*		
Geastrales	***Sphaerobolus***		*Sclerogaster*
Auriculariales	***Auricularia***		
Sebacinales			*Craterocolla*
Cantharellales	***Sistotrema***		***Botryobasidium***

[a]Bold = genome sequence of at least one species completed; ? indicates uncertainty. Bacteriovores, algal and cyanobacterial parasites, and animal pathogens are not included. For references, see this text and Hibbett and Thorn (2001). Some endophyte observations by R. Gazis and R. Martin (unpublished)

[b]This broad category includes saprotrophs on soil, litter, dung, and keratinic substrates, as well as wood decayers of uncertain rot type or that do not correspond to classical white or brown rot definitions

Agaricomycetes is presented in Table 14.1a, b. Saprotrophic taxa occur in all orders of Agaricomycetes, and ECM taxa occur in at least 13 orders. As many as 37 independent ECM lineages have been estimated to occur in Agaricomycetes (Tedersoo et al. 2010; Rinaldi et al. 2008). It has been proposed that the ancestor of Agaricomycetes might have been ECM, based on the occurrence of ECM taxa in Sebacinales (Weiß et al. 2004b), but molecular clock analyses (see below) suggest that the group is much older than potential ECM hosts, including Pinaceae (Floudas et al. 2012; Hibbett and Matheny 2009). Phylogenomic analyses suggest that the common ancestor of Agaricomycetes possessed multiple ligninolytic class II fungal peroxidases (PODs) and other plant cell wall (PCW)-decaying enzymes, implying that it was capable of producing a white rot (in which both the lignin and cellulose components of PCWs are degraded) (Floudas et al. 2012; Ruiz-Duenas et al. 2013). Wood decayers with multiple PODs and enzymes capable of attacking crystalline cellulose occur in diverse lineages of Agaricomycetes, and these may have retained the plesiomorphic white-rot mode of saprotrophy. Multiple origins of brown rot (in which lignin is not appreciably removed) in Polyporales, Boletales, and Gloeophyllales and the evolution of the ECM condition in *Laccaria bicolor* (Agaricales) seem to be associated with repeated losses of PODs and other PCW-degrading enzymes. Ongoing genomic comparisons are showing that some saprotrophic Agaricomycetes do not conform to the typical models of either white rot or brown rot, including wood decayers, such as *Schizophyllum commune* and *Fistulina hepatica*, and soil, litter, and dung fungi, such as *Agaricus bisporus* and *Coprinopsis cinerea* (Morin et al. 2012; Ohm et al. 2010; Stajich et al. 2010). Similarly, increased sampling of ECM genomes is revealing considerable diversity in genes encoding PCW-degrading enzymes among independently evolved symbiotic lineages (F. Martin and colleagues, unpublished).

Table 14.1b Nutritional modes in Agaricomycete orders, with examplar genera (Biotrophs)

	ECM	Orchid	Endophyte	Mycoparasite	Insect symbionts	Lichenized	Lichenicolous	Bryophilous	Plant pathogens	Nematode-trappers
Agaricales	***Laccaria***	***Armillaria***	*Coprinellus*	*Asterophora*	***Termitomyces***	*Dictyonema*		*Tephrocybe*	***Armillaria***	***Pleurotus***
Boletales	***Paxillus***			*Pseudoboletus*			*Leucogyrophana*			
Amylocorticiales	*Podoserpula?*								*Athelia rolfsii*	
Atheliales	***Piloderma***		*Athelia*		*Fibulorhizoctonia*		*Atheliaarachnoidea*		***Athelia***	
Lepidostromatales						*Lepidostroma*				
Polyporales		*Microporus*	***Phlebia***	*Lenzites?*	*Cerrena*					
Russulales	*Russula*		*Peniophora*		*Entomocorticium*				***Heterobasidion***	
Thelephorales	*Tomentella*									
Gloeophyllales										
Jaapiales										
Corticiales						*Marchandiomphalina*	*Marchandiomyces*		*Erythricium*	
Hymenochaetales	*Coltricia*	*Phellinus*	*Rigidoporus*			*Resinicium?*		***Rickenella***	***Fomitiporia***	*Peniophorella*
Trechisporales	*Trechispora?*								*"Trechispora alnicola"*	
Hysterangiales	*Hysterangium*									
Gomphales	*Ramaria*									
Phallales	*Protubera?*									
Geastrales	*Geastrum?*									
Auriculariales			*Elmerina*							
Sebacinales	***Sebacina***		***Piriformospora***							
Cantharellales	*Cantharellus*	***Tulasnella***	*Sistotrema*			*Multiclavula*	*Burgella*		***Rhizoctonia***	

Environmental studies are expanding our concepts of the ecological roles and diversity of Agaricomycetes (Hibbett et al. 2011). ECM and soil communities have been studied intensively for many years (Horton and Bruns 2001; Peay et al. 2008), but molecular environmental surveys are demonstrating the occurrence of diverse Agaricomycetes in other, often surprising, habitats. For example, a small number of freshwater, marine, and mangrove-inhabiting Agaricomycetes are known from cultures and fruiting bodies (Binder et al. 2006; Hibbett and Binder 2001; Frank et al. 2010; Jones and Fell 2012; Yamaguchi et al. 2008), but recent studies using molecular approaches have detected Agaricomycetes in marine planktonic communities (Gao et al. 2010) and in corals, which seem to harbor species of Agaricales, Auriculariales, Boletales, Corticiales, Hymenochaetales, Polyporales, and Russulales (Amend et al. 2012). The functional biology of these marine taxa, known only from DNA sequences, remains obscure. Numerous species of Agaricomycetes have also been discovered as endophytes (Oses et al. 2008; Rungjindamai et al. 2008; Thomas et al. 2008; Weiß et al. 2011). For example, culture-based studies of the foliar and sapwood endophytes of the rubber tree *Hevea brasiliensis* have detected many species of Polyporales (almost all white rot taxa), as well as Agaricales, Atheliales, Auriculariales, Cantharellales, Hymenochaetales, and Russulales (R. Gazis and R. Martin, unpublished). Most of the sapwood endophytes are closely related to known wood-decay species, suggesting that the endophytes may exist as latent saprotrophs. Other ecological associations of Agaricomycetes that have received significant attention recently include lichenized and lichenicolous forms (DePriest et al. 2005; Diederich et al. 2011; Diederich and Lawrey 2007; Lawrey et al. 2007) and insect symbionts (Aanen et al. 2002; Mueller et al. 2005; Nobre et al. 2011; Slippers et al. 2003). The latter group includes *Fibulorhizoctonia* (Atheliales), which produces sclerotia that mimic the eggs of its termite symbionts (Matsuura 2006; Matsuura et al. 2009).

C. Fossils and Molecular Clock Dating

Molecular clock studies have yielded diverse age estimates for Fungi, with the origin of the Basidiomycota inferred to be anywhere from 450 million years ago (mya) to over 1 billion years ago (Berbee and Taylor 2010; Blair 2009; Douzery et al. 2004; Gueidan et al. 2011; Hedges et al. 2004; Taylor and Berbee 2006). A genome-based molecular clock analysis (Floudas et al. 2012) estimated the age of the Agaricomycetes at ca. 290 million years (with a 95 % highest posterior density interval of 222–372 million years). Other molecular clock studies using rRNA genes, alone or in combination with selected protein-coding genes, have focused on groups within Agaricomycetes, such as Boletales (Skrede et al. 2011; Wilson et al. 2012), Agaricales (Matheny et al. 2009; Ryberg and Matheny 2012), and brown-rot lineages (Garcia-Sandoval et al. 2011). Taxon sampling in these analyses has been very divergent, and their results have often been inconsistent. For example, an analysis focused on Inocybaceae (Agaricales) (Matheny et al. 2009) suggested that the group arose 143 (99–191) mya, while another study that focused on Boletales but included diverse Agaricales (Skrede et al. 2011) suggested that the common ancestor of Inocybaceae and Crepidotaceae existed ca. 45 (30–60) mya.

New genome sequences are providing a wealth of data for molecular clock analyses in Agaricomycetes, but the paucity of reliably identified **fossils** continues to be a limiting factor. Basidiomycetous hyphae with clamp connections are known from the Pennsylvanian (ca. 330 mya) (Dennis 1970, 1976; Krings et al. 2011; see Taylor et al., Chap. 10, Vol. VII, Part B), but the earliest fossils that are clearly Agaricomycetes do not occur until the Cretaceous. The oldest, *Quatsinoporites cranhamii*, is a fragment of a poroid hymenophore from the lower Cretaceous (130–125 mya) that has simple (nonclamped) septate hyphae and hymenial elements that resemble setae, suggesting that it may be a member of Hymenochaetales (Smith et al. 2004). Two gilled mushrooms that are probably Agaricales are known from somewhat younger deposits, including *Archaeomarasmius leggetti*, from New Jersey amber (ca. 90–94 mya) (Hibbett et al. 1995, 1997a), and *Palaeoagaricites antiquus*, from Burmese amber (ca. 100 mya) (Poinar and Buckley 2007). Another fossil from Burmese amber, *Palaeoclavaria burmites* (Poinar and Brown 2003), was originally interpreted as a clavarioid member of the so-called Aphyllophorales (this name refers to a polyphyletic taxon and is no longer in use), but there are insufficient characters visible to determine its taxonomic placement. Eocene fossils of Agaricomycetes include ectomycorrhizae associated with pine roots that were interpreted as Suillaceae (Boletales) (LePage et al. 1997) and *Appianoporites vancouverensis*, a poroid fruiting body fragment similar to *Q. cranhamii* (Smith et al. 2004). Dominican amber from the Miocene–Oligocene (ca. 15–30 mya) has yielded several well-preserved mushrooms that resemble extant Agaricales (Hibbett et al. 1997a, 2003; Poinar and Singer 1990).

Of the fossils listed previously, several have been repeatedly used as calibration points in molecular clock analyses, including *Q. cranhamii*, *A. leggetti*, and the putative suilloid ectomycorrhiza (Floudas et al. 2012; Gueidan et al. 2011; Skrede et al. 2011). Additional fossils will surely be discovered, but it seems unlikely that they will ever provide numerous rigorously identified calibration points for the major clades of Agaricomycetes. Other sources of evidence that have the potential to address ages of diverse lineages of Agaricomycetes include vicariant events and fossils of obligate symbionts, such as ECM hosts (Hibbett 2001; Hibbett and Matheny 2009; Matheny et al. 2009; Wilson et al. 2012) and the arthropods associated with taxa such as *Termitomyces*, *Amylostereum*, and attine ant cultivars (Mikheyev et al. 2010; Nobre et al. 2011; Slippers et al. 2003).

II. Phylogenetic Diversity

A. Cantharellales

Overview: Cantharellales is a small order, comprising about 260 described and currently recognized species (Kirk et al. 2008), and is represented on all continents. Effused, skinlike fruiting bodies characterize roughly half of the species, e.g., in *Sistotrema*, *Botryobasidium*, and *Tulasnella*, some of which are extremely delicate and inconspicuous. Stipitate-hydnoid and stipitate-veined fruiting bodies occur in the edible genera *Hydnum* and *Cantharellus*, respectively, while coralloid fruiting structures are found in *Clavulina* and *Multiclavula* (Fig. 14.3e, f). The hymenophore is mostly smooth but sometimes hydnoid or poroid, while truly gilled structures are lacking.

With the possible exception of the genus *Tulasnella* (Rogers 1932), species in Cantharellales have a unique type of basidia called stichic, characterized by a longitudinal orientation of the spindle during meiosis, in contrast to the chiastic type with transversely oriented spindle present in all other Agaricomycetes. While the presence of four-spored basidia constitutes an almost universal condition within Agaricomycetes, it is not so in Cantharellales. Basidia with two sterigmata are found in, for example, *Clavulina* and *Membranomyces*, and six or eight sterigmata predominate in *Botryobasidium* and *Sistotrema*. Many species in *Cantharellus* have predominantly five-sterigmate basidia. The explanation for this variation in sterigma number is not known, but a connection to the unique mode of meiosis is perhaps not unlikely.

Species in Cantharellales also show variation in septal pore morphology. *Botryobasidium* and *Tulasnella* are examples of genera with imperforate parenthesomes, while species in *Cantharellus* and *Sistotrema* and at least some species in *Ceratobasidium* have perforate parenthesomes (van Driel et al. 2009).

Ecological diversity: most resupinate species in Cantharellales seem to be saprotrophs. However, most species in *Botryobasidium* and *Tulasnella* and the majority of species in *Sistotrema* are capable of growing on common malt agar, a likely indication of the presence of cellulolytic enzymes. On the other hand, no species in Cantharellales occur as primary decayers, and they do not develop an extensive mycelium within logs. Forthcoming genomes of *Botryobasidium botryosum*, *Sistotrema brinkmannii*, *Tulasnella calospora*, and others should provide clues to ecological capabilities within the order.

Symbiotic relationships are widespread within the order and occur in all families accepted here. ECM lineages include *Cantharellus/Craterellus*, *Clavulina/Membranomyces*, *Hydnum/Sistotrema* sensu stricto, *Ceratobasidium/Thanatephorus*, and *Tulasnella* (Tedersoo et al. 2010 and references therein). Another type of symbiosis is present in the lichenized *Multiclavula* species that always grow associated with unicellular green algae (Lawrey et al. 2007).

A parasitic lifestyle occurs in *Ceratobasidium*, where a common anamorph stage known as *Rhizoctonia solani* is a widespread and troublesome crop pest, seemingly capable of infecting a wide range of hosts (Mosquera-Espinosa et al. 2013; Parmeter 1970; Sneh et al. 1996; Veldre et al. 2013). The anamorph genera *Burgoa* and *Minimedusa*, both related to *Sistotrema*, are reported as lichen parasites (Diederich and Lawrey 2007).

Systematics: the first comprehensive, multiple-gene phylogeny of Cantharellales was presented by Moncalvo et al. (2006). Veldre et al. (2013) is the most recent phylogeny with coverage of the whole order. These studies support a division of Cantharellales into four families, which may be defined by septal pore structure and secondary spore production.

Ceratobasidiaceae (eight genera) includes species with thin, resupinate fruiting bodies developing on various kinds of fine woody debris and other plant remains, but also on living plants. Hyphae are broad and without clamps. With the exception of the type species of *Ceratobasidium*, all species studied so far have perforate parenthesomes (van Driel et al. 2009; Weiß and Oberwinkler 2001). Basidia are short, with 2–4 long sterigmata. Basidiospores are capable of forming secondary spores through the development of a functional sterigma from the primary spore. The formation of a secondary spore has been interpreted as a second chance to send propagules into the air. This ability is common also in Tulasnellaceae, Auriculariales, and Sebacinales but not known from other orders in Agaricomycotina. A recent molecular study of Ceratobasidiaceae suggests that only two or, perhaps, three genera should be recognized. *Ceratobasidium* is reduced to the type species, and most other species are referred to *Rhizoctonia* (Oberwinkler et al. 2013a).

Tulasnellaceae (three genera) is characterized by a unique basidium morphology. The young basidium is globose to club-shaped and develops four globose sterigma initials that at maturity become onion-shaped before developing spores, which are forcibly discharged. The unusual sterigmata have been interpreted as monosporic epibasidia. All species can form secondary spores in the same way as described for Ceratobasidiaceae. Hyphal septa have imperforate parenthesomes. Members of the family form thin resupinate basidiomata or develop a loose mycelium within fruiting bodies of other resupinate fungi, apparently without any interaction. Species seem to be saprotrophs or mutualists capable of forming orchid mycorrhizae (Cruz et al. 2011; Preussing et al. 2010) or ectomycorrhizae (Bidartondo et al. 2003). The nuclear ribosomal genes of large parts of the genus *Tulasnella* are inexplicably deviant from those of other fungi and often require tailored polymerase chain reaction primers for amplification (Taylor and McCormick 2008).

Botryobasidiaceae includes, as far as is known, a single genus, *Botryobasidium*. It is characterized by basidia that in most cases produce six or eight spores. A few species have spiny spores and four sterigmata. They were earlier referred to *Botryohypochnus*, but molecular data place all species examined firmly within *Botryobasidium* (Binder et al. 2005). Secondary spore formation has not been observed in the family, and septal pore parenthesomes are nonperforate. The basidiomata are very delicate, and hyphae are wide with a characteristic cruciate branching on subicular hyphae. Many species have an anamorph stage referred to the form genus *Haplotrichum*. They usually develop as separate, often brownish colonies sometimes integrated, however, with the teleomorph. Saprotrophy has been the assumed nutritional strategy, but a recent study detected orchid symbionts that, on the basis of DNA sequences, belong to *Botryobasidium* (Yukawa et al. 2009). There is no comprehensive molecular phylogeny for the family.

Hydnaceae (syn. Cantharellaceae, Clavulinaceae, Sistotremataceae; nine genera) is the largest family in Cantharellales in terms of the number of constituent genera and the most diverse in terms of described species. As in many other cases, genera dominated by corticioid species seem to represent the ancestral condition, and lineages with erect fruiting bodies seem to have evolved from such species (Moncalvo et al. 2006). The corticioid species belong to *Sistotrema* and *Membranomyces*. *Sistotrema* is a polyphyletic and ecologically diverse genus. The type species forms a stipitate fruiting body with a weakly hydnoid hymenophore and is closely related to *Hydnum*. Other species related to the type have resupinate basidiomata with a poroid or hydnoid hymenophore. They all seem to form ectomycorrhiza and share this strategy with the stipitate-hydnoid genus *Hydnum* (Nilsson et al. 2006). *Membranomyces* largely shares the micromorphological characteristics, ECM habit, and phylogenetic placement with the coralloid genus *Clavulina*. *Cantharellus* and *Craterellus* seem to make up a monophyletic group, and together they form a third lineage within Hydnaceae with ECM capacity. Fruiting bodies within this lineage are of the cantharelloid type, viz. more or less funnel-shaped and with a smooth, veined, or coarsely semigilled hymenophore. Secondary spore production is not known within this family, and the species examined have septa with perforate parenthesomes. Recent molecular studies in South America and Africa have unearthed a considerable number of new species and lineages in the family (Buyck et al. 2013a, b; Henkel et al. 2011; Tibuhwa et al. 2012; Uehling et al. 2012a, b). However, a comprehensive phylogeny for the family is still lacking. Several species in Hydnaceae (*Cantharellus cibarius*, *Craterellus tubaeformis*, and *Hydnum repandum*) are highly prized as culinary mushrooms, yet attempts to keep these fungi in culture and to grow mushrooms from those cultures have largely proved unsuccessful. As with *Tulasnella*, the nuclear ribosomal genes of the genus *Cantharellus* are very deviant from those of other fungi (Moncalvo et al. 2006).

B. Sebacinales

Overview: Sebacinales (Weiß et al. 2004b), one of the basal clades in Agaricomycetes, presently includes 8 genera with ca. 30 described species (Kirk et al. 2008). In contrast to these figures from the taxonomic literature, recent molecular phylogenetic studies have revealed a huge amount of cryptic species in this group (e.g., Riess et al. 2013; Selosse et al. 2007; Weiß et al. 2011) and also suggest that generic concepts will have to be revised in the future to yield monophyletic taxa. Morphological key features of Sebacinales include the ability of basidiospores to form ballistoconidia (secondary spores), dolipores with continuous parenthesomes, longitudinally septate basidia, septa without clamp connections, and often thickened hyphal walls in substrate hyphae (Weiß et al. 2004b; Wells and Bandoni 2001; Wells and Oberwinkler 1982); unique apomorphies are not known for this group. There is a remarkable range of basidiome shapes in Sebacinales, from taxa that completely lack macroscopically visible basidiomes (*Serendipita*) to forms with corticioid (*Sebacina*) (Fig. 14.3a), pustulate-confluent (*Efibulobasidium*), cushion-shaped (*Craterocolla*) (Fig. 14.3b), coralloid (*Tremellodendron*), or even stereoid (*Tremellostereum*) to infundibuliform (*Tremelloscypha*) appearance. Sebacinalean basidiomes most often have a gelatinous consistency. Known anamorphs in the Sebacinales comprise the species of *Piriformospora*, pycnidial conidiomata in *Craterocolla cerasi*, and coremioid stages of species of *Efibulobasidium* (Kirschner and Oberwinkler 2009; Wells and Bandoni 2001). Sebacinales species have a worldwide distribution and are even known from Antarctica (Newsham and Bridge 2010). A comprehensive review of this group was recently published by Oberwinkler et al. (2013b).

Ecological diversity: over the past decade, Sebacinales has received much attention because of the exceptionally wide spectrum and the ubiquity of mutualistic associations with plant roots in which members of this group are involved. Sebacinalean mycobionts have been detected in ectomycorrhizae (Glen et al. 2002; Tedersoo and Smith 2013; Urban et al. 2003) and orchid mycorrhizae (Selosse et al. 2002; Warcup 1988), as well as in ericoid (Allen et al. 2003; Selosse et al. 2007), arbutoid (Hynson et al. 2013), and cavendishioid (Setaro et al. 2006) mycorrhizae, and even in jungermannialean mycothalli (associations with liverworts) (Kottke et al. 2003). No other fungal group is known to have a broader spectrum of mycorrhizal types. In addition, members of Sebacinales have recently been shown to occur abundantly as endophytes in plant roots (Selosse et al. 2009; Weiß et al. 2011). A few Sebacinales strains have been studied in vitro for their impact on host plants in endophytic associations. Most of these studies used the anamorphic strain *Piriformospora indica* and reported significant increases in growth and yield and improved resistance of the plant hosts to abiotic and biotic stress (Qiang et al. 2012), rendering the members of Sebacinales promising bioagents for organic plant production. The genome sequence of *Piriformospora indica* (Zuccaro et al. 2011) facilitates functional studies. Because of the richness of their mutualistic associations with land plants, Sebacinales is a model group for studying the evolution of plant–fungal interactions.

Though there is an increasing body of evidence on the importance of Sebacinales as mutualistic mycobionts of plant roots in terrestrial ecosystems, some species seem to have a saprotrophic lifestyle (*Craterocolla*, *Efibulobasidium*). Since *P. indica* and members of the morphospecies *Serendipita vermifera* grow axenically in standard media, it can be assumed that many, if not all, species of Sebacinales Group B (see below) have saprotrophic abilities.

Systematics: the monophyly of the Sebacinales has been demonstrated in molecular phylogenetic analyses (Weiß and Oberwinkler 2001; Weiß et al. 2004b). All comprehensive analyses of phylogenetic relationships within Sebacinales have been based on nuclear-encoded rRNA genes, including the internal transcribed spacers (ITS) and partial large subunit (nuc-lsu) regions. Most of the sequences analyzed have come from environmental sources; multilocus data derived from fruiting bodies or cultures are needed to solidify the systematics of Sebacinales. The available molecular phylogenetic analyses indicate that Sebacinales is divided into two monophyletic subgroups, informally known as Group A and Group B.

Group A: Species forming macroscopically visible basidiomes have only been reported from **Group A**. The types of interactions with plant roots are not uniformly distributed over Groups A and B; most of the reported taxa known to be involved in ectomycorrhizae belong to Group A, whereas sebacinalean mycobionts of ericoid mycorrhizae have only been reported from Group B.

Group B: The vast majority of **Group B** taxa are known only from environmental sequences. *S. vermifera* is the only known teleomorph in this group, yet it has been shown that this morphospecies is in fact a broad complex of cryptic species (Weiß et al. 2004b), all of which may lack macroscopic basidiomes, that produce exceptionally long vermiform basidiospores and are very poor in distinctive microscopic characters. It is possible that all teleomorphic species in Sebacinales Group B belong to this morphospecies and that the anamorphic genus *Piriformospora*, with two currently described species, evolved within this group from a *S. vermifera*-like ancestor that lost the ability to reproduce sexually.

C. Auriculariales

Overview: Auriculariales in its current concept includes ca. 30 genera with ca. 200 described species (Kirk et al. 2008; Weiß et al. 2004a). It comprises wood-decaying fungi with a broad spectrum of basidiome shapes, including effused (*Exidiopsis*, *Basidiodendron*), effuso-reflex (*Eichleriella*), odontoid (*Stypella*), hydnoid (*Pseudohydnum*) (Fig. 14.3c), and infundibuliform (*Tremiscus*) basidiomes. Basidiomes of some species even have a poroid or daedaleoid habit (*Elmerina* [including *Aporpium* and *Protodaedalea*], *Protomerulius*) (Zhou and Dai 2013). All known species cause a white rot, some are regularly found on buried wood (*Tremiscus helvelloides*). Key characters of the Auriculariales include dolipores with continuous parenthesomes and the ability of basidiospores to form ballistoconidia (secondary basidiospores). With the exception of *Hyaloria pilacre*, all species are ballistosporic.

Most of the known species in the Auriculariales have longitudinally septate basidia, but there are also species with transversely (*Auricularia*) or obliquely septate (*Patouillardina*) and even nonseptate (*Oliveonia*) or apically partially septate (*Tremellodendropsis*) basidia. In some genera (e.g., *Myxarium*, *Protodontia*, *Pseudohydnum*, *Pseudomerulius*, *Stypella*, *Tremiscus*), basidia have a plasma-devoid "stalk" (*myxarioid*, *sphaeropedunculate* basidia), which probably represents a taxonomically relevant character (Weiß and Oberwinkler 2001; Wells and Bandoni 2001). An explanation of this peculiar morphology was given by Bandoni (1984), who interpreted the basidial compartments themselves as intrabasidial meiotic products (endospores) that in germination break the outer basidial wall and develop a single conidium, the "basidiospore" in common terminology. In the myxarioid members of the Auriculariales the endospores do not fill the complete basidium but leave the characteristic stalk.

Many investigated species show clamps with characteristic retrorse projections (so-called spurred clamps) (Bandoni and Wells 1992). Another characteristic microscopic feature reported from many species of Auriculariales is the ability of basidiospores to produce mostly crescent-shaped microconidia on short sterigmalike projections (Ingold 1982a, b). Little is known about other anamorphs in Auriculariales. From recent reports about sporodochial, synnematous, bulbilliferous, and possibly also pycnidial examples in this group

(Kirschner 2010; Kirschner and Chen 2004; Kirschner et al. 2010, 2012) we can extrapolate that there might be a rich diversity of forms still to be detected. Their ecological function in nature is still unknown.

Most species of Auriculariales are so-called jelly fungi. As in Tremellomycetes, Dacrymycetes, and Sebacinales, the basidiomes of most members of Auriculariales have a gelatinous consistency and are able to experience drought conditions in a state of cryptobiosis, where the water content of the basidiomes is drastically reduced and the basidiomes revive and continue growing and sporulating when soaked again (Wells 1994). The *Auricularia auricula-judae* complex includes the broadly distributed wood-ear (mu-err) fungus, which occurs on dead wood and is one of the most important edible mushroom species of the world, particularly in Asia (Chang and Wasser 2012). The group contains other edible mushrooms, for example, *Tremiscus helvelloides*; however, these are only sporadically collected in the field and not produced on an industrial scale.

Systematics: the monophyly of a core Auriculariales (excluding *Ceratosebacina* and *Exidiopsis gloeophora*) has been suggested by the molecular phylogenetic analysis of Weiß and Oberwinkler (2001), who included the broadest sampling of species of this group to date. Some prior analyses using only rRNA genes focused on Agaricomycetes (Binder et al. 2005; Hibbett and Binder 2002) resolved the Auriculariales as a paraphyletic grade, but support for these topologies has never been strong, and other analyses show the group to be monophyletic. Multigene or phylogenomic analyses are still lacking for this group. Given that many taxa of Auriculariales have not yet been sequenced and the monophyly of the group is still tentative, an infraordinal classification is not yet available. Elements of a future classification may include a family, Auriculariaceae, comprising *Auricularia*, *Eichleriella*, *Elmerina* [including *Aporpium* and *Protodaedalea*], *Exidia*, *Exidiopsis*, and *Heterochaete*; a clade comprising *Myxarium* (including *Hyaloria* with gasteroid sporulation) and the sporodochial anamorph *Helicomyxa everhartioides*; a clade including *Heterochaetella*, *Protodontia piceicola*, *Protomerulius*, and possibly *Tremellodendropsis*; and a clade comprising *Basidiodendron*, *Bourdotia*, *Ductifera*, and the cyphelloid bulbilliferous anamorph *Ovipoculum* (Weiß and Oberwinkler 2001; Zhou and Dai 2013).

D. Phallomycetidae

The group informally labeled the gomphoid-phalloid clade (Hibbett and Thorn 2001) has been classified as the subclass Phallomycetidae, with four orders: Geastrales, Phallales, Gomphales, and Hysterangiales (Hosaka et al. 2006) (Fig. 14.4).

1. Geastrales

Overview: this group is represented by earthstars (*Geastrum*) (Fig. 14.4b), cannonball fungi (*Sphaerobolus*), and false truffles (*Radiigera*, *Sclerogaster*, and *Schenella*). Taxa with nonsequestrate fruit bodies possess an exoperidium that opens in a stellate manner as it matures, exposing the endoperidium with one (*Geastrum*) or multiple stomata (*Myriostoma*) (Sunhede 1989). Most taxa, except *Sclerogaster* and *Sphaerobolus*, have a brownish to blackish gleba, which becomes powdery at maturity. Basidiospores of most taxa, including *Sclerogaster*, are globose with a warty to spiny ornamentation. The fruiting body structure of *Sphaerobolus* is unique for Geastrales in having a single peridiole instead of a powdery gleba. The mechanism of forcible ejection of peridioles was described in detail by Ingold (1972).

Ecological diversity: the ecological characters of this group have rarely been investigated. Many species of the order grow on soil but without obvious ECM plants nearby. In addition, some species of *Geastrum*, *Sclerogaster*, and *Sphaerobolus* often fruit on rotten wood or wood chips (Hosaka and Castellano 2008), and *Sphaerobolus* fruits abundantly on artificial media (Geml et al. 2005). Several species of *Geastrum* favor semiarid to arid environments, for example, well-drained sandy soils of coasts and deserts (Kasuya et al. 2011). Such evidence suggests that most, if not all, species in the order are saprotrophic, as suggested by several authors (Kreisel 1969; Sunhede 1989). Most trufflelike fungi are believed to form ectomycorrhizae; *Sclerogaster* is an exception. One species of *Geastrum*, *G. fimbriatum*, has been described as forming ectomycorrhizae with *Fagus* (Agerer and Beenken 1998), but their observation indicated the absence of a Hartig net. The ecological roles of Geastrales species warrant further investigation.

Systematics: most taxa in Geastrales have been treated in the order Lycoperdales, along with puffballs (*Lycoperdon*) (Fischer 1900; Miller and Miller 1988; Zeller 1949), but early molecular studies (Hibbett et al. 1997a) demonstrated that *Geastrum* and *Lycoperdon*, both of which possess a powdery gleba at maturity, are only distantly related. Kreisel (1969) first segregated Geastrales from Lycoperdales but did not provide a Latin diagnosis. In addition, Kreisel (1969) included only two genera, *Geastrum* and *Myriostoma*, in the order. Hosaka et al. (2006) formally described Geastrales with a broader concept, including several previously unrecognized taxa in the order.

Geastrales, which as a whole is moderately supported as monophyletic, is divided into four families—Geastraceae, Sclerogastraceae, Schenellaceae, and Sphaerobolaceae—that are all strongly supported as clades (Hosaka and Castellano 2008; Hosaka et al. 2006). Geastraceae, Sclerogastraceae, and Schenellaceae form a clade, with an ambiguous relationship among families (Hosaka and Castellano 2008). Within Geastraceae, *Myriostoma*, the only taxon possessing multiple stomata, represents the earliest branch, suggesting that the evolutionary trend is reduction from multiple stomata to a single stoma. The early-diverging taxa within the order, Sphaerobolaceae and Schenellaceae, form basidiospores in peridioles, and this may be the ancestral character state.

A total of 7 genera and 64 species are currently recorded in the order (Kirk et al. 2008), but a number of undescribed species have been discovered for *Geastrum* and *Sclerogaster* (Hosaka and Castellano 2008; Kasuya et al. 2012). Furthermore, Kasuya et al. (2012) demonstrated that *Geastrum triplex*, which was recorded from all continents except Antarctica, should be separated into multiple species. Therefore, a significantly higher number of species may be recognized in the future.

2. Phallales

Overview: this order is famous for its stinkhorns (Phallaceae) and lattice stinkhorns (Clathraceae) (Fig. 14.4a), but recent molecular phylogenetic studies have shown that a number of sequestrate taxa are also included. Most taxa in the Phallales have fruiting bodies with a gelatinous layer and a gelatinous to mucilaginous gleba. Fruiting bodies of epigeous stinkhorns are often brightly colored (white and yellow to bright red) and composed of a pseudoparenchymatous receptacle with multiple arms (Fig. 14.4a). The fruiting bodies of most sequestrate taxa contain thick gelatinous layers, and their gleba remains gelatinous to mucilaginous. However, *Gastrosporium* and *Calvarula* have a powdery gleba at maturity (Domínguez de Toledo and Castellano 1997). Spores of most taxa are small, ellipsoid, smooth, and without ornamentation, but a few taxa, such as *Kjeldsenia* and *Gastrosporium*, have warty spore surfaces (Colgan et al. 1995; Domínguez de Toledo and Castellano 1997).

Ecological diversity: most taxa are thought to be saprotrophic due to their lignicolous habit, but at least one species (*Protubera canescens*) has been reported to be ECM (Malajczuk 1988). This report, however, is suspect because *P. canescens* has recently been confirmed as an immature form of *Ileodictyon* (May et al. 2010). It is likely that all members of the order are saprotrophic, but further investigation is necessary.

Phallales represents one of the prime examples of interactions of Fungi with arthropods (Nouhra and Domínguez de Toledo 1994). The fruiting bodies of epigeous stinkhorns possess a gleba that becomes slimy and malodorous at maturity. The odor of a mature gleba attracts a variety of mycophagous arthropods, especially flies, that disperse the basidiospores (Tuno 1998). Unlike spores of many sequestrate fungi, those of Phallales (including sequestrate taxa) are rarely documented from mammal feces. It is possible that spore dispersal of Phallales is entirely dependent on arthropods.

Systematics: Phallales was described by Fischer (1900) with two families, Phallaceae and Clathraceae. A third family, Claustulaceae, was added to the order (Cunningham 1931; Jülich 1981; Zeller 1949), and this concept has been accepted for a long time. Miller and Miller (1988) further expanded the ordinal concept by including Protophallaceae in the order, but they also included Hysterangiaceae (now in Hysterangiales). Currently the order contains six families (Phallaceae, Clathraceae, Lysuraceae, Protophallaceae, Claustulaceae, and Trappeaceae), and the monophyly of the order and each family is strongly supported by multigene phylogenetic analyses (Hosaka et al. 2006).

The basal grades of the order are composed of Protophallaceae, Claustulaceae, and Trappeaceae, all of which exhibit an exclusively sequestrate habit, indicating that stinkhornlike fruit bodies are derived morphologies in Phallales (Hosaka et al. 2006). Among phylogenies of Agaricomycetes, Phallales represents the sole example of an unambiguous transition from sequestrate to nonsequestrate forms.

A total of 29 genera and ca. 100 species are currently recorded in the order (Kirk et al. 2008), but some genera, such as *Protubera* and *Trappea*, are polyphyletic, with species in both Phallales and Hysterangiales (Hosaka et al. 2006), and require further taxonomic revision. Some new genera and species have been described recently (Cabral et al. 2012; Desjardin and Perry 2009). Because the center of diversity of this order probably lies in the tropics (Miller and Miller 1988) and many such areas have not been extensively investigated, the number of taxa in this group will be significantly higher in the future.

3. Gomphales

Overview: the fungi in Gomphales (Jülich 1981; from the Greek *pluglike*) have long been recognized as a distinct, highly variable clade of Agaricomycetes (Bruns et al. 1998; Hibbett and Thorn 2001; Hosaka et al. 2006; Pine et al. 1999). They are characterized by a wide range of fruiting body morphologies, from stalked ramarioid/clavarioid (e.g., *Ramaria*, *Phaeoclavulina*, and Lentariaceae) to club (Clavariadelphaceae), gilled (*Gloeocantharellus*), cantharelloid-gomphoid (*Gomphus*, *Phaeoclavulina*, and *Turbinellus*) (Fig. 14.4d), tooth (*Beenakia*), resupinate-odontoid (*Kavinia*) (Fig. 14.4e), all the way to sequestrate fungi (Gauteriaceae) (Giachini et al. 2010; Hosaka et al. 2006; Humpert et al. 2001).

Ecological diversity: members of Gomphales show heterogeneity in their ecological characters. Most species in Beenakiaceae, Lentariaceae, Kaviniaceae, *Gloeocantharellus*, and *Phaeoclavulina* and some species of *Ramaria* (e.g., *R. moelleriana*, *R. stricta*, and *R. circinans*) grow and fruit on woody debris, a trait that has led to their general categorization as saprotrophs. The other taxa of the order are generally considered ECM, and while the nutritional status of many species of Gomphales is still unknown, the formation of ectomycorrhizae by *Turbinellus*, *Gomphus*, and some *Ramaria* species has been confirmed (Agerer 1996a, b, c, d; Agerer and Iosifidou 2004; Agerer et al. 1998; Castellano 1988; Griffiths et al. 1991; Masui 1926, 1927; Miller and Miller 1988; Nouhra et al. 2005; Rinaldi et al. 2008).

Systematics: the taxonomy and systematics of the Gomphales has traditionally relied on morphological characters now known to be subject to parallel evolution and phenotypic plasticity (Moncalvo et al. 2000). As a family, Gomphaceae has traditionally been classified within the Aphyllophorales, along with distantly related taxa such as Cantharellaceae, Ganodermataceae, and Polyporaceae (Donk 1964). The phylogenetic relationships of members of the order Gomphales, including its monophyly, have been estimated using molecular data (Giachini et al. 2010; Hosaka et al. 2006; Humpert et al. 2001), which revealed that gomphoid fungi are closely related to taxa in Geastrales, Hysterangiales, and Phallales in the subclass Phallomycetidae (Colgan et al. 1997; Giachini et al. 2010; Hibbett et al. 1997a; Hosaka et al. 2006; Humpert et al. 2001; Pine et al. 1999). Currently, the order encompasses six well-supported families, namely

Beenakiaceae (*Beenakia*, *Kavinia* and *Ramaricium*), Clavariadelphaceae (*Clavariadelphus*), Gautieriaceae (*Gautieria*), Gomphaceae (*Gloeocantharellus*, *Gomphus*, *Phaeoclavulina*, and *Turbinellus*), Lentariaceae (*Lentaria*), and Ramariaceae (*Ramaria*), distributed within 18 genera and ca. 336 species (Kirk et al. 2008).

Despite their macromorphological variation, the members of the order share a number of microscopic and macrochemical characters, including cyanophilic spore ornamentation, chiastic basidia, similar hyphal construction, and positive hymenial reaction to ferric sulfate (Donk 1961, 1964; Eriksson 1954; Humpert et al. 2001; Petersen 1971b; Villegas et al. 1999, 2005). In the studies of Hosaka et al. (2006), both Bayesian and parsimony analyses showed strong support for the monophyly of the Phallomycetidae. Even though no definitive synapomorphies have been identified for this gomphoid-phalloid clade, some potential synapomorphic characters, including rhizomorph morphology (presence of ampullate hyphae and acanthohypha), pistillarin content, and structures of septal pore cap, have been proposed (Agerer and Iosifidou 2004; Hibbett and Thorn 2001). In addition, some members of the gomphoid-phalloid clade, such as *Gautieria*, *Hysterangium*, *Ramaria*, and *Geastrum*, are known to produce thick hyphal mats in soil (Agerer and Iosifidou 2004; Nouhra et al. 2005; Sunhede 1989). Although most of these characters are not exclusive to the gomphoid-phalloid fungi, the yellowish filled acanthocystidia and associated "exuded drops of pigments" have been reported only from the gomphoid-phalloid fungi [e.g., *Geastrum*, *Gomphus*, *Phallogaster*, and *Ramaria* (Agerer and Iosifidou 2004)].

4. Hysterangiales

Overview: this group has long been considered a sequestrate (trufflelike) relative of stinkhorns (Phallales). The ordinal status was accepted by some authors (Hosaka et al. 2006; Jülich 1981; Zeller 1939), but others have included the group in the order Phallales (Kirk et al. 2008; Miller and Miller 1988). The order contains exclusively sequestrate taxa with hypogeous fruit bodies (e.g., *Hysterangium*, *Mesophellia*, *Austrogautieria*) (Fig. 14.4c), but taxa of epigeous habit (e.g., *Gallacea*, *Phallogaster*) are also known, and they often expose a gleba at maturity (Castellano and Beever 1994). Most taxa are characterized as having gelatinous to cartilaginous glebae of greenish to brownish tint, except Mesophelliaceae, which has a powdery gleba at maturity.

Ecological diversity: most taxa form ectomycorrhizae with various host trees, including Pinaceae and Fagaceae in the Northern Hemisphere and Myrtaceae (mostly *Eucalyptus* and *Leptospermum*) and Nothofagaceae in the Southern Hemisphere. In addition, some recent studies have extended the range of hosts to Caesalpiniaceae, Phyllanthaceae, and Dipterocarpaceae (Castellano et al. 2000; Henkel et al. 2012). Phallogastraceae is the only family in the order with a saprotrophic habit. Some species of *Hysterangium* form dense perennial hyphal mats, which change the soil chemistry and microorganism biomass (Griffiths et al. 1994). Fruiting bodies of Hysterangiaceae and Mesophelliaceae are consumed by small mammals and marsupials, and they often make up a significant portion of the animals' diet (Claridge 2002; Lehmkuhl et al. 2004).

Systematics: this order was proposed by Zeller (1939), whose treatment was followed by those of Locquin (1974) and Jülich (1981). However, a Latin diagnosis was not provided until Hosaka et al. (2006) formally described the order. The order is strongly supported as being monophyletic by multigene phylogenetic studies (Hosaka et al. 2006, 2008), but its relationships to other orders in Phallomycetidae are not well supported.

Hysterangiales is divided into four families, Hysterangiaceae, Mesophelliaceae, Gallaceaceae, and Phallogastraceae, all of which are strongly supported as being monophyletic (Hosaka et al. 2006, 2008). Hysterangiaceae and Mesophelliaceae form a clade, to which Gallaceaceae is the sister family (Hosaka et al. 2006). Phallogastraceae, the only saprotrophic member of the order, represents the earliest branch, which is separated from the remaining Hysterangiales (Hosaka et al. 2008), suggesting that the ECM habit was gained only once within the order.

Fifteen genera and ca. 110 species are currently recorded in the order (Kirk et al. 2008), but a number of new species have recently been discovered, mainly from the Southern Hemisphere (Henkel et al. 2011; Hosaka et al. 2008). It has been demonstrated that many genera in this order are polyphyletic. For example, *Hysterangium* spp. are placed in both Hysterangiaceae and Mesophelliaceae (Hosaka et al. 2008). Kirk et al. (2008) included Trappeaceae as the fourth family of the order, but because the genus *Trappea* is also polyphyletic and the type species, *Trappea darkeri*, belongs to Phallales (Hosaka et al. 2006), Trappeaceae should not be included in Hysterangiales.

The biogeography of the order was extensively studied by Hosaka et al. (2008), who demonstrated that the ECM lineages (Hysterangiaceae, Mesophelliaceae, and Gallaceaceae) originated in the Southern Hemisphere (presumably east Gondwana), with a few range expansions to the Northern Hemisphere. Although some area relationships can be explained by vicariance, many sister-group relationships, such as those of taxa from Australia and New Zealand separated by short branches, can only be explained by long-distance dispersal, suggesting that trufflelike fungi are capable of crossing ocean barriers.

E. Trechisporales

Overview: Trechisporales K. H. Larss. (2007) is a relatively small order with ca. 100 species and 8–13 genera. It was described only recently (Hibbett et al. 2007), after DNA studies confirmed it as a distinct clade (Binder et al. 2005; Larsson 2004; Matheny et al. 2007). The majority of species in the order belong to the genus *Trechispora* (including *Cristelloporia*, *Scytinopogon*), a highly diverse genus of mostly corticioid fungi. The other genera contain only corticioid fungi, with the exception of the monotypic polypore genus *Porpomyces*. A number of species in the order have an anamorphic stage: *Aegerita tortuosa* for *Subulicystidium* and *Osteomorpha* for *Trechispora*. Considering that almost all species in the order form inconspicuous fruiting bodies that rarely get collected and identified, the known species number is likely a fraction of the true diversity.

Fruiting body morphology ranges from clavarioid (*Scytinopogon*), stipitate hydnoid (*Trechispora thelephora*), and resupinate polyporoid (*Porpomyces*, *Trechispora*) to corticioid (Fig. 14.5f–h). Most species either have small spines (aculei) covering their hymenophore or are completely smooth. Fruiting-body-associated rhizomorphs are common, and all species have light-colored fruiting bodies that produce hyaline spores. Some dimitic species are found in *Cristelloporia*, *Fibrodontia*, and *Trechispora*, but most species are monomitic and bear clamps on all septa. Spore morphology is very variable, from very long and narrow spores of *Subulicystidium* to tiny ellipsoid spores of *Porpomyces* and spinose spores in most species of *Trechispora*. Conspicuous subulate cystidia are found in *Subulicystidium* and *Tubulicium*. Calcium oxalate crystals are common on subicular hyphae of *Trechispora* and have been shown to be species-specific in form (Larsson 1994).

Ecological diversity: most species in the genus appear to be white-rot wood-inhabiting (e.g., *Sistotremastrum*) or soil-inhabiting (e.g., *Porpomyces*, *Trechispora*) saprotrophs. *Trechispora* species are difficult to grow with standard culturing techniques for wood-decay fungi, whereas *Sistotremastrum* spp. pose no difficulties. Dunham et al. (2007) reported root-associated mycelial mats formed by *Trechispora*, indicating a possible mycorrhizal association, but further work is needed to confirm this inference. One species (misidentified as *T. alnicola*) has been reported as a grass parasite (Wilkinson 1987).

Systematics: Larsson (2007b) divides the order into two families: Hydnodontaceae Jülich 1982 (=Subulicystidiaceae Jülich 1982) with *Fibrodontia*, *Luellia*, *Porpomyces*, *Subulicystidium*, *Trechispora* (including *Cristelloporia*, *Hydnodon*), *Tubulicium*, and possibly *Subulicium*; and *Sistotremastrum* in its own, yet formally unnamed, family. Telleria et al. (2013) produced a phylogeny of the order and confirmed that *Brevicellicium* is also part of the Hydnodontaceae. Larsson et al. (2011)

and Birkebak et al. (2013) found that the clavarioid genus *Scytinopogon* was nested within *Trechispora*.

Only a few DNA-based species-level papers on Trechisporales have been published: Albee-Scott and Kropp (2010) on *Trechispora*, Telleria et al. (2012) on *Sistotremastrum*, and Telleria et al. (2013) on *Brevicellicium*. Other genera that may belong to the order include the corticioid *Brevicellopsis*, *Dextrinocystis*, and *Dextrinodontia*. A major open question in the systematics of the order is whether *Trechispora* should be divided or kept together so that it includes *Cristelloporia*, *Echinotrema*, *Hydnodon*, and *Scytinopogon*, as most authors currently do.

F. Hymenochaetales

Overview: Hymenochaetales Oberw. 1977 is one of the larger orders of basidiomycetes, with over 900 species and ca. 75 currently recognized genera. The order is dominated by wood-inhabiting polypores and bracket fungi (e.g., *Phellinus*, *Trichaptum*), as well as stereoid and corticioid fungi with a smooth or hydnoid hymenophore (e.g., *Hyphodontia*, *Resinicium*). A few coralloid fungi (*Alloclavaria*, *Clavariachaete*) and moss-associated agarics (e.g., *Rickenella*) are found in the order (Fig. 14.5a–e). The largest known fruiting body belongs to *Phellinus ellipsoideus* (Dai and Cui 2011).

Most Hymenochaetales species that have been studied have dolipore septa with continuous (imperforate) parenthesomes, in contrast to most other polypores, agarics, and corticioid fungi (van Driel et al. 2009). *Peniophorella praetermissa*, a corticioid fungus, is the only species in the order reported with perforate parenthesomes, but much of the diversity in the order remains unstudied in this respect.

Economically important pathogens of trees include *Rigidoporus microporus* in rubber and other tropical tree plantations (Farid et al. 2009) and *Phellinus sulphurascens* on temperate conifers (Lim et al. 2005). The fruiting bodies of *Inonotus sanghuan*, many *Phellinus* spp., and cankers of *I. obliquus* are used in herbal medicine and are reported to have anticancer properties (Dai et al. 2010; Ju et al. 2010; Wu et al. 2012).

Ecological diversity: the order exhibits a wide variety of different ecological strategies. Most species of Hymenochaetales are white-rot fungi. They are everywhere a major, and often the dominant, part of the wood-rot communities (e.g., species of *Hyphodontia*, *Phellinus*, *Trichaptum*). Two polypore genera (*Coltricia* and *Coltriciella*) form ectomycorrhizae (Tedersoo et al. 2007), and a number of species in other genera are parasites or pathogens of woody plants (e.g., many species of *Inonotus*, *Phellinus* s.l., and *Oxyporus*). Several species of *Peniophorella* have specialized organs for catching invertebrates, apparently an adaptation to a nitrogen-deprived environment (Tzean and Liou 1993). A peculiar ecological group of mostly agarics are moss-associated. Whether the association is parasitic or mutualistic is not clear (Larsson et al. 2006; Redhead 1981).

Systematics: Hymenochaetales as a clade is well supported, but its internal structure is largely unresolved. Larsson et al. (2006) provided the only broad phylogenetic overview of the order. They defined six clades using nuc-lsu rRNA sequences. Some of those clades were not corroborated in other studies using nrDNA, and the branching order of the groups varies from one analysis to another (Ghobad-Nejhad and Dai 2010; Larsson 2007b; Miettinen and Larsson 2010). The genome of *Fomitiporia mediterranea* has been published (Floudas et al. 2012), and two additional genomes (*Rickenella mellea*, *Trichaptum abietinum*) were produced in the US Department of Energy Joint Genome Institute in 2013.

Hymenochaetaceae: this family contains 60 % of described Hymenochaetales species, mostly polypores, a number of stereoid fungi, and a few hydnoid fungi. All species in this family are characterized by brown pigments that turn black in KOH (xanthochroic reaction). Hyphae are simple-septate and parenthesomes imperforate in all species studied so far. Many species have characteristic brown cystidia, setae, in the hymenium (which explains the name of the type genus, *Hymenochaete*). Due to morphological similarities, this order has long been recognized in the literature in a way that corresponds to the current concept: *Phellinus* s.l. (dimitic polypores), *Inonotus* s.l. (monomitic polypores), *Hymenochaete* s.l. (corticioid and hydnoid), and *Asterodon* (hydnoid). All species are wood inhabiting, and many species are parasites of living trees and bushes. A strange exception is *Phylloporia parasitica*, which reportedly grows on living leaves (Wagner and Ryvarden 2002). Whether the ECM *Coltricia* and *Coltriciella*, traditionally assigned to Hymenochaetaceae, belong here is unclear (see below). Recent regional morphology-based treatments of Hymenochaetaceae include Nuñez and Ryvarden (2000) for East Asia, Ryvarden (2004) for the Neotropics, and Dai (2010) for China. The basis for DNA-based assessments of generic concepts in the group comes from studies of Wagner and Fischer (2001, 2002a, b). Later family-level studies have been few and based on one or two ribosomal genes only (Jeong et al. 2005; Larsson et al. 2006; Parmasto et al. 2013). Many of the newly defined genera have received attention in more focused phylogeny papers: *Inonotus* (Tian et al. 2013), *Phellinus* s.s. (Cui and Decock 2012; Decock et al. 2006; Fischer and Binder 2004; Tomšovský et al. 2010b), *Phellinopsis* (Zhou and Qin 2013), *Phellopilus* (Keller and Hohn 1997), *Porodaedalea* (Brazee and Lindner 2013; Tomšovský et al. 2009), and *Phylloporia* (Valenzuela et al. 2010; Wagner and Ryvarden 2002; Zhou and Dai 2012). The genus *Fomitiporia* in particular has been the subject of many studies (Amalfi and Decock 2013; Amalfi et al. 2010, 2012; Decock et al. 2005, 2007; Fischer 2002; Fischer et al. 2005; Vlasák and Kout 2010). The corticioid genus *Hymenochaete* has turned out to be polyphyletic, and part of the species belongs to a new genus, *Pseudochaete* (He and Dai 2012; He and Li 2013; Parmasto et al. 2013; Wagner and Fischer 2002a). The genus now also includes poroid *Cyclomyces* spp. and hydnoid *Hydnochaete* spp. The position of the *Hymenochaete*-like coralloid genus *Clavariachaete* has not been studied, but morphologically it is a typical member of the Hymenochaetaceae (Parmasto 2010).

Schizoporaceae: this species-rich clade contains the bulk of corticioid *Hyphodontia*, now classified in the genus *Xylodon*, one of the largest genera of wood-rotting fungi (Hjortstam and Ryvarden 2007, 2009). Most poroid *Hyphodontia* or *Schizopora* belong here, and micromorphologically the clade is relatively homogeneous. Larsson et al. (2006) included the genus *Coltricia* here and called it the *Coltricia* clade. Here we consider the position of *Coltricia* unresolved. No phylogenetic overview of *Hyphodontia* s.l. or this clade exists. Paulus et al. (2000) studied species phylogeny in *Schizopora*.

Tubulicrinacae: initially called the *Hyphodontia* clade, Larsson (2007b) used the family name Tubulicrinaceae for this clade. It contains inconspicuous wood-rotting corticioid fungi with a variable micromorphology. The largest genus is *Tubulicrinis*.

***Kneiffiella* clade**: *Kneiffiella* is another segregate genus of *Hyphodontia* sensu lato with many hydnoid species. Most species in the clade have characteristic tubular tramal cystidia.

***Oxyporus* clade**: this small clade contains the polypore genus *Oxyporus* and poroid-hydnoid *Botryodontia* (Sell et al. 2013). The extent of the genus *Oxyporus* and delimitation against *Rigidoporus* is unclear.

***Rickenella* clade**: this is the most diverse in terms of morphology and ecological strategies of all the Hymenochaetales clades and includes moss-associated agarics (e.g., *Rickenella*), stereoid (*Cotylidia*), clavarioid (*Alloclavaria*), poroid (*Sidera*), and many wood-rotting corticioid (*Peniophorella*, *Resinicium*) species. Larsson (2007b) and Miettinen and Larsson (2010) found this clade to be paraphyletic, and clearly multigene data sets are needed to resolve the structure in this part of the fungal tree. It may be basal to the order, but again the current DNA data do not permit strong statements in this regard. A handful of studies have been conducted at the genus level in this group: Moncalvo et al. (2002) and Redhead et al. (2002) dealt with agarics, Dentinger and McLaughlin (2006) with *Alloclavaria*, Sjökvist et al. (2012) with *Cotylidia* and *Muscinupta*, Larsson (2007b) with *Peniophorella* and other *Hyphoderma* s. l., Miettinen and Larsson (2010) with *Sidera*, and Nakasone (2007, 2012) with *Resinicium* and *Tsugacorticium*. Hallenberg et al. (2007) studied the *Peniophorella praetermissa* species complex.

Lineages of uncertain position: *Coltricia* and *Coltriciella* are the only ECM genera in the order. Larsson et al. (2006) included them in the same clade (named *Coltricia* clade) with parts of *Hyphodontia* (the Schizoporaceae here). The nrDNA of *Coltricia* is highly divergent from that of other Hymenochaetales and occupies a long branch in nrDNA-based analysis, jumping around in phylogenies. Considering the long branch and divergent ecology and morphology, the position of *Coltricia* and *Coltriciella* within the order is in need of further study. Tedersoo et al. (2007) provide the only DNA-based study of the group. The wood-rotting polypore and corticioid genera *Trichaptum*, *Basidioradulum*, *Cyanotrama*, *Fibricium*, and *Poriodontia* belong in the vicinity of Hymenochaetaceae, but not within it. These genera seem to be somewhat closely related; they do not seem to form a monophyletic group (Binder et al. 2005; Ghobad-Nejhad and Dai 2010; Larsson et al. 2006; Miettinen and Larsson 2010). The polypore genus *Bridgeoporus* is a segregate of *Oxyporus* and has been shown to belong to Hymenochaetales (Redberg et al. 2003). It does not seem to be very closely related to *Oxyporus*, and its position in the order is open.

G. Polyporales

Overview: Polyporales (Gäumann 1926) includes approximately 1,800 described species (Kirk et al. 2008), making it one of the larger orders of Agaricomycetes. However, new species are continually being described, even in relatively well-studied areas such as western Europe (Bernicchia et al. 2010; Spirin et al. 2012; Vampola and Vlasak 2012), and the number of species known only from environmental sampling and studies of endophytic communities has increased dramatically in recent years (Fröhlich-Nowoisky et al. 2009, 2012; Hallenberg et al. 2008). Polyporales contains conspicuous bracket fungi, including perennial "conks" (e.g., *Ganoderma applanatum*, *Fomes fomentarius*), as well as more cryptic effused (resupinate) forms, which often fruit on the undersides of logs (Fig. 14.6b–d). Other species have pileate-stipitate fruiting bodies or multiple flabelliform lobes (e.g., *Sparassis*, *Hydnopolyporus*) (Fig. 14.6a, e, f). The hymenophore is frequently poroid (e.g., *Polyporus*) but can also be hydnoid (*Steccherinum*), lamellate (*Trametes* [*Lenzites*] *betulina*), merulioid (*Phlebia*), or smooth (*Phanerochaete*). No gasteroid taxa are known, but *Lentinus tigrinus* has a naturally occurring secotioid form in addition to the typical agaricoid form (Hibbett et al. 1994). A few species produce underground sclerotia (e.g., *Lignosus*, *Polyporus*, *Wolfiporia*). The order has varied hyphal anatomy, including monomitic forms (with only generative hyphae, e.g., *Ceriporia*), as well as dimitic and trimitic forms (with thick-walled skeletal or binding hyphae) (Gilbertson and Ryvarden 1986). No morphological synapomorphy characterizes the Polyporales, and the most common morphological types described previously also occur in other orders of Agaricomycetes.

Ecological diversity: along with members of Hymenochaetales and Russulales, members of this order dominate wood-decay communities in terrestrial ecosystems. A few species act as plant pathogens, causing timber damage (e.g., species of *Ganoderma*, *Fomitopsis*, and *Wolfiporia*), and others are major decay agents of structural timber (e.g., *Antrodia*). Wood decayers in Polyporales can be divided into two major groups: white-rot species, which are able to decay both lignin and cellulosic compounds, and brown-rot species, which remove cellulose and hemicellulose without significant lignin degradation (Worrall et al. 1997). *P. chrysosporium* and *P. placenta*, which are the model systems for white-rot and brown-rot biochemistry (respectively), are both in the Polyporales (Martinez et al. 2004, 2009). No mycorrhizal taxa are known in the order. Many members of Polyporales are commonly isolated as part of the endophytic communities in woody tissues and roots, and though several ecological roles have been proposed for these fungi, from latent saprotrophs to protective agents, their true function remains largely unknown (Porras-Alfaro et al. 2011).

Systematics: approximately 150 genera and 40 legitimate family names are available for use in Polyporales (Larsson 2007b; Ryvarden 1991), but there is no broadly accepted consensus infraordinal classification. Recent monographs on Polyporales include those of Nuñez and Ryvarden (2000) on East Asian polypores, Ryvarden (2004) on neotropical polypores (Ganodermataceae), Niemelä (2005) and Bernicchia (2005) on European polypores, and Bernicchia et al. (2010) on European corticioid fungi.

The monophyly of Polyporales was not well supported in analyses of rRNA gene sequences (Binder et al. 2005; Larsson 2007b). However, analyses adding single-copy protein-coding genes (Garcia-Sandoval et al. 2011; Justo and Hibbett 2011; Matheny et al. 2007; Miettinen et al. 2012; Sjökvist et al. 2012) and genome-based phylogenetic analyses (Binder et al. 2013; Floudas et al. 2012) have strongly supported the monophyly of the order. The sister group of Polyporales is not known with confidence; in the studies just mentioned, Corticiales, Gloeophyllales, Russulales, and Thelephorales usually seem to be closely related to Polyporales, but relationships between these orders remain in need of further study.

The studies of Binder et al. (2005, 2013) provide broad overviews of the major lineages of Polyporales based on taxon-rich sampling of rRNA genes in combination with *rpb1*, *rpb2*, and *tef1* sequences, as well as gene-dense phylogenomic analyses. Four major groups have been informally labeled as the *Antrodia*, core polyporoid, phlebioid, and "residual" clades, the latter being more a mixed bag of taxa that did not fit in the other clades. Larsson (2007b) and Miettinen et al. (2012) sought to apply existing family names to parts of the clades recognized in phylogenetic analyses.

Residual polyporoid clade: the monophyly of this clade remains uncertain. All taxa in this group produce a white rot, which is probably plesiomorphic for the Polyporales (Floudas et al. 2012), but morphologically they are very diverse and include poroid (*Rigidoporus*), agaricoid (*Panus*), corticioid (*Hyphoderma*), resupinate-hydnoid (*Steccherinum*), and stipitate-steroid forms (*Podoscypha*). Representative taxa that have been the subject of recent phylogenetic studies include *Antrodiella* (Miettinen et al. 2012), *Cerrena* (Lee and Lim 2009), *Hyphoderma* (Larsson 2007a), *Hypochnicium* (Telleria et al. 2010), *Pseudolagarobasidium* (Hallenberg et al. 2008), *Podoscypha* (Sjökvist et al. 2012), and *Steccherinum* (Miettinen et al. 2012). The group exemplifies the numerous transitions in hymenophore types and microscopic characters (e.g., cystidia and hyphal types) that have occurred repeatedly during the evolution of the Polyporales (Miettinen et al. 2012).

Phlebioid clade: largely dominated by corticioid forms, much of the taxonomy of this diverse group revolves around two large, highly polyphyletic genera, *Phlebia* and *Phanerochaete*, and their limits and relations with respect to several smaller genera. A number of polypore genera are also found in the clade (e.g., *Bjerkandera*, *Ceriporia* and *Irpex*). Taxa that have been the subject of phylogenetic studies include *Ceriporia* (Jia et al. 2013), *Ceriporiopsis* (Tomšovský et al. 2010a), *Phanerochaete* (De Koker et al. 2003; Greslebin 2004; Wu et al. 2010), and *Trametopsis* (Tomšovský 2008). *Leptoporus*, a close relative of *Ceriporia*, is often considered a brown-rot fungus (Gilbertson and Ryvarden 1986) and would be the only brown rotter in this lineage of white-rot taxa (Lindner and Banik 2008).

Antrodia clade: this lineage includes exclusively species that produce a brown-rot type of decay. The majority of all known brown-rot fungi belong to this clade. Pileate and resupinate polypores are predominant, along with a few corticioid taxa (e.g., *Dacryobolus*, possibly *Crustoderma*). Several genera have received attention in phylogenetic studies, including *Antrodia* sensu lato (Bernicchia et al. 2010; Rajchenberg et al. 2011; Spirin et al. 2013; Yu et al. 2010), *Daedalea* (Lindner et al. 2011), *Fomitopsis* (Kim et al. 2007), *Laetiporus* (Lindner and Banik 2008), *Postia* sensu lato (Pildain and Rajchenberg 2013), and *Sparassis* (Dai et al. 2006; Wang et al. 2004). A general overview of the clade is given by Ortiz-Santana et al. (2013), who showed that generic delimitation remains highly problematic, with most of the traditionally recognized genera being poly- or paraphyletic.

Core polyporoid clade: this group roughly corresponds to the families Polyporaceae and Ganodermataceae in the sense of Ryvarden (1991) and includes mostly polypores with a trimitic hyphal system. Some corticioid (*Epithele*, *Lopharia*) and agaricoid taxa (*Lentinus*) are also nested in the clade. This is the best sampled lineage of Polyporales, both in terms of taxa and genes, and the only one with a well-supported internal structure. Three major lineages were recognized by Justo and Hibbett (2011), termed the *Dentocorticium*, trametoid, and *Polyporus* clades. Representative genera with recent phylogenetic studies include *Lentinus* (Grand et al. 2010), *Megasporoporia* (Li and Cui 2013), *Melanoderma* (Cui et al. 2011), *Perenniporia* s. lato (Decock and Ryvarden 2003; Robledo et al. 2009; Zhao et al. 2013), *Polyporus* s. lato (Krüger 2008, 2010; Krüger and Gargas 2004; Sotome et al. 2008, 2013), and *Trametes* (Justo and Hibbett 2011; Tomšovský 2008; Welti et al. 2012).

Lineages of uncertain position: three relatively small lineages of white-rot polypores seem to be closely related to the *Antrodia* or core polyporoid clades, but they apparently do not belong to either group (Binder et al. 2013; Miettinen and Rajchenberg 2011). These lineages include the genus *Grifola*, the *Tyromyces* clade (*Piloporia*, *Skeletocutis*, *Tyromyces*), and the *Cinereomyces/Gelatoporia* clade (*Cinereomyces*, *Gelatoporia*, *Obba*, *Sebipora*). Resolving the position of these lineages and the phylogenetic structure of the residual polyporoid clade are two major issues for the higher-level taxonomy of Polyporales. Improving the internal resolution in the phlebioid, antrodia, and core polyporoid clades is necessary to move forward in the family-level and generic taxonomy of these groups.

H. Thelephorales

Overview: Thelephorales is a strongly supported clade that currently includes ca. 18 genera and 269 described species (Kirk et al. 2008). The group is morphologically diverse and contains corticioid (*Tomentella*), cantharelloid (*Polyozellus multiplex*), clavarioid (*Thelephora*), and pileate forms (*Hydnellum*) (Fig. 14.7c, d). Hymenophores of pileate taxa may be poroid (*Boletopsis*), toothed (*Hydnellum*, *Sarcodon*), smooth to wrinkled or tuberculate (*Thelephora*), or lamellate (*Lenzitopsis*). It was once suggested that the pileate-stipitate agaric *Horakia* (=*Verrucospora*) was related to Thelephoraceae based on spore morphology (Oberwinkler 1975), but molecular data place it in Agaricales (Matheny et al. 2006), as had been suggested by Singer (1986). Basidiospores are mostly dark, ornamented, and with a distinctive angular outline but may also be subglobose and spinose (*Bankera* and *Phellodon*). Thelephoric acid (a terphenyl quinone, similar to atrotomentin) is found in *Bankera*,

Boletopsis, *Hydnellum*, *Phellodon*, *Polyozellus*, *Pseudotomentella*, *Sarcodon*, and *Thelephora*. This compound also occurs in Boletales and other orders but nonetheless seems to be a distinguishing feature of the group (Bresinsky and Rennschmid 1971).

Ecological diversity: most members of Thelephorales are ECM and are often dominant components of mycorrhizal communities (Bruns et al. 1998; Tedersoo et al. 2010). However, *Lenzitopsis* produces fruiting bodies on wood of junipers and is reported to produce a white rot (Zhou and Kõljalg 2013). *Amaurodon* is also reported to grow on wood of living trees (U. Kõljalg, unpublished) and is presumably nonmycorrhizal. *Amaurodon* has been placed as the sister group to the remaining Thelephorales, but *Lenzitopsis* is nested within the group, closely related to *Tomentellopsis*, suggesting that there have been multiple transitions between nutritional modes (Larsson 2007b; Zhou and Kõljalg 2013).

Systematics: the current classification of Thelephorales (Kirk et al. 2008) includes two families, Thelephoraceae and Bankeraceae. Donk (1964, p. 247) thought that the similarity of the Bankeraceae to certain Thelephoraceae was "an example of extreme convergence," but other authors suggested that the two families were closely related (Jülich 1981; Stalpers 1993), and this has been repeatedly supported by molecular data (Binder et al. 2005; Bruns et al. 1998; Larsson 2004, 2007b; Zhou and Kõljalg 2013). The analysis of Zhou and Kõljalg (2013) resolved two nonsister clades corresponding to Bankeraceae (one including *Bankera* and *Phellodon* and another containing *Hydnellum*, *Sarcodon*, and *Boletopsis*) and a paraphyletic assemblage of taxa corresponding to Thelephoraceae. However, internal support for many deep nodes was weak. An in-depth multigene phylogenetic analysis is needed to assess the classification of the order.

I. Corticiales

Overview: Corticiales K.H. Larsson is a small order established to accommodate basidiomycetes recognized in recent molecular phylogenetic studies and included previously in the Vuilleminiales (Boidin et al. 1998), the *Dendrocorticium* clade (Binder and Hibbett 2002), and the corticioid clade (Binder et al. 2005; Larsson et al. 2004). It has been represented by a single family, Corticiaeae Herter, made up of ca. 29 genera and 136 species (Kirk et al. 2008), but the molecular phylogenetic study of Ghobad-Nejhad et al. (2010) recognized three families and several new genera (discussed subsequently). The order is made up of mostly resupinate species that produce smooth hymenophores, a monomitic hyphal system with or without clamps, and smooth basidiospores, often with pink walls. Many species produce pink or red basidiomata with a catahymenium in which young basidia do not form a palisade but are formed deep within a layer of hyphidia and then elongate to reach the hymenial surface, and some have dendrohyphidia; however, there is no morphological synapomorphy that characterizes the entire order.

Some species are known only from asexual stages. One genus of these is the anamorph-typified *Marchandiomyces*, which was commonly included as a core genus in recent molecular studies. The genus was originally established for asexual lichen parasites (Diederich 1990; Etayo and Diederich 1996), but sexual forms are also now known for the group. These include *Marchandiobasidium aurantiacum* as the teleomorph of *Marchandiomyces aurantiacus* (Diederich et al. 2003) and *Marchandiopsis quercina*, a species previously assigned to *Laeticorticium* or *Vuilleminia* that was found to be nested among asexual *Marchandiomyces* species by Ghobad-Nejhad et al. (2010). The clade containing *Marchandiomyces* also includes described plant pathogens in the teleomorph-typified genera *Laetisaria* and *Limonomyces*, indicating that sexual–asexual relationships among these species will require more study. The type species of the order, *Corticium roseum*, may form a bulbil-like anamorph known as *Hyphelia rosea*; these bulbils are similar to those of *Laetisaria* and *Marchandiomyces* (Eriksson and Ryvarden 1976).

Ecological diversity: fungi in Corticiales exhibit a remarkable range of ecologies, including saprotrophs, plant pathogens, lichen pathogens, and lichenized species (Lawrey et al. 2008). The basal position of saprotrophic species in Vuilleminiaceae and Punctulariaceae would indicate that this is the ancestral condition for the order, but the most derived clade, Corticiaceae, contains a complex mixture of ecological forms, which suggests an unusual tendency for ecological transitions (Ghobad-Nejhad et al. 2010; Lawrey et al. 2008). Ecology can sometimes be used to characterize genera, but most clades in this order have mixtures of nutritional modes. Entirely saprotrophic genera include *Corticium*, *Giulia*, and *Galzinia*, but *Erythricium*, which is mostly saprotrophic, also includes the plant pathogen *Erythricium salmonicolor*.

Most species of *Limonomyces*, *Laetisaria*, and *Waitea* are plant pathogens or endophytes but are able to persist in the field as saprotrophs (Andjic et al. 2005; Burdsall 1979; Burdsall et al. 1980; Stalpers and Loerakker 1982). However, *Laetisaria* also includes the lichen parasite *L. lichenicola* (Diederich et al. 2011). *Marchandiomyces*, originally described for lichen parasites (Diederich 1990; Etayo and Diederich 1996), is now known to include saprotrophic and foliicolous species (Diederich and Lawrey 2007; Lawrey et al. 2007, 2008). There is also one lichen-forming species, *Marchandiomphalina foliacea* (Lawrey et al. 2008; Palice et al. 2005), which seems to be most closely related to the lichen parasite *Marchandiobasidium aurantiacum* and the saprotrophic *Erythricium laetum*. Lichen mutualisms have evolved independently in Basidiomycota at least five times (Diederich and Lawrey 2007; Diederich et al. 2003, 2011; Ertz et al. 2008; Fischer et al. 2007; Hodkinson et al. 2012; Lawrey et al. 2007, 2008, 2009; Nelsen et al. 2007; Redhead et al. 2002). *Marchandiomphalina foliacea* is the only described lichen species in Corticiales. It forms a foliose thallus structure and asexual goniocysts resembling soredia, but no sexual stages are known (Jørgensen 1989).

Systematics: the monophyly of Corticiales is strongly supported by molecular phylogenies, mostly based on rRNA gene sequences (Binder et al. 2005; Boidin et al. 1998; DePriest et al. 2005; Diederich et al. 2011; Ghobad-Nejhad et al. 2010; Hibbett et al. 2007; Langer 2002; Larsson 2007b; Larsson et al. 2004; Lawrey et al. 2008). The sister group of Corticiales seems, on the basis of many of these studies, to be the order Gloeophyllales. One species of Corticiales, *Punctularia strigoso-zonata*, has been subject to whole-genome sequencing; phylogenomic analyses suggest that it is in a clade that also includes Gloeophyllales and Jaapiales (Fig. 14.1). A recent attempt to produce an infraordinal classification of Corticiales is based on a molecular phylogeny using nuc-lsu rRNA sequences (Ghobad-Nejhad et al. 2010). This analysis resolved three groups that were recognized at the family level using the existing names, Vuilleminiaceae, Punctulariaceae, and Corticiaceae. Until more sequences become available and more specimens sequenced, this represents the best current hypothesis for a family classification of the order.

Vuilleminiaceae Maire ex Lotsy: this clade contains saprotrophic species that develop dendrohyphidia and produce clamps and generally allantoid spores and gelatinous fruiting bodies. It includes the genera *Vuilleminia* and *Cytidia* and a new genus, *Australovuilleminia* (for *Vuilleminia coccinea*). Based on their molecular phylogeny and incompatibility crossing tests, Ghobad-Nejhad et al. (2010) found that the so-called core *Vuilleminia* species (*V. macrospora*, *V. pseudocystidia*, *V. alni*, *V. comedens*, *V. megalospora*) form a monophyletic group. These species are all decorticating, produce a gelatinous fruiting body, and exhibit a unique 13 bp insertion in the ITS2. Other described *Vuilleminia* species (*V. cystidiata* and *V. macrospora*) were recovered in the *Vuilleminia* clade but outside of the core *Vuilleminia* clade. Other species were recovered outside of the *Vuilleminia* clade and reassigned to new genera, including *V.* (*Punctulariopsis*) *obduscens* and *V.* (*Punctulariopsis*) *subglobispora*, which were recovered in the *Punctularia* clade, and *Vuilleminia* (*Marchandiopsis*) *quercina*, which was recovered in the *Corticium* clade.

Punctulariaceae Donk: the family introduced by Donk (1964) was intended to separate *Punctularia* species from other corticioid fungi in the Aphyllophorales, but few classifications recognized it, placing most corticioid species in the Corticiaeae. In the nuc-lsu rRNA phylogeny of Ghobad-Nejhad et al. (2010), this is a clade of saprotrophic species that produces clamps and ellipsoid spores and includes the genera *Punctularia* and *Dendrocorticium* and a new genus, *Punctulariopsis* (for *Vuilleminia subglobispora* and *Vuilleminia obducens*). These species cause a vigorous white rot compared to species in *Corticium*.

Corticiaceae Herter: this family, originally conserved against Vuilleminiaceae (Pouzar 1985) to represent a much broader circumscription than has emerged in recent molecular-based classifications, is now viewed by Ghobad-Nejhad et al. (2010) as a well-supported clade containing species with and without clamps, including the type species of *Corticium*, and a variety of sexual and asexual genera with diverse nutritional modes (*Erythricium*, *Galzinia*, *Giulia*, *Laetisaria*, *Limonomyces*, *Marchandiobasidium*, *Marchandiomphalina*, *Marchandiomyces*, *Marchandiopsis*, and *Waitea*). As mentioned by these authors, Corticiaceae is by far the most diverse family in Corticiales, both morphologically and ecologically. It also contains several polyphyletic genera in need of revision. The single most problematic clade, containing *Marchandiomyces*, *Marchandiopsis*, *Limonomyces*, and *Laetisaria*, is also the most interesting ecologically. Improving the internal resolution in this clade will not only resolve the generic taxonomy of these groups but also help to clarify some of the most interesting ecological transitions in Agaricomycetes.

J. Jaapiales

Overview: Jaapiales Manfr. Binder, K.H. Larss. & Hibbett (Binder et al. 2010) is the smallest order of Agaricomycetes, with a single genus of just two species, *Jaapia argillacea* and *J. ochroleuca*. Fruiting bodies of both species are resupinate, at first patchy, then thinly effused and monomitic, with thin-walled hyphae and frequent clamp connections. Basidiospores are narrowly fusoid (boletinoid) and cyanophilous.

Ecological diversity: both species of *Jaapia* fruit on wet, rotting wood on the margins of lakes and streams and are collected infrequently (Eriksson and Ryvarden 1976). Only *J. argillacea* is known in culture, and it is unreactive in tests for laccase, peroxidases, or tyrosinase (Stalpers 1978). Thus, *Jaapia* species might be brown-rot saprotrophs or ECM, but the biology of this group is unknown. BLAST searches using the sequences of *J. argillacea* or *J. ochroleuca* yield very few matches among environmental sequences, a rarity in the Agaricomycetes. At present, the only sequence matches are to a few ITS sequences of uncultured fungi from permafrost. It is likely that woody substrates in aquatic habitats have been undersampled for sequences of Basidiomycota, as well as for their fruiting bodies.

Systematics: the genus *Jaapia* Bres. was referred to the Coniophoraceae (Boletales) (Eriksson and Ryvarden 1976; Nannfeldt and Eriksson 1953) but later recognized as distinct from Boletales in rDNA analyses of Binder et al. (2005) and Larsson (2007a). The latter study, based on analyses of sequences of 5.8S and nuclsu rRNA, confirmed that *J. ochroleuca* is a member of the same lineage as the type species, *J. argillacea*. A 6-gene phylogeny placed *J. argillacea* as a sister group to Agaricomycetidae (Atheliales, Boletales, Amylocorticiales, and Agaricales) (Binder et al. 2010), but phylogenomic analyses place Jaapiales in a well-supported clade with Corticiales and Gloeophyllales (Fig. 14.1).

K. Gloeophyllales

Overview: Gloeophyllales Thorn is an odd taxon with no morphological or ecological characters that unite the 6 genera and perhaps 40 species (Hibbett et al. 2007; Kirk et al. 2008). The type genus, *Gloeophyllum*, is a bracket fungus with resupinate, effused-reflexed, or pileate fruiting bodies and poroid, daedaleoid, or lamellate hymenophores (Gilbertson and Ryvarden 1986). *Boreostereum*, *Chaetodermella*, and *Veluticeps* (including *Columnocystis*) are corticioid to stereoid, with resupinate to effused-reflexed fruiting bodies having a smooth or rugose-wrinkled hymenophore (Chamuris 1988; Eriksson and Ryvarden 1973; Nakasone 1990b) (Fig. 14.7b). *Neolentinus* and *Heliocybe* produce agaricoid fruiting bodies with a central or eccentric stipe, a convex to upturned pileus, and adnexed to decurrent lamellae (Redhead and Ginns 1985) (Fig. 14.7a). Most taxa are dimitic, but some are monomitic, and others trimitic. Simple clamp connections are constant (most taxa), rare (some *Veluticeps* in culture), or absent (*Boreostereum*). The context is pallid in *Heliocybe* and *Neolentinus* but brown in most other taxa and browning in KOH (turning green with KOH in *Boreostereum*) (Hibbett et al. 2007).

Sexuality ranges from homothallic (*Boreostereum*, several *Veluticeps*) to heterothallic and bipolar (*Gloeophyllum*, *Heliocybe*, *Neolentinus*) or tetrapolar (*V. berkeleyi*) (Ginns and Lefebvre 1993).

Ecological diversity: members of Gloeophyllales are wood-decay fungi mostly causing a brown rot of conifers, occasionally angiosperms, and frequently in wood in service. *Neolentinus lepideus* is known as the "train-wrecker" for its propensity to decay wooden railway trestles in bygone days (Redhead and Ginns 1985), and species of *Gloeophyllum* commonly decay outdoor wooden structures such as decks, playground equipment, and picnic tables, and sometimes wooden joists and timbers in homes (Gilbertson and Ryvarden 1986). *Chaetodermella* and *Veluticeps* cause brown rots of conifers, *Heliocybe* on angiosperms, often in quite dry situations such as fence posts or rails and exposed, decorticated logs. However, *Boreostereum*, which seems to be a sister group to the remainder of the order in a 6-gene phylogenetic analysis (Garcia-Sandoval et al. 2011), is associated with white rot of fire-charred coniferous or angiosperm wood, although spot tests for laccases, peroxidases, and tyrosinase in culture have been equivocal (Chamuris 1988; Nakasone 1990a).

Systematics: the core of Gloeophyllales was recognized by Kim and Jung (2000) as Chaetodermataceae, and the link between *Gloeophyllum*, *Heliocybe*, and *Neolentinus* was made by Thorn et al. (2000). Studies by Binder et al. (2005) strongly supported early suggestions (Hibbett and Donoghue 1995; Hibbett et al. 1997b) that *Gloeophyllum* was set apart from the true polypores and formed the basis for describing the Gloeophyllales (Hibbett et al. 2007). The polypore *Donkioporia*, which causes a white rot of conifer wood in service, was included in the order when it was first described (Hibbett et al. 2007) on the basis of its clustering in the *Gloeophyllum* clade in analyses of nuc-lsu rRNA sequence data by Kim and Jung (2000, 2001), but it can now be excluded as a member of the core polyporoid clade (Garcia-Sandoval et al. 2011). In addition to the six genera known to belong to the order on the basis of molecular studies, *Campylomyces* and *Pileodon*, which are segregates of *Veluticeps* (Nakasone 1990b), and *Mycothele* have been referred here, but no sequence data are available. *Mycobonia*, from which *Mycothele* was segregated, has also been suggested as belonging in Gloeophyllales, but analyses of rRNA gene sequences place it in the core polyporoid clade of Polyporales (Krüger and Gargas 2004).

Garcia-Sandoval et al. (2011) presented a 6-gene phylogenetic analysis of 18 species representing the 6 genera accepted in Gloeophyllales. Their results suggest that *Gloeophyllum* consists of at least two clades, one containing the type species, *G. sepiarium*, as well as *G. striatum*, *G. subferrugineum*, and *G. trabeum* (all species known from wood in service), and the other containing the type species of *Osmoporus*, *O. odoratus*, as well as *Osmoporus protractus* (both on exposed conifer wood in boreal-subarctic environments) (Garcia-Sandoval et al. 2011). *Gloeophyllum mexicanum* and *Gloeophyllum carbonarium* were basal to *Osmoporus*, and each might represent segregate genera upon further study. In addition, the type species of *Veluticeps* (*V. berkeleyi*), *Columnocystis* (*C. abietina*), and *Chaetodermella* (*C. luna*) formed a weakly supported clade for which the oldest generic name is *Veluticeps*. A sequence of *V. fimbriata* was placed on a long branch that split the genus *Neolentinus*, but other analyses placed it together with the remaining species of *Veluticeps*, and *Heliocybe* was recovered as sister to *Neolentinus*, leaving the possibility of its synonymy with *Neolentinus* undecided (Garcia-Sandoval et al. 2011). Thus, for the moment we advocate recognition of *Boreostereum*, *Gloeophyllum*, *Heliocybe*, *Neolentinus*, *Osmoporus*, and *Veluticeps* in Gloeophyllales.

L. Russulales

Overview: Russulales currently includes more than 1,700 described species (Kirk et al. 2008). This high number corresponds to an equally astonishing diversity of fruiting body morphologies (Fig. 14.8) and life strategies. From a phylogenetic perspective, the dominant life form is the skinlike, effused, and resupinate basidioma, often developing out of sight at the underside of decaying wood on the ground (Larsson and Larsson 2003). Examples include *Asterostroma*, *Gloeocystidiellum*, and *Boidinia*. From ancestors with such inconspicuous basidiomata, elaborate fruiting structures have developed, for example, coralloid in *Lachnocladium*, *Hericium*, and *Artomyces* (Fig. 14.8c); reflexed and bracketlike as in *Echinodontium*, *Stereum*, *Laurilia* (Fig. 14.8b), and (some species of) *Lentinellus*; pileate-stipitate mushrooms as in *Russula*, *Lactarius* (Fig. 14.8a), and *Albatrellus*; or sequestrate as in *Macowanites* (Fig. 14.8d, e) and *Leucophleps*. The hymenophore is most often smooth (e.g., *Peniophora*, *Stereum*) or hydnoid (e.g., *Auriscalpium*, *Hericium*), while a poroid hymenophore configuration is comparatively rare (e.g., *Albatrellus*, *Heterobasidion*, *Wrightoporia*). A lamellate hymenophore is known from Auriscalpiaceae (*Lentinellus*) and Russulaceae only. Many species have basidiospores with an amyloid reaction of the spore wall, and for most of them the amyloidity is combined with an ornamented outline of the wall.

There is no obvious morphological synapomorphy for Russulales, but the presence of gloeoplerous hyphae or gloeocystidia with contents rich in sesquiterpenes has not been demonstrated in any other basidiomycete order (Larsson and Larsson 2003). In a few cases such structures seem to have been secondarily lost (e.g., in *Byssoporia*) or transformed into homologous structures such as the lactiferous hyphae in *Stereum*. The term *gloeocystidium* refers to enclosed, bladderlike structures in fruiting bodies (Clémençon 2004). Structures termed *gloeocystidia* have been reported in many orders, but the specific type present in Russulales is associated with unique vesicles with tubular invaginations, which may constitute a synapomorphy for the order (McLaughlin et al. 2008). In Russulales the gloeoplerous hyphae and gloeocystidia have been suggested to serve as a chemical defense system against mycophagy (Sterner et al. 1985).

Ecological diversity: the dominant nutritional strategy in Russulales is saprotophic decay of organic matter, primarily wood. In Russulales only white rot has been documented. It can be highly intense, for example, by species in *Stereum* and *Scytinostroma*, and sometimes characteristic like the white pocket rot produced by *Conferticium* and *Xylobolus* spp. (Otjen and Blanchette 1984). Some species are capable of infecting living trees and perform decay in roots or heartwood. *Heterobasidion annosum* is considered the most severe forest pathogen in conifer forests in the Northern Hemisphere, causing economic losses of $1 billion annually in the USA alone (Woodward et al. 1998). Other species with a potentially pathogenic behavior in managed forests are *Echinodontium tinctorium*, *Hericium erinaceus*, *Scytinostroma galactinum*, and *Stereum sanguinolentum*.

Another threat to forestry is caused by wood wasps from the family Siricidae living in symbiosis with members of *Amylostereum* (Slippers et al. 2003). The wasp female transfers conidia of the fungus when she places eggs inside the wood of stressed trees using her needlelike ovipositor organ. Larvae then feed on the fungus while mining through fungus-infested wood. Infection by *Amylostereum areolatum* and its vector, *Sirex noctilio*, normally does not cause much damage within its native range in Europe and Asia. However, when the wasp was accidentally introduced into the Southern Hemisphere and in North America, infections in both exotic pine plantations and native American pine stands have become severe (Nielsen et al. 2009; Slippers et al. 2001).

ECM associations have developed independently within two Russulales lineages, in Russulaceae and in Albatrellaceae. Molecular phylogenies suggest that in Russulaceae the evolution proceeded from a saprotrophic to a mycorrhizal nutritional strategy and coincides with the development of erect fruiting bodies from effused ancestors (Larsson and Larsson 2003). Another trend confined to the same families is gasteromycetization (Albee-Scott 2007), which involves the evolution of closed fruiting bodies adapted for a dryer climate and animal dispersal. It is likely that change in lifestyle has driven the development of erect and closed fruiting structures since both are better adapted for a soil-oriented life than the effused structure typical for wood decayers.

Systematics: the potential presence of a russuloid lineage with the wide circumscription accepted here was first discussed by Donk (1971). Oberwinkler (1977) elaborated on Donk's ideas and used Russulales as an example of a higher-order group that contained multiple fruiting body morphologies. Molecular data have confirmed that Russulales sensu Oberwinkler is a monophyletic group (Bruns et al. 1998; Hibbett et al. 1997b, 2000; Larsson 2007b; Larsson et al. 2004; Binder et al. 2005). The most comprehensive phylogenies for Russulales have been published by Larsson and Larsson (2003) and Miller et al. (2006). Larsson and Larsson (2003) identified ten well-supported lineages that can be understood as representing families; the main lineages are briefly discussed in what follows. For many genera relationships are still not resolved.

Stereaceae (14 genera) is dominated by corticioid forms with a smooth hymenophore and smooth basidiospores. Most species fruit in exposed places like living or recently dead trunks and branches and have morphological adaptations for resisting drought. Examples include *Stereum*, *Aleurodiscus* sensu lato, *Aleurocystidiellum*, and *Xylobolus*. Only *Aleurodiscus* has been the subject of a detailed phylogenetic study (Wu et al. 2001).

Peniophoraceae (16 genera) includes mainly corticioid species with a smooth hymenophore and smooth spores, not always with an amyloid reaction of the spore wall. Some genera are characterized by branched, dextrinoid skeletal hyphae (e.g., *Scytinostroma*, *Vararia*). These genera are often referred to a separate family Lachnocladiaceae, but molecular data do not support its recognition. Also in this family, many species grow rather exposed and decay dead but still attached branches (*Peniophora*, *Scytinostroma*). Closely related to *Peniophora* is *Entomocorticium*, known as symbionts of bark beetles of the genus *Dendroctonus* that cause great damage to pine forests (Harrington 2005). Morphological studies of single genera are available, but a comprehensive phylogeny for the family is lacking.

Russulaceae (six genera) is the most species-rich family due to the high diversity seen in the mushroom genera *Lactarius* and *Russula*. Recent phylogenetic studies have shown that *Russula* is monophyletic, but *Lactarius* in a traditional sense is not (Buyck et al. 2008). The latter group is now divided into *Lactarius*, *Lactifluus*, and *Multifurca*. The many genera with sequestrate species recognized earlier (e.g., *Arcangeliella*, *Macowanites*) are now considered examples of adaptations to a dry habitat and animal dispersal that has repeatedly taken place within *Lactarius* and *Russula*, respectively (Lebel and Tonkin 2007; Miller et al. 2001; Nuytinck et al. 2004). Basal lineages are composed of saprotrophic taxa with corticioid fruiting bodies (*Boidinia*, *Gloeopeniophorella*) (Larsson and Larsson 2003). All species in Russulaceae have ornamented spores.

Albatrellaceae (six genera) is the second family where a mycorrhizal life strategy predominates. Species are stipitate-poroid (*Albatrellus*, *Polyporoletus*), effused-poroid (*Byssoporia*), or sequestrate (e.g., *Mycolevis*). Smith et al. (2013) present a phylogeny for the family.

Hericiaceae (three genera) species have coralloid or effused basidiocarps that are mostly strongly hydnoid. *Hericium erinaceus* (lion's mane, monkey's head) has been much used in Chinese folk medicine, and modern studies have demonstrated the presence of many medically active substances in this and related species (Lindequist et al. 2005; Mizuno 1999). A phylogeny for the genus is presented by Hallenberg et al. (2013).

Auriscalpiaceae (four genera) includes lamellate (*Lentinellus*), stipitate-hydnoid (*Auriscalpium*), and effused-hydnoid basidiocarps (*Dentipratulum*, *Gloiodon*).

Bondarzewiaceae (four genera) is a small family of hard and robust wood-decaying species with either a poroid (*Bondarzewia*, *Heterobasidion*) or a hydnoid hymenophore (*Echinodontium*).

Incertae sedis: the clavarioid genus *Artomyces* is sometimes included in Auriscalpiaceae, but that arrangement is not unambiguously supported by molecular data. *Amylostereum* could be placed in Bondarzewiaceae but also recognized as a separate family (Binder et al. 2005).

M. Agaricomycetidae

Agaricomycetidae contains two large and well-known orders, Agaricales and Boletales, as well as three small groups containing mostly corticioid forms, Atheliales, Amylocorticiales, and Lepidostromatales. Russulales seems to be the sister group of Agaricomycetidae. The clade containing Agaricomycetidae and Russulales largely corresponds to Agaricales sensu Singer (1986), which included four suborders, Agaricineae, Boletineae, Russulineae, and Polyporineae. The latter included agaricoid forms in *Lentinus* and other genera that are now known to be distributed among Polyporales, Gloeophyllales, and Agaricales.

1. Atheliales and Lepidostromatales

Overview: Atheliales Jülich in a broad sense includes 22 genera with 110 described species (Kirk et al. 2008). However, several genera of Atheliales are polyphyletic (e.g., some *Athelia*, *Athelopsis*, and *Leptosporomyces* species) with species in Agaricales, Amylocorticiales, Cantharellales, and Polyporales (Binder et al. 2005, 2010; Ertz et al. 2008; Larsson 2007b; Larsson et al. 2004; Matheny et al. 2006; Oberwinkler 2012). Many athelioid species produce loosely connected resupinate fruiting bodies lacking conspicuous morphological differentiation on various substrates, including branches, wooden debris, and mosses (Fig. 14.9d) (Larsson et al. 2004). *Stereopsis vitellina* forming stipitate-stereoid basidiocarps was recently separated from Cantharellales and placed in Atheliales (Sjökvist et al. 2012). The lichenized Lepidostromataceae was originally placed in a sister-group relationship to Atheliales and included three species that produce clavarioid basidiocarps similar to *Multiclavula* spp. in Cantharellales (Ertz et al. 2008).

Ecological diversity: Atheliales is not a species-rich group, but it is pervasive in terrestrial ecosystems. Some *Athelia* species parasitize cyanobacteria, green algae, and lichens (Oberwinkler 1970; Yurchenko and Golubkov 2003), and it has been suggested that the lifestyle of the lichen-forming Lepidostromataceae can be considered a similar form of interaction (Oberwinkler 2012). Other *Athelia*, *Athelopsis*, and *Tretomyces* spp. produce white rot on various trees, debris, leaf litter, grasses, and ferns (Eriksson and Ryvarden 1973; Kotiranta and Saarenoksa 2005; Kotiranta et al. 2011). Brown rot is absent in Atheliales (Binder et al. 2010). The *Athelia* anamorph *Fibulorhizoctonia* forms symbioses with termites by producing sclerotia that mimic termite eggs (Matsuura et al. 2000). *Amphinema*, *Byssocorticium*, *Piloderma*, and *Tylospora* spp. are ECM with Pinaceae and Fagaceae and are often dominant in ECM fungal communities (Erland and Taylor 1999; Lilleskov et al. 2004).

Systematics: Atheliales currently includes Atheliaceae as a single family. Lepidostromataceae had been formally left in Agaricomycetidae *incertae sedis* based on mixed support values from rDNA analyses (Ertz et al. 2008). In a recent multigene study this family was recognized as the order Lepidostromatales, including the genera *Lepidostroma*, *Sulzbacheromyces*, and *Ertzia* (Hodkinson et al. 2013). A taxonomic revision of Atheliaceae on the generic level is needed because new taxa are being described (Kotiranta et al. 2011) and previously unknown lineages are being added to the family (Sjökvist et al. 2012).

2. Amylocorticiales

Overview: Amylocorticiales K.H. Larss., Manfr. Binder & Hibbett is a recently described order (Binder et al. 2010) that includes roughly 70 species. Taxonomic concepts at the generic level are still in flux (Binder et al. 2005, 2010; Buyck et al. 2012; Gorjón et al. 2011; Larsson 2007b; Niemelä et al. 2007; Zmitrovich and Spirin 2002). Species of the nine genera accepted in Amylocorticiales usually form corticioid and resupinate fruiting bodies and produce smooth, merulioid, or sometimes poroid hymenophores (*Amylocorticiellum*, *Amylocorticium*, *Anomoloma*, *Anomoporia*, *Ceraceomyces*, *Serpulomyces*) (Fig. 14.9b). Others have evolved more elaborate fruiting bodies, including multistoried pileate-stipitate structures (*Podoserpula pusio*, the pagoda fungus), pendant fan-shaped fruiting bodies with wrinkled gill-like hymenophores (*Plicaturopsis crispa*), or hydnoid hymenophores (*Irpicodon pendulus*). Anatomical characters in Amylocorticiales are also diverse. The basidiospores are either thin- or thick-walled, smooth, ellipsoid, cylindrical, or allantoid, and most react positively (amyloid) to Melzer's reagent. All hyphal systems are monomitic (i.e., consist of generative hyphae only) and nodose septate; however, this character combination is not synapomorphic for Amylocorticiales and occurs in other groups (e.g., Polyporales). Cystidia are rare in Amylocorticiales.

Ecological diversity: species placed in Amylocorticiales are predominantly saprotrophic or, rarely, biotrophic. The modes of wood decay include brown rot (e.g., *Amylocorticium*, *Anomoporia*, *Podoserpula*) and white rot (e.g., *Anomoloma, Irpicodon, Plicaturopsis*). *Hypochniciellum molle* (*Leucogyrophana mollis*) is of economic importance as a causal agent of brown rot in timber (Mattsson et al. 2010; Niemelä et al. 2007). The decay strategy of *Serpulomyces* has not been studied in detail. ECM forms are seemingly absent in Amylocorticiales, but it has been suggested that *Anomoloma flavissimum* and *Podoserpula pusio* may represent transitions to ECM symbioses (Bougher and Syme 1998; Niemelä et al. 2007). "*Athelia*" *rolfsii* (anamorph *Sclerotium rolfsii*) is a serious soilborne pathogen, also known as Southern blight, that infects more than 500 plant species, including peanut, potato, and tomato (Punja 1985).

Systematics: Amylocorticiaceae Jülich is the single family in Amylocorticiales to date, and a major taxonomic revision on the generic level is needed. *Amyloathelia crassiuscula, Amyloxenasma allantosporum, Anomoporia kamtschatica, Athelia rolfsii, Athelopsis lacerata, Leptosporomyces septentrionalis*, and *Hypochniciellum molle* are distinct lineages in Amylocorticiales, but they do not represent the generic types.

3. Boletales

Overview: Boletales E.-J. Gilbert is one of the larger orders of fleshy Agaricomycetes, including 17 families, 88 genera, and roughly 1,400 species (Binder and Hibbett 2006; Kirk et al. 2008). The typical fruiting body of a bolete is pileate-stipitate with a tubular (e.g., *Boletus, Suillus*) (Fig. 14.9e) or sometimes gilled hymenophore (*Paxillus, Phylloporus*). Gasteroid forms (*Scleroderma, Rhizopogon, Astraeus*) have evolved several times independently from this morphology (Binder and Bresinsky 2002; Binder and Hibbett 2006; Bruns et al. 1989; Thiers 1984; Wilson et al. 2011). Roughly 77 % of the described species produce pileate-stipitate fruiting bodies (Binder et al. 2010). In addition, there are resupinate forms with smooth or warted (*Coniophora, Serpula*) (Fig. 14.9f), merulioid (*Leucogyrophana*), and toothed (*Gyrodontium*) hymenophores, and it has been suggested that pileate-stipitate fungi with gilled hymenophores, such as *Tapinella* or *Hygrophoropsis*, have evolved at least five times from resupinate ancestors (Binder et al. 2005, 2010). *Bondarcevomyces taxi* is the only species developing polyporelike basidiocarps (Larsson 2007b), but no coralloid or clavarioid forms are known in the order. The fruiting bodies of Boletales are specifically attacked by the ascomycete anamorph genus *Sepedonium* (teleomorph *Hypomyces*), suggesting some degree of coevolution between parasites and hosts (Douhan and Rizzo 2003; Sahr et al. 1999).

The morphological characters of Boletales have been studied intensively (Agerer 1999; Arpin and Kühner 1977; Both 1993; Corner 1972; Horak 2004, 2011; Moser 1983; Pegler and Young 1981; Singer 1986; Smith and Thiers 1971; Watling 1970), but there is no synapomorphic trait for the order as a whole. Boletales species produce unique pigments and colorless compounds during secondary metabolism, and the terphenyl quinone atromentin plays an essential role as building block for the synthesis of derivatives, including pulvinic acids (e.g., variegatic acid and xerocomic acid), cyclopentenones, grevillins, and other substances (Besl and Bresinsky 1977, 1997; Besl et al. 1986; Bresinsky 1974; Bresinsky and Orendi 1970; Gill and Steglich 1987). An atromentin pathway has evolved independently in Thelephorales, but it produces only simple terphenyl quinones, not the structurally more complex pigments (Besl and Bresinsky 1997).

Ecological diversity: members of Boletales have a worldwide distribution in forest ecosystems, with biodiversity hot spots in Southeast Asia and North America (Corner 1972; Singer 1965; Smith and Thiers 1971). The major nutritional modes of Boletales include brown-rot saprotrophy, ECM symbioses, and mycoparasitism; biotrophic plant pathogens and white-rot fungi are not known (Binder and Hibbett 2006). It has been suggested that brown rot is the ancestral lifestyle of Boletales, having a single evolutionary origin in the early branching lineages (Binder and Hibbett 2006). Based on the unique capability of brown-rot-producing Boletales to selectively depolymerize microcrystalline cellulose, which weakens the strength of wood, this form of wood decay has also been called Coniophoraceae-type rot (Kämmerer et al. 1985; Nilsson and Ginns 1979) to separate it from other brown-rot types. Most brown-rot-causing species contribute to carbon sequestration in conifer forests (*Tapinella*, *Pseudomerulius*), but a few have specialized on human-built timber environments. The so-called cellar fungus *Coniophora puteana* and especially the so-called dry rot fungus *Serpula lacrymans* cause significant damage in wooden building structures (Schmidt and Kebernik 1989; Schmidt et al. 2002), and the ecological diversification and structure of geographical lineages of these aggressive decayers have been studied in detail (Eastwood et al. 2011; Kauserud et al. 2007a, b; Skrede et al. 2011; Watkinson and Eastwood 2012). Serpulaceae is also a prime example of transitions from brown-rot to ectomycorrhiza associated with major morphological changes. *Austropaxillus* species (pileate-stipitate fruiting bodies with gilled hymenophores) and *Gymnopaxillus* species (gasteroid) are derived from within *Serpula* and form ectomycorrhizae with *Eucalyptus* and *Nothofagus* (Bresinsky et al. 1999; Jarosch 2001; Skrede et al. 2011).

Approximately 90 % of species in Boletales are involved in ECM symbioses, particularly with Betulaceae, Caesalpiniaceae, Dipterocarpaceae, Fagaceae, Myrtaceae, Nothofagaceae, Pinaceae, and Salicaceae, or in arbutoid mycorrhiza with Ericaceae (Newman and Reddell 1987; Rinaldi et al. 2008; Tedersoo et al. 2010). *Boletus* and *Leccinum* spp. show an increased tendency to associate with specific hosts; for example, *Leccinum scabrum* forms ectomycorrhizae with *Betula* (Singer 1967). *Suillus*, *Gomphidius*, *Chroogomphus*, and *Rhizopogon* spp. in the suborder Suillineae are almost exclusively associated with Pinaceae, which is probably the oldest clade of ECM partners for Boletales (Hibbett and Matheny 2009). Most ECM species are placed among the Boletaceae (roughly 400 plus species), which include highly prized edibles such as *Boletus edulis* (porcini).

Mycoparasites in Boletales represent transitions from the ECM lifestyle (Binder and Hibbett 2006) and have evolved at least twice independently. *Gomphidius* and *Chroogomphus* spp. are capable of parasitizing the established ectomycorrhizae of the closely related *Suillus* and *Rhizopogon* by penetrating their rhizomorphs (Agerer 1990, 1999; Miller 1964; Olsson et al. 2000), thereby circumventing competition for host plants (Binder and Hibbett 2006). *Pseudoboletus parasiticus* in Boletaceae produces its fruiting bodies on *Scleroderma citrinum* (a gasteroid bolete) while eroding the gleba of the host (Binder and Hibbett 2006). Other putative mycoparasites from the basal lineages of Boletaceae include the sister taxa *Buchwaldoboletus* and *Chalciporus* (Nuhn et al. 2013).

Systematics: Boletales includes five suborders that have been described based on disparate characteristics and methods: Boletineae, Suillineae, Sclerodermatineae, Tapinellineae, and Coniophorineae (Binder and Hibbett 2006). Boletineae was first introduced by Gilbert (1931) using fruiting body morphology and spore shape as distinctive characteristics. This suborder included all species with tubular hymenophores at that time and a few species with gilled hymenophores. Suillineae was separated later from Boletineae based on unique pigments occurring in this group (Besl and Bresinsky 1997). In addition, numerous resupinate and paxilloid taxa producing a brown rot were known to be closely related to Boletales based on their pigments (Besl et al. 1986), but they remained *incertae sedis*. The morphology of belowground rhizomorphs helped to formally place these taxa in Tapinellineae and Coniophorineae (Agerer 1999), which was supported by early major studies using DNA sequences (Bruns et al. 1998; Kretzer and Bruns 1999). Sclerodermatineae was described based on nuc-lsu rRNA sequences (Binder and Bresinsky 2002) integrating the gasteroid Sclerodermatales into Boletales.

Resupinate taxa are still a source of taxonomic uncertainty in Tapinellineae and Coniophorineae, particularly concerning the polyphyletic genus *Leucogyrophana* (Binder et al. 2010; Jarosch and Besl 2001). Coniophorineae, including three clades of *Leucogyrophana*, has been resolved as a monophyletic group, but without significant statistical support (Binder et al. 2010). Together, Boletineae, Sclerodermatineae, and Suillineae form the largest clade, including the overwhelming majority of ECM taxa (except *Austropaxillus* and *Gymnopaxillus*) and mycoparasites. *Hydnomerulius pinastri* (formerly *Leucogyrophana*), a single species that is not placed in any of the suborders, is sister to the remaining Boletineae members (Jarosch and Besl 2001). Current research is focused on the ecology of Sclerodermatineae (Wilson et al. 2007, 2012) and the taxonomic structure of Boletineae, especially the Boletaceae. Monographic work has led to a better definition of *Boletus* and its being restricted to the *B. edulis* group (Dentinger et al. 2010; Nuhn et al. 2013), the revision of gilled boletes in *Phylloporus* (Neves et al. 2012), and *Xerocomus*, which has been split into several new genera (Šutara 2008). Since 2007, 14 new genera have been described in Boletaceae (Desjardin et al. 2008, 2009; Halling et al. 2007, 2012a, b; Hosen et al. 2013; Lebel et al. 2012; Li et al. 2011; Orihara et al. 2010; Šutara 2008; Trappe et al. 2013; Zeng et al. 2012).

4. Agaricales

Overview: Agaricales (Underwood 1899) includes over 13,000 described species (Kirk et al. 2008), making it the largest order of Agaricomycetes. Despite being one of the most conspicuous and comparatively better studied groups of fungi, an immense amount of the actual diversity remains undescribed, hiding under commonly used names that molecular data have revealed to be clusters of morphologically cryptic species (e.g., *Amanita muscaria*) (Geml et al. 2006) and part of hyperdiverse lineages with over 2,000 estimated species such as *Cortinarius* (e.g., Harrower et al. 2011).

Agaricales is dominated by pileate-stipitate forms with lamellate hymenophores (e.g., *Amanita*, *Agaricus*, *Coprinus* s.l., *Entoloma*, *Lepiota*, *Tricholoma*) (Fig. 14.9a), but there is wide variation on this basic fruit-body morphology regarding characters such as the size of the basidiocarp, the presence of veils (universal and partial), gill attachment, and spore-print color (white, brown, purple-brown, black, pink). Microscopically, there is also a great diversity of characters, including spore size, shape, ornamentation, and chemical reactions; the arrangement of the covering layers of the fruit body (pileipellis, stipitipellis) and the hymenophoral trama; and the presence of specialized structures (cystidia, setae) (Clémençon 2004; Reijnders and Stalpers 1992; Singer 1986). All these characters have played a central role in defining the approximately 400 genera and 30 families in the order, but much of the taxonomy of the order is currently in flux as data from molecular phylogenies become incorporated. A promising pool of micromorphological features that may be useful as future systematic markers in Agaricales is the complex of characters related to conidiogenesis in anamorphic stages (Walther et al. 2005). Since monosporic cultures obtained from basidiospores are usually needed to study these characters, this complex of features has clearly been understudied.

The second major morphological component of Agaricales are the secotioid and gasteroid forms, including false truffles, puffballs, and bird-nest fungi, that have evolved repeatedly in different lineages within the order (e.g., *Lycoperdon*) (Fig. 14.9c). Additional morphologies that can be found in the order include resupinate (e.g., *Cylindrobasidium*), coralloid (e.g., *Clavaria*), cyphelloid (e.g., *Henningsomyces*), pileate with poroid (e.g., *Favolaschia*), or tubular hymenophores (e.g., *Fistulina*). There is no morphological synapomorphy that unites the Agaricales, and the typical pileate-stipitate gilled mushroom morphology that dominates the order also occurs in other orders of Agaricomycetes.

Ecological diversity: two ecological roles characterize the majority of Agaricales species: saprotrophy and ECM symbiosis. Saprotrophs can be broadly subdivided into soil/litter/dung fungi (e.g., *Agaricus*, *Coprinopsis*, *Gymnopus*) and wood decayers (e.g., *Pholiota*, *Pleurotus*), but the exact roles and capabilities of members of both ecological guilds remain largely unknown, although the emerging field of fungal genomics is bringing new insights into these aspects (e.g., Morin et al. 2012). There have been at least ten independent and asynchronous origins of the ECM symbiosis in Agaricales, involving associations with a great variety of vascular plants (Matheny et al. 2006; Ryberg and Matheny 2012; Tedersoo et al. 2012). There are many additional ecological roles in Agaricales, including plant pathogens (e.g., *Moniliophthora/Crinipellis*) (Meinhardt et al. 2008), mycoparasites (e.g., *Squamanita*) (Matheny and Griffith 2010), basidiolichens (e.g., *Lichenomphalia* and *Dictyonema*) (Dal-Forno et al. 2013; Lawrey et al. 2009), attine ant cultivars (*Leucoagaricus/Leucocoprinus* clade) (Mueller et al. 2005), or termite cultivars (*Termitomyces*) (Aanen et al. 2002). The ecology of many Agaricales species remains poorly understood, and new insights from molecular and isotopic data are challenging long-standing views even in relatively well-studied groups. For example, the so-called waxcaps (*Hygrocybe* sensu lato) have historically been considered saprotrophs but now are thought to be involved in some kind of biotrophic association (Lodge et al. 2013; Seitzman et al. 2011).

Systematics: three landmark papers published in recent years have redefined the taxonomic organization of Agaricales species and their closest relatives: (1) Moncalvo et al. (2002) presented the first broad phylogeny of the order, based on nuc-lsu rRNA, which resolved 117 clades and outlined conflicts with traditional morphologically defined groups; (2) Matheny et al. (2006) presented the first major multilocus overview of the order, including the protein-coding genes *rpb1*, *rpb2*, and *tef1-alpha*, and defined six major infraordinal clades: agaricoid, tricholomatoid, marasmioid, hygrophoroid, pluteoid, and plicaturopsidoid clades, although support for some of these groupings was weak; (3) Binder et al. (2010) built on the data set of Matheny et al. (2006) and formally recognized the plicaturopsidoid clade as the order Amylocorticiales.

Recent monographs on Agaricales include the titles of the *Fungi Europaei* series on *Agaricus* (Parra-Sánchez 2008), *Amanita* (Neville and Poumarat 2004), *Conocybe-Pholiotina* (Hausknecht 2009), and the family Strophariaceae (Noordeloos 2011). Outside Europe monographic work at the continental scale is rare, a situation that may change with large-scale cataloging efforts that include the use of molecular data, such as the North American Mycoflora project (Bruns 2012). Modern global monographs are lacking for all major genera of Agaricales.

Agaricoid clade: this clade is well supported in the study of Matheny et al. (2006) and is dominated by gilled pileate-stipitate forms with pigmented spores (brown, purple-brown, black). Exceptions to this general pattern include (1) white-spored taxa in Agaricaceae (*Lepiota* and allied genera), Cystodermateae, and Hydnangiaceae (e.g., *Laccaria*); (2) secotioid and gasteroid taxa that have repeatedly evolved in different lineages (e.g., Agaricaceae, *Cortinarius*); and (3) cyphelloid forms in the genera *Pellidiscus* and *Phaeosolenia* (Bodensteiner et al. 2004). At least six ECM lineages are included in this clade: *Cortinarius*, *Descolea* (and allied sequestrate taxa), Inocybaceae, *Laccaria* (and allied sequestrate taxa), *Phaeocollybia*, and some groups of Hymenogastraceae (*Alnicola*, *Hebeloma*). Most other members of the agaricoid clade are saprotrophs associated with the litter layer and similar substrates (e.g., *Agaricus*, *Coprinopsis*) or wood decayers (e.g., *Hypholoma*). Taxa with recent phylogenetic studies include Agaricaceae (Vellinga 2004; Vellinga et al. 2011), Bolbitiaceae (Tóth et al. 2013), Cortinariaceae (Frøslev et al. 2005; Garnica et al. 2005; Peintner et al. 2004), Crepidotaceae (Aime et al. 2005), Cystodermateae (Saar et al. 2009), Gymnopileae (Guzmán-Dávalos et al. 2003), Hebelomateae (Boyle et al. 2006; Moreau et al. 2006), Inocybaceae (Matheny 2005; Ryberg et al. 2010), Nidulariaceae (Zhao et al. 2007), Psathyrellaceae (Nagy et al. 2012), Strophariaceae (Ramírez-Cruz et al. 2013), Tubarieae (Gulden et al. 2005), and several secotioid/gasteroid taxa usually nested within agaricoid relatives (Larsson and Jeppson 2008; Lebel and Syme 2012).

Tricholomatoid clade: pileate-stipitate gilled mushrooms with white or pink spores dominate this lineage, which includes four traditionally recognized families: Entolomataceae, Lyophyllaceae, Mycenaceae, and a restricted version of Tricholomataceae. A fifth lineage includes the ECM *Catathelasma* and the saprotrophic *Callistosporium*. Additional ECM origins have also occurred in *Entoloma*, *Lyophyllum*, and Tricholomataceae. The clade also includes soil/litter saprotrophs (e.g., *Clitocybe*), mycoparasites (e.g., *Asterophora*), wood decayers associated with white rot (e.g., *Mycena*) or brown rot (e.g., *Ossicaulis*), and termite cultivars (*Termitomyces*). Entolomataceae remains one of the most distinct groups in Agaricales because of the pinkish spores that are warted, ridged, or angular. Co-David et al. (2009) recognize two broadly defined genera in the family (*Entoloma* and *Clitopilus*), while other authors prefer to recognize several narrowly defined genera (Baroni and Matheny 2011; Largent et al. 2011). The Lyophyllaceae has also been reviewed using molecular phylogenies (Hofstetter et al. 2002). The taxonomy of many traditionally recognized genera (e.g., *Tricholoma*, *Clitocybe*, *Lepista*, *Mycena*) is still in flux.

Marasmioid clade: this lineage is dominated by white-spored saprotrophic species associated with wood or leaf-litter substrates (e.g., *Lentinula*, *Marasmius*, *Xerula*), but important plant pathogens also belong here (e.g., *Armillaria*, *Moniliophthora*). Schizophyllaceae (*Fistulina*, *Schizophyllum*) and Lachnellaceae, dominated by cyphelloid forms, were recovered as part of the marasmioid clade by Matheny et al. (2006) but as independent lineages in Binder et al. (2010). Resupinate genera (e.g., *Chondrostereum*) also occur in the marasmioid clade. Representative taxa with recent phylogenetic studies include a general overview of marasmioid/gymnopoid fungi (Wilson and Desjardin 2005), *Omphalotus* (Kirchmair et al. 2004), *Rhodocollybia* (Mata et al. 2004), *Marasmius*, and *Crinipellis* (Kerekes and Desjardin 2009; Wannathes et al. 2009), and genera in the *Xerula/Oudemansiella* complex (Petersen and Hughes 2010).

Hygrophoroid clade: in the analyses of Matheny et al. (2006) this clade includes an expanded version of Hygrophoraceae as traditionally defined plus the club and coralloid fungi in the families Pterulaceae and Typhulaceae. However, these two families were later resolved as a separate lineage in Agaricales (Binder et al. 2010). The family Hygrophoraceae was recently extensively studied and redefined by Lodge et al. (2013) and now includes 18 genera, including the ECM *Hygrophorus*, several segregates from *Hygrocybe* sensu lato, and a diverse clade of basidiolichens such as *Dictyonema* s. l. (Dal-Forno et al. 2013). All members of the family are now assumed to be involved in some kind of biotrophic relation, but its exact nature remains obscure in most cases (Seitzman et al. 2011).

Pluteoid clade: this clade received weak support in Matheny et al. (2006), and its limits and composition require further study. The grouping of the Pluteaceae and *Melanoleuca* is well supported in most phylogenies (Justo et al. 2011; Matheny et al. 2006). The aquatic gasteromycete *Limnoperdon* is also part of this core pluteoid group in some analyses (Matheny et al. 2006). Amanitaceae, Pleurotaceae, and the genus *Macrocystidia* are, in some topologies, recovered as closely related to the core pluteoid genera but not always with statistical support or in a consistent position. Important taxonomic revisions include articles on the Pluteaceae (Justo et al. 2011), *Melanoleuca* (Sánchez-García et al. 2013; Vizzini et al. 2012), Pleurotaceae (Thorn et al. 2000), and sequestrate forms in *Amanita* (Justo et al. 2009). The iconic genus *Amanita* has received considerable attention in relation to biogeography (Geml et al. 2006), invasive species (Pringle et al. 2009), and transitions from saprotrophic to ECM nutrition (Wolfe et al. 2012).

III. Conclusions

The previous edition of *The Mycota* included a preliminary phylogenetic outline of Homobasidiomycetes, with eight informally named clades, that was based on 25 published and unpublished analyses (Hibbett and Thorn 2001). The present chapter cites nearly 300 phylogenetic studies, many combining rRNA and protein-coding genes, and a handful of phylogenomic analyses. Twenty strongly supported, mutually exclusive clades of Agaricomycetes are recognized as orders. Numerous studies, most of which are not cited here, have addressed species- and genus-level relationships within these groups. Nevertheless, the classification of Agaricomycetes is far from complete. There are weakly supported nodes throughout the phylogeny, and the catalog of described species is thought to be a tiny fraction of the actual diversity in the group (Blackwell 2011).

An overarching challenge of fungal systematics is to capture and integrate the massive volumes of data flowing from taxonomy, phylogenetics, genomics, and molecular ecology. Unfortunately, some common practices make it difficult to combine the products of different areas of research. For example, curated sequence databases are important for the identification of environmental sequences (Kõljalg et al. 2013), so it is unfortunate that many recent species descriptions have been published without sequences (Hibbett et al. 2011). It is also unfortunate that only about 17 % of published phylogenies, including those from fungal studies, are available in electronic form (not graphics files, but treefiles, in Newick or other formats), which limits efforts to assemble maximally inclusive phylogenies and combine them with taxonomic hierarchies (Collins et al. 2013; Drew et al. 2013). To approach a comprehensive phylogenetic classification of Agaricomycetes and other Fungi, it will be necessary to increase the pace of taxon discovery, encourage researchers to generate and deposit sequences, alignments, trees, and associated metadata (Hyde et al. 2013), and create new bioinformatics tools to synthesize the prodigious output of the fungal systematics community.

Acknowledgements The authors are grateful to Sigisfredo Garnica, Romina Gazis, Igor Grigoriev, Christiane Karasch-Wittmann, Urmas Kõljalg, Francis

Martin, Rachael Martin, Franz Oberwinkler, and Robert Riley, who provided access to specimens and unpublished results, and Yu-Cheng Dai, Paul Diederich, Doris Haas, M. Jeppson, Lasse Kosonen, Jaroslav Maly, O. Manninen, Guillermo Muñoz, Michael Wood (http://www.mykoweb.com), and Nourou Yourou, who provided images.

References

Aanen DK, Eggleton P, Rouland-Lefevre C, Guldberg-Froslev T, Rosendahl S, Boomsma JJ (2002) The evolution of fungus-growing termites and their mutualistic fungal symbionts. Proc Natl Acad Sci USA 99:14887–14892

Agerer R (1990) Studies on ectomycorrhizae XXIV. Ectomycorrhizae of *Chroogomphus helveticus* and *C. rutilus* (Gomphidiaceae, Basidiomycetes) and their relationship to those of *Suillus* and *Rhizopogon*. Nova Hedwigia 50:1–63

Agerer R (1996a) *Ramaria aurea* (Schaeff.:Fr.) Quel. + *Fagus sylvatica* L. Descr Ectomycorrhizae 1: 107–112

Agerer R (1996b) *Ramaria largentii* Man & D.E. Stuntz +*Picea abies* (L.) Karst. Descr Ectomycorrhizae 1:113–118

Agerer R (1996c) *Ramaria spinulosa* (Fr.) Quel. + *Fagus sylvatica* L. Descr Ectomycorrhizae 1:119–124

Agerer R (1996d) *Ramaria subbotrytis* (Coker) Corner +*Quercus robur* L. Descr Ectomycorrhizae 1:125–130

Agerer R (1999) Never change a functionally successful principle: the evolution of Boletales s. l. (Hymenomycetes, Basidiomycota) as seen from below-ground features. Sendtnera 6:5–91

Agerer R, Beenken L (1998) *Geastrum fimbriatum* Fr. + *Fagus sylvatica* L. Descr Ectomycorrhizae 3:13–18

Agerer R, Iosifidou P (2004) Rhizomorph structures of Hymenomycetes: a possibility to test DNA-based phylogenetic hypotheses? In: Agerer R, Piepenbring M, Blanz P (eds) Frontiers in basidiomycote mycology. IHW, Eching, Germany, pp 249–302

Agerer R, Beenken L, Christian J (1998) *Gomphus clavatus* (Pers.: Fr.) S. F. Gray + *Picea abies* (L.) Karst. Descr Ectomycorrhizae 3:25–29

Aime MC, Vilgalys R, Miller OK Jr (2005) The Crepidotaceae (Basidiomycota, Agaricales): phylogeny and taxonomy of the genera and revision of the family based on molecular evidence. Am J Bot 92:74–82

Albee-Scott SR (2007) Does secotoid inertia drive the evolution of false-truffles? Mycol Res 111:1030–1039

Albee-Scott SR, Kropp BR (2010) A phylogenetic study of *Trechispora thelephora*. Mycotaxon 114:395–399

Allen TR, Millar T, Berch SM, Berbee ML (2003) Culturing and direct DNA extraction find different fungi from the same ericoid mycorrhizal roots. New Phytol 160:255–272

Amalfi M, Decock C (2013) *Fomitiporia castilloi* sp. nov. and multiple clades around *F. apiahyna* and *F. texana* in Meso- and South America evidenced by multiloci phylogenetic inferences. Mycologia 105:873–887

Amalfi M, Yombiyeni P, Decock C (2010) *Fomitiporia* in sub-Saharan Africa: morphology and multigene phylogenetic analysis support three new species from the Guineo-Congolian rainforest. Mycologia 102:1303–1317

Amalfi M, Raymundo T, Valenzuela R, Decock C (2012) *Fomitiporia cupressicola* sp. nov., a parasite on *Cupressus arizonica*, and additional unnamed clades in the southern USA and northern Mexico, determined by multilocus phylogenetic analyses. Mycologia 104:880–893

Amend AS, Barshis DJ, Oliver TA (2012) Coral-associated marine fungi form novel lineages and heterogeneous assemblages. ISME J 6:1291–1301

Andjic V, Cole ALC, Klena JD (2005) Taxonomic identity of the Sterile Red Fungus inferred using nuclear rDNA ITS 1 sequences. Mycol Res 109:200–204

Arpin N, Kühner R (1977) Les grandes lignes de la classification des Boletales. Bulletin de la Société linnéenne de Lyon 46:83–108, 181–208

Bandoni RJ (1984) The Tremellales and Auriculariales: an alternative classification. Trans Mycol Soc Jpn 25:489–530

Bandoni RJ, Wells K (1992) Clamp connections and classification of the Auriculariales and Tremellales. Trans Mycol Soc Jpn 33:13–19

Baroni TJ, Matheny PB (2011) A re-evaluation of gasteroid and cyphelloid species of Entolomataceae from eastern North America. Harv Pap Bot 16:293–310

Benjamin DR (1995) Mushrooms: poisons and panaceas. W. H. Freeman, New York, NY

Berbee ML, Taylor JW (2010) Dating the molecular clock in fungi – how close are we? Fung Biol Rev 24:1–16

Bernicchia A (2005) Polyporaceae s.l. Fungi Europaei, vol 10. Candusso, Alassio

Bernicchia A, Gorjón SP, Vampola P, Ryvarden L, Prodi A (2010) A phylogenetic analysis of *Antrodia* s.l. based on nrDNA ITS sequences, with emphasis on rhizomorphic European species. Mycol Prog 11:93–100

Besl H, Bresinsky A (1977) Notizen über Vorkommen und systematische Bewertung von Pigmenten in Höheren Pilzen (2). Zeitschrift für Pilzkunde 43:311–322

Besl H, Bresinsky A (1997) Chemosystematics of Suillaceae and Gomphidiaceae (suborder Suillineae). Plant Syst Evol 206:223–242

Besl H, Bresinsky A, Kämmerer A (1986) Chemosystematik der Coniophoraceae. Zeitschrift für Mykologie 52:277–286

Bidartondo MI, Bruns TD, Weiß M, Sérgio C, Read DJ (2003) Specialized cheating of the ectomycorrhizal symbiosis by an epiparasitic liverwort. Proc R Soc B Biol Sci 270:835–842

Binder M, Bresinsky A (2002) Derivation of a polymorphic lineage of Gasteromycetes from boletoid ancestors. Mycologia 94:85–98

Binder M, Hibbett DS (2002) Higher-level phylogenetic relationships of Homobasidiomycetes (mushroom-forming fungi) inferred from four rDNA regions. Mol Phylogenet Evol 22:76–90

Binder M, Hibbett DS (2006) Molecular systematics and biological diversification of Boletales. Mycologia 98:971–981

Binder M, Hibbett DS, Larsson K-H, Larsson E, Langer E, Langer G (2005) The phylogenetic distribution of resupinate forms across the major clades of mushroom-forming fungi (Homobasidiomycetes). Syst Biodiv 3:113–157

Binder M, Hibbett DS, Wang Z, Farnham WF (2006) Evolutionary relationships of *Mycaureola dilseae* (Agaricales), a basidiomycete pathogen of a subtidal rhodophyte. Am J Bot 93:547–556

Binder M, Larsson K-H, Matheny PB, Hibbett DS (2010) Amylocorticiales ord. nov. and Jaapiales ord. nov.: early diverging clades of Agaricomycetidae dominated by corticioid forms. Mycologia 102:865–880

Binder M, Justo A, Riley R, Salamov A, Lopez-Giraldez F, Sjokvist E, Copeland A, Foster B, Sun H, Larsson E, Larsson K-H, Townsend JP, Grigoriev I, Hibbett D (2013) Phylogenetic and phylogenomic overview of the Polyporales. Mycologia 105:1350–1373

Birkebak JM, Mayor JR, Ryberg KM, Matheny PB (2013) A systematic, morphological and ecological overview of the Clavariaceae (Agaricales). Mycologia 105:896–911

Blackwell M (2011) The fungi: 1, 2, 3 ... 5.1 million species? Am J Bot 98:426–438

Blair JE (2009) Fungi. In: Hedges SB, Kumar S (eds) The timetree of life. Oxford University Press, Oxford, pp 215–219

Bodensteiner P, Binder M, Moncalvo J-M, Agerer R, Hibbett D S (2004) Phylogenetic relationships of cyphelloid homobasidiomycetes. Mol Phylogenet Evol 33:501–515

Boidin J, Mugnier J, Canales R (1998) Taxonomie moleculaire des Aphyllophorales. Mycotaxon 66:445–491

Both EE (1993) The Boletes of North America. A compendium. Buffalo Museum of Science, Buffalo, NY

Bougher NL, Syme K (1998) Fungi of southern Australia. University of Western Australia Press, Perth

Boyle H, Zimdars B, Renker C, Buscot F (2006) A molecular phylogeny of *Hebeloma* species from Europe. Mycol Res 110:369–380

Brazee NJ, Lindner DL (2013) Unravelling the *Phellinus pini* s.l. complex in North America: a multilocus phylogeny and differentiation analysis of *Porodaedalea*. Forest Pathol 43:132–143

Bresinsky A (1974) Zur Frage der taxonomischen Relevanz chemischer Merkmale bei Höheren Pilzen. Bulletin de la Société linnéenne de Lyon, numéro spécial, pp 61–84

Bresinsky A, Orendi P (1970) Chromatographische Analyse von Farbmerkmalen der Boletales und anderer Makromyzeten auf Dünnschichten. Zeitschrift für Pilzkunde 36:135–169

Bresinsky A, Rennschmid A (1971) Pigmentmerkmale, Organisationsstufen und systematische Gruppen bei höheren Pilzen. Berichte der Deutschen Botanischen Gesellschaft 84:313–329

Bresinsky A, Jarosch M, Fischer M, Schönberger I, Wittmann-Bresinsky B (1999) Phylogenetic relationships within *Paxillus* s. l. (Basidiomycetes, Boletales): separation of a southern hemisphere genus. Plant Biol 1:327–333

Bruns TD (2012) The North American Mycoflora project – the first steps on a long journey. New Phytol 196:972–974

Bruns TD, Fogel R, White TJ, Palmer JD (1989) Accelerated evolution of a false-truffle from a mushroom anchestor. Nature 339:140–142

Bruns TD, Szaro TM, Gardes M, Cullings KW, Pan JJ, Horton TR, Kretzer A, Garbelotto M, Li Y (1998) A sequence database for identification of ectomycorrhizal basidiomycetes by phylogenetic analysis. Mol Ecol 7:257–272

Burdsall HH Jr (1979) *Laetisaria* (Aphyllophorales, Corticiaceae), a new genus for the teleomorph of *Isaria fuciformis*. Trans Br Mycol Soc 72:419–422

Burdsall HH Jr, Hoch HC, Boosalis MG, Setliff EC (1980) *Laetisaria arvalis* (Aphyllophorales, Corticiaceae): a possible biological control agent for *Rhizoctonia solani* and *Pythium* species. Mycologia 72:728–736

Burdsall HH, Volk TJ, Ammirati JF (1996) *Bridgeoporus*, a new genus to accommodate *Oxyporus nobilissimus* (Basidiomycotina, Polyporaceae). Mycotaxon 60:387–395

Buyck B, Hofstetter V, Eberhardt U, Verbeken A, Kauff F (2008) Walking the thin line between *Russula* and *Lactarius*: the dilemma of *Russula* sect. *Ochricompactae*. Fungal Divers 28:15–40

Buyck B, Duhem B, Eyssartier G, Ducousso M (2012) *Podoserpula miranda* sp. nov. (Amylocorticiales, Basidiomycota) from New Caledonia. Cryptogamie Mycologie 33:453–461

Buyck B, Kauff F, Cruaud C, Hofstetter V (2013a) Molecular evidence for novel *Cantharellus* (Cantharellales, Basidiomycota) from tropical African miombo woodland and a key to all tropical African chanterelles. Fungal Divers 57:281–298

Buyck B, Kauff F, Eyssartier G, Couloux A, Hofstetter V (2013b) A multilocus phylogeny for worldwide *Cantharellus* (Cantharellales, Agaricomycetidae). Fungal Divers 64:101–121. doi:10.1007/s13225-013-0272-3

Cabral TS, Marinho P, Goto BT, Baseia IG (2012) *Abrachium*, a new genus in the Clathraceae, and *Itajahya* reassessed. Mycotaxon 119:419–429

Castellano MA (1988) The taxonomy of the genus *Hysterangium* (Basidiomycotina, Hysterangiaceae) with notes on its ecology. PhD Dissertation, Oregon State University, Corvallis, OR. 227 p

Castellano MA, Beever RE (1994) Truffle-like Basidiomycotina of New Zealand: *Gallacea, Hysterangium, Phallobata*, and *Protubera*. NZ J Bot 32:305–328

Castellano MA, Verbeken A, Walleyn R, Thoen D (2000) Some new or interesting sequestrate Basidiomycota from African woodlands. Karstenia 40:11–21

Chamuris GP (1988) The non-stipitate stereoid fungi in the northeastern United States and adjacent Canada. Mycologia Memoir No. 14. J. Cramer, Berlin

Chang ST, Wasser SP (2012) The role of culinary-medicinal mushrooms on human welfare with a pyramid model for human health. Int J Med Mushrooms 14:95–134

Claridge AW (2002) Ecological role of hypogeous ectomycorrhizal fungi in Australian forests and woodlands. Plant Soil 244:291–305

Clémençon H (2004) Cytology and plectology of the Hymenomycetes. Bibl Mycol 199:1–488

Co-David D, Langeveld D, Noordeloos ME (2009) Molecular phylogeny and spore evolution of Entolomataceae. Persoonia 23:147–176

Colgan W, Castellano MA, Bougher NL (1995) NATS truffle and truffle-like fungi 2: *Kjeldsenia aureispora* gen. et sp. nov., a truffle-like fungus in the Cortinariaceae. Mycotaxon 55:175–178

Colgan W, Castellano MA, Spatafora JW (1997) Systematics of the Hysterangiaceae (abstract). Inoculum 48:7

Collins T, Kearney M, Maddison D (2013) The ideas lab concept, assembling the tree of life, and AVAToL. PLoS Curr Tree Life 1. doi:10.1371/currents.tol.0fdb85e1619f313a2a5a2ec3d7a8df9e

Corner EJH (1972) *Boletus* in Malaysia. Government Printing Office, Singapore

Cruz D, Suárez JP, Kottke I, Piepenbring M, Oberwinkler F (2011) Defining species in *Tulasnella* by correlating morphology and nrDNA ITS-5.8S sequence data of basidiomata from a tropical Andean forest. Mycol Prog 10:229–238

Cui B-K, Decock C (2012) *Phellinus castanopsidis* sp. nov. (Hymenochaetaceae) from southern China, with preliminary phylogeny based on rDNA sequences. Mycol Prog 12:341–351

Cui B-K, Zhao C-L, Dai Y-C (2011) *Melanoderma microcarpum* gen. et sp. nov. (Basidiomycota) from China. Mycotaxon 116:295–302

Cunningham GH (1931) The Gasteromycetes of Australasia. XI. Linn Soc New South Wales 56:182–200

Dai Y-C (2010) Hymenochaetaceae (Basidiomycota) in China. Fungal Divers 45:131–343

Dai Y-C, Cui B-K (2011) *Fomitiporia ellipsoidea* has the largest fruiting body among the fungi. Fungal Biol 115:813–814

Dai YC, Wang Z, Binder M, Hibbett DS (2006) Phylogeny and a new species of *Sparassis* (Polyporales, Basidiomycota): evidence from mitochondrial *atp6*, nuclear rDNA and *rpb2* genes. Mycologia 98:584–592

Dai Y-C, Zhou L-W, Cui B-K, Chen Y-Q, Decock C (2010) Current advances in *Phellinus* sensu lato: medicinal species, functions, metabolites and mechanisms. Appl Microbiol Biotechnol 87:1587–1593

Dal-Forno M, Lawrey JD, Sikaroodi M, Bhattarai S, Gillevet PM, Sulzbacher M, Lücking R (2013) Starting from scratch: evolution of the lichen thallus in the basidiolichen *Dictyonema* (Agaricales: Hygrophoraceae). Fungal Biol 117:584–598

De Koker TH, Nakasone KK, Haarhof J, Burdsall HH, Janse BJH (2003) Phylogenetic relationships of the genus *Phanerochaete* inferred from the internal transcribed spacer region. Mycol Res 107:1032–1040

Decock C, Ryvarden L (2003) *Perenniporiella* gen. nov. segregated from *Perenniporia*, including a key to neotropical *Perenniporia* species with pileate basidiomes. Mycol Res 107:93–103

Decock C, Bitew A, Castillo G (2005) *Fomitiporia tenuis* and *Fomitiporia aethiopica* (Basidiomycetes, Hymenochaetales), two undescribed species from the Ethiopian highlands: taxonomy and phylogeny. Mycologia 97:121–129

Decock C, Figueroa SH, Robledo G, Castillo G (2006) *Phellinus caribaeo-quercicolus* sp. nov., parasitic on *Quercus cubana*: taxonomy and preliminary phylogenetic relationships. Mycologia 98:265–274

Decock C, Figueroa SH, Robledo G, Castillo G (2007) *Fomitiporia punctata* (Basidiomycota, Hymenochaetales) and its presumed taxonomic synonyms in America: taxonomy and phylogeny of some species from tropical/subtropical areas. Mycologia 99:733–752

Dennis RL (1970) A middle-Pennsylvanian basidiomycete mycelium with clamp connections. Mycologia 62:578–584

Dennis RL (1976) *Palaeosclerotium*, a Pennsylvanian age fungus combining features of modern ascomycetes and basidiomycetes. Science 192:66–68

Dentinger BTM, McLaughlin DJ (2006) Reconstructing the Clavariaceae using nuclear large subunit rDNA sequences and a new genus segregated from *Clavaria*. Mycologia 98:746–762

Dentinger BT, Ammirati JF, Both EE, Desjardin DE, Halling RE, Henkel TW, Moreau PA, Nagasawa E, Soytong K, Taylor AF, Watling R, Moncalvo JM, McLaughlin DJ (2010) Molecular phylogenetics of porcini mushrooms (*Boletus* section *Boletus*). Mol Phylogenet Evol 57:1276–1292

DePriest PT, Sikaroodi M, Lawrey JD, Diederich P (2005) *Marchandiomyces lignicola* sp. nov. shows recent and repeated transition between a lignicolous and a lichenicolous habit. Mycol Res 109:57–70

Desjardin DE, Perry BA (2009) A new species of *Phallus* from São Tomé, Africa. Mycologia 101:545–547

Desjardin DE, Wilson AW, Binder M (2008) *Durianella*, a new gasteroid genus of boletes from Malaysia. Mycologia 100:956–961

Desjardin DE, Binder M, Roekring S, Flegel T (2009) *Spongiforma*, a new genus of gastroid boletes from Thailand. Fungal Divers 37:1–8

Diederich P (1990) New or interesting lichenicolous fungi. 1. Species from Luxembourg. Mycotaxon 37:297–330

Diederich P, Lawrey JD (2007) New lichenicolous, muscicolous, corticolous and lignicolous taxa of *Burgoa* s. l. and *Marchandiomyces* s. l. (anamorphic Basidiomycota), a new genus for *Omphalina foliacea*, and a catalogue and a key to the non-lichenized, bulbilliferous basidiomycetes. Mycol Prog 6:61–80

Diederich P, Schultheis B, Blackwell M (2003) *Marchandiobasidium aurantiacum* gen. sp. nov., the teleomorph of *Marchandiomyces aurantiacus* (Basidiomycota, Ceratobasidiales). Mycol Res 107:523–527

Diederich P, Lawrey JD, Sikaroodi M, Gillevet PM (2011) A new lichenicolous teleomorph is related to plant pathogens in *Laetisaria* and *Limonomyces* (Basidiomycota, Corticiales). Mycologia 103:525–533

Domínguez de Toledo LS, Castellano MA (1997) First report of *Gastrosporium simplex* (Gasteromycetes) from South America. Mycotaxon 64:443–448

Donk MA (1961) Four new families of Hymenomycetes. Persoonia 1:405–407

Donk MA (1964) A conspectus of the families of Aphyllophorales. Persoonia 3:199–324

Donk MA (1971) Progress in the study of the classification of the higher Basidiomycetes. In: Petersen RH (ed) Evolution in the higher Basidiomycetes. The University of Tennessee Press, Knoxville, TN, pp 3–25

Douhan GW, Rizzo DM (2003) Host-parasite relationships among bolete infecting *Hypomyces* species. Mycol Res 107:1342–1349

Douzery EJ, Snell EA, Bapteste E, Delsuc F, Philippe H (2004) The timing of eukaryotic evolution: does a relaxed molecular clock reconcile proteins and fossils? Proc Natl Acad Sci USA 101:15386–15391

Drew BT, Gazis R, Cabezas P, Swithers KS, Deng J, Rodriguez R, Katz LA, Crandall KA, Hibbett DS, Soltis DE (2013) Lost branches on the tree of life. PLoS Biol 11:e1001636

Dunham SM, Larsson KH, Spatafora JW (2007) Species richness and community composition of mat-forming ectomycorrhizal fungi in old- and second-growth Douglas-fir forests of the HJ Andrews Experimental Forest, Oregon, USA. Mycorrhiza 17:633–645. doi:10.1007/s00572-007-0141-6

Eastwood DC, Floudas D, Binder M, Majcherczyk A, Schneider P, Aerts A, Asiegbu FO, Baker SE, Barry K, Bendiksby M, Blumentritt M, Coutinho PM, Cullen D, de Vries RP, Gathman A, Goodell B, Henrissat B, Ihrmark K, Kauserud H, Kohler A, LaButti K, Lapidus A, Lavin JL, Lee YH, Lindquist E, Lilly W, Lucas S, Morin E, Murat C, Oguiza JA, Park J, Pisabarro AG, Riley R, Rosling A, Salamov A, Schmidt O, Schmutz J, Skrede I, Stenlid J, Wiebenga A, Xie X, Kües U, Hibbett DS, Hoffmeister D, Högberg N, Martin F, Grigoriev IV, Watkinson SC (2011) The plant cell wall-decomposing machinery underlies the functional diversity of forest fungi. Science 333:762–765

Eriksson J (1954) *Ramaricium* n. gen., a corticoid member of the *Ramaria* group. Svensk Bot Tidskr 48:188–198

Eriksson J, Ryvarden L (1973) The Corticiaceae of North Europe: Aleurodiscus to Confertobasidium, vol 2. Fungiflora, Oslo

Eriksson J, Ryvarden L (1976) The Corticiaceae of North Europe: Hyphodermella to Mycoacia, vol 4. Fungiflora, Oslo

Erland S, Taylor AFS (1999) Resupinate ectomycorrhizal fungal genera. In: Cairney JM (ed) Ectomycorrhizal fungi: key genera in profile. Springer, Berlin, pp 347–363

Ertz D, Lawrey JD, Sikaroodi M, Gillevet PM, Fischer E, Killmann D, Serusiaux E (2008) A new lineage of lichenized basidiomycetes inferred from a two-gene phylogeny: the Lepidostromataceae with three species from the tropics. Am J Bot 95:1548–1556

Etayo J, Diederich P (1996) Lichenicolous fungi from the western Pyrenees, France and Spain. II. More deuteromycetes. Mycotaxon 60:415–428

Farid AM, Lee SS, Maziah Z, Patahayah M (2009) Pathogenicity of *Rigidoporus microporus* and *Phellinus noxius* against four major plantation tree species in Peninsular Malaysia. J Trop Forest Sci 21:289–298

Fischer E (1900) Phallineae, Hymenogastrineae, Lycoperdineae, Nidulariineae, Plectobasidiineae. In: Engler A, Prantl K (eds) Die Natürlichen Pflanzenfamilien, Teil I, Abt. 1, W Engelmann, Leipzig, pp 276–346

Fischer M (2002) A new wood-decaying basidiomycete species associated with esca of grapevine: *Fomitiporia mediterranea* (Hymenochaetales). Mycol Prog 1:315–324

Fischer M, Binder M (2004) Species recognition, geographic distribution and host-pathogen relationships: a case study in a group of lignicolous basidiomycetes, *Phellinus* s.l. Mycologia 96:799–811

Fischer M, Edwards J, Cunnington JH, Pascoe IG (2005) Basidiomycetous pathogens on grapevine: a new species from Australia – *Fomitiporia australiensis*. Mycotaxon 92:85–96

Fischer E, Ertz D, Killmann D, Sérusiaux E (2007) Two new species of *Multiclavula* (lichenized basidio-

mycetes) from savanna soils in Rwanda (East Africa). Bot J Linn Soc 155:457–465
Floudas D, Binder M, Riley R, Barry K, Blanchette RA, Henrissat B, Martinez AT, Otillar R, Spatafora JW, Yadav JS, Aerts A, Benoit I, Boyd A, Carlson A, Copeland A, Coutinho PM, de Vries RP, Ferreira P, Findley K, Foster B, Gaskell J, Glotzer D, Gorecki P, Heitman J, Hesse C, Hori C, Igarashi K, Jurgens JA, Kallen N, Kersten P, Kohler A, Kues U, Kumar TKA, Kuo A, LaButti K, Larrondo LF, Lindquist E, Ling A, Lombard V, Lucas S, Lundell T, Martin R, McLaughlin DJ, Morgenstern I, Morin E, Murat C, Nagy LG, Nolan M, Ohm RA, Patyshakuliyeva A, Rokas A, Ruiz-Duenas FJ, Sabat G, Salamov A, Samejima M, Schmutz J, Slot JC, St John F, Stenlid J, Sun H, Sun S, Syed K, Tsang A, Wiebenga A, Young D, Pisabarro A, Eastwood DC, Martin F, Cullen D, Grigoriev IV, Hibbett DS (2012) The Paleozoic origin of enzymatic lignin decomposition reconstructed from 31 fungal genomes. Science 336:1715–1719
Frank JL, Coffan RA, Southworth D (2010) Aquatic gilled mushrooms: *Psathyrella* fruiting in the Rogue River in southern Oregon. Mycologia 102:93–107
Fröhlich-Nowoisky J, Pickersgill DA, Despres VR, Poschl U (2009) High diversity of fungi in air particulate matter. Proc Natl Acad Sci USA 106:12814–12819
Fröhlich-Nowoisky J, Burrows SM, Xie Z, Engling G, Solomon PA, Fraser MP, Mayol-Bracero OL, Artaxo P, Begerow D, Conrad R, Andreae MO, Després VR, Pöschl U (2012) Biogeography in the air: fungal diversity over land and oceans. Biogeosciences 9:1125–1136
Frøslev TG, Matheny PB, Hibbett DS (2005) Lower level relationships in the mushroom genus *Cortinarius* (Basidiomycota, Agaricales): a comparison of RPB1, RPB2, and ITS phylogenies. Mol Phylogenet Evol 37:602–618
Gao Z, Johnson ZI, Wang G (2010) Molecular characterization of the spatial diversity and novel lineages of mycoplankton in Hawaiian coastal waters. ISME J 4:111–120
Garcia-Sandoval R, Wang Z, Binder M, Hibbett DS (2011) Molecular phylogenetics of the Gloeophyllales and relative ages of clades of Agaricomycotina producing a brown rot. Mycologia 103:510–524
Garnica S, Weiß M, Oertel B, Oberwinkler F (2005) A framework for a phylogenetic classification in the genus *Cortinarius* (Basidiomycota, Agaricales) derived from morphological and molecular data. Can J Bot 83:1457–1477
Gäumann E (1926) Vergleichende Morphologie der Pilze. Gustav Fischer, Jena, Germany
Geml J, Davis DD, Geiser DM (2005) Phylogenetic analyses reveal deeply divergent species lineages in the genus *Sphaerobolus* (Phallales: Basidiomycota). Mol Phylogenet Evol 35:313–322
Geml J, Laursen GA, O'Neill K, Nusbaum HC, Taylor DL (2006) Beringian origins and cryptic speciation events in the fly agaric (*Amanita muscaria*). Mol Ecol 15:225–239
Ghobad-Nejhad M, Dai YC (2010) *Diplomitoporus rimosus* is found in Asia and belongs to the Hymenochaetales. Mycologia 102:1510–1517
Ghobad-Nejhad M, Nilsson RH, Hallenberg N (2010) Phylogeny and taxonomy of the genus *Vuilleminia* (Basidiomycota) based on molecular and morphological evidence, with new insights into the Corticiales. Taxon 59:1519–1534
Giachini AJ, Hosaka K, Nouhra ER, Spatafora JW, Trappe JM (2010) Phylogenetic relationships of the Gomphales based on nuc-25S-rDNA, mit-12S-rDNA and mit-atp6-DNA combined sequences. Fungal Biol 114:224–234
Gilbert JE (1931) Les Bolets. Les livres du mycologue, vol 3. E Le Francois, Paris, pp 1–254
Gilbertson RL, Ryvarden L (1986) North American polypores, vol 1. Fungiflora, Oslo
Gill M, Steglich W (1987) Pigments of fungi (Macromycetes). Prog Chem Org Nat Prod 51:1–317
Ginns J, Lefebvre MNL (1993) Lignicolous corticioid fungi (Basidiomycota) of North America: systematics, distribution, and ecology. Mycologia Memoir No. 19. American Phytopathological Society, St. Paul, MN
Glen M, Tommerup IC, Bougher NL, O'Brien PA (2002) Are Sebacinaceae common and widespread ectomycorrhizal associates of *Eucalyptus* species in Australian forests? Mycorrhiza 12:243–247
Gorjón SP, Greslebin AG, Rajchenberg MP (2011) Notes on *Amylocorticiellum* (Amylocorticiales, Basidiomycota), with some new combinations. Mycotaxon 116:283–293
Grand EA, Hughes KW, Petersen RH (2010) Relationships within *Lentinus* subg. *Lentinus* (Polyporales, Agaricomycetes), with emphasis on sects. *Lentinus* and *Tigrini*. Mycol Prog 10:399–413
Greslebin A (2004) *Rhizochaete*, a new genus of phanerochaetoid fungi. Mycologia 96:260–271
Griffiths RP, Castellano MA, Caldwell BA (1991) Hyphal mats formed by two ectomycorrhizal fungi and their association with Douglas-fir seedlings: a case study. Plant Soil 134:255–259
Griffiths RP, Baham JE, Caldwell BA (1994) Soil solution chemistry of ectomycorrhizal mats in forest soil. Soil Biol Biochem 26:331–337
Grigoriev IV, Nordberg H, Shabalov I, Aerts A, Cantor M, Goodstein D, Kuo A, Minovitsky S, Nikitin R, Ohm RA, Otillar R, Poliakov A, Ratnere I, Riley R, Smirnova T, Rokhsar D, Dubchak I (2012) The genome portal of the Department of Energy Joint Genome Institute. Nucl Acids Res 40(Database issue):D26–D32
Gueidan C, Ruibal C, de Hoog GS, Schneider H (2011) Rock-inhabiting fungi originated during periods of dry climate in the late Devonian and middle Triassic. Fungal Biol 115:987–996

Gulden G, Stensrud O, Shalchian-Tabrizi K, Kauserud H (2005) *Galerina* Earle: a polyphyletic genus in the consortium of dark-spored agarics. Mycologia 97:823–837

Guzmán-Dávalos L, Mueller GM, Cifuentes J, Miller AN, Santerre A (2003) Traditional infrageneric classification of *Gymnopilus* is not supported by ribosomal DNA sequence data. Mycologia 95:1204–1214

Hallenberg N, Nilsson RH, Antonelli A, Wu SH, Maekawa N, Norden B (2007) The *Peniophorella praetermissa* species complex (Basidiomycota). Mycol Res 111:1366–1376

Hallenberg N, Ryberg M, Nilsson RH, Wood AR, Wu S-H (2008) *Pseudolagarobasidium* (Basidiomycota): on the reinstatement of a genus of parasitic, saprophytic, and endophytic resupinate fungi. Botany 86:1319–1325

Hallenberg N, Nilsson RH, Robledo G (2013) Species complexes in *Hericium* (Russulales, Agaricomycota) and a new species - *Hericium rajchenbergii* - from southern South America. Mycol Prog 12:413–420

Halling RE, Baroni TJ, Binder M (2007) A new genus of Boletaceae from eastern North America. Mycologia 99:310–316

Halling RE, Nuhn M, Fechner NA, Osmundson TW, Soytong K, Arora D, Hibbett DS, Binder M (2012a) *Sutorius*: a new genus for *Boletus eximius*. Mycologia 104:951–961. doi:10.3852/11-376

Halling RE, Nuhn M, Osmundson TW, Fechner NA, Trappe JM, Soytong K, Arora D, Hibbett DS, Binder M (2012b) Affinities of the *Boletus chromapes* group to *Royoungia* and the description of two new genera, *Harrya* and *Australopilus*. Aust Syst Bot 25:418–431

Harrington TC (2005) Ecology and evolution of mycophagous bark beetles and their fungal partners. In: Vega FH, Blackwell M (eds) Insect-fungal associations: ecology and evolution. Oxford University Press, Oxford, pp 257–291

Harrower E, Ammirati J, Cappuccino A, Ceska O, Kranabetter J, Kroeger P, Lim S, Taylor T, Berbee M (2011) *Cortinarius* species diversity in British Columbia and molecular phylogenetic comparison with European specimen sequences. Botany 89:799–810

Hausknecht A (2009) *Conocybe-Pholiotina*. Fungi Europaei, vol 11. Candusso, Alassio

He S-H, Dai Y-C (2012) Taxonomy and phylogeny of *Hymenochaete* and allied genera of Hymenochaetaceae (Basidiomycota) in China. Fungal Divers 56:77–93

He S-H, Li H-J (2013) *Pseudochaete latesetosa* and *P. subrigidula* spp. nov. (Hymenochaetales, Basidiomycota) from China based on morphological and molecular characters. Mycol Prog 12:331–339

Hedges SB, Blair JE, Venturi ML, Shoe JL (2004) A molecular timescale of eukaryote evolution and the rise of complex multicellular life. BMC Evol Biol 4:2

Heim R, Wasson RG (1958) Les champignons hallucinogenes du Mexique. Museum National d'Histoire Naturelle, Paris

Henkel TW, Aime MC, Uehling JK, Smith ME (2011) New species and distribution records of *Clavulina* (Cantharellales, Basidiomycota) from the Guiana Shield. Mycologia 203:883–894

Henkel TW, Aime MC, Chin MML, Miller SL, Vilgalys R, Smith ME (2012) Ectomycorrhizal fungal sporocarp diversity and discovery of new taxa in *Dicymbe* monodominant forests of the Guiana Shield. Biodiv Conserv 21:2195–2220

Hibbett DS (2001) Shiitake mushrooms and molecular clock: historical biogeography of *Lentinula*. J Biogeogr 28:231–241

Hibbett DS (2004) Trends in morphological evolution in homobasidiomycetes inferred using maximum likelihood: a comparison of binary and multistate approaches. Syst Biol 53:889–903

Hibbett DS (2006) A phylogenetic overview of the Agaricomycotina. Mycologia 98:917–925

Hibbett DS (2007) After the gold rush, or before the flood? Evolutionary morphology of mushroom-forming fungi (Agaricomycetes) in the early 21st century. Mycol Res 111:1001–1018

Hibbett DS, Binder M (2001) Evolution of marine mushrooms. Biol Bull 201:319–322

Hibbett DS, Binder M (2002) Evolution of complex fruiting-body morphologies in homobasidiomycetes. Proc R Soc B Biol Sci 269:1963–1969

Hibbett DS, Donoghue JD (1995) Progress towards a phylogenetic classification of the Polyporaceace through parsimony analysis of mitochondrial ribosomal DNA sequences. Can J Bot 73:S853–S861

Hibbett DS, Matheny PB (2009) The relative ages of ectomycorrhizal mushrooms and their plant hosts estimated using Bayesian relaxed molecular clock analyses. BMC Biol 7:13

Hibbett DS, Thorn RG (2001) Basidiomycota: homobasidiomycetes. In: McLaughlin DJ, McLaughlin EG, Lemke PA (eds) The Mycota, vol 7A, Systematics and evolution. Springer, Berlin, pp 121–168

Hibbett DS, Tsuneda A, Murakami S (1994) The secotioid form of *Lentinus tigrinus*: genetics and development of a fungal morphological innovation. Am J Bot 81:466–478

Hibbett DS, Grimaldi D, Donoghue MJ (1995) Cretaceous mushrooms in amber. Nature 377:487

Hibbett DS, Grimaldi D, Donoghue MJ (1997a) Fossil mushrooms from Cretaceous and Miocene ambers and the evolution of homobasidiomycetes. Am J Bot 84:981–991

Hibbett DS, Pine EM, Langer E, Langer G, Donoghue JD (1997b) Evolution of gilled mushrooms and puffballs inferred from riboomal DNA sequences. Proc Natl Acad Sci USA 94:12002–12006

Hibbett DS, Gilbert L-B, Donoghue MJ (2000) Evolutionary instability of ectomycorrhizal symbioses in basidiomycetes. Nature 407:506–508

Hibbett DS, Binder M, Wang Z (2003) Another fossil agaric from Dominican amber. Mycologia 95:685–687

Hibbett DS, Binder M, Bischoff JF, Blackwell M, Cannon PF, Eriksson OE, Huhndorf S, James T, Kirk PM, Lucking R, Thorsten Lumbsch H, Lutzoni F, Matheny PB, McLaughlin DJ, Powell MJ, Redhead S, Schoch CL, Spatafora JW, Stalpers JA, Vilgalys R, Aime MC, Aptroot A, Bauer R, Begerow D, Benny GL, Castlebury LA, Crous PW, Dai YC, Gams W, Geiser DM, Griffith GW, Gueidan C, Hawksworth DL, Hestmark G, Hosaka K, Humber RA, Hyde KD, Ironside JE, Koljalg U, Kurtzman CP, Larsson KH, Lichtwardt R, Longcore J, Miadlikowska J, Miller A, Moncalvo JM, Mozley-Standridge S, Oberwinkler F, Parmasto E, Reeb V, Rogers JD, Roux C, Ryvarden L, Sampaio JP, Schussler A, Sugiyama J, Thorn RG, Tibell L, Untereiner WA, Walker C, Wang Z, Weir A, Weiß M, White MM, Winka K, Yao YJ, Zhang N (2007) A higher-level phylogenetic classification of the Fungi. Mycol Res 111:509–547

Hibbett DS, Ohman A, Glotzer D, Nuhn M, Kirk P, Nilsson RH (2011) Progress in molecular and morphological taxon discovery in Fungi and options for formal classification of environmental sequences. Fung Biol Rev 25:38–47

Hibbett DS, Stajich JE, Spatafora JW (2013) Toward genome-enabled mycology. Mycologia 105:1339–1349

Hjortstam K, Ryvarden L (2007) Studies in corticioid fungi from Venezuela III (Basidiomycotina, Aphyllophorales). Synopsis Fungorum 23:56–107

Hjortstam K, Ryvarden L (2009) A checklist of names in *Hyphodontia* sensu stricto – sensu lato and *Schizopora* with new combinations in *Lagarobasidium*, *Lyomyces*, *Kneiffiella*, *Schizopora*, and *Xylodon*. Synopsis Fungorum 26:33–55

Hodkinson BP, Uehling JK, Smith ME (2012) *Lepidostroma vilgalysii*, a new basidiolichen from the New World. Mycol Prog 11:827–833

Hodkinson BP, Moncada B, Lücking R (2013) Lepidostromatales, a new order of lichenized fungi (Basidiomycota, Agaricomycetes), with two new genera, *Ertzia* and *Sulzbacheromyces*, and one new species, *Lepidostroma winklerianum*. Fungal Divers 64:165–179. doi:10.1007/s13225-013-0267-0

Hofstetter V, Clémençon H, Vilgalys R, Moncalvo JM (2002) Phylogenetic analyses of the Lyophylleae (Agaricales, Basidiomycetes) based on nuclear and mitochondrial rDNA sequences. Mycol Res 106:1043–1059

Horak E (2004) Röhrlinge und Blätterpilze in Europa. Elsevier, Spektrum, Munich

Horak E (2011) Revision of Malaysian species of *Boletales* s. l. (*Basidiomycota*) described by E. J. H. Corner (1972, 1974). Malayan Forest Rec 51:1–283

Horton TR, Bruns TD (2001) The molecular revolution in ectomycorrhizal ecology: peeking into the black-box. Mol Ecol 10:1855–1871

Hosaka K, Castellano MA (2008) Molecular phylogenetics of Geastrales with special emphasis on the position of *Sclerogaster*. Bull Nat Mus Nat Sci B 34:161–173

Hosaka K, Bates ST, Beever RE, Castellano MA, Colgan W, Dominguez LS, Nouhra ER, Geml J, Giachini AJ, Kenney SR, Simpson NB, Spatafora JW, Trappe JM (2006) Molecular phylogenetics of the gomphoid-phalloid fungi with an establishment of the new subclass Phallomycetidae and two new orders. Mycologia 98:949–959

Hosaka K, Castellano MA, Spatafora JW (2008) Biogeography of Hysterangiales (Phallomycetidae, Basidiomycota). Mycol Res 112:448–462

Hosen MI, Feng B, Wu G, Zhu XT, Li YC, Yang ZL (2013) *Borofutus*, a new genus of Boletaceae from tropical Asia: phylogeny, morphology and taxonomy. Fungal Divers 58:215–226

Humpert AJ, Muench EL, Giachini AJ, Castellano MA, Spatafora JW (2001) Molecular phylogenetics of *Ramaria* (Gomphales) and related genera: evidence from nuclear large subunit and mitochondrial small subunit rDNA sequences. Mycologia 93:465–477

Hyde KD, Udayanga D, Manamgoda DS, Tedersoo L, Larsson E, Abarenkov K, Bertrand YJK, Oxelman B, Hartman M, Kauserud H, Ryberg M, Kristiansson E, Nilsson RH (2013) Incorporating molecular data in fungal systematics: a guide for aspiring researchers. Curr Res Environ Appl Mycol 3:1–32

Hynson NA, Weiß M, Preiss K, Gebauer G, Treseder KK (2013) Fungal host specificity is not a bottleneck for the germination of Pyroleae species (Ericaceae) in a Bavarian forest. Mol Ecol 22:1473–1481

Ingold CT (1972) *Sphaerobolus*: the story of a fungus. Trans Br Mycol Soc 58:179–195

Ingold CT (1982a) Basidiospore germination and conidium development in *Auricularia*. Trans Br Mycol Soc 78:161–166

Ingold CT (1982b) Basidiospore germination and conidium formation in *Exidia glandulosa* and *Tremella mesenterica*. Trans Br Mycol Soc 79:370–373

James TY, Kauff F, Schoch CL, Matheny PB, Hofstetter V, Cox CJ, Celio G, Gueidan C, Fraker E, Miadlikowska J, Lumbsch HT, Rauhut A, Reeb V, Arnold AE, Amtoft A, Stajich JE, Hosaka K, Sung G-H, Johnson D, O'Rourke B, Crockett M, Binder M, Curtis JM, Slot JC, Wang Z, Wilson AW, Schüßler A, Longcore JE, O'Donnell K, Mozley-Standridge S, Porter D, Letcher PM, Powell MJ, Taylor JW, White MM, Griffith GW, Davies DR, Humber RA, Morton JB, Sugiyama J, Rossman AY, Rogers JD, Pfister DH, Hewitt D, Hansen K, Hambleton S, Shoemaker RA, Kohlmeyer J, Volkmann-Kohlmeyer B, Spotts RA, Serdani M, Crous PW,

Hughes KW, Matsuura K, Langer E, Langer G, Untereiner WA, Lücking R, Büdel B, Geiser DM, Aptroot A, Diederich P, Schmitt I, Schultz M, Yahr R, Hibbett DS, Lutzoni F, McLaughlin DJ, Spatafora JW, Vilgalys R (2006) Reconstructing the early evolution of Fungi using a six-gene phylogeny. Nature 443:818–822

Jarosch M (2001) Zur molekularen Systematik der Boletales: Coniophorineae, Paxillineae und Suillineae. Bibl Mycol 191:1–158

Jarosch M, Besl H (2001) *Leucogyrophana*, a polyphyletic genus of the order Boletales (Basidiomycetes). Plant Biol 3:443–448

Jeong WJ, Lim YW, Lee JS, Jung HS (2005) Phylogeny of *Phellinus* and related genera inferred from combined data of ITS and mitochondrial SSU rDNA sequences. J Microbiol Biotechnol 15:1028–1038

Jia B-S, Zhou L-W, Cui B-K, Rivoire B, Dai Y-C (2013) Taxonomy and phylogeny of *Ceriporia* (Polyporales, Basidiomycota) with an emphasis of Chinese collections. Mycol Prog 13:81–93. doi:10.1007/s11557-013-0895-5

Jones EBG, Fell JW (2012) Basidiomycota. In: Jones EBG, Pang K-L (eds) Marine fungi. De Gruyter, Berlin, pp 48–61

Jørgensen PM (1989) *Omphalina foliacea*, a new basidiolichen from America. Nord J Bot 9:89–95

Ju HK, Chung HW, Hong S-S, Park JH, Lee J, Kwon SW (2010) Effect of steam treatment on soluble phenolic content and antioxidant activity of the Chaga mushroom (*Inonotus obliquus*). Food Chem 119:619–625

Jülich W (1981) Higher taxa of basidiomycetes. J. Cramer, Vaduz

Justo A, Hibbett D (2011) Phylogenetic classification of Trametes (Basidiomycota, Polyporales) based on a five-marker dataset. Taxon 60:1567–1583

Justo A, Morgenstern I, Hallen-Adams HE, Hibbett DS (2009) Convergent evolution of sequestrate forms in *Amanita* under Mediterranean climate conditions. Mycologia 102:675–688

Justo A, Vizzini A, Minnis AM, Menolli N Jr, Capelari M, Rodríguez O, Malysheva E, Contu M, Ghignone S, Hibbett DS (2011) Phylogeny of the Pluteaceae (Agaricales, Basidiomycota): taxonomy and character evolution. Fungal Biol 115:1–20

Kämmerer A, Besl H, Bresinsky A (1985) Omphalotaceae fam. nov. und Paxillaceae, ein chemotaxonomischer Vergleich zweier Pilzfamilien der Boletales. Plant Syst Evol 150:101–117

Kasuya T, Hosaka K, Sakamoto H, Uchida A, Hoshino T, Kakishima M (2011) New records of *Geastrum* from Japanese sand dunes. Mycotaxon 118:1–15

Kasuya T, Hosaka K, Uno K, Kakishima M (2012) Phylogenetic placement of *Geastrum melanocephalum* and polyphyly of *Geastrum triplex*. Mycoscience 53:411–426

Kauserud H, Shalchian-Tabrizi K, Decock C (2007a) Multi-locus sequencing reveals multiple geographically structured lineages of *Coniophora arida* and *C. olivacea* (Boletales) in North America. Mycologia 99:705–713

Kauserud H, Svegården IB, Sætre G-P, Knudsen H, Stensrud Ø, Schmidt O, Doi S, Sugiyama T, Högberg N (2007b) Asian origin and rapid global spread of the destructive dry rot fungus *Serpula lacrymans*. Mol Ecol 16:3350–3360

Keller N, Hohn T (1997) Metabolic pathway gene clusters in filamentous fungi. Fung Genet Biol 21:17–29

Kerekes J, Desjardin DE (2009) A monograph of the genera *Crinipellis* and *Moniliophthora* from Southeast Asia including a molecular phylogeny of the nrITS region. Fungal Divers 37:101–152

Kim SY, Jung HS (2000) Phylogenetic relationships of the Aphyllophorales inferred from sequence analysis of nuclear small subunit ribosomal DNA. J Microbiol 38:122–131

Kim SY, Jung HS (2001) Phylogenetic relationships of the Polyporaceae based on gene sequences of nuclear small subunit ribosomal RNAs. Mycobiology 29:73–79

Kim KM, Lee JS, Jung HS (2007) *Fomitopsis incarnatus* sp. nov. based on generic evaluation of *Fomitopsis* and *Rhodofomes*. Mycologia 99:833–841

Kirchmair M, Morandell S, Stolz D, Pöder R, Sturmbauer C (2004) Phylogeny of the genus *Omphalotus* based on nuclear ribosomal DNA-sequences. Mycologia 96:1253–1260

Kirk PM, Cannon P, Stalpers J (eds) (2008) Dictionary of the fungi, 10th edn. CABI, Wallingford

Kirschner R (2010) The synnematous anamorph of *Exidia saccharina* (Auriculariales, Basidiomycota): morphology, conidiogenesis and association with bark beetles. Polish Bot J 55:335–342

Kirschner R, Chen C-J (2004) *Helicomyxa everhartioides*, a new helicosporous sporodochial hyphomycete from Taiwan with relationships to the Hyaloriaceae (Auriculariales, Basidiomycota). Stud Mycol 50:337–342

Kirschner R, Oberwinkler F (2009) Supplementary notes on *Basidiopycnis hyalina* (Basidiomycota, Atractiellales) and its anamorph. Mycotaxon 109:29–38

Kirschner R, Yang ZL, Zhao Q, Feng B (2010) *Ovipoculum album*, a new anamorph with gelatinous cupulate bulbilliferous conidiomata from China and with affinities to the Auriculariales (Basidiomycota). Fungal Divers 43:55–65

Kirschner R, Lee I-S, Piepenbring M (2012) A new pycnidial fungus with clamped hyphae from Central America. Mycol Prog 11:561–568

Kõljalg U, Nilsson RH, Abarenkov K, Tedersoo L, Taylor AFS, Bahram M, Bates ST, Bruns TD, Bengtsson-Palme J, Callaghan TM, Douglas B, Drenkhan T, Eberhardt U, Dueñas M, Grebenc T, Griffith GW, Hartmann M, Kirk PM, Kohout P, Larsson E, Lindahl BD, Lücking R, Martín MP,

Matheny PB, Nguyen NH, Niskanen T, Oja J, Peay KG, Peintner U, Peterson M, Põldmaa K, Saag L, Saar I, Schüßler A, Scott JA, Senés C, Smith ME, Suija A, Taylor DL, Telleria MT, Weiß M, Larsson K-H (2013) Towards a unified paradigm for sequence-based identification of fungi. Mol Ecol 22:5271–5277

Kotiranta H, Saarenoksa R (2005) The genus *Athelopsis* (Aphyllophorales, Corticiaceae) in Finland. Annales Botanici Fennici 42:335–342

Kotiranta H, Larsson K-H, Saarenoksa R, Kulju M (2011) *Tretomyces* gen. novum, *Byssocorticium caeruleum* sp. nov., and new combinations in *Dendrothele* and *Pseudomerulius* (Basidiomycota). Annales Botanici Fennici 48:37–48

Kottke I, Beiter A, Weiß M, Haug I, Oberwinkler F, Nebel M (2003) Heterobasidiomycetes form symbiotic associations with hepatics: Jungermanniales have sebacinoid mycobionts while *Aneura pinguis* (Metzgeriales) is associated with a *Tulasnella* species. Mycol Res 107:957–968

Kreisel H (1969) Grundzüge eines natürlichen Systems der Pilze. J. Cramer, Jena, Germany

Kretzer AM, Bruns TD (1999) Use of *atp6* in fungal phylogenetics: an example from the Boletales. Mol Phylogenet Evol 13:483–492

Krings M, Dotzler N, Galtier J, Taylor TN (2011) Oldest fossil basidiomycete clamp connections. Mycoscience 52:18–23

Krüger D (2008) Notes on the molecular phylogeny of the '*Polyporellus*' group within *Polyporus*: identity of collections from Canada and Ecuador, and relationships with Lentinus. Sydowia 60:213–233

Krüger D (2010) Unusual polypore fungi – a taxonomic emendation of *Polyporus* (Basidiomycotina) after ribosomal spacer characters. Cryptogamie Mycologie 31:389–401

Krüger D, Gargas A (2004) The basidiomycete genus *Polyporus* – an emendation based on phylogeny and putative secondary structure of ribosomal RNA molecules. Feddes Repertorium 115:530–546

Kühner R (1984) Some mainlines of classification in the gill fungi. Mycologia 76:1059–1074

Langer E (2002) Phylogeny of non-gilled and gilled basidiomycetes: DNA sequence inference, ultrastructure and comparative morphology. Universität Tübingen, Tübingen

Largent DL, Abell-Davis SE, Cummings GA, Ryan KL, Bergemann SE (2011) Saxicolous species of *Claudopus* (Agaricales, Entolomataceae) from Australia. Mycotaxon 116:253–264

Larsson K-H (1994) Poroid species in *Trechispora* and the use of calcium oxalate crystals for species identification. Mycol Res 98:1153–1172

Larsson K-H (2004) High phylogenetic diversity among corticioid homobasidiomycetes. Mycol Res 108:983–1002

Larsson K-H (2007a) Molecular phylogeny of *Hyphoderma* and the reinstatement of *Peniophorella*. Mycol Res 111:186–195

Larsson K-H (2007b) Re-thinking the classification of corticioid fungi. Mycol Res 111:1040–1063

Larsson E, Jeppson M (2008) Phylogenetic relationships among species and genera of Lycoperdaceae based on ITS and LSU sequence data from north European taxa. Mycol Res 112:4–22

Larsson E, Larsson K-H (2003) Phylogenetic relationships of russuloid basidiomycetes with emphasis on aphyllophoralean taxa. Mycologia 95:1037–1065

Larsson K-H, Parmasto E, Fischer M, Langer E, Nakasone KK, Redhead SA (2006) Hymenochaetales: a molecular phylogeny for the hymenochaetoid clade. Mycologia 98:926–936

Larsson K-H, Læssøe T, Yorou NS, Ryvarden L (2011) The phylogenetic position of *Hydnodon* and *Scytinopogon*. Inoculum 62:28

Lawrey JD, Binder M, Diederich P, Molina MC, Sikaroodi M, Ertz D (2007) Phylogenetic diversity of lichen-associated homobasidiomycetes. Mol Phylogenet Evol 44:778–789

Lawrey JD, Diederich P, Sikaroodi M, Gillevet PM (2008) Remarkable nutritional diversity of basidiomycetes in the Corticiales, including a new foliicolous species of *Marchandiomyces* (anamorphic Basidiomycota, Corticiaceae) from Australia. Am J Bot 95:816–823

Lawrey JD, Lücking R, Sipman HJ, Chaves JL, Redhead SA, Bungartz F, Sikaroodi M, Gillevet PM (2009) High concentration of basidiolichens in a single family of agaricoid mushrooms (Basidiomycota: Agaricales: Hygrophoraceae). Mycol Res 113:1154–1171

Lebel T, Syme A (2012) Sequestrate species of *Agaricus* and *Macrolepiota* from Australia: new species and combinations and their position in a calibrated phylogeny. Mycologia 104:496–520

Lebel T, Tonkin JE (2007) Australasian species of *Macowanites* are sequestrate species of *Russula* (Russulaceae, Basidiomycota). Aust Syst Bot 20:355–381

Lebel T, Orihara T, Maekawa N (2012) The sequestrate genus *Rossbeevera* T. Lebel & Orihara gen. nov. (Boletaceae) from Australasia and Japan: new species and new combinations. Fungal Divers 52:49–71

Lee JS, Lim YW (2009) *Cerrena aurantiopora* sp. nov. (Polyporaceae) from eastern Asia. Mycologia 102:211–216

Lehmkuhl JF, Gould LF, Cazares E, Hosford DR (2004) Truffle abundance and mycophagy by northern flying squirrels in eastern Washington forests. Forest Ecol Manage 200:49–65

LePage BA, Currah RS, Stockey RA, Rothwell GW (1997) Fossil ectomycorrhizae from the Middle Eocene. Am J Bot 84:410

Li H, Cui B (2013) Taxonomy and phylogeny of the genus *Megasporoporia* and its related genera. Mycologia 105:368–383

Li YC, Feng B, Yang ZL (2011) *Zangia*, a new genus of Boletaceae supported by molecular and morphological evidence. Fungal Divers 49:125–143

Lilleskov EA, Bruns TD, Horton TR, Taylor D, Grogan P (2004) Detection of forest stand-level spatial structure in ectomycorrhizal fungal communities. FEMS Microbiol Ecol 49:319–332

Lim Y, Yeung Y, Sturrock R, Leal I, Breuil C (2005) Differentiating the two closely related species, *Phellinus weirii* and *P. sulphurascens*. Forest Pathol 35:305–314

Lindequist U, Niedermeyer THJ, Jülich W-D (2005) The pharmacological potential of mushrooms. Evid Based Complement Altern Med 2:285–299

Lindner DL, Banik MT (2008) Molecular phylogeny of *Laetiporus* and other brown rot polypore genera in North America. Mycologia 100:417–430

Lindner DL, Ryvarden L, Baroni TJ (2011) A new species of *Daedalea* (Basidiomycota) and a synopsis of core species in *Daedalea* sensu stricto. North American Fungi 6:1–12

Liu YJ, Whelen S, Hall BD (1999) Phylogenetic relationships among ascomycetes: evidence from an RNA polymerase II subunit. Mol Biol Evol 16:1799–1808

Locquin MV (1974) De Taxia Fungorum I. U.A.E. Mondedition, Paris

Lodge DL, Padamsee M, Matheny PB, Aime MC, Cantrell SA, Boertmann D, Kovalenko A, Vizzini A, Dentinger B, Kirk P, Ainsworth M, Moncalvo J-M, Vilgalys R, Larsson E, Lücking R, Griffith G, Smith M, Norvell LL, Desjardin DE, Redhead SA, Ovrebo CL, Lickey E, Ercole E, Hughes KW, Courtecuisse R, Young A, Binder M, Minnis A, Lindner DL, Ortiz-Sanatan B, Haight J, Læssøe T, Baroni TJ, Geml J, Hattori T (2013) Molecular phylogeny, morphology, pigment chemistry and ecology in Hygrophoraceae (Agaricales). Fungal Divers 64:1–99. doi:10.1007/s13225-013-0259-0

Malajczuk N (1988) Ecology and management of ectomycorrhizal fungi in regenerated ecosystems in Australia. In: Sylvia DM, Hung LL, Graham JH (eds) Mycorrhizae in the next decade. Practical applications and research priorities. University of Florida, Gainesville, FL, pp 118–120

Martinez D, Larrondo L, Putnam N, Gelpke M, Huang K, Chapman J, Helfenbein K, Ramaiya P, Detter J, Larimer F, Coutinho P, Henrissat B, Berka R, Cullen D, Rokhsar D (2004) Genome sequence of the lignocellulose degrading fungus *Phanerochaete chrysosporium* strain RP78. Nat Biotechnol 22:695–700

Martinez D, Challacombe J, Morgenstern I, Hibbett D, Schmoll M, Kubicek CP, Ferreira P, Ruiz-Duenas FJ, Martinez AT, Kersten P, Hammel KE, Vanden Wymelenberg A, Gaskell J, Lindquist E, Sabat G, Bondurant SS, Larrondo LF, Canessa P, Vicuna R, Yadav J, Doddapaneni H, Subramanian V, Pisabarro AG, Lavin JL, Oguiza JA, Master E, Henrissat B, Coutinho PM, Harris P, Magnuson JK, Baker SE, Bruno K, Kenealy W, Hoegger PJ, Kues U, Ramaiya P, Lucas S, Salamov A, Shapiro H, Tu H, Chee CL, Misra M, Xie G, Teter S, Yaver D, James T, Mokrejs M, Pospisek M, Grigoriev IV, Brettin T, Rokhsar D, Berka R, Cullen D (2009) Genome, transcriptome, and secretome analysis of wood decay fungus *Postia placenta* supports unique mechanisms of lignocellulose conversion. Proc Natl Acad Sci USA 106:1954–1959

Masui K (1926) A study of the mycorrhiza of *Abies firma*, S. et Z., with special reference to its mycorrhizal fungus, *Cantharellus floccosus*, Schw. Memoirs of the College of Science, Kyoto Imperial University, Series B 2:1–84

Masui K (1927) A study of the ectotrophic mycorrhizas of woody plants. Memoirs of the College of Science, Kyoto Imperial University, Series B 3:149–279

Mata JL, Halling RE, Hughes KW, Petersen RH (2004) *Rhodocollybia* in neotropical montane forests. Mycol Prog 3:337–352

Matheny PB (2005) Improving phylogenetic inference of mushrooms with RPB1 and RPB2 nucleotide sequences (*Inocybe*; Agaricales). Mol Phylogenet Evol 35:1–20

Matheny PB, Griffith GW (2010) Mycoparasitism between *Squamanita paradoxa* and *Cystoderma amianthinum* (Cystodermateae, Agaricales). Mycoscience 51:456–461

Matheny PB, Curtis JM, Hofstetter V, Aime MC, Moncalvo JM, Ge ZW, Slot JC, Ammirati JF, Baroni TJ, Bougher NL, Hughes KW, Lodge DJ, Kerrigan RW, Seidl MT, Aanen DK, DeNitis M, Daniele GM, Desjardin DE, Kropp BR, Norvell LL, Parker A, Vellinga EC, Vilgalys R, Hibbett DS (2006) Major clades of Agaricales: a multilocus phylogenetic overview. Mycologia 98:982–995

Matheny PB, Wang Z, Binder M, Curtis JM, Lim YW, Nilsson RH, Hughes KW, Hofstetter V, Ammirati JF, Schoch CL, Langer E, Langer G, McLaughlin DJ, Wilson AW, Frøslev T, Ge ZW, Kerrigan RW, Slot JC, Yang Z-L, Baroni TJ, Fischer M, Hosaka K, Matsuura K, Seidl MT, Vauras J, Hibbett DS (2007) Contributions of *rpb2* and *tef1* to the phylogeny of mushrooms and allies (Basidiomycota, Fungi). Mol Phylogenet Evol 43:430–451

Matheny PB, Aime MC, Bougher NL, Buyck B, Desjardin DE, Horak E, Kropp BR, Lodge DJ, Soytong K, Trappe JM, Hibbett DS (2009) Out of the Palaeotropics? Historical biogeography and diversification of the cosmopolitan ectomycorrhizal mushroom family Inocybaceae. J Biogeogr 36:577–592

Matsuura K (2006) Termite-egg mimicry by a sclerotium-forming fungus. Proc R Soc B Biol Sci 273:1203–1209

Matsuura K, Tanaka C, Nishida T (2000) Symbiosis of a termite and a sclerotium forming fungus: Sclerotia mimic termite eggs. Ecol Res 15:405–414

Matsuura K, Yashiro T, Shimizu K, Tatsumi S, Tamura T (2009) Cuckoo fungus mimics termite eggs by producing the cellulose-digesting enzyme beta-glucosidase. Curr Biol 19:30–36

Mattsson J, Anne-Cathrine Flyen A-C, Nunez M (2010) Wood-decaying fungi in protected buildings and structures on Svalbard. Agarica 29:5–14

May TW, Sinnott N, Sinnott A (2010) The truffle-like *Protubera canescens* is an early developmental stage of the cage fungus *Ileodictyon*. Victorian Nat 127:49–54

McLaughlin DJ, Celio GJ, Padamsee M, Dentinger BTM (2008) Cystidial structure in two genera of the Russulales. Botany 86:545–550

Meinhardt LW, Rincones J, Bailey BA, Aime MC, Griffith GW, Zhang D, Pereira GA (2008) *Moniliophthora perniciosa*, the causal agent of witches' broom disease of cacao: what's new from this old foe? Mol Plant Pathol 9:577–588

Miettinen O, Larsson K-H (2010) *Sidera*, a new genus in Hymenochaetales with poroid and hydnoid species. Mycol Prog 10:131–141

Miettinen O, Rajchenberg M (2011) *Obba* and *Sebipora*, new polypore genera related to *Cinereomyces* and *Gelatoporia* (Polyporales, Basidiomycota). Mycol Prog 11:131–147

Miettinen O, Larsson E, Sjökvist E, Larsson K-H (2012) Comprehensive taxon sampling reveals unaccounted diversity and morphological plasticity in a group of dimitic polypores (Polyporales, Basidiomycota). Cladistics 28:251–270

Mikheyev AS, Mueller UG, Abbot P (2010) Comparative dating of attine ant and lepiotaceous cultivar phylogenies reveals coevolutionary synchrony and discord. Am Nat 175:E126–E133

Miller OK Jr (1964) Monograph of *Chroogomphus* (Gomphidiaceae). Mycologia 56:526–549

Miller OK Jr, Miller HH (1988) Gasteromycetes: morphological and developmental features with keys to orders, families, and genera. Mad River, Eureka, CA

Miller SL, McClean TM, Walker JF, Buyck B (2001) A molecular phylogeny of the Russulales including agaricoid, gasteroid and pleurotoid taxa. Mycologia 93:344–354

Miller SL, Larsson E, Larsson KH, Verbeken A, Nuytinck J (2006) Perspectives in the new Russulales. Mycologia 98:960–970

Mizuno T (1999) Bioactive substances in *Hericium erinaceus* (Bull.:Fr.) Pers. (Yamabushitake), and its medicinal utilization. Int J Med Mushrooms 1:105–119

Moncalvo JM, Lutzoni FM, Rehner SA, Johnson J, Vilgalys R (2000) Phylogenetic relationships of agaricoid fungi based on nuclear large subunit ribosomal DNA sequences. Syst Biol 49:278–305

Moncalvo JM, Vilgalys R, Redhead SA, Johnson JE, James TY, Catherine Aime M, Hofstetter V, Verduin SJW, Larsson E, Baroni TJ, Greg Thorn R, Jacobsson S, Clémençon H, Miller OK Jr (2002) One hundred and seventeen clades of euagarics. Mol Phylogenet Evol 23:357–400

Moncalvo J-M, Nilsson RH, Koster B, Dunham SM, Bernauer T, Matheny PB, Porter TM, Margaritescu S, Weiß M, Garnica S, Danell E, Langer G, Langer E, Larsson E, Larsson K-H, Vilgalys R (2006) The cantharelloid clade: dealing with incongruent gene trees and phylogenetic reconstruction methods. Mycologia 98:937–948

Moreau PA, Peintner U, Gardes M (2006) Phylogeny of the ectomycorrhizal mushroom genus *Alnicola* (Basidiomycota, Cortinariaceae) based on rDNA sequences with special emphasis on host specificity and morphological characters. Mol Phylogenet Evol 38:794–807

Morin E, Kohler A, Baker AR, Foulongne-Oriol M, Lombard V, Nagy LG, Ohm RA, Patyshakuliyeva A, Brun A, Aerts AL, Bailey AM, Billette C, Coutinho PM, Deakin G, Doddapaneni H, Floudas D, Grimwood J, Hildén K, Kües U, Labutti KM, Lapidus A, Lindquist EA, Lucas SM, Murat C, Riley RW, Salamov AA, Schmutz J, Subramanian V, Wösten HA, Xu J, Eastwood DC, Foster GD, Sonnenberg AS, Cullen D, de Vries RP, Lundell T, Hibbett DS, Henrissat B, Burton KS, Kerrigan RW, Challen MP, Grigoriev IV, Martin F (2012) Genome sequence of the button mushroom *Agaricus bisporus* reveals mechanisms governing adaptation to a humic-rich ecological niche. Proc Natl Acad Sci USA 109:17501–17506

Moser M (1983) Die Röhrlinge und Blätterpilze (Polyporales, Boletales, Agaricales, Russulales). Kleine Kryptogamenflora IIb/2. Basidiomyceten, 5th edn. Gustav Fischer, Stuttgart

Mosquera-Espinosa AT, Bayman P, Prado GA, Gómez-Carabali A, Otero JT (2013) The double life of *Ceratobasidium* orchid mycorrhizal fungi and their potential for biocontrol of *Rhizoctonia solani* sheath blight of rice. Mycologia 205:141–150

Mueller UG, Gerardo NM, Aanen DK, Six DL, Schultz TR (2005) The evolution of agriculture in insects. Annu Rev Ecol Evol Syst 36:563–595

Nagy LG, Házi J, Szappanos B, Kocsubé S, Bálint B, Rákhely G, Vágvölgyi C, Papp T (2012) The evolution of defense mechanisms correlate with the explosive diversification of autodigesting *Coprinellus* mushrooms (Agaricales, Fungi). Syst Biol 61:595–607

Nakasone KK (1990a) Cultural studies and identification of wood-inhabiting Corticiaceae and selected Hymenomycetes from North America. Mycologia Memoir No. 15. J. Cramer, Berlin

Nakasone KK (1990b) Taxonomic study of *Veluticeps* (Aphyllophorales). Mycologia 82:622–641

Nakasone KK (2007) Morphological and molecular studies on *Resinicium* s. str. Can J Bot 85:420–436

Nakasone KK (2012) *Tsugacorticium kenaicum* (Hymenochaetales, Basidiomycota), a new corticioid genus and species from Alaska. North Am Fungi 7:1–9. doi:10.2509/naf2012.007.001

Nannfeldt JA, Eriksson J (1953) On the hymenomycetous genus *Jaapia* Bres. and its taxonomical position. Svensk Bot Tidskr 47:177–189

Nelsen MP, Lücking R, Umaña L, Trest MT, Will-Wolf S, Chaves JL, Gargas A (2007) *Multiclavula ichthyiformis* (Fungi: Basidiomycota: Cantharellales: Clavulinaceae), a remarkable new basidiolichen from Costa Rica. Am J Bot 94:1289–1296

Neves MA, Binder M, Halling RE, Hibbett DS, Soytong K (2012) The phylogeny of selected *Phylloporus* species, inferred from NUC-LSU and ITS sequences, and descriptions of new species from the Old World. Fungal Divers 55:109–123

Neville P, Poumarat S (2004) Amaniteae (*Amanita, Limacella, Torrendia*). Fungi Europaei, vol 9. Candusso, Alassio

Newman EI, Reddell P (1987) The distribution of mycorrhizas among families of vascular plants. New Phytol 106:745–751

Newsham KK, Bridge PD (2010) Sebacinales are associates of the leafy liverwort *Lophozia excisa* in the southern maritime Antarctic. Mycorrhiza 20:307–313

Nielsen C, Williams DW, Hajek AE (2009) Putative source of the invasive *Sirex noctilio* fungal symbiont, *Amylostereum areolatum*, in the eastern United States and its association with native siricid woodwasps. Mycol Res 113:1242–1253

Niemelä T (2005) Käävät, puiden sienet. Polypores, lignicolous fungi. Norrlinia 13:1–320

Niemelä T, Larsson K-H, Dai Y-C, Larsson E (2007) *Anomoloma*, a new polypore genus separated from *Anomoporia* on the basis of decay type and molecular phylogenetic data. Mycotaxon 100:305–318

Nilsson T, Ginns J (1979) Cellulolytic activity and the taxonomic position of selected brown-rot fungi. Mycologia 71:170–177

Nilsson RH, Larsson K-H, Larsson E, Kõljalg U (2006) Fruiting body-guided molecular identification of root-tip mantle mycelia provides strong indications of ectomycorrhizal associations in two species of *Sistotrema* (Basidiomycota). Mycol Res 110:1426–1432

Nobre T, Kone NA, Konate S, Linsenmair KE, Aanen DK (2011) Dating the fungus-growing termites' mutualism shows a mixture between ancient codiversification and recent symbiont dispersal across divergent hosts. Mol Ecol 20:2619–2627

Noordeloos ME (2011) Strophariaceae s.lato. Fungi Europaei, vol 13. Candusso, Alassio

Nouhra ER, Domínguez de Toledo LS (1994) Interacción entre Phallales (Basidiomycotina) e insectos (Coleópteros y Dípteros). Boletín de la Sociedad Argentina de Botánica 30:21–24

Nouhra ER, Horton TR, Cazares E, Castellano MA (2005) Morphological and molecular characterization of selected *Ramaria* mycorrhizae. Mycorrhiza 15:55–59

Nuhn ME, Binder M, Taylor AFS, Halling RE, Hibbett DS (2013) Phylogenetic overview of Boletineae. Fungal Biol 117:479–511

Nuñez M, Ryvarden L (2000) East Asian Polypores, volume 1 – Ganodermataceae and Hymenochaetaceae. Synopsis Fungorum 13:1–168

Nuytinck J, Verbeken A, Delarue S, Walleyn R (2004) Systematics of European sequestrate lactarioid Russulaceae with spiny spore ornamentation. Belgian J Bot 136:145–153

O'Donnell K, Lutzoni F, Ward TJ, Benny GL (2001) Evolutionary relationships among mucoralean Fungi (Zygomycota): evidence for family polyphyly on a large scale. Mycologia 93:286–297

Oberwinkler F (1970) Die Gattungen der Basidiolichenen. Vorträge aus dem Gesamtgebiet der Botanik, Neue Folge 4:139–169

Oberwinkler F (1975) Eine agaricoide Gattung der Thelephorales. Sydowia 28:359–361

Oberwinkler F (1977) Das neue System der Basidiomyceten. In: Frey W, Hurka H, Oberwinkler F (eds) Beiträge zur Biologie der niederen Pflanzen. Gustav Fischer, Stuttgart, pp 59–105

Oberwinkler F (2012) Basidiolichens. In: Hock B (ed) Fungal associations. The Mycota IX, 2nd edn. Springer, Berlin

Oberwinkler F (2014) Dacrymycetes. In: McLaughlin DJ, Spatafora JW (eds) The Mycota: Systematics and evolution: Part B. Springer, Heidelberg

Oberwinkler F, Riess K, Bauer R, Kirschner R, Garnica S (2013a) Taxonomic re-evaluation of the *Ceratobasidium-Rhizoctonia* complex and *Rhizoctonia butinii*, a new species attacking spruce. Mycol Prog 12:763–776

Oberwinkler F, Riess K, Bauer R, Selosse MA, Weiss M, Garnica S, Zuccaro A (2013b) Enigmatic Sebacinales. Mycol Prog 12:1–27

Ohm RA, de Jong JF, Lugones LG, Aerts A, Kothe E, Stajich JE, de Vries RP, Record E, Levasseur A, Baker SE, Bartholomew KA, Coutinho PM, Erdmann S, Fowler TJ, Gathman AC, Lombard V, Henrissat B, Knabe N, Kües U, Lilly WW, Lindquist E, Lucas S, Magnuson JK, Piumi F, Raudaskoski M, Salamov A, Schmutz J, Schwarze FWMR, vanKuyk PA, Horton JS, Grigoriev IV, Wosten HAB (2010) Genome sequence of the model mushroom *Schizophyllum commune*. Nat Biotechnol 28:957–963

Olsson PA, Münzenberger B, Mahmood S, Erland S (2000) Molecular and anatomical evidence for a three-way association between *Pinus sylvestris* and the ectomycorrhizal fungi *Suillus bovinus* and *Gomphidius roseus*. Mycol Res 104:1372–1378

Orihara T, Sawada F, Ikeda S, Yamato M, Tanaka C, Shimomura N, Hashiya M, Iwase K (2010) Taxonomic reconsideration of a sequestrate fungus, *Octaviania columellifera*, with the proposal of a new genus, *Heliogaster*, and its phylogenetic relationships in the Boletales. Mycologia 102:108–121

Ortiz-Santana B, Lindner DL, Miettinen O, Justo A, Hibbett D (2013) A phylogenetic overview of the *Antrodia* clade (Basidiomycota, Polyporales). Mycologia 105:1391–1411

Oses R, Valenzuela S, Freer J, Sanfuentes E, Rodrigues J (2008) Fungal endophytes in xylem of healthy Chilean trees and their possible role in early wood decay. Fungal Divers 33:77–86

Otjen L, Blanchette RA (1984) *Xylobolus frustulatus* decay of oak: patterns of selective delignification and subsequent cellulose removal. Appl Environ Microbiol 47:670–676

Padamsee M, Kumar TKA, Riley R, Binder M, Boyd A, Calvo AM, Furukawa K, Hesse C, Hohmann S, James TY, LaButti K, Lapidus A, Lindquist E, Lucas S, Miller K, Shantappa S, Grigoriev IV, Hibbett DS, McLaughlin DJ, Spatafora JW, Aime MC (2012) The genome of the xerotolerant mold *Wallemia sebi* reveals adaptations to osmotic stress and suggests cryptic sexual reproduction. Fung Genet Biol 49:217–226

Palice Z, Schmitt I, Lumbsch HT (2005) Molecular data confirm that *Omphalina foliacea* is a lichen-forming basidiomycete. Mycol Res 109:447–451

Parmasto E (2010) Clavariachaetaceae, a family of neotropical Hymenochaetales (Basidiomycota) including clavarioid, pileate and resupinate species. Folia Cryptogamica Estonica 47:51–57

Parmasto E, Saar I, Larsson E, Rummo S (2013) Phylogenetic taxonomy of *Hymenochaete* and related genera (Hymenochaetales). Mycol Prog 12:1–10

Parmeter JR (ed) (1970) *Rhizoctonia solani*: biology and pathology. University of California Press, Berkeley, CA

Parra-Sánchez LA (2008) *Agaricus* L. – *Allopsalliota*. Fungi Europaei, vol 1. Candusso, Alassio

Paulus B, Hallenberg N, Buchanan PK, Chambers GK (2000) A phylogenetic study of the genus *Schizopora* (Basidiomycota) based on ITS DNA sequences. Mycol Res 104:1155–1163

Peay KG, Kennedy PG, Bruns TD (2008) Fungal community ecology: a hybrid beast with a molecular master. BioScience 58:799–810

Pegler DN, Young TWK (1981) A natural arrangement of the Boletales, with reference to spore morphology. Trans Br Mycol Soc 76:103–146

Peintner U, Moncalvo JM, Vilgalys R (2004) Toward a better understanding of the infrageneric relationships in *Cortinarius* (Agaricales, Basidiomycota). Mycologia 96:1042–1058

Petersen RH (1971a) Evolution in the higher basidiomycetes. University of Tennessee Press, Knoxville, TN

Petersen RH (1971b) The genera *Gomphus* and *Gloeocantharellus* in North America. Nova Hedwigia 21:1–118

Petersen JH (2012) The kingdom of fungi. Princeton University Press, Princeton, NJ

Petersen RH, Hughes KW (2010) The *Xerula/Oudemansiella* complex. Nova Hedwigia Beih 137:1–625

Pildain MB, Rajchenberg M (2013) The phylogenetic position of *Postia* s.l. (Polyporales, Basidiomycota) from Patagonia, Argentina. Mycologia 105:357–367

Pine EM, Hibbett DS, Donoghue MJ (1999) Phylogenetic relationships of cantharelloid and clavarioid Homobasidiomycetes based on mitochondrial and nuclear rDNA sequences. Mycologia 91:944–963

Poinar GO Jr, Brown AE (2003) A non-gilled hymenomycete in Cretaceous amber. Mycol Res 107:763–768

Poinar GO Jr, Buckley R (2007) Evidence of mycoparasitism and hypermycoparasitism in Early Cretaceous amber. Mycol Res 111:503–506

Poinar GO Jr, Singer R (1990) Upper Eocene gilled mushroom from the Dominican Republic. Science 248:1099–1101

Porras-Alfaro A, Lipinski K, Herrera J, Natvig DO, Sinsabaugh RL (2011) Diversity and distribution patterns of soil fungal communities in a semiarid grassland. Mycologia 103:10–21

Pouzar Z (1985) Proposals for the conservation of five family names of fungi. Taxon 34:709–712

Preussing M, Nebel M, Oberwinkler F, Weiß M (2010) Diverging diversity patterns in the *Tulasnella* (Basidiomycota, Tulasnellales) mycobionts of *Aneura pinguis* (Marchantiophyta, Metzgeriales) from Europe and Ecuador. Mycorrhiza 20:147–159

Pringle A, Adams RI, Cross HB, Bruns TD (2009) The ectomycorrhizal fungus *Amanita phalloides* was introduced and is expanding its range on the west coast of North America. Mol Ecol 18:817–833

Punja ZK (1985) The biology, ecology, and control of *Sclerotium rolfsii*. Annu Rev Phytopathol 23:97–127

Qiang X, Weiß M, Kogel K-H, Schäfer P (2012) *Piriformospora indica* – a mutualistic basidiomycete with an exceptionally large plant host range. Mol Plant Pathol 13:508–518

Rajchenberg M, Gorjón SP, Pildain MB (2011) The phylogenetic disposition of *Antrodia* s.l. (Polyporales, Basidiomycota) from Patagonia, Argentina. Aust Syst Bot 24:111–120

Ramírez-Cruz V, Guzmán G, Villalobos-Arámbula AR, Rodríguez A, Matheny PB, Sánchez-García G-DL (2013) Phylogenetic inference of the psychedelic mushroom genus *Psilocybe* sensu lato. Botany 91:573–591

Raper JR, Miles PG (1958) The genetics of *Schizophyllum commune*. Genetics 43:530–546

Redberg GL, Hibbett DS, Ammirati JF, Rodriguez RJ (2003) Phylogeny and genetic diversity of *Bridgeoporus nobilissimus* inferred using mitochondrial and nuclear rDNA sequences. Mycologia 95:836–845

Redhead SA (1981) Parasitism of bryophyes by agarics. Can J Bot 59:63–67

Redhead SA, Ginns J (1985) A reappraisal of agaric genera associated with brown rots of wood. Trans Mycol Soc Japan 26:349–381

Redhead SA, Lutzoni F, Moncalvo JM, Vilgalys R (2002) Phylogeny of agarics: partial systematics solutions for core omphalinoid genera in the Agaricales (Euagarics). Mycotaxon 83:19–57

Reijnders AFM, Stalpers JS (1992) The development of the hymenophoral trama in the Aphyllophorales and the Agaricales. Stud Mycol 34:1–109

Riess K, Oberwinkler F, Bauer R, Garnica S (2013) High genetic diversity at the regional scale and possible speciation in *Sebacina epigaea* and *S. incrustans*. BMC Evol Biol 13:102

Rinaldi AC, Comandini O, Kuyper TW (2008) Ectomycorrhizal fungal diversity: separating the wheat from the chaff. Fungal Divers 33:1–45

Robledo GL, Amalfi M, Castillo G, Rajchenberg M, Decock C (2009) *Perenniporiella chaquenia* sp. nov. and further notes on *Perenniporiella* and its relationships with *Perenniporia* (Poriales, Basidiomycota). Mycologia 101:657–673

Rogers DP (1932) A cytological study of *Tulasnella*. Botanical Gazette 94:86–105

Ruiz-Dueñas FJ, Martínez AT (2009) Microbial degradation of lignin: how a bulky recalcitrant polymer is efficiently recycled in nature and how we can take advantage of this. Microb Biotechnol 2:164–177

Ruiz-Duenas FJ, Lundell T, Floudas D, Nagy L, Barrassa JM, Hibbett D, Martínez AT (2013) Lignin-degrading peroxidases in Polyporales: an evolutionary survey based on ten sequenced genomes. Mycologia 105:1428–1444

Rungjindamai N, Pinruan U, Choeyklin R, Hattori T, Jones EBG (2008) Molecular characterization of basidiomycetous endophytes isolated from leaves, rachis and petioles of the oil palm, *Elaeis guineensis*, in Thailand. Fungal Divers 33:139–161

Ryberg M, Matheny PB (2012) Asynchronous origins of ectomycorrhizal clades of Agaricales. Proc R Soc B Biol Sci 279:2003–2011

Ryberg M, Larsson E, Jacobsson S (2010) An evolutionary perspective on morphological and ecological characters in the mushroom family Inocybaceae (Agaricomycotina, Fungi). Mol Phylogenet Evol 55:431–442

Ryvarden L (1991) Genera of polypores. Nomenclature and taxonomy. Synopsis Fungorum 5:1–363

Ryvarden L (2004) Neotropical polypores, part 1 – introduction, Ganodermataceae and Hymenochaetaceae. Synopsis Fungorum 19:1–228

Saar I, Põldmaa K, Kõljalg U (2009) The phylogeny and taxonomy of genera *Cystoderma* and *Cystodermella* (Agaricales) based on nuclear ITS and LSU sequences. Mycol Prog 8:59–73

Sahr T, Ammer H, Besl H, Fischer M (1999) Infrageneric classification of the boleticolous genus *Sepedonium*: species delimitation and phylogenetic relationships. Mycologia 91:935–943

Sánchez-García M, Cifuentes-Blanco J, Matheny PB (2013) Revisión taxonómica del género *Melanoleuca* en México y descripción de especies nuevas. Revista Mexicana de Biodiversidad 84:S111–S127

Schmidt O, Kebernik U (1989) Characterization and identification of the dry rot fungus *Serpula lacrymans* by polyacrylamide gel electrophoresis. Holzforschung 43:195–198

Schmidt O, Grimm K, Moreth U (2002) Molecular identity of species and isolates of the *Coniophora* cellar fungi. Holzforschung 56:563–571

Seitzman BH, Ouimette A, Mixon RL, Hobbie EA, Hibbett DS (2011) Conservation of biotrophy in Hygrophoraceae inferred from combined stable isotope and phylogenetic analyses. Mycologia 103:280–290

Sell I, Kotiranta H, Miettinen O, Põldmaa K (2013) Molecular analysis confirms that *Botryodontia millavensis* and *Oxyporus philadelphi* are conspecific. Mycol Prog 13:65–74. doi:10.1007/s11557-013-0893-7

Selosse M-A, Weiß M, Jany J-L, Tillier A (2002) Communities and populations of sebacinoid basidiomycetes associated with the achlorophyllous orchid *Neottia nidus-avis* (L.) L.C.M. Rich. and neighbouring tree ectomycorrhizae. Mol Ecol 11:1831–1844

Selosse M-A, Setaro S, Glatard F, Richard F, Urcelay C, Weiß M (2007) Sebacinales are common mycorrhizal associates of Ericaceae. New Phytol 174:864–878

Selosse M-A, Dubois M-P, Alvarez N (2009) Do Sebacinales commonly associate with plant roots as endophytes? Mycol Res 113:1062–1069

Setaro S, Weiß M, Oberwinkler F, Kottke I (2006) Sebacinales form ectendomycorrhizas with *Cavendishia nobilis*, a member of the Andean clade of Ericaceae, in the mountain rain forest of southern Ecuador. New Phytol 169:355–365

Sigler L, de la Maza LM, Tan G, Egger KN, Sherburne RK (1995) Diagnostic difficulties caused by a nonclamped *Schizophyllum commune* isolate in a case of fungus ball of the lung. J Clin Microbiol 33:1979–1983

Singer R (1965) Die Röhrlinge. Teil I. Boletaceae (ohne Boletoideae). J. Klinkhardt, Bad Heilbrunn, Germany

Singer R (1967) Die Röhrlinge. Teil II. Die Boletoideae und Strobilomycetaceae. J. Klinkhardt, Bad Heilbrunn, Germany

Singer R (1986) The Agaricales in modern taxonomy, 4th edn. Koeltz, Königstein, Germany

Sjökvist E, Larsson E, Eberhardt U, Ryvarden L, Larsson K-H (2012) Stipitate stereoid basidiocarps have evolved multiple times. Mycologia 104:1046–1055

Skrede I, Engh I, Binder M, Carlsen T, Kauserud H, Bendiksby M (2011) Evolutionary history of Serpulaceae (Basidiomycota): molecular phylogeny, historical biogeography and evidence for a single transition of nutritional mode. BMC Evol Biol 11:230

Slippers B, Wingfield MJ, Coutinho TA, Wingfield BD (2001) Population structure and possible origin of *Amylostereum areolatum* in South Africa. Plant Pathol 50:206–210

Slippers B, Coutinho TA, Wingfield BD, Wingfield MJ (2003) A review of the genus *Amylostereum* and its association with woodwasps. S Afr J Sci 99:70–74

Smith AH, Thiers HD (1971) The boletes of Michigan. University of Michigan Press, Ann Arbor, MI

Smith SY, Currah RS, Stockey RA (2004) Cretaceous and Eocene poroid hymenophores from Vancouver Island, British Columbia. Mycologia 96:180–186

Smith ME, Schell KJ, Castellano MA, Trappe MJ, Trappe JM (2013) The enigmatic truffle *Fevansia aurantiaca* is an ectomycorrhizal member of the *Albatrellus* lineage. Mycorrhiza 23:663–668

Sneh B, Jabaji-Hare S, Neate S, Dijst G (eds) (1996) *Rhizoctonia* species: taxonomy, molecular biology, ecology, pathology and disease control. Kluwer, Dordrecht

Sotome K, Hattori T, Ota Y, To-anun C, Salleh B, Kakishima M (2008) Phylogenetic relationships of *Polyporus* and morphologically allied genera. Mycologia 100:603–615

Sotome K, Akagi Y, Lee SS, Ishikawa NK, Hattori T (2013) Taxonomic study of *Favolus* and *Neofavolus* gen. nov. segregated from *Polyporus* (Basidiomycota, Polyporales). Fungal Divers 58:245–266

Spirin V, Miettinen O, Pennanen J, Kotiranta H, Niemelä T (2012) *Antrodia hyalina*, a new polypore from Russia, and *A. leucaena*, new to Europe. Mycol Prog 12:53–61

Spirin V, Vlasák J, Niemelä T, Miettinen O (2013) What is *Antrodia* sensu stricto? Mycologia 105:1555–1576

Stajich JE, Wilke SK, Ahren D, Au CH, Birren BW, Borodovsky M, Burns C, Canback B, Casselton LA, Cheng CK, Deng J, Dietrich FS, Fargo DC, Farman ML, Gathman AC, Goldberg J, Guigo R, Hoegger PJ, Hooker JB, Huggins A, James TY, Kamada T, Kilaru S, Kodira C, Kues U, Kupfer D, Kwan HS, Lomsadze A, Li W, Lilly WW, Ma LJ, Mackey AJ, Manning G, Martin F, Muraguchi H, Natvig DO, Palmerini H, Ramesh MA, Rehmeyer CJ, Roe BA, Shenoy N, Stanke M, Ter-Hovhannisyan V, Tunlid A, Velagapudi R, Vision TJ, Zeng Q, Zolan ME, Pukkila PJ (2010) Insights into evolution of multicellular fungi from the assembled chromosomes of the mushroom *Coprinopsis cinerea* (*Coprinus cinereus*). Proc Natl Acad Sci USA 107:11889–11894

Stalpers JA (1978) Identification of wood-inhabiting Aphyllophorales in pure cuture. Stud Mycol 16:1–248

Stalpers JA (1993) The aphyllophoraceous fungi. I. Keys to the species of the Thelephorales. Stud Mycol 35:1–168

Stalpers JA, Loerakker WM (1982) *Laetisaria* and *Limonomyces* species (Corticiaceae) causing pink diseases in turf grasses. Can J Bot 60:529–537

Sterner O, Bergman R, Kihlberg J, Wickberg B (1985) The sesquiterpenes of *Lactarius vellereus* and their role in a proposed chemical defense system. J Nat Prod 48:279–288

Sunhede S (1989) Geastraceae (Basidiomycotina): morphology, ecology, and systematics with special emphasis on the North European species. Synopsis Fungorum 1. Fungiflora, Oslo

Šutara J (2008) *Xerocomus* s. l. in the light of the present state of knowledge. Czech Mycol 60:29–62

Taylor JW, Berbee ML (2006) Dating divergences in the fungal tree of life: review and new analyses. Mycologia 98:838–849

Taylor DL, McCormick MK (2008) Internal transcribed spacer primers and sequences for improved characterization of basidiomycetous orchid mycorrhizas. New Phytol 177:1020–1033

Taylor TN, Krings M, Taylor EL (2014) Fungal diversity in the fossil record. In: McLaughlin DJ, Spatafora JW (eds) The Mycota, vol 7B, Systematics and evolution. Springer, Berlin

Tedersoo L, Smith ME (2013) Lineages of ectomycorrhizal fungi revisited: foraging strategies and novel lineages revealed by sequences from belowground. Fung Biol Rev 27:83–99

Tedersoo L, Suvi T, Beaver K, Saar I (2007) Ectomycorrhizas of *Coltricia* and *Coltriciella* (Hymenochaetales, Basidiomycota) on Caesalpiniaceae, Dipterocarpaceae and Myrtaceae in Seychelles. Mycol Prog 6:101–107

Tedersoo L, May TW, Smith ME (2010) Ectomycorrhizal lifestyle in fungi: global diversity, distribution, and evolution of phylogenetic lineages. Mycorrhiza 20:217–263

Tedersoo L, Bahram M, Toots M, Diédhiou AG, Henkel TW, Kjøller R, Morris MH, Nara K, Nouhra E, Peay KG, Põlme S, Ryberg M, Smith ME, Kõljalg U (2012) Towards global patterns in the diversity and community structure of ectomycorrhizal fungi. Mol Ecol 21:4160–4170

Telleria MT, Duenas M, Melo I, Hallenberg N, Martin MP (2010) A re-evaluation of *Hypochnicium* (Polyporales) based on morphological and molecular characters. Mycologia 102:1426–1436

Telleria MT, Melo I, Dueñas M, Salcedo I, Beltrán-Tejera E, Rodríguez-Armas JL, Martín M (2012) *Sistotremastrum guttuliferum*: a new species from the Macaronesian islands. Mycol Prog 12:687–692

Telleria MT, Melo I, Dueñas M, Larsson K-H, Martin MP (2013) Molecular analyses confirm *Brevicellicium* in Trechisporales. IMA Fungus 4:21–28

Thiers HD (1984) The secotioid syndrome. Mycologia 76:1–8

Thomas SE, Crozier J, Aime MC, Evans HC, Holmes KA (2008) Molecular characterisation of fungal endophytic morphospecies associated with the indigenous forest tree, *Theobroma gileri*, in Ecuador. Mycol Res 112:852–860

Thorn RG, Moncalvo J-M, Reddy CA, Vilgalys R (2000) Phylogenetic analyses and the distribution of nematophagy support a monophyletic Pleurotaceae within the polyphyletic pleurotoid-lentinoid fungi. Mycologia 92:241–252

Tian X-M, Yu H-Y, Zhou L-W, Decock C, Vlasák J, Dai Y-C (2013) Phylogeny and taxonomy of the *Inonotus linteus* complex. Fungal Divers 58:159–169

Tibuhwa DD, Savić S, Tibell L, Kivaisi AK (2012) *Afrocantharellus* gen. stat. nov. is part of a rich diversity of African Cantharellaceae. IMA Fungus 3:25–38

Tomšovský M (2008) Molecular phylogeny and taxonomic position of *Trametes cervina* and description of a new genus *Trametopsis*. Czech Mycol 60:1–11

Tomšovský M, Sedlák P, Jankovský L (2009) Species recognition and phylogenetic relationships of European *Porodaedalea* (Basidiomycota, Hymenochaetales). Mycol Prog 9:225–233

Tomšovský M, Menkis A, Vasaitis R (2010a) Phylogenetic relationships in European *Ceriporiopsis* species inferred from nuclear and mitochondrial ribosomal DNA sequences. Fungal Biol 114:350–358

Tomšovský M, Vampola P, Sedlák P, Byrtusová Z, Jankovský L (2010b) Delimitation of central and northern European species of the *Phellinus igniarius* group (Basidiomycota, Hymenochaetales) based on analysis of ITS and translation elongation factor 1 alpha DNA sequences. Mycol Prog 9:431–445

Tóth A, Hausknecht A, Krisai-Greilhuber I, Papp T, Vágvölgyi C, Nagy LG (2013) Iteratively refined guide trees help improving alignment and phylogenetic inference in the mushroom family Bolbitiaceae. PLoS One 8:e56143

Trappe JM, Castellano MA, Halling RE, Osmundson TW, Binder M, Fechner N, Malajczuk N (2013) Australasian sequestrate fungi. 18: *Solioccasus polychromus* gen. & sp. nov., a richly colored, tropical to subtropical, hypogeous fungus. Mycologia 105:888–895

Tuno N (1998) Spore dispersal of *Dictyophora* fungi (Phallaceae) by flies. Ecol Res 13:7–15

Tzean SS, Liou JY (1993) Nematophagous resupinate basidiomycetous fungi. Phytopathol 83:1015–1020

Uehling JK, Henkel TW, Aime MC, Vilgalys R, Smith ME (2012a) New species of *Clavulina* (Cantharellales, Basidiomycota) with resupinate and effused basidiomata from the Guiana Shield. Mycologia 104:547–556

Uehling JK, Henkel TW, Aime NC, Vilgalys R, Smith ME (2012b) New species and distribution records for *Clavulina* (Cantharellales, Basidiomycota) from the Guiana Shield, with a key to the lowland neotropical taxa. Fungal Biol 116:1263–1274

Underwood LM (1899) Moulds, mildews and mushrooms: a guide to the systematic study of the Fungi and Mycetozoa and their literature. Henry Holt, New York, Ny

Urban A, Weiß M, Bauer R (2003) Ectomycorrhizae involving sebacinoid mycobionts. Mycol Res 107:3–14

Valenzuela R, Raymundo T, Cifuentes J, Castillo G, Amalfi M, Decock C (2010) Two undescribed species of *Phylloporia* from Mexico based on morphological and phylogenetic evidence. Mycol Prog 10:341–349

Vampola P, Vlasak J (2012) *Rigidoporus pouzarii*, a new polypore species related to *Rigidoporus crocatus*. Czech Mycol 64:3–11

van Driel KGA, Humbel BM, Verkleij AJ, Stalpers J, Müller WH, Boekhout T (2009) Septal pore complex morphology in the Agaricomycotina (Basidiomycota) with emphasis on the Cantharellales and Hymenochaetales. Mycol Res 113:559–576

Veldre V, Abarenkov K, Bahram M, Martos F, Selosse M-A, Tamm H, Kõljalg U, Tedersoo L (2013) Evolution of nutritional modes of Ceratobasidiaceae (Cantharellales, Basidiomycota) as revealed from publicly available ITS sequences. Fungal Ecol 6:256–268

Vellinga EC (2004) Genera in the family Agaricaceae: evidence from nrITS and nrLSU sequences. Mycol Res 108:354–377

Vellinga EC, Sysouphanthong P, Hyde KD (2011) The family Agaricaceae: phylogenies and two new white-spored genera. Mycologia 103:494–509

Villegas M, De Luna E, Cifuentes J, Estrada-Torres A (1999) Phylogenetic studies in Gomphaceae sensu lato (Basidiomycetes). Mycotaxon 70:127–147

Villegas M, Cifuentes J, Torres AE (2005) Sporal characters in Gomphales and their significance for phylogenetics. Fungal Divers 18:157–175

Vizzini A, Para R, Fontenla R, Ghignone S, Ercole E (2012) A preliminary ITS phylogeny of *Melanoleuca* (Agaricales), with special reference to European taxa. Mycotaxon 118:361–381

Vlasák J, Kout J (2010) Pileate *Fomitiporia* species in the USA. New combinations *Fomitiporia calkinsii* and *F. bakeri*. Mycol Prog 10:445–452

Wagner T, Fischer M (2001) Natural groups and a revised system for the European poroid Hymenochaetales (Basidiomycota) supported by nLSU rDNA sequence data. Mycol Res 105:773–782

Wagner T, Fischer M (2002a) Classification and phylogenetic relationships of *Hymenochaete* and allied genera of the Hymenochaetales, inferred from rDNA sequence data and nuclear behaviour of vegetative mycelium. Mycol Prog 1:93–104

Wagner T, Fischer M (2002b) Proceedings towards a natural classification of the worldwide taxa *Phellinus* s.l. and *Inonotus* s.l., and phylogenetic relationships of allied genera. Mycologia 94:998–1016

Wagner T, Ryvarden L (2002) Phylogeny and taxonomy of the genus *Phylloporia* (Hymenochaetales). Mycol Prog 1:105–116

Walther G, Garnica S, Weiß M (2005) The systematic relevance of conidiogenesis modes in the gilled Agaricales. Mycol Res 109:525–544

Wang Z, Binder M, Dai YC, Hibbett DS (2004) Phylogenetic relationships of *Sparassis* inferred from nuclear and mitochondrial ribosomal DNA and

RNA polymerase sequences. Mycologia 96:1015–1029

Wannathes N, Desjardin DE, Hyde KD, Perry BA, Lumyong S (2009) A monograph of *Marasmius* (Basidiomycota) from Northern Thailand based on morphological and molecular (ITS sequences) data. Fungal Divers 37:209–306

Warcup JH (1988) Mycorrhizal associations of isolates of *Sebacina vermifera*. New Phytol 110:227–231

Watkinson SC, Eastwood DC (2012) *Serpula lacrymans*, wood and buildings. In: Laskin I, Sariaslani S, Gadd GM (eds) Advances in applied microbiology, vol 78. Academic, Burlington, pp 121–149

Watling R (1970) Boletaceae: Gomphidiaceae: Paxillaceae. In: Henderson DM, Orton PD, Watling R (eds) British fungus flora. Agarics and boleti I. Royal Bot Garden, Edinburgh

Weiß M, Oberwinkler F (2001) Phylogenetic relationships in Auriculariales and related groups – hypotheses derived from nuclear ribosomal DNA sequences. Mycol Res 105:403–415

Weiß M, Bauer R, Begerow D (2004a) Spotlights on heterobasidiomycetes. In: Agerer R, Piepenbring M, Blanz P (eds) Frontiers in Basidiomycote mycology. IHW, Eching, pp 7–48

Weiß M, Selosse M-A, Rexer K-H, Urban A, Oberwinkler F (2004b) Sebacinales: a hitherto overlooked cosm of heterobasidiomycetes with a broad mycorrhizal potential. Mycol Res 108:1003–1010

Weiß M, Sýkorová Z, Garnica S, Riess K, Martos F, Krause C, Oberwinkler F, Bauer R, Redecker D (2011) Sebacinales everywhere: previously overlooked ubiquitous fungal endophytes. PLoS One 6: e16793

Weiß M, Bauer R, Sampaio JP, Oberwinkler F (2014) Tremellomycetes and related groups. In: McLaughlin DJ, Spatafora JW (eds) The Mycota: Systematics and evolution: Part B. Springer, Heidelberg

Wells K (1994) Jelly fungi, then and now! Mycologia 86:18–48

Wells K, Bandoni R (2001) Heterobasidiomycetes. In: McLaughlin DJ, McLaughlin EG, Lemke PA (eds) The Mycota. Part B: systematics and evolution, vol VII. Springer, Berlin, pp 85–120

Wells K, Oberwinkler F (1982) *Tremelloscypha gelatinosa*, a species of a new family Sebacinaceae. Mycologia 74:325–331

Welti S, Moreau P-A, Favel A, Courtecuisse R, Haon M, Navarro D, Taussac S, Lesage-Meessen L (2012) Molecular phylogeny of *Trametes* and related genera, and description of a new genus *Leiotrametes*. Fungal Divers 55:47–64

Wilkinson HT (1987) Association of *Trechispora alnicola* with yellow ring disease of Poa pratensis. Can J Bot 65:150–153

Wilson AW, Desjardin DE (2005) Phylogenetic relationships in the gymnopoid and marasmioid fungi (Basidiomycetes, euagarics clade). Mycologia 97:667–679

Wilson AW, Hobbie EA, Hibbett DS (2007) The ectomycorrhizal status of *Calostoma cinnabarinum* determined using isotopic, molecular, and morphological methods. Can J Bot 85:385–393

Wilson AW, Binder M, Hibbett DS (2011) Effects of fruiting body morphology on diversification rates in three independent clades of fungi estimated using binary state speciation and extinction analysis. Evolution 65:1305–1322

Wilson AW, Binder M, Hibbett DS (2012) Diversity and evolution of ectomycorrhizal host associations in the Sclerodermatineae (Boletales, Basidiomycota). New Phytol 194:1079–1095

Wolfe BE, Tulloss RE, Pringle A (2012) The irreversible loss of a decomposition pathway marks the single origin of an ectomycorrhizal symbiosis. PLoS One 7:e39597

Woodward S, Stenlid J, Karjalainen R, Hüttermann A (eds) (1998) *Heterobasidion annosum*: biology, ecology, impact and control. CABI, Wallingford

Worrall JJ, Anagnost SE, Zabel RA (1997) Comparison of wood decay among diverse lignicolous fungi. Mycologia 89:199–219

Wu S-H, Hibbett DS, Binder M (2001) Phylogenetic analyses of *Aleurodiscus* s.l. and allied genera. Mycologia 93:720–731

Wu S-H, Nilsson HR, Chen C-T, Yu S-Y, Hallenberg N (2010) The white-rotting genus *Phanerochaete* is polyphyletic and distributed throughout the phleboid clade of the Polyporales (Basidiomycota). Fungal Divers 42:107–118

Wu S-H, Dai Y-C, Hattori T, Yu T, Wang D, Parmasto E, Chang H, Shih S (2012) Species clarification for the medicinally valuable 'sanghuang' mushroom. Bot Stud 53:135–149

Yamaguchi K, Degawa Y, Nakagiri A (2008) An aero-aquatic fungus, *Peyronelina glomerulata*, is shown to have teleomorphic affinities with cyphelloid basidiomycetes. Mycoscience 50:156–164

Yu Z-H, Wu S-H, Wang D-M, Chen C-T (2010) Phylogenetic relationships of *Antrodia* species and related taxa based on analyses of nuclear large subunit ribosomal DNA sequences. Bot Stud 51:53–60

Yukawa T, Ogura-Tsujita Y, Shefferson RP, Yokoyama J (2009) Mycorrhizal diversity in *Apostasia* (Orchidaceae) indicates the origin and evolution of orchid mycorrhiza. Am J Bot 96:1997–2009

Yurchenko EO, Golubkov VV (2003) The morphology, biology, and geography of a necrotrophic basidiomycete *Athelia arachnoidea* in Belarus. Mycol Prog 2:275–284

Zeller SM (1939) New and noteworthy Gasteromycetes. Mycologia 31:1–32

Zeller SM (1949) Keys to the orders, families, and genera of the Gasteromycetes. Mycologia 41: 36–58

Zeng NK, Cai Q, Yang ZL (2012) *Corneroboletus*, a new genus to accommodate the southeastern Asian *Boletus indecorus*. Mycologia 104:1420–1432
Zhao RL, Jeewon R, Desjardin DE, Soytong K, Hyde KD (2007) Ribosomal DNA phylogenies of *Cyathus*: is the current infrageneric classification appropriate? Mycologia 99:385–395
Zhao C-L, Cui B-K, Dai Y-C (2013) New species and phylogeny of *Perenniporia* based on morphological and molecular characters. Fungal Divers 58:47–60
Zhou L-W, Dai Y-C (2012) Phylogeny and taxonomy of *Phylloporia* (Hymenochaetales): new species and a worldwide key to the genus. Mycologia 104:211–222
Zhou L-W, Dai Y-C (2013) Phylogeny and taxonomy of poroid and lamellate genera in the Auriculariales (Basidiomycota). Mycologia 105:1219–1230
Zhou L-W, Kõljalg U (2013) A new species of *Lenzitopsis* (Thelephorales, Basidiomycota) and its phylogenetic placement. Mycoscience 54:87–92
Zhou L-W, Qin W-M (2013) Phylogeny and taxonomy of the recently proposed genus *Phellinopsis* (Hymenochaetales, Basidiomycota). Mycologia 105:689–696
Zmitrovich IV, Spirin VA (2002) A contribution to the taxonomy of corticioid fungi II. The genera *Serpula*, *Serpulomyces* gen. nov., *Amylocorticiellum* gen. nov. Mikologya Fitopatologya 36:11–26
Zuccaro A, Lahrmann U, Guldener U, Langen G, Pfiffi S, Biedenkopf D, Wong P, Samans B, Grimm C, Basiewicz M, Murat C, Martin F, Kogel KH (2011) Endophytic life strategies decoded by genome and transcriptome analyses of the mutualistic root symbiont *Piriformospora indica*. PLoS Pathog 7: e1002290

Biosystematic Index

Systematics and Evolution, 2nd Edition
The Mycota VII Part A
D.J. McLaughlin and J.W. Spatafora (Eds.)

Subject Index

Systematics and Evolution, 2nd Edition
The Mycota VII Part A
D.J. McLaughlin and J.W. Spatafora (Eds.)

MIX
Papier aus verantwortungsvollen Quellen
Paper from responsible sources
FSC® C105338

If you have any concerns about our products,
you can contact us on
ProductSafety@springernature.com

In case Publisher is established outside the EU,
the EU authorized representative is:
Springer Nature Customer Service Center GmbH
Europaplatz 3, 69115 Heidelberg, Germany

Printed by Libri Plureos GmbH
in Hamburg, Germany